DATE DUE

Sewall Wright
EVOLUTION
Selected Papers

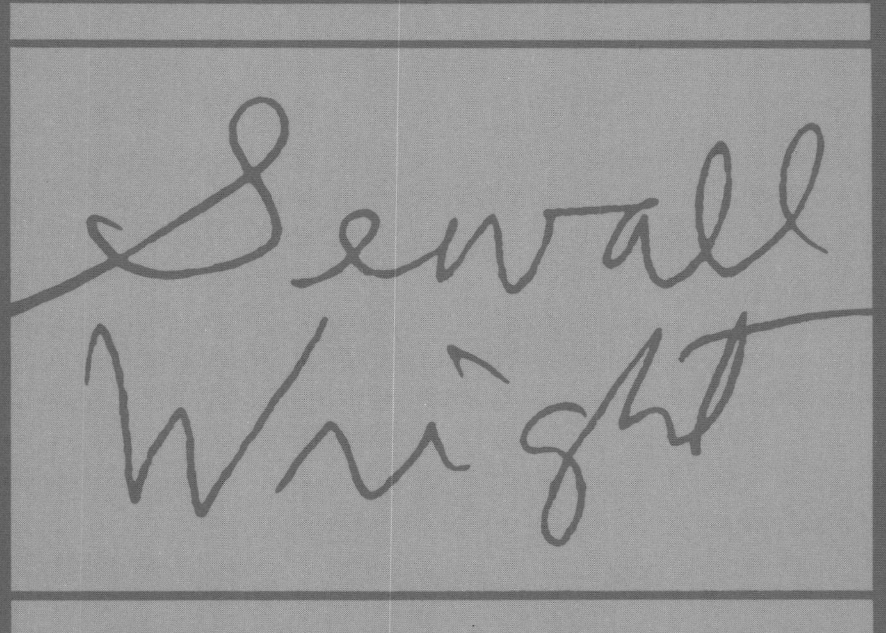

Edited and with Introductory Materials by

WILLIAM B. PROVINE

EVOLUTION

EVOLUTION
Selected Papers

Sewall Wright

Edited and with Introductory Materials by
William B. Provine

The University of Chicago Press
Chicago and London

SEWALL WRIGHT is the Ernest D. Burton Distinguished Service Professor Emeritus of zoology at the University of Chicago and professor emeritus of genetics at the University of Wisconsin. He has received the National Medal of Science, the Darwin Medal of the Royal Society, the Balzan Prize, and many other honors during his extraordinary career. His four-volume work, *Evolution and the Genetics of Populations,* is also published by the University of Chicago Press.

WILLIAM B. PROVINE is professor in the Department of History and in the Division of Biological Sciences, section of ecology and systematics, at Cornell University. He is the author of many books, including *The Origin of Theoretical Population Genetics,* and *Sewall Wright and Evolutionary Biology,* also published by the University of Chicago Press.

The University of Chicago Press, Chicago 60637
The University of Chicago Press, Ltd., London

Printed in the United States of America

95 94 93 92 91 90 89 88 87 86   5 4 3 2 1

Library of Congress Cataloging-in-Publication Data

Wright, Sewall, 1889–
Evolution: selected papers:

Bibliography: p.
Includes index.
1. Evolution. I. Provine, William B. II. Title.
QH360.W75 1986 575 86-11327
ISBN 0-226-91053-9
ISBN 0-226-91054-7 (pbk.)

William B. Provine
and
Sewall Wright
join in dedicating this volume to

James F. Crow

Contents

Preface

I have designed this volume as a companion to *Sewall Wright and Evolutionary Biology* and to Wright's four-volume treatise, *Evolution and the Genetics of Populations*. In his treatise, Wright constantly referred to his earlier publications on evolutionary biology. He did not, however, simply summarize these papers, but instead generally began his discussion from where they left off. Much of *Sewall Wright and Evolutionary Biology* (hereafter referred to as *SW&EB*) is devoted to an analysis of the background and content of Wright's papers on evolutionary biology. Thus a volume containing a judicious selection of his published papers on evolutionary biology would be a very useful companion to these two works. Wright is one of the most influential evolutionary biologists of the twentieth century, and this volume might also be useful for those merely wishing to have a collection of his papers on that subject.

Another compelling reason for publishing a volume of Wright's papers, especially those before about 1950, is that they were little understoood at the time of publication, even if widely read. Still Wright enjoyed much prestige as a quantitative population geneticist. The primary reasons for this curious situation are easily seen. Evolutionary biologists in general had very little training in mathematics or specifically in statistics, or in quantitative reasoning generally. Moreover, Wright was rather insensitive to the inability of his audience to follow his quantitative reasoning. Even those with some mathematical training had much difficulty following Wright's idiosyncratic method of path coefficients. Inability to read much more than his introductions and conclusions did not, however, prevent many evolutionists from admiring Wright's quantitative work. As Ernst Mayr and John A. Moore have pointed out on many occasions in letters to me, evolutionary biologists were extremely susceptible to the attractions of quantitative models whose derivations they could not understand.

Now, however, the situation has changed dramatically. Most graduate students in evolutionary biology are expected to develop at least minimal sophistication in quantitative reasoning, and many have mathematical skills that equal or often surpass the sophistication of any used by Wright in his papers. Thus the time is ripe for a rereading of Wright's papers.

This volume emphasizes the period before 1950. During this period Wright thought of himself primarily as an experimental physiological geneticist rather than as a theoretical population geneticist. His work from 1912 to 1950 comprised substantial contributions to physiological genetics, including theories of gene action; mammalian coat color genetics; developmental genetics including teratology; tissue transplants; animal breeding, including the quantitative analysis of inbreeding and hybrid vigor; separating the effects of heredity and environment (including human IQ); and statistical methodology. He was the leading physiological geneticist in the United States, more trusted in the genetics community than Richard Goldschmidt and admired by such newcomers as George Wells Beadle and Tracy Sonneborn. The experimental organism that he worked with on a daily basis from 1912 to 1955 was the guinea pig, which he never saw in the wild or used for any studies in evolutionary biology. Wright considered his papers on evolutionary biology as a sideline to his major interests until after his retirement from the University of Chicago in 1954.

Included in this volume are all of Wright's published papers on evolutionary biology up to 1950 and a selection of those published after that, including two important papers published after the last volume of his treatise (1978; the treatise is hereafter referred to as *E&GP*). I have included all of the earlier papers because in *SW&EB* most of my discussion of Wright's published papers concerns the period before 1950; the papers published after 1950 tend to be repetitious of earlier papers or of the later treatise and, being older, the earlier papers are more difficult to locate. I have excluded from this category only the five papers that Wright wrote in collaboration with Theodosius Dobzhansky and published in Dobzhansky's famous series, "The Genetics of Natural Populations," which has recently been republished in its entirety with extensive introductions (Lewontin et al. 1981). I included the two very recent papers because one is Wright's own account of the origins of his shifting balance theory of evolution in nature and the other his analysis of speciation and the theory of punctuated equilibrium, popularized by Niles Eldredge and Stephen Jay Gould (Eldredge and Gould 1972). Wright received more requests for reprints of this last paper than any other he ever published (he was ninety-two when it appeared in print).

There was another reason for this pattern of selection of the papers. Continuity versus change is a fundamental theme in the development of an individual scientist's thought and influence. This theme is especially important in Wright's case because he has strongly emphasized the continuity in his thinking about mechanisms of evolution from the mid-1920s to the present. While agreeing with Wright that his thought about evolution has been in many ways remarkably consistent over the years, I argue in *SW&EB* that in some crucial ways he changed his mind about the mechanisms of evolution, especially concerning the problem of adaptation in relation to random drift and selection. By presenting Wright's papers on evolution in their entirety up to 1950, a comparison of his earlier and later views is possible simply by examining this volume and his *E&GP*, an exercise that I warmly recommend to anyone interested in the development of Wright's views and their influence in evolutionary biology.

Both this introduction and those for the individual papers are very brief because most of the historical background is already in *SW&EB*; later extensions of the papers are detailed in *E&GP*. Thus aside from the barest information about each paper, the primary purpose of the introductions is to refer the reader to the pertinent sections of *SW&EB* and *E&GP*, where often extensive background may be found.

With a few exceptions, the order of presentation of papers in this volume is chronological. For obvious reasons, I have placed first Wright's 1978 paper on the origins of his shifting balance theory. Otherwise, the only departure from strict chronological order comes when papers sharing a basic theme are introduced together, as in the case of Wright's three papers on isolation by distance.

All but a few of the papers in this volume were photocopied from reprints that Wright sent out to other biologists. Occasional typesetting errors appeared in the published journals, and Wright corrected these by hand in the reprints he sent to others. Wright's corrections thus appear in these photocopies. The print quality of the original reprints from which the reproductions are taken varies considerably. Many academic journals in the first half of the century operated on minimal budgets; low-quality print was one way to save money. Every effort has been made in this volume to maintain the highest quality in photoreproduction. The variation in the results is therefore a function of variation in the originals. The advantages to scholars of having the originals is enormous, well worth the cost of some variation in print quality.

Sewall Wright deserves the credit for this volume. He wrote all of the papers and told me about the origins of each one. He also wrote a detailed critique of the draft version of my introductions to the papers. Our conversations about the background of the papers are recorded on audiotape (and are partially transcribed) and are available at the Library of the American Philosophical Society. I am enormously grateful to the library for providing funds for the transcriptions. James F. Crow provided constant support for this volume and made a great many thoughtful suggestions for the introductions and choice of papers. I also thank Michael J. Wade for a careful reading of the introductions and many helpful suggestions. I gratefully acknowledge the support of the National Science Foundation, Section on the History and Philosophy of Science (Ronald Overman, director).

1
The Relation of Livestock Breeding to Theories of Evolution

Journal of Animal Science 46, no. 5 (1978):1192–1200

Introduction

A more accurate title for this paper would have been "Relation of Livestock Breeding to My Shifting Balance Theory of Evolution in Nature." Wright's purpose in writing the paper (shortly after he had completed the fourth and final volume of his treatise in 1977) was clearly to explain the origins of his shifting balance theory, which was the backbone of the treatise and, indeed, of all of his work in evolutionary theory.

In this paper Wright argued that his shifting balance theory grew out of his synthesis of four separate (Wright terms them "very diverse") research projects. These were (1) William Castle's selection experiment with hooded rats (*SW&EB*, chap. 2; Provine 1971, 109–14, 126–29); (2) Wright's thesis research on the physiological genetics of color inheritance in guinea pigs (Wright 5; *SW&EB*, chap. 3); (3) Wright's analysis of inbreeding in guinea pigs at the United States Department of Agriculture, including his invention of methods for calculating the degree of inbreeding (Wright 33, 34, 35; *SW&EB*, chaps. 4, 5); and (4) the analysis by Wright and H. C. McPhee of the evolution of the Shorthorn breed of cattle (Wright 38, 39, 44; *SW&EB*, chap. 5). The first two of these research projects are treated in extensive detail in *SW&EB* and are not represented in this volume. The third is represented by Wright's theoretical paper on coefficients of inbreeding and relationship, and the fourth by the complete set of three papers by Wright and McPhee on evolution in the Shorthorns.

It is significant that Wright's shifting balance theory of evolution in nature grew directly out of his theory of evolution in domestic breeds. Darwin's theory of evolution by natural selection was similarly based upon his great understanding of animal and plant breeding. In the early twentieth century, almost all geneticists grounded their views of evolution in nature upon what they knew of evolution in domesticated animals and plants. Wright, who worked at the Animal Husbandry Division of the USDA, J.B.S. Haldane, who worked for many years at the John Innes Horticultural Institution, and R. A. Fisher, who worked during the same years at the Rothamsted Experimental Station, all mostly derived their evolutionary theories from their knowledge of domesticated animals and plants.

At present, evolutionary theory is based primarily upon studies of the genetics and ecology of natural populations. But it should be clearly recognized that these careful studies of evolution in natural populations are recent (beginning in the late 1930s and reaching large numbers only in the 1960s).

Thus one reason why evolutionary geneticists based their theories of evolution in nature on evolution in domesticated species is because they knew very little about the genetics of natural populations. Most young evolutionists are unaware that many of the intense controversies that rage among evolutionary biologists today were initiated by evolutionists whose knowledge of natural populations was minimal. The robustness of these controversies is a tribute to the real and fundamental similarities of evolution in domestic breeds and in natural populations. Population geneticists who concentrate exclusively upon natural populations still have a great deal to learn from animal and plant breeders.

Wright was almost ninety when he wrote this paper, and one might expect a heavy dose of hindsight rather than historical accuracy. After an extended examination of Wright's work, I think his recollection of the origins of his shifting balance theory is remarkably accurate. My only reservation about historical accuracy in this paper concerns Wright's labeling of his famous diagrams of fitness surfaces redrawn from Wright 70 (1932). Here Wright changed the parameters of the surface from his original conception of gene combinations (each point on the fitness surface an individual fitness) to each axis representing the frequency of a gene (where each point on the surface is the mean fitness of an entire population with a specified set of gene frequencies). I have argued (*SW&EB*, 307–17) that the 1932 diagrams are unintelligible with either set of parameters.

THE RELATION OF LIVESTOCK BREEDING TO THEORIES OF EVOLUTION[1]

Sewall Wright

University of Wisconsin[2]*, Madison 53706*

INTRODUCTION *

The debt owed by quantitative geneticists to today's guest speaker is immeasurable. We quantitative geneticists are much concerned with the covariances among relatives, so I thought this introduction should include information on near relatives as well as individual performance data. For pedigree evaluation purposes, I thought it necessary to tell you a bit about Dr. Wright's father. Philip Green Wright was born in 1861, and took a Master's Degree at Harvard in Economics. Most of his academic career was spent at Lombard College in Galesburg, IL. There he taught Economics, Mathematics, Astronomy and English, and was Director of the Gymnasium. He was a minor poet and would probably have to be judged unsuccessful in this endeavor, except that one of his student's, Carl Sandburg, enjoyed a bit of success. Another important relationship is that of full-sibs. Dr. Wright's two brothers were successful and able men. Quincy was a Professor of Political Science at the University of Chicago. Theodore was Director of the Aircraft Resources Office, 1942 to 1944, and had primary responsibility of overcoming the lead of the Axis powers in military airplanes. Later he became Vice-President-in-Charge-of-Research at Cornell University. Positive assortative mating occurs in humans, thus information on a spouse is of interest. Mrs. Wright was also a geneticist with a Master's Degree in Zoology. She taught at Smith College for 3 years, during which time she started the first course in Genetics there. For progeny test information, I offer that the Wrights had three children – two sons and a daughter – all of whom took advanced degrees.

Finally, we must consider individual performance. Professor Wright took a B.S. degree at Lombard College in 1911, having successfully passed the course in Calculus and Analytical Geometry given by his father. His M.S. degree was taken at the University of Illinois in 1912 where he studied the microscopic anatomy of a flatworm. Upon completion of his Master's Degree, he studied genetics under Dr. William E. Castle and received his Doctor of Science in 1915. The next 10 years were spent as Senior Animal Husbandman with the U.S. Department of Agriculture in Washington. From there he went to the University of Chicago, and was a member of that faculty from 1926 to 1954. In 1955, he accepted an appointment as the L. J. Cole Professor of Genetics here at Wisconsin, and, since 1960 has been an Emeritus Professor of Genetics. Dr. Wright has received honorary degrees from a number of prestigious universities and many awards, the foremost of which is the National Medal of Science.

His classical series of papers on Systems of Mating, published in 1921, provides the theoretical foundation for plant and animal breeding, our main interest here. But we should also be cognizant of his contributions in other areas of knowledge. His method of analysis of correlated variables, quite familiar to us here, is now widely used in the social sciences. His work on pigment genetics led him to a theory of genes as producers of enzymes, thus anticipating in many ways the work of Beadle and Tatum which initiated modern biochemical genetics. Every student of evolution theory is familiar with his recognition that stochastic changes and the effect of subdivision of populations can be important evolutionary determinants. He has emphasized many times that the evolutionary path of a population is determined by a balance among the various simultaneously-acting forces that change gene frequency. Dr. Wright's first scientific contribution was in 1912. Now some 65 years later, we quantitative geneticists are eagerly awaiting Volume 4 of his encyclopedic treatise – *Evolution and the Genetics of*

[1] Invited paper presented at the 69th Annual Meeting of the American Society of Animal Science, Madison, WI, July 25, 1977.

[2] Department of Genetics.

* Introduction by J. J. Rutledge

Populations.

Dr. Wright was a member of the American Society of Animal Production for some 25 years and made several contributions to the annual Proceedings. Dr. Wright, although the Society may have changed its name since you were last a member, we sincerely want to welcome you home.

THEORIES OF EVOLUTION

It is appropriate to begin with Darwin's theory of evolution by natural selection (1859). Direct evidence from nature was not available. The only direct evidence of its possibility was from the results that had been obtained from artificial selection by British livestock breeders during the preceding century. Darwin discussed this at considerable length.

He also discussed whether advances had occurred typically by abrupt steps (as in polled breeds of cattle and in the short-legged Ancon sheep) or had been built up gradually by selection of slight quantitative variants. He decided on the latter—"without variability nothing can be effected: Slight individual differences, however, suffice and are probably the chief or sole means in the production of species."

The opposite alternative was adopted in extreme form by deVries near the beginning of this century (1901–1903). He held that species arise from single mutations that are responsible both for the character changes and for the reproductive isolation that define a new species.

The mutations on which he largely based his theory, those that he observed in the American plant, Oenothera lamarckiana, that had escaped from cultivation in the Netherlands, turned out later to be chromosome aberrations, mostly the presence of an extra chromosome. This proved to be transmissible only by the ovules, so that the mutants existed only as segregants without the reproductive isolation of a true species.

It came to be recognized that certain balanced chromosome changes (tetraploids and translocations) play an important role in the origin of species by leading to the essential reproductive isolation, but are of little significance in character change. The unbalanced ones cause drastic character change, but are in general unfixable and soon lost. I will be concerned here with the transformation of characters within species, not with the origin of species.

After the rediscovery of Mendelian heredity in 1900, most geneticists, led by Bateson (1909) in England and Morgan (1932) in America, thought of species as fixed with respect to "wild type" alleles at most loci. They held that evolution consisted in the replacement of such alleles by very rare mutations that happened to be favorable.

The course of fixation was worked out by Castle in 1903 in a special case. The process was worked out systematically by J. B. S. Haldane under a great variety of genetic conditions in a series of papers from 1924 summarized in his book, "The Causes of Evolution," in 1932. He also dealt with quantitative variation and such complications as gene interaction, but emphasized rare major mutations, probably from considerations of available time. He recognized the principle, painfully familiar to livestock breeders, that selection for any one character restricts the possibility of selection for any other because of limited reproductive excess. He estimated later (1957) that the process of fixation of any gene, major or minor, required some 300 generations. Even geologic time was not enough to fix a very large number of genes with very slight effects. Figure 1 illustrates the course of fixation of a recessive mutation under two conditions with respect to heritability.

At the time of the rediscovery of Mendelian heredity, Darwinian evolution from quantitative variability was being advocated in England by the biometricians under the leadership of Karl Pearson. They strenuously rejected Mendelian heredity in favor of Galton's law of

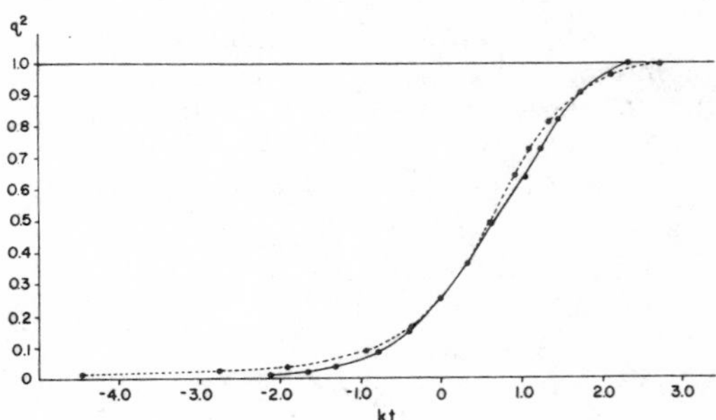

Figure 1. Comparison of the courses of change of the frequencies of recessives (q^2) under directional selection, where variability is entirely genetic (solid line) or is so nearly entirely nongenetic that the selection intensity is treated as constant (broken line). For ease of comparison, the rates are taken as the same as $q = .5$ ($q^2 = .25$). From Wright (1977, figure 6.2).

ancestral heredity (1889) and thus rejected Bateson's (1909) ideas on evolution. The acrimonious debate that ensued probably delayed experimental population genetics for several decades in England, that is, until Mather (1941) began his studies of what he called polygenic heredity. This was in spite of papers by the biometrician, Yule, in 1902 and 1906, showing that there was no irreconcilable difference between Galton's purely statistical law, essentially a multiple regression equation, and a physiological law such as Mendel's.

The situation was wholly different in America. Pearson's biometry (1901) was brought to America by Davenport, Harris and Pearl, but merely as a statistical tool. Castle not only published the first experimental Mendelian results from this country (1903a,b) but also soon began experimenting with inbreeding (Castle *et al.*, 1906) and selection, experimental population genetics. deVries had maintained that selection of quantitative variability had no permanent effect. This was probably because his artificial selection happened to be opposed by such strong natural selection that any progress was soon reversed on cessation of the artificial selection. This became something of a dogma, however, among many early Mendelians. Castle, with an agricultural background, challenged this dogma in an extensive experiment with black and white hooded rats, of which I was his assistant from 1912 to 1915. He selected toward self white and toward self black and was approaching these goals when high mortality and low fecundity put an end to both strains. Figure 2 shows the results of some 20 generations of selection in each direction.

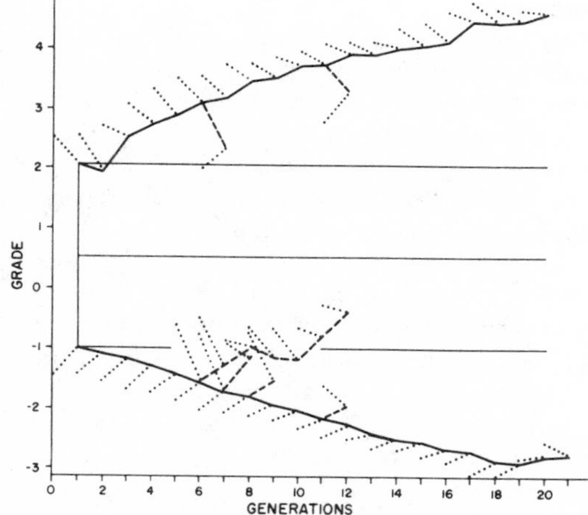

Figure 2. Courses of plus and minus selection (solid lines) and of reverse selection (broken lines) for amount of color in coats of hooded rats. The dotted lines show the regressions from the grades of the selected parents. From data of Castle and Phillips (1914) and Castle (1916).

Meanwhile Nilsson-Ehle (1909) in Sweden, G. H. Shull (1908) and E. M. East (1910) in this country were giving extensive experimental support to the interpretation of quantitative variability as due to multiple minor Mendelian differences. Castle had originally held that his successful selection experiments depended on variability of the major spotting factor of the rat, and in one case clearly demonstrated a mutation at this locus. Later, however, he made crucial tests that convinced him that the results depended largely on multiple independent modifiers (1919).

Darwin's theory that evolution is due primarily to mass selection of quantitative variability was put in mathematical form by Fisher in 1930 in his book, "The Genetical Theory of Natural Selection." According to his "fundamental theorem of natural selection": "the rate of increase in fitness of an organism at any time is equal to its genetic variance in fitness at that time." Evolution under given conditions according to this theorem, slows down to the exceedingly low rate supported by mutation as the genetic variance becomes exhausted. Long continued evolution at a higher rate depends on continued renewal of the genetic variance by changing conditions, which favor different alleles. Since he defined "genetic variance" as the additive component, dominance and gene interaction, the nonadditive components, merely reduce the rate. This theorem depended on Fisher's assumption that species are essentially homogeneous except in so far as different local environments may bring about selective differentiation. Fisher thus held that accidents of sampling, so called sampling drift, play no significant role apart from occasionally raising the frequency of a more favorable mutation to a safe level.

The shifting balance theory of evolution that I arrived at while in the Animal Husbandry Division of the U.S. Bureau of Animal Industry in Washington was very different from both Haldane's and Fisher's theories of the same period. It was based on my own experience in four very diverse research projects.

The first was Prof. Castle's selection experiment with hooded rats to which I have already referred.

The second was my thesis project (1916), studies of the genetics of several continuous series of variations in Prof. Castle's guinea pig colony; the intensities of the eumelanic and phaeomelanic colors, and patterns of hair direction, ranging from smooth to one of multiple small rosettes. There were independent loci in each case with allelic differences of various magnitudes, sets of multiple alleles, and different background heredities in different strains that showed apparent blending heredity, due presumably to multiple minor factors.

In the course of this study, I became fascinated with the frequently unpredictable effects of combinations. I had originally gone into genetics from interest in how genes act. The interaction effects seemed to give a basis for deducing the chains of gene action. Figure 3 shows the combination effects of the major genes on the eumelanic and phaeomelanic colors. Figure 4 shows the complex patterns of interaction of these genes and the multifactorial background heredities that seemed indicated. Figure 5 shows similarly those deduced in the case of hair direction.

The third project was the study of numerous closely inbred strains of guinea pigs which I

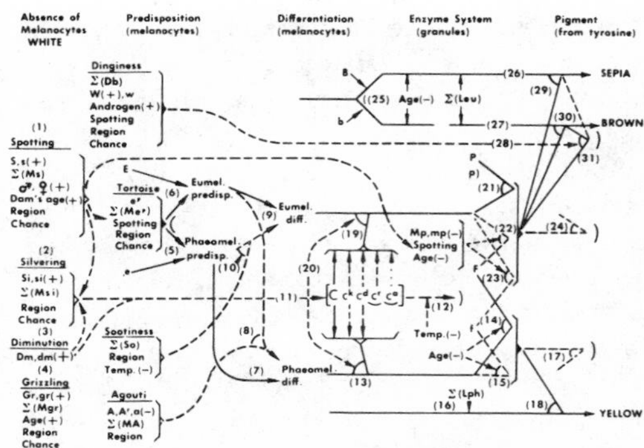

Figure 4. Factor interactions in the determination of coat color of the guinea pig. (Wright, 1968, figure 5.11).

took up on going to the Animal Husbandry Division in 1915. George Rommel, Chief of the Division, had started many strains from pairs in 1906 and had them maintained along lines of exclusive brother-sister mating. Twenty-three of these persisted long enough to yield significant data. Seventeen were still on hand when I was brought in to analyze the data and continue the experiment.

All strains exhibited the traditional inbreeding depression. There was not, however, a uniform decline in general vigor. Each strain had a unique combination of relative strength or weakness in size of litter and regularity in producing litters, mortality at birth and later, and in weight at birth and in later gains. Each strain had a particular combination of known color factors and its own background heredity with respect to the spotting pattern and intensity of color. There was also much differentiation in conformation. The very large animals of one strain (No. 13) (figure 6) had such short legs that they seemed to glide on the floor like

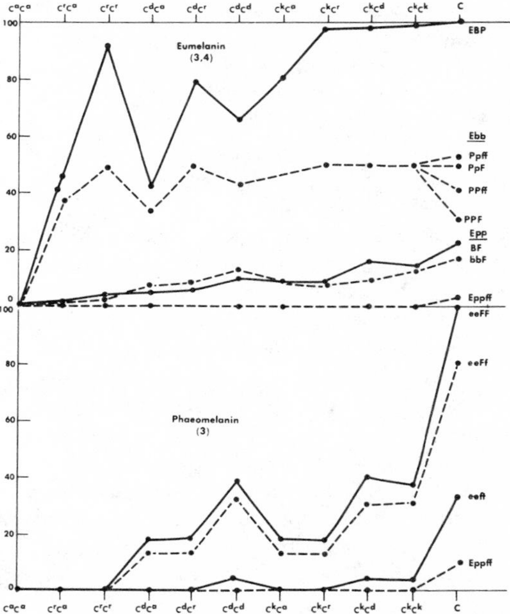

Figure 3. Relative amounts of pigment (eumelanin above, and phaeomelanin below) in coats of guinea pigs of various genotypes. (Wright, 1968, figure 5.8).

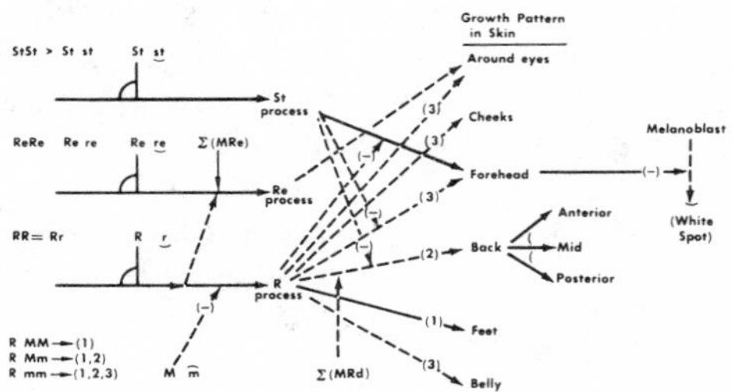

Figure 5. Diagram of factor interactions in the determination of hair direction and of a pleiotropic effect on a white forehead spot in the guinea pig. (Wright, 1968, figure 5.13).

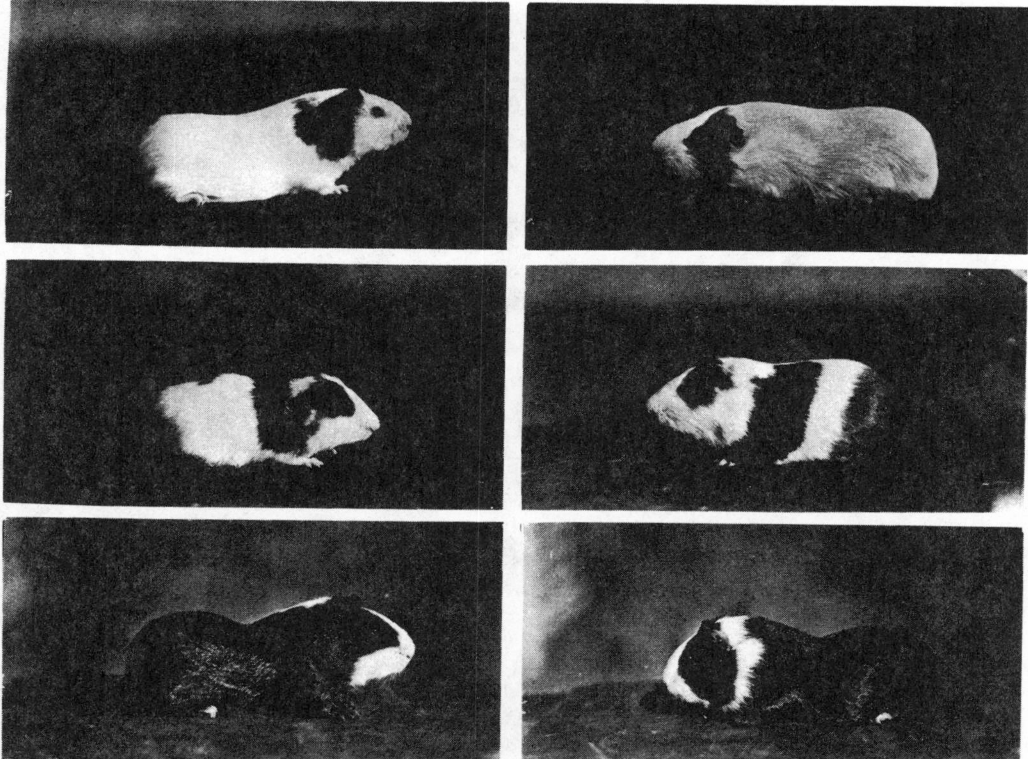

Figure 6. Illustration of varying color patterns and conformation in closely inbred strains of guinea pigs. Males (left) and females (right) of inbred strains 13, 2 and 39 (top to bottom) in generations 18, 12 and 13, respectively, of brother-sister mating. From Wright 1922c, plates II, I and V.

oversized planarians. The small animals of strain No. 2 had legs as long or longer than the preceding and ran well off the floor. Those of strain No. 13 had rounded noses and bent ears. Those of No. 2 had pointed noses and erect ears. Those of No. 2 had pointed noses and erect ears. Those of strains No. 39 had notably swayed backs. Strain 35 had protruding eyes; strain 13 sunken ones. The internal organs such as thyroid, adrenals and spleen differed strikingly in size and shape (Strandskov 1939, 1942). Some of the strains produced high frequencies of particular abnormalities such as otocephaly in No. 13, anophthalmia in No. 38. An atavistic little toe was common in No. 35, wholly absent in most others. There were notable differences temperament. The pigs of strain 13 could be picked up like sacks of meal while those of strains 2 and 35 would struggle and kick a hole in one's wrist unless picked up properly. Experiments by Dr. P. A. Lewis of the Phipps and later the Rockefeller Institute (Wright and Lewis, 1921), revealed striking differences in resistance to inoculations of tuberculosis in the order 35, 2, 13 and 32, 39 from high to low. Ones by Dr. Leo Loeb of Washington University, St. Louis, showed that each strain had its characteristic array of histocompatibility genes (Loeb and Wright, 1927).

My general project in the Animal Husbandry Division was clarification of the roles of inbreeding and selection in the breeding of livestock. This was the fourth of the projects referred to. I developed an inbreeding coefficient (the theoretical correlation between uniting gametes) that could be shown to measure the decrease in heterozygosis from that in the foundation stock. It could easily be derived, not only for any regular system of mating (Wright, 1921), but from the irregular ones of livestock pedigrees (Wright, 1922a) and by an approximation method, for whole breeds (Wright and McPhee, 1925; McPhee and Wright, 1925). Figure 7 shows the pedigree of the foundation Shorthorn bulls, Favourite and Comet. I calculated the coefficients for all of the 64 cows of Bates' famous Duchess line and their sires (Wright, 1923). Figure 8 shows the

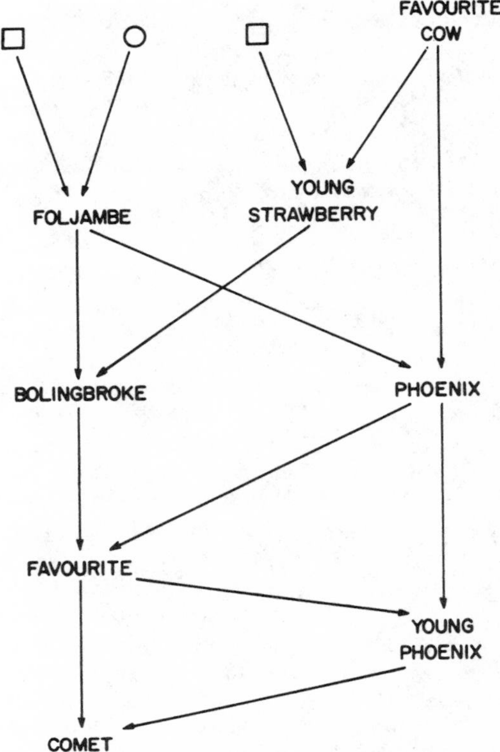

Figure 7. Pedigree of the Shorthorn bulls Favourite (252) and Comet (115). (Wright, 1977, figure 16.1).

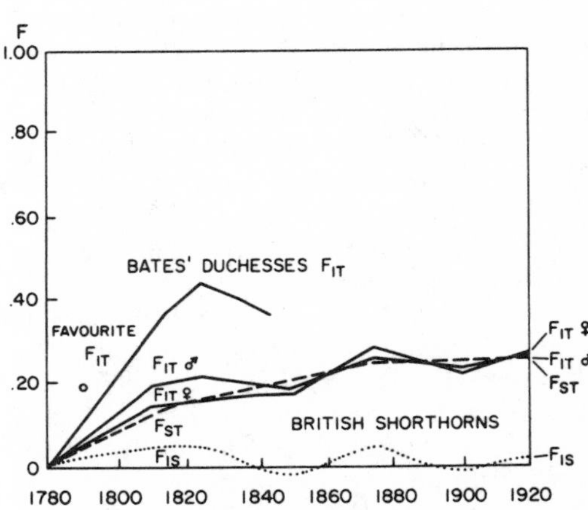

Figure 8. Inbreeding coefficients of British Shorthorn cattle from the foundation period 1780 to 1920. Inbreeding of bulls and cows relative to the foundation stock, F_{IT}, of the breed, and of the Bates' Duchesses, are in solid lines. Those for the hypothetical offspring of randomly mated animals, F_{ST}, are in a broken line. Those for individuals relative to the contemporary breed, F_{IS}, where $1 - F_{IT} = (1 - F_{IS})(1 - F_{ST})$ are in a dotted line. The inbreeding coefficient of the foundation bull, Favourite, is indicated by a circle. (Wright, 1977, figure 16.2).

results for the whole Shorthorn breed at intervals from about 1780, the foundation period, to 1920. This includes the coefficient F_{IT} (males and females separately) for inbreeding relative to the foundation stock, F_{ST}, derived by matching random two-line pedigrees of random sires and dams to find the cumulative inbreeding that would persist in spite of random breeding in the last generation, and F_{IS}, a derived coefficient, that measures current consanguine mating. The figure shows the strong inbreeding of Charles Colling's bull Favourite, born 1793, to which random animals in 1920, and doubtless 1977, were more closely related (.55) than offspring to sire under random mating. It also shows the extraordinarily high inbreeding of Bates' Duchesses, from the leading herd of the early 19th century. The relatively high values of current consanguine mating, F_{IS}, in the early years was followed by negative values in 1850. There was renewed excess inbreeding in 1875, but another dip at about 1900. These waves in F_{IS} undoubtedly represent alternate periods of development of strongly inbred herds and of prevailing crossing of animals as little related as possible within the breed.

I have described briefly the four rather unrelated lines of research that led me to a viewpoint on evolution very different from those propounded by Haldane and Fisher at about the same time, the 1920's. Let me indicate briefly, what it was that I derived from each of these lines.

From assisting Prof. Castle, I learned at firsthand the efficacy of mass selection in changing permanently a character subject merely to quantitative variability. Because of this and a distaste for miracles in science, I started with full acceptance of Darwin's contention that evolution depends mainly on quantitative variability rather than on favorable major mutations. Thus, I have assumed that species are typically heterallelic in tens of thousands of loci in which the leading alleles differ only slightly in effect, a situation that is maintained in a continually shifting state of near-equilibrium by the opposing pressures of recurrent mutation, diffusion and weak selection. Only a few loci at any time can show fairly rapid changes in allelic frequencies from strong selection.

My recognition of the efficacy of mass selection was, however, qualified by recognition that this process is likely to lead to deleterious side effects because of the interference of selection in any one respect with selection of the others.

From my studies of gene combinations, the second line, I recognized that an organism must never be looked upon as a mere mosaic of "unit characters," each determined by a single gene, but rather as a vast network of interaction systems. The indirectness of the relations of genes to characters insures that gene substitutions often have very different effects in different combinations and also multiple (pleiotropic) effects in any given combination. The latter consideration gives another reason for the tendency of mass selection to lead to deterioration.

The dependence of favorable interaction systems on multiple loci implies that if a selective value is attributed to each of the millions upon millions of possible sets of gene frequencies, the "surface" of selective values will have many peaks, each separated from the neighboring ones by saddles (figure 9). The elementary evolutionary process becomes passage from control by a relatively low selective peak to control by a higher one across a saddle, at first against the pressure of weak selection but under pressure of selection toward the new peak after the saddle has been passed. All of these considerations indicate that natural selection must somehow operate on combinations of interacting genes as wholes to be most effective. This raises a serious problem, however, since mass selection can operate only according to the net effects of substitutions under biparental reproduction (except for combinations of genes that are so closely linked that the combinations behave almost as if alleles).

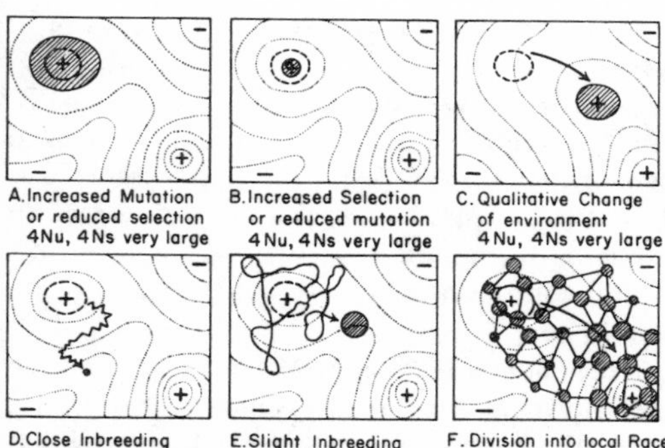

Figure 9. Hypothetical multidimensional field of gene frequencies (represented in two dimensions) with fitness contours. Field initially occupied by a population indicated by heavy broken contour except in case of multiple subdivisions in F. Field occupied later in A to E indicated by cross hatched area. Courses indicated in C, D, E, and F by arrows. Effective population number N (total), n (local). Coefficients u (mutation), s (selection), m (immigration). (Wright, 1977, figure 13.1, redrawn from Wright, 1932).

My studies of closely inbred lines of guinea pigs revealed the profound differentiations brought about in all respects by the cumulative accidents of sampling expected at all heterallelic loci under such inbreeding. Random fixation against the pressure of even rather strong selection under brother-sister mating accounts for the deterioration in random aspects. It shows, however, how a saddle leading to a higher selective peak may occasionally be crossed against the pressure of weak selection in small populations in which inbreeding is much less than under brother-sister mating. This suggested a solution to the difficulty of selection for interaction systems as wholes. There may often be subdivision of the species into small local populations that permit wide stochastic variability (but not fixation) at all nearly neutral loci and thus favor occasional local crossing of saddles leading to higher selective peaks. The firm establishment at this peak may then be brought about by mass selection. The spreading of this interaction system to neighboring localities may then be brought about by excess diffusion, followed again by mass selection after these have been pulled across the same saddle, and so on. The process may go on indefinitely as long as complete fixation does not occur. A major gene substitution and an array of modifiers that remove its deleterious side effects may also be established in this way.

It would have been very desirable to be able to point to studies of natural populations for evidence that such a process is actually occurring, or at least that the actual population structures are often favorable for its occurrence. Unfortunately there were few studies of differentiation within species except at the subspecies level at the time when I published the theory (1929, 1931). I had, however, been much impressed by Gulick's studies (1905) of what he called nonutilitarian differentiation in species of land snails in the numerous mountain valleys of Oahu in the Hawaiian Islands.

It was apparent, however, from my studies of the breeding history of Shorthorn cattle (the fourth line of research) that their improvement had actually occurred essentially by the shifting balance process rather than by mere mass selection. There were always many herds at any given time, but only a few were generally perceived as distinctly superior; those of Charles and Robert Collins near the end of the 18th century, those of Thomas Bates and the Booths in the first half of the 19th century, and that of Amos Cruickshank later in the century (figure 10). These herds successively made over the whole breed by being principal sources of sires.

My 1931 paper on evolution did not go into the factual side: the evidence for the multiple factor theory, for the importance of gene interaction and pleiotrophy or for the differentiation of inbred lines. Neither did I go into the exemplification on the process that I advocated in the history of breeds of livestock. I had discussed these elsewhere. The paper was almost wholly mathematical.

A

B

Figure 10. Prominent 19th Century Shorthorn sires. A: Cruickshank bull "Field Marshall" (47870). B: Bates' bull "Earl of Oxford, 3rd" (51186).

I attempted to introduce parameters for all of the factors into a single expression for the change of gene frequency, at least in token form, leading to an expression for the equilibrium frequency. Haldane had derived expressions for the change of gene frequency under various kinds of selection and integrated them to obtain curves describing the course toward fixation followed by the favorable allele under various conditions. Fisher's "fundamental theorem of natural selection" implies steady progress in exact accord with the steadily decreasing additive genetic variance as long as conditions remain unchanged. I did not derive any such expressions since under my view, evolution advanced by irregular, wholly unpredictable steps – the occasional occurrence off a peak-shift in some locality at some time, followed by relatively rapid spread throughout the species.

The mathematical emphasis in this paper was thus not on the course but on the momentary states of balance. The final summing up began as follows:

"Evolution as a process of cumulative change depends on a proper balance of the conditions which at each level of organization – gene, chromosome, cell, individual, local race – make for genetic homogeneity or genetic heterogeneity of the species."

After referring to the extensive field of variability always present in the state of balance at most loci, I went into the unlikelihood, under random mating, of evolution in very small populations or very large ones, or even in ones of any intermediate size. I stressed the great importance of subdivision into partially isolated local populations. I concluded in the final sentence that "conditions in nature are often such as to bring about the state of poise among opposing tendencies on which an indefinitely continuing evolutionary process depends."

I may have used the word "poise" in the final sentence partially because I had rather overworked the word "balance" (but not wholly). Balance may be maintained so firmly that further evolution is prevented. It itself becomes a product of evolution, an end result, as in most of the conspicuous polymorphisms. The word "poise" implies a lightly held state of balance at each moment, implying continual readiness to shift to a superior state of balance which is the essence of the shifting balance theory of evolution.

LITERATURE CITED

Bateson, W. 1909. Mendel's Principles of Heredity. Cambridge University Press, Cambridge.

Castle, W. E. 1903a. Mendel's Law of Heredity. Proc. Amer. Acad. Arts and Sci. 38.

Castle, W. E. 1903b. The laws of heredity of Galton and Mendel, and some laws governing race improvement by selection. Proc. Amer. Acad. Sci. 39:233.

Castle, W. E. 1916. Studies of inheritance in guinea-pigs and rats. Pub. No. 241. Carnegie Institution of Washington.

Castle, W. E. 1919. Piebald rats and selection: a correction. Amer. Naturalist 53:370.

Castle, W. E., F. W. Carpenter, A. H. Clark, S. O. Mast and W. M. Barrows. 1906. The effects of inbreeding, crossbreeding and selection upon the fertility and variability of Drosophila. Proc. Amer. Acad. Arts and Sci. 41:731.

Castle, W. E. and J. C. Phillips. 1914. Piebald rats and selection. Pub. No. 195. Carnegie Institution of Washington.

Darwin, C. 1859. The Origin of Species by Means of Natural Selection. John Murray, London.

East, E. M., 1910. A Mendelian interpretation of variation that is apparently continuous. Amer. Naturalist 44:65.

Fisher, R. A. 1930. The Genetical Theory of Natural Selection. Clarendon Press, Oxford.

Galton, F. 1889. Natural Inheritance. Macmillan and Co., London.

Gulick, J. T. 1905. Evolution, racial and habitudinal. Pub. No. 25. Carnegie Institution of Washington.

Haldane, J. B. S. 1924. A mathematical theory of natural and artificial selection. Part I. Trans. Cambridge Phil. Soc. 23:19.

Haldane, J. B. S. 1932. The Causes of Evolution. Longman, Green and Co., London.

Haldane, J. B. S. 1957. The Cost of Natural Selection. Genet. 55:511.

Loeb, Leo and S. Wright. 1927. Transplantation and individuality differentials in inbred strains of guinea pigs. Amer. J. Pathol. 31:251.

Mather, K., 1941. Variation and selection of polygenic characters. J. Genet. 41:159.

McPhee, H. C. and S. Wright. 1925. Mendelian analysis of the pure breeds of livestock, III The Shorthorns. J. Hered. 16:205.

Morgan, T. H. 1932. The Scientific Basis of Evolution. W. W. Norton and Co., Inc., New York.

Nilsson-Ehle, H. 1909. Kreuzungsuntersuchungen au Hafer und Weizen. Lunds Univ. Aerskr. N.F. 5:2:1–122.

Pearson, K. 1901. Mathematical contributions to the theory of evolution. VIII. On 'the inheritance of characters not capable of exact quantitative measurement. Part I. Introductory Part II. On the inheritance of coat-colour in horses. Part III. On the inheritance of eye-colour in man. Phil. Trans. Roy. Soc. London A 195:79.

Shull, G. H. 1908. The composition of a field of maize. Rep. Amer. Breeders Assoc. 4:290.

Strandskov, H. H. 1939. Inheritance of internal organ differences in guinea pigs. Genetics 24:722.

Strandskov, H. H. 1942. Skeletal variations in guinea pigs and their inheritance. J. Mammalogy 23:65.

Vries, Hugo de 1901–03. Die Mutationstheorie. 2 vol. Leipzig: Veit and Co.

Wright, S. 1916. An intensive study of the inheritance of color and other coat characters in guinea pigs with special reference to graded variation. Carnegie Institution of Washington: Pub. No. 241:59.

Wright, S. 1921. Systems of mating. Genetics 6:111.

Wright, S. 1922a. Coefficients of inbreeding and relationship. Amer. Naturalist 56:330.

Wright, S. 1922b. The effects of inbreeding and crossbreeding on guinea pigs. I. Decline in vigor. USDA Bull. 1090, p. 1. Washington, DC.

Wright, S. 1922c. Ibid. II. Differentiation among inbred families. USDA Bull. 1090, p. 37. Washington, DC.

Wright, S. 1922d. Ibid. III. Crosses between highly inbred families. USDA Bull. 1121. Washington, DC.

Wright, S. 1923. Mendelian analysis of the pure breeds of livestock. II. The Duchess family of Shorthorns as bred by Thomas Bates. J. Hered. 14:379.

Wright, S. 1929. Evolution in a Mendelian population. Anat. Rec. 44:87 (Abstr.).

Wright, S. 1931. Evolution in Mendelian populations. Genetics 16:97.

Wright, S. 1932. The roles of mutation, inbreeding, crossbreeding, and selection in evolution. Proc. VI. Intern. Cong. Genetics 1:356.

Wright, S. 1968. Evolution and the Genetics of Populations. Vol. I. Genetic and Biometric Foundations. Univ. of Chicago Press, Chicago, IL.

Wright, S. 1977. Evolution and the Genetics of Populations. Vol. III. Experimental Results and Evolutionary Deductions. Univ. of Chicago Press, Chicago, IL.

Wright, S. and P. A. Lewis. 1921. Factors in the resistance of guinea pigs to tuberculosis with especial regard to inbreeding and heredity. Amer. Naturalist 55:20.

Wright, S. and H. C. McPhee. 1925. An approximate method of calculating coefficients of inbreeding and relationship from livestock pedigrees. J. Agr. Res. 31:377.

Yule, G. U. 1902. Mendel's laws and their probable relation to intraracial heredity. New Phytol. 1:192, 222.

Yule, G. U. 1906. On the theory of inheritance of quantitative compound characters and the basis of Mendel's law: A preliminary note. Proc. III. Intern. Cong. Genetics, p. 140.

2
Coefficients of Inbreeding and Relationship

American Naturalist 56 (July–August 1922):330–38

Introduction

This paper set the theoretical framework for the "Mendelian Analysis of the Pure Breeds of Livestock" series that follows in this volume. Measuring the degree of inbreeding in a domesticated population had been recognized as an important problem by animal breeders for centuries, but only after the rise of Mendelian genetics in the early twentieth century was a quantitative measure feasible. Wright was unhappy with Raymond Pearl's measure of inbreeding, which was based upon a ratio of the actual smaller number of ancestors behind an inbred individual to the larger maximum possible number. The basic problem with Pearl's measure was that it could give the same degree of inbreeding with different systems of inbreeding that had very different biological consequences. Thus Wright set out to devise an inbreeding coefficient that was closer to the actual biological effects of inbreeding. He reasoned that the biological effects of inbreeding were caused by increasing homozygosity and that the best measure of inbreeding was therefore based upon the percentage decrease in heterozygosis.

The proper background for a full understanding of this paper requires an appreciation of Wright's method of path coefficients and its application to breeding pedigrees, no matter how irregular. I have addressed these issues in *SW&EB*, chapter 5; I would also recommend the following earlier papers by Wright: 7, 22–24, 27–31.

The quantitative measurement of inbreeding was crucial to Wright's theory of evolution in domestic breeds and evolution in nature. This paper is therefore a foundation stone of his theory of evolution. His use of the inbreeding coefficient *f* in this paper he later developed into his famous *F*-statistics, which provided a more detailed quantitative description of inbreeding in populations (see especially Wright 179, 196) and which proved to be extremely useful to animal and plant breeders as well as important for the elaboration of his shifting balance theory of evolution.

Although a generation of animal breeders used the method in this paper to compute inbreeding and relationship coefficients, Wright later (Wright 140) streamlined the method to make the computation of coefficients of inbreeding easier, especially with X-linked loci.

COEFFICIENTS OF INBREEDING AND RELATIONSHIP

DR. SEWALL WRIGHT

BUREAU OF ANIMAL INDUSTRY, UNITED STATES DEPARTMENT OF AGRICULTURE

IN the breeding of domestic animals consanguineous matings are frequently made. Occasionally matings are made between very close relatives—sire and daughter, brother and sister, etc.—but as a rule such close inbreeding is avoided and there is instead an attempt to concentrate the blood of some noteworthy individual by what is known as line breeding. No regular system of mating such as might be followed with laboratory animals is practicable as a rule.

The importance of having a coefficient by means of which the degree of inbreeding may be expressed has been brought out by Pearl[1] in a number of papers published between 1913 and 1917. His coefficient is based on the smaller number of ancestors in each generation back of an inbred individual, as compared with the maximum possible number. A separate coefficient is obtained for each generation by the formula

$$Z_n = 100\left(1 - \frac{q_{n+1}}{p_{n+1}}\right) = 100\left(1 - \frac{q_{n+1}}{2^{n+1}}\right)$$

where $q_{n+1}/2^{n+1}$ is the ratio of actual to maximum possible ancestors in the $n+1$st generation. By finding the ratio of a summation of these coefficients to a similar summation for the maximum possible inbreeding in higher animals, *viz.*, brother-sister mating, he obtains a single coefficient for the whole pedigree.

This coefficient has the defect, as Pearl himself pointed

[1] AMERICAN NATURALIST, 1917, 51: 545–559; 51: 636–639.

out, that it may come out the same for systems of breeding which we know are radically different as far as the effects of inbreeding are concerned. For example, in the continuous mating of double first cousins, an individual has two parents, four grandparents, four great grandparents and four in every generation, back to the beginning of the system. Exactly the same is true of an individual produced by crossing different lines, in each of which brother-sister mating has been followed. Yet in the first the individual will be homozygous in all factors if the system has been in progress sufficiently long; in the second he will be heterozygous in a maximum number of respects.

In order to overcome this objection Pearl has devised a partial inbreeding index which is intended to express the percentage of the inbreeding which is due to relationship between the sire and dam, inbreeding being measured as above described. A coefficient of relationship is used in this connection. These coefficients have been discussed by Ellinger[2] who suggests certain alterations and extensions by means of which the total inbreeding coefficient, a total relationship coefficient and a total relationship-inbreeding index for a given pedigree can be compared on the same scale.

An inbreeding coefficient to be of most value should measure as directly as possible the effects to be expected on the average from the system of mating in the given pedigree.

There are two classes of effects which are ascribed to inbreeding: First, a decline in all elements of vigor, as weight, fertility, vitality, etc., and second, an increase in uniformity within the inbred stock, correlated with which is an increase in prepotency in outside crosses. Both of these kinds of effects have ample experimental support as average (not necessarily unavoidable) consequences of inbreeding. The best explanation of the decrease in vigor is dependent on the view that Mendelian

[2] AMERICAN NATURALIST, 1920, 54: 540–545.

factors unfavorable to vigor in any respect are more frequently recessive than dominant, a situation which is the logical consequence of the two propositions that mutations are more likely to injure than improve the complex adjustments within an organism and that injurious dominant mutations will be relatively promptly weeded out, leaving the recessive ones to accumulate, especially if they happen to be linked with favorable dominant factors. On this view it may readily be shown that the decrease in vigor on starting inbreeding in a previously random-bred stock should be directly proportional to the increase in the percentage of homozygosis. Numerous experiments with plants and lower animals are in harmony with this view. Extensive experiments with guinea-pigs conducted by the Bureau of Animal Industry are in close quantitative agreement. As for the other effects of inbreeding, fixation of characters and increased prepotency, these are of course in direct proportion to the percentage of homozygosis. Thus, if we can calculate the percentage of homozygosis which would follow on the average from a given system of mating, we can at once form the most natural coefficient of inbreeding. The writer[3] has recently pointed out a method of calculating this percentage of homozygosis which is applicable to the irregular systems of mating found in actual pedigrees as well as to regular systems. This method, it may be said, gives results widely different from Pearl's coefficient, in many cases even as regards the relative degree of inbreeding of two animals.

Taking the typical case in which there are an equal number of dominant and recessive genes (A and a) in the population, the random-bred stock will be composed of 25 per cent. AA, 50 per cent. Aa and 25 per cent. aa. Close inbreeding will tend to convert the proportions to 50 per cent. AA, 50 per cent. aa, a change from 50 per cent. homozygosis to 100 per cent. homozygosis. For a natural coefficient of inbreeding, we want a scale which

[3] *Genetics*, 1921, 6: 111–178.

runs from 0 to 1, while the percentage of homozygosis is running from 50 per cent. to 100 per cent. The formula $2h-1$, where h is the proportion of complete homozygosis, gives the required value. This can also be written $1-2p$ where p is the proportion of heterozygosis. In the above-mentioned paper it was shown that the coefficient of correlation between uniting egg and sperm is expressed by this same formula, $f=1-2p$. We can thus obtain the coefficient of inbreeding f_b for a given individual B, by the use of the methods there outlined.

The symbol r_{bc}, for the coefficient of the correlation between B and C, may be used as a coefficient of relationship. It has the value 0 in the case of two random individuals, .50 for brothers in a random stock and approaches 1.00 for individuals belonging to a closely inbred subline of the general population.

In the general case in which dominants and recessives are not equally numerous, the composition of the random-bred stock is of the form x^2 AA, $2xy$ Aa, y^2 aa. The percentage of homozygosis is here greater than 50 per cent. The rate of increase, however, under a given system of mating, is always exactly proportional to that in the case of equality. The coefficient is thus of general application.

If an individual is inbred, his sire and dam are connected in the pedigree by lines of descent from a common ancestor or ancestors. The coefficient of inbreeding is obtained by a summation of coefficients for every line by which the parents are connected, each line tracing back from the sire to a common ancestor and thence forward to the dam, and passing through no individual more than once. The same ancestor may of course be involved in more than one line.

The path coefficient, for the path, sire (S) to offspring (O), is given by the formula $p_{o.s}=\frac{1}{2}\sqrt{(1+f_s)/(1+f_o)}$, where f_s and f_o are the coefficients of inbreeding for sire and offspring, respectively. The coefficient for the path, dam to offspring, is similar.

In the case of sire's sire (G) and individual, we have $p_{o.g} = p_{o.s}\, p_{s.g} = \frac{1}{4}\sqrt{(1+f_g)/(1+f_o)}$, and for any ancestor ($A$) we have for the coefficient pertaining to a given line of descent $p_{o.a} = (\frac{1}{2})^n\sqrt{(1+f_a)/(1+f_o)}$, where n is the number of generations between them in this line.

The correlation between two individuals (r_{bc}) is obtained by a summation of the coefficients for all connecting paths.

Thus

$$\begin{aligned} r_{bc} &= \Sigma p_{ba} p_{ca} \\ &= \Sigma \left(\frac{1}{2}\right)^{n+n'} \frac{1+f_a}{\sqrt{(1+f_b)(1+f_c)}}, \end{aligned}$$

where n and n' are the number of generations in the paths from A to B and from A to C, respectively.

The formula for the correlation between uniting gametes, which is also the required coefficient of inbreeding, is

$$f_o = \tfrac{1}{2} r_{sd}\sqrt{(1+f_s)(1+f_d)},$$

where r_{sd} is the correlation between sire and dam and f_s and f_d are coefficients of inbreeding of sire and dam. Substituting the value of r_{sd} we obtain

$$f_o = \Sigma\, (\tfrac{1}{2})^{n+n'+1}(1+f_a).$$

If the ancestor (A) is not inbred, the component for the given path is simply $(\frac{1}{2})^{n+n'+1}$ where n and n' are the number of generations from sire and dam respectively to the ancestor in question. If the common ancestor is inbred himself, his coefficient of inbreeding (f_a) must be worked out from his pedigree.

This formula gives the departure from the amount of homozygosis under random mating toward complete homozygosis. The percentage of homozygosis (assuming 50 per cent. under random mating) is $\frac{1}{2}(1+f_o) \times 100$.

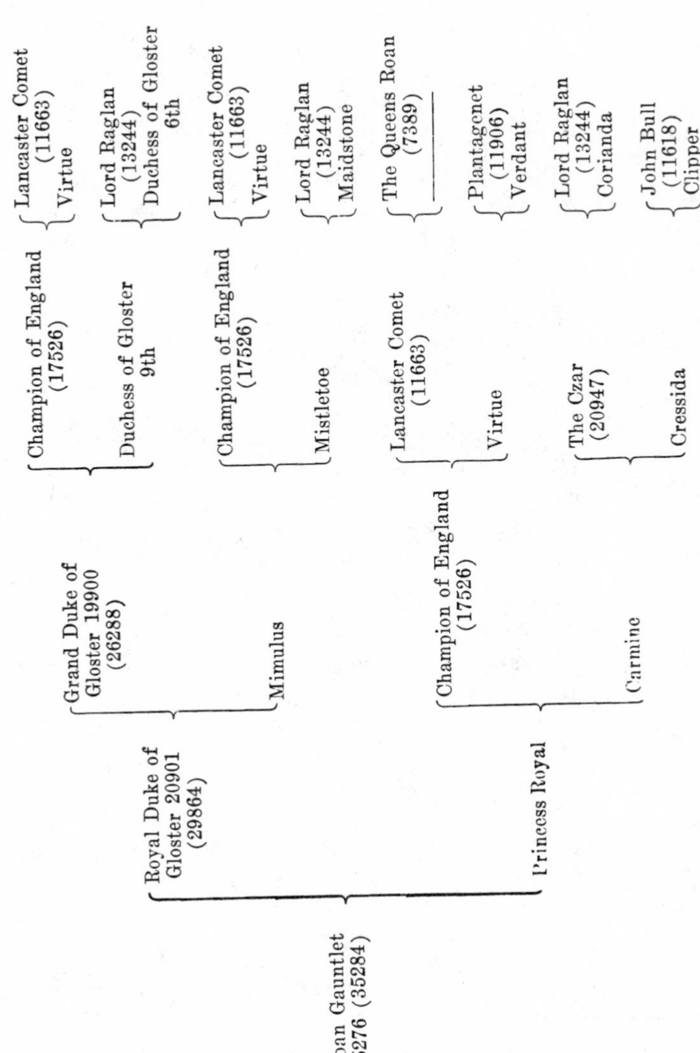
Roan Gauntlet
45276 (35284)
Royal Duke of
Gloster 20901
(29864)
Princess Royal
Grand Duke of
Gloster 19900
(26288)
Mimulus
Champion of England
(17526)
Carmine
Champion of England
(17526)
Duchess of Gloster
9th
Champion of England
(17526)
Mistletoe
Lancaster Comet
(11663)
Virtue
The Czar
(20947)
Cressida
Lancaster Comet
(11663)
Virtue
Lord Raglan
(13244)
Duchess of Gloster
6th
Lancaster Comet
(11663)
Virtue
Lord Raglan
(13244)
Maidstone
The Queens Roan
(7389)
Plantagenet
(11906)
Verdant
Lord Raglan
(13244)
Corianda
John Bull
(11618)
Clipper

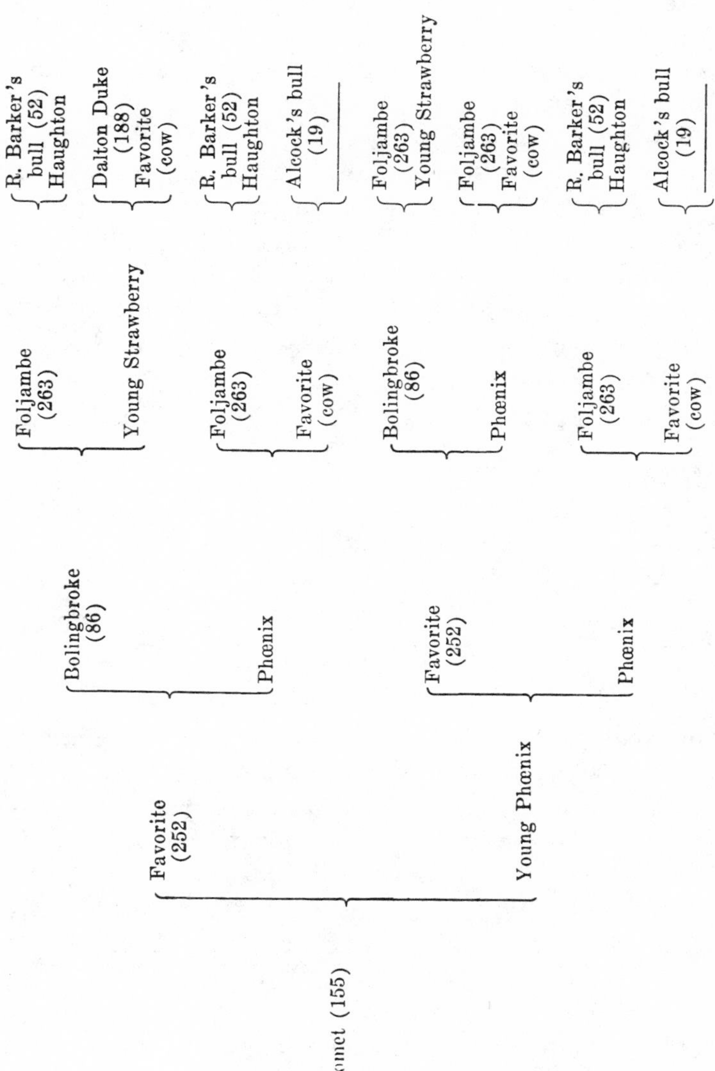

Comet (155)
Favorite (252)
Young Phœnix
Bolingbroke (86)
Phœnix
Favorite (252)
Phœnix
Foljambe (263)
Young Strawberry
Foljambe (263)
Favorite (cow)
Bolingbroke (86)
Phœnix
Foljambe (263)
Favorite (cow)
R. Barker's bull (52)
Haughton
Dalton Duke (188)
Favorite (cow)
R. Barker's bull (52)
Haughton
Alcock's bull (19)
Foljambe (263)
Young Strawberry
Foljambe (263)
Favorite (cow)
R. Barker's bull (52)
Haughton
Alcock's bull (19)

By this means the inbreeding in an actual pedigree, however irregular the system of mating, can be compared accurately with that under any regular system of mating.

As an illustration, take the pedigree of Roan Gauntlet, a famous Shorthorn sire, bred by Amos Cruickshank. This bull traces back in every line to a mating of Champion of England with a daughter or granddaughter of Lord Raglan. For the present purpose we will assume that these bulls were not at all inbred themselves and not related to each other. Since the sire traces twice to Champion of England and twice to Lord Raglan and the dam once to each bull, there are in all four lines by which the sire and dam are connected.

Individual	Common Ancestors of Sire and Dam	f_a	n	n'	$(\frac{1}{2})^{n+n'+1} \times (1+f_a)$
Roan Gauntlet 45,276 (35,284)	Champion of England (17,526)...........	0	2	1	.062500
			2		.062500
	Lord Raglan (13,244)..	0	3	3	.007812
			3		.007813
					.140625

The coefficient of inbreeding comes out 14.1 per cent., a rather low figure when compared to such systems as brother-sister mating (one generation 25 per cent., two generations 37.5 per cent., three generations 50 per cent., ten generations 88.6 per cent.) or parent-offspring mating, (one generation 25 per cent., two generations 37.5 per cent., three generations 43.8 per cent., approaching 50 per cent. as a limit).

As an example of closer inbreeding, take the pedigree of Charles Collings' bull, Comet. The sire was the bull Favorite and the dam was from a mating of Favorite with his own dam. As Favorite was himself inbred to some extent, it is necessary to calculate first his own coefficient of inbreeding.

Individual	Common Ancestors of Sire and Dam	f_a	n	n'	$(\frac{1}{2})^{n+n'+1} \times (1+f_a)$
Favorite (252)	Foljambe (263).......	0	1	1	.1250
	Favorite (cow)........	0	2	1	.0625
					.1875
Comet (115)	Favorite (252)........	.1875	0	1	.2969
	Phœnix..............	0	1	1	.1250
	Foljambe.............	0	2	2	.0312
	Favorite (cow)........	0	3	2	.0156
					.4687

In the case of Comet, Foljambe and Favorite (cow) each appears twice in the pedigree of the sire and three times in the pedigree of the dam. However, only those pedigree paths which connect sire and dam and which do not pass through the same animal twice are counted. The listing of Favorite (252) and Phœnix as common ancestors eliminates all but one path in each case as regards Foljambe and Favorite cow. The remaining paths are those due to the common descent of Bolingbroke, the sire's sire and Phœnix as the dam's dam from the above two animals.

By tracing the pedigrees back to the beginning of the herd book, the coefficients of inbreeding are slightly increased. This meant going back to the seventh generation for one common ancestor of the sire and dam of Favorite. The coefficient in the case of Favorite becomes .192 instead of .188 and that of Comet .471 instead of .469. Remote common ancestors in general have little effect on the coefficient. It will be noticed that Comet has a degree of inbreeding almost equal to three generations of brother-sister mating or an indefinite amount of sire-daughter mating where the sire is not himself inbred.

3

Mendelian Analysis of the Pure Breeds of Livestock I: The Measurement of Inbreeding and Relationship

Journal of Heredity 14, no. 8 (1923):339–48

4

II: The Duchess Family of Shorthorns as Bred by Thomas Bates

Journal of Heredity 14, no. 9 (1923):405–22

5

III: The Shorthorns

Journal of Heredity 16, no. 6 (1925):205–15

Introduction

In this series of three papers Wright presented his theory of the evolution of the Shorthorn breed of cattle. The series is pivotally important because Wright applied basically the same theory to evolution in natural populations. The third paper of the series appeared in the June 1925 issue of *Journal of Heredity*; by this time, Wright had already begun to write his long paper on evolution in natural populations, although it was not published until 1931.

From Castle's selection experiment on hooded rats, Wright had learned that mass selection (selection applied to all members of a randomly breeding population) was effective in changing the genetic constitution of a population, but was an extremely slow process. From the inbreeding experiment on guinea pigs, Wright learned that the process of inbreeding (in this case by brother-sister mating) produced a series of families that differed by distinctive characters because different interactive gene systems had become fixed in each. Although heterozygosity had decreased from the inbreeding, the visible variation between families, upon which selection could act, had been greatly increased. Wright concluded that the really effective breeder trying to create a new breed should practice a judicious amount of inbreeding (enough to fix different gene combinations in different subgroups, but not enough to reduce fecundity to dangerously low levels), followed by selection for the desired gene combinations. Transformation of the entire breed was then accomplished by the exportation of bulls from the founder herd.

The first test of Wright's theory of evolution in domestic breeds came from examination of the breeding records of Shorthorn cattle carried out by Wright and his assistant, Hugh Clyde McPhee. Using the method for calculating inbreeding coefficients that Wright had developed in the previous paper in this volume, the results of the analysis strongly supported Wright's theory. It should be noted, however, that later similar analyses of different breeds of hogs and other domestic animals by Jay L. Lush and his associates at Iowa State University revealed significantly lower levels of inbreeding during the foundation period than Wright and McPhee had discovered in Shorthorn cattle. Nevertheless, Lush, who was an extremely influential theorist of animal breeding, was strongly influenced by Wright's theory of evolution in domestic breeds (see *SW&EB*, 317–26).

MENDELIAN ANALYSIS OF THE PURE BREEDS OF LIVESTOCK

I. The Measurement of Inbreeding and Relationship

SEWALL WRIGHT

Bureau of Animal Industry, United States Department of Agriculture, Washington, D. C.

THE pure breeds of livestock which we have today are the result of many years of patient effort. It seems likely that they will furnish the material for further improvement for many years to come. It should thus be of value to study the methods used in their development and attempt to express in terms of modern genetics what these methods should have accomplished, and what, in consequence, is the present genetic status of the breeds. Before dealing directly with the work of certain of the leading breeders it will be well to consider briefly the successive phases in the history of livestock breeding, the light which discoveries in genetics has thrown on the earlier breeding methods, and finally, the analytic methods by which the data in livestock pedigrees can be related to Mendelian theory.

Successive Phases in Livestock Breeding

Livestock breeders did not have to wait for the development of a science of breeding to accomplish a great deal in the improvement of livestock. From the first, indeed, there must have been modification of the wild types through the retention of those animals which were most tractable, and it was merely necessary for the early shepherds and herdsmen to come to a realization of the great fact of heredity, for the conscious molding of animal forms and function toward greater usefulness to man to commence.

Primitive man naturally made no distinction between innate differences between animals and those due to differences in care and feeding, and believing that all were equally transmissible, concluded that good care and liberal feeding were short cuts to livestock improvement. Numerous other beliefs, partially true or wholly false, such as those concerning the injurious effects of matings between close relatives, the effect of maternal impressions, telegony, and so forth, contributed to the traditional lore of breeding. The realization of the fact of heredity, aided indirectly by some of the other beliefs, was undoubtedly enough to lead to a very considerable improvement.

Intensive efforts toward the improvement of local types of cattle and sheep began in England early in the eighteenth century. After the leaders in this movement had reached a certain degree of success, they began to find difficulty in securing animals from other herds and flocks which did not have a detrimental influence. They began cautiously to breed within their own stocks. Such experiments were conducted most boldly by Robert Bakewell,[1]* who discovered that he could practice close and continued inbreeding not only without necessarily causing degeneration of his stock, but with a rapid fixation of the types for which he was selecting. Those seeing his Longhorn cattle and Leicester sheep could not but see the difference between them and the stock of his neighbors. Their uniformity of type brought

*For numbered references, see *Literature Cited* at end of article.

clearly to the eye their features of superiority.

Bakewell's example was followed with enthusiasm by other breeders and closely bred strains of the best local types of animals began to appear in all parts of England. The Colling brothers began following this system with the shorthorned cattle of the north of England. Their work was carried on by other breeders, notably by Thomas Booth and his descendants and by Thomas Bates, with whose methods we are to deal in a succeeding paper.

While many of these early breeders had notable success with close breeding, injurious effects seem to have been encountered with such increasing frequency in later years as to discourage the practice. The superior types which had been developed in each region, were, however, maintained and developed into the pure breeds as we know them. Herd and flock books were established for the recording of pedigrees. Efforts at improvement took the form of selection within these pure breeds and in the grading up of common stock by the continued use of purebred sires. This is essentially the status of the art of livestock breeding today.

During the last quarter of a century a real science of breeding has been developing about the principles of heredity discovered by Mendel. Mendelian heredity has been found to be the regular mode of inheritance for all sorts of characters and in all sorts of organisms. We have come to believe that it is the general law of heredity under sexual reproduction.

With accurate knowledge of the principles of heredity, the hereditary characteristics of many plants and of the small and rapidly breeding laboratory animals can be controlled with something of the precision of the chemist working with non-living materials. Livestock breed relatively slowly, however, and the large generations obtainable from laboratory animals are not feasible. Genetic analysis must thus be slow. Nevertheless, genetics has an important contribution to practical breeding in the insight which it gives into the results of the long-known mass methods of breeding: assortative and disassortative mating, selection and culling, inbreeding and outcrossing. This is particularly true in the case of inbreeding and outcrossing.

The Effects of Inbreeding and Crossbreeding

The principal effect of inbreeding, we find, is in automatically making homozygous some combination of the factors which were heterozygous in the original random-bred stock. Immediately related to this is the increase in uniformity in an inbred stock, which makes it possible to recognize genetic differences which would otherwise be overlooked. Increased prepotency in outside crosses is another direct consequence of increase in homozygosis. A usual but not necessary decline in vigor in all respects is explained as due to a tendency for recessive factors, brought to light by increased homozygosis, to be more frequently deleterious than dominant factors. This tendency in turn is explained as due merely to the greater rapidity with which natural selection can eliminate the deleterious dominant variations that arise in a random-bred stock. These conclusions and their interpretations have been tested with diverse kinds of plants and animals. Among mammals, Miss King's experiments[2] at the Wistar Institute with inbreeding and selection in a line of rats have shown clearly that inbreeding does not necessarily lead to deterioration. The experiments of the Bureau of Animal Industry[3] with twenty-three different inbred lines of guinea pigs have also demonstrated that even twenty-five generations of brother-sister mating may not cause any obvious degeneration. They have, however, demonstrated that some decline with inbreeding is the usual result in such characters as weight, fertility and vitality. They have also brought out a conspicuous differentiation among different inbred lines in characters of the above kinds in which it has been almost impossible to demonstrate heredity otherwise. Another result has been the recovery of full vigor on crossing different inbred lines, explained as due to the complementary nature of these lines, each in general, supplying the particular dominant factors for vigor which had been lost in the other.

Characteristics differ greatly in the extent to which they are determined by heredity. Some, like most coat colors, are almost wholly hereditary. Others, like fertility and length of life are largely environmental as far as the individual is concerned. Inbreeding is as effective in making homozygous such heredity as there is in the latter case as in the former. Selection, on the other hand, is effective only in proportion to the importance of the hereditary element. The number of independent factors which effect the character also play a part in determining the effectiveness of selection methods With knowledge of the effects to be expected theoretically from the various methods of breeding, in dealing with characters determined to any given extent by heredity, and showing any degree of segregation in the second hybrid generation as compared with the first generation, and showing any degree of dominance, it should become possible to lay plans on a sound basis for the best system of mating to follow to obtain the results which are desired with any particular character.

The application of even these methods to livestock improvement must necessarily, however, be a rather slow and expensive process. Meanwhile it is important to check them as far as possible by study of the methods actually used by those breeders who have been recognized as having been most successful and find out in terms of modern genetics just what it was that they did.

Importance of Skill in Judging

That the breeders who laid the foundations for the pure breeds were exceptional among their contemporaries as judges of livestock and thus exceptionally skillful in making matings between animals which were really superior, we may take for granted as a big element in their success. The importance which Thomas Bates attributed to his ability in this direction may be inferred from a quotation from a letter, "A hundred men may be found to make a Prime Minister to one fit to judge of the real merits of Shorthorns." We can form some idea of the ideals which these men strove for from the descriptions and pictures of the noted animals which they bred. Recognizing the importance of their ability as judges, it is nevertheless difficult to put a quantitative measure on what they accomplished by selection.

Measurement of Inbreeding

We can, however, discover by study of pedigrees how far they practiced inbreeding and how far they maintained a general resemblance to particular worthy animals by concentrating their blood. We need to use measures of inbreeding and relationship which shall make it possible to relate the methods of these breeders directly to the modern theory of genetics. Such coefficients have been described in a previous paper.[4]

The coefficient of inbreeding (F) depends primarily on the number and closeness of the ancestral connections between the sire and dam and secondarily on the degree of inbreeding of the common ancestors of the latter. Every chain of generations in the pedigree by which one may trace back from the sire to a common ancestor and then forward to the dam, passing through no animal more than once (within the given chain), contributes to the inbreeding an amount equal to one-half used as a factor one more time than there are generations in the chain, with the qualification that this must be multiplied by a corrective term $(1 + F_A)$, in case the common ancestor (A) is himself inbred.*

It was demonstrated that this coefficient measures accurately the percentage departure from the number of homozygous factors in the random-bred stock toward complete homozygosis. If there were for example, 60 per cent homozygosis in the random-bred ancestral stock, a coefficient of inbreeding of 50 per cent means that the individual in question should be homozygous in 80 per cent of his factors.

The method of calculation may be made clearer by a few examples. Let us take first the simple case in which a mating is made between half brother and sister (Figure 1a). The sire and dam are connected in only one way. There are two generations in the connecting chain, *S-A-D*. As the common ancestor (A) is not himself inbred, the formula is simply $(\frac{1}{2})^3$. The coefficient of inbreeding is thus 12.5 per cent.

If the sire is a grandson and the dam a daughter of the common ancestor (Figure 1b), there are three generations between sire and dam and the coefficient is $(\frac{1}{2})^4$, or 6.25 per cent.

If now the mating is between full brother and sister (Figure 1c), the lattter are connected by two chains, each containing two generations. The formula requires us to find the sum of the contributions of each chain. We have then $(\frac{1}{2})^3 + (\frac{1}{2})^3 = \frac{1}{4}$ or 25 per cent as the coefficient of inbreeding. With two generations of brother-sister mating (Figure 2a), the reader will readily see that sire and dam are connected

*Letting F_X and F_A be coefficients for the individual and for a representative common ancester of his sire and dam, and letting n and n' be the number of generations between the sire and dam respectively and their common ancester, we have as the general formula:

$$F_X = \Sigma_A \left[\left(\frac{1}{2}\right)^{n+n'+1} (1+F_A) \right]$$

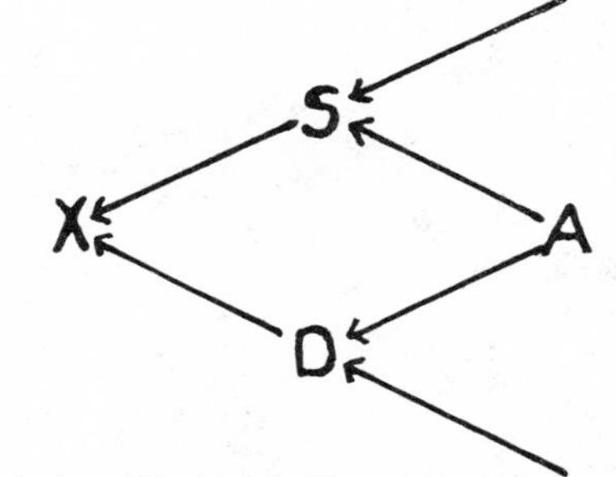

(a) A mating of half brother and sister
$F_X = 12.5\%$, $R_{SD} = 25\%$,
$R_{XA} = 47.1\%$

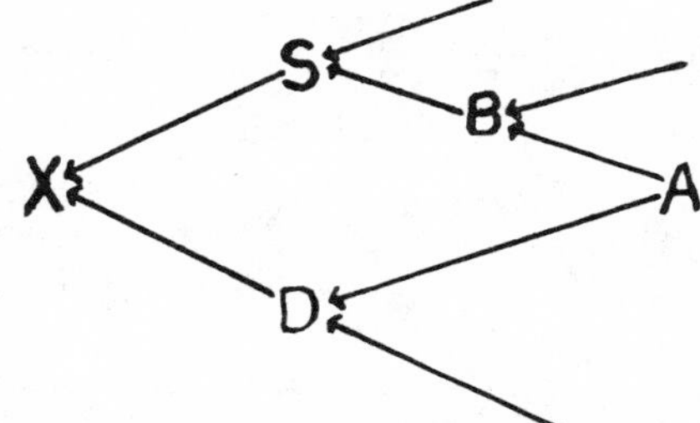

(b) A mating between grandson and daughter
$F_X = 6.25\%$, $R_{SD} = 12.5\%$,
$R_{XA} = 36.4\%$

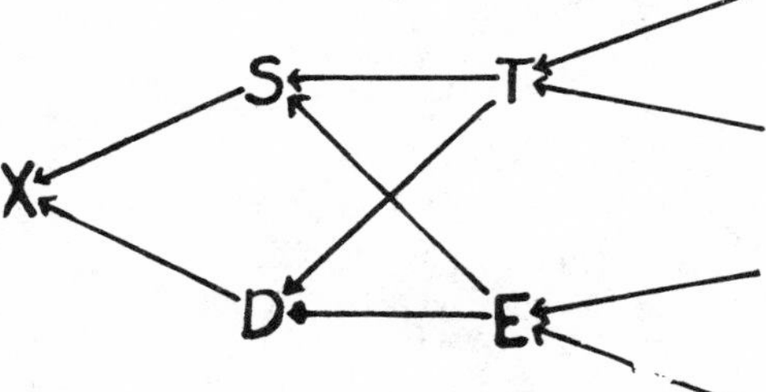

(c) A mating between full brother and sister
$F_X = 25\%$, $R_{SD} = 50\%$,
$R_{XT} = 44.7\%$

INBREEDING AND RELATIONSHIP

FIGURE 1. The chart shows three cases of inbreeding: mating of half brother and sister, mating between grandson and daughter, and mating of full brother and sister. The coefficient of inbreeding (F_X) of the resulting individual is given, in each case, also the coefficient of relationship between his sire and dam (R_{SD}), and between X and the ancestor to whom the inbreeding is due (R_{XA}). A coefficient of relationship of 50 per cent represents that existing between ordinary brothers and sisters.

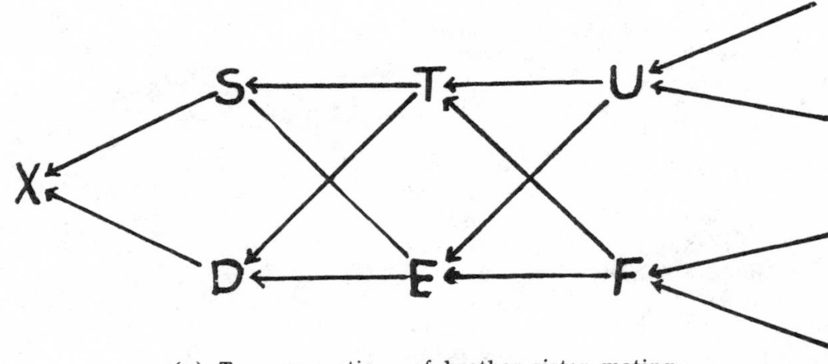

(a) Two generations of brother-sister mating

F_X=37.5%, R_{SD}=60%,
R_{XU}=42.7%

(b) Three generations of brother-sister mating

F_X=50%, R_{SD}=72.7%,
R_{XV}=40.8%

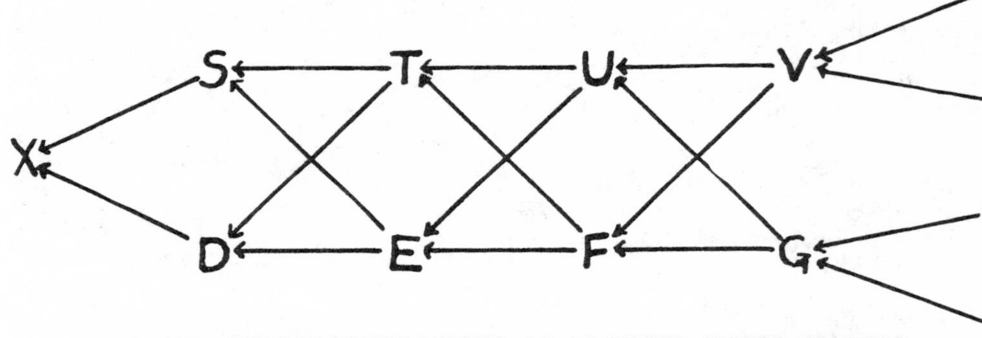

TWO AND THREE GENERATIONS OF BROTHER-SISTER MATING

Figure 2. For explanation of symbols see Figure 1. The effects of continued close breeding on homozygosis and resemblance due to relationship are analyzed in the text with the help of these diagrams.

through six independent chains, *S-T-D*, *S-E-D*, *S-T-U-E-D*, *S-T-F-E-D*, *S-E-U-T-D*, *S-E-F-T-D*. The formula is thus $2(\frac{1}{2})^3 + 4(\frac{1}{2})^5 = \frac{3}{8}$ or 37.5 per cent.

With three generations of brother-sister mating (Figure 2b) we encounter for the first time ancestors of the sire and dam which are inbred. There are as before two chains of two generations each by which sire and dam are connected through their parents, and in addition, four chains of four generations each, by which they are connected through their grandparents. Similarly we find eight chains of six generations each tracing to their great-grandparents. The parents of the sire and dam are themselves 25 per cent inbred, having one generation of inbreeding back of them. The formula thus becomes $2(\frac{1}{2}^3 \times 1.25) + 4(\frac{1}{2})^5 + 8(\frac{1}{2})^7 = \frac{1}{2}$. The individual *X* is 50 per cent inbred. The reader will have no difficulty in carrying the results to further generations and discovering that the limit under indefinitely continued inbreeding is 1, complete homozygosis, as it should be according

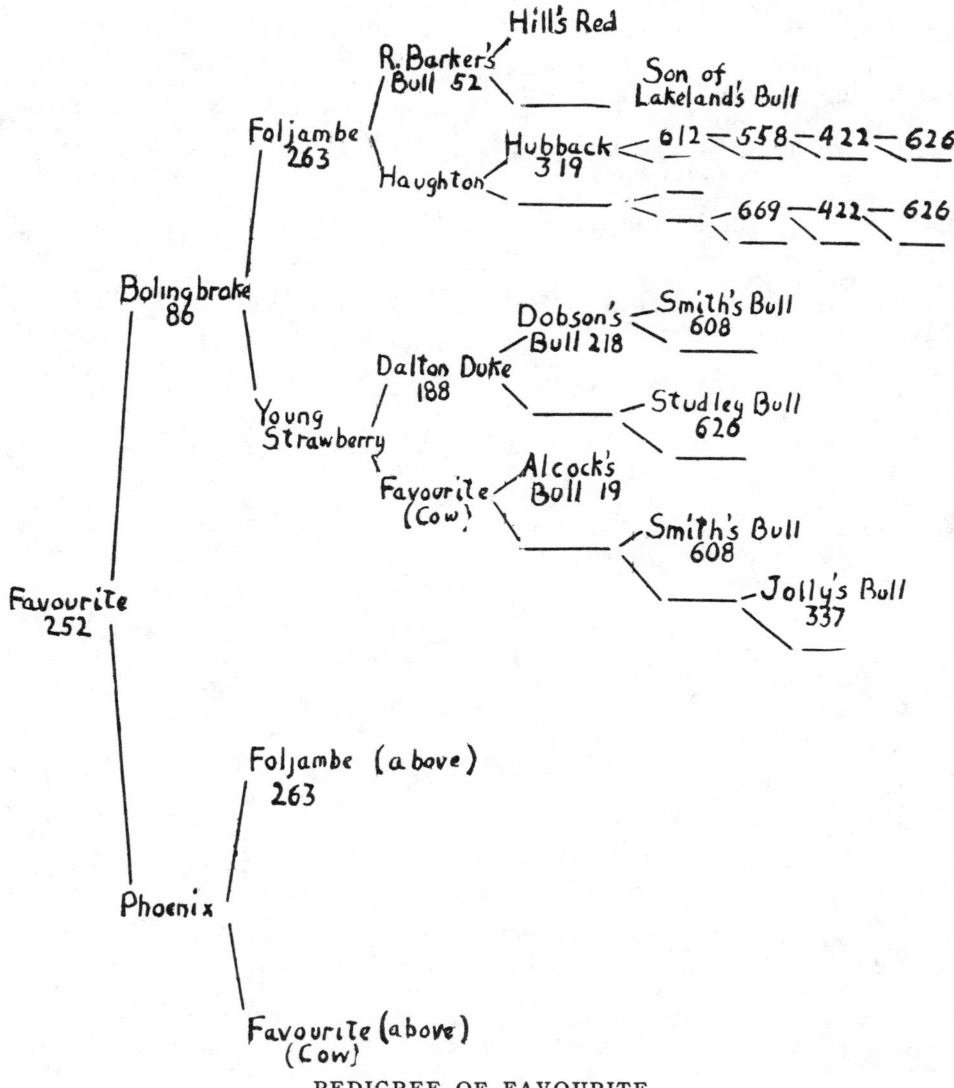

PEDIGREE OF FAVOURITE

FIGURE 3. Charles Colling's bull, Favourite 252, might be considered the foundation bull of the Shorthorns, as his blood became more widely distributed in the developing breed than that of any other bull. Favourite was inbred to a considerable extent, 19.2 per cent, which means that he was 19.2 per cent less heterozygous than the random-bred foundation stock. As an illustration of his influence in the early history of the breed it may be mentioned that forty years after Favourite's death there was a closer relationship between him and the cows of the famous Dutchess strain, bred by Thomas Bates, than between ordinary parent and offspring.

to theory. The results in regular systems of breeding, such as continued brother-sister mating, it should be said, can be worked out much more simply by special methods.[5] The case is merely brought up here to show the application of the general formula which must be used in dealing with the irregular systems encountered in ordinary pedigrees.

The Pedigree of Favourite

As an example of such an irregular case, let us take the pedigree of the bull Favourite (Figure 3). None of the ancestors of the sire and dam (Bolingbroke and Phoenix) are shown as inbred, so that we merely need to count the generations in each path connecting sire and dam, add one and use as an exponent of ½ to obtain the contribution of each path to the coefficient.

We find that Foljambe was the sire of both Bolingbroke and Phoenix. The contribution due to this connection is thus $(½)^3$ or 12.5 per cent. Foljambe as Bolingbroke's sire is not connected with the dam of Phoenix. This disposes of all possible connections through Foljambe as the sire's sire. The sire's dam, Young Strawberry, is connected twice with Foljambe as the dam's sire through common descent from the Studley Bull (626). The latter is four generations back of Bolingbroke and seven back of Phoenix in each of two lines through her sire. These contributions are thus $(½)^{12} + (½)^{12}$ or 0.05 per cent, an almost negligible quantity. We have finally to consider the connections between sire's dam (Young Strawberry) and dam's dam (Favourite Cow). First we note that Favourite Cow was herself the dam of Young Strawberry. We have here a contribution of $(½)^4$ or 6.25 per cent. The sire of Young Strawberry is connected with Favourite Cow through descent from Smith's Bull. The contribution is $(½)^8$ or 0.39 per cent. This disposes of all connections. The work may be arranged as follows:

Inbreeding of Favourite (252)

Common Ancestor		Generation from		
Name	Inbreeding	Bolingbroke	Phoenix	Contribution
Foljambe	0	1	1	12.50
Studley Bull	0	4	7,7	0.05
Favourite Cow	0	2	1	6.25
Smith's Bull	0	4	3	0.39
Total				19.19

The total percentage of inbreeding is 19.2. It will be noticed that this would not be appreciably modified by omission of the remote connections through descent from Studley Bull and Smith's Bull. This coefficient means that Favourite departs 19.2 per cent from the genetic heterogeneity of the foundation Shorthorn stock in the direction of complete genetic homogeneity in other words has 19.2 per cent less heterozygosis.

Comparison with Other Measurements of Inbreeding

As the coefficient of inbreeding is derived from theoretical considerations, it will be well to consider for a moment how it is related to the degree of inbreeding in the popular sense. This is a somewhat difficult question to answer since inbreeding in the popular sense is a rather vague term. Pearl[6] has defined inbreeding on the basis of the ratio of the number of ancestors in each ancestral generation to the maximum possible number. The smaller the ratio the greater the inbreeding. This is a very different conception from the writer's, which is based on the reduction of genetic heterogeneity present in the foundation stock and is hence in the main a function of the ancestral connections between the parents. The contrast in the two conceptions can be seen very clearly by comparing the two pedigrees of Figure 4. The upper represents a mating be-

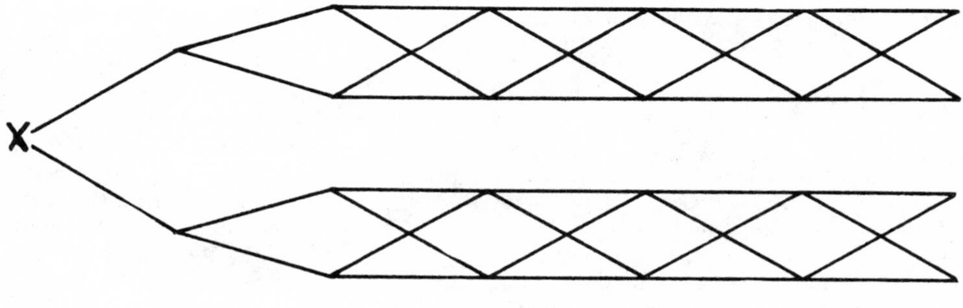

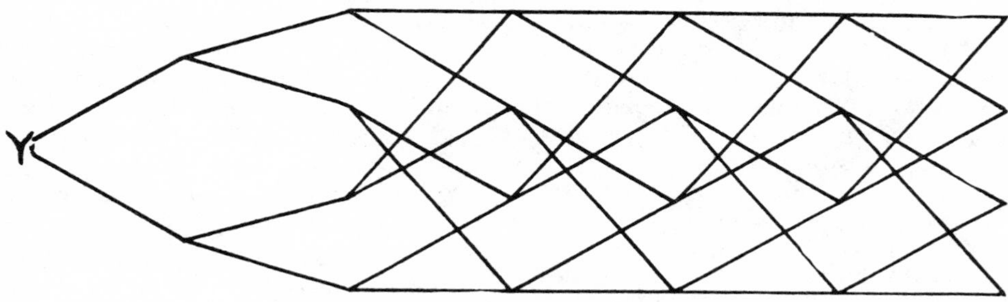

TWO METHODS OF BREEDING

FIGURE 4. Two pedigrees in which the number of ancestors in each generation remains the same, but in which the coefficient of inbreeding (F) of the individual X equals 0, while that of Y approaches 100 per cent. This is in accord with the results of experiments in inbreeding, where crossing two weakened inbred lines results in the return of normal vigor. Methods of calculating inbreeding that are based only on the ratio of the number of ancestors in each generation to the greatest possible number would not bring out this difference.

tween two individuals, each of which comes from a line in which brother-sister matings have been made generation after generation. The other comes from a line in which double first cousin matings have been made for generations. In each case there are two parents, four grandparents, four great grandparents, and four in every ancestral generation to the point where inbreeding commenced. They both, therefore, have the same very high coefficient of inbreeding according to Pearl's formula. According to the writer's formula, on the other hand, they are nearly as different as possible. The former has a coefficient of 0, the latter of nearly 1. The former is heterozygous in all respects in which the ancestral lines differ; the latter approaches perfect homozygosis.

With respect to vigor, the experiments of the Bureau of Animal Industry with guinea pigs, already referred to, demonstrate that the crossing of two weakened inbred lines results in a recovery of the vigor that has been lost. Thus generally speaking, the animal X (Figure 4), would be a vigorous animal; the animal Y would be weak in various respects. The experimental evidence is thus in accord with the coefficients 0 and 1 while as far as the writer knows there is no experimental evidence to indicate that the number of ancesters *per se* has anything to do with the uniformity, vigor, prepotency or other characteristics.

Even if the measurement of inbreeding is based on the ancestral connections between sire and dam, there remain many possible formulae which could be suggested.

Perhaps the first thought would be to base the coefficient on the percentage of common blood in sire and dam. A little consideration will show, however, the unreliability of such a measure. Full brother and sister have 100 per cent common blood, but we know that much greater effects can be produced by continued brother-sister mating than appear in the first generation. Indeed, there might be 100 per cent common blood with no common ancestors of sire and dam for an indefinite number of generations. Let us suppose that the sire and dam trace to the same 1,024 individuals ten generations back, but have no common ancestors in later generations. Common sense and experiment tell us that their progeny could hardly be considered to be inbred at all. The coefficient of inbreeding of their progeny comes out only $1024 \times (\frac{1}{2})^{21} = (\frac{1}{2})^{11} = 0.05$ per cent in spite of the 100 per cent common blood.

Pearl[6] has attempted to separate from the amount of inbreeding as found by his formula a portion due only to relationship between the parents, this relationship being based on the number of animals common to the pedigrees of sire and dam in proportion to the greatest possible number of common ancestors. This comes closer to the writer's conception of inbreeding but in practice gives very different results from that given by the coefficient F. For example, continued mating of single first cousins gives results rapidly approaching 100 per cent according to this or any other of Pearl's coefficients, while the percentage of heterozygosis is decreased only by about 25 per cent in ten generations ($F = 25.2\%$). As far as the writer knows there is no purely experimental evidence to determine whether such a system as continued mating of single first cousins really approaches the effect of brother-sister mating in a few generations or whether it has the relatively slight effects indicated by the value of F. The writer, however, has sufficient confidence in the generality of Mendelian inheritance to believe that it is an advantage to use a coefficient which measures directly the consequences logically to be expected from it as well as measuring accurately the results of experiments in inbreeding and crossbreeding as far as these have been carried.

Measurement of Relationship

The measurement of relationship* is naturally closely related to measurement of inbreeding. In order to meas-

*Following is the formula where n and n' are the generations from the individuals X and Y to a given common ancestor A, and F_A, F_X and F_Y are the coefficients of inbreeding of A, X and Y respectively:

$$R_{XY} = \frac{\Sigma\left[\left(\frac{1}{2}\right)^{n+n'}(1+F_A)\right]}{\sqrt{(1+F_X)(1+F_Y)}}$$

In the important case of sire and dam of the individual X we have:

$$R_{SD} = \frac{2F_X}{\sqrt{(1+F_S)(1+F_D)}}$$

Thus if the sire and dam are related but not themselves inbred, the coefficient of relationship is just twice the coefficient of inbreeding of their progeny.

ure the relationship between animals in such a way that direct comparison can be made with Mendelian theory, the most satisfactory method seems to be to find the coefficient of correlation to be expected with respect to a character determined wholly by heredity with no dominance. Such conditions give the maximum correlation and any other conditions give results strictly proportional. The correlation between parent and offspring or between full brothers should be 0.50 in such cases, according to Mendelian theory, in close agreement with many actual determinations. The correlation to be expected when the relationship is more complex can readily be calculated.[5]

Literature Cited

[1] The statements in regard to the early history of the Shorthorn cattle are largely drawn from *Shorthorn Cattle* by Alvin H. Sanders and *Farm Livestock of Great Britain* by Robert Wallace.

[2] King, Helen Dean, 1919. Studies on Inbreeding. The Wisner Institute of Anatomy and Biology, 175 pages. Reprinted from the *Journal of Experimental Zoology*. 26:1-98; 27:1-35, and 29:134-135.

[3] Wright, S., 1922. The Effects of Inbreeding and Crossbreeding on Guinea Pigs. *Bulletins* 1090 and 1121. U. S. Department of Agriculture.

[4] Wright, S., 1922. Coefficients of Inbreeding and Relationship. *American Naturalist*, 56:330-338.

[5] Wright, S., 1921. Systems of Mating. *Genetics*, 6:111-178.

[6] Pearl, R., 1917. Studies on Inbreeding. *American Naturalist*, 51:545-559; 51:636-639.

MENDELIAN ANALYSIS OF THE PURE BREEDS OF LIVESTOCK

II. The Duchess Family of Shorthorns As Bred By Thomas Bates

SEWALL WRIGHT

Bureau of Animal Industry, United States Department of Agriculture, Washington, D. C.

A VARIETY of cattle with short horns and red, roan, and white as the prevalent colors had long been established in parts of Durham, Yorkshire and surrounding counties in England and had been bred to a fair degree of excellence by the middle of the eighteenth century. They have been described from contemporary accounts as "generally wide-backed, well-framed cows, deep in their fore-quarters, soft and mellow in their hair and handling and possessing, with average milking qualities, a remarkable disposition to fatten. Their horns were rather longer than those of their descendants of the present day and inclining upward. The defects were those of an undue prominence of the hip and shoulder point, a want of length in the hind quarters, of width in the floor of the chest, of fullness generally before and behind the shoulders, as well as of flesh upon the shoulder itself. They had a somewhat disproportionate abdomen, were too long in the legs and showed a want of substance, indicative of delicacy, in the hide. They failed also in the essential requisite of taking on their flesh evenly and firmly over the whole frame, which frequently gave them an unlevel appearance. There was, moreover, a general want of compactness in their conformation." This foundation stock was thus decidedly open to improvement. In fact, Robert Bakewell of Dishley, whose leadership in the improvement of cattle and sheep we have already mentioned[2]* is said to have kept a few of the old sort merely to set off his improved Longhorns.

The foundations for the improvement of the Shorthorns were laid by Charles Colling of Ketton farm, who made a prolonged study of Bakewell's methods at Dishley in 1783. In the following year he bought, among other cattle, a massive, short-legged, wide-backed cow named Duchess, who became the progenitress of the family which we are to study. Meanwhile Robert Colling, a brother of Charles, purchased the bull Hubback, who came to be considered the best Shorthorn bull of the time. One of his grandsons was the noted bull Foljambe. Charles Colling seems to have been sufficiently impressed with the merit of the produce of Foljambe to begin efforts at fixing their qualities by inbreeding. A daughter, Phoenix, was bred to a son, Bolingbroke, and produced (in 1793) the bull Favourite, whose pedigree we have already considered. Favourite was inbred to a very appreciable extent (F = 19.2 per cent), not only through Foljambe but through his granddams, one of which was the dam of the other. He is described as "a large massive bull of good constitution with a fine bold eye, remarkably good loins and long level hind quarters." Mr. Colling was so satisfied with Favourite that he began breeding him to his daughters and granddaughters, in some cases even for five or six generations. Through the demand for

*For numbered references see "Literature Cited" at end of article.

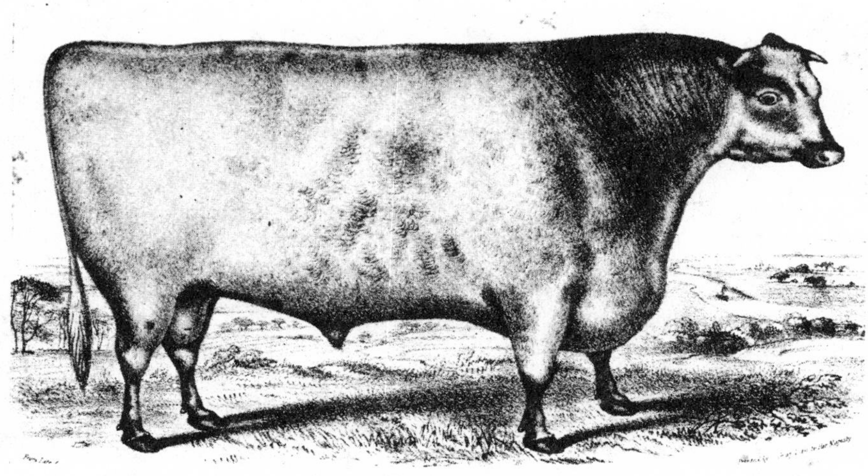

DUKE OF NORTHUMBERLAND

FIGURE 13. Bates is considered to have won his greatest triumph as a breeder in producing the Duke of Northumberland (1940) from a mating of Duchess 34th with her own sire, Belvidere.* Duke of Northumberland won first prize at the Oxford Royal Show of 1839 and was generally conceded to be the best bull in England in his time. Without taking the old lithographs too seriously, we may infer from them that Bates was aiming at a compact build, smoothness, and great fineness of bone, and that he came near achieving his ideal in Duke of Northumberland.

---

*Correct spelling **Belvedere.** Error in spelling this name noted just before going to press, and too late to change.

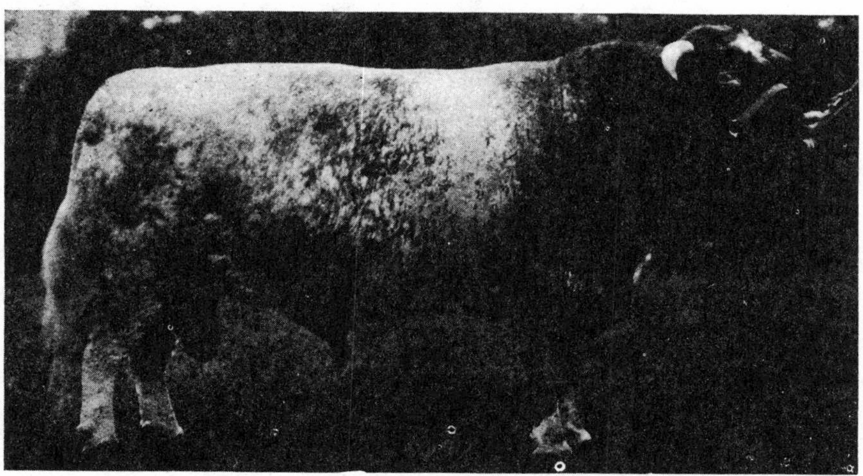

A PRESENT DAY SHORTHORN CHAMPION

FIGURE 14. A comparison between the photograph of this bull and the lithographs of the Bates' cattle brings out the greater ruggedness of type that has come into favor, even after due allowance is made for the obvious exaggeration of the old illustrations.

bulls of Colling's improved stock, his blood undoubtedly became more widely distributed in the developing breed than that of any other one animal. Darwin[1] cites his case as an illustration of prepotency. As we shall see, Bates maintained a coefficient of relationship to Favourite of about 60 per cent in his Duchess family for forty years after Favourite's death (in 1809).

At the dispersion sale of Charles Colling's herd, in 1810, forty-seven head, all by Favourite or his son Comet or their get, sold for £7,115, an unprecedented amount. The record price, 1,000 guineas, was paid for Comet, a son of Favourite, dam by Favourite and out of Favourite's dam. This remarkably inbred animal ($F = 47.1$ per cent) was declared by Mr. Colling to be the best bull he ever bred or saw. For years after, it seems to have been the general opinion that no Shorthorn bull had ever been quite the equal of Comet. Mr. Colling's stock are described by a contemporary, as of great size and substance, fine long hind quarters, space from hip to rib long and counteracted by a broad back and high round ribs.

The Duchesses

One of the bidders at this dispersion sale was a young breeder, Thomas Bates, who had made a careful study of cattle pedigrees. He purchased a rather "shabby" cow called Young Duchess, largely, it appears, on the strength of her pedigree, her top sires being Comet, Favourite, Daisy Bull (a son of Favourite), Favourite again, and Hubback. She was a descendent in the straight female line of the cow Duchess purchased by Charles Colling in 1784. Following the custom of naming families by the female line, Bates developed a Duchess family from this cow, which he renamed Duchess I. Up to the time of his death in 1849 he had bred sixty-three cows in the family which he named Duchess 2 to Duchess 64. Two other females are recorded as having died. Forty-five males are recorded as dropped by Duchess cows, twenty-nine of which were named. The family was not a prolific one. They won, however, an extraordinary reputation. The Duchess bull, Duke of Northumberland, which won first at the Oxford Royal Show of 1839, was conceded to be the best bull in England in his time. Bates had other notable successes in the show yard although an opponent of the system of specially fitting for the shows.

This reputation extended to America and Bates-bred cattle, especially bulls of the Duchess family, played a notable part in the improvement of American Shorthorns.

After Bates' death, a line of Duchesses was maintained without outcrossing. These became the aristocrats of the cattle world. The family had never been a prolific one and became increasingly difficult to maintain as a pure strain. This, however, does not seem to have been held as a detriment but rather the reverse, since it resulted in enhanced values due to scarcity. The climax came in a sale at New York Mills, near Utica, N. Y., in 1873. The "pure" line of Duchesses had become extinct in England and all in America had come into the hands of one man. The sale at New York Mills developed into an international competition for the "pure" Duchesses. One cow sold for $40,600. The average of the eleven Duchess cows was $21,705, that for three bulls was $7,866.

While these prices were largely speculative and the real merit of the family was falling off, in the hands of the speculators as the prices mounted higher, the fact remains that as bred by Bates, the Duchess cattle were a most notable achievement of breeding skill. Through crosses with other lines they have had a conspicuous part in the improvement of the Shorthorn breed.

Description of Tables

The sixty-three Duchesses bred by Bates, and their ancestors in the straight female line are listed in Table II with

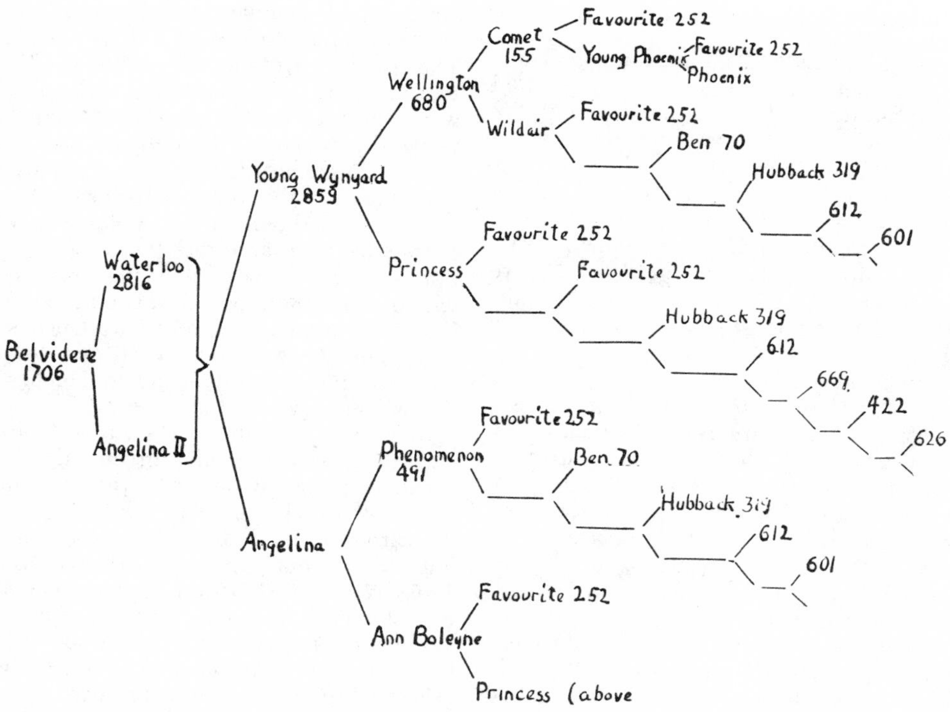

THE PEDIGREE OF BELVIDERE

FIGURE 15. The purchase of Belvidere (1706) from Mr. Stevenson is considered to mark the turning point in Bates' career. Note that Belvidere's sire and dam were full brother and sister, making possible condensation of the pedigree as indicated.

their date of birth and the dam and sire of each. The coefficient of inbreeding has been calculated as described[2], for all of these animals and is given in the next three columns. The relationship of the sires and dams of the Duchesses is given in the next column. Finally it is of interest to find how far there was relationship to Favourite, the bull to which the greater part of the inbreeding is due. The last three columns are devoted to this purpose, dealing with the individual and her sire and dam respectively.

All pedigrees have been traced to the beginning of the Coates' herd books in making these calculations. They naturally become very complex after a few generations. The labor in calculating the coefficients is not so great as might be imagined, however, since the calculation of the contribution to inbreeding or relationship made by two animals on opposite sides of a pedigree can be used in all other pedigrees in which these same animals appear on opposite sides. In the course of the work a list was made of the contributions due to the various important sires with each other and with the more important cows. The parentage of the sires is given in Table I. It will be noticed that all but four of the twenty-five of them were from sires that appear earlier in the list and that eleven of them were from Duchess cows. The pedigrees can be practically completed by reference to the notes and to the pedigrees of Favourite[2], 2nd Hubback, Belvidere, Gambier and Norfolk (Figures 15-18).

Relationship of Sires and Dams

The results are summed up in Table III by generations and are presented graphically in Figures 21, 24 and 25. Duchess 59 and 62 were eight generations from Duchess I, and thirteen generations from the original Duchess purchased by Charles Colling in 1784.

The coefficients of relationship between sire and dam are given in Figure 21. For three generations there was little or no relationship. For Duchess by Daisy Bull in the next generation the coefficient rises to 39 per cent, both sire and dam having been progeny of Favourite. The parents of Duchess by Favourite and Duchess I (by Comet) in the next two generations show coefficients of 53 per cent and 59 per cent respectively. These figures are to be compared with coefficients of 50 per cent between ordinary brothers and sisters or between parent and offspring in a random-bred stock. They measure the actual correlation in appearance in characters which are wholly determined by nondominant genetic factors. They are doubtless about what the reader would expect from a cursory examination of the pedigrees.

It is not so obvious, however, what the correlations in later generations should be. The general opinion has been that Bates began his career with very intensive inbreeding and on encountering deterioration was obliged to make outcrosses. Darwin[1] states, "For thirteen years he bred most closely in-and-in; but during the next seventeen years, although he had the most exalted notion of his own stock, he thrice infused fresh blood into his herd; it is said that he did this not to improve the form of the animals, but on account of their lessened fertility."

The calculation of the actual relationships of the sires and dams, however, shows no such history. The sire and dam of Bates' purchase, Duchess I, were correlated about 0.59 ($R = 59$ per cent) on account of relationship. During the eight generations bred by Bates, he practically maintained the same level of relationship, the coefficients fluctuating about 60 per cent. The parents of the ten cows of the second generation after Duchess I were correlated 0.67, the high point, while the parents of the nine cows of the fifth generation were correlated only 0.52 at the other extreme. The average for the next generation, however, was 0.64 and the eighth generation averaged 0.63. Turning to individuals, the high point was reached in the mating which produced Duchess 21, where the coefficient is 87 per cent. The low point was reached in Duchess 41, where it was 42 per cent.

The purchase of the "Princess" bull Belvidere is generally given as the turning point in Bates' career, the point at which he found it necessary to introduce fresh blood and the point with which his greatest success began. It is true that Belvidere had no Duchess blood. He was, however, rich in the blood of Favourite as may be seen from his pedigree (Figure 15). His relationship with Favourite was about 65 per cent. In fact his relationship with Duchess 19 and with Duchess 29, the cows of the Duchess family with which he was first mated, were as high as 45 per cent and 47 per cent respectively.

Previous to the use of Belvidere, outside blood was introduced through 2nd Hubback. However, while Red Rose I, his dam, had no Duchess blood, she was closely related genetically (see pedigree, Figure 16). Moreover the sire of 2nd Hubback was the Duchess bull, The Earl. Second Hubback should have shown a correlation of about .65 with Favourite as far as all purely hereditary characters were concerned and correlations .49 and .62 with the first Duchesses with which he was mated.

Fresh blood was introduced later through the bulls Gambier and Norfolk, bred by Jonas Whitaker, which were without Duchess blood on either side of their pedigrees (Figures 17 and 18). Nevertheless both of these bulls were closely related to Favourite (65 per

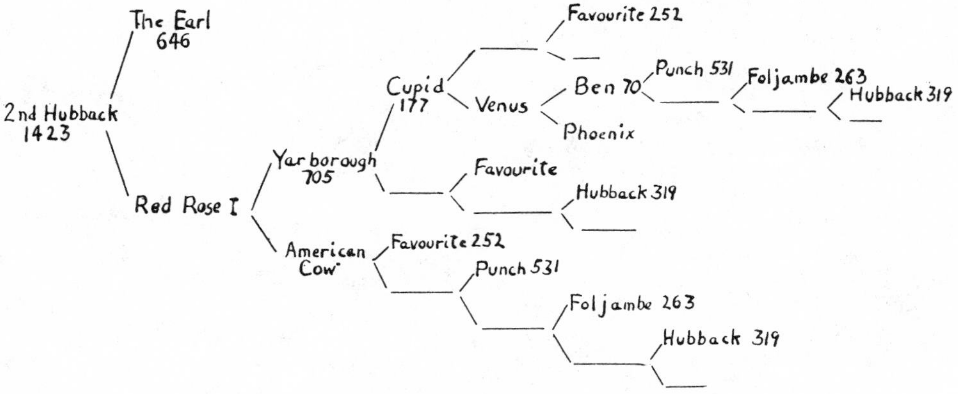

The Earl 646
2nd Hubback 1423
Red Rose I
Cupid 177
Yarborough 705
American Cow
Venus
Favourite 252
Ben 70
Phoenix
Punch 531
Foljambe 263
Hubback 319
Favourite
Hubback 319
Favourite 252
Punch 531
Foljambe 263
Hubback 319

THE PEDIGREE OF 2ND HUBBACK

FIGURE 16. Second Hubback (1423) was one of the bulls used by Bates to keep the degree of inbreeding in his Duchess herd from rising too high. The pedigree of 2nd Hubback's sire is given in Table I and is not repeated here.

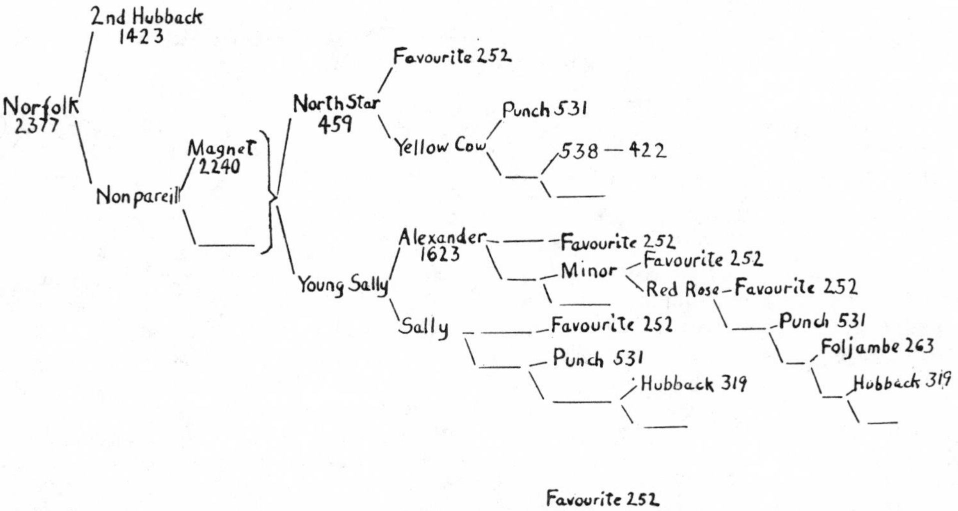

2nd Hubback 1423
Norfolk 2377
Nonpareil
Magnet 2240
North Star 459
Young Sally
Favourite 252
Yellow Cow
Punch 531
538 — 422
Alexander 1623
Sally
Favourite 252
Minor
Favourite 252
Red Rose — Favourite 252
Favourite 252
Punch 531
Punch 531
Foljambe 263
Hubback 319
Hubback 319
Favourite 252

PEDIGREE OF NORFOLK

FIGURE 17. The bull Norfolk (2377), bred by Mr. Whitaker, had no Duchess blood, but was closely related to Favourite, as the pedigree shows. The pedigree of the sire, 2nd Hubback, is given above, and in Table I and hence is not repeated.

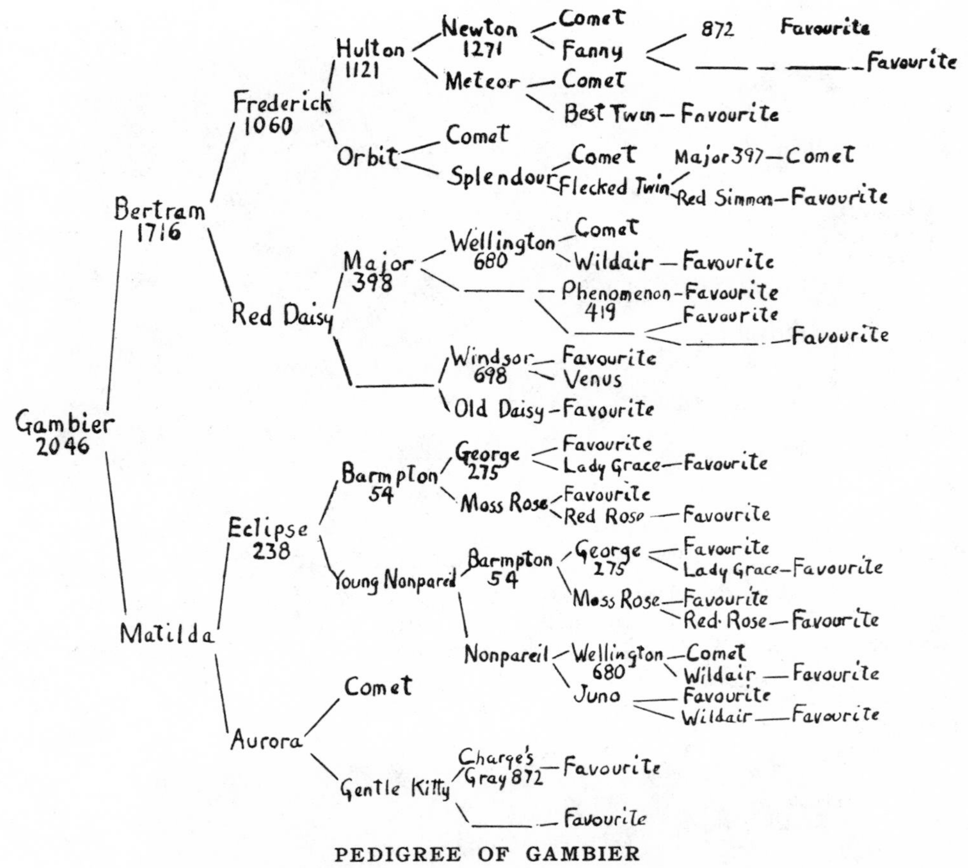

PEDIGREE OF GAMBIER

FIGURE 18. The bull Gambier (2046) was bred by Jonas Whitaker. The numerous lines tracing to Favourite (252) and Comet (155), the son of Favourite, are shown. There are, however, other lines early in the pedigree which trace to the bulls, Foljambe, Dalton Duke, Hubback, Ben, and Punch, and to the cow Phoenix (Dam of Favourite), which are not shown to avoid undue complexity, but are taken account of in calculating coefficients of inbreeding and relationship.

cent and 55 per cent respectively), and both were genetically almost as closely related to the Duchess with which they were mated as ordinary full brothers and sisters (47 per cent and 49 per cent).

The most important "outcrosses" in the later years were those with the descendants of the Matchem Cow. The latter was not closely related to the Duchesses or even to Favourite ($R = 22$ per cent), but as her blood was introduced into the Duchess line only through her sons, the Cleveland Lads, sired by the Duchess bull, Shorttail, and through her grandson, 2nd Duke of Oxford, with two Duchess top crosses, the result was merely that Bates kept the degree of inbreeding from rising above its previous level by introducing this dash of outside blood.

We must conclude that Bates simply maintained from the beginning to the end of his breeding career a certain average degree of relationship between the animals he mated (about 60 per cent), a degree distinctly higher than that between an ordinary brother and

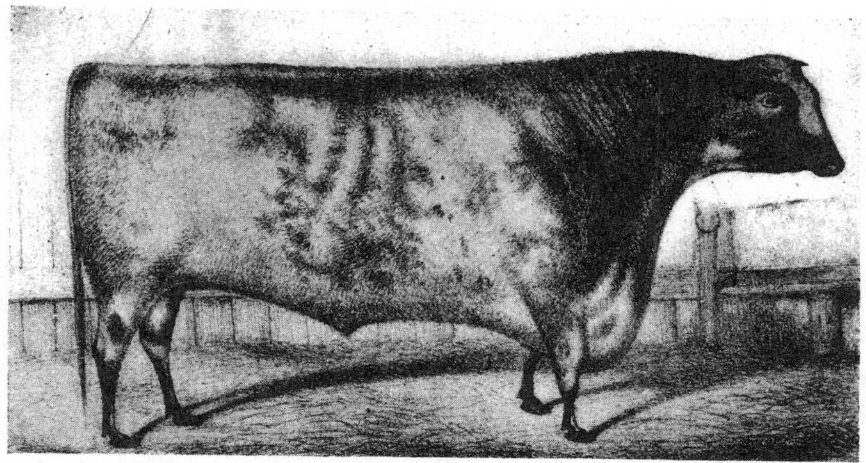

BELVIDERE

FIGURE 19. Bates' greatest success began with the use of the "Princess" bull, Belvidere (1706), which, although lacking in Duchess blood, was closely inbred to the foundation sire, Favourite, and through the latter was as closely related to the Duchess cows as an ordinary full brother.

OLD BROKEN-LEG

FIGURE 20. Daughter of Belvidere and dam of Duke of Northumberland, Old Broken Leg (Duchess 34th) was one of Bates' most famous Duchesses. Her merits as an individual were established by her notable victory over John Booth's cow, Necklace, at the York show in 1842.

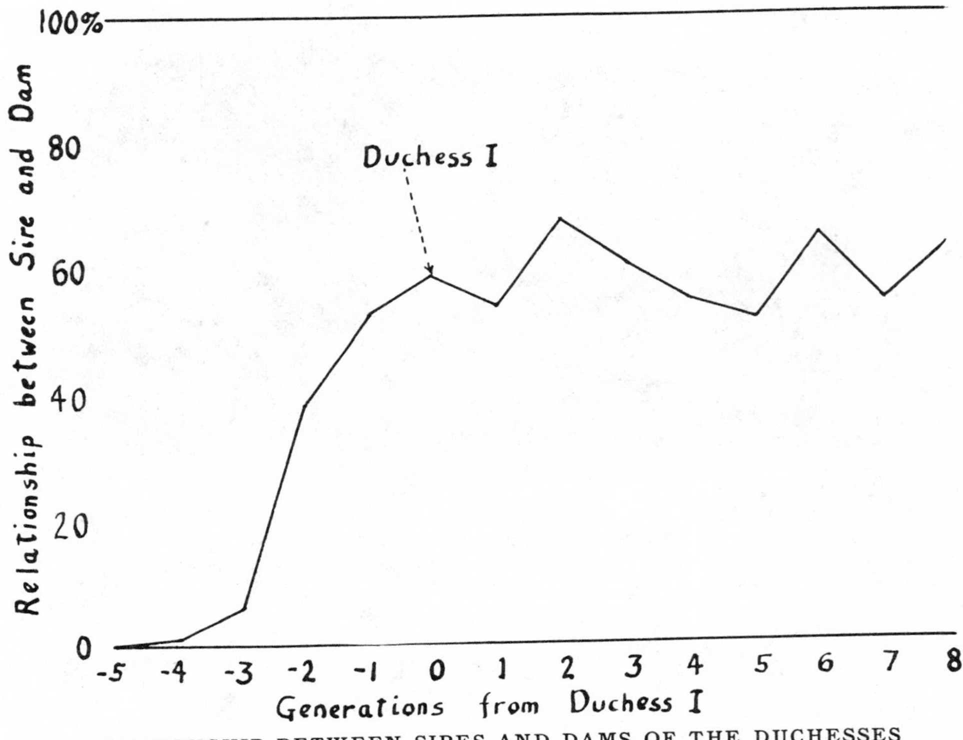

RELATIONSHIP BETWEEN SIRES AND DAMS OF THE DUCHESSES

FIGURE 21. The degrees of relationship between the sires and dams of the eight generations of Duchesses bred by Thomas Bates and six earlier generations in the straight female line. The coefficients of relationship are theoretically equal to the coefficients of correlation with respect to characters which are wholly hereditary and lacking in dominance.

sister. He constantly introduced fresh blood but only to such an extent as to prevent the relationship from rising above this level.

The Degree of Inbreeding Used by Bates

The coefficients of inbreeding of the thirteen generations of Duchesses are shown in the solid line in Figure 24. That of their dams (also, of course, Duchesses, but in a different order) and of their sires are given in dotted and broken lines respectively. The striking feature of this diagram is the similarity of the three lines at a constant level of about 40 per cent beginning with Duchess I. Duchess by Daisy Bull, two generations earlier, is the first Duchess which shows an appreciable amount of inbreeding (20 per cent). Duchess by Favourite in the next generation rises to 32 per cent. Duchess I, bred by Colling and purchased by Bates, rises to 41 per cent. From this point on no generation rises above 47 per cent or falls below 36 per cent. The coefficient for the last generation, 43 per cent, is practically the same as that for Duchess I and for the average of the first two generations which Bates himself bred. This level of inbreeding is approximately equivalent to two generations of straight brother-sister mating, something over four generations of pure breeding within a herd (half brother-sister mating) and six generations of double first-cousin mating. In maintaining it, Bates used bulls, wheth-

COMET

FIGURE 22. The bull Comet (155), bred by Charles Colling, was one of the most important foundation sires of the Shorthorn breed. He was the sire of Bates' Duchess I. In spite of his close inbreeding, 47.1 per cent, he was a remarkably vigorous animal and considered the best bull of his time. At Charles Colling's dispersion sale he sold for the record price of 1,000 guineas.

NORFOLK

FIGURE 23. Norfolk (2377), bred by James Whitaker, was one of the bulls used by Bates to introduce a dash of fresh blood into his herd. Many of the later day Duchesses descended from him. Like the other bulls used by Bates for this purpose, he was closely related to the Duchesses genetically through his descent in many lines from Favourite. Bates' so-called "outcrosses" merely kept the degree of inbreeding from rising above the level—about 40 per cent—with which he started.

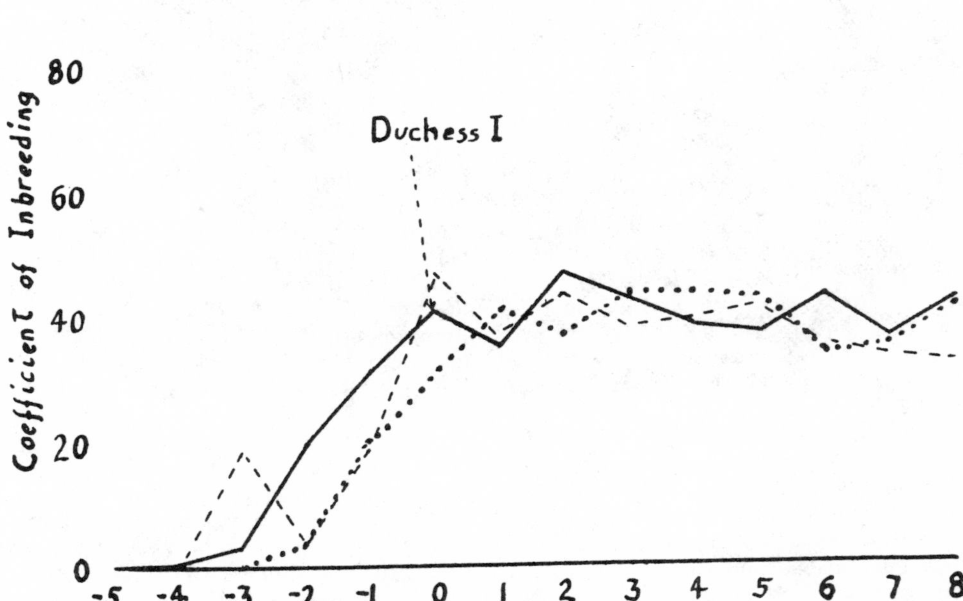

DEGREE OF INBREEDING OF THE DUCHESSES

FIGURE 24. The degrees of inbreeding of the Duchesses (solid line), of their sires (broken line), and dams (dotted line). The original Duchess was purchased by Charles Colling in 1784. Duchess I, five generations later, was purchased from Colling by Bates, who bred eight generations, including sixty-three cows, before his death in 1849. The coefficient of inbreeding measures the approach toward complete homozygosis. The coefficient of 40 per cent maintained by Bates means 40 per cent less heterozygosis than was present in the foundation stock.

er of his own or other breeding, which were inbred to substantially the same extent, 40 per cent, as the cows and related to them as we have seen, about 60 per cent. With regard to factors in which the original Shorthorns were only 50 per cent homozygous, the sires used by Bates and his whole Duchess family were about 70 per cent homozygous from the automatic effect of inbreeding alone.

Relationship to Favourite

Finally it is interesting to see how far Bates maintained relationship to the bull, Favourite, about which as we have seen the Collings centered their efforts at the improvement of Shorthorns. Figure 25 shows the coefficients for the generations of Duchesses in the solid line and those for their dams and sires in dotted and broken lines respectively. Favourite himself appears as a sire of two of the early Duchesses bred by Colling. Bates' Duchess I shows a relationship of 76 per cent. During the following generations bred by Bates, the relationship slowly falls from 69 per cent to about 57 per cent. It is noteworthy that the sires, the majority of which were bred by others, show practically the same coefficients as the Bates-bred cows with which they were bred.

While these lines show a gradual decline, it is a surprising result to find that Bates maintained a strain for forty years after the death of Favourite in which there was a distinctly closer relationship to the latter and hence presumably a closer resemblance than between an ordinary parent and offspring.

The Significance of Bates' Methods

Bates' own view of inbreeding is contained in a statement handed on by Darwin[1] "to breed in and in from a bad stock was ruin and devastation yet that the practice may be safely followed within certain limits when the parents so related are descended from first-rate animals."

The striking feature of his actual practices as brought out by the above diagrams is their uniformity throughout his whole career. He did not inbreed at the closest possible rate for a few generations and then make violent outcrosses. Neither did he concentrate the blood of one bull for a few generations and then turn to a wholly different line. Whatever the basis in his own mind, he actually pursued a steady policy—maintaining a relationship of nearly 60 per cent between the animals he mated, maintaining a coefficient of inbreeding of something over 40 per cent and maintaining a relationship to the foundation bull, Favourite, falling only slowly from 76 per cent to 57 per cent in eight generations. While our figures do not bring it out, it is well known that he also steadily selected for a certain type. The uniform degree of inbreeding was doubtless a somewhat unconscious result of a balancing between his desire to inbreed to maintain his type and his constant watchfulness over the characteristics of his animals, leading to prompt recognition of the need for a dash of fresh blood.

Through this system he must have been able to maintain without effort a close resemblance to Favourite in numerous features of quality and conformation. The fixation of characters (40 per cent), was not so great, however, but that there was variation and room for constant selection in which he could strengthen those features of the Ketton stock which he favored and get rid of those which he disliked. That he was able to mold the conformation into a distinctive type is the universal testimony of his contemporaries. One of these speaks of seeing his herd driven across country from Ridley Hall to Kirklevington in 1830, being "50 cows and heifers by 2nd Hubback, all as alike as beans and leaving a great impression wherever they passed." The Booths also founded their herds on Ketton stock and their animals would probably show relationship to Favourite of the order of those shown by the Duchesses. Indeed calculation would probably indicate a fairly close relationship between the Bates and Booth herds. Yet Bates and Booth types were recognized as distinctly different. This difference must have been due largely to selection.

Livestock breeders like to compare their work to that of one who molds figures in clay, as suggested above. The successful breeder is often spoken of as molding the conformation of his animals to the ideal type which he has in mind. If clay is to be worked into shape it must have just the right plasticity. Similarly with livestock. If Bates had not maintained close relationship between the animals which he mated, the relatively high degree of inbreeding, and close relationship to one animal (Favourite) his material would probably have been too plastic. The simultaneous variation in all characters would have been more than he could have contended with. If on the other hand he had bred wholly within his herd and between full brother and sister as far as possible, his material would soon not have been plastic enough to mold into shape. Undesirable characters, moreover, would almost certainly have become ineradicably fixed. As it was, a low level of fertility seems to have become fixed and to have doomed the efforts to maintain

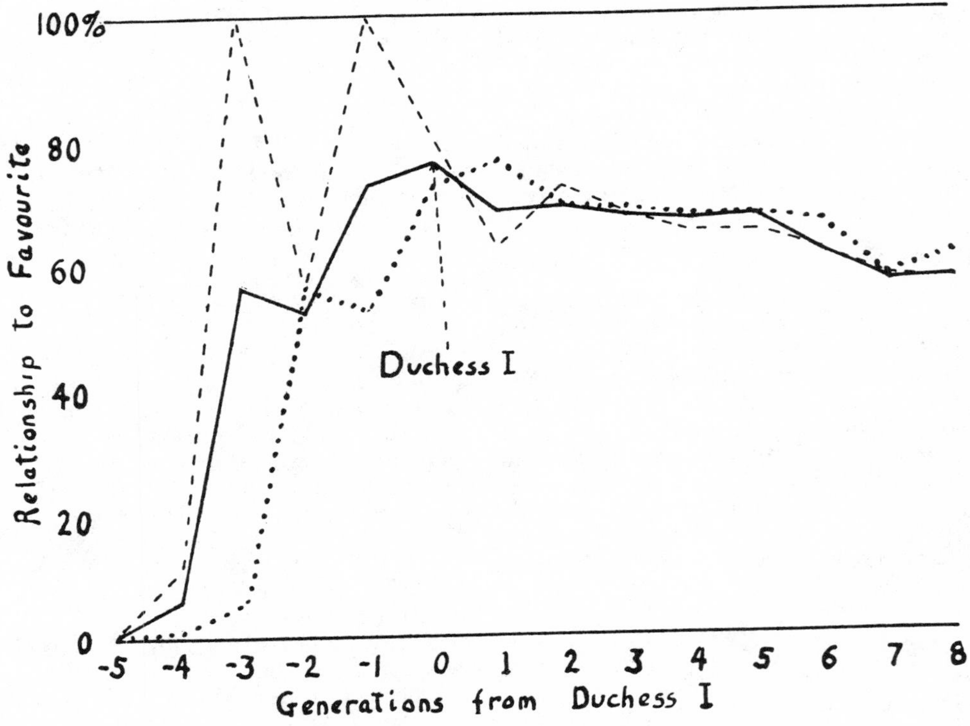

RELATIONSHIP OF THE DUCHESS COWS TO FAVORITE

FIGURE 25. The degrees of relationship to the bull Favourite of the Duchesses (solid line), of their sires (broken line), and of their dams (dotted line). For eight generations Bates maintained a closer resemblance to Favourite than between parent and offspring (50 per cent) in a stock not inbred.

a "pure" Duchess strain after Bates' death. On the whole it must be conceded that Bates managed to maintain a happy medium with respect to the plasticity of his stock.

Let us compare a little more closely the methods followed by Bates and those suggested by Mendelian theory. In combining inbreeding and selection there are several methods which may logically be followed depending on the genetic complexity of the characters, the importance of environmental variation and such factors as the extent of the operations and the amount of risk to be undertaken.

The first step in any case should be selection of a vigorous foundation, approaching as closely as possible to the desired type. This was the step taken by the Collings in purchasing the original Duchess, Favourite Cow, the bull Hubback, and so forth.

With such a foundation stock, one might practice the most intensive inbreeding in a large number of distinct lines, knowing that most lines would inevitably deteriorate greatly, but trusting that a few would be found in which desirable qualities would become fixed, and in which the deterioration in any vital respect would be so slight that they could be maintained successfully. By crossing such lines which have withstood this acid test of inbreeding, one might reasonably hope to recover more than the original vigor and retain those characters which had been fixed. Such a method is especially indicated where the characters are of a kind determined so slightly by heredity that genetic differences can be recognized only on comparing lines which have been kept distinct and free from outside blood. This method, an alternation of intensive inbreeding with selection and crossbreeding of the few successful lines must naturally be done on a large scale and with the undertaking of considerable risk. It is a method adapted rather to experiment stations than to private individuals. It is, however, an important method and has some parallel in the general history of the breeds. Many of the early breeders practiced close inbreeding. Only a few like the Collings were notably successful. The strains developed by these successful breeders were intercrossed to found the present pure breeds.

For the individual breeder, however, theory as well as practice indicate that the most reliable method is the maintenance of a steady level in closeness of breeding coupled with persistent selection toward the desired type, the requisite closeness of breeding depending, naturally, on the heterogeneity of the foundation animals and the breeder's skill as a judge of livestock.

Our analysis indicates that this was the method pursued by Bates. In view of Bates' success we may infer that the degree of inbreeding practiced by him, 40 per cent, represents about the right amount in the hands of an exceptionally able judge of cattle, working with a material as heterogeneous as the original Shorthorns.

Summary

It is shown that in establishing the famous Duchess family of Shorthorns, Thomas Bates started with Colling-bred stock already about 40 per cent inbred (i. e., which was 40 per cent less heterozygous than the original Shorthorns). During the eight generations which he bred himself, through a period of about forty years, he maintained substantially the same level of inbreeding by constantly introducing just the right amount of fresh blood to keep the percentage from rising above 40 per cent. He used bulls, whether of his own or other breeding, which averaged about 40 per cent inbred. The relationship between the Duchess cows and the bulls with which they were mated, whether bred by himself or others, was kept at such point that a correlation of about + .60 would be present in purely hereditary nondominant characters, throughout the eight generations. Finally during these eight generations a high correlation, falling gradually from .76 to .57 was maintained with Colling's bull, Favourite.

It is suggested that these levels of inbreeding and relationship yielded the proper balance between the extreme plasticity of the original heterogeneous Shorthorn stock and the more complete fixation of characters which would have resulted from closer inbreeding, to enable Bates, with his great skill as a judge of cattle to maintain a high degree of vigor and to mold a new type according to his ideals on the basis of the type represented by Charles Colling's bull, Favourite.

Literature Cited

[1] Darwin, Charles. Animals and Plants Under Domestication, 1868.

[2] Wright, S. Mendelian Analysis of the Pure Breeds of Livestock. I. The Measurement of Inbreeding and Relationship. *Journal of Heredity* 14:339-348. 1923.

TABLE I—*The parentage of the bulls used by Bates in developing his Duchess family.*

Bull	Sire	Dam
Favourite (252)	Bolingbroke (86)	Phoenix
Daisy Bull (186)	Favourite (252)	By Punch[1] (531) out of daughter of Hubback (319)
Comet (156)	Favourite (252)	Young Phoenix by Favourite out of Phoenix (his dam)
Ketton (709)	Favourite (252)	Duchess by Daisy Bull
Ketton II (710)	Ketton (709)	By son of Favourite out of daughter of J. Brown's bull[2] (97)
Ketton III (349)	Ketton II (710)	Duchess 3
Marske (418)	Favourite (252)	By Favourite out of daughter of Favourite[3]
Cleveland (146)	Ketton III (349)	Duchess 1
Young Marske (419)	Marske (418)	Duchess 4
The Earl (646)	Duke[4] (226)	Duchess 3
2nd Earl (1511)	The Earl (646)	Duchess 3
3rd Earl (1514)	The Earl (646)	Duchess 8
2nd Hubback (1423)	The Earl (646)	Red Rose I (see Figure 16)
		Angelina (see Figure 15)
Belvidere (1706)	Waterloo (2816)	Matilda (see Figure 18)
Gambier (2046)	Bertram (1716)	Nonpareil (see Figure 17)
Norfolk (2377)	2nd Hubback (1423)	Duchess 32
Shorttail (2621)	Belvidere (1706)	By 2nd Hubback out of daughter of 2nd Hubback[5]
Holkar (4041)	Belvidere (1706)	Duchess 34
Duke of Northumberland (1940)	Belvidere (1706)	Duchess 34
2nd Duke of Northumberland (3646)	Belvidere (1706)	Duchess 34
4th Duke of Northumberland (3649)	Shorttail (2621)	
Cleveland Lad (3407)	Shorttail (2621)	Matchem Cow[6]
2nd Cleveland Lad (3408)	Shorttail (2621)	Matchem Cow[6]
Lord Barrington (9308)	2nd Duke of Northumberland (3646)	By Cleveland Lad out of daughter of Belvidere[7]
2nd Duke of Oxford (9046)	Duke of Northumberland (1940)	Oxford 2nd (full sister of the Cleveland Lads)

[1] Punch was by Broken Horn (95) out of his own daughter. Broken Horn in turn was both son and grandson of Hubback (319), who thus appears four times in the ancestry of Punch within four generations.

[2] J. Brown's Bull (97) was a great grandson of the Studley Bull (626).

[3] The more remote top sires of Marske were Punch, Hubback, Snowden's Bull (612) and Masterman's Bull (422). The last three appear in the pedigree of Favourite.

[4] Duke (226) was by Comet out of Duchess by Favourite and was hence a full brother of Duchess I.

[5] The great great granddam of Holkar (female line) was by a son of Hubback.

[6] The Matchem Cow introduces more outside blood than any other animal in the table. Her sire, Matchem (2281) had four top crosses of bulls without any blood of Favourite or closely related animals as far as known certainly. Favourite appears as the fifth top sire and again as the eighth. Bolingbroke was the seventh, Foljambe the ninth and Hubback the tenth. Only the maternal grandsire of the Matchem Cow, Young Wynyard (2859), had much Favourite blood. He appears in the pedigree of Belvidere as both maternal and paternal grandsire.

[7] The fourth top sire was a grandson of Favourite and had other Favourite blood; the fifth and sixth top sires also had much Favourite blood while the seventh was Favourite himself.

TABLE II—*The coefficients of inbreeding of the Bates Duchesses and of their sires and dams, the degree of relationship between the latter and the degrees of relationship to the foundation bull, Favourite.*

Individual	Born	Generation	Parents: Sire	Parents: Dam	Inbreeding: Ind.	Inbreeding: Sire	Inbreeding: Dam	Relationship: Sire and Dam	Relationship To Favourite: Ind.	Relationship To Favourite: Sire	Relationship To Favourite: Dam
A (Duchess)	....	1	J. Brown's Bull (97)	....	0	0	0	0	0.3	0.5	....
B (Unnamed)	....	2	Hubback (319)	A	0.2	0	0	0.4	6.0	11.8	0.3
C (Unnamed)	....	3	Favourite (252)	B	3.3	19.2	0.2	6.0	56.7	100.0	6.0
D (Duchess)	1800	4	Daisy Bull (186)	C	19.9	3.4	3.3	38.6	52.6	56.8	56.7
E (Duchess)	....	5	Favourite (252)	D	31.5	19.2	19.9	52.6	72.7	100.0	52.6
Duchess 1	1808	6	Comet (155)	E	40.8	47.1	31.5	58.7	76.3	80.5	72.7
" 2	1812	7	Ketton (709)	1*	43.2	31.5	40.8	63.5	72.7	72.7	76.3
" 3	1815	7	Ketton (709)	1	43.2	31.5	40.8	63.5	72.7	72.7	76.3
" 4	1816	7	Ketton 2nd (710)	1	27.9	11.8	40.8	44.5	64.5	52.4	76.3
" 5	1817	7	Ketton 2nd (710)	1	27.9	11.8	40.8	44.5	64.5	52.4	76.3
" 6	1819	8	Ketton 3rd (349)	4	43.5	33.3	27.9	66.6	60.2	61.6	64.5
" 7	1820	8	Marske (418)	3	42.5	45.7	43.2	58.8	76.8	79.9	72.7
" 8	1820	8	Marske (418)	2	42.5	45.7	43.2	58.8	76.8	79.9	72.7
" 9	1821	8	Marske (418)	2	42.5	45.7	43.2	58.8	76.8	79.9	72.7
" 10	1821	8	Cleveland (146)	4	46.3	42.4	27.9	68.7	63.6	67.8	64.5
" 11	1822	8	Young Marske (419)	5	40.6	35.6	27.9	61.6	66.4	72.7	64.5
" 12	1822	8	The Earl (646)	4	43.3	49.2	27.9	62.7	67.6	72.6	64.5
" 13	1823	9	The Earl (646)	9	47.9	49.2	42.5	65.6	74.2	72.6	76.8
" 14	1823	9	The Earl (646)	6	44.4	49.2	43.5	60.7	66.9	72.6	60.2
" 15	1824	9	The Earl (646)	8	47.9	49.2	42.5	65.6	74.2	72.6	76.8
" 16	1824	8	The Earl (646)	3	60.4	49.2	43.2	82.6	69.4	72.6	72.7
" 17	1825	9	3rd Earl (1514)	11	44.0	47.9	40.6	61.0	70.4	74.2	66.4
" 18	1825	9	2nd Hubback (1423)	6	33.2	27.0	43.5	49.1	62.9	64.9	60.2
" 19	1825	9	2nd Hubback (1423)	12	41.8	27.0	43.3	61.9	64.7	64.9	67.6
" 20	1825	9	2nd Earl (1511)	8	48.5	60.4	42.5	64.1	73.7	69.4	76.8
" 21	1825	8	2nd Earl (1511)	3	66.0	60.4	43.2	87.1	67.9	69.4	72.7
" 22	1826	9	2nd Hubback (1423)	9	37.8	27.0	42.5	56.2	70.2	64.9	76.8
" 23	1826	9	2nd Earl (1511)	11	43.9	60.4	40.6	58.4	69.5	69.4	66.4
" 24	1826	9	2nd Hubback (1423)	6	33.2	27.0	43.5	49.1	62.9	64.9	60.2
" 25	1826	9	2nd Hubback (1423)	8	37.8	27.0	42.5	56.2	70.2	64.9	76.8
" 26	1826	8	2nd Hubback (1423)	3	43.5	27.0	43.2	64.5	66.8	64.9	72.7
" 27	1827	9	2nd Hubback (1423)	16	47.2	27.0	60.4	66.0	66.4	64.9	69.4
" 28	1827	9	2nd Hubback (1423)	6	33.2	27.0	43.5	49.1	62.9	64.9	60.2
" 29	1829	10	2nd Hubback (1423)	20	42.5	27.0	48.5	61.9	68.2	64.9	73.7
" 30	1830	10	2nd Hubback (1423)	20	42.5	27.0	48.5	61.9	68.2	64.9	73.7
" 31	1830	9	2nd Hubback (1423)	26	53.5	27.0	43.5	79.3	61.8	64.9	66.8
" 32	1831	10	2nd Hubback (1423)	19	52.6	27.0	41.8	78.5	60.8	64.9	64.7

—*Continued*

Table II—*Continued*

Individual	Born	Generation	Parents: Sire	Parents: Dam	Inbreeding: Ind.	Inbreeding: Sire	Inbreeding: Dam	Relationship: Sire and Dam	Relationship to Favourite: Ind.	Relationship to Favourite: Sire	Relationship to Favourite: Dam
Duchess 33	1832	10	Belvidere (1706)	19	32.8	52.3	41.8	44.6	68.1	64.8	64.7
" 34	1832	11	Belvidere (1706)	29	34.5	52.3	42.5	46.8	69.6	64.8	68.2
" 35	1833	10	Gambier (2046)	19	31.9	32.3	41.8	46.5	66.4	65.5	64.7
" 36	1834	10	Belvidere (1706)	19	32.8	52.3	41.8	44.6	68.1	64.8	64.7
" 37	1834	11	Belvidere (1706)	30	34.5	52.3	42.5	46.8	69.6	64.8	68.2
" 38	1835	11	Norfolk (2377)	33	31.2	19.5	32.8	49.5	60.7	55.4	68.1
" 39	1835	11	Belvidere (1706)	30	34.5	52.3	42.5	46.8	69.6	64.8	68.2
" 40	1835	10	Belvidere (1706)	19	32.8	52.3	41.8	44.6	68.1	64.8	64.7
" 41	1835	11	Belvidere (1706)	32	32.1	52.3	52.6	42.2	67.5	64.8	60.8
" 42	1837	11	Belvidere (1706)	30	34.5	52.3	42.5	46.8	69.6	64.8	68.2
" 43	1837	12	Belvidere (1706)	34	55.3	52.3	34.5	77.3	64.5	64.8	69.6
" 44	1838	12	Shorttail (2621)	37	48.5	32.1	34.5	72.8	65.0	67.5	69.6
" 45	1838	11	Shorttail (2621)	30	42.9	32.1	42.5	62.5	66.5	67.5	68.2
" 46	1838	12	Shorttail (2621)	34	48.5	32.1	34.5	72.8	65.0	67.5	69.6
" 47	1839	12	Shorttail (2621)	37	48.5	32.1	34.5	72.8	65.0	67.5	69.6
" 48	1839	11	Shorttail (2621)	30	42.9	32.1	42.5	62.5	66.5	67.5	68.2
" 49	1839	11	Shorttail (2621)	30	42.9	32.1	42.5	62.5	66.5	67.5	68.2
" 50	1839	12	Duke of Northumberland (1940)	38	40.1	55.3	31.2	56.2	63.3	64.5	60.7
" 51	1840	12	Cleveland Lad (3407)	41	32.9	11.6	32.1	54.2	55.3	47.4	67.5
" 52	1841	12	Holkar (4041)	38	36.0	24.0	31.2	56.4	58.8	60.7	60.7
" 53	1842	12	Duke of Northumberland (1940)	41	51.3	55.3	32.1	71.7	64.2	64.5	67.5
" 54	1844	12	2nd Cleveland Lad (3408)	49	32.6	11.6	42.9	51.6	56.3	47.4	66.5
" 55	1844	12	4th Duke of Northumberland (3649)	38	40.7	48.5	31.2	58.3	62.7	64.9	60.7
" 56	1844	13	2nd Duke of Northumberland (3646)	51	41.7	55.3	32.9	58.1	60.5	64.5	55.3
" 57	1845	13	2nd Cleveland Lad (3408)	50	28.8	11.6	40.1	46.1	59.0	47.4	63.3
" 58	1846	13	Lord Barrington (9308)	54	34.5	36.8	32.6	51.3	57.2	58.1	56.3
" 59	1847	14	2nd Duke of Oxford (9046)	56	42.8	32.1	41.7	62.6	57.4	56.7	60.5
" 60	1848	13	2nd Duke of Oxford (9046)	54	36.3	32.1	32.6	54.9	55.7	56.7	56.3
" 61	1848	13	2nd Duke of Oxford (9046)	51	37.8	32.1	32.9	57.1	54.9	56.7	55.3
" 62	1848	14	2nd Duke of Oxford (9046)	56	42.8	32.1	41.7	62.6	57.4	56.7	60.5
" 63	1848	13	2nd Duke of Oxford (9046)	54	36.3	32.1	32.6	54.9	55.7	56.7	56.3
" 64	1849	13	2nd Duke of Oxford (9046)	55	39.1	32.1	40.7	57.4	57.6	56.7	62.7

TABLE III—*A summary by generations (female line) of the coefficients of inbreeding and relationship for the 64 Duchesses, including Duchess I, bred by Charles Colling and purchased by Bates and the 8 generations bred by Bates.*

Generation from Duchess I	No. of Cows	Inbreeding: Individual	Inbreeding: Sire	Inbreeding: Dam	Relationship: Sire and Dam	Relationship To Favourite: Individual	Relationship To Favourite: Sire	Relationship To Favourite: Dam
0 (= Duchess I)	1	40.8	47.1	31.5	58.7	76.3	80.5	72.7
1 (Bates)	4	43.5	21.6	40.8	54.0	68.6	62.5	76.3
2 "	10	47.1	43.4	37.1	67.0	69.2	72.1	69.4
3 "	14	42.4	38.0	43.9	60.2	67.9	67.8	68.7
4 "	7	38.3	38.6	43.7	54.7	66.8	64.9	67.3
5 "	9	36.7	41.9	42.5	51.8	67.3	64.7	67.4
6 "	10	43.4	35.5	33.9	64.4	62.0	61.7	66.2
7 "	7	36.4	33.2	34.9	54.3	56.7	56.7	57.9
8 "	2	42.8	32.1	41.7	62.6	57.4	56.7	60.5
Total	64	40.9	37.5	39.6	59.2	65.6	65.1	67.2

MENDELIAN ANALYSIS OF THE PURE BREEDS OF LIVESTOCK

III. The Shorthorns

HUGH C. MCPHEE and SEWALL WRIGHT

Bureau of Animal Industry, United States Department of Agriculture

MUCH has been written about the role of inbreeding in the development of the pure breeds of livestock but the illustrations have been confined largely to pedigrees of certain individuals chosen in most cases merely as notorious examples of close inbreeding. Similarly, the importance of "prepotent" sires has been popularly discussed in breed literature but again for the most part without any definite measures of their influence. After reading the glowing accounts of the achievements of early breeders one often has an uneasy feeling that all of this may have had to do merely with a temporary aristocracy of show ring winners and have had little to do with the many thousands of animals which make up the rank and file of the breed, or even to any permanent extent with the more improved section of the breed.

Pearl, Gowen and others[2,3,4] have published studies dealing with the extent to which inbreeding is being practiced at present in the Jersey and Holstein-Friesian breeds of cattle. They find very low figures, especially for the latter breed. These results, as will be seen, agree with those which we have obtained for the Shorthorns. It may be well to point out here that such a result as regards current inbreeding is not incompatible with a high degree of inbreeding relative to the foundation stock. In the present paper we shall be concerned more with questions of breed origin and development than with current practices.

A noteworthy numerical analysis of the question of breed composition is that recently published by D. F. Malin[2], who deals with the Shorthorn, Hereford and Aberdeen-Angus breeds of cattle and the Poland China and Duroc Jersey breeds of swine. His method is the determination of the percentages of the blood of outstanding sires contained in the ancestry of the better stock of today as represented by a carefully selected list of the leading sires of prize winners and by a number of other animals of recognized importance. We shall have frequent occasion to compare his results with those obtained by the methods used here.

The preceding paper of the present series[8] presented the results of a detailed study of the breeding of a particular noted strain of Shorthorn cattle, Bates' Duchess family, in terms intended to be readily interpretable on a Mendelian basis. It should be said, however, that while this family has been of great importance in the development of the breed, it should not be considered as typical.

In order to obtain a proper perspective it is desirable to learn the situation in the breed as a whole. The purpose of the present paper is to give in broad outline the history of the British Shorthorns with respect to the use of inbreeding, the influence of particular sires, and the degree of homegeneity of the breed. The Shorthorn breed is especially adapted to such a study because of the relative completeness of the records of the foundation period. While the first herd book was published in 1822, almost forty years after the Collings began their work, a great many of the pedigrees extend back to animals of the latter period. Since the above date, the records are practically complete. All but two or three per cent of lines of ancestry of modern animals, chosen at random, can be traced back this far.

Methods

The methods of analysis are essentially those described in the first paper[7] of this series. The coefficient of inbreeding measures the correlation between the uniting germ cells with respect to their genetic constitution, relative, of course, to the diversity of the foundation stock[8]. This coefficient measures also the percentage decrease in heterozygosis relative to that in the foundation stock. The coefficient of relationship is simply the correlation between the genetic constitutions of the individuals concerned. It represents the correlation to be expected in appearance with respect to non-dominant characteristics which are determined entirely by heredity. Since the formulae and method of calculating these coefficients have been previously discussed in detail[9] they need not be repeated here. It may be well to call attention to the fact, however, that they are simple deductions from the Mendelian theory of heredity.

It should be emphasized that the coefficient of inbreeding cannot be expected to yield significant information concerning the approach toward homozygosity of particular factors in individual animals. It should give, however, the average approach toward homozygosity when a large number of factors are considered even as applied to individuals and, of course, with greater reliability as applied in large populations. A similar caution applies to the interpretation of the coefficient of relationship.

In the study of Bates' Duchesses the pedigrees were traced complete to the foundation animals. The coefficients calculated from such pedigrees should measure, therefore, the relative degree of homozygosis of the individuals as well as the actual degrees of resemblance to each other in characteristics involving a sufficiently large number of factors to eliminate the vagaries of random sampling.

The present study is an analysis of a breed rather than of individuals. In such a study it is, of course, impossible to consider all the animals of the breed and the analysis must be confined, therefore, to a random sample. In our study not only have the individuals been chosen at random from the appropriate herd books, which represent the registered breed at particular times, but also a strictly random selection of the ancestral lines of the parents of these animals has been made. The first step in obtaining the random sample has been to ascertain the number of pages in the herd book occupied by registrations in the year in question, and then to divide this range into such regular intervals as would yield the desired number of individuals on taking the first animal at the top of the page. In tabulating the lines of ancestry of the parents of these animals, a random sequence of sires and dams has been obtained by following the sequence of heads (sires) and tails (dams) in coin tossing experiments, a different sequence being used in each case. The coefficients depend in the main upon the percentage of occurrence of connections between the chosen ancestral lines of sire and dam of the individuals.

Whether the result obtained from a certain sized sample represents accurately the condition of the entire breed is purely a question of random sampling. A probable error may be obtained by the usual methods. A detailed discussion of the methods is now in press[9].

The period for which samples have been taken, the size of sample, and the number of ancestral lines tabulated in each are shown in Table I. It will be noted that the samples taken from the year 1810 are larger than those of other periods. This condition has resulted from the selection of all of the registered bulls and cows born in the years 1809, 1810 and 1811. Since the number born in 1810 was too small for the purpose of this study it was decided to select also all those born in the year preceding and following. Four ancestral lines, two back of the sire and two back of the dam, have been tabulated for each of the groups of bulls, except those of the 1920 period, for which only two lines have been tabulated. In the case of the groups of cows four lines have been tabulated only for those born in the 1810 period, all the others being two line pedigrees. The method of using four line pedigrees was used at first because of the greater amount of information furnished for the same amount of work. A change was made to the two line pedigree on recognition of its more satisfactory character from the standpoint of random sampling. It did not seem necessary to repeat the work done by the four line method since practically it should give as reliable results as the other unless the breed showed much more breaking up into distinct lines than our data indicate.

In addition to the study of the entire breed, coefficients of inbreeding and relationship have been calculated for the twenty leading Shorthorn prize-winning sires at the International Livestock Exposition during the years 1918-22. These were taken from Malin's book. Malin has determined the relative importance of each of these sires by assigning different weights to each as sires, grandsires or great grandsires of 1st, 2d and 3d prizewinners. The list of total weights assigned is given in Table II.

The History of the Breed

The Shorthorn breed originated in northeastern England. The original stock was probably a rather heterogenous mixture of native British, Saxon and possibly some Norman cattle. A rather extensive use of imported Dutch bulls upon such native cows took place during the sixteenth and seventeenth centuries and is credited by Wallace[5] with imparting to the Shorthorn breed "its most distinctive characteristics, large size, short horns, broken colour and fine dairy qualities." The genetic homogeneity of this stock relative to the general stock of England at that time cannot be estimated but doubtless it was slight. Outcrosses with other cattle were common during the early period but became rare after the middle of the eighteenth century.

The names and achievements of several prominent breeders of the middle and latter part of the eighteenth century stand out in history as testimony of a period of enthusiastic effort toward the improvement of the Shorthorn breed within itself. The best of the products of the earlier breeders seem to have been brought together in the herds of Charles and Robert Colling about 1780. Probably the greatest individual product of their breeding was the famous bull Favourite (252), dropped in 1793. His ancestry has been given in a previous paper[7]. The Collings bred him to his daughters generation after generation with notable success. Such families as the Duchesses discussed in a previous paper[8] were founded largely on his blood. Through the efforts of Thomas Bates (1775-1849), the Booths, and many others, his blood became diffused to such an extent that he may almost be said to have sired the entire breed.

What may be designated as a second period of breed improvement has been the development of the so-called Scotch type of Shorthorns, based in general on the herd of Amos Cruickshank (1808-1895) at Sittyton, Aberdeen, Scotland, and especially on his noted bull, Champion of England (17526), dropped in 1859. The herd had its origin in animals from many sources selected without regard to blood lines but with particular attention to vigor, early maturity, compactness, short legs and ability to lay on flesh rapidly. This type is somewhat of a contrast to that of some of the earlier herds, such as that of Bates, in which milk production was given considerable attention. The Scotch type predominates in the Shorthorn breed today.

Photograph by Courtesy of Robert Wallace.

THE CRUICKSHANK TYPE OF SHORTHORN

Figure 8

The rugged type and coarse bone of the Cruickshank bred stock is well illustrated by the above photograph of the Shorthorn bull "Field Marshal," (47870). A comparison with the photograph of "Earl of Oxford 3rd" shows the contrast between the type bred by Bates. The Cruickshank type predominates in the breed today.

Average Inbreeding

The problem of the present paper is to obtain numerical measures of the genetic situation in the breed throughout its history. First, consideration will be given to the degree of inbreeding. Figure 10 shows the coefficients of inbreeding of bulls and cows of the British Shorthorns during the period 1810-1920. The previously obtained coefficients of Bates' Duchesses are given for comparison. The figures for the whole breed, as represented in Coates' Herd Book, show a rather abrupt rise in the coefficient from 0 in 1790 to about 17 per cent in 1810. By 1825 it had risen to 20 per cent. It has changed relatively little to the present time, the figure for 1920 being 26 per cent. It is likely, however, that the figure for 1810 is somewhat too high to be representative of the whole breed at that time. Probably only the best bred animals of this period were recorded in the first volume of Coates' Herd Book which was published twelve years later. It should also be mentioned here that the choice of 1790 to represent the period of the foundation stock, relative to which all the coefficients are to be interpreted, is somewhat arbitrary. It was chosen as a date just preceding the birth of Favourite.

The relatively high level of inbreeding which was undoubtedly reached before 1825 and to a large extent before 1810, must be attributed to the direct influence of the Colling bred stock. Bates' Duchesses represent an extreme example of the concentration of this blood but the early herd books show that many other breeders of the time

Photograph by Courtesy of Robert Wallace.

THE BATES TYPE OF SHORTHORN

Figure 9

This photograph of "Earl of Oxford 3rd," (51186) illustrates the compact build, smoothness, and the fineness of bone which was characteristic of the latter day Bates type Shorthorns.

were following the same course to some extent. The high level of such families as the Duchesses must, however, have been balanced by a low level in other families.

The graph shows a slight drop if anything in inbreeding between 1825 and 1850. This was a period in which the influence of the best bred herds was diffusing through the breed, as will be brought out later in the discussion of the coefficients of relationship. Such a grading up process would tend to be accompanied by less intensive inbreeding. During the next period, however, a genuine rise of inbreeding is indicated from a study of both the bulls and cows. The change from 18.0 in 1850 to 27.4 in 1875 is more than five times its probable error. This indicates renewed attention to the building up of closely bred families. These families seem to have been based largely on Bates and Booth bred stock. The graph shows that Bates maintained a level of 40 per cent inbreeding in the Duchesses practically up to 1850. On the other hand, little if any of the increase in inbreeding at this time (1850-1875) can be attributed to Champion of England. His blood can hardly have spread far beyond the Sittyton herd by 1875. Paradoxical as it may seem, it is likely that the diffusion of Champion of England blood was responsible for the drop in inbreeding shown by both bulls and cows between 1875 and 1900 (difference 2.6 times its probable error). The first effect of this diffusion was something like that of an outcross with respect to the Bates and Booth topped herds of the time.

We come now to the level of inbreeding in the breed today, viz., 26 per cent. A slight rise over 1900 is shown. As far as any significance can be at-

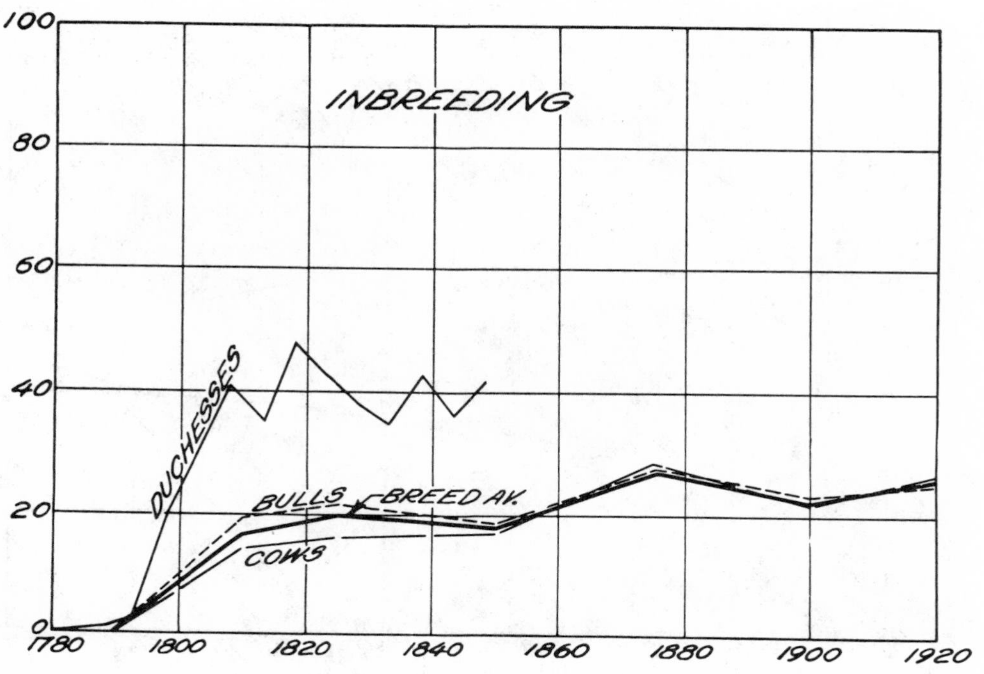

VARIATION IN INBREEDING

Figure 10

The degree of inbreeding of Shorthorn bulls (dotted line), Shorthorn cows (broken line), breed average (heavy solid line), with Bates' Duchesses (light solid line) for comparison. The origin at 1790 is somewhat arbitrary. The coefficient of inbreeding measures the approach toward complete homozygosis. The coefficient of 26 per cent in 1920 means that the breed at that time was 26 per cent less heterozygous than the foundation stock of 1790.

tributed to this rise (difference 1.4 times its probable error) it is to be assigned to the influence of Champion of England.

Twenty-six per cent inbreeding means a correlation of 0.26 between the uniting germ cells in their genetic constitution, relative, of course, to the condition existing in the foundation stock of about 1790. It means also that there has been a loss of 26 per cent in heterozygosis due to inbreeding alone besides that which has probably been accomplished through selection. The real significance of this coefficient can be made clear by comparing it with the results of certain regular systems of mating. If, for example, an entire breed were descended from a single pair of animals chosen at random from the foundation stock, the coefficient of inbreeding would be 25 per cent or practically the same as that of the present day Shorthorns. Again, an entire breed descended from a herd in which an individual sire had been used exclusively for two generations would have an inbreeding coefficient of 28.1 per cent in all later generations, if there were no further inbreeding. In the light of this result the coefficient of 26 per cent for the year 1920 is remarkably high especially considering the large number of animals that are registered every year. How nearly this figure represents the condition of the Shorthorn cattle of the entire world cannot be stated but it is believed that the American, Canadian and Argentinian sections differ but little in this respect from the British whence they were derived.

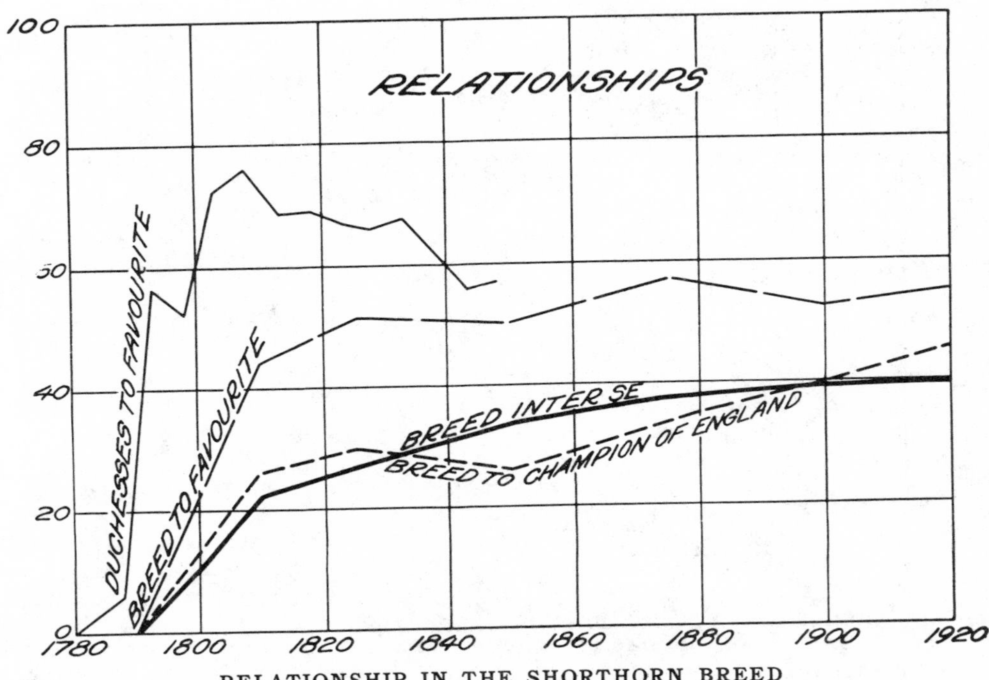

RELATIONSHIP IN THE SHORTHORN BREED

Figure 11

The relationships between the random individuals of the breed, the breed and Favourite, and the breed and Champion of England at different periods. The relationship between the Bates' Duchesses and Favourite (252) are shown for comparison.

Relationship to Favourite

The evident extensive influence of Colling stock on the breed in this early period makes it important to study intensively the degree of relationship to Charles Colling's most famous sire, Favourite, in all periods up to the present. As shown in Figure 11, the relationship of the whole breed to this bull rises from the basic 0 per cent in 1790 to 44 per cent in 1810 and 51 per cent in 1825. From that time it has shown only minor fluctuations from 50 per cent, being 55 per cent in 1920. These are remarkable figures when it is recalled that 50 per cent is the coefficient which applies to the relationship between parent and offspring in a random bred stock. It means that in so far as unselected characters are concerned the Shorthorn breed of today resembles Favourite slightly more than the average resemblance between sire and son among the old Teeswater stock of his time.

As already noted, this relationship had already risen above 50 per cent by 1825. At this date there was not a uniform diffusion, however, as indicated by the much higher coefficients in the Duchess family, balanced presumably by low coefficients in other strains. Even the Duchesses had practically fallen to the breed level, however, by 1850. It may safely be assumed that by this time such a uniform diffusion of Favourite blood had taken place that no further change is possible.

Relationship to Champion of England

Champion of England, who was dropped in 1859, had an inbreeding coefficient of 18.4 ± 0.56 per cent, or about the same as that of the entire breed at that time (18.0). His relationship to Favourite was about 49 per cent as compared with 50 per cent for the breed average. The practical identity in this respect is important as showing that no amount of concentration of the blood of Champion of England could appreciably change the breed's relationship to Favourite. His relationship to the breed at the time of his birth was about 26 per cent which is distinctly less than that of two random animals of the breed to each other in the periods preceding or following his birth (34 per cent and 38 per cent, respectively). In the years following, the figure was steadily increased until it was 46 per cent in 1920, an increase of 20 per cent since 1850.

In this connection it is interesting to compare the relationship of Champion of England to the breed with his average relationship to the 20 leading Shorthorn prizewinning sires at the International Livestock Exposition, 1918-22. The latter relationship has been found to be about 47 per cent or only about one per cent higher than the relationship between Champion of England and the breed. Since the coefficient of relationship between an ancestor and his descendants is about the same as percentage of blood, the increase of 20 per cent in relationship to Champion of England since 1850 may be considered as giving approximately the percentage of Champion of England blood in the breed as a whole today. Likewise, the average for the 20 prizewinners may be taken as about 21 per cent. These results are in harmony with Malin's calculation of 22.5 per cent Champion of England blood for these same animals, using complete pedigrees. These figures indicate then that there is no significant difference between the amount of Champion of England blood carried by the prizewinners and the average for the breed as a whole. In other words, this means, as Malin has already suggested, that Champion of England blood has now become so thoroughly diffused throughout the breed that a further concentration is impossible.

The preceding results indicate that Champion of England has been the most potent sire since Favourite in affecting the character of the breed.

The Homogeneity of the Breed

A very important question connected with a breed is the degree of homogeneity. Is it a well knit unit or a collection of diverse tribes? The coefficient of inbreeding does not answer this question. A breed might become split up into several such tribes each closely inbred within itself but having little relationship to each other. Such an assemblage might well show a high coefficient of inbreeding. To take an extreme illustration, a group of pedigrees picked at random from the record books of Shorthorn cattle, Holstein cattle, Percheron horses, Duroc-Jersey hogs, Shropshire sheep and Persian cats might yield a high coefficient of inbreeding with no homogeneity whatever. Returning to the Shorthorns, high coefficients of relationship to particular animals, such as to Favourite and to Champion of England, go farther in indicating breed homogeneity but do not give a complete answer. The most direct method of determining homogeneity is to calculate the degree of correlation between random animals. For this purpose pedigree lines of different animals for a particular year were matched at random using no line more than once. There were 150 such comparisons in 1810, 150 in 1825, and 100 in the later years. The calculated coefficients of relationship are shown in Table III (Breed inter se) and in the heavy line of Figure 11. This figure shows a consistent rise in breed homogeneity. The most rapid rise occurs before 1850 in agreement with the rise in the coefficient of inbreeding.

With such high coefficients of relationship as shown by the table, purely random mating within the breed should give fairly high coefficients of inbreeding. It is easy to calculate the amount of inbreeding due to random mating. The formula for the interrelationship at a given time in the symbols used in the preceding papers of this series is $R = \frac{\Sigma[(\frac{1}{2})^{n+n^1}(1+F_A)]}{1+F}$

The formula for inbreeding is $F = \Sigma[(\frac{1}{2})^{n+n^1+1}(1+F_A)]$.

The relations $R = \frac{2F}{1+F}$ and $F = \frac{R}{2-R}$ follow immediately. Applying the latter formula we obtain the following figures for the degree of inbreeding resulting merely from random mating within the breed.

Coefficients of Inbreeding

Date	*Due to Random Mating*	*Observed*	*Difference*
1810	12.4	16.6	+4.2
1825	15.5	19.9	+4.4
1850	20.4	18.0	—1.6
1875	23.3	27.4	+4.1
1900	24.4	22.9	—1.5
1920	24.6	26.0	+1.4

These figures indicate that there has been little active inbreeding in the Shorthorns since the foundation period. Some active inbreeding is indicated by the excess in the observed inbreeding of 1810 and 1825, and again in 1875 in agreement with the previous conclusions. As for the breed today, the high correlation between random animals (nearly .40) and the practical identity of the observed inbreeding and that due to random mating for 1900 and 1920, indicates that the breed is now a rather homogeneous unit.

A partial analysis of the ancestry of the 20 leading American prizewinning sires taken from Malin[2] indicates that they do not form a specialized group. Their average inbreeding came out 25 per cent as compared with 26 per cent for the breed average. Their relationship to each other was 39.8 ± 3.4. The average relationship between the breed of 1920 and these sires was 33.0±2.8 per cent. Champion of England and Favourite were related to the prizewinning sires 47.0 and 45 per cent respectively.

Summary

The present study is an attempt to interpret the history of the Shorthorn on the basis of the values of certain Mendelian coefficients. The coefficients of inbreeding measure the decrease in heterozygosis relative to the foundation stock. The changes in the coefficients at different times indicate the prevailing system of mating. The changes in the coefficients of relationship to particular animals, Favourite and Champion of England, measure the influence of these animals on the characteristics of the breed. The correlation between random individuals of the breed measures the amount of homogeneity. Comparisons with the coefficients in a noted early family, the Bates' Duchesses, give suggestions as to the means by which the present condition of the breed was attained.

Prior to 1810, herds with an intense concentration of the blood of one animal, Favourite (252) were built up by the Collings. This concentration of blood was carried on for many years in particular families of which Bates' Duchesses may be taken as an extreme representative. The widespread use of bulls of this stock rapidly raised the whole breed to such a point that there came to be a general resemblance to Favourite equal to that between parent and offspring in the foundation stock, and a correlation between random individuals as great as between half brothers of that stock. This was accompanied by an elimination of 15 to 20 per cent of the heterozygosis originally present. Most of these changes took place before 1825. By 1850 the breed was a fairly homogeneous unit, with the blood of Favourite uniformly diffused. There seems to have followed a period of renewed inbreeding in particular lines, based probably on Bates and Booth stock. Meanwhile, the blood of a bull, Champion of England, equal to the average of the breed in heredity and relationship to Favourite, but otherwise somewhat apart from the breed as a whole, was being concentrated by Amos Cruickshank. The diffusion of the blood of this bull through the breed between 1875 and 1900, having to a slight extent the effect of a cross, brought down the degree of inbreeding. The homogeneity of the breed as measured by the relationship of random animals remained practically stationary as Champion of England blood, increasing in amount, gradually displaced Bates and Booth blood. The increase in relationship to Champion of England has gone on to the present, until in 1920, the coefficient of 46 per cent is 20 per cent greater than the original relationship to this bull and approaches the present relationship to Favourite of 55 per cent. The relationship of random animals to each other, however, approaches 40 per cent. The present coefficient of inbreeding, 26 per cent, indicates an elimination of more than one-fourth of the heterozygosis of the foundation stock. The indications are that the breed is practically in equilibrium in all of these respects, an equilibrium which can only be disturbed by the formation and diffusion of the influence of a new, closely bred family based on the excellence of a new Favourite or a new Champion of England. The increased size of the breed makes the difficulty of exerting an influence comparable to that of these earlier bulls enormously greater even than it was in Cruickshank's time.

TABLE I. The coefficients of inbreeding of Shorthorn sires and dams at each period.

Year	*No. Animals*	*No. Ancestral Lines*	*Inbreeding*	*Coefficient*
1810	75 bulls	300	19.1 ± 1.0	16.6 ± 0.7
1810	85 cows	340	14.3 ± 0.9	
1825	50 bulls	200	21.9 ± 1.4	19.9 ± 1.2
1825	50 cows	100	16.1 ± 2.5	
1850	50 bulls	200	18.4 ± 1.3	18.0 ± 1.2
1850	50 cows	100	16.9 ± 2.5	
1875	50 bulls	200	25.9 ± 1.4	27.4 ± 1.3
1875	50 cows	100	29.0 ± 2.6	
1900	50 bulls	200	23.2 ± 1.4	22.9 ± 1.2
1900	50 cows	100	22.6 ± 2.6	
1920	50 bulls	100	25.4 ± 2.7	26.0 ± 2.0
1920	50 cows	100	26.7 ± 2.8	

TABLE II. The leading sires of shorthorn winners at the International, 1918-1922.

	Points	Number Winners		Points	Number Winners
Avondale 245144	490	34	Rodney 753273	136	7
Whitehall Sultan 163573 ...	380	38	Cumberland's Last 229822 .	122	15
Villager 295884	280	22	Fair Acres Sultan 354154..	122	8
Revolution 388359	280	22	Dale Clarion 385195	120	6
Master Ruby 446601	244	9	Sultan Supreme 367161	118	7
Sultan Stamp 334974	214	17	Superb 300054	116	7
Double Dale 337156	182	17	Village Beau 295883	112	14
Archer's Hope 402425	170	9	SanquharDreadnaught680399	108	9
Lespedeza Sultan 406929 ..	156	7	Village Supreme 423865 ..	108	5
Master Bapton 556804	152	3	Maxwalton Pride 367542 ..	96	2

TABLE III. The coefficients of relationship between Favourite and the breed, Champion of England and the breed and between random individuals at each of the periods studied.

Date	*Breed to Favourite*	*Breed to Champion of England*	*Breed Inter Se*
1810	44.3 ± 1.2	26.3 ± 1.2	22.0 ± 2.8
1825	51.3 ± 1.9	29.9 ± 1.6	26.8 ± 2.5
1850	50.1 ± 1.9	26.1 ± 1.6	33.9 ± 3.1
1875	57.6 ± 1.8	32.9 ± 1.5	37.8 ± 3.2
1900	52.1 ± 1.9	39.2 ± 1.6	39.3 ± 3.2
1920	55.2 ± 2.3	45.5 ± 2.3	39.5 ± 3.2

Literature Cited

1. GOWEN, JOHN W., and MILDRED R. COVELL, 1921. Studies in Milk Secretion. XII. Transmitting Qualities of Holstein-Friesian Sires for Milk Yield, Butterfat Percentages and Butterfat. *Maine Agri. Exp. Sta. Bull.* 301.

2. MALIN, DONALD F., 1923. The Evolution of Breeds. Wallace Publishing Co., Des Moines, Iowa.

3. PEARL, RAYMOND, 1915. Report of Progress of Animal Husbandry Investigations in 1915. *Paper No.* 92 from the Biological Laboratory of the Maine Agri. Exp. Sta.

4. PEARL, RAYMOND, JOHN W. GOWEN and JOHN RICE MINER, 1919. Studies in Milk Secretion. VII. Transmitting Qualities of Jersey Sires for Milk Yield, Butterfat Percentage and Butterfat. *Maine Agri. Exp. Sta. Bull.* 281.

5. WALLACE, ROBERT, 1923. Farm Live Stock of Great Britain.

6. WRIGHT, S., 1922. Coefficients of Inbreeding and Relationship. *Amer. Nat.* 56: 330-338.

7. WRIGHT, S., 1923. Mendelian Analysis of Pure Breeds of Livestock. I. The Measurement of Inbreeding and Relationship. *Journal of Heredity.* 14:339-348.

8. WRIGHT, S., 1923. Mendelian Analysis of the Pure Breeds of Livestock. II. The Duchess Family of Shorthorns as Bred by Thomas Bates. *Journal of Heredity,* 14:405-422.

9. WRIGHT, S. and HUGH C. MCPHEE. An Approximate Method of Calculating Coefficients of Inbreeding and Relationship from Livestock Pedigrees. *Journal of Agricultural Research.* In Press.

6
Fisher's Theory of Dominance

American Naturalist 63 (May–June 1929):274–79

Introduction

Wright wrote this paper in direct response to two papers by R. A. Fisher on the evolution of dominance (Fisher 1928a, 1928b). Fisher's conception of evolution in nature at this time was that it was dominated by very small rates of natural selection operating upon very large random breeding populations, a process he had likened to the general laws governing the behavior of gases (Fisher 1922, 321–22). Faced by the observation that most mutations are completely or almost completely recessive instead of exhibiting intermediate dominance, Fisher theorized that natural selection had made them that way from an original state of intermediate dominance. His argument was that deleterious mutations occurred at a small but definite rate and eventually reached an equilibrium frequency between adverse selection and recurrent mutation. Since homozygous mutants were extremely rare, natural selection operated primarily upon the heterozygotes, causing them to resemble the wild type, thus resulting in dominance. Fisher thought that the hypothesized selection rates would be "extremely slow," but effective in the long run.

Wright was very skeptical about Fisher's general theory of evolution in nature and doubted in particular the efficacy of Fisher's proposed mechanism for the evolution of dominance. Wright calculated that the selection rate was never more than half of the recurrent mutation rate from mutant to wild type. (Fisher soon showed that the correct figure was really twice the recurrent mutation rate, a figure Wright immediately accepted). Because Wright believed that genes had multiple physiological effects (pleiotropic effects), he thought that selection pressures upon the gene came from all directions and that the extremely small selection rate hypothesized by Fisher was unlikely to control the fate of the gene.

Drawing upon his background in physiological genetics and belief that genes acted by producing catalysts, Wright suggested that the prevailing recessivity of mutations resulted from the inactivation of gene products. Fisher, Ford, and others thought Wright meant that dominance was a physiological fact that could not be changed by selection. Actually, Wright had reversed dominance in one of his guinea pig strains and published the results only two years earlier and was well aware that selection could alter dominance relations. His criticism of Fisher's theory of the evolution of dominance was that the postulated selective forces were too weak, not that dominance was unmodifiable by selection.

It is worth noticing that Wright did not use his shifting balance theory or its primary elements in this first paper on the evolution of dominance. Subdivision of populations and random genetic drift are nowhere visible.

When Fisher read this paper, he was stimulated to answer Wright by letter, thus initiating a significant correspondence that lasted for more than two years. This correspondence is reproduced in its entirety in *SW&EB*, chapters 8 and 9. The evolution of dominance question is treated in *SW&EB*, 243–60, 299–303. Wright reassessed theories of the evolution of dominance in *E&GP*, volume 3, chapter 15.

The corrections on page 276 (of the original) in this reprint were made by Wright in the reprint he sent to Edward Murray East, whose copy was used for reproduction here. These were not corrections of printer's errors, but corrections that Wright made in response to Fisher's criticisms.

FISHER'S THEORY OF DOMINANCE

THE phenomenon of dominance in Mendelian heredity has from the first presented certain puzzling features. There was at one time a tendency to look upon dominance as a principle of heredity on a par with the principles of gene autonomy, the orderly segregation of allelomorphs and the assortment of non-allelomorphs. It was soon discovered that dominance is often merely a superficial appearance which disappears on careful examination and that even superficial dominance is far from invariable. It became clear that in any case dominance is a phenomenon of the physiology of development to be associated with the various types of epistatic relationships among factors rather than with the more fundamental genetic principles cited above.

The reasons for the undoubted frequency of real dominance, and for the also undoubted tendency of the type to be dominant over the mutant, are not obvious. A possible explanation has been developed by R. A. Fisher in two recent papers (2, 3) according to which these tendencies are statistical consequences of natural selection rather than of anything inherent in the physiology of gene action. Assuming that the pristine character of heterozygotes is intermediate, his suggestion is, in brief, that the observed gene mutations, however rare in laboratory experience, have nevertheless had opportunity to recur a very large number of times in the evolutionary history of the species; that such mutations are usually deleterious; that natural selection plays with much greater force on the heterozygotes than on the homozygous mutants (because of the enormously greater relative abundance of the former); that the character of the heterozygote is subject to the action of modifying factors, and that in consequence of all these facts there will be a gradual drift of the heterozygote toward the wild type, not necessarily involving the homozygous mutant type to any great extent. In the course of geologic time this may bring about complete dominance of the wild type or dominance so nearly complete that the heterozygote is no longer subject to adverse selection. He concludes that "with mutation rates of the order of one in a million, the corresponding selection in a state of nature, though extremely slow, can not be safely neglected in the case of heterozygotes."

This suggestion merits careful consideration, not only because of its interest on the evolutionary side, but especially because of its bearing on any deduction as to the nature of gene action which might be based on dominance. Fisher does not present his calculation of the selection rate due to this cause. The present writer reaches a figure so "extremely slow" as to make its efficacy seem highly questionable.

Suppose that a large population is in approximate equilibrium between the pressure of recurrent mutation from type factor A to a and adverse selection of heterozygotes as well as of homozygous mutants. It will be assumed that these are autosomal genes, as more favorable to the theory than sex-linked ones. Let q be the proportion of A's and $(1-q)$ that of a's in the initial constitution of a generation and assume that individuals of the three phases AA, Aa and aa are successful in attaining parenthood in the proportions $1:1-hs:1-s$, respectively. The initial zygotic proportions $(1-q)^2\,aa+2q\,(1-q)\,Aa+q^2\,AA$ become at the time of gametogenesis:

$$\frac{(1-s)\,(1-q)^2\,aa+2\,(1-hs)\,q\,(1-q)\,Aa+q^2\,AA}{1-2hsq\,(1-q)-s\,(1-q)^2}$$

The proportion of A's increases slightly:

$$q_1=q\left[\frac{1-hs\,(1-q)}{1-2hsq\,(1-q)-s\,(1-q)^2}\right]$$

The initial zygotic constitution of the next generation would show this increase were it not for a tendency for A to mutate to a, which at equilibrium just balances it. Let u be this mutation rate per gamete. Then $q_1-q=uq$, giving $q^2(1-2h)-q(2-3h+hu)+1-h-\frac{u}{s}+hu=0$.

For values of h larger than $\sqrt{\frac{u}{s}}$ this yields for the equilibrium value of q approximately $q=1-\frac{u}{hs}$

Since $\sqrt{\frac{u}{s}}$ is in general a small fraction, this formula holds until selection against the heterozygote has almost ceased. The value for complete dominance (at least with respect to selection effects) is $q=1-\sqrt{\frac{u}{s}}$, and is approached only as h becomes smaller than $\sqrt{\frac{u}{s}}$. These values agree approximately with those reached by Fisher for the equilibrium conditions.

Suppose that there is another factor, M, which modifies the selection against the heterozygote, giving it a reproductive rate with the ratio $1-h's$ to that of the wild type instead of $1-hs$. The reproductive ratios for mm and M – will be as follows:

$$\text{mm} \quad \overset{AA}{q^2} + \overset{Aa}{2\,(1-hs)\,q\,(1-q)} + \overset{aa}{(1-s)\,(1-q)^2} = 1-2u \text{ approximately}$$

$$\text{M}- \quad q^2 + 2\,(1-h's)\,q\,(1-q) + (1-s)\,(1-q)^2 = 1-2\frac{h'}{h}u \text{ approximately}$$

Thus the selection against *Aa* (and *aa*), however intense, will carry with it a selection in favor of M – as compared with mm of only about $(1+2u(1-\frac{h'}{h}):1)$.

The advantage of the modifier is thus decidedly small. Even if it is able at once to bring the heterozygote *Aa* to equality with the wild type in reproductive rate ($h'=0$) the pressure per generation toward its fixation (which is $2up(1-p)$, where p is the proportion of m's) is never more than half that which would result from recurrent mutation from m to M at the same rate as that from A to a (*i.e.*, up). It may be doubted whether mutation pressure by itself has been important in fixing factors.

It might be suggested that mutation pressure is the factor which reduces useless vestigial organs beyond the stage at which they can exert any conceivable deleterious effect, *e.g.*, in the frequently cited case of the whale's hind legs; and that another evolutionary force of the same order of magnitude can not be considered negligible. It is common experience that mutations tend to reduce development of parts. It is only necessary to compare the number of mutations which reduce wing size or venation, size of eye or eye pigmentation, number and size of bristles, etc., in Drosophila as compared with those with the opposite effects to find an illustration of this principle. The explanation usually advanced, that random changes in a gene are more likely to lead to injury to developmental processes than stimulation, seems a reasonable one. Mutation pressure would thus, no doubt, tend to reduce the development of useless characteristics. It is probable, however, that there is a more important factor in this case. It has been shown that genes often have multiple effects and it is not unlikely in view of results such as those of Dobzhansky (1) that in general any given gene has some effect on nearly all parts of the organism. Thus the evolutionary changes in a system of genes, some bringing increased development and others reduced development to those parts of the organism which are under direct selection, should have a net effect on *indifferent* parts in the direction of reduced development. The reduction of indifferent vestigial organs need not therefore be looked upon as the direct consequence of mutation pressure but as a by-product of multiple selection pressures of a higher order of magnitude. In the case of modifiers of dominance there seems to be no reason why increase rather than decrease of resemblance to wild type should come about as a by-product of the general selection process.

It will be seen that the hypothesis that a selection pressure, of the order calculated here, can be the *general* factor making for dominance of wild type, depends on the assumption that modifiers of dominance (assumed to be sufficiently abundant) are in general so nearly indifferent to selection on their own account that a force of the order of mutation pressure is the *major* factor controlling their fate. With the prevalence of multiple effects in mind it seems doubtful to the present writer whether there are many such genes.

It should be emphasized, however, that this conclusion rests to a large extent on the low frequency of even heterozygous mutants, to be expected where mutation is balanced by adverse selection of a higher order. If for any reason the proportion of heterozygous mutants reaches the same order as that of the type, selection of modifiers of dominance approaches the order of direct selection in its effects and might well become of evolutionary importance. Fisher's argument for the importance of this factor under certain conditions of artificial selection seems valid and must be taken into account in interpretation of dominance in domestic animals, especially, perhaps, in poultry. It seems unlikely that similar conditions would occur in nature except in special cases.

If this interpretation of the dominance of wild type allelomorphs is not available, what alternative is there? Probably most geneticists would hold that dominance in general has some immediate physiological explanation. Bateson long ago suggested that pairs of allelomorphs represent the presence and absence of something and that it was to be expected that one dose of an entity would give a result more like that of two doses than like complete absence. There are many reasons which have led to the general abandonment of the presence and absence hypothesis in its literal form. There is still much to be said, however, for the idea that the commonest type of change in a gene is one which partially or completely inactivates it in one or more respects. The effects of mutant genes, and especially of intermediate mutant allelomorphs, in combination with known deficiencies indicate differences from wild type of a negative rather than a positive sort.

On the view that genes act as catalysts and largely through bringing about the production of catalysts of a second order, it is easy to show that increase in the activity of a gene should soon lead to a condition in which even doubling of its immediate effect brings about little or no increase in the ultimate effects. Under such conditions, there would be dominance of the active phase. Lower levels of activity of the gene would show more or less imperfect dominance, corresponding to the usual case among intermediate allelomorphs in multiple series. Increased activity of other genes affecting transformation of the same substrate would also reduce dominance. Conversely, reduction of activity of such modifiers gives increase in dominance and thus the kind of effect required in Fisher's theory. There seem to be only rather special conditions (*e.g.*, certain threshold effects) under which it is to be expected that heterozygotes will be closer in effect to the inactivated than the active gene, making the latter recessive. Without going into further detail, it seems that in the hypotheses that mutations are most frequently in the direction of inactivation and that for physiological reasons inactivation should generally behave as recessive, at least among factors with major effects, may be found the explanation of the prevalence of recessiveness among observed mutations.

There remains the difficulty emphasized by Fisher that whatever the status of the genes in a species at a given time evolution through replacement of type genes by mutant allelomorphs, in general more or less completely recessive, would ultimately bring it to a condition in which the type genes would no longer be prevailingly dominant, were there no process continually at work increasing the dominance of new type genes. This argument, however, seems to involve the assumption that the array of observed gene mutations is a fair sample of those which are seized upon and fixed in the evolutionary process. This can hardly be the case if most such mutations are of the nature of inactivation.

Probably few geneticists, including Bateson himself, have ever taken very seriously the latter's suggestion that evolution may have taken place wholly by losses from a primitive complex. Presumably a species remains about the same for long periods of geologic time in the average degree of physiological activity of its genes. The mutations which actually are *fixed* must thus as frequently involve increases in activity as decreases. Under such an evolution, the species may continue indefinitely in a condition in which the most frequently *occurring* major mutations are recessive.

SEWALL WRIGHT

UNIVERSITY OF CHICAGO

LITERATURE CITED

1. Dobzhansky, Th.
   1927. "Studies on the Manifold Effects of Certain Genes in *Drosophila melanogaster.*" *Zeit. f. ind. Abst. u. Ver.*, 43: 330–388.

2. Fisher, R. A.
   1928. "The Possible Modification of the Response of the Wild Type to Recurrent Mutations." AMER. NAT., 62: 115–126.

3. Fisher, R. A.
   1928. "Two Further Notes on the Origin of Dominance." AMER. NAT., 62: 571–574.

7
The Evolution of Dominance: Comment on Dr. Fisher's Reply

American Naturalist 63 (November–December 1929):556–61

Introduction

Fisher replied to Wright's paper above with a note to the *American Naturalist* in which he argued that a tiny selection rate operating over a sufficiently long time was just as effective as a large selection rate operating for a correspondingly shorter time. In addition, he argued that since most genes in a given population were almost selectively neutral at any given time, then very small selection rates could determine their fates.

This paper was Wright's reply to those two arguments. He first reiterated his position that selection rates of the order of mutation rates could hardly be expected to control the frequencies of genes with pleiotropic effects. Then, for the first time in print, Wright began to use the elements of his shifting balance theory of evolution in nature. He answered both of Fisher's arguments by presenting (without derivation) his statistical distribution of genes in a population (to be contrasted with that of Fisher 1922) and demonstrated that effective population size was a crucially important factor. If effective size were sufficiently small, then random genetic drift could become a significant factor. Wright argued that there were many reasons to think that effective population sizes in nature were small enough for random drift to be important. He calculated that if effective population sizes were anything less than one million, then the selection rates hypothesized by Fisher as sufficient to account for the evolution of dominance would be swamped by random genetic drift. To indicate the power of random genetic drift, Wright suggested that the nonadaptive differences observed by systematists between "local races, subspecies and even species" could have been caused by random drift. If effective population sizes were as small as Wright supposed, then Fisher's theory of the evolution of dominance was impossible unless heterozygotes were more numerous than Fisher had originally supposed (a view Fisher and Ford later adopted).

After reading this paper, Fisher wrote to Wright to say that he believed effective population size to be a crucial variable, but that he considered most species to be panmictic (random breeding) throughout their ranges, even if the range was the entire world. Wright disagreed strongly.

THE EVOLUTION OF DOMINANCE

Comment on Dr. Fisher's Reply

Dr. Fisher's objection to my statement[1] that the pressure per generation toward fixation of a modifier of dominance "is never more than half that which would result from recurrent mutation from m to M at the same rate as that for A to a" is undoubtedly well taken. If account be taken of the effect of the frequency of the modifier on the frequency of the heterozygote, which must be done if p is to depart much from the value 1, my formula for the relative reproductive rates of the individuals with and without the modifier becomes $1:1-2up^2$ instead of $1:1-2u$. Similarly, the change in the value of p per generation (selection pressure) becomes $\delta p = 2u\,(1-p)$ (until p is nearly 0, when δp itself rapidly approaches 0). This agrees with Fisher's result for i "selection intensity" defined as $\frac{\delta p}{p\,(1-p)}$.

It still appears to me, however, to be proper to compare $\delta p = 2u\,(1-p)$, the change in frequency due to the selection, with $\delta p = up$, the change in frequency in the same direction due to mutation at rate u; or better, with the change due to opposing mutation at this rate, which is $u\,(1-p)$. In the latter case, the selection pressure, being always just twice the mutation pressure (until dominance is nearly complete), would in time practically fix the modifier against such mutation. But, as Dr. Fisher notes, the case was chosen as the most favorable to the theory which could be expected in nature. The selection pressure on a gene which merely increases the dominance of another by a small fraction becomes a third order phenomenon, if direct selection and mutation be considered as of the first and second order respectively. Dr. Fisher stresses increasing speed of selection as dominance becomes more complete but the speed can never reach twice the mutation rate of the mutant, unless the frequency of the mutant heterozygote is made greater than its value at equilibrium. The latter remains rather low ($\sqrt{\frac{u}{s}}$) even when dominance is complete, because of the selection against the homozygous mutant. In the laboratory one can readily pick out the hetero-

[1] Amer. Nat., 1929, 63: 274–279.

zygotes for breeding, *i.e.*, practice artificial selection *against* wild type, and thereby enormously increase the natural selection in favor of heterozygotes which *resemble* the wild type, but such a process of opposing selections would seem to have no counterpart in nature.

As there seems to be no essential disagreement on this matter of rate, we may come to the real point at issue: can a selection pressure of this order produce any appreciable evolutionary effect, however long it may continue? There seems to be no disagreement on the view that the fate of a factor is determined by the net effect of the opposing evolutionary forces and that practically this is determined by only the most important of these. Dr. Fisher's comparison of the influence of a second order factor with that of a continuous wind in controlling the destination of a railroad train expresses well the idea. The question at issue reduces then to whether there are genes so neutral in relation to all other evolutionary forces that selections which change their frequencies at rates of the order of mutation pressure are the most important forces acting on them, and whether such genes are sufficiently numerous to give a basis for such a common phenomenon as dominance. Dr. Fisher holds on the whole to the affirmative and I am still skeptical, admitting that questions are involved on which we know very little.

I have not raised the question of insufficiency of geologic time, to which Dr. Fisher refers, although a factor which at best requires a number of generations greater than the reciprocal of the natural mutation rate per locus probably has little to spare in this respect.

My first point was that, in the majority of cases, the selection rate of Fisher's hypothesis would be unable to fix the modifier even against an average mutation pressure. With mutation acting in the same direction as selection, the gene would be fixed but the selection would merely be playing the rôle of a following wind in the case of the railroad train.

It seemed unlikely to me, however, that mutation itself is a factor of appreciable importance in fixing genes. I raised the probability that all genes have multiple effects and, through one or other of these, each in general is subject to direct selection which takes precedence in controlling its fate. The recent estimates of the total number of genes in Drosophila as only a few thousand are suggestive in this connection.

I do not hold, however, that even the most important selective action on a gene is necessarily the controlling factor. This brings up considerations relating to the stability of gene frequencies, which I did not discuss in my previous paper but which are raised by Dr. Fisher as favorable to his view. Both of us used the conditions of equilibrium in connection with the ordinary mutations whose characteristic recessiveness constitutes the problem. Such factors are in general rare in nature, and probably always have been rare. Though continually mutating down the ages, they are kept from rising appreciably above the estimated figure for equilibrium by powerful negative selection. The equilibrium may be considered stable for the present purpose. On the other hand, factors which are almost neutral to all other evolutionary forces should be highly unstable with respect to gene frequency. I can not accept the view that this instability is favorable to the success of feeble selection. In fact, I would say that it sets a lower limit below which selection is not effective.

In a freely interbreeding population of limited size (n) gene frequency shows random variation with a standard deviation $\sqrt{\frac{q(1-q)}{2n}}$. Being random, such variations largely neutralize each other, but there is a second order drift which can not be ignored. Dr. Fisher himself[2] has calculated the probability distribution resulting from this cause under various conditions. Thus with no selection and very low mutation rate, he found the distribution $df = \frac{1}{2\pi} \operatorname{sech} \frac{1}{2} z dz$, where $z = \log \frac{q}{1-q}$. Translated into terms of q, this becomes $\frac{1}{\pi} q^{-\frac{1}{2}}(1-q)^{-\frac{1}{2}} dq$, which is a U shaped distribution. The factor is usually close to fixation one way or the other but may drift from one extreme to the other. There can hardly be said to be any equilibrium point. He showed that feeble selection does not alter the situation in this respect. I have made a similar calculation by a different method and have reached a general formula (unpublished) which agrees roughly with Dr. Fisher's results. I find the probability of different values of q to be proportional to $e^{2nsq} q^{4nv-1}(1-q)^{4nu-1}$, where s measures the selection favoring the gene whose frequency is q,

[2] *Proc.* Roy. Soc. Edinburgh, 1922.

u the mutation rate of this gene and v that of reverse mutation. The approximation is quantitatively valid only for weak selections and does not, of course, give the frequencies of the terminal, temporarily fixed classes, which are $\frac{f_1}{4nv}$ and $\frac{f_{2n-1}}{4nu}$ respectively, where f_1 and f_{2n-1} are the frequencies of the subterminal classes as given approximately by the formula. The form of the curve under different conditions indicates whether selection (s), isolation effect $(\frac{1}{2n})$, or mutation (u, v) is the factor which controls the fate of the gene and thus in a sense adjudicates between the principles of Darwin, Wagner and de Vries, respectively.

Selection controls the situation if s is larger than $\frac{1}{2n}$ but is of little importance below this figure. In small inbred populations ($\frac{1}{2n}$ large) even vigorous selection is ineffective in keeping injurious factors from drifting into fixation, and in the histories of a number of such lines one can follow in the laboratory the course of evolution under extreme isolation. In the case of Fisher's modifiers of dominance with selection coefficients at best of the order of mutation rate, the latter must be greater than $\frac{1}{2n}$ if the gene is not to drift back and forth in the course of geologic time from one state of approximate fixation to the other and practically as freely in the face of the selection pressure as with it.

Unfortunately it is difficult to estimate *n* in animal and plant populations. In the calculations, it refers to a population breeding at random, a condition not realized in natural species as wholes. In most cases random interbreeding is more or less restricted to small localities. These and other conditions such as violent seasonal oscillation in numbers may well reduce *n* to moderate size, which for the present purpose may be taken as anything less than a million. If mutation rate is of the order of one in a million per locus, an interbreeding group of less than a million can show little effect of selection of the type which Dr. Fisher postulates even though there be no more important selection process and time be unlimited.

The non-adaptive nature of the differences which usually seem to characterize local races, subspecies and even species of the same genus indicates that this factor of isolation is in fact of first importance in the evolutionary origin of such groups, a point on which field naturalists (*e.g.*, Wagner, Gulick, Jordan, Kellogg, Osborn and Crampton) have long insisted. With nearly complete local isolation and correspondingly large value of $\frac{1}{2n}$, one might expect to find the entire probability array of values of q exhibited in different localities contemporaneously in the case of factor pairs of little selective significance. It seems likely then that, in many cases, factors which are not fixed by rather strong selection will be subject to this effect and thus irresponsive to feeble selection of the order of effectiveness of mutation pressure.

Both Dr. Fisher and I have dealt only with the case of pairs of allelomorphs. If mutations are occurring in many directions in each gene, as seems likely, a complete analysis would require consideration of an indefinitely extended multiple allelomorphic series. This complicates the matter, but it seems to me that here also those genes which are not controlled by moderately strong selections would ordinarily drift at random through the multiple dimensional system of gene frequencies, regardless of any second or third order evolutionary pressures, and that consequently the explanation of the frequency of dominance as a Mendelian phenomenon must be sought elsewhere. The suggestion that mutations most frequently represent inactivations of genes, and that, for simple physiological reasons, inactivation should generally behave as recessive, still seems adequate as a positive alternative hypothesis.

SEWALL WRIGHT

UNIVERSITY OF CHICAGO

8
Evolution in a Mendelian Population

Anatomical Record 44 (1929):287

Introduction

The long manuscript on evolution in nature that Wright had written in 1925 was still not in print by the time of the AAAS annual meeting in December of 1929. Wright presented his shifting balance theory at the meeting and the abstract of that paper was published (along with many other abstracts) in the *Anatomical Record*. This is the first published account of his shifting balance theory.

Wright, S. 1929. Evolution in a mendelian population. Anatomical Record, 44: 287.

The frequency of a given gene in a population is affected by mutation, selection, migration and chance variation. The pressure exerted by these factors (excluding chance) and the position of equilibrium between opposing pressures are easily found. Gene frequency fluctuates about this equilibrium in a distribution curve, determined by size of population and the various pressures. The mean and variability of characters, correlation between relatives and the evolution of the population, depend on these distributions. In too small a population, there is nearly complete random fixation, little variation, little effect of selection and thus a static condition, modified occasionally by chance fixation of a new mutation, leading to degeneration and extinction. In too large a freely interbreeding population, there is great variability, but such a close approach of all gene frequencies to equilibrium that there is no evolution under static conditions. Changed conditions cause a usually slight and reversible shift of the gene frequencies to new equilibrium points. With intermediate size of population, there is continual random shifting of gene frequencies and consequent alteration of all selection coefficients, leading to relatively rapid, indefinitely continuing, irreversible and large fortuitous but not degenerative changes even under static conditions. The absolute rate, however, is slow, being limited by mutation pressure. Finally, in a large but subdivided population, there is continually shifting differentiation among the local races, even under uniform static conditions, which through intergroup selection, brings about indefinitely continuing, irreversible, adaptive and much more rapid evolution of the species as a whole.

9

Review of *The Genetical Theory of Natural Selection* by R. A. Fisher

Journal of Heredity 21, no. 8 (1930):349–56

Introduction

Wright always wrote substantive book reviews. This one pointed out the agreement that Wright and Fisher had reached (though reached by quite different methods) on quantitative details in their statistical distributions of genes in a population. It also explored in considerable depth the qualitative differences in their conceptions of the evolutionary process.

Although Fisher disagreed with Wright's shifting balance theory of evolution, he was much pleased by Wright's thoughtful review of the book and wrote a long letter to Wright in response to the review. This review is important for understanding the differences between the evolutionary views of Fisher and Wright in the early 1930s.

Fisher, with his poor eyesight, complained to Wright about the way his handwritten formulas were reduced to such small size by the *Journal of Heredity*. The formulas were difficult for anyone with good eyesight to read, and Fisher's well-justified complaint is ironic only because his own handwriting was frequently so small that others, including Wright, used a magnifying glass to read it.

I have discussed Wright's reaction to Fisher's *The Genetical Theory of Natural Selection* in *SW&EB*, 260–76.

THE GENETICAL THEORY OF NATURAL SELECTION

A Review

SEWALL WRIGHT

Department of Zoology, University of Chicago

DURING the latter part of the nineteenth century, increasing difficulty was felt in accepting Darwin's conception of the evolutionary process as one in which variation merely plays the subordinate (though necessary) rôle of providing a field of potentialities, through which the actual direction of advance is determined by natural selection. Theories were developed according to which the "origin of species" was to be sought more directly in the "origin of variation." Most of these were Lamarckian, others, of which de Vries' theory was most important, were not. The rediscovery of Mendelian heredity was a direct consequence of the mutation theory of the origin of species and was naturally seized upon as supporting this view. Only gradually has it become apparent that the real implications of Mendelian heredity are exactly the opposite and that in fact, it supplies the answer to some of the main difficulties felt with Darwin's theory. Dr. Fisher has played a leading part in developing the statistical consequences of Mendelian heredity and here brings together his views in a unified form.* It is a book which is certain to take rank as one of the major contributions to the theory of evolution.

The first chapter is concerned with a comparison of the consequences of blending and particulate heredity. A consequence of blending heredity, which Dr. Fisher shows was well understood by Darwin, and which was felt by him, and others, as a major difficulty with his theory, is the fact that under such heredity, the variability of a population tends to be greatly reduced in each successive generation. The portion of the variance lost per generation is one-half (if there is no assortative mating) and after ten generations only one-tenth of one per cent is left. Thus Darwin felt constrained to believe that an enormous amount of new variation appears in each generation, the differences among brothers being of this sort. This variability must be seized upon at once by natural selection or it will be lost. With even a slight departure from randomness in its occurrence, direction of mutation, rather than natural selection becomes the guiding principle of evolution.

All of this was changed with the demonstration of particulate inheritance and orderly segregation. The frequencies of zygotes of the types *aa, Aa* and *AA* tend to remain indefinitely in the proportions of a binomial square $p^2+2pq+q^2$ where p and q are the proportion in which alternative genes are represented in the population. In a population of limited size, to be sure, there is some variability of gene frequency, due to the accidents of sampling from generation to generation, but this brings about only a very low rate of reduction of variance. As to the actual rate of reduction of variance (and of heterozygosis), Fisher here confirms the figure which I had obtained by the method of path coefficients, viz. $\frac{1}{2n}$ per generation, where n is the effective size of the breeding population. The modern geneticist may get an appreciation of the difficulties which confronted

*The Genetical Theory of Natural Selection, by R. A. Fisher Sc. D., F. R. S., Price $6.00. Oxford University Press, New York. 1930.

Darwin, in attempting to account for natural variability and to apply selection as a guiding principle, by considering the case of a self-fertilized line, in which the loss of variance actually is 50% per generation, the same as with a random breeding population under blending heredity. The difference that with any initial variability, the inbred line tends to split up into many diverse lines, while the population under blending heredity becomes fixed as of one type , further emphasizes the difficulty. That pure lines actually show very little genetic variability, Fisher points out, is convincing evidence that substantially all inheritance is Mendelian. A quotation will bring out his conclusions with regard to mutation and selection:

> For mutations to dominate the trend of evolution it is thus necessary to postulate mutation rates immensely greater than those which are known to occur and of an order of magnitude which in general would be incompatible with particulate inheritance. * * * The whole group of theories which ascribe to hypothetical physiological mechanisms, controlling the occurrence of mutations, a power of directing the course of evolution, must be set aside once the blending theory of inheritance is abandoned. The sole surviving theory is that of natural selection and it would appear impossible to avoid the conclusion that if any evolutionary phenomenon appears to be inexplicable on this theory it must be accepted at present merely as one of the facts which in the present state of knowledge seems inexplicable.

I may state at this point that I am in accord with Dr. Fisher on the rôle of mutation, except that I would perhaps allow occasional significance to chromosome aberration, and to hybridization, as direct species forming agencies. It appears to me, however, that in this statement and throughout the book, he overlooks the rôle of inbreeding as a factor leading to nonadaptive differentiation of local strains, through selection of which, adaptive evolution of the species as a whole may be brought about more effectively than through mass selection of individuals.

Distribution of Gene Frequencies

The central problem in the analysis of the statistical consequences of Mendelian heredity is that of determining the distribution of gene frequencies under the pressures of mutation, selection, migration, etc., and not least important, as affected by size of population. Under given conditions, what proportion of the genes will be fixed? How many will have frequencies in the neighborhood of 50%? How many 99%? How rapidly will new mutations attain fixation under favorable selection? Two of the chapters (IV, V), are devoted to a mathematical investigation of such questions. As I have recently presented certain results in this field,* it may be of interest to bring out the points of agreement and disagreement.

My approach to the subject was from a different angle than Dr. Fisher's in being through the problem of inbreeding. I found that the decrease in heterozygosis, to be expected under inbreeding (but ignoring new mutations and selection) could be obtained by an application of the method of path coefficients.[4] The method could be applied to complex pedigrees encountered in livestock, and studies of the history of the Shorthorn breed of cattle have been made by means of it by Dr. McPhee and myself[5] of Clydesdale horses by Calder,[1] and of Jersey cattle by Buchanan Smith.[3] In the case of random mating in a population of N_m males and N_f females, it gave as a close approximation $\frac{1}{8N_m}+\frac{1}{8N_f}$ as the rate of loss of heterozygosis (and hence of variance) per generation. With an equal number of males and females in a total breeding population of n this reduces to the $\frac{1}{2n}$ referred to above. Fisher, studying the problem of evolution of large popula-

*These results were presented at the 1929 meeting of the A. A. A. S. An abstract appeared in the *Anatomical Record* (Vol. 44, p. 287, 1929). The full paper is to appear in *Genetics*.

tions, made the first attempt to find the actual distribution of gene frequencies under various conditions.[2] He reached a solution for the case of unselected genes not replenished by mutation, which indicated loss of variance at the rate of $\frac{1}{4n}$ per generation, just half of the rate indicated by my method. His formula for the distribution of gene frequencies was expressed on a scale of the logarithms of the ratio of alternative gene frequencies $\log\frac{q}{1-q}$, a scale which has the advantage of stretching the important regions close to 0% and 100% and also of making the effect of simple selection uniform at all points. It is interesting to note, however, that on transforming his formula to the simple scale of percentage frequencies it indicates an equal number of genes at all frequencies ($y = 1$). He also obtained a solution for the case in which decrease in heterozygosis is just balanced by mutation.

On noting the discrepancy between his result and mine for decrease in the rate of heterozygosis, I was not able to correct a questionable point in his derivation, but was able to reach a formula for the distribution of gene frequencies in a different way. The result agreed with his solution in form ($y = 1$), but with the rate of decline as $\frac{1}{2n}$ per generation. In the case of loss of variance, balanced by mutation, the distribution differed considerably in form, being $y = \frac{1}{2n[.577 + \log(2n)]q(1-q)}$ instead of his $\frac{1}{\pi\sqrt{q(1-q)}}$. It appeared further from this method that a selective advantage such that genes A and a reproduce in the ratio $1:1-S$ introduced an exponential term e^{2snq} into the formula. This is valid, however, only for irreversible mutation and then only for extremely small values of the selection coefficient. It now appears that for reversible mutation and in any case for values considerably larger than $\frac{1}{2n}$ it should be e^{4snq}. Appreciable rates of recurrence of mutation (u) and reverse mutation (v) were stated* to throw the formula into the form $y = Ce^{4nsq}q^{4nv-1}(1-q)^{4nu-1}$ a curve which for high mutation rates (relative to $\frac{1}{2n}$) approaches the form of a probability curve and indicates a random drifting of gene frequency about an equilibrium point. The case which has seemed most important to me is that of the effects of migration in a population which is a sub-group of a large one. The formula is similar mathematically to that for mutation. It is given below in a revised form.

These results were communicated to Dr. Fisher, who now finds on reexamination of his method, that the addition of a term which had seemed unimportant gives a confirmation of my formulae in the first two cases. He obtains on the other hand, a somewhat different form for the effect of selection, viz., $y = \frac{2}{pq}\left(\frac{1-e^{-4anq}}{1-e^{-4an}}\right)$ where his a is my s, except for a change of sign. The exact case which he deals with, is not one which I had considered, a fact which reflects our differences in viewpoint on the general problem. His formula refers to flux equilibrium with respect to an inexhaustible supply of irreversible mutations. On solving for it by my method I get results substantially identical with his as long as the selection coefficient is less than $\frac{1}{2n}$. Above this there is rapid divergence. His formula is undoubtedly a better approximation, and in fact, I may say that on reexamination, I find that I, in turn, have here neglected terms which should be taken account of. The general formula for a partially isolated population by my method, as now revised, is as follows: $y = Ce^{4nsq}q^{4n(mq_m+v)-1}(1-q)^{4n[m(1-q_m)+u]-1}$ where m is the rate of population exchange with the species as a whole, q_m is the gene frequency in the latter and s measures the differential selection of the group as compared with the species as a whole. If v is actually zero (completely irreversible mutation from an inexhaustible supply of genes), and no im-

*Presented without proof in *American Nat.* 63:556-561 (1929).

migration is assumed, the formula takes a somewhat different form and in fact reduces to Dr. Fisher's result, identically. Summing up, our mathematical results on the distribution of gene frequencies are now in complete agreement as far as comparable, although based on very different methods of attack. He has not yet checked my conclusions as to the effects of recurrent and reversible mutation and of immigration by his method.

Differences in Interpretation

There are, however, important differences in interpretation. Dr. Fisher is interested in the figure $\frac{1}{2n}$, measuring decrease in variance, only because of its extreme smallness, from which he argues that the effects of random sampling are negligible in evolution (except as bearing on the chances of loss of a recently originated gene). I, on the contrary, have attributed to the inbreeding effect, measured by this coefficient, an essential rôle in the theory of evolution, arguing that the effective breeding population, represented by n of the formula may after all be relatively small compared with the actual size of the population. In this view I have been encouraged by the rather high coefficients of inbreeding found even in entire breeds of livestock. Calder, for example, finds a rate of increase of the inbreeding coefficient in Scotch Clydesdale horses of nearly 1% per generation which let it be emphasized again is a direct determination of the value of $\frac{1}{2n}$ for this large breed, assuming as seems to be justified, that there is no important subdivision into local strains.

The core of Dr. Fisher's theory of selection is given in Chapter II. He reaches a formula on which he lays great emphasis as "the fundamental theorem of natural selection." "The rate of increase in fitness of any organism at any time is equal to its genetic variance in fitness at that time."

This is given as exact for idealized populations in which fortuitous fluctuations in genetic composition have been excluded *i. e.,* in indefinitely large populations. He calculates the standard error of the rate of advance in fitness, due to such fluctuations, and concludes that this is negligibly small; even over a single generation, in populations of the order of size of natural species. This means that the small random fluctuations in the frequencies of individual genes balance each other in their effect on a selected character to such an extent that irregularities in evolutionary advance are of the second order with respect to the rate of advance. He compares this principle to the regular increase of entropy in a physical system. The only effective offset to undeviating increase in fitness, which he recognizes, is change of environment, living or non-living, which he points out must usually be for the worse. The net effect of natural selection, and change of environment is registered in an increase or decrease in numbers and a somewhat winding course of evolution.

The splitting of species, he attributes to differences in the direction of selection in different parts of the range. The process may be facilitated by geographic (or other) isolation, but he holds that it may also be brought about wholly by selection, the primary selection tending to set up secondary processes (including especially preferential mating *i. e.,* sexual selection), which in the end may lead to complete fission of the species.

It will be seen that Dr. Fisher's conception of evolution is pure Darwinian selection. The extent to which he carries the principle is well illustrated in his theory of dominance (chapter III) in which he attempts to account for the prevalent dominance of type genes, over mutant genes by the natural selection of modifiers of dominance.* I

---

*This theory was first elaborated by Fisher in papers which appeared in *The American Naturalist* (62:115-126, 571-574, 1928). In a criticism of it (*American Naturalist* 63:274-

have pointed out elsewhere and he has agreed, that the selection pressure on the modifiers is here of the second order, compared with the rate of mutation of the primary gene. It seemed probable to me that such a minute selection pressure would ordinarily be of the second order compared with other selection pressures acting on the same gene, and therefore negligible. Dr. Fisher on the other hand, adheres to the effectiveness of selection in this case.

In order to bring out the point at which we part company with respect to the efficacy of selection, it will be necessary to return to Dr. Fisher's fundamental theorem: "The rate of increase in fitness of any organism at any time is equal to its genetic variance in fitness at that time." One's first impression is that the genetic variance in fitness must in general be large and that hence if the theorem is correct the rate of advance must be rapid. As Dr. Fisher insists, however, the statement must be considered in connection with the precise definition which he gives of the terms. He uses "genetic variance" in a special sense. It does not include all variability due to differences in genetic constitution of individuals. He assumes that each gene is assigned a constant value, measuring its contribution to the character of the individual (here fitness) in such a way that the sums of the contributions of all genes will equal as closely as possible the actual measures of the character in the individuals of the population. Obviously there could be exact agreement in all cases only if dominance and epistatic relationships were completely lacking. Actually, dominance is very common and with respect to such a character as fitness, it may safely be assumed that there are always important epistatic effects. Genes favorable in one combination, are, for example, extremely likely to be unfavorable in another. Thus allelomorphs which are held in equilibrium by a balance of opposing selection tendencies (possibilities of which are discussed in Chapter V) may contribute a great deal to the total genetically determined variance but not at all to the genetic variance in Fisher's special sense, since at equilibrium there is no difference in their contributions. The formula itself seems to need revision in the case of another important class of genes, ones slightly deleterious in effect but maintained at a certain equilibrium in frequency by recurrent mutation (or migration). These contribute to the genetic variance of the species, but not to the increase in fitness. Terms involving mutation (and migration) rates seem to be omitted in the formula as given.

Mutational Flux as a Factor

Consider now the case of a population so large that fortuitous variation of gene frequency is negligible. According to my view, such a population is one in which all mutations which can occur will recur at measurable rates. All genes which are not fixed will be held in equilibrium by opposing selections, or by selection opposed by mutation, the cases just discussed. Thus while there may be a great deal of genetically determined variance, there will be no movement of gene frequencies and hence no evolution as long as external conditions remain constant. This state of equilibrium may be upset by change of external conditions, bringing changes in the direction and intensity of selection. All gene frequencies may then be expected to shift in an orderly fashion until the equilibrium consistent with the new conditions is attained. On return to the old conditions, all gene frequencies should shift back to the old positions. It may be granted that an irregular sequence

297, 1929), I proposed an alternative directly physiological interpretation of the phenomenon. Further discussion may be found in Fisher's reply to this criticism (*American Naturalist* 63:553-556, a counter-reply *ibid* 556-561 and a paper by J. B. S. Haldane, *American Naturalist* 64:87-90 1930).

of environmental condition would result occasionally in irreversible changes (because of epistatic relationships) thus giving a real, if very slow, evolutionary process; but this is not Dr. Fisher's scheme under which evolution should proceed under constant external conditions. He would have the system of equilibria of gene frequencies kept in motion by a steady flux of novel mutations. These to be effective must be advantageous practically from the first, since non-recurrent, unfavorable mutations would be lost (in an indefinitely large population) before they could reach such a frequency as to have any appreciable effect on the situation. Even those advantageous at once would also usually be lost within a few generations of their appearance. They would, however, as Dr. Fisher shows, have a finite chance of reaching high frequencies and ultimately fixation. In their progress, they may be expected to unsettle the equilibria of other genes by creating new favorable (or unfavorable) combinations. Thus the entire system of gene frequencies is thrown into motion and may yield the steady adaptive advance of the theory.

As noted above, this scheme appears to depend on an inexhaustible flow of new favorable mutations. Dr. Fisher does not go into this matter of inexhaustibility but presumably it may be obtained by supposing that each locus is capable of an indefinitely extended series of multiple allelomorphs, each new gene becoming a potential source of genes which could not have appeared previously. The greatest difficulty, seems to be in the posited favorable character of the mutations. Dr. Fisher, elsewhere, presents cogent reasons as to why the great majority of all mutations should be deleterious. He shows that all mutations affecting a metrical character "unless they possess countervailing advantages in other respects will be initially disadvantageous." He shows that in any case the greater the effect, the less the chance of being adaptive. Add to this the point that mutations as a rule probably have multiple effects, and that the sign of the net selection pressure is determined by the greater effects, and it will be seen that the chances of occurrence of new mutations, advantageous from the first are small indeed.

Partial Isolation as a Factor

I would not deny the possibility of very slow evolutionary advance through this mechanism but it has seemed to me that there is another mechanism which would be much more effective in preventing the system of gene frequencies from settling into a state of equilibrium, than the occurrence of new immediately favorable mutations. If the population is not too large, the effects of random sampling of gametes in each generation brings about a random drifting of the gene frequencies about their mean positions of equilibrium. In such a population we can not speak of single equilibrium values but of probability arrays for each gene, even under constant external conditions. If the population is too small, this random drifting about leads inevitably to fixation of one or the other allelomorph, loss of variance, and degeneration. At a certain intermediate size of population, however (relative to prevailing mutation and selection rates), there will be a continuous kaleidescopic shifting of the prevailing gene combinations, not adaptive itself, but providing an opportunity for the occasional appearance of new adaptive combinations of types which would never be reached by a direct selection process. There would follow thorough-going changes in the system of selection coefficients, changes in the probability arrays themselves of the various genes and in the long run an essentially irreversible adaptive advance of the species. It has seemed to me that the conditions for evolution would be more favorable here than in the indefinitely large population of Dr. Fisher's scheme. It would, however, be very slow, even in terms of geologic time, since it can be shown to be limited by mutation rate. A much more favorable condition would be that of a large population, broken up into imperfectly isolated local strains. The probability array for genes within such a local strain has been given on a previous page. The rate of evolutionary change depends primarily on the balance between the effective size of population in the local strain (n) and the amount of interchange of individuals with the species as a whole (m) and is therefore not limited by mutation rates. The consequence would seem to be a rapid differentiation of local strains, in itself non-adaptive, but permitting selective increase or decrease of the numbers in different strains and thus leading to relatively rapid adaptive advance of the species as a whole. Thus I would hold that a condition of subdivision of the species is important in evolution not merely as an occasional precursor of fission, but also as an essential factor in its evolution as a single group. Between the primary gene mutations, gradually carrying each locus through an endless succession of allelomorphs, and the control of the major trends of evolution by natural selection, I would interpolate a process of largely random differentiation of local strains. As to the existence of such strain differences, the situations described in the herring by Heinke, in Zoarces and Lebistes by J. Schmidt, and in deer mice by Sumner, as well as the situation in man, may be called to mind.

Sexual Selection and Mimicry

To the general biological reader, the later chapters of the book dealing with concrete applications of the selection principle may prove most attractive. A well sustained attempt to rehabilitate Darwin's theory of sexual selection has already been noted. Another chapter deals with mimicry. The validity of both Batesian and Müllerian mimicry is accepted and the possibility of accounting for the origin of each sort by natural selection is developed after careful analysis of opposing arguments which have been widely accepted, especially in the case of Müllerian mimicry. The author naturally applies his theory of direct progress through mass selection. It appears to me, however, that these cases fall at least equally well under the viewpoint which I have developed, which does not require such a minutely continuous path of selective advantage between the original pattern of the species and that ultimately reached.

Evolution In Man

More than one-third of the book is devoted to discussion of the trend of evolution in man. This portion deserves the most careful consideration by all interested in problems of Eugenics. The course of the argument may be summarized briefly as follows: One might expect to find that civilization once started on the earth would give such an advantage that its history would be an uninterrupted succession of triumphs. Instead of this, we find that every civilization, after a period of prosperity, has fallen into decay, and succumbed to the onslaughts of numerically weak, barbarous peoples. The cause of this decay, he finds reason to believe, is genetical rather than social. Evidence indicates that differences in fertility are in part hereditary, whether dependent on physical or mental qualities. The bulk of the evidence from civilized communities, ancient and modern, indicates that fertility is lowest in the upper classes of the population, where qualities which make for individual ability and leadership are most frequent. The reason for this inversion of the normal relation is seen in the tendency (first pointed out by Galton) for infertility as well as ability to rise in the social scale. The result is a tendency to extinction of ability, applying to all classes in society. Examination of conditions in more primitive societies organized on the clan basis, lead to the conclusion that the play of natural selection is here exactly the opposite. The evolution of individual qualities he believes reaches its climax just before civilization begins.

The final chapter deals with the conditions necessary for a permanent civilization. The author holds that only a wage system definitely designed to remove the present severe social penalty on fertility and indeed tending to promote fertility would adequately oppose the present tendency toward racial deterioration.

Literature Cited

1. Calder, A., 1927. *Proc. Roy. Soc. Edinburgh.* 47:118-140.

2. Fisher, R. A., 1922. *Proc Roy. Soc. Edinburgh.* 42:321-341.

3. Smith, A. D. B., 1926. *Eugenics Review.* 14:189-204. 1928. *Report of Brit. Ass. Adv. Science,* 649-655.

4. Wright, S., 1921. *Genetics.* 6:111-178.

5. Wright, S., 1922. *Amer. Nat.* 61:330-338. 1923, *Jour. Her.* 14:339-348, 405-422. McPhee, H. C., and S. Wright. 1925-26. *Jour. Hered.* 16:205-215, 17:397-401.

10
Statistical Theory of Evolution

Journal of the American Statistical Association 26, suppl. (March 1931):201–8

Introduction

This paper was a brief summary of Wright's long manuscript on evolution, which by the time this paper was published (March 1931) had been in press for fourteen months. The paper seems to have been little read at the time, perhaps because the much larger paper from which it was drawn was also published in the same month. Still, this paper is of considerable interest as representing what Wright thought statisticians might wish to know about his theory of evolution.

STATISTICAL THEORY OF EVOLUTION

By Sewall Wright

A recent writer on evolution, Dr. R. A. Fisher, has made an interesting comparison between the position of the evolutionary principle in biology and the second law of thermodynamics in the physical sciences. He quotes Eddington to the effect that the law that entropy increases holds the supreme position among the laws of nature and notes that the principle of evolution holds a similar position among the biological sciences. Both describe irreversible processes and thus mark a direction in time, the law of increase of entropy, according to Eddington, being unique among the physical sciences in so doing. Both are statistical laws. Dr. Fisher notes, however, a remarkable contrast between them. The operation of the second law of thermodynamics brings about a disorganization of the systems concerned, a passage from less probable to more probable states. Evolution, on the other hand, is nearly always described in terms of progress, a passage from simple to complex organization, from more probable to less probable states. Fitness takes the place of entropy in the formulation.

Whether evolution is a mere eddy in a general process of running downhill in the universe, as Eddington doubtless would hold, or whether the developmental side of nature so conspicuous in the biological sciences is an aspect of reality more basic than increase of entropy in physical systems, or whether time is essentially without direction one way or the other—as G. N. Lewis holds, at least with respect to the physical sciences—are philosophical questions which I shall not attempt to discuss. The comparison sufficiently brings out the difficulty of accounting for the evolutionary process on the same basis, statistical theory, as that which leads to the law of increase of entropy in the physical sciences. Yet it seems the only course open to scientific analysis.

The first attempts at explanation, to be sure, interepreted the process as directly physiological rather than statistical. Lamarck assumed that the physiological adaptations or organisms to varying environments produced parallel changes in the heredity which they transmitted. Experimental study of heredity and of development have shown that this interpretation is not available. The observed properties of variation, for one thing, are as far as possible from those postulated by this theory.

Darwin was the first to present effectively a sketch of a statistical interpretation, the play of natural selection upon random hereditary variability. But practically nothing was known in Darwin's time of the physical mechanism of heredity. He merely assumed that there was a tendency toward persistence of type, qualified by small random variations. Since the rediscovery of Mendelian heredity in 1900, the subject of heredity has changed from one which was a plaything of every speculative writer, biological or otherwise, to one in which we have as much exact knowledge as, perhaps, in any other field of biology. This knowledge has raised some difficulties of which Darwin was not aware. It includes, however, mathematically expressed rules whose statistical consequences, in populations, can be worked out with something like the confidence that they correspond to realities, which we find in the physical sciences.

The basic fact of modern genetics is, of course, that heredity is composed of units, "genes," whose most essential property is that of duplicating themselves with most extraordinary precision (as determined by effects under controlled genetic and environmental conditions), quite regardless of the characteristics of the organism in whose cells they are carried. The effects on the developmental process, dependent on the general genetic and environmental situation, constitute secondary properties. It is the property of autosynthesis of a doubtless highly complex material which makes possible the apparently disentropic aspect of evolution. Certain highly "improbable" states of organization are hereby multiplied instead of being dissipated as in ordinary thermodynamic systems.

Absolute precision of gene duplication is, however, incompatible with evolution. The exceptions, so-called "gene mutations," have been much studied of late. Their properties at first sight seem as far as possible from those required for progressive evolution. The typical rate of mutation for individual genes can hardly be more than 1 per million per generation. Direction of mutation has no relation to external conditions, although at least one agent, X-rays, greatly increases the rate. Mutations are practically never adaptive. They are usually definitely injurious, although sometimes apparently indifferent, especially if very slight in effect. The low rate of mutation observed is probably about as much as species can stand in view of the prevailingly injurious effects. The observed time rate of lethal mutation in the vinegar fly, with two weeks between generations, would bring about extinction of the human species in one generation.

I shall pass rapidly over two factors of the greatest significance in making progressive evolution a conceivable process upon such unfavorable material as described above. First is the aggregation of numerous genes in the same cell, and the evolution of a mechanism of exact equational division of such an aggregation, the mechanism of mitosis. This is apparently absent in bacteria and blue-green algae but present in all higher plants and all animals, and is doubtless necessary for any high degree of organization. Associated with mitosis is sexual reproduction, involving the union of half samples of the parental heredities. Biparental heredity makes it possible for a not too injurious mutation to be tried out in combination with all mutations carried by the species, and since it is really the combination and not the individual gene which is injurious or adaptive, it becomes possible that an initially injurious gene may find a place in an adaptive combination. Under uniparental reproduction, each mutation adds only one new type to the species. The occurrence of 100 different mutations means only 101 types. Under biparental reproduction each new mutation doubles the number of possible gene combinations. One hundred different mutations means 2^{100} or about 10^{30} potential types. Compare this with 101 and you will appreciate the enormously greater field of variation presented under biparental reproduction for the play of natural selection than under uniparental reproduction. The problem is for the species to hold its slightly injurious mutations until it can work its way in some way through the nearly infinite field of gene combinations to the particular combination which will mark an advance. Whether this can be brought about by natural selection alone is a moot question. Dr. R. A. Fisher, to whom I referred at the beginning, has made a mathematical investigation which leads to the conclusion that natural selection is enough, that such selection must inevitably lead the species along the road of increasing fitness even in the minutest detail, assuming that the environment does not deteriorate. I have been led to somewhat different conclusions.

In considering the problem, it is necessary to start with a conception of the differences between species different from that which is perhaps most usual. We are likely to think of a natural species as an assemblage of individuals all homozygous for the same genes except for rare mutants. According to this view two species differ in that certain genes of one are replaced by allelomorphs in the other and the elementary evolutionary process is looked upon as the replacement of one gene by a mutation. It corresponds better with observation to assume that a species is made up of individuals no two of which are alike. Mutations have been occurring for millions of years, and each series of allelomorphs is typically represented in the species by more than one gene. What characterizes a species is a certain ratio in each series of allelomorphs. The symmetry of the Mendelian mechanism insures the constancy of such ratios in large populations, unless disturbed by evolutionary pressure. The elementary evolutionary process, according to this view, is merely change of gene frequency. It is conceivable that two species may have all of their genes in common. A difference in the frequencies of a large number of genes could well bring about such a differentiation that the probability would be indefinitely small that any individuals of one group could be mistaken for the other.

The effects of various evolutionary pressures on gene frequency are easily deduced. In the case of a gene which is mutating with measurable frequency, the rate of change is, of course, directly proportional to the frequency, giving a straight line on a graph. Selection can have no effect on the frequency of a gene which is completely lost or fixed. The rate of change rises to a maximum at some point between, depending on dominance. In the case of unfavorable recurrent mutations, the mutation pressure, tending to shift gene frequency in one direction, is opposed to selection, tending to shift it in the other. At a certain point, the two lines intersect. This is a point of equilibrium. Whatever the initial constitution of the population, there will be change until this equilibrium is reached. After this, evolution ceases as long as the conditions lead to the same mutation pressure and the same selection. The effects of migration can be treated similarly. Given constant conditions, evolution should occur only until every series of allelomorphic genes has reached the equilibrium determined by the various evolutionary pressures. Change in conditions should be followed by systematic changes in all gene frequencies until all have reached the appropriate new positions of equilibrium. Return to the old conditions should be followed by return to the old equilibria. We have here reached a theory of specific stability, amid infinite individual variability, rather than a theory of evolution. And even the changes brought about by changes in conditions—more severe selection, for example—being reversible, are scarcely to be dignified by the term evolution.

Our analysis so far, however, has certain limitations. We have really treated only the relations between a gene and a single allelomorph. If, as is probable, each gene is capable of mutation through an indefinitely extended series of allelomorphic conditions, new ones may appear sufficiently favorable from the first to upset the equilibrium, to make possible new combinations, to alter all of the selection coefficients and thus bring about a continuous, essentially irreversible process even under constant conditions. This seems to be Fisher's scheme. The difficulty is the extreme rarity of new mutations favorable from the start.

It has seemed to me that another factor should be much more important in keeping the system of gene frequencies from settling into equilibrium. This is the effect of random sampling in a breeding population of limited size. The gene frequencies of one generation may be expected to differ a little from those of the preceding merely by chance. In the course of generations this may bring about important changes, although the farther the drift from the theoretical equilibrium, the greater will be the pressure toward return. The resultant of these tendencies is a certain frequency distribution, or probability curve, for gene frequencies in place of a single equilibrium value.

The most general solution which I have reached for the formula of this probability curve is as follows:

$$y = Ce^{4Nsq}q^{4N(mq_m+v)-1}(1-q)^{4N[m(1-q_m)+u]-1}$$

It includes terms representing effective size of population (N), mutation rate from (u) and to (v) the gene in question, amount of exchange with other populations than that in question (m) and selection coefficient (s such that the gene and its array of allelomorphs reproduce at the rate $1{:}1-s$ respectively). Gene frequency is represented by q (abscissa in charts) and that of the species as a whole by q_m. The form of the curve and the consequent statistical situation in the population vary greatly, depending on the relative magnitudes of the coefficients.

The bearing on evolution can perhaps best be brought out by comparing certain extreme and intermediate cases. Chart I represents the situation in a small population. The probability array for genes is approximately of the form, $y = Cq^{-1}(1-q)^{-1}$. Most genes drift at random into complete fixation or loss, bringing the well known effects of close inbreeding. The population is extremely uniform. Only very rarely is an old gene replaced by a new one. Even severe selection has little effect, and the fixation process, being random, is in general injurious. The end can only be extinction for a group, permanently reduced below a certain size of population (in relation to other evolutionary factors).

Thus a small population is not favorable to evolution. Consider the opposite extreme, a very large population (see Chart III). Here random variation of gene frequencies is a negligible factor and we have the situation considered first. Each gene is held in equilibrium at a certain frequency determined by selection, mutation, etc. Although the variability of the population may be great, the genes of different series combining in different ways probably in every individual, the average condition remains the same as long as conditions are constant,

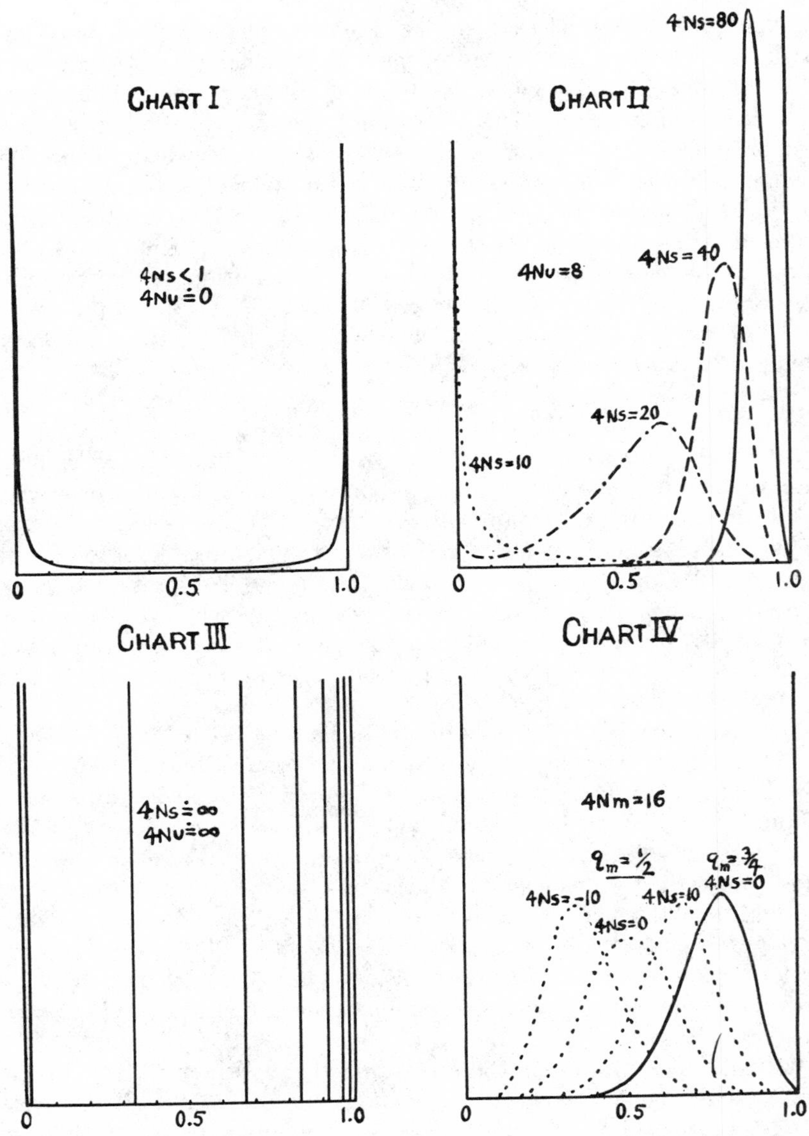

subject to the possibility, already discussed, that wholly novel favorable mutations may disturb the situation. A change in conditions, such as more severe selection, may rapidly change the average of the population, but the change is at the expense of the store of variability of the species and compromises evolutionary advance for a long time following, since there is no escape from fixation except through the slow process of mutation. As previously noted, such change is of an essentially reversible sort and thus not really of an evolutionary character.

Thus it seems that neither a small nor a large freely interbreeding population offers an adequate basis for a continuing evolutionary process. How is it with a population of intermediate size, defined as one in which the reciprocal of population size, the selection coefficient and mutation rate are all of about the same order. As shown in Chart II, gene frequencies drift at random about their equilibrium values, but not to the point of fixation as in a smaller population. Just because the direction of drift is accidental, the result is a kaleidoscopic shifting of the average characters of the population through predominant types which practically are never repeated. The selection coefficients cannot be expected to be constant under such conditions. It is the organism as a whole that is selected, not the individual genes, and a gene favored in one combination may be unfavorable in another. Thus the probability arrays themselves will be constantly changing—some moving to the right and closing up, others to the left and opening out, some to the extreme left and loss. A continuous and irreversible, though primarily non-adaptive, evolutionary process will take place even under constant conditions. The rate, however, is limited by mutation rate and hundreds of thousands of generations are required for important evolutionary changes. Nevertheless this case seems the most rapid to be considered so far.

We have not dealt wholly fairly with the case of large species. As size of population increases, the tendency to spread out and break up into partially isolated groups increases. Each sub-group has a system of frequency arrays for the genes in which they drift about at random and at rates determined, not by the size of the whole species and mutation pressure, but by the size of the sub-group and migration pressure. (See Chart IV.) The rate of decrease in heterozygosis due to size of population is $1/2N$, where N is the effective size of the breeding population. The result is a geologically rapid drifting apart of the various sub-groups, even under uniform conditions. This is a non-adaptive radiation, but, on the average, not such as to lead to appreciable deterioration. Exceptionally favorable combinations of genes may come to predominate in some of the sub-groups. These may be expected to expand their ranges while others dwindle. This process of intergroup selection may be very rapid as compared with mass selection of individuals, among whom favorable combinations are broken up by the reduction-fertilization mechanism in the next generation after formation. With partial isolation and differentiation accompanying expansion of the successful sub-groups, the process may go on indefinitely. In short this seems from statistical considerations to be the only mechanism which offers an adequate basis for a continuous and progressive evolutionary process. It may be added that when tested by observation, it accords excellently with the actual situation found among natural species. It agrees well with the views reached by many field biologists.

The final conclusion to which this analysis leads seems to be as follows: The conditions favorable to progressive evolution as a process of cumulative change are neither extreme mutation, extreme selection, extreme hybridization nor any other extreme, but rather a certain balance between conditions which make for genetic homogeneity and genetic heterogeneity. Such a situation means on the one hand the retention of a great store of variability in the population and on the other hand a random drifting of the mean grade of all characters, leading, occasionally by chance, to the attainment of exceptionally favorable gene combinations. In particular, a state of sub-division of a sexually reproducing population into small, incompletely isolated groups provides the most favorable condition, not merely for branching of the species, but also for its evolution as a single group.

11
Evolution in Mendelian Populations

Genetics 16, no. 2 (1931):97–159

12
The Roles of Mutation, Inbreeding, Crossbreeding, and Selection in Evolution

Proceedings of the Sixth International Congress of Genetics, 1 (1932):356–66

Introduction

These are Wright's two most seminal early papers on evolutionary theory. The first resulted in his admission to the National Academy of Science at a young age, and the second was probably the most influential paper he ever published.

I have devoted chapter 9 of *SW&EB* to discussion and analysis of these two papers and the differences between them; no brief summary is feasible here.

In a real sense, the statistical distribution of genes that Wright derived in the 1931 paper was the basis for all of his later work in mathematical population genetics theory. Throughout his later career, he attempted to elaborate and apply this statistical distribution of genes. The four volumes of Wright's *E&GP* are, more than anything else, his later elaboration of these two papers as he attempted to achieve greater generality and simplicity for his distribution of genes and its relations with his qualitative shifting balance theory of evolution in nature.

EVOLUTION IN MENDELIAN POPULATIONS

SEWALL WRIGHT

University of Chicago, Chicago, Illinois

Received January 20, 1930

TABLE OF CONTENTS

THEORIES OF EVOLUTION

One of the major incentives in the pioneer studies of heredity and variation which led to modern genetics was the hope of obtaining a deeper insight into the evolutionary process. Following the rediscovery of the Mendelian mechanism, there came a feeling that the solution of problems of evolution and of the control of the process, in animal and plant breeding and in the human species, was at last well within reach. There has been no halt in the expansion of knowledge of heredity but the advances in the field of evolution have, perhaps, seemed disappointingly small. One finds the subject still frequently presented in essentially the same form as before 1900, with merely what seems a rather irrelevant addendum on Mendelian heredity.

The difficulty seems to be the tendency to overlook the fact that the evolutionary process is concerned, not with individuals, but with the species, an intricate network of living matter, physically continuous in space-time, and with modes of response to external conditions which it appears can be related to the genetics of individuals only as statistical consequences of the latter. From a still broader viewpoint (compare LOTKA 1925) the species itself is merely an element in a much more extensive evolving pattern but this is a phase of the matter which need not concern us here.

The earlier evolutionists, especially LAMARCK, assumed that the somatic effects of physiological responses of individuals to their environments were transmissible to later generations, and thus brought about a directed evolution of the species as a whole. The theory remains an attractive one to certain schools of biologists but the experimental evidence from genetics is so overwhelmingly against it as a general phenomenon as to render it unavailable in present thought on the subject.

DARWIN was the first to present effectively the view of evolution as primarily a statistical process in which random hereditary variation merely furnishes the raw material. He emphasized differential survival and fecundity as the major statistical factors of evolution. A few years later, the importance of another aspect of group biology, the effect of isolation, was brought to the fore by WAGNER. Systematic biologists have continued to insist that isolation is the major species forming factor. As with natural selection, a connection with the genetics of individuals can be based on statistical considerations.

There were many attempts in the latter part of the nineteenth century to develop theories of direct evolution in opposition to the statistical viewpoint. Most of the theories of orthogenesis (for example, those of EIMER and of COPE) implied the inheritance of "acquired characters." NÄGELI postulated a slow but self contained developmental process within protoplasm; practically a denial of the possibility of a scientific treatment of the problem. Differing from these in its appeal to experimental evidence and from the statistical theories in its directness, was DE VRIES' theory of the abrupt origin of species by "mutations." A statistical process, selection or isolation, was indeed necessary to bring the new species into predominance, but the center of interest, as with Lamarckism, was in the physiology of the mutation process.

The rediscovery of Mendelian heredity in 1900 came as a direct consequence of DE VRIES' investigations. Major Mendelian differences were naturally the first to attract attention. It is not therefore surprising that the phenomena of Mendelian heredity were looked upon as confirming DE VRIES' theory. They supplemented the latter by revealing the possibilities of hybridization as a factor bringing about an extensive recombination of mutant changes and thus a multiplication of incipient species, a phase emphasized especially by LOTSY. JOHANNSEN'S study of pure lines was interpreted as meaning that DARWIN'S selection of small random variations was not a true evolutionary factor.

A reaction from this viewpoint was led by CASTLE, who demonstrated the effectiveness of selection of small variations in carrying the average of a stock beyond the original limits of variation. This effectiveness turned out to depend not so much on variability of the principal genes concerned as on residual heredity. As genetic studies continued, ever smaller differences were found to mendelize, and any character, sufficiently investigated, turned out to be affected by many factors. The work of NILSSON-EHLE, EAST, SHULL, and others established on a firm basis the multiple factor hypothesis in cases of apparent blending inheritance of quantitative variation.

The work of MORGAN and his school securely identified Mendelian heredity with chromosomal behavior and made possible researches which further strengthened the view that the Mendelian mechanism is the general mechanism of heredity in sexually reproducing organisms. The only exceptions so far discovered have been a few plastid characters of plants. That differences between species, as well as within them, are Mendelian, in the broad sense of chromosomal, has been indicated by the close parallelism between the frequently irregular chromosome behavior and the genetic phenomena of species crosses (FEDERLEY, GOODSPEED and CLAUSEN, etc.). Most of DE VRIES' mutations have turned out to be chromosome aberrations, of occasional evolutionary significance, no doubt, in increasing the number of genes and in leading to sterility of hybrids and thus isolation, but of secondary importance to gene mutation as regards character changes. As to gene mutation, observation of those which have occurred naturally as well as of those which MULLER, STADLER, and others have recently been able to produce wholesale by X-rays, reveals characteristics which seem as far as possible from those required for a directly adaptive evolutionary process. The conclusion nevertheless seems warranted by the present status of genetics that any theory of evolution must be based on the properties of Mendelian factors, and beyond this, must be concerned largely with the statistical situation in the species.

VARIATION OF GENE FREQUENCY

Simple Mendelian equilibrium

The starting point for any discussion of the statistical situation in Mendelian populations is the rather obvious consideration that in an indefinitely large population the relative frequencies of allelomorphic genes remain constant if unaffected by disturbing factors such as mutation, migration, or selection. If $[(1-q)a+qA]$ represents the frequencies of two allelomorphs, (a, A) the frequencies of the zygotes reach equilibrium according to the expansion of $[(1-q)a+qA]^2$ within at least two generations,[1] whatever the initial composition of the population (HARDY 1908). Combinations of different series are in equilibrium when these are combined at random, but as WEINBERG (1909) and later, in more detail, ROBBINS (1918) have shown, equilibrium is not reached at once but is approached asymptotically through an infinite number of generations. Linkage slows down the approach to equilibrium but has no effect on the ultimate frequencies.

Mutation pressure

The effects of different simple types of evolutionary pressure on gene frequencies are easily determined. Irreversible mutation of a gene at the rate u per generation changes gene frequency (q) at the rate $\Delta q = -uq$. With reverse mutation at rate v the change in gene frequency is $\Delta q = -uq+v(1-q)$. In the absence of other pressures, an equilibrium is reached between the two mutation rates when $\Delta q = 0$, giving $q = v/(u+v)$.

Migration pressure

The frequency of a gene in a given population may be modified by migration as well as by mutation. As an ideal case, suppose that a large population with average frequency q_m for a particular gene, is composed of subgroups each exchanging the proportion m of its population with a random sample of the whole population. For such a subgroup, $\Delta q = -m(q-q_m)$.

The conditions postulated above are rather artificial since, in an actual species, subgroups exchange individuals with neighboring subgroups rather

[1] This statement assumes that there is no overlapping of generations which may bring about some delay in the attainment of equilibrium.

than with a random sample of the whole species and the change in q will be only a fraction of that given above. The fraction is the average degree of departure of the neighboring subgroups toward the population average. The formula may be retained by letting q_m stand for the gene frequency of immigrants rather than of the whole species.

Selection pressure

Selection, whether in mortality, mating or fecundity, applies to the organism as a whole and thus to the effects of the entire gene system rather than to single genes. A gene which is more favorable than its allelomorph in one combination may be less favorable in another. Even in the case of cumulative effects, there is generally an optimum grade of development of the character and a given plus gene will be favorably selected in combinations below the optimum but selected against in combinations above the optimum. Again the greater the number of unfixed genes in a population, the smaller must be the average effectiveness of selection for each one of them. The more intense the selection in one respect, the less effective it can be in others. The selection coefficient for a gene is thus in general a function of the entire system of gene frequencies. As a first approximation, relating to a given population at a given moment, one may, however, assume a constant net selection coefficient for each gene. Assume that the genes a and A tend to be reproduced in the ratio $(1-s):1$ per generation. The gene array $[(1-q)a+qA]$ becomes $[(1-s)(1-q)a+qA]/[1-s(1-q)]$. The change in the frequency of A is $\Delta q=[sq(1-q)]/[1-s(1-q)]$ or with sufficiently close approximation $\Delta q=sq(1-q)$ if the selection coefficient is small.

A second approximation may be obtained by considering the zygotic frequencies. Assume that the types aa, Aa, and AA reproduce in the ratio $(1-s'):(1-hs'):1$ per generation. The change in the frequency of A to a sufficiently close approximation is $\Delta q=s'q(1-q)[1-q+h(2q-1)]$ (WRIGHT 1929). In the case of selection for or against a complete recessive ($h=0$, s' negative or positive respectively), $\Delta q=s'q(1-q)^2$.

The case of no dominance ($h=\frac{1}{2}$) is the same as the case of genic selection except that the selection against the gene is $s'/2$ instead of s.

The two factor case in which the phenotypes aabb, aaB–, A–bb and A–B– reproduce at the rates $(1-s_{ab}):(1-s_a):(1-s_b):1$ respectively yields (for low values of the selection coefficients):

$$\Delta q_A = q_A(1-q_A)^2[s_a+(s_{ab}-s_a-s_b)(1-q_B)^2].$$

The frequency of A depends on the frequency and selection of B, becoming independent only if $s_{ab}=s_a+s_b$, that is, if the two series of genes are cumulative with respect to selection. It does not seem profitable to pursue this subject further for the purpose of the present paper, since in the general case, each selection coefficient is a complicated function of the entire system of gene frequencies and can only be dealt with qualitatively. Attention may, however, be called to HALDANE's (1924–1927) studies of selection rates and of the consequent number of generations required for unopposed selection to bring about any required change in gene frequency under various assumptions with respect to mode of inheritance and system of mating.

Equilibrium under selection

There may be equilibrium between allelomorphs as a result wholly of selection, namely, selection against both homozygotes in favor of the heterozygous type. Putting $\Delta q=s'q(1-q)[1-q+h(2q-1)]=0$ gives $q=[1-h]/[1-2h]$ as the condition.[2] This includes the case of selection against both homozygotes and also that in favor of them, but examination of the signs of Δq above and below the equilibrium point shows that only the former is in stable equilibrium in agreement with FISHER (1922). The linkage of a favorable dominant with an unfavorable recessive of another series is a case in which selection would be against both homozygotes as JONES (1917) has pointed out, and stressed as a factor in the vigor of heterozygosis. In a population produced by the intermingling of types in which different deleterious recessives have become fixed, there will be a temporary selection in favor of the heterozygotes even without any linkage at all. Unless the linkage is very strong, however, this effect does not persist long enough to have much effect on gene equilibrium. The extreme case of equilibrium of the sort discussed here is, of course, that of balanced lethals, found in nature in Oenothera.

In the two factor case, discussed in the preceding section,

$$\Delta q_A = 0 \text{ if } q_B = 1 - \sqrt{\frac{s_a}{s_a + s_b - s_{ab}}}$$

$$\text{and } \Delta q_B = 0 \text{ if } q_A = 1 - \sqrt{\frac{s_b}{s_a + s_b - s_{ab}}}.$$

[2] The condition can be expressed in a more symmetrical form by using a different form of statement of the selection coefficients. Assume that the rates of reproduction of the three types aa, Aa and AA are as $(1-s_a):1:(1-s_A)$. The value of q at equilibrium comes out $q=\frac{s_a}{s_A+s_a}$ with stable equilibrium only for positive values of the two selection coefficients.

There may be equilibrium here, if s_a and s_b are alike in sign, and s_{ab} is either opposite in sign or of the same sign and smaller, but it is an unstable equilibrium. Of more general importance, perhaps, is the equilibrium reached by a deleterious mutant gene. For mutation opposed by genic selection $\Delta q = -uq + sq(1-q) = 0$, $q = 1 - u/s$. For mutation opposed by zygotic selection (aa, Aa and AA reproducing at rates $(1-s'):(1-hs'):1$ it is easily shown that $q = 1 - u/hs'$ (WRIGHT 1929), unless h approaches 0. Thus with no dominance, $q = 1 - 2u/s'$, and for selection against a dominant mutation, $q = 1 - u/s'$. The important case of selection against a recessive is that in which $h = 0$. The formula becomes $q = 1 - \sqrt{u/s'}$. All of these cases are illustrated in figure 1 in which the ordinates show the selection pressure as related to factor frequency, under the different conditions of selection. The intersections with the straight line representing mutation pressure give the points of equilibrium. If deleterious dominant and recessive mutations occur with equal frequency and are subject to the same degree of selection, the frequency of the recessive mutant genes will be greater than that of the dominant ones in the ratio $\sqrt{u/s'}$ to u/s'. The corresponding figure for factors lacking dominance is $2u/s'$, where s' is the selection against the homozygote. These considerations alone should lead to a marked correlation in nature between recessiveness and deleterious effect. This correlation is further increased by the greater frequency of recessive mutation which seems to be a general phenomenon. It is this correlation which gives the theoretical basis for the immediate degeneration which usually accompanies inbreeding, a process which increases the proportion of recessive phenotypes.

The amount to which gene frequency in a subgroup may depart from the species average as a result of local selection held in check by population interchange with other regions may be calculated by solving the quadratic $\Delta q = sq(1-q) - m(q - q_m) = 0$. If the local selection coefficient is much greater than the proportion of migration ($s > m$), $q = 1 - \frac{m}{s}(1 - q_m)$ or $-mq_m/s$ depending on the direction of selection, formulae analogous to those for the equilibrium between mutation and selection. If, on the other hand, selection is weak compared with migration ($s < m$), the departure from q_m is small and $q = q_m[1 + \frac{s}{m}(1 - q_m)]$. This case is doubtless the more important in nature. Large subgroups living under different selection pressures should show gene frequencies clustering about the average according to this expression. The effect of small size of the subgroups in bringing about random deviation in this and other cases is not here considered. The case in which s and m are of the same order of magnitude may be illustrated by the case of exact equality. Here $q=\sqrt{q_m}$ or $1-\sqrt{1-q_m}$ depending on the direction of selection.

Multiple allelomorphs

The foregoing discussion has dealt formally only with pairs of allelomorphs, a wholly inadequate basis for consideration of the evolutionary pro-

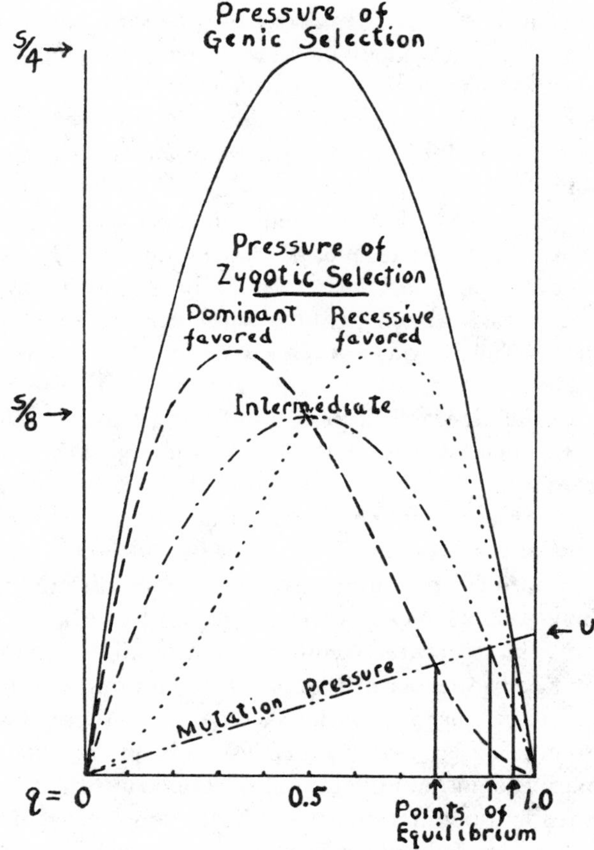

FIGURE 1.—Rate of change of gene frequency under selection or mutation. Genic selection (A, a reproducing at rates $1:1-s$); Zygotic selection: dominant (B-, bb at rates $1:1-s$), recessive (cc, C— at rates $1:1-s$), intermediate (DD, Dd, dd at rates $1:1-\frac{1}{2}s:1-s$), Mutation such that $u=-0.05$ s. Intersections of line of mutation pressure with those for selection pressure determine the equilibrium frequencies.

cess unless extension can be made to multiple allelomorphs. Among the laboratory rodents some 40 percent of the known series of factors affecting coat color are already known to be multiple. The number of multiple series is large in other organisms, for example, Drosophila (MORGAN, STURTEVANT and BRIDGES 1925). It is not unlikely that further study will indicate that all series are potentially multiple. In this case, each gene has a history which is not a mere oscillation between approximate fixation of two conditions but a real evolutionary process in its internal structure. Presumably any particular gene of such an indefinitely extended series can arise at a step from only a few of the others[3] and in turn mutate to only a few. Since genes as a rule have multiple effects and change in one effect need not involve others, it is probable that in time a gene may come to produce its major effects on wholly different characters than at first. Continuing this line of thought, it indeed seems possible that all genes of all organisms may ultimately be traced to a common source, mitotic irregularities furnishing the basis for multiplication of genes.

The relative frequencies of all allelomorphs in a series tend, of course, to remain constant in the absence of disturbing forces. The zygotes reach the equilibrium of random combination of the genes in pairs by the second generation from any initial constitution of the population. The effects of the various kinds of evolutionary pressure on the frequency of each gene may be treated as before by contrasting each gene with the totality of its allelomorphs. In the binomial expression $[(1-q)a+qA]$, A may be understood as representing any gene, and a as including all others of its series. Such treatment, however, requires further qualification with regard to the constancy of the various coefficients. It may still be assumed that the rate (u) of mutational breakdown of the gene in question (A) is reasonably constant, but its rate (v) of mutational origin from allelomorphs must be expected to change. This may be expected to rise to a maximum, as genes closely allied to A in structure become frequent, and to fall off to zero as changes accumulate in the locus. Even at its maximum, however, the rate of formation should in general be of the second order compared with the rate of change to something else, simply because it is one and its alternatives many. Moreover, there is an indication that the genes which become more or less established in a population are not a random sample of the types of mutations which occur. It has been the common experience that mutations are usually recessive. Recessiveness is most simply interpreted physiologically as due to inactivation which may well be the commonest type of mutational change. But the evolutionary process presumably involves in-

[3] Those most closely related genetically, however, need not always be closest in effect. The complete inactivation of a gene in a particular respect may for example occur more freely than a partial inactivation.

crease in activity of genes at least as frequently as inactivation with the consequence that the rate of formation (v) of genes of evolutionary significance becomes negligibly small in comparison with rate of breakdown (u) of such genes. It should be said that FISHER has advanced an alternative hypothesis according to which genes originally without dominance become dominant through a process of selection of modifiers (FISHER 1928, 1929, WRIGHT 1929, 1929a).

The selection coefficient, s, relating to a gene A cannot be expected to be constant if the alternative term a includes more than one gene. The coefficient should rise to a maximum positive value as A replaces less useful genes but should fall off and ultimately become negative as the group of allelomorphs comes to include still more useful genes. But as already discussed, even if A has only one allelomorph, the dependence of the selection coefficient on the frequencies and selection coefficients of non-allelomorphs keeps it from being constant. The existence of multiple allelomorphs merely adds another cause of variation.

Random variation of gene frequency

There remains one factor of the greatest importance in understanding the evolution of a Mendelian system. This is the size of the population. The constancy of gene frequencies in the absence of selection, mutation or migration cannot for example be expected to be absolute in populations of limited size. Merely by chance one or the other of the allelomorphs may be expected to increase its frequency in a given generation and in time the proportions may drift a long way from the original values. The decrease in heterozygosis following inbreeding is a well known statistical consequence of such chance variation. The extreme case is that of a line propagating by self fertilization which may be looked upon as a self contained population of one. In this case, 50 percent of the factors with equal representation of two allelomorphs (that is, in which the individual is heterozygous) shift to exclusive representation of one of the allelomorphs in the following generation merely as a result of random sampling among the gametes. From the series of fractions given by JENNINGS (1916) for the change in heterozygosis under brother-sister mating (population of two) it may be deduced that the rate of loss in this case is a little less than 20 percent per generation. A general method for determining the decrease in heterozygosis under inbreeding has been presented in a previous paper (WRIGHT 1921). It can be shown from this that there is a rate of loss of about 1/2N in the case of a breeding population of N individuals whether equally divided between males and females or composed of monoecious individuals, assuming pairs of allelomorphs. HAGEDOORN (1921) has urged the importance of such random fixation as a factor in evolution.

Another phase of this question was opened by FISHER (1922) who attempted to discover the distribution of gene frequencies ultimately reached in a population as a result of the above process. He studied a number of conditions relative to mutation and selection. He does not state the rate of decrease in heterozygosis (where any) which would follow from the solutions which he reached but this can be deduced very directly from them. It comes out 1/4N for a population of N breeding individuals in the absence of selection or mutation. This is just half the rate indicated by the method referred to above.

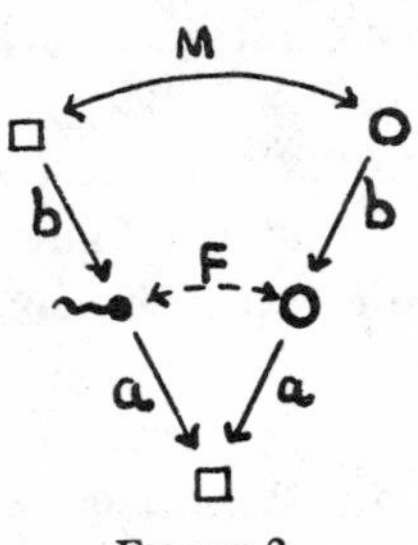

FIGURE 2.

Rate of decrease in heterozygosis

The following symbols and formulae were used in the previous paper in determining the consequences of systems of inbreeding. Primes were used to indicate the number of generations preceding the one in question. Only pairs of allelomorphs are considered here.

M correlation between genotypes of mates

b $(=\sqrt{\tfrac{1}{2}(1+F')})$ path coefficient, zygote to gamete

a $\left(=\sqrt{\dfrac{1}{2(1+F)}}\right)$ path coefficient, gamete to fertilized egg

F $(=b^2M)$ correlation between uniting egg and sperm, also, total proportional change in heterozygosis.

P $(=2q(1-q)(1-F))$ proportion of heterozygosis.

The general formula for the correlation between uniting gametes is easily deduced and has been used as a coefficient of inbreeding in dealing with complex livestock pedigrees (WRIGHT 1922, 1923, 1925, MCPHEE and WRIGHT 1925, 1926),

$$F = \Sigma\left[(\tfrac{1}{2})^{n_s+n_d+1}(1 + F_A)\right].$$

Here F_A is the coefficient of inbreeding of any common ancestor that makes the connecting link between a line of ancestry tracing back from the sire and one tracing back from the dam. The numbers of generations from sire and dam to such a common ancestor are designated n_s and n_d respectively. The contribution of a particular tie between the pedigrees of sire and dam is $(\frac{1}{2})^{n_s+n_d+1}(1+F_A)$ and the total coefficient is simply the sum of all such contributions. This formula makes it possible to compare quantitatively the statistical situation in actual populations with that in ideal populations.

In dealing with regular systems of mating the method of analysis consists in expressing the correlation between mated individuals in terms of path coefficients and correlations pertaining to the preceding generation ($M=\phi(a,'b,'M')$) and from this obtaining expression for F in terms of the F's of the preceding generations.

Consider a population in which there are N_m breeding males and N_f breeding females, and random mating. The proportion of matings between full brother and sister will be $1/(N_mN_f)$, that between half brother and sister $(N_m+N_f-2)/(N_mN_f)$, and that between less closely related individuals $(N_m-1)(N_f-1)/(N_mN_f)$. The correlation between mated individuals may be written as follows, giving due weight to these three possibilities:

$$M = a'^2b'^2\left[\frac{1}{N_mN_f}(2+2M') + \frac{N_m+N_f-2}{N_mN_f}(1+3M') + \frac{(N_m-1)(N_f-1)}{N_mN_f}4M'\right]$$

This leads to the following formula for proportional change in heterozygosis since the foundation period:

$$F = F' + \left(\frac{N_m+N_f}{8N_mN_f}\right)(1-2F'+F'').$$

The proportion of heterozygosis may be written, relative to that of preceding generations:

$$P = P' - \left(\frac{N_m+N_f}{8N_mN_f}\right)(2P'-P'').$$

It is to be expected that the proportional change per generation will reach approximate constancy. This rate may be found by equating P/P' to P/P''

$$-\frac{\Delta P}{P'} = \frac{1}{2}\left(1+\frac{N_m+N_f}{4N_mN_f}\right) - \frac{1}{2}\sqrt{1+\left(\frac{N_m+N_f}{4N_mN_f}\right)^2}.$$

This gives $(1/8N_m+1/8N_f)(1-1/8N_m-1/8N_f)$ as a close approximation even for the smallest populations while for reasonably large ones the form $1/8N_m+1/8N_f$ is sufficiently accurate.

The simplest case is that of continued mating of brother with sister ($N_m=N_f=1$). The rate of loss of heterozygosis comes out $\frac{1}{4}(3-\sqrt{5})$ or 19.1 percent per generation. The formula for proportion of heterozygosis takes the form $P=\frac{1}{2}P'+\frac{1}{4}P''$ as given in the previous paper, with results in exact agreement with those derived by JENNINGS (1916) by working out in detail the consequences of every possible mating from generation to generation.

Another simple case is that in which one male is mated with an indefinitely large number of half-sisters. This is approximately the system of breeding continuously within one herd, headed always by just one male. In this case $N_m=1$, $N_f \doteq \infty$, with rate of loss of heterozygosis of 11.0 percent per generation in agreement with previous results (WRIGHT 1921).

With a relatively limited number of males but unlimited number of females, the rate becomes approximately $1/8N_m$.

An especially important case is that in which the population is equally divided between males and females. Here $N_m=N_f=\frac{1}{2}N$ and the rate of loss is approximately $1/2N$ (or somewhat more closely $1/(2N+1)$) where N is the total size of the breeding population.

It is not, perhaps, clear at first sight that a population of N monoecious organisms, in which self fertilization is prevented, should show a decrease in heterozygosis exactly equal to that in a population of the same size equally divided between males and females. The chance that uniting gametes come from full sisters is $2/[N(N-1)]$, the chance that they come from half sisters is $4(N-2)/[N(N-1)]$ while the chance that they come from less closely related individuals is $(N-2)(N-3)/[N(N-1)]$ giving

$$M=\frac{a'^2b'^2}{N(N-1)}[2(2+2M')+4(N-2)(1+3M')+(N-2)(N-3)4M']$$

$$P=P'-\frac{1}{2N}(2P'-P'')$$

exactly as in the preceding case.

The somewhat arbitrary case in which the gametes produced by N monoecious individuals unite wholly at random is that which can be compared directly with FISHER'S results. The gametes have a chance $1/N$ of coming from the same individual and of $(N-1)/N$ of coming from different individuals. The correlation between uniting gametes may thus be written

$$F = \frac{1}{N}b^2 + \left(\frac{N-1}{N}\right)4b^2a'^2F'$$

$$P = \frac{(2N-1)}{2N}P'.$$

As might be expected, the result does not differ appreciably from that of the preceding case. The rate of loss of heterozygosis is exactly 1/2N instead of merely approximately this figure. The simplest special case is, of course, continued self fertilization in which N = 1 and the formula gives the obviously correct result of 50 percent loss of heterozygosis per generation.

From the mode of analysis it might be thought that the loss in heterozygosis is wholly the consequence of the occasional matings between very close relatives. This, however, is not the case. If instead of random sampling of the gametes produced by the population it is assumed that all individuals reproduce equally and that inbreeding is consistently avoided as much as possible, the percentage of heterozygosis still falls off. The rate of loss is, however, only about half as rapid (approximately 1/4N) in a reasonably large population equally divided between males and females. The cases of N = 2, 4, 8 and 16 have been given previously (WRIGHT 1921).

In dealing with heterozygosis in the foregoing, it has been assumed for simplicity that each locus was represented by only two allelomorphs in the population in question and that either complete fixation or complete loss of a particular gene means homozygosis of all individuals with respect to the locus. But in any case beyond that of self fertilization, more than two allelomorphs may be present and complete loss of the gene no longer implies homozygosis of the locus. The initial rate of loss of heterozygosis in a large population may thus be only 1/4N with gradual approach to 1/2N as the number of loci with only two remaining allelomorphs increases. The rate of decay of the distribution of gene frequencies is 1/2N regardless of number of allelomorphs.

The population number

It will be well to discuss more fully, before going on, what is to be understood by the symbol N used here for population number. The conception is that of two random samples of gametes, N sperms and N eggs, drawn from the total gametes produced by the generation in question (N/2 males and N/2 females each with a double representation from each series of allelomorphs). Obviously N applies only to the breeding population and not to the total number of individuals of all ages. If the population fluctuates greatly, the effective N is much closer to the minimum number than to the maximum number. If there is a great difference between the number of mature males and females, it is closer to the smaller number than to the larger. In fact, as just shown, a population of N_m males and an indefinitely large number of females is approximately equivalent to a population of $4N_m$ individuals equally divided between males and females.

The conditions of random sampling of gametes will seldom be closely approached. The number of surviving offspring left by different parents may vary tremendously either through selection or merely accidental causes, a condition which tends to reduce the effective N far below the actual number of parents or even of grandparents. How small the effective N of a population may be is indicated by recent studies of SMITH and CALDER (1927) on the Clydesdale breed of horses in Scotland, in which they find a steady increase in the degree of inbreeding (Coefficient F) equivalent to that in population headed by only about a dozen stallions. Even more striking is the rapid increase in the coefficient of inbreeding in the early history of the Shorthorn breed of cattle (MCPHEE and WRIGHT 1925).

THE DISTRIBUTION OF GENE FREQUENCIES AND ITS IMMEDIATE CONSEQUENCES

No mutation, migration or selection

On making a cross between two homozygous strains a population is produced in which the members of each pair of allelomorphs in which the strains differ are necessarily equally numerous. The proportional frequency of each allelomorph in unfixed series is $q = 0.50$. In an indefinitely large population, there should be no change in this frequency in later generations (except by recurrent mutation or selection). In any finite population, however, some genes will come to be more frequent than their allelomorphs merely by chance. This means a decrease in heterozygosis, since the proportion of heterozygosis under random mating is $2q(1-q)$, and this quantity is maximum when $q = 0.50$. As time goes on, divergences in the frequencies of factors may be expected to increase more and more until at last some are either completely fixed or completely lost from the population. The distribution curve of gene frequencies should, however, approach a definite form if the genes which have been wholly fixed or lost are left out of consideration. This can easily be seen by considering a case opposite in a sense to that considered above. Suppose that a large number of different mutations occur in a previously pure line. The frequency ratio of mutant to type allelomorphs is initially $(1/2N):(2N-1)/2N$ where N is the number of individuals. The great majority of such mutations will be lost, by the chances of sampling, as Fisher (1922) points out. Those which persist are largely those for which there has been a chance increase in frequency. The distribution curve of frequencies of persisting mutations will thus continually spread toward higher frequencies. There must be a position of equilibrium as far as form is concerned between this situation and that first considered, although a uniform decline in absolute numbers.

As noted above, decrease in heterozygosis takes place in the early generations following a cross without any appreciable fixation or loss of genes. But after equilibrium has been reached in the form of the distribution curve, further loss in heterozygosis must be identical in rate with fixation plus loss.

In simple cases, the equilibrium distribution of gene frequencies can easily be worked out directly. Under brother-sister mating, for example, the following relative frequencies of the 4 possible types of mating involving unfixed factors are in equilibrium although the absolute frequencies of all are falling off 19.1 percent ($=\frac{1}{4}(3-\sqrt{5})$) each generation as new genes enter the fixed states, AA × AA or aa × aa.

Mating	*Relative Frequency*	
		Percent
AA×Aa	$7-3\sqrt{5}=$	29.2
Aa ×Aa	$-22+10\sqrt{5}=$	36.1
AA×aa	$9-4\sqrt{5}=$	5.6
Aa ×aa	$7-3\sqrt{5}=$	29.2
		100.1

Similarly in populations of 2 and 3 monoecious individuals with random union of gametes, the following relative frequencies are in equilibrium although the absolute frequencies are decreasing in each generation by exactly 25 percent and 16⅔ percent respectively verifying the 1/2N of theory.

Gene Frequency	*Class Frequency* Percent	
3A:1a	32	Case of 2 monoecious
2A:2a	36	individuals per gen-
1A:3a	32	eration

Gene Frequency	*Class Frequency* Percent	
5A:1a	18.3	
4A:2a	21.0	Case of 3 monoecious
3A:3a	21.4	individuals per gen-
2A:4a	21.0	eration
1A:5a	18.3	
	100.0	

In order to determine generally the distribution of gene frequencies, consider the way in which genes (A) with frequency q are distributed after one generation of random mating. In a population of N breeding individuals, each of the specified genes will have 2Nq representatives among the zygotes and their allelomorphs $2N(1-q)$. A random sample of the same size will be distributed according to the expression $[(1-q)a+qA]^{2N}$. The contribution of this sample to the frequency class with an allelomorphic ratio of $q_1:(1-q_1)$ will be in proportion to the $2Nq_1$'th term of the above expression and to the number of genes included in the contributing class (f). The sum of contributions from all such classes should give the $2Nq_1$'th term an absolute frequency smaller than its value in the preceding generation (f_1) by the amount $1/(2N+1)$ deduced above. Following is the equation to be solved for f as a function of q.

$$f_1\left(1 - \frac{1}{2N+1}\right) = \frac{\underline{|2N}}{\underline{|2Nq_1}\ \underline{|2N(1-q_1)}}\Sigma q^{2Nq_1}(1-q)^{2N(1-q_1)}f$$

Replacing summation by integration and letting $f=\phi(q)/2N=\phi(q)dq$ we have[4]

$$\frac{\phi(q_1)}{2N+1} = \frac{\underline{|2N}}{\underline{|2Nq_1}\ \underline{|2N(1-q_1)}}\int_0^1 q^{2Nq_1}(1-q)^{2N(1-q_1)}\phi(q)dq.$$

The cases of 2 and 3 monoecious individuals as worked out by simple algebra suggest an approach to a uniform distribution. As a trial let $\phi(q) = C$. It will be found that this makes the right and left members of the equation identical and is thus a solution.

$$\frac{C}{2N+1} = \frac{C\,\underline{|2N}}{\underline{|2Nq_1}\ \underline{|2N(1-q_1)}}\ \frac{\Gamma(2Nq_1+1)\Gamma(2N-2Nq_1+1)}{\Gamma(2N+2)} = \frac{C}{2N+1}.$$

The case of loss at rate $1/2N$ should not differ appreciably from that at rate $1/(2N+1)$. It would appear that after a cross, the gene frequencies will spread out from 50 percent toward fixation and loss until a practically uniform distribution is reached. The frequencies of all classes will then slump at a rate of about $1/2N$ as $1/4N$ of the genes become fixed and the same number lost per generation. Figure 3 is intended to illustrate this situation.

[4] f must be equated to $\phi(q)/2N$ here, rather than $\phi(q)/(2N-1)$, if the convenient limits 0 and 1 are to be used for integration in place of the limits $1/2N$ and $(2N-1)/2N$ of the summation with its $2N-1$ terms.

Before finally accepting this solution, however, it will be well to examine the terminal conditions. The amount of fixation at the extremes if N is large can be found directly from the Poisson series according to which the chance of drawing 0 where m is the mean number in a sample is e^{-m}. The contribution to the 0 class will thus be $(e^{-1}+e^{-2}+e^{-3}\cdots)f=$

$$\frac{e^{-1}}{1-e^{-1}}f, = 0.582f.$$

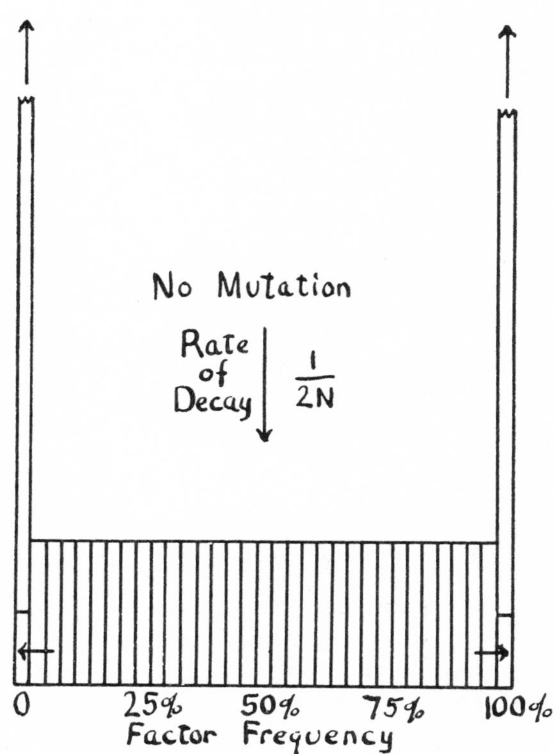

FIGURE 3.—Distribution of gene frequencies in an isolated population in which fixation and loss of genes each is proceeding at the rate 1/4N in the absence of appreciable selection or mutation. $y=L_0e^{-T/2N}$.

This is a little larger than the $\frac{1}{2}$f deduced above and indicates a small amount of distortion near the ends due to the element of approximation involved in substituting integration for summation. The nature and amount of this distortion are indicated by the exact distributions obtained in the extreme cases of only 2 and 3 monoecious individuals.

Letting L_0 be the initial number of unfixed loci (pairs of allelomorphs) and T the number of generations we have approximately

Unfixed loci in the T 'th generation $L_T = L_0 e^{-T/2N}$

An analogous formula holds for genes in multiple series, but in this case, as previously noted, the rate of fixation of loci is only half that given above.

The amount of genetic variation with respect to cumulative characters is easily calculated assuming for simplicity pairs of allelomorphs. The contribution of each factor pair to variance, in the case of no dominance, is $2a^2q(1-q)$ where a is the average difference in effect between plus and minus allelomorphs. The general formula for variance in this or any other distribution is thus $\sigma^2 = 2a^2\int_0^1 q(1-q)\phi(q)dq$. In the present case in which $\phi(q) = L$ this reduces to $\sigma^2 = \frac{1}{3}La^2$. PEARSON's β_2 comes out with a value 2.8 a slightly platykurtic distribution. Since the percentage of heterozygosis for a given factor frequency, q, is $2q(1-q)$, the formula for heterozygosis is the same as that for variance except that a^2 is to be omitted.

Similarly in the case of dominance, the contribution of a single factor pair to variance is $4a^2(1-q)^2(2q-q^2)$ where a is half the average difference in effect between dominant and recessive zygotes. The total variance with perfect dominance is thus in general

$$\sigma^2 = 4a^2 \int_0^1 (1 - q)^2(2q - q^2)\phi(q)dq.$$

In the case of a uniform distribution this gives $\sigma^2 = 8/15\ La.^2$

FISHER (1918) has emphasized the importance of a characteristic of the population which he calls the dominance ratio. He analyzes the variance of characters into three portions, that due to genetic segregation (τ^2) that due to dominance, as something which causes deviations of the phenotype from the closest possible linear relation with the genotype (ϵ^2), and that due to environment. Assuming environment constant, $\sigma^2 = \tau^2 + \epsilon^2$. The simple formulae for the correlations between relatives, to be found if there is random mating and no dominance, must be modified, if dominance is present depending on the value of the dominance ratio defined as ϵ^2/σ^2. Following are examples which he gives:[5]

[5] The author wishes here to correct an error in his 1921 paper which was written without knowledge of FISHER's results cited above. In this paper it was assumed that the correlation with no dominance needed merely to be multiplied by the squared correlation between genotype and phenotype, the same as FISHER's $\tau^2/\sigma^2 = (1-\epsilon^2/\sigma^2)$, to obtain that with dominance. This gives correct results (if there is no assortative mating) in the case of offspring with parents, all other ancestors and also in the case of collaterals where one of the individuals is related to the other through only one parent but it is more or less in error in other cases, the fraternal correlation being that most affected. The reasoning followed was not exact because a correlation in the deviations due to dominance in the cases indicated was overlooked.

	Correlation	
	No dominance	*Dominance*
Parent and offspring	$\frac{1}{2}$	$\frac{1}{2}(1-(\epsilon^2/\sigma^2))$
Brothers	$\frac{1}{2}$	$\frac{1}{2}(1-(\epsilon^2/2\sigma^2))$
Uncle and nephew	$\frac{1}{4}$	$\frac{1}{4}(1-(\epsilon^2/\sigma^2))$
Double first cousins	$\frac{1}{4}$	$\frac{1}{4}(1-(3\epsilon^2/4\sigma^2))$

FISHER has shown that the contribution of a single factor to ϵ^2, if there is complete dominance, may be written $\delta^2 = 4q^2(1-q)^2a^2$ where q is the frequency for dominant allelomorphs. Whether a particular dominant gene has a plus or minus effect on the character under consideration is immaterial. The contribution due to genetic segregation he gives as $\beta^2 = 8q(1-q)^3a^2$ thus

$$\frac{\epsilon^2}{\sigma^2} = \frac{\Sigma\delta^2}{\Sigma(\delta^2+\beta^2)} = \frac{\int_0^1 q^2(1-q)^2\phi(q)dq}{2\int_0^1 q(1-q)^3\phi(q)dq + \int_0^1 q^2(1-q)^2\phi(q)dq}$$

In the present case this reduces to $\frac{1}{4}$ as given by FISHER who also obtains a uniform distribution of factor frequencies for the case of no mutation or selection, although a different rate of decay.

Nonrecurrent mutation

If mutation is occurring, however low the rate, the decline in heterozygosis, following isolation of a relatively small group from a large population, cannot go on indefinitely. There will come a time when the chance elimination of genes will be exactly balanced by new genes arising by mutation. The equation to be solved is obviously as follows:

$$\frac{\phi(q_1)}{2N} = \frac{\underline{|2N}}{\underline{|2Nq_1}\ \underline{|2N(1-q_1)}}\int_0^1 q^{2Nq_1}(1-q)^{2N(1-q_1)}\phi(q)dq.$$

It may be found by trial that the expression $\phi(q) = C_1q^{-1}+C_2(1-q)^{-1}$ is a solution. The terminal condition, reduction of the class of fixed genes $(q=1)$ by an occasional mutation (contributing to the class $q=(2N-1)/2N$ necessarily involves the appearance of new genes (contributing to the class $q=1/2N$) and therefore means that only the symmetrical solution $\phi(q) = Cq^{-1}(1-q)^{-1}$ can be accepted as descriptive of the distribution of the entire array of genes at equilibrium (under the rather arbitrary postulated condition, no selection, no migration, no recurrence of the same mutations). Letting $f = (C/2N)\ q^{-1}(1-q)^{-1}$ and making $\sum f = 1$,

$$C = \frac{1}{2[(0.577 + \log(2N-1)]}$$

or approximately $C = 1/(2 \log 3.6N)$ (compare figure 5).

Before attainment of equilibrium with respect to heterozygosis the distribution will pass through phases of approximately the form $\phi(q) = C_1 q^{-1}(1-q)^{-1} + C_3$ in which the term C_1 gradually displaces C_3 as the number of temporarily fixed genes approaches equilibrium with mutation.

Each particular gene has a probability distribution for the future which spreads in time from the initial frequency in curves which are at first approximately normal in form but later (if the initial q was not too close to 1) become flat, the chances of complete fixation or complete loss each increasing by $1/4N$ each generation. As the chances of complete fixation increase, the chance of mutation must be taken into account. The distribution passes through phases of the type $C_2(1-q)^{-1} + C_3$, C_2 gradually displacing C_3, relatively, but itself beginning to decline as the chance of complete loss increases. With initial q equal (or close) to 1, equilibrium with mutation, and hence the hyperbolic distribution, is reached directly. The ultimate result in any case is complete loss of the gene in question (still assuming no recurrence of the same mutation and hence mutation *of* the gene but not *to* it). If there is reverse mutation, but at very low rate, a term $C_1 q^{-1}$ must be added to the formula, and an equilibrium will be reached in the form $Cq^{-1}(1-q)^{-1}$. This last formula means that in the long run (assuming no disturbances from selection, migration, etc.) the gene will usually be found either completely fixed or completely absent from the population (with frequencies proportional to the mutation rates to and from the gene respectively) but that occasionally fixation or absence will not be quite complete and that at extremely rare intervals the gene will drift from one state to the other.

The turnover among genes in equilibrium in the distribution $Cq^{-1}(1-q)^{-1}$ can be determined from consideration of the variance of q, and independently by application of the Poisson law.

Let $\sigma_q^2 = \Sigma(q-\frac{1}{2})^2 f/\Sigma f$ be the variance of q, excluding the terminal classes, the summation including $2N-1$ terms. This variance is increased in the following generation by the spreading out of each frequency class as a result of random sampling. The variance from the spreading of a single class is $q(1-q)/2N$ and the average is thus

$$\Delta\sigma_q^2 = \frac{\Sigma q(1-q)f}{2N\Sigma f} = \frac{1}{2N}\left(\frac{1}{4} - \sigma_q^2\right) = \frac{2N-1}{(2N)^2}C.$$

The sum $\sigma_q^2+\Delta\sigma_q^2$ includes the newly fixed factors whose contribution is $\frac{1}{4}$K where K is the rate of fixation, plus loss, but excludes mutation.

Digressing for a moment to the case of no mutation but equilibrium of form, we have at once

$$\sigma_q^2 + \Delta\sigma_q^2 = K\tfrac{1}{4} + (1 - K)\sigma_q^2$$

$\left(K-\frac{1}{2N}\right)\left(\sigma_q^2-\frac{1}{4}\right) = 0$ giving an independent demonstration that the rate of decay is 1/2N in this case.

Returning to the case of equilibrium under mutation, the contribution of new mutations to variance is $K(N-1)^2/(2N)^2$.

$$\sigma_q^2 + \Delta\sigma_q^2 - \tfrac{1}{4}K + K\left(\frac{N-1}{2N}\right)^2 = \sigma_q^2$$

$$K = C = \frac{1}{2 \log 3.6N}.$$

The proportion exchanged at each extreme is thus about half the corresponding subterminal class where N is large ($f_1 = f_{2N-1} = 2NC/(2N-1)$ by this method. This compares fairly well with the proportion as determined by the Poisson law, which is 0.46 times the subterminal class instead of 0.50.

The equilibrium frequencies can be worked out algebraically in simple cases. The figures below give the results in the case of a population of 3 monoecious individuals for comparison with the theoretical values deduced above. The rate of exchange at each extreme is actually 10.8 percent in comparison with 11.0 percent as $\frac{5}{12}\left(=\frac{2N-1}{4N}\right)$ the subterminal class, or 11.4 percent from the formula $\frac{1}{4(.577+\log 5)}$. The case of irreversible mutation is also given.

Gene frequency	Reversible Mutation: Exact equilibrium	Reversible Mutation: $Cq^{-1}(1-q)^{-1}$	Irreversible Mutation: Exact	Irreversible Mutation: $C(1-q)^{-1}$
5A:1a	27.5	26.3	47.7	43.8
4A:2a	15.4	16.4	20.6	21.9
3A:3a	14.1	14.6	14.1	14.6
2A:4a	15.4	16.4	10.2	10.9
1A:5a	27.5	26.3	7.3	8.8
Totals	99.9	100.0	99.9	100.0
Terminal exchange	10.8	11.0	18.0	18.25
Loss			3.6	3.65

The number of unfixed loci (L) which a given mutation rate per individual (μ) will support in a population is easily found, assuming only pairs of allelomorphs. The number of mutations is KL as well as $N\mu$. Therefore $L = N\mu/K = 2N\mu \log 3.6N$. The variance of cumulatively determined characters worked out as in the preceding case comes out $2N\mu a^2$ in the case of no dominance and $10/3\ N\mu a^2$ in the case of dominance, in both cases, directly proportional to the size of population[6] and to the mutation rate. In view of the piling up of new mutations, one might perhaps, expect to find a leptokurtic distribution for characters. This, however, turns out not to be the case: PEARSON'S β_2 comes out exactly 3 in the case of no dominance on substitution in the general formula

$$\beta_2 = 3 + \frac{\int_0^1 q(1-q)[1-6q(1-q)]\phi(q)dq}{\left[\int_0^1 q(1-q)\phi(q)dq\right]^2}$$

FISHER'S dominance ratio comes out 1/5 in this case.

The preceding results differ somewhat from those presented by FISHER (1922). The latter's analysis was based on a transformation of the scale of factor frequencies designed to make the variance due to random sampling uniform at all points. The variance at a given value of q is $q(1-q)/2N$. FISHER assumes that if the ratio of small differences on the q scale to the corresponding differences on a new θ scale be made proportional to the varying standard deviation of q, the standard deviation on the θ scale will be uniform. Letting $dq/d\theta = \sqrt{q(1-q)}$ leads to the transformation $\theta = \cos^{-1}(1-2q)$ with uniform variance of factor frequencies of $1/2N$. Letting $y = F(\theta)$ be the distribution of factor frequencies in one generation, he wrote that in the next as

$$y + \Delta y = \int_0^\pi \frac{1}{\sigma\sqrt{2\pi}} e^{-\delta\theta^2/2\sigma^2}\left(y + y'\delta\theta + \frac{\delta\theta^2}{\underline{|2}} y'' \cdots\right)$$

and measuring time in generations (T) he reached the equation

$$\frac{\partial y}{\partial T} = \frac{1}{4N}\frac{\partial^2 y}{\partial \theta^2}.$$

After noting that the solution for the symmetrical stationary case is

[6] These estimates of number of unfixed loci and of variance depend, of course, on the validity of the conditions on which the formula of the distribution curve is based. How far the mutation rate per locus can be considered negligibly small as size of population increases is discussed later.

$y = L/\pi$, he proceeded to derive the formulae for increasing and decreasing y. Considering the latter, $dy/dT = -Ky$ where K is the rate of decay, giving $1/4N \; d^2y/d\theta^2 = -Ky$ as the equation to be solved. In the symmetrical case this yields $y = C \cos [\sqrt{4NK}(\theta - \pi/2)]$ where $C = \sqrt{4NK}/[2 \sin (\frac{1}{2}\pi\sqrt{4NK})]$ in order to give a total frequency of unity and is to be multiplied further by L_0e^{-KT} to show change from the initial frequency of L_0.

The maximum value which K can take without giving negative frequencies within the range is obviously 1/4N and FISHER found reason for accepting this as the value in the case of no mutation. The formula for the distribution in this case reduces to $y = \frac{1}{2} \sin \theta$. FISHER transformed these equations to the scale $Z = \log [q/(1-q)]$ in which the case of no mutation becomes $y = \frac{1}{4} \operatorname{sech}^2 \frac{1}{2}Z$ and the case of loss balanced by mutation becomes $y = 1/2\pi \operatorname{sech} \frac{1}{2}Z$. This transformation brings the curves into an approach to the form of the normal probability curve. For our present purpose it is preferable to transform to the scale of actual factor frequencies. The case of steady decay becomes $y = 1$ with which my results are in agreement, although in disagreement as to rate of decay. In the case of loss balanced by mutation, FISHER's formula transforms into $y = 1/[\pi\sqrt{q(1-q)}]$ instead of $1/[2(\log 3.6N)q(1-q)]$ as developed in the present paper. FISHER obtained $\sqrt{\pi/2}N^{3/2}\mu$ for the number of unfixed factors, in contrast with $2N\mu \log 3.6N$; and $\sqrt{\frac{2}{\pi N}}$ for the factor turnover in contrast with $1/[2 \log 3.6N]$. It will be seen that FISHER'S solution leads to a smaller number of unfixed factors with more rapid turnover in very small populations (less than 81) but to a larger number of such factors with slower turnover in larger populations. In a breeding population of one million with one mutation per 1000 individuals, FISHER's formula gives 1,250,000 unfixed factors with a turnover of 0.08 percent while I get 30,000 unfixed factors with a turnover of 3.3 percent.

The exact harmonizing of the results of the two methods of attack has been a somewhat puzzling matter, but Doctor FISHER, on examination of the manuscript of the present paper, has written to me the following which I quote at his suggestion. ". . . . I have now fully convinced myself that your solution is the right one. It may be of some interest that my original error lay in the differential equation

$$\frac{\partial y}{\partial T} = \frac{1}{4N} \frac{\partial^2 v}{\partial \theta^2}$$

which ought to have been

$$\frac{\partial v}{\partial T} = \frac{1}{4N} \frac{\partial}{\partial \theta}(y \cot \theta) + \frac{1}{4N} \frac{\partial^2 y}{\partial \theta^2}$$

the new term coming in from the fact that the mean value of δp in any generation from a group of factors with gene frequency p is exactly zero,[7] and consequently the mean value of $\delta\theta$ is not exactly zero but involves a minute term $-1/4N \cot \theta$. With this correction, I find myself in entire agreement, with your value 2N, for the time of relaxation and with your corrected distribution for factors in the absence of selection."

Reversible recurrent mutation

It only requires a very moderate mutation rate in a large population for the number of unfixed loci to become enormous. This raises the question as to the effect of a limitation in the number of mutable loci, and recurrence of mutations.

Consider now the case of genes with uniform rates of recurrence of mutation and reverse mutation. Let u be the rate per generation for break down of the gene A and v that for origin from allelomorphs. A class of genes with frequency q (that of all allelomorphs, $1-q$) will be distributed in the following generation under random sampling according to the expansion of the expression

$$\{[(1-q) - v(1-q) + uq]a + [q + v(1-q) - uq]A\}^{2Nf}.$$

Equating the total contribution to a given class, to the frequency of this class in the parent generation, reduced by the proportion K, if there is a uniform rate of decay, gives as the equation to be solved:

$$\phi(q_1)\frac{1-K}{2N} = \frac{\underline{|2N}}{\underline{|2Nq_1}\ \underline{|2N(1-q_1)}}\int_0^1 [q(1-u-v)+v]^{2Nq_1}$$

$$[1-q(1-u-v)-v]^{2N(1-q_1)}\phi(q)dq.$$

It will be found by trial that the right and left members became identical in certain cases in which $\phi(q)$ is of the form $q^s(1-q)^t$

Let $x = q(1-u-v)+v$

$$q = \frac{x-v}{1-u-v} \qquad dq = \frac{dx}{1-u-v}$$

$$q^s = \frac{x^s - svx^{s-1}\ldots}{(1-u-v)^s} \qquad (1-q)^t = \frac{(1-x)^t - tu(1-x)^{t-1}}{(1-u-v)^t}.$$

[7] p is the q of the present paper. Since the above was written, FISHER has published this revision of his results in *The genetical theory of natural selection*, 1930.

The small amount of spread from a given class will justify retention of the untransformed limits of integration.

Noting that $\Gamma(c+s+1) = \underline{|c}\, c^s \left(1+\frac{s(1+s)}{2c}\right)$ approximately when c is an integer and s is small compared with c, and making use of the following derived relation

$$\frac{2N\underline{|2N}}{\underline{|2Nq_1}\;\underline{|2N(1-q_1)}}\int_0^1 x^{2Nq_1+s}(1-x)^{2N(1-q_1)+t}dx$$
$$= \frac{4N + s(s+1)q_1^{-1} + t(t+1)(1-q_1)^{-1}}{4N + (s+t+1)(s+t+2)} q_1^s(1-q_1)^t$$

the equation may be written as followsf or small values of s and t (compared with N) and values of u and v of a still lower order of size.

$$(1-K)q_1^s(1-q_1)^t$$
$$= \frac{1}{1-(u+v)(s+t+1)}\left[\frac{4N}{4N+(s+t+1)(s+t+2)} q_1^s(1-q_1)^t\right.$$
$$+\left(\frac{s(s+1)}{4N+(s+t+1)(s+t+2)} - \frac{4Nsv}{4N+(s+t)(s+t+1)}\right)q_1^{s-1}(1-q_1)^t$$
$$\left.+\left(\frac{t(t+1)}{4N+(s+t+1)(s+t+2)} - \frac{4Ntu}{4N+(s+t)(s+t+1)}\right)q_1^s(1-q_1)^{t-1}\right]$$

The coefficients of $q_1^{s-1}(1-q_1)^t$ and of $q_1^s(1-q_1)^{t-1}$ must equal 0 either under complete equilibrium or equilibrium merely in form of distribution. Neglecting small quantities:

$$s = 0 \quad \text{or} \quad s = 4Nv - 1$$
$$t = 0 \quad \text{or} \quad t = 4Nu - 1.$$

In the case of complete equilibrium (K = O), it turns out that the coefficients of $q_1^s(1-q_1)^t$ in the left and right members are also satisfied to a first approximation by $s = 4Nv - 1$, $t = 4Nu - 1$. They are also satisfied by letting $s = 0$, $t = 0$ provided that $u + v = 1/2N$. The relation between the fixed terminal and the unfixed subterminal classes, however, requires that $u = v = 1/4N$ in this case, which thus becomes merely a special case of the first solution. Similarly, the solutions $s = 0$, $t = 4Nu - 1$ and $s = 4Nv - 1$, $t = 0$ require that $v = 1/4N$ and $u = 1/4N$ respectively and thus also reduce to special cases of the first solution. It appears then that the distribution of gene frequencies in equilibrium under mutation and reverse mutation may be represented approximately by curves of PEARSON'S Type I,

$$\phi(q) = \frac{\Gamma(4Nu + 4Nv)}{\Gamma(4Nu)\Gamma(4Nv)} q^{4Nv-1}(1 - q)^{4Nu-1}.$$

The terminal conditions are of interest in this and other cases to be considered. The factor turnover at each extreme may be written

$$K_0 = \tfrac{1}{2}f_1 = 2Nvf_0$$

$$K_{2N} = \tfrac{1}{2}f_{2N-1} = 2Nuf_{2N}$$

where the subterminal classes have the frequencies

$$f_1 = \frac{1}{2N}\phi\left(\frac{1}{2N}\right)$$

$$f_{2N-1} = \frac{1}{2N}\phi\left(1 - \frac{1}{2N}\right).$$

In the present case, the terminal classes have the frequencies $f_0 = C/[4Nv(2N)^{4Nv}]$ and $f_{2N} = C/[4Nn(2N)^{4Nu}]$ where C is the coefficient in the expression for $\phi(q)$.

It will be seen that the form of the curve depends not only on the rates of mutation of the genes but also on the size of the breeding population. With small populations or rare recurrence of mutations, the distribution approaches the symmetrical form $y = 1/(2 \log 3.6N)\, q^{-1}(1-q)^{-1}$ already discussed (figure 5). The ratio of the class of temporarily fixed genes (f_{2N}) to the class of complete absence (f_0) must be approximately v:u in this case in order that the number of mutations at each extreme of the symmetrical distribution of unfixed factors may be equal.

With increase in size of the population, the gene frequencies tend in general to be distributed in asymmetrical U– or even I– or J–shaped curves. For example, if the size of population reaches 1/4u and v is much smaller than u, the distribution will be the hyperbola $\phi(q) = Cq^{-1}$ with a piling up of factors with few or no plus representatives.

With sufficient increase in the size of population, the distribution at length takes a form approaching that of the normal probability curve; centered about the point $\bar{q} = v/(u+v)$ which, indeed, is always the mean

$$\left(\bar{q} = \int_0^1 q\phi(q)dq = \frac{v}{u+v}\right).$$

The variance of gene frequencies, $\sigma_q^2 = \int_0^1 (q-\bar{q})^2\phi(q)dq$ is

$$\frac{\bar{q}(1 - \bar{q})}{4N(u + v) + 1}.$$

The amount of genetic variation of cumulative characters may be calculated as before. In the case of no dominance and paired allelomorphs

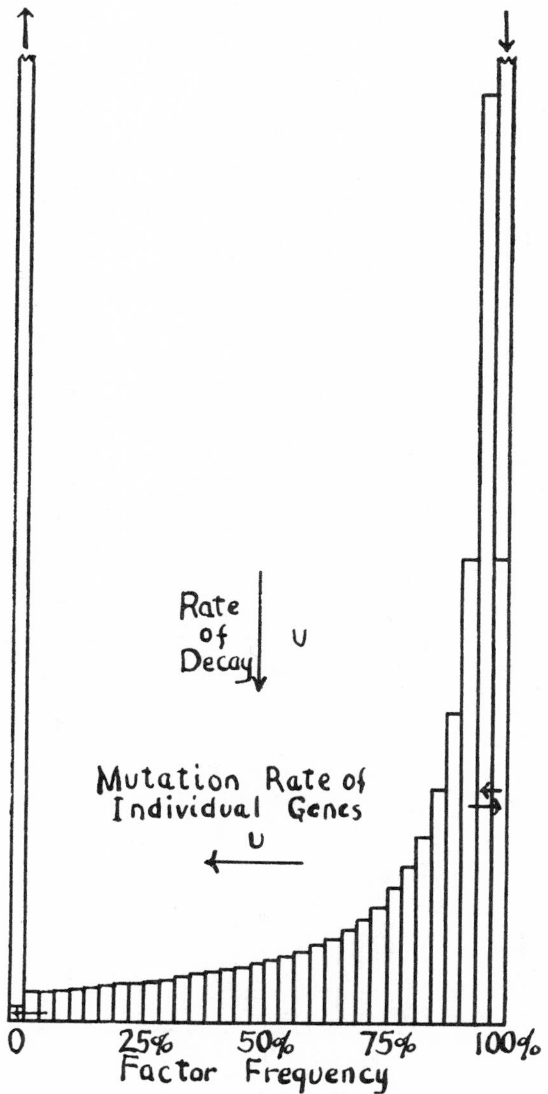

FIGURE 4.—Distribution of type genes in an isolated population in which equilibrium has been reached with destructive mutation but has not been approached with respect to formative mutation. $y = 4NuL_0e^{-uT}(1-q)^{4Nu-1}$ with 4Nu much smaller than 1 and the formula approximately $\frac{L_0(1-q)^{-1}}{\log 3.6N}$.

it is $\sigma^2 = 2La^2\ 4Nu\bar{q}/(4Nu+4Nv+1)$ or $2La^2[\bar{q}(1-q)-\sigma_q{}^2]$. Where u is much greater than v, this can be written approximately $\sigma^2 = 2La^2 4Nv/(4Nu+1)$ approaching $2La^2\bar{q}$ as a limit, as N increases and L comes to include all loci.

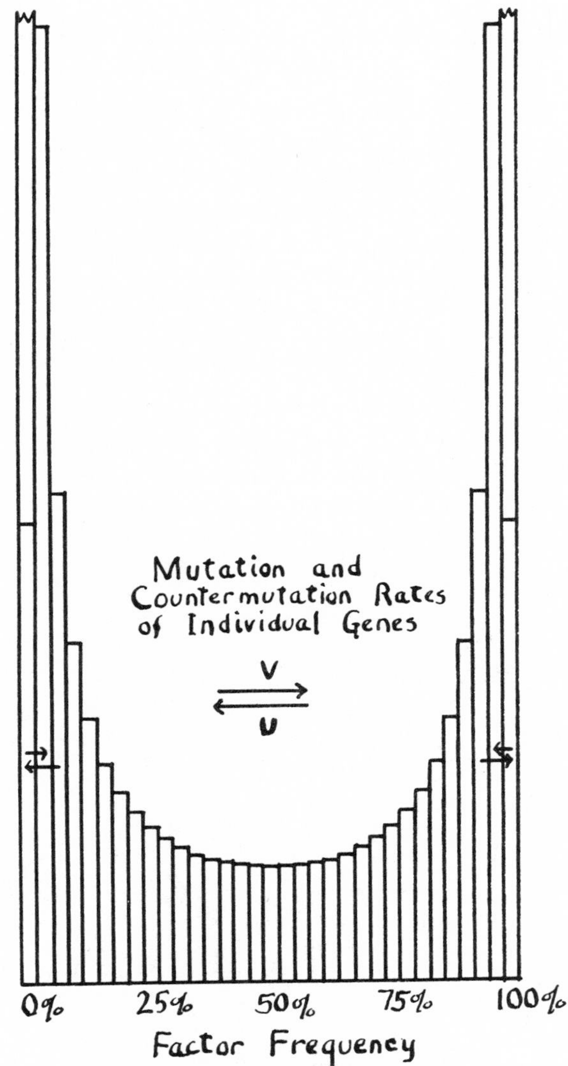

FIGURE 5.—Distribution of gene frequencies (or probability array of gene) where equilibrium with mutation has been attained. Population so small that the terms 4Nu and 4Nv are both much smaller than 1. $y = Cq^{4Nv-1}(1-q)^{4Nu-1}$, approximately $\dfrac{q^{-1}(1-q)^{-1}}{2 \log 3.6N}$.

As the formula for this case was derived on the assumption of small values of u and v, it is desirable to obtain an independent test of its applicability to large values. This can be done as follows: the increase in variance of q due to random sampling is

$$\frac{1}{2N}\int_0^1 q(1-q)\phi(q)\,dq=\frac{\bar{q}(1-\bar{q})-\sigma^2_q}{2N}.$$

Letting $\Delta q=-uq+v(1-q)$ be, as before, the change in q due to mutation, $q+\Delta q-\bar{q}=(q-\bar{q})(1-u-v)$. Thus the effect of mutation is as if all deviations from the mean were reduced in the proportion $(1-u-v)$. The decrease in the variance of q, due to mutation is therefore $\sigma_q{}^2[1-(1-u-v)^2]$. At equilibrium the increase in $\sigma_q{}^2$ due to random sampling is exactly balanced by the decrease due to mutation yielding:

$$\sigma^2_q=\frac{\bar{q}(1-q)}{4N(u+v)-2N(u+v)^2+1.}$$

The term $-2N(u+v)^2$ in the denominator is important only when $(u+v)$ has a large absolute value. Omitting this, the formula is identical with that deduced by the first method and thus gives an independent demonstration of its validity. As mutation approaches its maximum value $(u+v=1)$, the variance of q approaches $\bar{q}(1-\bar{q})/2N$, that due to random sampling alone.

Migration

The distribution of gene frequencies in an incompletely isolated subgroup of a large population can be obtained immediately from the preceding results. The change in gene frequency per generation under migration $\Delta q=-m(q-q_m)$ can be written $-m(1-q_m)q+mq_m(1-q)$ which is in the same form as the change of q under mutation, $\Delta q=-uq+v(1-q)$. We may write at once for the distribution under negligible mutation rates:

$$\phi(q)=\frac{\Gamma(4Nm)}{\Gamma[4Nmq_m]\Gamma[4Nm(1-q_m)]}q^{4Nmq_m-1}(1-q)^{4Nm(1-q_m)-1}.$$

The mutation terms 4Nu and 4Nv can be inserted, if mutation rates are not negligible.

Figure 6 shows how the form of the distribution changes with change in m or N. Where m is less than 1/2N there is a tendency toward chance fixation of one or the other allelomorph. Greater migration prevents such fixation. How little interchange would appear necessary to hold a large population together may be seen from the consideration that $m=1/2N$

means an interchange of only one individual every other generation, regardless of the size of the subgroup. However, this estimate must be much qualified by the consideration that the effective N of the formula is in general much smaller than the actual size of the population or even than the breeding stock, and by the further consideration that q_m of the formula refers to the gene frequency of actual migrants and that a further factor must be included if q_m is to refer to the species as a whole. Taking both of these into account, it would appear that an interchange of the order of thousands of individuals per generation between neighboring subgroups of a widely distributed species might well be insufficient to prevent a considerable random drifting apart in their genetic compositions. Of course,

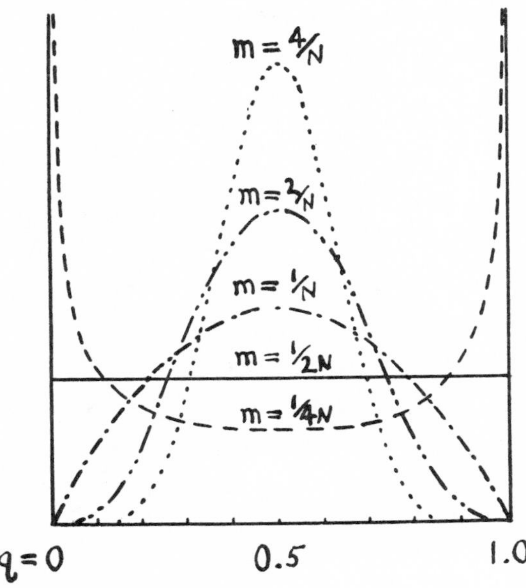

FIGURE 6.—Distribution of frequencies of a gene among subdivisions of a population in which $q_m = 1/2$ (or probability array of gene within a subdivision) under various amounts of intermigration. $y = Cq^{4Nmq_m - 1}(1-q)^{4Nm(1-q_m)-1}$.

differences in the condition of selection among the subgroups may greatly increase this divergence. It appears, however, that the actual differences among natural geographical races and subspecies are to a large extent of the nonadaptive sort expected from random drifting apart. An interesting example, apparently nonadaptive, is the racial distribution of the 3 allelomorphs which determine human blood groups (BERNSTEIN 1925).

The variance of distribution of values of q among subgroups (in the ideal case) is $\sigma_q^2 = q_m(1-q_m)/(4Nm+1)$ by substitution in the formula for the preceding case.

The zygotic distribution $[(1-q)a+qA]^2$ cannot be expected to hold in a population made up of isolated groups among which gene frequency varies. WAHLUND (1928) has shown that the proportions in each homozygous class are increased at the expense of the heterozygotes by the amount of the variance of the gene frequencies among the subgroups,[8] the proportions becoming $[(1-q)^2+\sigma_q^2]aa+[2q(1-q)-2\sigma_q^2]\ Aa+[q^2+\sigma_q^2]\ AA$. By substituting the expression for σ_q^2, given above, in WAHLUND's formula one might determine empirically the effective value of 4Nm for the population, except that it would be difficult to rule out the possibility that some of the variance of gene frequencies might be due to differences in the selection coefficients among the subgroups instead of merely to random drifting apart.

Irreversible recurrent mutation

The solution $s=0$, $t=4Nu-1$ for the equation reached in the case of recurrent mutation satisfies the conditions for equilibrium of form under irreversible mutation ($v=0$), with decay at rate $K=u$.

$$\phi(q) = 4NuL_0e^{-uT}(1-q)^{4Nu-1}.$$

The proportional frequency of the unfixed subterminal class which is not replenished by mutation is $f_1/(L_0e^{-uT})=2u$, twice the rate of decay and thus approximately satisfying the necessary terminal condition.

For values of u as small as $1/(2N \log 3.6N)$ the coefficient in the expression for ϕ (q) must be calculated to a closer approximation $\dfrac{4NuL_0e^{-uT}}{1-\left(\dfrac{1}{2N}\right)^{4Nu}}$ which approaches $\dfrac{L}{\log 3.6N}$ as u approaches zero.

The distribution of gene frequencies under irreversible mutation is illustrated in figure 4.

This case is of most interest as representing for a long time the distribution of type genes in a small group isolated from a large one in which all type genes are close to fixation. The release of deleterious mutation pressure from equilibrium with selection will result in approximate equi-

[8] The percentage of heterozygotes is $2\int_0^1 q(1-q)\phi(q)dq$ where $\phi(q)$ is the distribution of values of q among the subgroups. As shown above this reduces to $2\bar{q}(1-\bar{q})-2\sigma_q^2$, thus demonstrating WAHLUND's principle.

librium of the form described above. With decay at the rate u, it may be a very long time before effects of reverse mutation become appreciable and the final equilibrium $y=Cq^{-1}(1-q)^{-1}$ approached. Assuming that type genes are dominant, the dominance ratio in this case is 1/3.

Selection[9]

Using $\Delta q=sq(1-q)$ as the measure of the effect of genic selection, the class of genes with frequency $(1-q)a:qA$ is distributed after one generation according to the expression:

$$\{(1-q)(1-sq)a+q[1+s(1-q)]A\}^{2N}.$$

The distribution of gene frequencies which is in equilibrium may be obtained from the following equation which represents the total contribution to class q_1 after one generation, as equal to its previous frequency.

$$\frac{\phi(q_1)}{2N}=\frac{\lfloor 2N}{\lfloor 2Nq_1\ \lfloor 2N(1-q_1)}$$

$$\int_0^1 q^{2Nq_1}(1-q)^{2N(1-q_1)}(1+s(1-q))^{2Nq_1}(1-sq)^{2N(1-q_1)}\phi(q)dq$$

To a first approximation, the selection terms approach the value $e^{2Ns(q_1-q)}$. The introduction of a factor e^{2Nsq} into the previously reached formula for $\phi(q)$ gives a solution of the equation (for very small values of s) since it cancels the new term e^{-2Nsq} in the integral, and leaves e^{2Nsq_1} as a factor in $\phi(q_1)$. This was the basis for the formula published (WRIGHT 1929a) as $\phi(q)=Ce^{2Nsq}q^{4Nv-1}(1-q)^{4Nu-1}$ intended to exhibit in combination the effects of selection, mutation in both directions and size of population. Further consideration reveals that this solution is the correct one only for the case of irreversible mutation and then only when the selection coefficient is exceedingly small, less than 1/2N in fact. FISHER (1930) in his recently published revision of the results of his method of attack on this problem has given a formula for a special case of selection, equilibrium of flux from an inexhaustible supply of mutating genes. This is given as accurate as long as Ns^2 is small. Assuming one mutation per generation, he writes:

$$y=\frac{2dp(1-e^{-4anq})}{pq(1-e^{-4an})}.$$

[9] This and the following section have been rewritten since submission of the manuscript in order to take account of the correction of my formula, suggested by FISHER's results in *The genetical theory of natural selection*, 1930 as noted herein.

In this formula, $a(=-s)$ is the selection coefficient, $p(=1-q)$ is frequency of mutant genes and dp may be taken as $1/2N$ numerically. This agrees with my previous formula for irreversible mutation, $y = Ce^{2Nsq}(1-q)^{-1}$ only when s is less than $1/2N$, above which value my formula rapidly leads to impossible results. On reexamination of my method, however, I find that the same degree of approximation can be reached by it. The expansion of $[1+s(1-q)]^{2Nq_1}[1-sq]^{2N(1-q_1)}$ yields series of terms which condense into the expression $e^{2Ns(q_1-q)}\{1-Ns^2[q_1(1-q_1)+(q_1-q)^2]\}$ taking into account terms in Ns^2, N^2s^3, N^3s^4, N^4s^5 as well as those in which N and s have the same exponent. Since the random deviations of q have a variance of $q_1(1-q_1)/2N$ the term (q_1-q) is of the order $\sqrt{1/2N}$. A second order approximation should be obtainable by retaining the term $Ns^2q_1(1-q_1)$ while that in $Ns^2(q_1-q)^2$ may be dropped. The equation to be solved can now be written.

$$\phi(q_1) = \frac{2N\,\lfloor 2N}{\lfloor 2Nq_1\ \lfloor 2N(1-q_1)} e^{2Nsq_1}[1 - Ns^2q_1(1-q_1)] \int_0^1 q^{2Nq_1}(1-q)^{2N(1-q_1)}e^{-2Nsq}\phi(q)\,dq.$$

Let $\phi(q) = e^{2Nsq}q^{-1}(1-q)^{-1}(a+bq+cq^2+dq^3\cdots)$.

The exponential term in the integral being cancelled, it becomes possible to carry out the integration by means of the approximate formula already used in the case of mutation (page 122).

$$\frac{2N\,\lfloor 2N}{\lfloor 2Nq_1\ \lfloor 2N(1-q_1)} \int_0^1 q^{2Nq_1+z-1}(1-q)^{2N(1-q_1)-1}dq = \frac{4N + z(z-1)q_1^{-1}}{4N + z(z-1)} q_1^{z-1}(1-q_1)^{-1}.$$

The resulting coefficients of the powers of q_1 on the right side of the equation may now be equated separately to those of $\phi(q_1)$. To a sufficient approximation it turns out that $c = \frac{(2Ns)^2}{\lfloor 2}a$, $d = \frac{(2Ns)^2}{\lfloor 3}b$, $e = \frac{(2Ns)^4}{\lfloor 4}a$, $f = \frac{(2Ns)^4}{\lfloor 5}b$, $g = \frac{(2Ns)^6}{\lfloor 6}a$, etc.

Letting $C_1 = a/2$ and $C_2 = \frac{b}{4Ns}$

$$\phi(q) = 2e^{2Nsq}q^{-1}(1-q)^{-1}[C_1 \cosh 2Nsq + C_2 \sinh 2Nsq].$$

From considerations of symmetry, it is obvious that another solution may be obtained by replacing q by $(1-q)$ and s by $-s$. The full solution may be written in the form

$$\phi(q) = q^{-1}(1-q)^{-1}[C_1(e^{4Nsq}+1) + C_2(e^{4Nsq}-1) + C_3(1+e^{-4Ns(1-q)}) + C_4(1-e^{-4Ns(1-q)})].$$

The relative values of the coefficients in the case of equilibrium can be obtained by setting up the equation for the absence of flux. Each group of genes, $f=\phi(q)dq$ tends to be shifted by the amount $\Delta q = sq(1-q)$ in a generation. There is thus a total flux measured by $\int_0^1 \phi(q)\Delta q dq$ unless there is counterbalancing mutation. The amount of mutation in each direction (assuming the rates of recurrence to be very small compared with $1/4N$) is approximately half the respective subterminal classes, as demonstrated in the preceding cases.

$$f_1 = 2C_1 + 2sC_2 + (1+e^{-4Ns})C_3 + (1-e^{-4Ns})C_4$$
$$f_{2N-1} = (e^{4Ns}+1)C_1 + (e^{4Ns}-1)C_2 + 2C_3 + 2sC_4.$$

Since mutation moves genes from the fixed classes to the subterminal classes with gene frequencies of $1/2N$ and $\left(1-\frac{1}{2N}\right)$ respectively, it creates a net flux of $\frac{f_{2N-1}}{4N} - \frac{f_1}{4N}$ which at equilibrium should balance that due to selection

$$\int_0^1 \phi(q)\Delta q dq - \frac{f_{2N-1}}{4N} + \frac{f_1}{4N} = 0.$$

Substitution of the values given above leads to the condition $C_1 - C_2 + C_3 + C_4 = 0$. Under this condition the formula simplifies greatly, becoming for all values of s $\left(\text{of lower order than } \frac{1}{\sqrt{N}}\right)$

$$\phi(q) = Ce^{4Nsq}q^{-1}(1-q)^{-1}.$$

The effect of selection in this case is perhaps best exhibited in the ratio of the classes of alternative fixed genes in the highly artificial case of equality in the rates of mutation in opposite directions. This ratio is e^{4Ns}. More generally, $f_0 = \frac{C}{4Nv}$ and $f_{2N} = \frac{Ce^{4Ns}}{4Nu}$ where u and v, both assumed to be very small compared with 1/4N, are the opposing mutation rates. The number of unfixed loci (pairs of allelomorphs) takes the form

$$L = \frac{2N\mu}{f_1 + f_{2N-1}} = \frac{2N\mu}{C(e^{4Ns} + 1)}$$

where μ is the mutation rate per individual and C is chosen so that $\int_0^1 \phi(q)dq = 1$. The effect of selection on the variance of cumulative characters (pairs of allelomorphs) may be seen by comparing the formula

$$\sigma^2 = 2\mu a^2\left(\frac{e^{4Ns} - 1}{2s(e^{4Ns} + 1)}\right)$$

with the previously given form $2N\mu a^2$ which it approaches as s approaches 0.

In the case treated by FISHER, there is assumed to be irreversible mutation at the rate of one per generation from an inexhaustible supply. As each new gene becomes fixed, it may be considered as transferred to the type class, ready to mutate to new allelomorphs in its series. Thus in place of a return flux of $\frac{f_1}{4N}$, due to reversible mutation, we must write $\frac{f_1}{2}$ (if $v = 0$)

$$\int_0^1 \phi(q)\Delta q dq - \frac{f_{2N-1}}{4N} + \frac{f_1}{2} = 0.$$

This is solved if $C_1 = C_3 = C_4 = 0$ and

$$\phi(q) = C_2(e^{4Nsq} - 1)q^{-1}(1 - q)^{-1}.$$

In case the direction of mutation coincides with that of selection ($u = 0$), the mutational terms must be written $\frac{f_{2N-1}}{2} - \frac{f_1}{4N}$ giving the solution

$$\phi(q) = C_4(1 - e^{-4Ns(1-q)})q^{-1}(1 - q)^{-1}.$$

These are identical with FISHER's results on proper choice of the coefficient.

An interesting question which FISHER has discussed, is the chance of fixation of a single mutation. This is given by the ratio of the subterminal classes in the formulae just considered. Where selection opposes mutation, $\frac{f_1}{f_{2N-1}} = \frac{2s}{e^{4Ns} - 1}$, always less than 1/2N. In the case of favorable mutations, $\frac{f_{2N-1}}{f_1} = \frac{2s}{1 - e^{-4Ns}}$, or approximately 2s. FISHER also gives an independent derivation of the last figure.

General formula

It is of especial importance to assemble the effects of all evolutionary factors into a single formula. Unfortunately, the equation of equilibrium of class frequencies becomes rather complicated and has not yet been worked through. Presumably the form is given at least approximately by a formula of the type $Ce^{aq}q^{4Nv-1}(1-q)^{4Nu-1}$ in the case of reversible mutation. In order that there may be no flux, $\int_0^1 \phi(q)\Delta q dq = 0$. It is not necessary to consider the terminal classes in this case. Thus

$$C \int_0^1 e^{aq}q^{4Nv-1}(1 - q)^{4Nu-1}[- uq + v(1 - q) + sq(1 - q)]dq = 0.$$

Integration of the first term (that in $-uq$) by parts gives an expression which is immediately solved by letting $a = 4N$. Thus the selection term appears to be e^{4Nsq} regardless of the rates of mutation provided there is reversibility. It is approximately of this value in the case of irreversible mutation, discussed above, provided that s is considerably larger than 1/4N. The conclusions based on the previously presented value e^{2Nsq} still hold,[10] except that they should be applied to selection intensities just half as great.

The position of the mode of the I-shaped distribution curve given when u and s are greater than 1/4N can be found by equating the differential coefficient of the logarithm of the formula to zero.

$$4Ns + \frac{4Nv - 1}{q} - \frac{4Nu - 1}{1 - q} = 0.$$

When v is small but u and s are both large, q approaches the value $1 - \frac{u}{s}$ already given as the equilibrium point. The mean would be somewhat below this point, as expected from the curvilinear relation of selection

[10] These conclusions were presented at the meeting of the AMERICAN ASSOCIATION FOR THE ADVANCEMENT OF SCIENCE for 1929 and were summarized in the abstracts (WRIGHT 1929b).

pressure to gene frequency and in contrast with the case of equilibrium between opposing mutation pressures (but no selection) in which the mean is always the equilibrium point $\frac{v}{u + v}$ and the mode is more extreme than this figure, $\frac{4Nv - 1}{4N(u + v) - 2}$.

Migration pressure introduces no other complications. Combining all factors:

$$\phi(q) = Ce^{4Nsq}q^{4N(mq_m+v)-1}(1 - q)^{4N[m(1-q_m)+u]-1}.$$

The selection coefficient refers here to the difference between the selection in the group under consideration and that in the species as a whole, the effect of the latter being taken account of in the mean gene frequency of the species q_m.

The distribution curves

Some of the forms taken by the probability array of gene frequencies, in cases involving selection, are illustrated in figures 7 to 14. Figures 7 to 10 deal with the case in which the rates of mutation are negligibly small compared with 1/4N. The curves are thus all variants of the form $y = Ce^{4Nsq}q^{-1}(1 - q)^{-1}$. Figure 7 illustrates the relatively slight effect of selection below a certain relation to size of population. All conditions are the same in figure 8 except that the populations are four times as large as in figure 7. Thus while the absolute intensities of the selection coefficients are the same, the relations to size of population are altered.[11] The curves bring out the great effect of selection beyond the critical point, $s = \frac{1}{2N}$ (where mutation rates are low). Figures 9 and 10 are intended to show the effects of change in size of population where the intensity of selection remains constant (low in figure 9, four times as severe in figure 10). Up to a certain point $\left(N = \frac{1}{2s}\text{, figure 9}\right)$ increase in population raises the middle portion of the curve. Above this point (figure 10) increase in population depresses the middle portion. In the former case, the increase in unfixed factors brings increased variability of cumulative characters, in the latter there is little change of variability in relation to population size, the depression among middle frequencies being balanced by the accumulation of nearly but not quite fixed factors. All of these fig-

[11] The probability that increase in the number of unfixed genes would react on the individual gene selection coefficients, reducing them, is here ignored.

ures (7 to 10) may be taken as representative of conditions in small inbred populations which have been isolated sufficiently long to reach equilibrium in relation to mutation. It will be recalled that figures 3 and 4 represent successive stages preceding the attainment of such equilibrium.

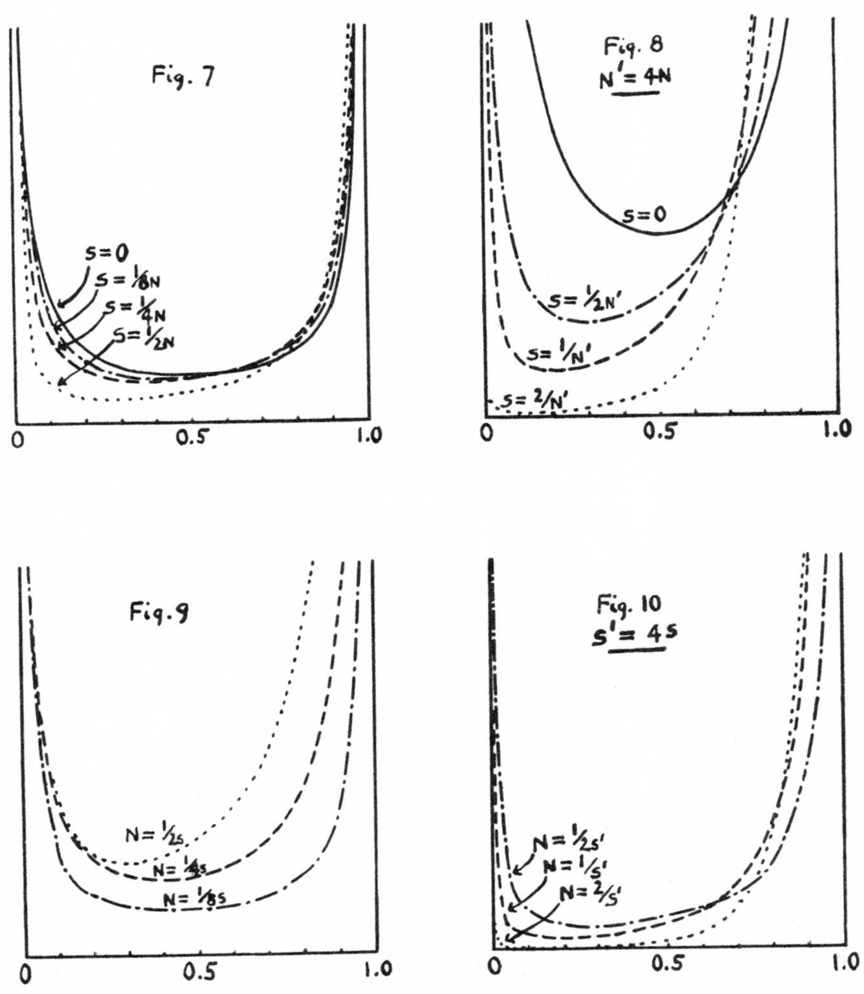

FIGURES 7, 8, 9, and 10—Distribution of gene frequencies in relation to size of population and intensity of selection where rates of mutation and migration are small compared with 1/4N. Formulae all of type $y = Cy^{4Nsq}q^{-1}(1-q)^{-1}$.

Figure 7. Small population, four degrees of selection. Figure 8. Population four times as large as in figure 7 under the same four (absolute) degrees of selection. Figure 9. Three sizes of population under given weak selection. Figure 10. Same three sizes of population as in figure 9, under selection four times as severe.

Figures 11 to 14 present exactly the same series of comparisons as figures 7 to 10, for small populations that are not completely isolated from the main body of a species.[12] In all cases the gene frequency (q_m) of the

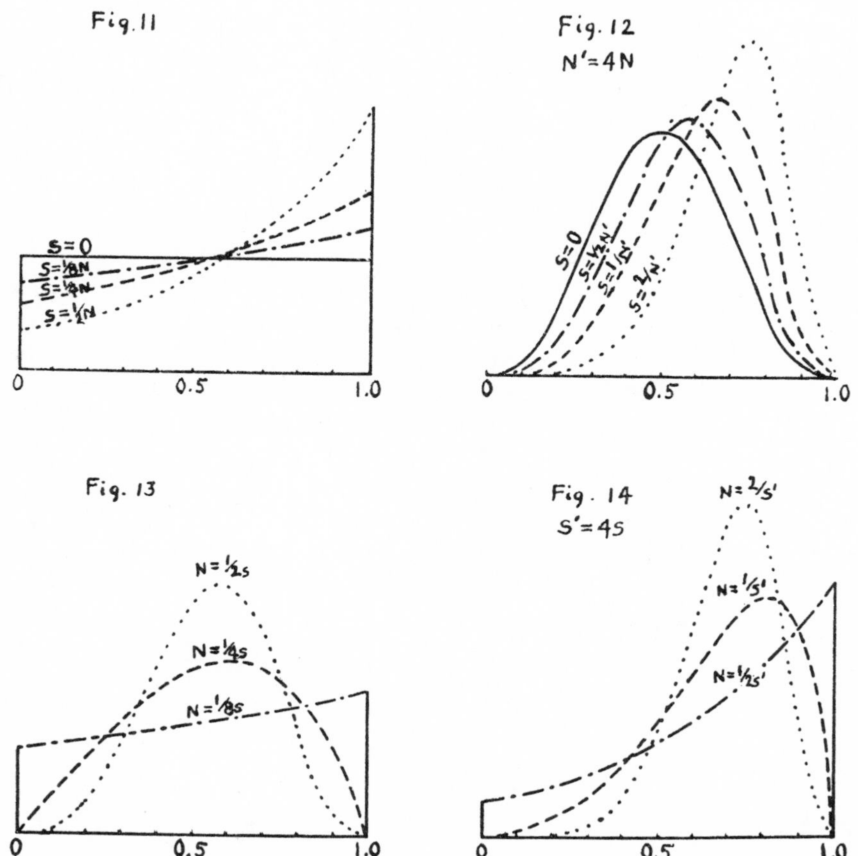

FIGURES 11 to 14.—Distribution of gene frequencies in subgroups of large population (mean frequency $q_m = 1/2$) in relation to size of population and intensity of selection. Formulae all of the type $y = Ce^{4Nsq}q^{4Nmq_m-1}(1-q)^{4Nm(1-q_m)-1}$ Same comparisons as in figures 7 to 10.

Figure 11. Small subgroups ($2Nm = 1$), four degrees of selection. Figure 12. Subgroups four times as large as in figure 11, under same four (absolute) degrees of selection. Figure 13. Subgroups of three sizes under given weak selection. Figure 14. Same three sizes as in figure 13 under selection four times as severe.

The figures may also be used to illustrate cases of equal mutation to and from a gene ($u = v$). $y = Ce^{4Nsq}q^{4Nv-1}(1-q)^{4Nu-1}$

[12] These figures also represent the distribution of gene frequencies in population in which mutation and reverse mutation are equally frequent, but this seems to be so exceptional a case especially under multiple allelomorphism, as to be of little importance.

whole species is assumed to be 1/2. The relations of migration to size of population are such that there is very little complete fixation of genes. In figure 11, $m=1/2N$ and the purely exponential curves show how increasing intensity of genic selection shifts a uniform distribution in the direction favored by the selection. The fourfold greater population of figure 12 brings about concentration, in curves approaching the normal in form. Figure 13 brings out the concentrating effect of increase in population in the case of weak selection while figure 14 does the same for the case of selection four times as severe.

The important case in which mutation is balanced by selection in a moderately large population (both s and u large compared with 1/4N) is illustrated in figure 19. The four curves represent four degrees of selection, rising by doubling of severity at each step from a case in which mutation pressure practically overwhelms the effect of selection to the reverse situation. The limiting condition in populations so large that 1/4N is very small compared with both s and u is that of concentration of factor frequency almost at a single value (figure 20, page 148).

Dominance ratio

The form of the distribution of the frequencies of the dominant genes affecting a character is of interest in connection with the dominance ratio. Since different genes have different mutation rates and selection coefficients, this distribution is a composite of curves of the types discussed. In small populations which have reached equilibrium, all of these arrays and hence their composite are of the type $Cq^{-1}(1-q)^{-1}$. The dominance ratio is 1/5 in so far as dominance is complete. FISHER gives the value as $3/13=0.23$ for the case "when in the absence of selection, sufficient mutation takes place to counteract the effect of random survival." The difference from the value 0.20 given above is due solely to the difference in the formula for the curve, discussed earlier.

In the case of the isolation of a small part of a large population, the dominance ratio takes the value 1/3 in so far as dependent on dominant type genes in equilibrium with recessive mutation but not with reverse mutation ($y=C/(1-q)$). Where following isolation both fixation and loss are substantially irreversible ($y=1$) the dominance ratio is 1/4 in agreement with FISHER'S result. In both of these cases, of course, the dominance ratio falls to 1/5 when equilibrium is finally attained.

The foregoing discussion applies practically only to very small completely isolated populations. In large populations where the distribution of gene frequencies, even in partially isolated subgroups, tends to approach the normal type, the dominance ratio comes to depend mainly on the mean gene frequency which depends on the relation of selection to mutation, or on selection against both homozygotes in favor of heterozygotes. In the extreme case in which the gene frequency is reduced to a single value, the dominance ratio is $q/(2-q)$. Values close to unity should not be uncommon, especially where gene frequency is controlled by the balance of selection and mutation. Such a dominance ratio has rather surprising effects on the correlation between relatives. The correlation between parent and offspring approaches 0 although that between brothers may remain as high as 0.25. However, the occurrence of an appreciable number of genes at lower frequencies, for example, held in equilibrium by selection favoring the heterozygote against both homozygotes would greatly lower the dominance ratio.

All of these figures are on the assumption that dominance is complete. Dominance, however, is frequently not complete. Among 22 heterozygotes in the guinea pig which have been studied with some care, at least 9 or about 40 percent are to some extent intermediate. Most of these have to do with color characters. It is not unlikely that incomplete dominance will be found to be even more frequent on careful study of size characters.

Fisher (1922) comes rather definitely to the conclusion that the dominance ratio is typically in the neighborhood of 1/3. This was based primarily on a distribution of factor frequencies which he reached for the case of selection against a recessive,[13] with which the results of the present study are not at all in agreement. He also finds, however, that the differences between fraternal and parent-offspring correlations in data which he analyzes indicate the same figure. The analysis of a large number of correlations of these sorts would undoubtedly furnish valuable information with regard to the statistical situation in populations. It is to be noted, however, that similarity in the environment of brothers as compared with parent and offspring may also contribute to a higher fraternal correlation and that in any case one cannot reason from the dominance ratio deduced from correlations to the distribution of factor frequencies without making some assumption as to the prevalence of dominance.

About all that seems justified by the present analysis, is the statement that for permanently small populations under low selection the value should

[13] This was given as $df \propto \frac{d\theta}{\sin \frac{1}{2}\theta \cos^3 \frac{1}{2}\theta}$ or $\phi(q) = Cq^{-1}(1-q)^{-2}$ on transformation of scale. Fisher does not discuss dominance ratio in connection with his recent revision of his results in *The genetical theory of natural selection*, 1930.

be less and probably considerably less than 0.20 but that this figure may be raised by severe selection (favoring dominants) and especially by increase in size of population. It may even approach unity in very large populations under severe selection, if complete dominance is the rule.

Mean and variability of characters

In the case of genes which are indifferent to selection (s less than u), the mean frequency $\bar{q} = v/(u+v)$ remains unchanged through all transformations from a U-shaped distribution in small populations to an I-shaped one in large populations. The variance, due to such genes, is small in small populations, rises in nearly direct proportion to size of population up to a certain critical point (about $N = 1/4u$) and then approaches a limiting value. For the case in which mutations in one direction (u) occur at a much greater rate than in the other (v), the general formula reduces to $\sigma^2 = \sigma_\infty^2 \left(\frac{4Nu}{1+4Nu}\right)$, in which $\sigma_\infty^2 (= 2La^2v/u)$ is the limiting value. This case is illustrated in figure 15. The dotted lines represent mean gene frequencies and the line of dashes the variance.

Actual changes in the size of a given population are not of course accompanied by instant adjustment of the distribution of gene frequencies. A decrease in size to a point well below the critical value is followed by decrease in heterozygosis and variance at a rate between $1/4N$ and $1/2N$ per generation depending on the number of allelomorphs. This may be a fairly rapid process in terms of geologic time but the recovery of heterozygosis through growth of the population to its original size occurs more slowly, since this depends on mutation pressure. On the other hand, the intercrossing of a number of isolated strains, in each of which the reduction of variance has occurred, is followed by immediate recovery of the original statistical situation (except with respect to factor combinations in which there is some delay).

In the opposite case of genes under vigorous selection (s much greater than u) mean frequency as well as variance is affected by size of population and by severity of selection. As in the preceding case, variance is small in small populations, rises in nearly direct proportion to growth of population[14] until a critical point is approached (here about $N = 1/4s$) and then rapidly approaches a limiting value $\sigma^2 = \sigma_\infty^2 \left(\frac{e^{4Ns} - 1}{e^{4Ns} + 1}\right)$ where

[14] As before, the probability that the increase in variance, due to growth of population, would react on the selection coefficients of the individual genes, reducing them, requires some qualification of this statement in application to actual populations.

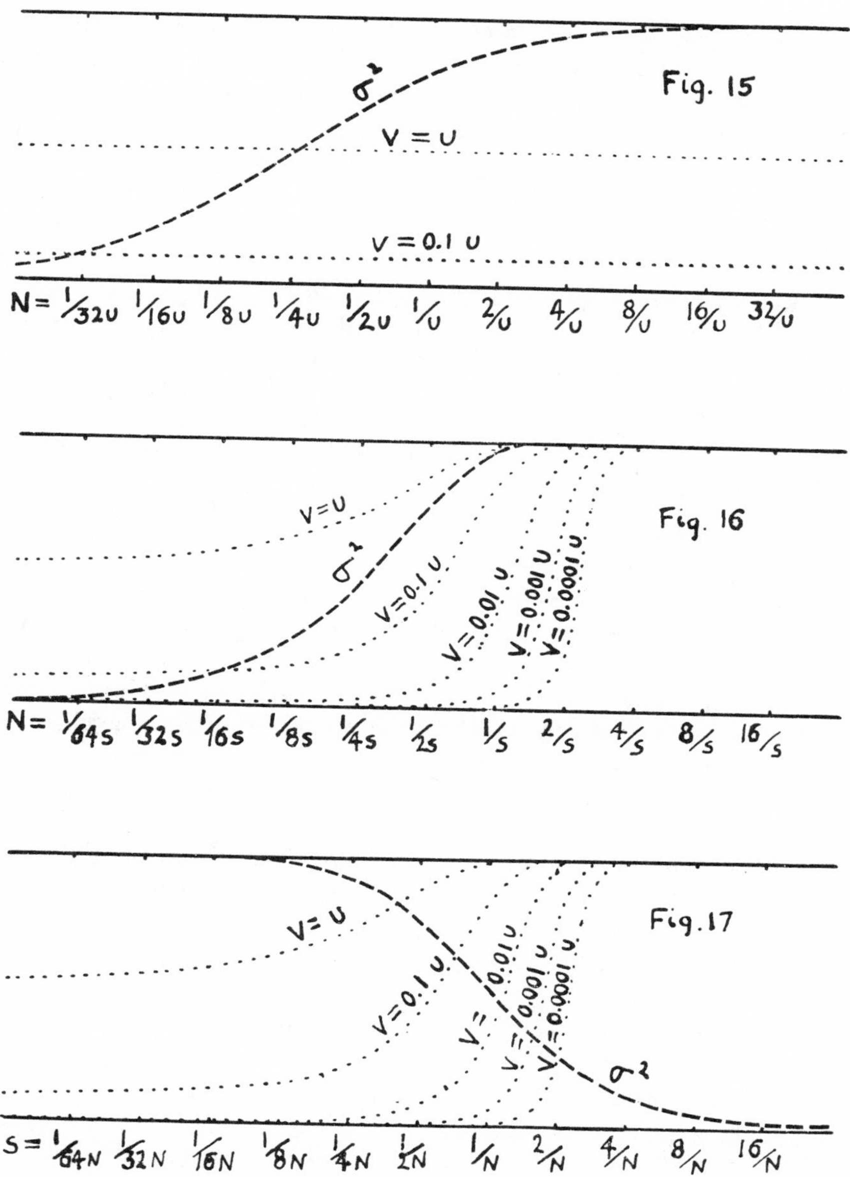

FIGURES 15 to 17.—The variance (σ^2) and mean gene frequencies (dotted lines) at equilibrium under various conditions of mutation, selection and size of population. Figure 15. Effects of increasing population where selection is negligible relative to mutation. Figure 16. Effects of increasing population where mutation rates are small compared with 1/4N. Figure 17. Effects of increasing selection where mutation rates are small compared with 1/4N.

$\sigma_\infty^2 = \mu a^2/s$. The mean factor frequency in large populations ($\bar{q} = 1 - u/s$) is close to 1, nearly complete fixation of the favorable gene. In small populations, on the other hand, the equilibrium point approaches that of the opposing mutation pressures ($\bar{q} = v/(u+v)$) and hence practically 0, with complete loss ($v = 0$) as the inevitable ultimate fate in an extended multiple allelomorphic series. Up to the point at which mutation pressure seriously disturbs the form of the distribution curve, the mean gene frequencies are simply the ratios of the chances of fixation at each extreme, namely, $ve^{4Ns}/(u + ve^{4Ns})$.

The relations of mean frequency and variance to size of population in this case are both shown in figure 16, the former for various relative values of u and v. Inspection of figures 9 and 10 may also be of assistance in understanding this situation.

As in the other case, actual change in size of population is not accompanied by immediate attainment of the new equilibrium. Decrease in population to a number well below the critical point is followed by decrease in heterozygosis at the rate described, bringing with it at the same rate the well known inbreeding effects, loss of variance and, in general, decline in vigor toward a new level. This immediate decline in vigor is not due to change in mean gene frequency, but merely to the greater proportion of recessive phenotypes as homozygosis increases, and thus comes to an end when the degree of homozygosis has reached equilibrium. The change in mean gene frequency proceeds more slowly since it depends on mutation pressure. Long continued isolation should thus involve two distinct degeneration processes, a rapid but soon completed process of fixation and a very slow process of accumulation of injurious genes. The recovery on increase in size of population is slow in both cases, depending on mutation pressure. The intercrossing of isolated lines, on the other hand, is followed by immediate return to the original status of the population if only the immediate inbreeding effect has occurred, but must wait on favorable mutations if there has been time for the slower process.

The effects of different intensities of selection on mean gene frequency and variance (population size constant) are illustrated in figure 17, still assuming that the selection coefficient is of higher order than mutation rate. Figures 7 and 8 showing the distribution of gene frequencies in this case may also be of assistance here. Selection has little effect on variability until it reaches about the value $1/8N$, about half the variance is eliminated when selection reaches $1/N$ and most of it at $s = 4/N$. The formula is $\sigma^2 = \sigma_0^2 \left(\frac{e^{4Ns} - 1}{2Ns(e^{4Ns} + 1)} \right)$. Selection, of course, affects the mean gene frequency, the formula being the same as that given above under the effect of size of population. On actual increase in the intensity of selection, the rate of change toward the new equilibrium both in mean and variance is controlled by selection pressure and may thus be fairly rapid in terms of geologic time in a large population. This is the case in which HALDANE'S formulae for progress under selection are most applicable. The increase in variance and in the proportion of unfavorable genes following relaxation of selection, on the other hand, are controlled by mutation pressure and thus approach equilibrium relatively slowly. A shift in gene frequency at rate uq may well mean no more than 0.000,001 per generation.

The type of result where the selection coefficient is of the same order of magnitude as mutation rate can be inferred, qualitatively, at least from the preceding extreme cases. Inspection of figure 19 may also be of assistance here.

THE EVOLUTION OF MENDELIAN SYSTEMS

Classification of the factors of evolution

In attempting to draw conclusions with respect to evolution one is apt, perhaps, to assume that factors which make for great variation are necessarily favorable while those which reduce variation are unfavorable. Evolution, however, is not merely change, it is a process of *cumulative* change: fixation in some respects is as important as variation in others. Live stock breeders like to compare their work to that of a modeller in clay. They speak of moulding the type toward the ideal which they have in mind. The analogy is a good one in suggesting that in both cases it is a certain intermediate degree of plasticity that is required.

The basic cumulative factor in evolution is the extraordinary persistence of gene specificity. This doubtless rests on a tendency to precise duplication of gene structure in the proper environment. The basic change factor is gene mutation, the occasional failure of precise duplication. Since the time of LAMARCK, a school of biologists have held that the primary changes in hereditary constitution must be adaptive in direction in order to account for evolutionary advance. Unfortunately, the results of experimental study have given no support to this view. Instead, the characteristics of actually observed gene mutations seem about as unfavorable as could be imagined for adaptive evolution. In the first place, is their fortuitous occurrence. No correlation has been found between external conditions and direction of mutation, and those few agents which have been found to affect the rate (X-ray, radium, and to a relatively unimportant extent, temperature) merely speed up the rate of random mutation. The great majority of mutations are either definitely injurious to the organism or produce such small effects as to be seemingly negligible. MULLER has graphically compared the range of mutations to a spectrum in which the nonlethal conspicuous mutations form a narrow field between broad regions of individually inconspicuous mutations on the one hand and of sublethal and lethal mutations on the other. In addition, the great majority of mutations are more or less completely recessive to the type genes from which they arise. These effects are easily understood if mutation is an accidental process. Random changes in a complex organization are more likely to injure than to improve it, and with respect to the immediate products of the gene, random changes are more likely to be of the nature inactivation (and hence probably recessive) than of increased activation. Finally is to be mentioned the extreme rarity of gene mutation. Even in Drosophila, mutation rates as high as $u = 10^{-5}$ per locus seem to be exceptional and 10^{-6} or less more characteristic. This infrequency seems unfavorable to rapid evolution, yet it is a necessary corollary of the usually injurious effect, if life is to persist at all. Moreover, the more advanced the evolution, the slower must be the time rate of mutation. In one-celled organisms, dividing several times a day, a rapid time rate of mutation will not prevent the production of sufficient normal offspring to maintain the species. The same time rate in Drosophila with an interval of some two weeks between generations would mean such an accumulation of lethals in every gamete that the species would come to an abrupt end. The time rate of lethal mutation in Drosophila (7 per 1000 chromosomes per month under ordinary conditions according to MULLER (1928)) would be quite impossible in the human species. The problem is to determine how an adaptive evolutionary process may be derived from such unfavorable raw material as the infrequent, fortuitous and usually injurious gene mutations.

It will be convenient here to classify factors of evolution according as they tend toward genetic homogeneity or heterogeneity of the species. They are grouped below in more or less definitely opposing pairs.

Factors of Genetic Homogeneity	*Factors of Genetic Heterogeneity*
Gene duplication	Gene mutation (u, v)
Gene aggregation	Random division of aggregate
Mitosis	Chromosome aberration
Conjugation	Reduction (meiosis)
Linkage	Crossing over
Restriction of population size ($1/2N$)	Hybridization (m)
Environmental pressure (s)	Individual adaptability
Crossbreeding among subgroups (m_1)	Subdivision of group ($1/2N_1$)
Individual adaptability	Local environments of subgroups (s_1)

The first pair have been discussed above. They enter into the formulae through the mutation rates u and v. MULLER has pointed out the necessary similarity, in order of size, of genes and of filterable viruses and has suggested the possibility that the latter may consist of single genes. If so, their evolution rests wholly on a not too high rate of mutation, and selection, which seems possible enough in organisms as simple as these presumably are, especially as in this case where the gene is the organism, the mutation of the gene need not be expected to be as fortuitously related to the activities of the organism as in more complex cases.

Presumably the first step toward higher organisms is the aggregation of such genes with multiplication of the aggregate by random division. Given occasional gene mutation, this leads to a new kind of variation, that in proportional abundance of the different kinds of genic material. The larger the aggregate, the less violent the variation. Large aggregates present a labile system capable of quantitative variation in response (perhaps physiologically as well as through selection) to changing conditions. As far as observation goes, the bacteria, and blue green algae have no mechanism of division beyond a random division of the protoplasmic constituents. Such apportionment of more or less autonomous materials may also be important in the differentiating cell lineages of multicellular organisms, but, except for a few plastid characters, seems to play no important role in heredity from generation to generation, as far as has been determined by experiment. There seems here an adequate basis for an evolutionary process in organisms so simple that the handing on of a few different protoplasmic constituents can determine all of the characteristics of the species but the conditions are not favorable for an extensive cumulative process.

Mitosis provides the mechanism by which an indefinitely large number of qualitatively different elements may be maintained in the same proportions. But it provides so perfectly for the persistence of complex organization that further change is difficult. Irregularities in mitosis provide a source of variation but of so violent a nature for the most part as to be of infrequent evolutionary importance, although the differences in chromosome numbers of related species demonstrate that they play a genuine rôle. Complete duplication (tetraploidy) is important in doubling the possible number of different kinds of genes. Other aberrations, especially translocations, are probably more important in isolating types, than for the character changes which they bring. Gene mutation remains the principal factor of variation, but seems inadequate as the basis of an evolutionary process under exclusively mitotic (asexual) reproduction.

The most important factor in transcending the evolutionary difficulties inherent in the characteristics of gene mutation is undoubtedly the attainment of biparental reproduction (EAST 1918). This involves two phases, conjugation, a factor which makes the entire interbreeding group a physiological unit in evolution, and meiosis, with its consequence, Mendelian recombination which enormously increases the amount of variability within the limits of the species. Each additional viable mutation in an asexually reproducing form merely adds one to the number of types subject to natural selection. The chance that two or more indifferent or injurious mutations may combine in one line to produce a possibly favorable change is of the second or higher order. Under biparental reproduction, each new mutation doubles the number of potential variations which may be tried out. The contrast is between $n+1$ and 2^n types from n viable mutations.

Biparental reproduction solves the evolutionary requirement of a rich field of variation. But by itself it provides rather too much plasticity. It makes a highly adaptable species, capable of producing types fitted to each of a variety of conditions, but a successful combination of characteristics is attained in individuals only to be broken up in the next generation by the mechanism of meiosis itself.

An excellent illustration of the principle that a balance between factors of homogeneity and of heterogeneity may provide a more favorable condition for evolution than iether factor by itself may be found in the effects of an alternation of a series of asexual generations with an occasional sexual generation. Evolution is restrained under exclusive asexual reproduction by the absence of sufficient variation, and under exclusive sexual reproduction by the noncumulative character of the variation, but, on alternating with each other, any variety in the wide range of combinations provided by a cross may be multiplied indefinitely by asexual reproduction. The selection of individuals is replaced by the much more effective selection of clones and leads to rapid statistical advance which, however, comes to an end with reduction to a single successful clone. On the other hand a new cross (before reduction to a single clone) may provide a new field of variation making possible a repetition of the process at a higher level. This method has been a favorite of the plant breeder and is perhaps the most successful yet devised for human control of evolution in those cases to which it can be applied at all. Under natural conditions, alternation of asexual and sexual reproduction is characteristic of many organisms and doubtless has played an important rôle in their evolution.

The demonstration of the evolutionary advantages of an alternation of the two modes of reproduction seems to prove too much. Asexual reproduction is practically absent in the most complex group of animals, the vertebrates, and is rather sporadic in its occurrence elsewhere. The purpose of the present paper has been to investigate the statistical situation in a population under exclusive sexual reproduction in order to obtain a clear idea of the conditions for a degree of plasticity in a species which may make the evolutionary process an intelligible one.

First may be mentioned briefly a modification of the meiotic mechanism which has been introduced only qualitatively into the investigation where at all. This is the aggregation of genes into more or less persistent systems, the chromosomes. Complete linkage cuts down variability by preventing recombination. Wholly random assortment gives maximum recombination but does not allow any important degree of persistence of combinations once reached. An intermediate condition permits every combination to be formed sooner or later and gives sufficient persistence of such combinations to give a little more scope to selection than in the case of random assortment. Close linkage, moreover, brings about a condition in which selection tends to favor the heterozygote against both homozygotes and so helps in maintaining a store of unfixed factors in the population.

Resrtiction of size of population, measured by $1/2N$, is a factor of homogeneity and conversely with increase of size. The effects of restricted size may also be balanced by those of occasional external hybridization, measured by m.

Environmental pressure on the species as a whole is a factor of homogeneity. It has been urged by some that because natural selection is a factor which reduces variability, and most conspicuously by eliminating extreme types, it cannot be the guiding principle in adaptive evolution. From the viewpoint of evolution as a moving equilibrium, however, the guiding principle may be found on the conservative as well as on the radical side. The selection coefficient, s, depends on the balance between environmental pressure and individual adaptability. High development of the latter permits the survival of genetically diverse types in the face of severe pressure.

Subdivision of a population into almost completely isolated groups, whether by prevailing self fertilization, close inbreeding, assortative mating, by habitat or by geographic barriers is a factor of heterogeneity with effects measured by $1/2N_1$, N_1 being here the size of the subgroup. This factor may be balanced by crossbreeding between such groups, measured by m_1.

It is interesting to note that restriction of population size is a factor of homogeneity or of heterogeneity for the species, depending on whether it relates to the species as a whole or to subgroups and conversely with the crossbreeding coefficient. Similarly the selection pressures of varied environments within the range of the species (s_1) constitute factors of heterogeneity, restrained from excessive genetic effect by the same individual adaptability which appears in the opposite column in relation to the general environment of the species. Individual adaptability is, in fact, distinctly a factor of evolutionary poise. It is not only of the greatest significance as a factor of evolution in damping the effects of selection and keeping these down to an order not too great in comparison with 1/4N and u, but is itself perhaps the chief object of selection. The evolution of complex organisms rests on the attainment of gene combinations which determine a varied repertoire of adaptive cell responses in relation to external conditions. The older writers on evolution were often staggered by the seeming necessity of accounting for the evolution of fine details of an adaptive nature, for example, the fine structure of all of the bones. From the view that structure is never inherited as such, but merely types of adaptive cell behavior which lead to particular structures under particular conditions, the difficulty to a considerable extent disappears. The present difficulty is rather in tracing the inheritance of highly localized structural details to a more immediate inheritance of certain types of cell behavior.

Lability as the condition for evolution

The statistical effects of the more important of these factors in a freely interbreeding population are brought together in the formula

$$y = Ce^{4Nsq}q^{-1}(1 - q)^{4Nu-1}.$$

The term 4Nv in the exponent of q is here assumed to be negligible and the terms applicable in case of external hybridization are also omitted.

Consider first the situation in a small population in which 1/4N is much greater than u and than s (figure 18). Nearly all genes are fixed in one phase or another. Even rather severe selection is without effect. There is no equilibrium for individual genes. They drift from one state of fixation to another in time regardless of selection, but the rate of transfer is extremely slow. Such evolution as there is, is random in direction and tends toward extinction of the group.

Consider next the opposite extreme, a very large undivided population under severe selection. Assume that s is in general much greater than u and that the latter is much greater than 1/4N. There is almost complete

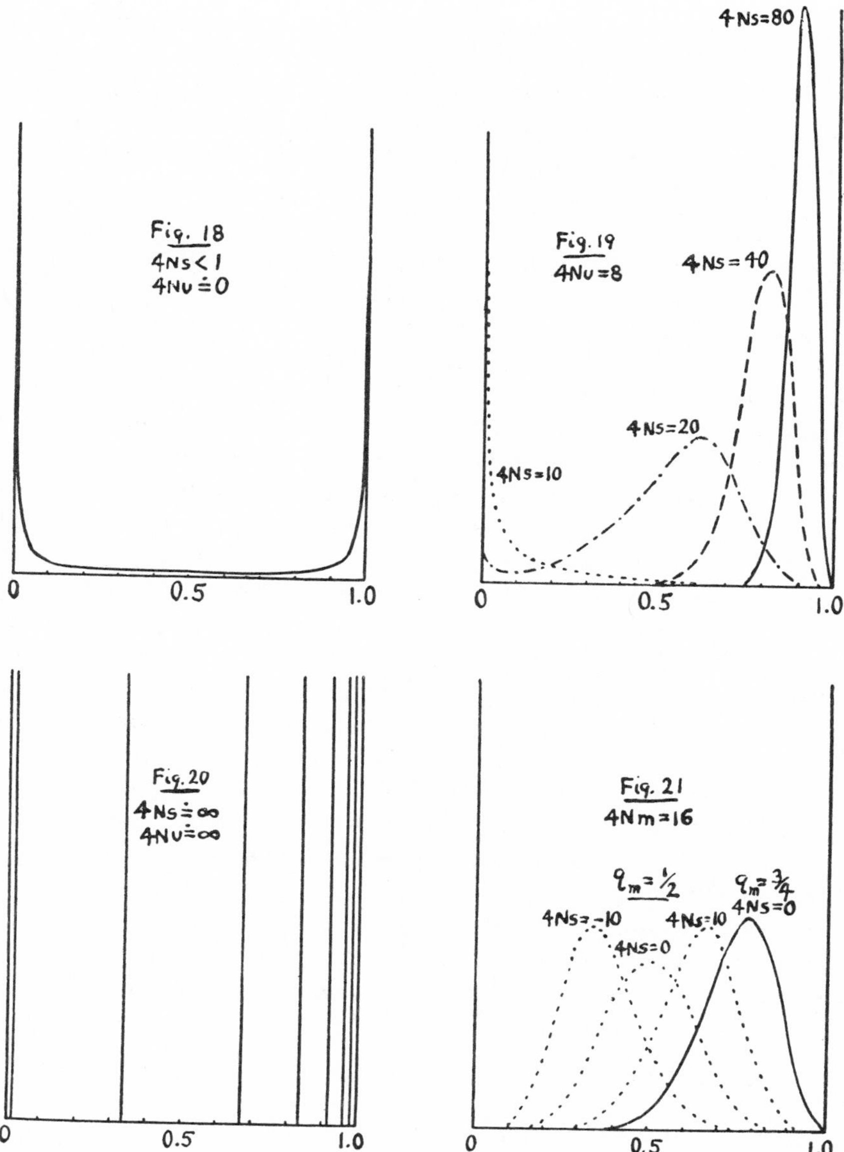

FIGURES 18 to 21.—Distributions of gene frequencies in relation to size of population, selection, mutation and state of subdivision. Figure 18. Small population, random fixation or loss of genes ($y=Cq^{-1}(1-q)^{-1}$. Figure 19. Intermediate size of population, random variaton of gene frequencies about modal values due to opposing mutation and selection ($y=Ce^{4Nsq}q^{-1}(1-q)^{4Nu-1}$. Figure 20. Large population, gene frequencies in equilibrium between mutation and selection ($q=1-u/s$, etc.). Figure 21. Subdivisions of large population, random variation of gene frequencies about modal values due to immigration and selection. ($y=Ce^{4Nsq}q^{4Nmq_m-1}(1-q)^{4Nm(1-q_m)-1}$.

fixation of the favored gene for each locus. Here also there is little possibility of evolution. There would be complete equilibrium under uniform conditions if the number of allelomorphs at each locus were limited. With an unlimited chain of possible gene transformations, new favorable mutations should arise from time to time and gradually displace the hitherto more favored genes but with the most extreme slowness even in terms of geologic time.[15]

Even if selection is relaxed to such a point that the selection coefficients of many of the genes are not much greater than mutation rates, the conditions are not favorable for a rapid evolution (figure 20). The amount of variability in the population may be great, maximum in fact, but if the distributions of gene frequencies are closely concentrated about single values, the situation approaches one of complete equilibrium and hence of complete cessation of evolution. At best an extremely slow, adaptive, and hence probably orthogenetic advance is to be expected from new mutations and from the effects of shifting conditions.

It should be added that a relatively rapid shift of gene frequencies can be brought about in this case by vigorous increase in the intensity of selection. The effects of unopposed selection of various sorts and in various relations of the genes has been studied exhaustively by HALDANE, with regard to the time required to bring about a shift of gene frequency of any required amount. The end result, however, is the situation previously discussed. The rapid advance has been at the expense of the store of variability of the species and ultimately puts the latter in a condition in which any further change must be exceedingly slow. Moreover, the advance is of an essentially reversible type. There has been a parallel movement of all of the equilibria affected and on cessation of the drastic selection, mutation pressure should (with extreme slowness) carry all equilibria back to their original positions. Practically, complete reversibility is not to be expected, and especially under changes in selection which are more complicated than can be described as alternately severe and relaxed. Nevertheless, the situation is distinctly unfavorable for a continuing evolutionary process.

Thus conditions are unfavorable for evolution both in very small and in very large, freely interbreeding, populations, and largely irrespective of severity of selection. We have next to consider the intermediate situa-

[15] This, nevertheless, seems to be the case which FISHER (1930) considers most favorable to evolution. The greatest difference between our conclusions seems to lie here. His theory is one of complete and direct control by natural selection while I attribute greatest immediate importance to the effects of incomplete isolation.

tion in which s is not much greater than u for many genes and the latter is not much greater than 1/4N. Such a case is illustrated in figure 19. The size of population is sufficient to prevent random fixation of genes, but insufficient to prevent random drifting of gene frequencies about their mean values, as determined by selection and mutation. It is to be supposed that the relations of the selection and mutation coefficients vary from factor to factor. The more indifferent ones drift about through a wide range of frequencies in the course of geologic time while those under more severe selection oscillate about positions close to complete fixation. In any case, all gene frequencies are continually changing even under uniform environmental conditions. But the selection coefficients themselves are in general to be considered functions of the entire array of gene frequencies and will therefore also be continually changing. The probability arrays of some genes will travel to the right and close up as their selection coefficients stiffen, while some of the genes which have been nearly fixed will come to be less severely selected and their probability arrays will shift to the left and open out or even move to the extreme left under displacement by another allelomorph. A continuous and essentially irreversible evolutionary process thus seems inevitable even under completely uniform conditions. The direction is largely random over short periods but adaptive in the long run. The less the variation of gene frequency about its mean value, the closer the approach to an adaptive orthogenesis. Complete separation of the species into large subspecies should be followed by rather slow more or less closely parallel evolutions, if the conditions are similar, or by adaptive radiation, under diverse conditions, while isolation of smaller groups would be followed by a relatively rapid but more largely nonadaptive radiation.

As to rate, since the process depends mainly on the value of 1/4N, assumed to be somewhat less than u (and s) the process cannot be as rapid as one due temporarily to either unopposed selection or unopposed mutation pressure. Hundreds of thousands of generations seem to be required at best for important nonadaptive evolutionary changes of the species as a whole; while adaptive advance, depending on the chance attainment of favorable combinations would be much slower. Even so the process is much the most rapid non-self-terminating one yet considered.

In reaching the tentative conclusion that the situation is most favorable for evolution in a population of a certain intermediate size, one important consideration has been omitted. This is the tendency toward subdivision into more or less completely isolated subgroups in widely distributed populations. Within each subgroup there is a distribution of gene frequencies dependent largely on its own size (N_1), amount of crossing with the rest of the species (m_1), selection due to local conditions (s_1) and the mean gene frequency of migrants from the rest of the species (q_m). It may be assumed that $1/4N_1$ is so much larger than u that mutation pressure (and also the average selection coefficient, s, for the whole species) can be ignored.

$$y = Ce^{4N_1s_1q}q^{4N_1m_1q_m-1}(1 - q)^{4N_1m_1(1-q_m)-1}.$$

Gene frequency in each subgroup oscillates about a mean value, which is that of the whole species only if conditions of selection are uniform. Figure 21 represents various cases. The random variations of gene frequency have effects similar to those described above within each group. The result is a partly nonadaptive, partly adaptive radiation among the subgroups. Those in which the most successful types are reached presumably flourish and tend to overflow their boundaries while others decline, leading to changes in the mean gene frequency of the species as a whole. In this case, the rate of evolution should be much greater than in the previous cases. The coefficients $1/4N_1$ and s_1 may be relatively large and bring about rapid differentiation of subgroups, while the competition between subgroups will bring about rapid changes in the gene frequencies of the species as a whole. The direction of evolution of the species as a whole will be closely responsive to the prevailing conditions, orthogenetic as long as these are constant, but changing with sufficiently long continued environmental change.

A question which requires consideration is the effect of alternation of conditions, large and small size of population, severe and low selection. The effects of changes in the conditions of selection have already been touched upon. Persistence of small numbers or of severe selection for such periods of time as to bring about extensive fixation of factors compromises evolution for a long time following, there being no escape from fixation except by mutation pressure. Many thousands of generations may be required after restoration to large size and not too severe selection, before evolutionary plasticity is restored. Short time oscillations in population number or severity of selection, on the other hand, probably tend to speed up evolutionary change by causing minor changes in gene frequency.

Control of evolution

With regard to control of the process, it is evident that little is possible either within a small stock or a freely interbreeding large one. Even drastic selection is of little effect in the former, and in the latter, while it may bring about a rapid immediate change in the particular respect selected, this must be at the expense of other characters, and in any case, soon leads to a condition in which further advance must wait on the occurrence of mutations more favorable than those fixed by the selection. The limitations in this case have been well brought out in a recent discussion by KEMP (1929). Maximum continuous progress in a homogeneous population requires an intensity of selection for each of the more indifferent genes not much greater than its mutation rate and also a certain size of population. Even so, the direction of advance is somewhat uncertain and the rate to be measured in geologic time.

If infrequency of mutation is the limiting factor here, it would seem that a considerable increase in the rate of evolution should be made possible by a speeding up of mutation, as by X-rays. There is a limit, however, imposed by the prevailingly injurious character of mutations. Even the most rigorous culling of individuals means in general, only a low selection coefficient (in absolute terms) for each of the presumably numerous unfixed genes, which are not in themselves lethal or sublethal in effect. Such culling would become insufficient to hold mutation pressure in check when the latter had increased beyond a certain point ($u>s$). Moreover, as the number of unfixed genes becomes greater under an increased mutation rate, the smaller becomes the separate gene selection coefficients, making it certain that mutation rate could not increase very much before the possibility of effective selection (in all respects at once) rather than infrequency of mutation would become the limiting factor. With respect to lethal mutations, it has already been noted that the observed natural time rate in Drosophila is such as would mean immediate extinction, if transferred to the human species. It is clear that an evolution in the direction of increased gene stability, rather than mutability, has been a necessary phase, in the evolution of the longer lived higher animals. This makes it unlikely that a general increase in mutation rate would increase the rate of evolutionary advance along adaptive lines.

The only practicable method of bringing about a rapid and non-self-terminating advance seems to be through subdivision of the population into isolated and hence differentiating small groups, among which selection may be practiced, but not to the extent of reduction to only one or two types (WRIGHT 1922a). The crossing of the superior types followed by another period of isolation, then by further crossing and so on *ad infinitum* presents a system by means of which an evolutionary advance through the field of possible combinations of the genes present in the original stock, and arising by occasional mutation, should be relatively rapid and practically unlimited. The occasional use of means for increasing mutation rate within limited portions of the population should add further to the possibilities of this system.

Agreement with data of evolution

We come finally to the question as to how far the characteristics of evolution in nature can be accounted for on a Mendelian basis. A review of the data of evolution would go far beyond the scope of the present paper. It may be suggested, however, that the type of moving equilibrium to be expected, according to the present analysis, in a population comparable to natural species in numbers, state of subdivision, conditions of selection, individual adaptability, etc. agrees well with the apparent course of evolution in the majority of cases, even though heredity depend wholly on genes with properties like those observed in the laboratory. Adaptive orthogenetic advances for moderate periods of geologic time, a winding course in the long run, nonadaptive branching following isolation as the usual mode of origin of subspecies, species and perhaps even genera, adaptive branching giving rise occasionally to species which may originate new families, orders, etc.; apparent continuity as the rule, discontinuity the rare exception, are all in harmony with this interpretation.

The most serious difficulties are perhaps in apparent cases of nonadaptive orthogenesis on the one hand and extreme perfection of complicated adaptations on the other. In so far as extreme degeneration of organs is concerned, there is little difficulty—this is to be expected as a by-product of other evolutionary changes. Because of their multiple effects, there can be no really indifferent genes, whatever may be true of organs which have been reduced beyond a certain size. Zero as the value of a selection coefficient is merely a mathematical point between positive and negative values. It is common observation that mutations are more likely to reduce the development of an organ than to stimulate it. It follows that evolutionary change in general will have as a by product the gradual elimination of indifferent organs. Nonadaptive orthogenesis of a positive sort, increase of size of organs to a point which threatens the species, constitutes a more difficult problem, if a real phenomenon. Probably many of the cases cited are cases in which the line of evolution represents the most favorable immediately open to a species doomed by competition with a form of of radically different type or else cases in which selection based on individual advantage leads the species into a cul-de-sac. The nonadaptive differentiation of small subgroups and the great effectiveness of subsequent selection between such groups as compared with that between individuals seem important factors in the origin of peculiar adaptations and the attainment of extreme perfection. It is recognized that there are specific cases which seem to offer great difficulty. This should not obscure the fact that the bulk of the data indicate a process of just the sort which must be occurring in any case to some extent as a statistical consequence of the known mechanism of heredity. The conclusion seems warranted that the enormous recent additions to knowledge of heredity have merely strengthened the general conception of the evolutionary process reached by DARWIN in his exhaustive analysis of the data available 70 years ago.

"Creative" and "emergent" evolution

The present discussion has dealt with the problem of evolution as one depending wholly on mechanism and chance. In recent years, there has been some tendency to revert to more or less mystical conceptions revolving about such phrases as "emergent evolution" and "creative evolution." The writer must confess to a certain sympathy with such viewpoints philosophically but feels that they can have no place in an attempt at scientific analysis of the problem. One may recognize that the only reality directly experienced is that of mind, including choice, that mechanism is merely a term for regular behavior, and that there can be no ultimate explanation in terms of mechanism—merely an analytic description. Such a description, however, is the essential task of science and because of these very considerations, objective and subjective terms cannot be used in the same description without danger of something like 100 percent duplication. Whatever incompleteness is involved in scientific analysis applies to the simplest problems of mechanics as well as to evolution. It is present in most aggravated form, perhaps, in the development and behavior of individual organisms, but even here there seems to be no necessary limit (short of quantum phenomena) to the extent to which mechanistic analysis may be carried. An organism appears to be a system, linked up in such a way, through chains of trigger mechanisms, that a high degree of freedom of behavior as a whole merely requires departures from regularity of behavior among the ultimate parts, of the order of infinitesimals raised to powers as high as the lengths of the above chains. This view implies considerable limitations in the synthetic phases of science, but in any case it seems to have reached the point of demonstration in the field of quantum physics that prediction can be expressed only in terms of probabilities, decreasing with the period of time. As to evolution, its entities, species and ecologic systems, are much less closely knit than individual organisms.

One may conceive of the process as involving freedom, most readily traceable in the factor called here individual adaptability. This, however, is a subjective interpretation and can have no place in the objective scientific analysis of the problem.

SUMMARY

The frequency of a given gene in a population may be modified by a number of conditions including recurrent mutation to and from it, migration, selection of various sorts and, far from least in importance, mere chance variation. Using q for gene frequency, v and u for mutation rates to and from the gene respectively, m for the exchange of population with neighboring groups with gene frequency q_m, s for the selective advantage of the gene over its combined allelomorphs and N for the effective number in the breeding stock (much smaller as a rule than the actual number of adult individuals) the most probable change in gene frequency per generation may be written:

$$\Delta q = v(1 - q) - uq - m(q - q_m) + sq(1 - q)$$

and the array of probabilities for the next generation as $[(1-q-\Delta q)a + (q+\Delta q)A]^{2N}$. The contribution of zygotic selection (reproductive rates of aa, Aa and AA as $1-s^1:1-hs^1:1$) is $\Delta q = s^1 q(1-q)[1-q+h(2q-1)]$. In interpreting results it is necessary to recognize that the above coefficients are continually changing in value and especially that the selection coefficient of a particular gene is really a function not only of the relative frequencies and momentary selection coefficients of its different allelomorphs but also of the entire system of frequencies and selection coefficients of non-allelomorphs. Selection relates to the organism as a whole and its environment and not to genes as such. The mutation rate to a gene (v) can usually be treated as of negligible magnitude assuming the prevalence of multiple allelomorphs.

In a population so large that chance variation is negligible, gene frequency reaches equilibrium when $\Delta q = 0$. Among special cases is that of opposing mutation rates $\left(q = \frac{v}{u+v}\right)$, of selection against both homozygotes $\left(q = \frac{1-h}{1-2h}\right)$, of mutation against genic selection $\left(q = 1 - \frac{u}{s}\right)$, of mutation against zygotic selection ($q = 1 - \frac{u}{hs^1}$ unless h approaches 0, when $q = 1 - \sqrt{\frac{u}{s}}$), of selection and migration ($q = 1 - \frac{m}{s}(1 - q_m)$) or

$-\frac{mq_m}{s}$ if s is much greater than m, $q = q_m\left(1 + \frac{s}{m}(1 - q_m)\right)$ if s is much smaller than m, while the values $q = \sqrt{q_m}$ or $1 - \sqrt{1 - q_m}$ when $s = \pm m$ illustrate the intermediate case).

Gene frequency fluctuates about the equilibrium point in a distribution curve, the form of which depends on the relations between population number and the various pressures. The general formula in the case of a freely interbreeding group, assuming genic selection, is

$$y = Ce^{4Nsq}q^{4N[mq_m+v]-1}(1 - q)^{4N[m(1-q_m)+u]-1}.$$

The correlation between relatives is affected by the form of the distribution of gene frequencies through FISHER's "dominance ratio." It appears that this is less than 0.20 in small populations under low selection but may even approach 1 in large populations under severe selection against recessives.

In a large population in which gene frequencies are always close to their equilibrium points, any change in conditions other than population number is followed by an approach toward the new equilibria at rates given by the Δq's. Great reduction in population number is followed by fixation and loss of genes, each at the rate $1/4N$ per generation, where N refers to the new population number. This applies either in a group of monoecious individuals with random fertilization or, approximately, in one equally divided between males and females (9.6 percent instead of 12.5 percent, however, under brother-sister mating, $N = 2$). More generally with an effective breeding stock of N_m males and N_f females, the rates of fixation and of loss are each approximately $(1/16N_m + 1/16N_f)$ until mutation pressure at length brings equilibrium in a distribution approaching first the form $y = C(1-q)^{-1}$ with decay at rate u and ultimately $Cq^{-1}(1-q)^{-1}$. The converse process, great increase in the size of a long inbred population, is followed by a slow approach toward the new equilibrium at a rate dependent in the early stages on mutation pressure.

With respect to genes which are indifferent to selection, the mean frequency is always $q = v/(u+v)$. The variance of characters, dependent on such genes, is proportional (at equilibrium) to population number up to about $N = 1/4u$. Beyond this, there is approach of variance to a limiting value.

In the presence of selection (s considerably greater than 2u) the mean frequency at equilibrium varies between approximate fixation of the favored genes ($q = 1 - u/s$) in large populations and approximate, if not complete, fixation of mutant allelomorphs ($q = v/(u+v)$) in small populations, the rate of change from one state to the other being the mutation rate (u). A consequence is a slow but increasing tendency to decline in vigor in inbred stocks, to be distinguished from the relatively rapid but soon completed fixation process, described above as occurring at rate 1/2N. The variance of characters in this as in the preceding case, is approximately proportional to population number up to a certain point (N less than 1/4s) and above this rapidly approaches a limiting value. Variance is inversely proportional to the severity of selection in large populations unless the selection is very slight but in small populations is little affected by selection unless the latter is very severe (s greater than 1/4N).

Evolution as a process of cumulative change depends on a proper balance of the conditions, which, at each level of organization—gene, chromosome, cell, individual, local race—make for genetic homogeneity or genetic heterogeneity of the species. While the basic factor of change—the infrequent, fortuitous, usually more or less injurious gene mutations, in themselves, appear to furnish an inadequate basis for evolution, the mechanism of cell division, with its occasional aberrations, and of nuclear fusion (at fertilization) followed at some time by reduction make it possible for a relat vely small number of not too injurious mutations to provide an extensive field of actual variations. The type and rate of evolution in such a system depend on the balance among the evolutionary pressures considered here. In too small a population (1/4N much greater than u and s) there is nearly complete fixation, little variation, little effect of selection and thus a static condition modified occasionally by chance fixation of new mutations leading inevitably to degeneration and extinction. In too large a freely interbreeding population (1/4N much less than u and s) there is great variability but such a close approach to complete equilibrium of all gene frequencies that there is no evolution under static conditions. Change in conditions such as more severe selection, merely shifts all gene frequencies and for the most part reversibly, to new equilibrium points in which the population remains static as long as the new conditions persist. Such evolutionary change as occurs is an extremely slow adaptive process. In a population of intermediate size (1/4N of the order of u) there is continual random shifting of gene frequencies and a consequent shifting of selection coefficients which leads to a relatively rapid, continuing, irreversible, and largely fortuitous, but not degenerative series of changes, even under static conditions. The rate is rapid only in comparison with the preceding cases, however, being limited by mutation pressure and thus requiring periods of the order of 100,000 generations for important changes.

Finally in a large population, divided and subdivided into partially isolated local races of small size, there is a continually shifting differentiation among the latter (intensified by local differences in selection but occurring under uniform and static conditions) which inevitably brings about an indefinitely continuing, irreversible, adaptive, and much more rapid evolution of the species. Complete isolation in this case, and more slowly in the preceding, originates new species differing for the most part in nonadaptive respects but is capable of initiating an adaptive radiation as well as of parallel orthogenetic lines, in accordance with the conditions. It is suggested, in conclusion, that the differing statistical situations to be expected among natural species are adequate to account for the different sorts of evolutionary processes which have been described, and that, in particular, conditions in nature are often such as to bring about the state of poise among opposing tendencies on which an indefinitely continuing evolutionary process depends.

LITERATURE CITED

BERNSTEIN, F., 1925 Zusammenfassende Betrachtungen über die erblichen Blutstrukturen des Menschen. Z. indukt. Abstamm.-u. VererbLehre **37:** 237–269.

CALDER, A., 1927 The role of inbreeding in the development of the Clydesdale breed of horse. Proc. Roy. Soc. Edinb. **47:** 118–140.

EAST, E. M., 1918 The role of reproduction in evolution. Amer. Nat. **52:** 273–289.

FISHER, R. A., 1918 The correlation between relatives on the supposition of Mendelian inheritance. Trans. Roy. Soc., Edinb. 52 part **2:** 399–433.

1922 On the dominance ratio. Proc. Roy. Soc., Edinb. **42:** 321–341.

1928 The possible modification of the response of the wild type to recurrent mutations. Amer. Nat., **62:** 115–126.

1929 The evolution of dominance; reply to Professor Sewall Wright: Amer. Nat. **63:** 553–556.

1930 The genetical theory of natural selection. 272 pp. Oxford: Clarendon Press.

HAGEDOORN, A. L., and HAGEDOORN A. C., 1921 The relative value of the processes causing evolution. 294 pp. The Hague: Martinus Nijhoff.

HALDANE, J. B. S., 1924–1927 A mathematical theory of natural and artificial selection. Part I. Trans. Camb. Phil. Soc. **23:** 19–41. Part II. Proc. Camb. Phil. Soc. (Biol. Sci) **1:** 158–163. Part III. Proc. Camb. Phil. Soc. **23:** 363–372. Part IV. Proc. Camb. Phil. Soc. **23:** 607–615. Part V. Proc. Camb. Phil. Soc. **23:** 838–844.

HARDY, G. H., 1908 Mendelian proportions in a mixed population. Science **28:** 49–50.

JENNINGS, H. S., 1916 The numerical results of diverse systems of breeding. Genetics **1:** 53–89.

JONES, D. F., 1917 Dominance of linked factors as a means of accounting for heterosis. Genetics **2:** 466–479.

KEMP, W. B., 1929 Genetic equilibrium and selection. Genetics **14:** 85–127.

LOTKA, A. J., 1925 Elements of Physical Biology. Baltimore: Williams and Wilkins.

MCPHEE, H. C., and WRIGHT S., 1925, 1926 Mendelian analysis of the pure breeds of livestock. III. The Shorthorns. J. Hered., **16:** 205–215. IV. The British dairy Shorthorns. J. Hered. **17:** 397–401.

MORGAN, T. H., BRIDGES, C. B., STURTEVANT, A. H., 1925 The genetics of Drosophila. 262 pp. The Hague: Martinus Nijhoff.

MULLER, H. J., 1922 Variation due to change in the individual gene. Amer. Nat. **56**: 32–50.

1928 The measurement of gene mutation rate in Drosophila, its high variability, and its dependence upon temperature. Genetics **13**: 279–357.

1929 The gene as the basis of life. Proc. Int. Cong. Plant Sciences **1**: 897–921.

ROBBINS, R. B., 1918 Some applications of mathematics to breeding problems. III. Genetics **3**: 375–389.

SMITH, A. D. B., 1926 Inbreeding in cattle and horses. Eugen. Rev. **14**: 189–204.

WAHLUND, STEN., 1928 Zusammensetzung von Populationen und Korrelationserscheinungen vom Standpunkt der Vererbungslehre aus betrachtet. Hereditas **11**: 65–106.

WEINBERG, W., 1909 Über Vererbungsgesetze beim Menschen. Z. indukt. Abstamm.-u. Vererb-Lehre **1**: 277–330.

1910 Weiteres Beiträge zur Theorie der Vererbung. Arch. Rass.-u. Ges. Biol. **7**: 35–49, 169–173.

WRIGHT, S., 1921 Systems of mating. Genetics **6**: 111–178.

1922 Coefficients of inbreeding and relationship. Amer. Nat. **61**: 330–338.

1922a The effects of inbreeding and crossbreeding on guinea-pigs. III. Crosses between highly inbred families. U. S. Dept. Agr. Bull. No. 1121.

1923 Mendelian analysis of the pure breeds of live stock. Part I. J. Hered. **14**: 339–348. Part II, J. Hered. **14**: 405–422.

1929 FISHER's theory of dominance. Amer. Nat. **63**: 274–279.

1929a The evolution of dominance. Comment on Doctor FISHER's reply. Amer. Nat. **63**: 556–561.

1929b Evolution in a Mendelian population. Anat. Rec. **44**: 287.

WRIGHT, S., and MCPHEE, H. C., 1925 An approximate method of calculating coefficients inbreeding and relationship from livestock pedigrees. J. Agric. Res. **31**: 377–383.

THE ROLES OF MUTATION, INBREEDING, CROSSBREEDING AND SELECTION IN EVOLUTION

Sewall Wright, University of Chicago, Chicago, Illinois

The enormous importance of biparental reproduction as a factor in evolution was brought out a good many years ago by East. The observed properties of gene mutation—fortuitous in origin, infrequent in occurrence and deleterious when not negligible in effect—seem about as unfavorable as possible for an evolutionary process. Under biparental reproduction, however, a limited number of mutations which are not too injurious to be carried by the species furnish an almost infinite field of possible variations through which the species may work its way under natural selection.

Estimates of the total number of genes in the cells of higher organisms range from 1000 up. Some 400 loci have been reported as having mutated in Drosophila during a laboratory experience which is certainly very limited compared with the history of the species in nature. Presumably, allelomorphs of all type genes are present at all times in any reasonably numerous species. Judging from the frequency of multiple allelomorphs in those organisms which have been studied most, it is reasonably certain that many different allelomorphs of each gene are in existence at all times. With 10 allelomorphs in each of 1000 loci, the number of possible combinations is 10^{1000} which is a very large number. It has been estimated that the total number of electrons and protons in the whole visible universe is much less than 10^{100}.

However, not all of this field is easily available in an interbreeding population. Suppose that each type gene is manifested in 99 percent of the individuals, and that most of the remaining 1 percent have the most favorable of the other allelomorphs, which in general means one with only a slight differential effect. The average individual will show the effects of 1 percent of the 1000, or 10 deviations from the type, and since this average has a standard deviation of $\sqrt{10}$ only a small proportion will exhibit more than 20 deviations from type where 1000 are possible. The population is thus confined to an infinitesimal portion of the field of possible gene combinations, yet this portion includes some 10^{40} homozygous combinations, on the above extremely conservative basis, enough so that there is no reasonable chance that any two individuals have exactly the same genetic constitution in a species of millions of millions of individuals persisting over millions of generations. There is no difficulty in accounting for the probable genetic uniqueness of each individual human being or other organism which is the product of biparental reproduction.

If the entire field of possible gene combinations be graded with respect to adaptive value under a particular set of conditions, what would be its nature? Figure 1 shows the combinations in the cases of 2 to 5 paired allelomorphs. In the last case, each of the 32 homozygous combinations is at one remove from 5 others, at two removes from 10, etc. It would require 5 dimensions to represent these relations symmetrically; a sixth dimension is needed to represent level of adaptive value. The 32 combina-

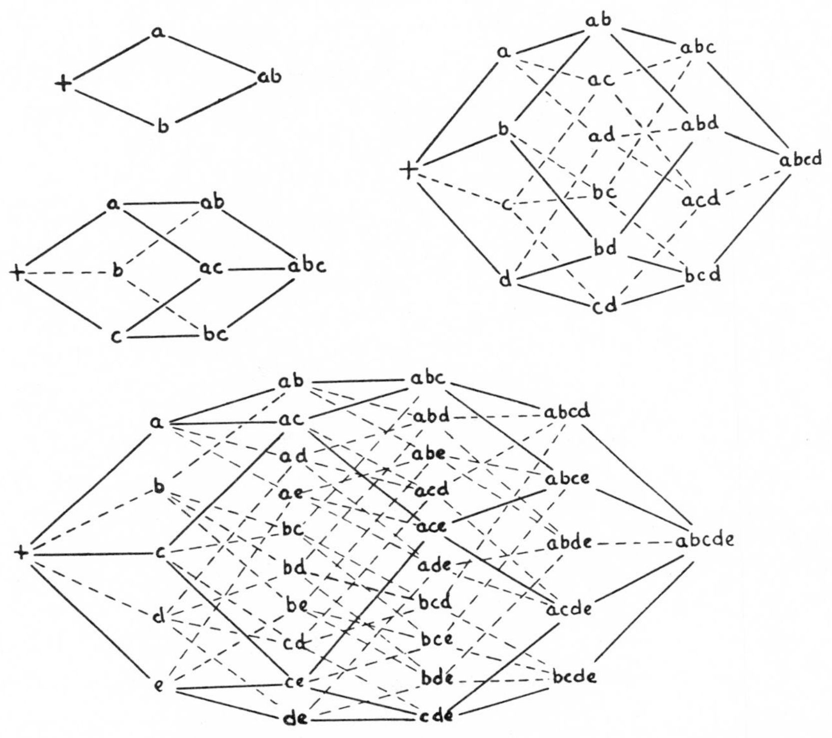

FIGURE 1.—The combinations of from 2 to 5 paired allelomorphs.

tions here compare with 10^{1000} in a species with 1000 loci each represented by 10 allelomorphs, and the 5 dimensions required for adequate representation compare with 9000. The two dimensions of figure 2 are a very inadequate representation of such a field. The contour lines are intended to represent the scale of adaptive value.

One possibility is that a particular combination gives maximum adaptation and that the adaptiveness of the other combinations falls off more or less regularly according to the number of removes. A species whose individuals are clustered about some combination other than the highest would move up the steepest gradient toward the peak, having reached which it would remain unchanged except for the rare occurrence of new favorable mutations.

But even in the two factor case (figure 1) it is possible that there may be two peaks, and the chance that this may be the case greatly increases with each additional locus. With something like 10^{1000} possibilities (figure 2) it may be taken as certain that there will be an enormous number of

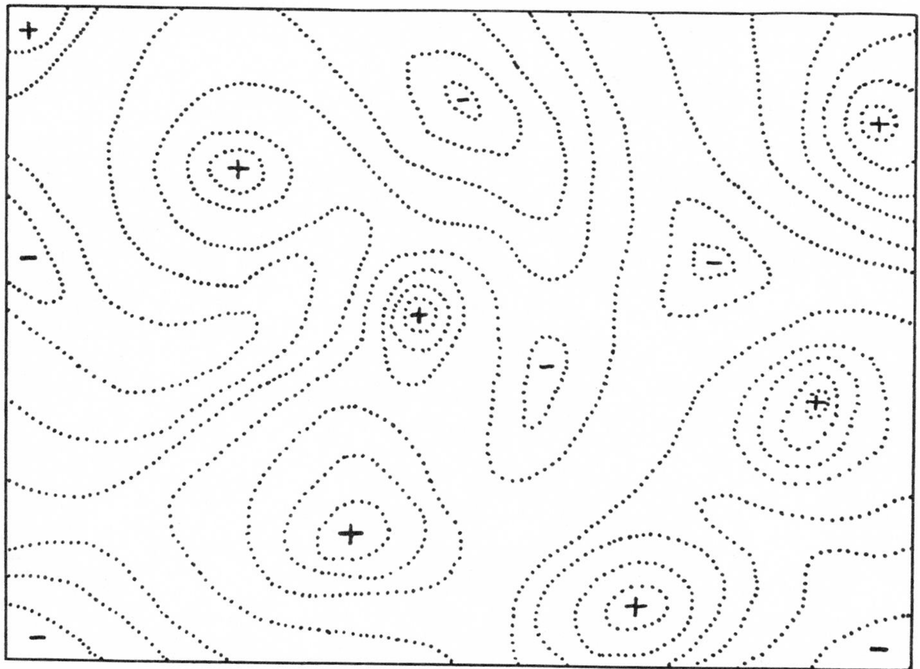

FIGURE 2.—Diagrammatic representation of the field of gene combinations in two dimensions instead of many thousands. Dotted lines represent contours with respect to adaptiveness.

widely separated harmonious combinations. The chance that a random combination is as adaptive as those characteristic of the species may be as low as 10^{-100} and still leave room for 10^{800} separate peaks, each surrounded by 10^{100} more or less similar combinations. In a rugged field of this character, selection will easily carry the species to the nearest peak, but there may be innumerable other peaks which are higher but which are separated by "valleys." The problem of evolution as I see it is that of a mechanism by which the species may continually find its way from lower to higher peaks in such a field. In order that this may occur, there must be some trial and error mechanism on a grand scale by which the species may explore the region surrounding the small portion of the field which it occupies. To evolve, the species must not be under strict control of natural selection. Is there such a trial and error mechanism?

At this point let us consider briefly the situation with respect to a single locus. In each graph in figure 3 the abscissas represent a scale of gene frequency, 0 percent of the type genes to the left, 100 percent to the right. The elementary evolutionary process is, of course, change of gene frequency, a

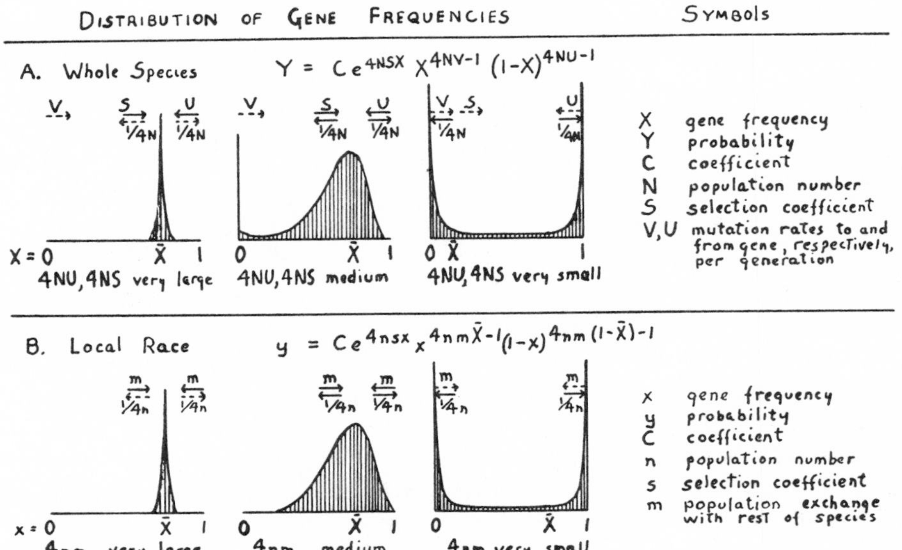

FIGURE 3.—Random variability of a gene frequency under various specified conditions.

practically continuous process. Owing to the symmetry of the Mendelian mechanism, any gene frequency tends to remain constant in the absence of disturbing factors. If the type gene mutates at a certain rate, its frequency tends to move to the left, but at a continually decreasing rate. The type gene would ultimately be lost from the population if there were no opposing factor. But the type gene is in general favored by selection. Under selection, its frequency tends to move to the right. The rate is greatest at some point near the middle of the range. At a certain gene frequency the opposing pressures are equal and opposite, and at this point there is consequently equilibrium. There are other mechanisms of equilibrium among evolutionary factors which need not be discussed here. Note that we have here a theory of the stability of species in spite of continuing mutation pressure, a continuing field of variability so extensive that no two individuals are ever genetically the same, and continuing selection.

If the population is not indefinitely large, another factor must be taken into account: the effects of accidents of sampling among those that survive and become parents in each generation and among the germ cells of these, in other words, the effects of inbreeding. Gene frequency in a given generation is in general a little different one way or the other from that in the preceding, merely by chance. In time, gene frequency may wander a long way from the position of equilibrium, although the farther it wanders the greater the pressure toward return. The result is a frequency distribution within which gene frequency moves at random. There is considerable spread even with very slight inbreeding and the form of distribution becomes U-shaped with close inbreeding. The rate of movement of gene frequency is very slow in the former case but is rapid in the latter (among unfixed genes). In this case, however, the tendency toward complete fixation of genes, practically irrespective of selection, leads in the end to extinction.

In a local race, subject to a small amount of crossbreeding with the rest of the species (figure 3, lower half), the tendency toward random fixation is balanced by immigration pressure instead of by mutation and selection. In a small sufficiently isolated group all gene frequencies can drift irregulary back and forth about their mean values at a rapid rate, in terms of geologic time, without reaching fixation and giving the effects of close inbreeding. The resultant differentiation of races is of course increased by any local differences in the conditions of selection.

Let us return to the field of gene combinations (figure 4). In an indefinitely large but freely interbreeding species living under constant conditions, each gene will reach ultimately a certain equilibrium. The species will occupy a certain field of variation about a peak in our diagram (heavy broken contour in upper left of each figure). The field occupied remains constant although no two individuals are ever identical. Under the above conditions further evolution can occur only by the appearance of wholly new (instead of recurrent) mutations, and ones which happen to be favorable from the first. Such mutations would change the character of the field itself, increasing the elevation of the peak occupied by the species. Evolutionary progress through this mechanism is excessively slow since the chance of occurrence of such mutations is very small and, after occurrence, the time required for attainment of sufficient frequency to be subject to selection to an appreciable extent is enormous.

The general rate of mutation may conceivably increase for some reason. For example, certain authors have suggested an increased incidence of cosmic rays in this connection. The effect (figure 4A) will be as a rule a spreading of the field occupied by the species until a new equilibrium is reached. There will be an average lowering of the adaptive level of the species. On the other hand, there will be a speeding up of the process discussed above, elevation of the peak itself through appearance of novel favorable mutations. Another possibility of evolutionary advance is that the spreading of the field occupied may go so far as to include another and

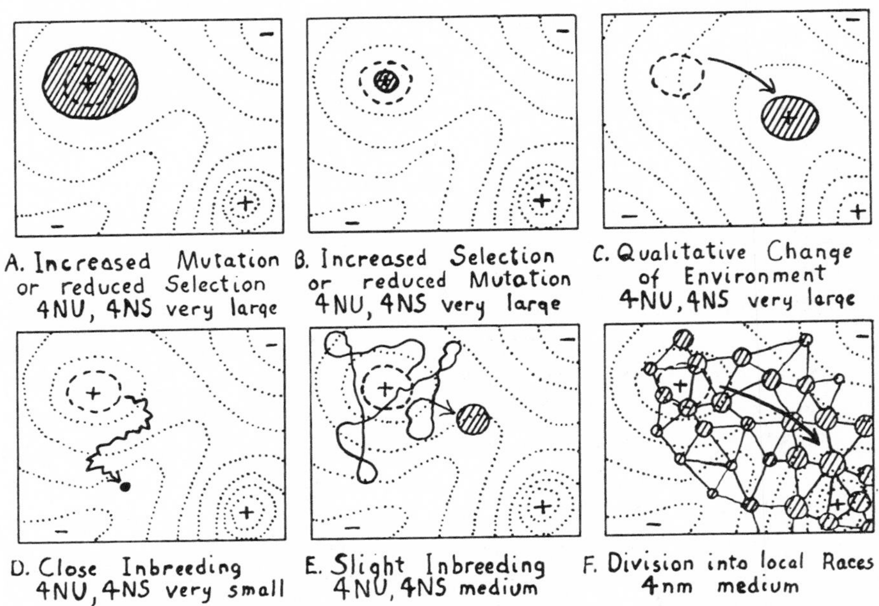

FIGURE 4.—Field of gene combinations occupied by a population within the general field of possible combinations. Type of history under specified conditions indicated by relation to initial field (heavy broken contour) and arrow.

higher peak, in which case the species will move over and occupy the region about this. These mechanisms do not appear adequate to explain evolution to an important extent.

The effects of reduced mutation rate (figure 4B) are of course the opposite: a rise in average level, but reduced variability, less chance of novel favorable mutation, and less chance of capture of a neighboring peak.

The effect of increased severity of selection (also 4B) is, of course, to increase the average level of adaptation until a new equilibrium is reached. But again this is at the expense of the field of variation of the species and reduces the chance of capture of another adaptive peak. The only basis for continuing advance is the appearance of novel favorable mutations which are relatively rapidly utilized in this case. But at best the rate is extremely slow even in terms of geologic time, judging from the observed rates of mutation.

Relaxation of selection has of course the opposite effects and thus effects somewhat like those of increased mutation rate (figure 4A).

The environment, living and non-living, of any species is actually in continual change. In terms of our diagram this means that certain of the high places are gradually being depressed and certain of the low places are becoming higher (figure 4C). A species occupying a small field under influence of severe selection is likely to be left in a pit and become extinct, the victim of extreme specialization to conditions which have ceased, but if under sufficiently moderate selection to occupy a wide field, it will merely be kept continually on the move. Here we undoubtedly have an important evolutionary process and one which has been generally recognized. It consists largely of change without advance in adaptation. The mechanism is, however, one which shuffles the species about in the general field. Since the species will be shuffled out of low peaks more easily than high ones, it should gradually find its way to the higher general regions of the field as a whole.

Figure 4D illustrates the effect of reduction in size of population below a certain relation to the rate of mutation and severity of selection. There is fixation of one or another allelomorph in nearly every locus, largely irrespective of the direction favored by selection. The species moves down from its peak in an erratic fashion and comes to occupy a much smaller field. In other words there is the deterioration and homogeneity of a closely inbred population. After equilibrium has been reached in variability, movement becomes excessively slow, and, such as there is, is nonadaptive. The end can only be extinction. Extreme inbreeding is not a factor which is likely to give evolutionary advance.

With an intermediate relation between size of population and mutation rate, gene frequencies drift at random without reaching the complete fixation of close inbreeding (figure 4E). The species moves down from the extreme peak but continually wanders in the vicinity. There is some chance that it may encounter a gradient leading to another peak and shift its allegiance to this. Since it will escape relatively easily from low peaks as compared with high ones, there is here a trial and error mechanism by which in time the species may work its way to the highest peaks in the general field. The rate of progress, however, is extremely slow since change of gene frequency is of the order of the reciprocal of the effective population size and this reciprocal must be of the order of the mutation rate in order to meet the conditions for this case.

Finally (figure 4F), let us consider the case of a large species which is subdivided into many small local races, each breeding largely within itself but occasionally crossbreeding. The field of gene combinations occupied by each of these local races shifts continually in a nonadaptive fashion (except in so far as there are local differences in the conditions of selection). The rate of movement may be enormously greater than in the preceding case since the condition for such movement is that the reciprocal of the population number be of the order of the proportion of crossbreeding instead of the mutation rate. With many local races, each spreading over a considerable field and moving relatively rapidly in the more general field about the controlling peak, the chances are good that one at least will come under the influence of another peak. If a higher peak, this race will expand in numbers and by crossbreeding with the others will pull the whole species toward the new position. The average adaptiveness of the species thus advances under intergroup selection, an enormously more effective process than intragroup selection. The conclusion is that subdivision of a species into local races provides the most effective mechanism for trial and error in the field of gene combinations.

It need scarcely be pointed out that with such a mechanism complete isolation of a portion of a species should result relatively rapidly in specific differentiation, and one that is not necessarily adaptive. The effective intergroup competition leading to adaptive advance may be between species rather than races. Such isolation is doubtless usually geographic in character at the outset but may be clinched by the development of hybrid sterility. The usual difference of the chromosome complements of related species puts the importance of chromosome aberration as an evolutionary process beyond question, but, as I see it, this importance is not in the character differences which they bring (slight in balanced types), but rather in leading to the sterility of hybrids and thus making permanent the isolation of two groups.

How far do the observations of actual species and their subdivisions conform to this picture? This is naturally too large a subject for more than a few suggestions.

That evolution involves nonadaptive differentiation to a large extent at the subspecies and even the species level is indicated by the kinds of differences by which such groups are actually distinguished by systematists. It is only at the subfamily and family levels that clear-cut adaptive differences become the rule (Robson, Jacot). The principal evolutionary mechanism in the origin of species must thus be an essentially nonadaptive one.

That natural species often are subdivided into numerous local races is indicated by many studies. The case of the human species is most familiar. Aside from the familiar racial differences recent studies indicate a distribution of frequencies relative to an apparently nonadaptive series of allelomorphs, that determining blood groups, of just the sort discussed above. I scarcely need to labor the point that changes in the average of mankind in the historic period have come about more by expansion of some types and decrease and absorption of others than by uniform evolutionary advance. During the recent period, no doubt, the phases of intergroup competition and crossbreeding have tended to overbalance the process of local differentiation, but it is probable that in the hundreds of thousands of years of prehistory, human evolution was determined by a balance between these factors.

Subdivision into numerous local races whose differences are largely nonadaptive has been recorded in other organisms wherever a sufficiently detailed study has been made. Among the land snails of the Hawaiian Islands, Gulick (sixty years ago) found that each mountain valley, often each grove of trees, had its own characteristic type, differing from others in "nonutilitarian" respects. Gulick attributed this differentiation to inbreeding. More recently Crampton has found a similar situation in the land snails of Tahiti and has followed over a period of years evolutionary changes which seem to be of the type here discussed. I may also refer to the studies of fishes by David Starr Jordan, garter snakes by Ruthven, bird lice by Kellogg, deer mice by Osgood, and gall wasps by Kinsey as others which indicate the role of local isolation as a differentiating factor. Many other cases are discussed by Osborn and especially by Rensch in recent summaries. Many of these authors insist on the nonadaptive character of most of the differences among local races. Others attribute all differences to the environment, but this seems to be more an expression of faith than a view based on tangible evidence.

An even more minute local differentiation has been revealed when the methods of statistical analysis have been applied. Schmidt demonstrated the existence of persistent mean differences at each collecting station in certain species of marine fish of the fjords of Denmark, and these differences were not related in any close way to the environment. That the differences were in part genetic was demonstrated in the laboratory. David Thompson has found a correlation between water distance and degree of differentiation within certain fresh water species of fish of the streams of Illinois. SUMNER's extensive studies of subspecies of Peromyscus (deer mice) reveal genetic differentiations, often apparently nonadaptive, among local populations and demonstrate the genetic heterogeneity of each such group.

The modern breeds of livestock have come from selection among the products of local inbreeding and of crossbreeding between these, followed by renewed inbreeding, rather than from mass selection of species. The recent studies of the geographical distribution of particular genes in livestock and cultivated plants by SEREBROVSKY, PHILIPTSCHENKO and others are especially instructive with respect to the composition of such species.

The paleontologists present a picture which has been interpreted by some as irreconcilable with the Mendelian mechanism, but this seems to be due more to a failure to appreciate statistical consequences of this mechanism than to anything in the data. The horse has been the standard example of an orthogenetic evolutionary sequence preserved for us with an abundance of material. Yet MATHEW's interpretation as one in which evolution has proceeded by extensive differentiation of local races, intergroup selection, and crossbreeding is as close as possible to that required under the Mendelian theory.

Summing up: I have attempted to form a judgment as to the conditions for evolution based on the statistical consequences of Mendelian heredity. The most general conclusion is that evolution depends on a certain balance among its factors. There must be gene mutation, but an excessive rate gives an array of freaks, not evolution; there must be selection, but too severe a process destroys the field of variability, and thus the basis for further advance; prevalence of local inbreeding within a species has extremely important evolutionary consequences, but too close inbreeding leads merely to extinction. A certain amount of crossbreeding is favorable but not too much. In this dependence on balance the species is like a living organism. At all levels of organization life depends on the maintenance of a certain balance among its factors.

More specifically, under biparental reproduction a very low rate of mutation balanced by moderate selection is enough to maintain a practically infinite field of possible gene combinations within the species. The field actually occupied is relatively small though sufficiently extensive that no two individuals have the same genetic constitution. The course of evolution through the general field is not controlled by direction of mutation and not directly by selection, except as conditions change, but by a trial and error mechanism consisting of a largely nonadaptive differentiation of local races (due to inbreeding balanced by occasional crossbreeding) and a determination of long time trend by intergroup selection. The splitting of species depends on the effects of more complete isolation, often made permanent by the accumulation of chromosome aberrations, usually of the balanced type. Studies of natural species indicate that the conditions for such an evolutionary process are often present.

LITERATURE CITED

Crampton, H. E., 1925 Contemporaneous organic differentiation in the species of Partula living in Moorea, Society Islands. Amer. Nat. **59**:5-35.

East, E. M., 1918 The role of reproduction in evolution. Amer. Nat. **52**:273-289.

Gulick, J. T., 1905 Evolution, racial and habitudinal. Pub. Carnegie Instn. **25**:1-269.

Jacot, A. P., 1932 The status of the species and the genus. Amer. Nat. **66**:346-364.

Jordan, D. S., 1908 The law of geminate species. Amer. Nat. **42**:73-80.

Kellogg, V. L., 1908 Darwinism, today. 403 pp. New York: Henry Holt and Co.

Kinsey, A. C., 1930 The gall wasp genus Cynips. Indiana Univ. Studies. **84-86**:1-577.

Mathew, W. D., 1926 The evolution of the horse. A record and its interpretation. Quart. Rev. Biol. **1**:139-185.

Osborn, H. F., 1927 The origin of species. V. Speciation and mutation. Amer. Nat. **49**:193-239.

Osgood, W. H., 1909 Revision of the mice of the genus Peromyscus. North American Fauna **28**:1-285.

Philiptschenko, J., 1927 Variabilität and Variation. 101 pp. Berlin.

Rensch, B., 1929 Das Prinzip geographischer Rassenkreise und das Problem der Artbildung. 206 pp. Berlin: Gebrüder Borntraeger.

Robson, G. C., 1928 The species problem. 283 pp. Edinburgh and London: Oliver and Boyd.

Ruthven, A. G., 1908 Variation and genetic relationships of the garter snakes. U. S. Nat. Mus. Bull. **61**:1-301.

Schmidt, J., 1917 Statistical investigations with *Zoarces viviparus* L. J. Genet. **7**:105-118.

Serebrovsky, A. S., 1929 Beitrag zur geographischen Genetic des Haushahns in Sowjet-Russland. Arch. f. Geflügelkunde, Jahrgang **3**:161-169.

Sumner, F. B., 1932 Genetic, distributional, and evolutionary studies of the subspecies of deer mice (Peromyscus). Bibl. genet. **9**:1-106.

Thompson, D. H., 1931 Variation in fishes as a function of distance. Trans. Ill. State Acad. of Sci. **23**:276-281.

Wright, S., 1931 Evolution in Mendelian populations. Genetics **16**:97-159.

13

Physiological and Evolutionary Theories of Dominance

American Naturalist 68 (January–February 1934):25–53

14

Professor Fisher on the Theory of Dominance

American Naturalist 68 (November–December 1934):562–65

Introduction

The interchange between Fisher and Wright in 1928–29 on the evolution of dominance stimulated tremendous interest among evolutionary biologists and geneticists, resulting in a host of publications on the subject. In 1933 Wright was known as a major participant in the controversy over the evolution of dominance, but he had never written a paper fully detailing his physiological theory of dominance or his explanation of the evolution of dominance. Thus Wright wrote the first of these two papers.

Fisher was unhappy with this paper and wrote a strongly worded rebuttal (Fisher 1934). Wright's reply to Fisher's rebuttal is the second paper. The source of disagreement between Fisher and Wright was far more than just different views about the evolution of dominance. At stake were their different theories of evolution in nature. If Fisher's theory of the evolution of dominance were correct, then Wright's shifting balance theory was wrong; and if Wright's objections to Fisher's theory of the evolution of dominance were correct, then Fisher's genetical theory of natural selection was wrong. The focus of their fundamental disagreement happened at this time to be the evolution of dominance; the focus would later change, but the fundamental disagreement over the process of evolution in nature would change very little.

The acrimonious exchange in 1934 between Wright and Fisher over the evolution of dominance marked the end of their previously cordial interchange of ideas. After 1934 they never again corresponded or even had a congenial conversation (stories about their occasional meetings abound), although they did continue to exchange reprints.

I examine these two papers, and the interaction between Wright and Fisher, in *SW&EB*, 299–303.

PHYSIOLOGICAL AND EVOLUTIONARY THEORIES OF DOMINANCE[1]

PROFESSOR SEWALL WRIGHT
DEPARTMENT OF ZOOLOGY, THE UNIVERSITY OF CHICAGO

INTRODUCTION

MENDEL found that one member of each of his seven pairs of alternative characters of the pea reappeared in the first cross-bred generation to the complete or nearly complete exclusion of the other. Although he attributed no great importance to this himself, there was some tendency, following the rediscovery in 1900, to consider a law of dominance as one of the fundamental principles of heredity. It has long fallen from this estate and it has been questioned whether careful measurements would not show complete dominance to be the exception rather than the rule. An approach to complete dominance is common enough, however, to present a number of interesting problems.

In the first place, it is clear that dominance has to do with the physiology of the organism and has nothing to do with the mechanism of transmission, *i.e.,* with heredity in the narrow sense. Studies of dominance bear on two different groups of problems. The accurate measurement of degree of dominance of particular genes under varying conditions furnishes one of the most available tools for carrying physiological analysis back to the ultimate controlling factors. On the other hand, statistical generalizations in regard to dominance may rest not only on the average consequences of physiological principles but also on general evolutionary trends and thus may throw light on the evolutionary process.

[1] Presented at the Symposium of the American Society of Naturalists at the Century of Progress meeting of the American Association for the Advancement of Science, Chicago, June 21, 1933. The mathematical portions have been extended.

PART I. STATISTICAL PRINCIPLES OF DOMINANCE

GENERALIZATIONS

Among the statistical generalizations which have been made are the following: (1) Dominance of one allelomorph is the rule. (2) The recessive genes are usually less advantageous to the species than their dominant allelomorphs. (3) The effect of the recessive is usually the absence of a positive property found in the dominant in those cases in which there seems to be an approach to knowledge of the primary gene effects. (4) The recessive genes are usually less abundant in natural population than their dominant allelomorphs. (5) In cases in which origin is known the mutant gene is usually recessive. All of these principles are subject to much qualification.

CONDITIONS IN NATURAL POPULATIONS

With regard to the last two, we must carefully distinguish between the properties of uncommon genes found in natural population and those of observed mutations. The evolutionary problems are somewhat different. Even if mutation of all sorts, dominant, semi-dominant and recessive, occurred at random, these properties would not remain distributed at random in a population. Most mutations are observed to be injurious as is to be expected if a mutation is of the nature of an accident to a gene. Injurious dominant and semi-dominant mutations are vigorously selected against and kept at low frequencies ($q = \frac{u}{hs}$ where u is rate of recurrence of the particular mutation and s is the selective disadvantage per generation of the homozygous mutant, hs that of the heterozygote). A mutation that happens to be completely recessive is largely protected from selection and will be kept at relatively high frequencies in the population ($q = \sqrt{u/s}$). With one per cent. selective discrimination against homozygous mutants ($s = .01$) and a rate of recurrence of one in a million per generation, deleterious recessive genes will be kept at 50 times the frequency of ones with exactly intermediate heterozygotes and 100 times that of equally deleterious dominants. Thus the first four of the above principles as applied to genes in natural populations are readily understandable (assuming that most mutations interfere with, rather than increase, the development of characters), without assuming any special tendency for mutation to be recessive.

There is, however, abundant evidence that this is not the whole story. Observed mutations in general are not only injurious and manifested generally in loss of characters, but are actually much more frequently recessive than dominant or semi-dominant.

The Presence and Absence Hypothesis

The first widely accepted hypothesis was Bateson and Punnett's theory of presence and absence. They assumed that the commonest type of mutation was simply loss of a positive entity, that loss of a gene brought loss of the corresponding character (3) that such loss was usually injurious (2), and usually recessive (5) on the principle that the presence of one dose of a positive entity should give results more like those of two doses than like none at all (a law of diminishing returns). The observed phenomena of dominance were thus accounted for on the basis of a physiological theory of the relative effects of presence and absence and a theory of the prevailing nature of mutation.

I need not dwell long on the difficulties with this theory that have rendered it unacceptable. Among these have been the discovery of numerous series of multiple allelomorphs, not always to be interpreted as quantitative series, the observation of dominant effects from known losses of genes (*e.g.*, Haplo IV. and Notch in Drosophila) and particularly the demonstration of reverse mutation from recessive mutant genes. There is also the difficulty of accounting for evolution, if mutation is wholly a matter of loss, to which Bateson himself drew attention.

SELECTION OF MODIFIED HETEROZYGOTES

R. A. Fisher (1928a, b, 1928, 1930, 1931) has attempted to account for the prevailing dominance of wild type over rare and deleterious mutations as the result of a long continued process of selection of heterozygotes toward resemblance to wild type. With respect to the necessary physiological basis, he merely assumes that intermediacy of the heterozygote is the most natural condition.

The physiological basis of dominance on this view is the assemblage of a very special group of modifying factors under pressure of natural selection. He assumes, doubtless correctly, that the mutations observed in the laboratory must have recurred innumerable times in the history of the species. He notes that heterozygotes, while uncommon, must always have been enormously more common than homozygous mutants and concludes that the selective advantage of those heterozygotes which most closely resemble wild type would result in the end in carrying them to identity of appearance (and physiology) with wild type.

I have criticized this theory on the ground that the selection pressure would be too small to be the controlling factor in the fixation of the postulated modifiers of dominance (1929a, b). I will take this opportunity to correct a slight error in the formulae.

Let A, a be the type and mutant genes respectively.
Let M be a modifier which causes Aa to approach AA, but which has no effect on AA.

It is convenient to assume dominance of M as the most favorable case.

Let q_M be the frequency of M, $(1-q_M)$ that of m.
Let s be the selective disadvantage of aa (compared with AA).
Let hs be the selective disadvantage of Aamm (compared with AA).
Let $h's$ be the selective disadvantage of AaM– (compared with AA).
Let $h''s$ be the net selective disadvantage of Aa (compared with AA).

$$h'' = (1 - q_M)^2 h + [2q_M(1 - q_M) + q_M^2]h'.$$

Let u be rate of mutation from A to a per generation.

The opposed pressures of mutation and selection would keep the frequency of gene a at the point $\frac{u}{h''s}$. The genotypes would be distributed according to the expansion of the following expression.

$$\left[\frac{u}{h''s}a + \left(1 - \frac{u}{h''s}\right)A\right]^2\left[(1 - q_M)m + q_M M\right]^2.$$

The selective disadvantage of the mm individuals as a group (compared with AA) is $2\frac{u}{h''s}\left(1 - \frac{u}{h''s}\right)hs + \left(\frac{u}{h''s}\right)^2 s$, or $\frac{2uh}{h''}$ ignoring second order terms. Similarly that of M— individuals is $\frac{2uh'}{h''}$ to a sufficient approximation. The advantage of M— over mm may be written $s_M = \frac{2u(h - h')}{h''}$.

The most favorable case is that in which the modifier is not only dominant but also produces complete dominance of A over a at a single step. In this case $h' = 0$, $h'' = (1 - q_M)^2 h$ and $s_M = \frac{2u}{(1 - q_M)^2}$.

The rate of change per generation of frequency of the dominant gene M is of the form

$$\Delta q_M = s_M\, q_M (1 - q_M)^2. \qquad \text{(Wright, 1931)}$$

Substituting,

$$\Delta q_M = 2uq_M. \qquad (1)$$

If gene M were mutating to m at a rate only twice that at which A is mutating to a, its frequency would change at the rate $\Delta q_M = -2uq_M$, exactly neutralizing the above selection pressure.

But this selection pressure is the maximum possible under Fisher's hypothesis (in the case of rare mutation).

In general, single modifiers of dominance could not be expected to produce complete dominance by themselves nor would they always be completely dominant. Thus in general, selection favoring the heterozygotes of rare mutations would tend to change the gene frequencies of modifiers much more slowly than would recurrent mutation at ordinary rates.

There is no disagreement over the magnitude of the effect of selection here. Fisher's contention is that the action of a steady selection pressure, however minute, would be effective if continued long enough. Yet he states elsewhere of mutation pressure "for mutations to dominate the trend of evolution, it is thus necessary to postulate mutation rates immensely greater than those which are known to occur." I quite agree with this statement and it has accordingly seemed to me that still smaller selection pressures would be practically negligible in determining which phase of the modifier approached fixation. There should always be other evolutionary pressures of greater magnitude acting in one direction or the other.

In particular, if the modifier has any effect on its own account which is subject to selection (*i.e.*, if the combination AAM— has any advantage or disadvantage relative to AAmm the pressure due to such selection is certain to take precedence over that due to its effect on the rare heterozygote. Haldane (1930a) has made essentially the same criticism. His form of statement brought out the likelihood that a modifier of the heterozygote would be for a similar reason a modifier of the much more abundant wild type, while I emphasized more the point that even in those cases in which the homozygote is not capable of being modified in the same respect as the heterozygote, the modifier is likely to have other effects subject to direct selection.

If this hypothesis is untenable what alternative is there? I believe that little progress can be made without development of a physiological theory. It must be remembered that whatever the evolutionary mechanism by which a particular gene complex has been reached, the state of dominance of all of the genes in the complex must always have a completely physiological explanation.

PART II. PHYSIOLOGICAL EXPLANATIONS OF DOMINANCE

The Factors of Development

The development of such a theory is of course a rather speculative matter. It involves the whole question of the mechanism of control of the developmental process. However, the present states of knowledge of embryology and physiology impose certain limitations on speculation. We are sure, for example, that development is an epigenetic process. The genes can not stand in the simple one to one relation to morphological characters of a preformationist theory. Genetic data are of course in harmony. Each character is affected by many genes and each gene affects many characters.

Any particular physiological process occurring at a particular time and place in the history of the organism is the resultant of a complex of factors (Fig. 1). These may conveniently be grouped in four categories.

First, is the chain of past events in the same region. These give in general a continually narrowing field of potentialities associated with an increasing capacity for self differentiation, as development proceeds.

Second, is the array of correlative influences emanating from other regions, by means of which the region in question comes to fill a place in the integrated pattern of the organism as a whole. Among these are relations of physiological dominance in the sense of Child, embryonic inductions, mechanical stresses, hormone effects, stimulation through the nervous system and perhaps mitogenetic rays.

The internal environment, thus constituted, must be sharply distinguished from a third class of factors, the direct influences from the external world. Little importance can be attributed to these in controlling the course of normal embryonic development, apart from the establishment of a primary simple gradient pattern by trivial differential stimuli, the maintenance of differential conditions at the surface, and of course the supply of the necessary conditions for life.

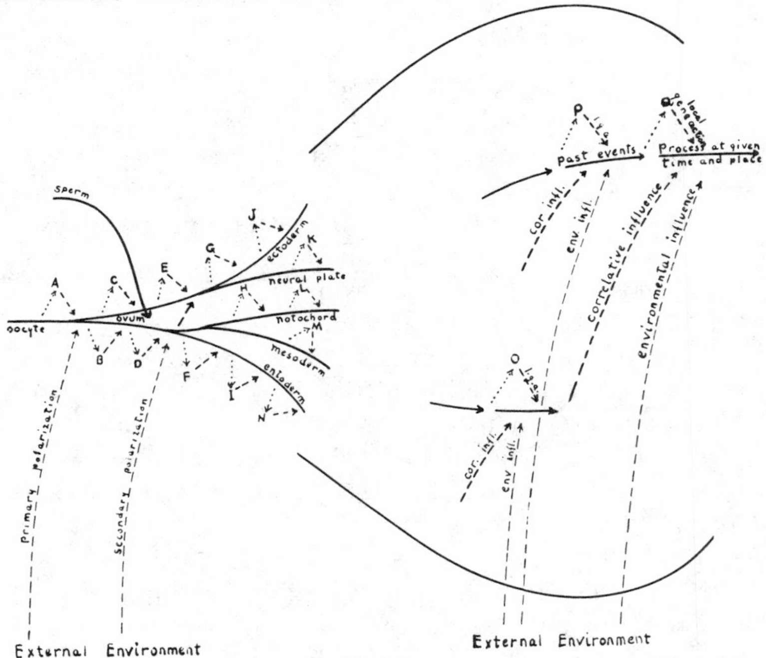

FIG. 1. Diagram illustrating the relation of factors to any developmental process of a vertebrate, occurring at a particular time and place. The letters A to Q represent genes which affect the process in question.

Finally an event in the place in question may be the result of current activity of genes in its own cells.

These are obviously not final terms, except in the case of external influences. On following the history of our region a step back, analysis can again be made in terms of these four classes of factors. Similarly each correlative influence traces to events in other parts of the body which are capable of the same sort of analysis. Ultimately all chains begin either in an external stimulus somewhere or a gene action somewhere. But gene action is not a final term. The gene is presumably called into play at a particular time and place by conditions external to it. These conditions may simply be ones that present it with its proper substrate for action, but it is possible that in some cases there is activation of a mutational nature.

It has been usual for geneticists to assume that all cells of an organism contain the same array of genes, as a necessary consequence of the equational character of mitosis, and that differentiation depends merely on differential responses of this uniform heredity to diverse local environments. But genes which mutate so frequently as to bring about genetic differentiation of the tissues of every individual in a strain were long ago discovered by Correns in variegated plants. Emerson and his coworkers have shown that mutation in variegated corn is not merely a result of chance. Demerec (1929) has shown that a similar type of mutation in *Drosophila virilis* is brought to some extent under control of local conditions (stable in gonads, highly mutable in wings) in the presence of specific genes. The resulting situation closely resembles that of the familiar spotting patterns of mammals ranging from the highly irregular piebald of guinea-pigs to the orderly one of hooded rats. At present, the most plausible explanation of these is that they depend on genes which mutate in relation to a more or less regular gradient pattern in the embryonic skins, although proof of the type given by Demerec is lacking because of the stability of the genes in the germ plasm (*e.g.*, through 30 generations of brother-sister mating in guinea-pigs). There is here a suggestion that diphasic genes, stable in the germ plasm, but regularly and irreversibly changing phase in parts of the soma in response to specific stimuli, may be of general importance in stabilizing otherwise labile embryonic patterns.

In any event, on continuing to trace back, the various paths should ultimately be found to be diverging from a relatively small number of gene activities in the egg and, even earlier, in the oocyte. Through such mechanisms, a primary merely quantitative pattern of metabolic activity, due to external relations (Child), may be elaborated along branching paths, possibly involving irreversible nuclear changes (perhaps related to the embryonic segregation of F. R. Lillie) to any degree of complexity as a largely self-contained system.

THE CHAIN OF REACTIONS BETWEEN GENE AND CHARACTER

If we isolate from this network a chain of processes connecting a single gene action with an observed character, we reach a representation of the type of Fig. 2.

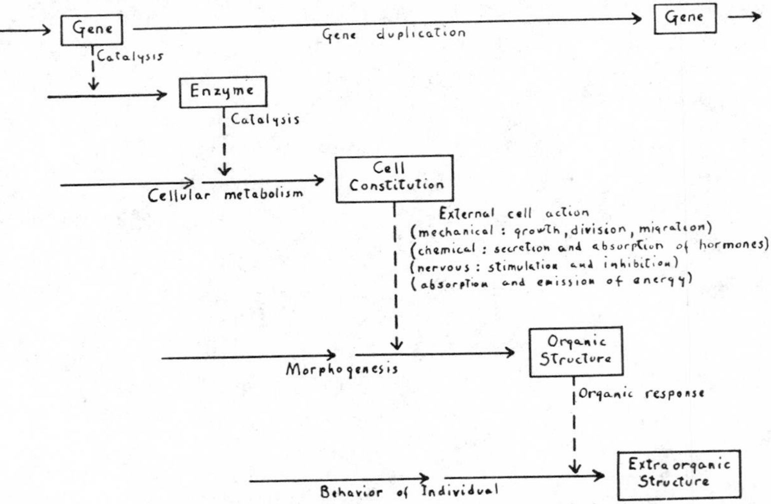

FIG. 2. Diagram illustrating the chain of processes relating the immediate physiological action of a gene to characters at different levels.

Since the genes are contained within the nuclei of cells, it is obvious that a genetic difference in any character must trace back to a difference in a cell character—rate of growth and division, orientation of mitotic spindle, movement in response to stimuli, production of specific substances, etc. In some cases the chain of developmental processes between gene and character may be fairly short. Danforth (1928, 1929), for example, has shown by transplantation experiments that the differences between breeds of poultry in color, in rate of growth of feathers (as affected by the well-known sex-linked genes, but not by another factor) and in the form of feathers of males of hen-feathered and cock-feathered breeds all depend on gene differences in the cells of the feather follicles themselves. The study of gynandromorphs and other mosaics in *Drosophila* reveals that with a few interesting exceptions the observed character differences depend on local gene action. The same is shown by somatic mutations to be true of most at least of the color variations of guinea-pigs (Wright and Eaton, 1926). On the other hand, numerous experiments show that the difference between male and female fowls in color and form of feathers do not depend on local genes but on genetic differences elsewhere, *i.e.*, the presence or absence of an ovarian hormone, probably tracing in normal birds to differential gene action in the cells of the embryonic gonad, although the chain may be still longer. Similarly, Smith and MacDowell (1930) have demonstrated that a type of dwarfism, shown by Snell (1929) to be a simple recessive, depends on a deficiency of the anterior pituitary. The gene difference between these mice and normals exists in all cells of the body, but differential action apparently occurs only in the pituitary. The chains of reactions may be even longer. The characteristic specific, generic and familial differences between the webs of spiders are as clearly hereditary as are the differences in arrangements of their eyes, but must trace to their genetic basis through somewhat longer chains of processes. It is important to recognize that there is no essential difference in the relation of character to gene in these cases. The form of the web is a resultant of the behavior of the organism as a whole, the arrangement of the eyes is a resultant of cellular behavior. The ultimate differential factor between species in both cases presumably resides in the nuclei of cells called into action somewhere in the organism.

As to the chains of intracellular processes, physiologists seem agreed that the specific syntheses to which differences in constitution and behavior of cells trace, depend on the presence of specific catalysts—enzymes. Cuénot, as early as 1903, suggested that the color genes of the mouse act in part through control of the production of specific enzymes. Since then the suggestion has been made repeatedly by both geneticists and physiologists that this is the characteristic mode of action of genes.

Enzyme differences have been demonstrated to be associated with gene differences in a considerable number of cases. One of the most interesting is Brink's demonstration of a reduction in amylase content in pollen grains containing the recessive gene wx of waxy corn as compared with those containing its normal allelomorph. Only slightly less direct, however, is the relation established between genetic constitution and the presence or absence of oxidases and of oxidase inhibitors in the skin of mammals (Onslow, Schultz, Kröning, Koller and others). In other cases it has been established by tracing the embryonic history, that a genetic character difference is first manifested as a difference in rate of reaction, strongly suggestive of enzyme control. Goldschmidt (1927) has presented a number of beautiful examples in Lepidoptera. Of especial importance is his evidence that differences which on first analysis depend on regulation of time of onset or cessation of a reaction, on further analysis can be shown to depend on rates. Ford and Huxley (1927) have given an analysis of variation in eye color of Gammarus in similar terms.

The genes themselves would seem to belong to the class of catalysts from mere consideration of the definitions of the two terms (Goldschmidt, 1916, Troland,

1917). Whatever reactions genes may bring about, it appears that they remain constant in the amount per cell, through duplication (autocatalysis) and separation in orderly relation to cell division. Brink's case is of interest here as one in which the same gene clearly produces physiological effects in the pollen grain without being destroyed thereby.

It is hardly likely, however, that the genes are themselves the enzymes immediately responsible for observed cell products: that Brink's extracted amylase or Onslow's tyrosinase, for example, actually consist of genes. The fact that the genes are in the nucleus while specific cell products (such as starch grains and pigment granules) are elaborated in the cytoplasm suggests that genes, as catalysts, act largely through control of the production of other catalysts, the extractible enzymes.

With the character in general at many additional removes from the primary gene action, it is obvious that no single theory can be expected to account for all cases of dominance.

As regards the mere question of frequency of dominance, however, it is obvious that a condition at any link in the long chain between gene and character, which causes identity of expression in the heterozygote and one of the homozygotes will thereby result in dominance at all later links. The chance of intermediacy persisting to the final character is thus only that for a typical link raised to a power as great as the number of links in the chain.

Mathematical Theory

As a starting point, let us consider the conditions for dominance at a single link in such a chain and since rate effects seem to be more fundamental than timing or duration effects, let us consider the relation between amounts of catalyst and the amount of product in a chain of practically irreversible transformations during a period in which the amounts of intermediary substances may be treated as in flux equilibrium. The transformations will be assumed to be such that they can be treated as monomolecular, and the rate of reaction will be assumed to depend jointly on the concentration of the substrate and of the catalyst.

Assume that the substance Y (Fig. 3) is supplied at a

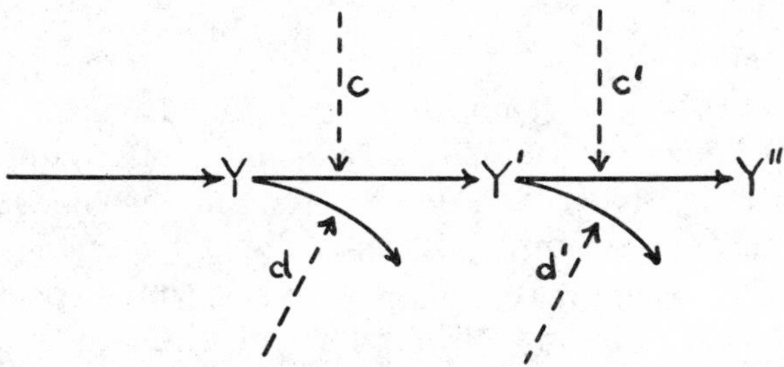

FIG. 3. A chain of monomolecular transformations.

constant rate C, that it is transformed into Y′ at a rate per unit mass c and that it is otherwise disposed of at a rate per unit mass d. The product Y′ is disposed of constructively (forming Y″) and destructively at the respective rates per unit mass c′ and d′. Y″ is treated as a final product. The rate of change of each substance (except Y″) is to be equated to zero, to represent the postulated state of flux equilibrium.

$$
\begin{aligned}
\frac{dY}{dt} &= C - (c+d)\,Y = 0 & \qquad Y &= C\left(\frac{1}{c+d}\right)\\
\frac{dY'}{dt} &= cY - (c'+d')\,Y' = 0 & \qquad Y' &= C\left(\frac{c}{c+d}\right)\left(\frac{1}{c'+d'}\right)\\
\frac{dY''}{dt} &= c'Y' & \qquad Y'' &= C\left(\frac{c}{c+d}\right)\left(\frac{c'}{c'+d'}\right)t
\end{aligned}
\tag{2}
$$

Thus the quantity of the product depends (apart from duration) on the product of a series of terms, one for each preceding link in the chain, and each consisting of the ratio of the constructive rate constant to the total rate constant for transformation of the substrate in question. In case the substance itself is being transformed or lost, the reciprocal of the total rate constant for these processes constitutes another term.

Consider now the relation of variation in a product (Y, dropping primes) to variation in a catalyst (X) responsible in part for one of the transformations. The rate constant x varies directly with the quantity of catalyst. Using c only for the rate constant additional to x,

$$Y = Y_{\infty}\left(\frac{x + c}{x + c + d}\right) \tag{3}$$

The curve expressing the relation of (Y) to (X) is a hyperbola, asymptotic at its upper limit ($Y\infty$). Doubling the quantity of x will less than double the amount of product. There should be some approach to dominance of the active phase of the gene and the greater the activity the closer the approach to complete dominance. It is convenient to measure dominance in this case by the ratio of the excess of the positive homozygote over the heterozygote to its excess over the other homozygote.

$$\frac{Y_{AA} - Y_{Aa}}{Y_{AA} - Y_{aa}} = \left(\frac{X_{AA} - X_{Aa}}{X_{AA} - X_{aa}}\right)\left(1 - \frac{X_{Aa} - X_{aa}}{X_{Aa} + c + d}\right) \tag{4}$$

From this it will be seen that there is an approach to a linear relation only if the increase in the rate constant due to one dose of the catalyst ($X_{Aa} - X_{aa}$) is small compared with the total rate constant ($X_{Aa} + c + d$) for disposal of the substrate, while there is an approach to complete dominance if the transformation largely depends on the positive phase of the gene (X_{aa}, c, d all small, compared with X_{Aa}).

If the gene (X) catalyzes a destructive process as far as the product in question is concerned

$$Y \propto \frac{c}{x + c + d} \tag{5}$$

The relation between character and catalyst is again represented by a hyperbola, but one descending toward a lower asymptote with increasing amount of the catalyst. The formula for degree of dominance comes out exactly the same as before. A gene with this mode of action will appear as an incompletely dominant inhibitor of the character in question.

If a second catalytic action intervenes between the primary catalyst and the product (Fig. 4), the relation is

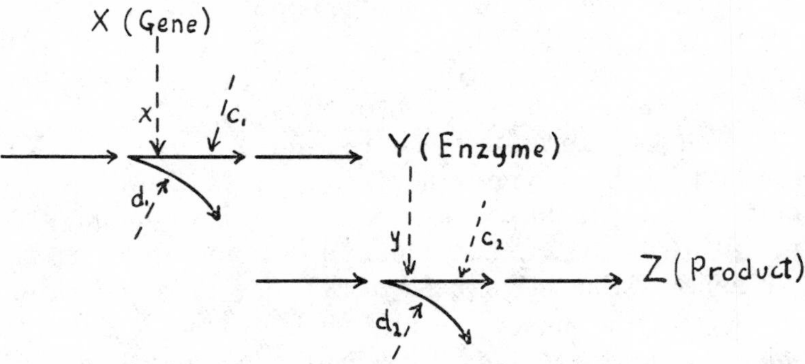

FIG. 4. Relation of gene action to amount of enzyme, as primary product, and of this to amount of secondary product.

still hyperbolic, irrespective of whether both catalytic actions are constructive, both destructive, or differ in this respect.

$$Z \propto \frac{y + c_2}{y + c_2 + d_2} \quad \text{or} \quad \frac{c_2}{y + c_2 + d_2}$$

depending on whether the action of Y is constructive or destructive in relation to Z.

$$Y \propto \frac{x + c_1}{x + c_1 + d_1} \quad \text{or} \quad \frac{c_1}{x + c_1 + d_1}$$

Measuring dominance as before

$$\frac{Z_{AA} - Z_{Aa}}{Z_{AA} - Z_{aa}} = \left(\frac{X_{AA} - X_{Aa}}{X_{AA} - X_{aa}}\right)\left(1 - \frac{Y_{Aa} - Y_{aa}}{Y_{Aa} + c_2 + d_2}\right)\left(1 - \frac{X_{Aa} - X_{aa}}{X_{Aa} + c_1 + d_1}\right) \qquad (6)$$

If X acts constructively on production of Y, Z exhibits greater dominance of the active phase of X than does Y. If, however, X acts destructively on Y, X exhibits less dominance than does Y, the most extreme case being that in which c_2 and d_2 approach zero. Even in this case the heterozygote can not fall below intermediacy in relation to the active phase of X.

A bimolecular reaction may intervene between gene and product. In Fig. 5, Z is a joint product of B and M.

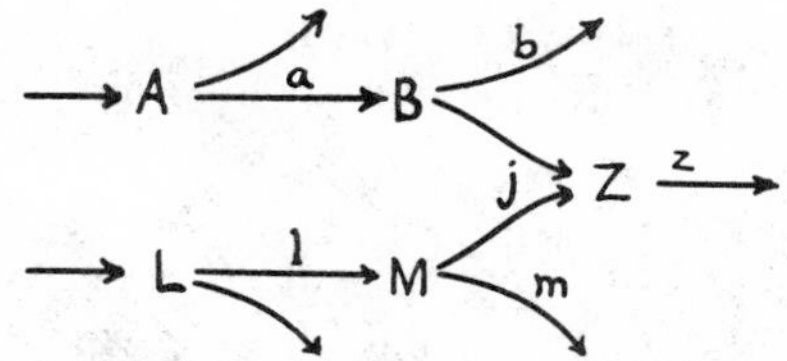

FIG. 5. Diagram of bimolecular reaction.

The rate constant (j) for this reaction being rate per unit mass of B and M jointly is of the dimensions of a reciprocal quantity.

Equations of flux equilibrium

$$\frac{dB}{dt} = aA - bB - jMB = 0$$

$$\frac{dM}{dt} = lL - mM - jMB = 0$$

$$\frac{dZ}{dt} = jMB - zZ = 0$$

Values for B, M and Z can be found directly in terms of A and L. Two extreme cases are of most interest. If B and M are largely disposed of in other ways than by formation of Z $\left(\frac{mb}{j}\right.$ large compared with aA and lL$\left.\right)$ the solutions are approximately

$$M = \frac{lL}{m}\left[1 - \frac{aAj}{mb}\right] \text{ or even } \frac{lL}{m} \qquad (7)$$

$$B = \frac{aA}{b}\left[1 - \frac{lLj}{mb}\right] \text{ or even } \frac{aA}{b} \qquad (8)$$

$$Z = \frac{aAlLj}{zmb} \qquad (9)$$

The dominance relations of B and M to genes acting on processes back of them are of course approximately as if there were no combination to form Z, while the dominance relations of Z to genes acting on either chain are approximately as if these were combined in one chain under the preceding theory.

If Z is, on the contrary, produced almost quantitatively from the more limited component $\left(\frac{mb}{j}\right.$ smaller than aA and lL) the approximate solution is as follows, taking $aA > lL$:

$$B = \frac{aA - lL}{b} \tag{10}$$

$$M = \frac{blL}{jaA} \tag{11}$$

$$Z = \frac{lL}{z} \tag{12}$$

The amounts of M and its products other than Z are negligible. For B and its products (other than Z) there is the interesting possibility of a reversal of dominance of active genes in the chain leading to B. The production of Z, limited as it is by that of M, subtracts from that of B, giving a threshold for the latter. If B falls below M in the heterozygote of such a gene, but exceeds M in the homozygote, the inactive phase would appear to be completely dominant over the active phase. The dominance of the gene for albinism (c^a) in certain combinations in the guinea-pig (c^dc^dff pale cream, c^dc^aff, c^ac^aff white) has been interpreted in this way (Wright, 1927).

Another interesting consequence of this mechanism is that inactivation of the more abundant component may release the products of the less abundant component. B behaves as an inhibitor for M and *vice versa*. There is here a possible mechanism for such antagonistic reactions as Goldschmidt finds indicated in his intersexes in Lymantria.

As for Z and later derivations, its quantity depends on that of the more limited component (M) as if it were the product of a monomolecular reaction and the dominance relation to genes in this series are thus the same as in the latter case. But while minor variations back of the more abundant component (B) are without effect, such a reduction at any link that aA becomes less than lL reverses the situation. Thus complete or nearly complete inactivation of a gene necessary to the progress of the reaction toward B will behave as a practically complete recessive to the active phase.

Several bimolecular reactions may intervene between gene and product. The active phase will completely dominate over the inactive phase if at any one of these reactions the gene in question is back of the more concentrated component. The increased tendency toward dominance of the active phase of genes through this mechanism much more than overweighs the possibility of reversal of dominance by the special condition discussed in a preceding paragraph.

So far as it has been assumed that rate of transformation varies jointly with the concentration of substrate and enzyme. This has been found to apply at low concentrations of both, but in general requires qualification (Waldschmidt-Leitz, 1929; Haldane, 1930b). It appears that the rate of transformation varies in such a way as to indicate dependence on the concentration of a compound of the substrate with the enzyme. The reaction must thus in general be treated as involving a bimolecular one between substrate and enzyme and a diagram such as the following (Fig. 6) should be substituted for Fig. 4.

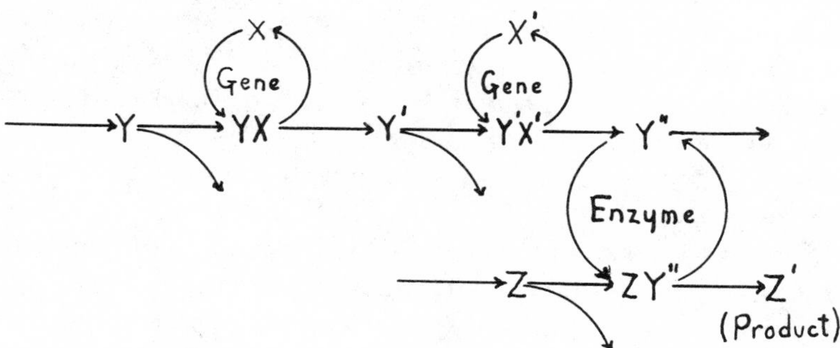

FIG. 6. Relations of gene to enzyme as primary product and of this to secondary product, treating gene and enzyme as entering into temporary combination with their substrates.

For our present purpose it is only necessary to consider the extreme cases. That in which the compound is low in amount compared with both components gives the theory already discussed. If the total catalyst (X + YX) is in excess, but most of the substrate is always in combination, further increase in the catalyst can have no effect and all genes necessary for the production of the catalyst should exhibit complete dominance above a certain level. If, on the other hand, the substrate is in excess while the gene itself as a catalyst, or a catalyst directly controlled by it, is kept continually in combination, the gene would exhibit intermediacy in its phases. Such a case as that of the gene affecting yellow endosperm in corn, demonstrated by Mangelsdorf and Fraps (1931) to exhibit exact proportionality of effect (measured as vitamin A content) to number of positive genes (yyy, Yyy, YYy and YYY) might come here. Conversely, a gene necessary for any step in the formation of the substrate would tend to exhibit complete dominance.

MULTIPLE ALLELOMORPHS

The interpretation of multiple allelomorphic series as often representing different grades of activation of a gene is an obvious extension of this theory of dominance. Fig. 7 illustrates the numerical relations under the two extreme forms of the theory.

The dominance phenomena are somewhat different. Under the theory of proportionality of rate to substrate and catalyst jointly, the higher allelomorphs should show increasing degrees of dominance over the completely inactive phase (or indeed over any particular lower allelomorph) but absolutely complete dominance is theoretically never reached. Under the theory of control of rate by the limiting factor, there is exact intermediacy of heterozygotes (up to a certain level), then a range of increasing dominance of the higher allelomorph until at a certain level complete dominance is attained. The general observation that the highest member of a series shows most complete dominance is in harmony with both theories, although the apparent completeness of dominance often found accords best with the limitation theory.

Considering the series of heterozygotes between a given higher member and successively lower members, the two theories give opposite expectations. With a hyperbolic relation, there is maximum dominance (by formula 4) over the lowest allelomorph. Under the limitation theory, there is *least* dominance in this case (if dominance is present at all, but is not complete). Cases are known which definitely accord with the limitation hypothesis in this respect. In guinea-pigs the gene C seems to be completely dominant over all its allelomorphs ex-

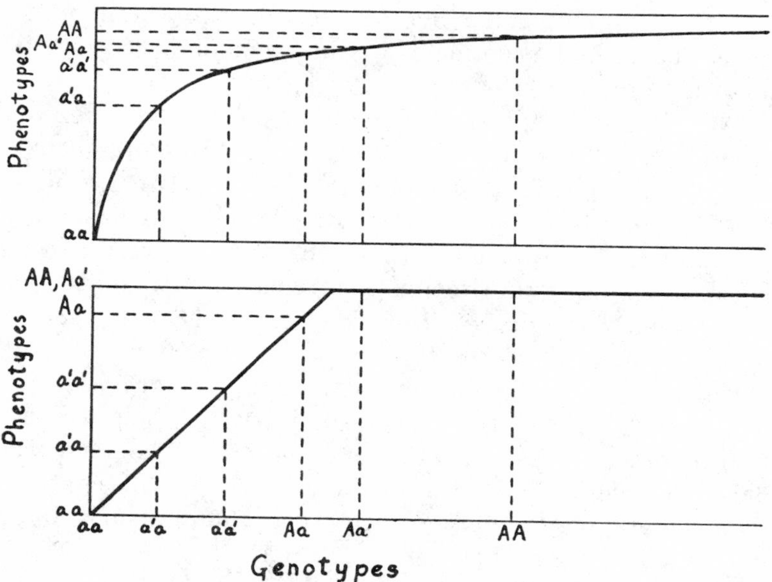

FIG. 7. Relation of gene activity to phenotypic effect under two extreme hypotheses: above, on theory that reaction varies jointly with gene activity and amount of substrate; below, on theory that the reaction varies only with that component (gene or substrate) which is most limited in amount.

cept the lowest, c^a, but Cc^a (in the presence of gene pp) shows intermediacy. The situation is the same in the mouse (Snell, 1931). Cases of partially dominant effects of complete loss (deficiency) may also be explained in this way.

It must be remembered in this connection that the theory of joint control and of limiting factors are not really separate theories but merely extreme cases of a single theory. Some examples may require one, some the other and yet others an intermediate case of the general theory.

All-or-None Reactions

The condition of flux equilibrium has been assumed as characteristic of vital phenomena over considerable periods of time. But developmental processes must have a beginning and an end. As already noted, it has been demonstrated in several cases that genes may affect a character by advancing one or the other of these processes and such mechanisms would seem of necessity to be of first importance in morphogenesis.

This does not necessarily wholly change the relation of dominance to gene activity, since as Goldschmidt especially has emphasized such effects probably trace to differences in rates. Graded effects of delay in onset or termination should thus be functions of underlying rates. However, mere quantitative differences in the duration of an elementary process are likely to have consequences of an all-or-none character, making dominance (if not already shown by the elementary process) a largely accidental matter in relation to gene activity. If the termination of one process occurs after another has begun with one homozygote, but occurs before, with the alternative homozygote, it is largely chance (from the present standpoint) whether the heterozygote occurs before or after.

Modifying Factors

The relative frequency of complete dominance in cases of genes with major and minor effects is a question of great importance. Most of the direct evidence of dominance naturally comes from genes with major effects. Genes with minor effects are largely a deduction from the statistical phenomena of quantitative variation following crosses.

In so far as genes are related to characters through catalyzed reactions whose rates vary jointly with both substrate and catalyst (hyperbolic theory) the smaller the difference in the catalyst, the less the dominance (formulae 4 and 6). Under this theory, minor factors should rarely exhibit dominance.

In the case in which reaction rate depends only on the limiting factor, the situation is more mixed. Minor variations back of the limiting factor exhibit intermediacy. On the other hand minor variations back of a component only slightly in excess are more likely to show complete recessiveness than larger variations. Where there is a larger factor of safety small variations, however, will have no effect.

In the more general case, between the above extremes, minor variations tend to have intermediate heterozygotes.

If an all-or-none reaction intervenes, the heterozygotes of major variations are likely to fall invariably on one or the other side of the threshold, giving complete dominance or recessiveness. The effects of minor variations tend to be confused by those of other modifying factors, including environment, and the effects of the alternative homozygotes are manifested merely in differences in percentage of occurrence, with heterozygotes showing an intermediate percentage. On the whole it appears that an approach to complete dominance or recessiveness should be expected more of factors with major than with minor effects.

The evidence from crosses between strain shown to differ in multiple minor factors is, I believe, in harmony with the view that lack of dominance is the rule for such factors. It may be noted that only slight departures from intermediacy of individual heterozygotes need be postulated to account for lack of intermediacy of F_1 or even for an excess development of F_1 over both parents (if these are not at opposite extremes and if there is any correlation between direction of deviation from exact intermediacy and direction of effect).

Direct Evidence in Nature of Dominance

The argument for the prevailing dominance of the active phase of factors with major effects has been based here largely on theoretical grounds. I will not go far into detail on the observations which indicate the actual occurrence of such a relation. The superficial correlation between dominance and positive effect is familiar and formed one of the bulwarks of the old presence and absence hypothesis. Thus presence of pigment is usually dominant over absence. But the occurrence of definite exceptions such as dominant whites was an embarrassing fact, It is, therefore, of especial significance that where physiological analysis of characters is pushed back a step, such exceptions tend to disappear. From the work on extracts from the skins of mammals of different genetic constitution it has come out that in those cases in which the dominant phase is more pigmented, the extract from the dominant contains an enzyme which is merely absent (or reduced in effectiveness) in the recessive phase. There is no antienzyme present in the latter. On the other hand, in those exceptional cases in which the dominant is less pigmented, the extract has been shown to possess the property of inhibiting the action of extracts from colored skin.

The blood groups of man (Bernstein, 1925) depend primarily on 3 allelomorphic genes of the interesting type which can not be arranged in a quantitative series with respect to effect. They also appear to have no adaptive significance. It is significant therefore to find that agglutinogen A is dominant over its absence (Group O), agglutinogen B is dominant over its absence (also Group O). The heterozygote between the two positive agglutinogens shows both. In the heterozygotes each gene dominates in its positive effect. This seems to be the rule in non-linear multiple allelomorphic series. A woman who carries the recessive gene for the "red blind" type of color blindness in one X chromosome and that for the "green blind" type in the other, has normal vision, but all her sons are of one or the other type (Waaler, 1927). A similar situation holds for the scute locus in *Drosophila* (Agol, 1931). The phenomena of pseudodominance of recessive genes in organisms heterozygous for deficiencies in the locus in question indicates that the dominant gene is a positive entity (Mohr, 1923). In *Drosophila,* allelomorphs of white eye, heterozygous for a deficiency in the same locus, resemble the combinations of these allelomorphs with white eye itself. "Exaggeration" of the recessive effect in heterozygous deficiencies calls to mind the tendency toward incomplete dominance of type over extreme inactivation discussed on a previous page.

Perhaps the most direct evidence comes from the experiments of Stern (1929) and of Muller, League and Offermann (1931). Stern showed that the piling up of Y chromosomes containing "bobbed bristle" causes, not more extreme development of the character bobbed bristles, but normal type. Muller and his coworkers, following x-ray treatment, caused an accumulation of small fragments of chromosomes carrying particular recessive genes. In general there was approach to wild type indicating that the genes (hypomorphs) behaved like type genes of low activity. In a few cases there was increasing divergence from type with increase in number of genes. But these "neomorphs" were largely semi-dominants. Their occurrence does not invalidate the hypothesis that the ordinary recessive mutations represent partial or complete inactivation.

PART III. REINTERPRETATION OF STATISTICAL PRINCIPLES

Statistical Results of Gene Physiology

Let us return now to the statistical phenomena of dominance. Such considerations as the above (as far as available) led me several years ago (1929a) to suggest that "in the hypothesis that mutations are most frequently in the direction of inactivation and that for physiological reasons, inactivation should generally behave as recessive, at least among factors with major effects, may be found the explanation of the prevalence of recessiveness among observed mutations.''

Ford (1930) has objected as follows: ''The only explanation of dominance previous to the publication of Dr. Fisher's theory was the long-discussed physical one of partial or complete loss (or inactivation) of genic material at mutation, and Professor Sewall Wright falls back upon this in his criticism of it. Not only does the loss hypothesis lead to an evolutionary impasse when accepted as a general principle, but it can not be too strongly urged that as such it is untenable in the face of the proved fact of reverse mutation.''

Ford here seems to imply that the loss and inactivation theories are equivalent. There are many ways of inactivating a complex mechanism such as a motor car (and probably also a gene). The momentary effect on progress is the same as that of complete loss, but the ultimate outlook is not so hopeless. More to the point perhaps is the fact that enzymes can be inactivated by addition of certain substances and completely reactivated by later treatment. I see no difficulty in holding that variegated corn possesses an inactivated gene which reverts to activity by mutation thousands of times in the same corn plant.

With respect to the supposed evolutionary impasse, the objection applies only if it be supposed that all mutation is of the nature of inactivation. This I explicitly avoided. There seems little reason to believe that the rare major mutations, whose prevailing recessiveness was the object of discussion, are representative of the mutations which are of evolutionary significance. The latter are probably as a rule ones with minor effects and thus often semidominant over the previous type gene, but still dominant over complete inactivation. In any case genes with increased or new activities are probably as likely to be seized upon as those with reduced activity, even though the latter be much more numerous.

Effects of Selection

However, I have no objection to an evolutionary process by which the dominance of wild type over mutations may be increased provided the pressure toward fixation is sufficient to be effective. Plunkett[2] (1932a, b) has recently pointed out that the direct selection of a type, subject to variation from numerous minor genetic and environmental factors, would tend to build up a factor of safety against such variation and automatically against mutations, irrespective of their past occurrence.

He calls attention to the well-known fact that wild type is much less susceptible to variation from such factors (*e.g.*, temperature) than are mutations and explains this as due to "differences in the distance of the developmental process from its asymptote at the time it ends: processes controlled by wild type genes usually closely approach their asymptotes, while those modified by mutant genes may be terminated by the effects of other developmental processes while still very incomplete." This theory seems quite in harmony with that which I presented. From the standpoint of the present paper, we are probably concerned here with those genes which have to do with the timing of processes whose dominance was found to be largely a matter of chance from the physiological standpoint and by this very fact easily subject to natural selection. Any one who has worked with genes whose effects are subject to threshold effects (*e.g.*, white spotting) is familiar with the ease with which the dominance of these can be changed by introduction of modifiers.

Evolutionary Implications

From the standpoint of the theory of dominance it may seem of little importance which mechanism is accepted if it be granted that selection has been an important factor. This is not at all the case, however, with the impli-

[2] Essentially the same view-point was also presented by Muller at a meeting of the 6th International Genetics Congress in 1932, but unfortunately did not reach publication until after the above was written.

cation of Fisher's and Plunkett's selection theories, for the theory of evolution. Fisher used the observed frequency of dominance as evidence for his conception of evolution as a process under complete control of selection pressure, however small the magnitude of the latter.

My interest in his theory of dominance was based in part on the fact that I had reached a very different conception of evolution (1931) and one to which his theory of dominance seemed fatal if correct. As I saw it, selection could exercise only a loose control over the momentary evolutionary trend of populations. A large part of the differentiation of local races and even of species was held to be due to the cumulative effects of accidents of sampling in populations of limited size. Adaptive advance was attributed more to intergroup than intragroup selection.

Plunkett's theory of dominance as an incident to the building up of a factor of safety of wild type against variation in general seems wholly compatible with such a view of evolution. It may be noted that intergroup as well as intragroup selection would be effective here.

GENERAL CONCLUSIONS

As a rule the most active phase of a gene with major effects is dominant over less active phases, as a result of physiological relations, allied to the law of diminishing returns, applying to its own activity or to that of primary or secondary products.

Incomplete dominance implies that variations of the gene and its successive products leading to the observed character *all* affect the limiting components in the reactions into which they enter or at least components which are not in excess. Genes with minor effects (so called modifiers) are especially likely to fall in this class.

If there is still incomplete dominance at a stage in the reaction chain at which an all-or-none reaction intervenes, dominance of a gene comes to depend on the effects of independent modifiers and in this case is without relation to its primary activity. On the other hand, dominance in this case is likely to depend on evolutionary history, specifically to the building up of a factor of safety for the most adaptive type of the species.

The measurement of dominance as an element in the study of the effects of gene substitutions in every possible genetic and environmental background furnishes one of the most useful tools in tracing the genetic physiology of characters.

LITERATURE CITED

I. J. Agol.
1931. *Genetics,* 16: 254–266.

F. Bernstein
1925. *Zeit. f. ind. Abst. u. Vererb.,* 37: 273–269.

R. A. Brink
1929. *Genetics,* 14: 569–590.

C. M. Child
1921. "The Origin and Development of the Nervous System." The University of Chicago Press.

L. Cuénot
1903. *Arch. Zool. Exp. et Gén.* (4) 1, Notes et revue, pp. 33–41.

C. H. Danforth
1928. *Proc. Soc. Exp. Biol. and Med.,* 26: 86–87.
1929. *Genetics,* 14: 256–269.

C. H. Danforth and Frances Foster
1929. *Jour. Exp. Zool.,* 52: 443–470.

M. Demerec
1929. *Proc. Nat. Acad. Sci.,* 15: 834–838.

R. A. Fisher
1928a. AMER. NAT., 62: 115–126.
1928b. AMER. NAT., 62: 571–574.
1929. AMER. NAT., 63: 553–556.
1930. "The Genetical Theory of Natural Selection." Oxford, Clarendon Press, 272 pp.
1931. *Biol. Reviews,* 6: 345–368.

E. B. Ford
1930. AMER. NAT., 64: 560–566.

E. B. Ford and J. S. Huxley
1927. *Brit. Jour. Exp. Biol.,* 5: 112–134.

R. Goldschmidt
1916. *Science,* N. S. 43: 98–100.
1927. "Physiologische Theorie der Vererbung." Berlin. Julius Springer. 247 pp.

J. B. S. Haldane
1930a. AMER NAT., 64: 87–90.

1930b. "Enzymes. Monographs on Biochemistry." Longmans, Green and Co. 235 pp.

P. Koller

1930. *Jour. Genet.*, 22: 103–109.

F. Kröning

1930. *Arch. f. Entw. d. Org.*, 121: 470–484.

F. R. Lillie

1929. *Arch f. Entw. d. Org.*, 118: 499–533.

P. C. Mangelsdorf and G. S. Fraps

1931. *Science*, 73: 241–242.

O. L. Mohr

1923. *Zeit. f. ind. Abst. u. Vererb.*, 32: 208–232.

H. J. Muller, B. B. League and C. A. Offermann

1931. *Anat. Rec.*, 51: 110.

H. Onslow

1915. *Proc. Roy. Soc. London B.*, 89: 36–58.

C. R. Plunkett

1932a. *Proc.*, 6th Internat. Congress of Genetics, 2: 160–162.

1932b. AMER. NAT., 67: 27–28.

W. Schultz

1925. *Arch. f. Entw. d. Org.*, 105: 677–710.

P. E. Smith and E. C. MacDowell

1930. *Anat. Rec.*, 46: 249–257.

G. D. Snell

1929. *Proc. Nat. Acad. Sci.*, 15: 733–734.

1931. *Genetics*, 16: 42–74.

C. Stern

1929. *Biol. Zentralbl.*, 49: 261–290.

L. T. Troland

1917. AMER. NAT., 321–350.

G. H. M. Waaler

1927. *Zeit. f. ind. Abst. u. Vererb.*, 45: 279–333.

E. Waldschmidt-Leitz

1929. "Enzyme Actions and Properties." Translated and extended by Robert P. Walton. John Wiley and Sons. 255 pp.

Sewall Wright

1927. *Genetics*, 12: 530–569.

1929a. AMER. NAT., 63: 274–279.

1929b. AMER. NAT., 63: 556–561.

1931. *Genetics*, 16: 97–159.

Sewall Wright and O. N. Eaton

1926. *Genetics*, 11: 333–351.

PROFESSOR FISHER ON THE THEORY OF DOMINANCE

Dr. Fisher's recent note (Fisher, 1934) on the theory of dominance seems to require some comment on my part as he makes several complaints of my treatment of his work.

To begin with, he complains that "Professor Sewall Wright, who had perhaps overlooked or misunderstood the calculations in my paper, put forward some calculations of his own expressed in a different notation which for the general case gave a result identical with mine (Fisher, 1928)." If Dr. Fisher will reread the paper (Wright, 1929a) to which he refers he will find explicit recognition of agreement up to a certain point.

He also makes the following complaint: "Wright has asserted that he is discussing a case especially favorable to my views, while in fact restricting himself to the minimal postulates in its favor." I can only say that it still seems to me that a factor which in a single dose gives complete dominance of wild type would be subject to more rapid net selection than any other sort of modifier of dominance contemplated by Fisher's theory. Perhaps I should have excepted the production of superdominance, but I was not aware that this was postulated in the theory. My treatment dealt explicitly with the case of rare factors held in equilibrium by the opposing pressures of mutation and selection, the case at issue and that to which complaint No. 1 referred. No geneticist would question that dominance can often be shifted by direct selection of heterozygotes. I accepted this in my first paper (Wright, 1929a) on Fisher's theory and had in fact previously reported on such modifications of dominance in two cases in the guinea pig (albinism (Wright, 1927), white spotting (Wright, 1928)).

Again Fisher complains—and here I think with more justice—that in my recent paper (Wright, 1934) I present corrections of certain of my formulae, tracing to a correction made by him in 1929, without referring to this history. I did, however, acknowledge Dr. Fisher's correction at the time (Wright, 1929b). On rereading my recent paper, I see that I should have repeated the acknowledgment there. The correction, I may add, was too small to affect the argument against the dominance theory.

I trust that Dr. Fisher will acknowledge that he also has not always been careful in historical matters. In 1922 (Fisher, 1922) he attempted to determine the distribution of gene frequencies in certain special cases. One of his results (decay of variability under inbreeding) was inconsistent with results which I already had. On attacking the problem by a different method (involving integral instead of differential equations) I reached a general formula for the distribution of gene frequencies in populations of limited size under the simultaneous pressures of selection, mutation and migration as well as ones for simpler cases. It appeared that in the simplest case Fisher's form of distribution was correct but only through the compensatory effects of two errors. It was in this case that his rate of decay was wrong. All of his other cases were wrong. I sent my manuscript to him in 1929 in order to iron out our differences as rapidly as possible. On returning it, Dr. Fisher stated that he had found the error in his work and that he confirmed my conclusions in the first two cases. He authorized me to present this correction in my paper, which, however, owing to the saturated condition of *Genetics,* did not reach publication until 1931 (Wright, 1931). Meanwhile in his book on the "Genetical Theory of Natural Selection" and in a technical paper, both published in 1930, Dr. Fisher acknowledged my correction with respect to rate of decay of variance under inbreeding but went on to state that this had led him to a more exact examination of the whole problem. He thereupon derived corrected results for the distribution of gene frequencies in 3 special cases. There is nothing to indicate that I had obtained any results on the distribution of gene frequencies and accordingly no indication that in 2 of his 3 cases his results were merely confirmations of mine. In the third case he undoubtedly made an improvement in my formula, as given in the manuscript. The latter, while superior to his original utterly erroneous one, admittedly (Wright, 1929b) applied only to weak selection rates. I was later (Wright, 1931) able to obtain by my method the same degree of approximation not only for this special case but also in the more general cases which Fisher has never treated. The reader may compare Fisher's formulae in his publications of 1930 (Fisher, 1930a, 1930b) not only with those in my detailed presentation sent to him in 1929 but delayed in publication to 1931, but also with a general formula published in 1929 in my second paper on his theory of dominance (Wright, 1929b). Perhaps he has overlooked this, since I find it nowhere listed in his references.

Returning to the theory of dominance, Fisher's recent paper (Fisher, 1934) goes on to expose various alleged fallacies in my reasoning. I, in turn, am unable to follow his reasoning but am willing to leave the matter at this point to the reader who wishes to compare his arguments with those which I have expressed in preceding papers or with those of J. B. S. Haldane (1930) who independently made similar criticisms of the theory. Fisher concludes with a reference to the physiological theory of dominance which I developed. "If, however, Professor Wright's views can only be made plausible by the exclusion of all alternatives he must find other objections to the selection theory more weighty than those he has revived." The reader of my paper (Wright, 1934) will find that while I contend that dominance must always have a complete immediate explanation in terms of gene physiology, I also accept selection theories as important with respect to the statistical phenomena of dominance. On pages 25 and 26, I sketched some rather obvious selective mechanisms of this sort of which I had recorded my acceptance at least as early as 1921 (Wright, 1921). On pages 50 and 51, the reader will note my acceptance of the form of selection theory advanced independently by Muller (1932) and by Plunkett (1932) as one under which dominance of wild type may increase irrespective of the past occurrence of mutants required by Fisher's theory. It was merely Fisher's particular selection theory that I excluded.

Finally in view of the tone of Dr. Fisher's note let me say here that I have the greatest admiration for his contributions to statistical genetics and biometry. If I have devoted considerable space to their criticism it is only because his views have seemed to me to be worthy of exhaustive study.

SEWALL WRIGHT

THE UNIVERSITY OF CHICAGO

LITERATURE CITED

Fisher, R. A.
1922. *Proc. Roy. Soc. Edinburgh,* 42: 321–341.
1928. AMER. NAT., 62: 115–126.
1930a. "The Genetical Theory of Natural Selection." Clarendon Press.
1930b. *Proc. Roy. Soc. Edinburgh,* 50: 205–220.
1934. AMER NAT., 68: 370–374.

Haldane, J. B. S.
1930. AMER, NAT., 64: 87–90.
Muller, H. J.
1932. *6th Internat. Genetics Cong.*, 1: 213–255.
Plunkett, C. R.
1932. *6th Internat. Genetics Cong.*, 2: 158–160.
Wright, S.
1921. *Bull.* No. 1121, U. S. Dept. Agr.
1927. *Genetics,* 12: 530–569.
1928. *Genetics,* 13: 508–531.
1929a. AMER. NAT., 63: 274–279.
1929b. AMER. NAT., 63: 556–561.
1931. *Genetics,* 16: 97–159.
1934. AMER. NAT., 68: 24–53.

15
The Analysis of Variance and the Correlations between Relatives with Respect to Deviations from an Optimum

Journal of Genetics 30, no. 2 (1935):243–56

16
Evolution in Populations in Approximate Equilibrium

Journal of Genetics 30, no. 2 (1935):257–66

Introduction

Originally written as one, these two papers represent Wright's first attempt to increase the sophistication of his statistical distribution of genes presented in detail in "Evolution in Mendelian Populations" (1931). The greatest problem with Wright's statistical distribution of genes was that it was confined to a very simplified set of assumptions (such as one locus with two alleles, one level of intermediate dominance, one kind of selection in the selection term). This distribution did not incorporate the effects of interaction between loci, which was central to Wright's qualitative shifting balance theory of evolution. One of Wright's lifelong tasks was to bring his quantitative theory of evolution closer to his qualitative theory.

Wright's approach in these two papers was to examine measurable characters for which the selective optimum was at or near the mean, a situation that Wright and others (Fisher 1930, 104–11; Haldane 1932, 196–98) considered to occur frequently in nature. His next step was to assess the drop in selective values according to the square of the deviations above or below the mean. The structure of the approach guaranteed that much genic interaction would occur with respect to selection of the character. Thus this approach enabled Wright to indirectly incorporate genic interaction and other effects such as dominance or environmental circumstances into the calculation of correlations between relatives or into his statistical distribution of genes. In the first paper, Wright applied this approach to the correlations between relatives, and in the second to the problem of evolution in populations.

In the second paper, Wright showed that it was possible, with certain simplifying assumptions, to connect intermediate optima for a character with underlying sets of gene frequencies determining the character. This held out a tantalizing prospect. One of the weakest links in his statistical distribution of genes was the selection term, into which it was almost impossible to incorporate gene interaction or dominance. But if he could replace the selection term with the average adaptive value term, then at one time he could connect his fitness surface (whose axes were sets of gene frequencies, not the gene combinations version of 1932) with his statistical distribution of genes, thus tying together the qualitative and quantitative aspects of his evolutionary theory.

I discuss these papers in *SW&EB*, 304–7. Both are difficult reading and, while little read at the time of publication, are well worth the effort now.

THE ANALYSIS OF VARIANCE AND THE CORRELATIONS BETWEEN RELATIVES WITH RESPECT TO DEVIATIONS FROM AN OPTIMUM.

By SEWALL WRIGHT.

(*Department of Zoology, The University of Chicago.*)

CONTENTS.

INTRODUCTION.

NATURAL selection is often thought of as favouring one extreme of each measurable character. Doubtless there is such a process under exceptional conditions, but it is certainly more often the case that the best adapted individuals are those nearest the average in every respect. In such cases, any gene with a constant effect on a measurable character increases the total fitness of the organism in combinations below the average but decreases fitness above the average. The analysis of variability in fitness into components due to additive gene effects, dominance, and epistasis, and the determination of the correlations to be expected between relatives, are questions of considerable importance in statistical genetics.

There are also probably many characters more easily measurable than total fitness, whose grade of development depends more on the harmony among a number of elementary processes than on the absolute magnitudes of the latter. Here also the direction and degree of effect of each gene substitution depends on the array of other genes with which it is associated and there are complications in the analysis of variability and in the correlations between relatives similar to those in the case of fitness.

In the present paper it is assumed that the grade of a "primary" character is determined by a number of independent pairs of genes whose effects combine additively (no epistasis). The cases of no dominance and of complete dominance will be treated in succession. A "secondary" character is assumed to depend on the deviation of the primary from a certain optimum grade. It is convenient to use the squared deviations in order to bring deviations above and below the optimum to a common sign. It is also more natural in that it avoids an abrupt change at the optimum. Such a scale has been used by Fisher in dealing with the effects of selection on "simple metrical characters" (1930, p. 104) and by Haldane (1932, p. 196). It will be assumed at first that there is no environmental variability.

SYMBOLISM.

Frequencies.

$q_1 \dots q_n$	Frequencies of genes $A_1 \dots A_n$ (*plus* on primary scale, case of no dominance) in contrast with $(1-q_1) \dots (1-q_n)$ the frequencies of their minus allelomorphs.
$p_1 \dots p_n$	Frequencies of genes $a_1 \dots a_n$ (*recessive* on primary scale, case of complete dominance) in contrast with $(1-p_1) \dots (1-p_n)$ the frequencies of the dominant allelomorphs.
f	Frequency of genotype.

Primary scale.

$\alpha_1 \dots \alpha_n$	Effects of gene replacements, a_1 by A_1, etc.
S	Grade of primary character.
$\mu_T(S)$	Tth moment of S $(=\Sigma(S-\bar{S})^T f)$.
M	Mean on primary scale $(=\bar{S})$.
O	Optimum on primary scale.

Secondary scale.

H	Grade of secondary character (zero at optimum).
G	Least square approximation to H in particular population on assigning additive effects (γ's) to each gene replacement.
$\gamma_1 \dots \gamma_n$	Net effects of gene replacement, a_1 by A_1, etc., in particular population. $\gamma_1 = G_{A_1A_1} - G_{A_1a_1} = G_{A_1a_1} - G_{a_1a_1}$.

$\delta_{A_1A_1}$ etc.	Dominance deviations of gene pairs.
D	Deviation from G due to dominance deviations ($=\Sigma\delta$).
$\epsilon_{A_1A_1A_2A_2}$ etc.	Epistatic deviations of sets of two gene pairs.
E	Total epistatic deviation of genotype ($=\Sigma\epsilon$).

$$H=G+D+E.$$

σ_H^2	The variance of H to be analysed into three portions, σ_G^2, σ_D^2 and σ_E^2, due respectively to additive gene effects, dominance deviations and epistatic deviations. The contributions of single gene pairs (or two in case of epistasis) are σ_γ^2, σ_δ^2, and σ_ϵ^2.

Analysis of variance due to one gene pair.

It will be convenient to consider first the general one-factor case. The method followed here is essentially that given by Fisher (1918), and the results are in agreement.

σ_δ^2 $(=\Sigma\,(H-G)^2 f)$ is to be minimised, noting that $G_{AA}=2G_{Aa}-G_{aa}$

$$\frac{\partial\sigma_\delta^2}{\partial G_{Aa}}=0,\quad \frac{\partial\sigma_\delta^2}{\partial G_{aa}}=0. \qquad \ldots\ldots(1)$$

The solution is as follows, letting

$$\lambda=\frac{H_{AA}-2H_{Aa}+H_{aa}}{\left[\dfrac{1}{f_{AA}}+\dfrac{4}{f_{Aa}}+\dfrac{1}{f_{aa}}\right]}:$$

$$\left.\begin{aligned}\delta_{AA}&=H_{AA}-G_{AA}=\frac{\lambda}{f_{AA}}\\ \delta_{Aa}&=H_{Aa}-G_{Aa}=-\frac{2\lambda}{f_{Aa}}\\ \delta_{aa}&=H_{aa}-G_{aa}=\frac{\lambda}{f_{aa}}\end{aligned}\right\}, \qquad \ldots\ldots(2)$$

$$\sigma_\delta^2=\lambda^2\left[\frac{1}{f_{AA}}+\frac{4}{f_{Aa}}+\frac{1}{f_{aa}}\right], \qquad \ldots\ldots(3)$$

$$\sigma_\gamma^2=\gamma^2\,[f_{AA}f_{Aa}+4f_{AA}f_{aa}+f_{Aa}f_{aa}]. \qquad \ldots\ldots(4)$$

Under random mating, which will be assumed in what follows,

$$f_{AA}=q^2,\quad f_{Aa}=2q\,(1-q)\ \text{ and }\ f_{aa}=(1-q)^2.$$

$$\lambda=[H_{AA}-2H_{Aa}+H_{aa}]\,q^2\,(1-q)^2, \qquad \ldots\ldots(5)$$

$$\gamma=[H_{AA}-H_{Aa}]\,q+[H_{Aa}-H_{aa}]\,(1-q), \qquad \ldots\ldots(6)$$

$$\sigma_\delta^2=\frac{\lambda^2}{q^2\,(1-q)^2}, \qquad \ldots\ldots(7)$$

$$\sigma_\gamma^2=2\gamma^2\,q\,(1-q). \qquad \ldots\ldots(8)$$

CASE OF NO DOMINANCE IN PRIMARY CHARACTER.

Mean and variance of squared deviations from optimum.

Assuming now that H depends on the squared deviation from the optimum of a primary scale, we have the following relations:

$$H = -(S-O)^2 = -[(S-M)+(M-O)]^2, \qquad \ldots\ldots(9)$$

$$\bar{H} = -[\mu_2(S)+(M-O)^2], \qquad \ldots\ldots(10)$$

$$\begin{aligned}\sigma_H{}^2 &= \Sigma H^2 f - \bar{H}^2 \\ &= \mu_4(S) + 4(M-O)\,\mu_3(S) + 4(M-O)^2\,\mu_2(S) - [\mu_2(S)]^2. \qquad \ldots\ldots(11)\end{aligned}$$

If there is no dominance on the primary scale, the moments due to a single pair are those of the squared binomial $[(1-q)\,a+qA]^2$:

$$\begin{aligned}M_1 &= 2\alpha q,\\ \mu_2 &= 2\alpha^2 q\,(1-q),\\ \mu_3 &= 2\alpha^3 q\,(1-q)\,(1-2q),\\ \mu_4 &= 2\alpha^4 q\,(1-q).\end{aligned}$$

The totals for the first three moments are given by simple summation and for the fourth by the formula $\mu_4(S) = \Sigma\mu_4 + 3\,[\Sigma\mu_2]^2 - 3\Sigma(\mu_2)^2$:

$$M = 2\Sigma\alpha q, \qquad \ldots\ldots(12)$$

$$\mu_2(S) = 2\Sigma\alpha^2 q\,(1-q), \qquad \ldots\ldots(13)$$

$$\mu_3(S) = 2\Sigma\alpha^3 q\,(1-q)\,(1-2q), \qquad \ldots\ldots(14)$$

$$\mu_4(S) = 2\Sigma\alpha^4 q\,(1-q) + 12\,[\Sigma\alpha^2 q\,(1-q)]^2 - 12\Sigma\alpha^4 q^2\,(1-q)^2. \qquad \ldots\ldots(15)$$

Returning to the secondary scale and substituting (12)–(15) in (10) and (11):

$$\bar{H} = -[2\Sigma\alpha^2 q\,(1-q) + (M-O)^2], \qquad \ldots\ldots(16)$$

$$\begin{aligned}\sigma_H{}^2 &= 8\,[\Sigma\alpha^2 q\,(1-q)]^2 + 2\Sigma\alpha^4 q\,(1-q) - 12\Sigma\alpha^4 q^2\,(1-q)^2 \\ &\quad + 8(M-O)\,\Sigma\alpha^3 q\,(1-q)\,(1-2q) + 8\,(M-O)^2\,\Sigma\alpha^2 q\,(1-q). \qquad \ldots(17)\end{aligned}$$

Analysis of variance.

The rate at which the mean of a character ($\bar{H}$) changes in relation to changes in the frequency (q_i) of one of the genes (A_i) must equal twice the net effect of substituting A_i for a_i (twice because change of q_i from 0 to 1 replaces two a_i's):

$$\frac{\partial \bar{H}}{\partial q_i} = 2\gamma_i. \qquad \ldots\ldots(18)$$

Applying this to the present case

$$\gamma_i = -\alpha_i\,[\alpha_i\,(1-2q_i) + 2\,(M-O)], \qquad \ldots\ldots(19)$$

$$\sigma_G^2 = 2\Sigma\gamma^2 q(1-q) = 2\Sigma\alpha^4 q(1-q)(1-2q)^2 + 8(M-O)\Sigma\alpha^3 q(1-q)(1-2q) + 8(M-O)^2\Sigma\alpha^2 q(1-q) \qquad ......(20)$$

from (8).

Next let O' be the deviation on the primary scale of the optimum from the grade of a_1a_1 when associated with a particular complex of the other genes. The grades (H) of A_1A_1, A_1a_1 and a_1a_1 on the secondary scale are then $-(2\alpha_1-O')^2$, $-(\alpha_1-O')^2$ and $-O'^2$ respectively. The value of λ can now be found from (5):

$$\lambda_1 = -2\alpha_1^2 q_1^2 (1-q_1)^2. \qquad ......(21)$$

But this is independent of O' and is therefore the same for all associated gene complexes. Thus with a given gene frequency and gene effect, the gene pair A_1, a_1 makes a constant contribution,

$$\sigma_{\delta_1}^2 = 4\alpha_1^4 q_1^2 (1-q_1)^2$$

by (7), to the portion of the total variance due to dominance deviations.

$$\sigma_D^2 = 4\Sigma\alpha^4 q^2 (1-q)^2. \qquad ......(22)$$

The contribution of epistatic deviations can now be obtained by subtracting (20) and (22) from (17). It will be desirable, however, to indicate an independent derivation. The epistatic deviation in the combined effects of two gene pairs may be found by subtracting the deviations from the mean due to each separately, from the deviation due to the combination. These means are easily found from the definition $H = -(S-O)^2$:

$$\epsilon_{A_1A_1A_2A_2} = (\bar{H}_{A_1A_1A_2A_2} - \bar{H}) - (\bar{H}_{A_1A_1} - H) - (\bar{H}_{A_2A_2} - \bar{H}). \qquad ......(23)$$

This yields the following results:

Genotype	ϵ	f
$A_1A_1A_2A_2$	$-8\alpha_1\alpha_2(1-q_1)(1-q_2)$	$q_1^2q_2^2$
A_2a_2	$-4\alpha_1\alpha_2(1-q_1)(1-2q_2)$	$2q_1^2q_2(1-q_2)$
a_2a_2	$+8\alpha_1\alpha_2(1-q_1)q_2$	$q_1^2(1-q_2)^2$
$A_1a_1A_2A_2$	$-4\alpha_1\alpha_2(1-2q_1)(1-q_2)$	$2q_1(1-q_1)q_2^2$
A_2a_2	$-2\alpha_1\alpha_2(1-2q_1)(1-2q_2)$	$4q_1(1-q_1)q_2(1-q_2)$
a_2a_2	$+4\alpha_1\alpha_2(1-2q_1)q_2$	$2q_1(1-q_1)(1-q_2)^2$
$a_1a_1A_2A_2$	$+8\alpha_1\alpha_2q_1(1-q_2)$	$(1-q_1)^2q_2^2$
A_2a_2	$+4\alpha_1\alpha_2q_1(1-2q_2)$	$2(1-q_1)^2q_2(1-q_2)$
a_2a_2	$-8\alpha_1\alpha_2q_1q_2$	$(1-q_1)^2(1-q_2)^2$

......(24)

$$\bar{\epsilon} = 0, \qquad ......(25)$$

$$\sigma_\epsilon^2 = 16\alpha_1^2\alpha_2^2 q_1(1-q_1)q_2(1-q_2). \qquad ......(26)$$

With n gene pairs, the total variance due to epistasis is the sum of the $\frac{n(n-1)}{2}$ contributions of the above type from each possible two gene pairs:

$$\sigma_E^2 = 8\,[\Sigma\alpha^2 q\,(1-q)]^2 - 8\Sigma\alpha^4 q^2\,(1-q)^2. \qquad \text{......(27)}$$

Equations (20), (22) and (27) complete the analysis of variance. Their sum agrees with (17).

The correlations between relatives.

The correlations between parent (p) and offspring (o) and between two offspring involve the correlations between the dominance deviations and also those between epistatic deviations. The joint frequencies, under random mating, are as follows:

$p \backslash o$	AA	Aa	aa
AA	q^3	$q^2(1-q)$	0
Aa	$q^2(1-q)$	$q(1-q)$	$q(1-q)^2$
aa	0	$q(1-q)^2$	$(1-q)^3$
Total	q^2	$2q(1-q)$	$(1-q)^2$

......(28)

$o \backslash o$	AA	Aa	aa
AA	$\frac{1}{4}q^2(1+q)^2$	$\frac{1}{2}q^2(1-q^2)$	$\frac{1}{4}q^2(1-q)^2$
Aa	$\frac{1}{2}q^2(1-q^2)$	$q(1-q)(1+q-q^2)$	$\frac{1}{2}q(1-q)^2(2-q)$
aa	$\frac{1}{4}q^2(1-q)^2$	$\frac{1}{2}q(1-q)^2(2-q)$	$\frac{1}{4}(1-q)^2(2-q)^2$
Total	q^2	$2q(1-q)$	$(1-q)^2$

......(29)

The correlation between the dominance deviations (δ's equation (2)) comes out zero in the case of parent and offspring (frequencies of (28)) but $\frac{1}{4}$ in the case of two offspring (frequencies of (29)), as given by Fisher (1918).

$$r_{\delta_p\delta_o} = 0, \quad r_{D_pD_o} = 0, \qquad \text{......(30)}$$

$$r_{\delta_o\delta_{o'}} = \tfrac{1}{4}, \quad r_{D_oD_{o'}} = \tfrac{1}{4}. \qquad \text{......(31)}$$

The correlation between epistatic deviations of parent and offspring (two gene pairs) is the sum of 81 terms, each consisting of the product of the joint frequency, $f_{(po)_1}$, for A_1, a_1 (as given in (28)), the joint frequency, $f_{(po)_2}$, for A_2, a_2 from a table similar to (28), the epistatic deviation of parent (24) and the epistatic deviation (24) of the offspring, divided by σ_ϵ^2 (26). The correlation between the epistatic deviations of two offspring is similarly derived, using table (29) instead of (28). Both of these come out $\frac{1}{4}$.

$$r_{\epsilon_p\epsilon_o} = \frac{1}{\sigma_\epsilon^2}\,\Sigma\epsilon_p\epsilon_o f_{(po)_1} f_{(po)_2} = \tfrac{1}{4}, \qquad \text{......(32)}$$

$$r_{\epsilon_o \epsilon_{o'}} = \frac{1}{\sigma_\epsilon^2} \Sigma \epsilon_o \epsilon_{o'} f_{(oo')_1} f_{(oo')_2} = \tfrac{1}{4}. \qquad \text{......(33)}$$

Since these are constant,

$$r_{E_p E_o} = \tfrac{1}{4} \qquad \text{......(34)}$$

and

$$r_{E_o E_{o'}} = \tfrac{1}{4}. \qquad \text{......(35)}$$

The parent-offspring and fraternal correlations are therefore as follows:

$$r_{po} = \tfrac{1}{2} \frac{\sigma_G^2}{\sigma_H^2} + \tfrac{1}{4} \frac{\sigma_E^2}{\sigma_H^2}, \qquad \text{......(36)}$$

$$r_{oo} = \tfrac{1}{2} \frac{\sigma_G^2}{\sigma_H^2} + \tfrac{1}{4} \frac{\sigma_D^2}{\sigma_H^2} + \tfrac{1}{4} \frac{\sigma_E^2}{\sigma_H^2}. \qquad \text{......(37)}$$

It appears that r_{oo} ranges only between $\frac{1}{4}$ and $\frac{1}{2}$, while r_{po} can be zero if all variability consists of dominance deviations (case of 1 factor, $\gamma = 0$). In the case of populations in which the optimum on the primary scale is far from the mean, the largest component of σ_H^2 is σ_G^2. Both r_{po} and r_{oo} are then close to $\frac{1}{2}$. Such a population is in course of rapid transformation.

Of more interest is the case in which the optimum and mean coincide. Consider the case of n genes, equal in their effects ($\alpha_1 = \alpha_2 \ldots = \alpha_n$) and frequencies all q or $(1-q)$.

$$\sigma_G^2 = 2nq\,(1-q)\,(1-2q)^2\alpha^4, \qquad \text{......(38)}$$

$$\sigma_D^2 = 4nq^2\,(1-q)^2\alpha^4, \qquad \text{......(39)}$$

$$\sigma_E^2 = 8n\,(n-1)\,q^2\,(1-q)^2\alpha^4, \qquad \text{......(40)}$$

$$\sigma_H^2 = 2nq\,(1-q)\,[1+(4n-6)\,q\,(1-q)]\,\alpha^4. \qquad \text{......(41)}$$

With multiple factors, epistatic deviations contribute much more than dominance deviations. ($\sigma_E^2 = 2(n-1)\sigma_D^2$.) There can, therefore, be little difference between r_{po} and r_{oo}. Both are close to $\frac{1}{4}$ if there are many gene pairs of which neither allelomorph is rare but approach $\frac{1}{2}$ if variability is largely due to rare genes.

A case of special interest is that in which the tendency to fixation is balanced by mutation in a large random breeding population. Assume that mutation is occurring at the rate u per gene. The selective advantage of one allelomorph over the other is proportional to γ and thus may be written $s\gamma = s\alpha^2(2q-1)$ still assuming that the optimum coincides with the mean. The rate of change of gene frequency due to mutation $(-uq)$ is balanced by that due to selection $(s\gamma q(1-q))$. The condition at equilibrium may thus be represented by the equation:

$$\Delta q = s\alpha^2(2q-1)q(1-q) - uq = 0, \qquad \text{......(42)}$$

$$q = \tfrac{1}{4}\left(3 \pm \sqrt{1 - \frac{8u}{s\alpha^2}}\right) \text{ or } 0. \qquad \text{......(43)}$$

There is no equilibrium if $8u > s\alpha^2$ (except $q=0$). Otherwise there are two positions of equilibrium (besides $q=0$), one stable and one unstable. Where $\frac{8u}{s\alpha^2}$ is a small fraction, the position of stable equilibrium is approximately $1-\frac{u}{s\alpha^2}$, and for unstable equilibrium $\frac{1}{2}+\frac{u}{s\alpha^2}$. Letting $q=1-\frac{u}{s\alpha^2}$ in equations (38)–(41) and substituting in (36) and (37), the approximate value of either r_{po} or r_{oo} comes out

$$\frac{s\alpha^2+2nu}{2s\alpha^2+8nu}. \qquad \text{......(44)}$$

This approaches $\frac{1}{2}$ under severe selection, low mutation rate, and in cases of variability dependent on a relatively small number of genes with large effects, while it approaches $\frac{1}{4}$ under the opposite conditions.

It was assumed above that there was no variation of q about its position of equilibrium. Where the effective size of population (N) is sufficiently small $\left(N \text{ much less than } \frac{1}{2Ns\alpha^2} \text{ and also than } \frac{1}{4u}\right)$, gene frequencies become distributed in the form $y=\frac{C}{q(1-q)}$ (Wright (1929, 1931) confirmed by a different method by Fisher (1930)). The number of unfixed loci maintained by such a population was found to be

$$L=2N\mu \log 3{\cdot}6N,$$

where μ is the mutation rate for all loci per individual. In the above formula, $C=\frac{L}{2 \log 3{\cdot}6N}=N\mu$. Take all α's as equal, and mean at optimum:

$$\sigma_G{}^2=2\alpha^4\int_0^1(1-2q)^2q(1-q)y\,dq=\tfrac{2}{3}N\mu\alpha^4, \qquad \text{......(45)}$$

$$\sigma_D{}^2=4\alpha^4\int_0^1 q^2(1-q)^2y\,dq=\tfrac{2}{3}N\mu\alpha^4, \qquad \text{......(46)}$$

$$\sigma_E{}^2=8\alpha^4\left[\int_0^1 q(1-q)y\,dq\right]^2-2\sigma_D{}^2=\tfrac{2}{3}N\mu\alpha^4(12N\mu-2). \quad \text{...(47)}$$

In this case $\sigma_D{}^2=\sigma_G{}^2$ giving $r_{po}=\frac{1}{4}$ irrespective of the relative importance of $\sigma_E{}^2$. The fraternal correlation does not appreciably exceed $\frac{1}{4}$ where the number of loci is sufficiently great to approximate the assumed distribution. It appears that both the correlation between parent and offspring, and that between two offspring is typically about 0·25 under the assumed conditions.

Case of complete dominance.

Analysis of variance, general formulae.

It is interesting to compare the preceding results with those which hold when there is complete dominance. It is convenient in this case to use $p\ (=1-q)$ as the frequency of the recessive gene irrespective of direction of effect. $\bar{H}_A$ and $\bar{H}_a$ will be used for the mean phenotypes $A-$ and aa respectively, $\bar{H}_{Ab}$ for A–bb etc. No assumptions will be made at first with respect to the nature of the epistatic relations.

The contribution of epistasis to variance in the general 2-factor case can be found directly by minimising $\sigma_\epsilon^2\ (=\Sigma(H-J)^2f)$, where $J=G+D$ and $J_{A_1A_2}=J_{A_1a_2}+J_{a_1A_2}-J_{a_1a_2}$. This requires solution of the following three simultaneous equations:

$$\frac{\partial\sigma_\epsilon^2}{\partial J_{A_1a_2}}=0,\quad \frac{\partial\sigma_\epsilon^2}{\partial J_{a_1A_2}}=0,\quad \frac{\partial\sigma_\epsilon^2}{\partial J_{a_1a_2}}=0. \qquad \text{......(48)}$$

The solutions for $\epsilon\ (=H-J)$ and for σ_ϵ^2 are as follows, letting

$$\kappa=\frac{(H_{A_1A_2}-H_{A_1a_2}-H_{a_1A_2}+H_{a_1a_2})}{\Sigma(1/f)}.$$

$$\left.\begin{aligned}\epsilon_{A_1A_2}&=+\kappa/f_{A_1A_2}\\ \epsilon_{A_1a_2}&=-\kappa/f_{A_1a_2}\\ \epsilon_{a_1A_2}&=-\kappa/f_{a_1A_2}\\ \epsilon_{a_1a_2}&=+\kappa/f_{a_1a_2}\end{aligned}\right\}, \qquad \text{......(49)}$$

$$\sigma_\epsilon^2=\kappa^2\Sigma(1/f). \qquad \text{......(50)}$$

With n gene pairs, σ_E^2 is the sum of the $\frac{n\,(n-1)}{2}$ terms of the type σ_ϵ^2 which can be made by taking every possible set of two gene pairs:

$$\sigma_E^2=\Sigma\sigma_\epsilon^2. \qquad \text{......(51)}$$

Under random mating, which will be assumed hereafter,

$$f_{A_1A_2}=(1-p_1^2)(1-p_2^2),\ f_{A_1a_2}=(1-p_1^2)p_2^2,$$
$$f_{a_1A_2}=p_1^2(1-p_2^2),\ f_{a_1a_2}=p_1^2p_2^2, \qquad \text{......(52)}$$

$$\Sigma(1/f)=\frac{1}{p_1^2p_2^2(1-p_1^2)(1-p_2^2)}, \qquad \text{......(53)}$$

from (6),

$$\gamma=-p(\bar{H}_A-\bar{H}_a), \qquad \text{......(54)}$$

from (8),

$$\sigma_G^2=2\Sigma\gamma^2p(1-p), \qquad \text{......(55)}$$

from (7).

$$\sigma_D^2=\Sigma\gamma^2(1-p)^2, \qquad \text{......(56)}$$

The correlations between relatives.

The parent-offspring and fraternal correlation tables relative to one gene pair condense into 2×2 form:

$p \backslash o$	$A-$	aa	
$A-$	$(1-p)(1+p-p^2)$	$p^2(1-p)$	......(57)
aa	$p^2(1-p)$	p^3	
Total	$1-p^2$	p^2	

Joint frequencies: parent and offspring.

$o \backslash o'$	$A-$	aa	
$A-$	$\frac{1}{4}(1-p)(4+4p-3p^2-p^3)$	$\frac{1}{4}p^2(1-p)(3+p)$	...(58)
aa	$\frac{1}{4}p^2(1-p)(3+p)$	$\frac{1}{4}p^2(1+p)^2$	
Total	$1-p^2$	p^2	

Joint frequencies: two full brothers.

The correlations between the epistatic deviations of combinations of two gene pairs can be found by adding the 16 terms of the following summations and dividing by $\sigma_\epsilon{}^2$. These reduce to the simple expressions at the right:

$$r_{\epsilon_p\epsilon_o} = \frac{1}{\sigma_\epsilon{}^2} \Sigma \epsilon_p \epsilon_o f_{(po)_1} f_{(po)_2} = \frac{p_1 p_2}{(1+p_1)(1+p_2)}, \quad \text{......(59)}$$

$$r_{\epsilon_o\epsilon_{o'}} = \frac{1}{\sigma_\epsilon{}^2} \Sigma \epsilon_o \epsilon_{o'} f_{(oo')_1} f_{(oo')_2} = \frac{(1+3p_1)(1+3p_2)}{16(1+p_1)(1+p_2)}. \quad \text{......(60)}$$

For n gene pairs, the correlations between the epistatic deviations depend on the summation of $\frac{n(n-1)}{2}$ terms taking once every possible set of two gene pairs:

$$r_{E_pE_o} = \frac{1}{\sigma_E{}^2} \Sigma (\sigma_\epsilon{}^2 r_{\epsilon_p\epsilon_o}), \quad \text{......(61)}$$

$$r_{E_oE_{o'}} = \frac{1}{\sigma_E{}^2} \Sigma (\sigma_\epsilon{}^2 r_{\epsilon_o\epsilon_{o'}}). \quad \text{......(62)}$$

The parent-offspring and fraternal correlations are as follows:

$$r_{po} = \frac{1}{2} \frac{\sigma_G{}^2}{\sigma_H{}^2} + r_{E_pE_o} \frac{\sigma_E{}^2}{\sigma_H{}^2}, \quad \text{......(63)}$$

$$r_{oo'} = \frac{1}{2} \frac{\sigma_G{}^2}{\sigma_H{}^2} + \frac{1}{4} \frac{\sigma_D{}^2}{\sigma_H{}^2} + r_{E_oE_{o'}} \frac{\sigma_E{}^2}{\sigma_H{}^2}. \quad \text{......(64)}$$

It is to be noted that these formulae (48–64) involve no assumptions as to the nature of the epistatic relations.

Application to squared deviations from optimum.

If now it be assumed that the value of the character (H) depends on the squared deviations from the optimum on the primary scale, its mean and variance can be obtained from (10) and (11). The moments on the primary scale can be deduced from those of the binomial

$$[p^2{}_{aa}+(1-p^2)\,A-]:$$

$$M=\Sigma\alpha(1-p^2), \qquad \text{......(65)}$$

$$\mu_2(S)=\Sigma\alpha^2p^2(1-p^2), \qquad \text{......(66)}$$

$$\mu_3(S)=\Sigma\alpha^3p^2(1-p^2)(2p^2-1), \qquad \text{......(67)}$$

$$\mu_4(S)=\Sigma\alpha^4p^2(1-p^2)[1-6p^2(1-p^2)]+3[\Sigma\alpha^2p^2(1-p^2)]^2. \qquad \text{......(68)}$$

The various statistics of the secondary scale can now be calculated:

$$\bar{H}=-[\Sigma\alpha^2p^2(1-p^2)+(M-O)^2], \qquad \text{......(69)}$$

$$\gamma_1=-\alpha_1p_1[\alpha_1(2p_1{}^2-1)+2(M-O)], \qquad \text{......(70)}$$

$$\kappa=2\alpha_1\alpha_2p_1{}^2p_2{}^2(1-p_1{}^2)(1-p_2{}^2), \qquad \text{......(71)}$$

$$\sigma_\epsilon{}^2=4\alpha_1{}^2\alpha_2{}^2p_1{}^2p_2{}^2(1-p_1{}^2)(1-p_2{}^2), \qquad \text{......(72)}$$

$$\sigma_E{}^2=2[\Sigma\alpha^2p^2(1-p^2)]^2-2\Sigma\alpha^4p^4(1-p^2)^2. \qquad \text{......(73)}$$

It will be found that $(\sigma_G{}^2+\sigma_D{}^2+\sigma_E{}^2)$ as given by (55), (56), (70) and (73) agrees with $\sigma_H{}^2$ as calculated by substitution of the moments in (11).

Haldane (1932, p. 196) gives the rate of change of the gene frequency *ratio* $u\left(=\frac{1-p}{p}\right)$ in the case in which mean and optimum coincide as $\Delta u=\frac{c\alpha^2u(u^2+2u-1)}{(u+1)^3}$. In my terminology the rate of change of gene frequency is always $s\gamma p\,(1-p)$, which in the present case becomes $\Delta p=s\alpha^2p^2(1-p)(1-2p^2)$. This agrees with Haldane's result, on making the substitutions $\Delta u=-\frac{\Delta p}{p^2}$, $u=\frac{1-p}{p}$, $c=-s$ in the latter.

For comparison with the case of no dominance, we will consider the case in which mean and optimum coincide and variability is due to n pairs of genes, all with the same frequency (p) of the recessive and all with an equal effect, α (in one direction or the other), on the primary scale:

$$\sigma_G{}^2=2np^3(1-p)(1-2p^2)^2\alpha^4, \qquad \text{......(74)}$$

$$\sigma_D{}^2=np^2(1-p)^2(1-2p^2)^2\alpha^4, \qquad \text{......(75)}$$

$$\sigma_E{}^2=2n(n-1)p^4(1-p^2)^2\alpha^4, \qquad \text{......(76)}$$

$$\sigma_H{}^2=np^2(1-p^2)[1+(2n-6)p^2(1-p^2)]\alpha^4, \qquad \text{......(77)}$$

$$r_{E_pE_o} = \left(\frac{p}{1+p}\right)^2, \qquad \text{......(78)}$$

$$r_{E_oE_o'} = \left[\frac{1+3p}{4(1+p)}\right]^2. \qquad \text{......(79)}$$

Where p is very small (all recessive allelomorphs rare), σ_G^2, σ_D^2 and σ_E^2 are in the approximate ratio $2p : 1 : 2(n-1)p^2$. Thus nearly all of the variability is due to dominance deviations (unless n is enormously great) and r_{po} is nearly zero and r_{oo} is close to $\frac{1}{4}$.

Where $(1-p)$ is very small (all dominant allelomorphs rare), σ_G^2, σ_D^2 and σ_E^2 are in the approximate ratio $1 : \frac{1}{2}(1-p) : 4(n-1)(1-p)$. Here very little of the variability is due to dominance deviations, and with large n both r_{po} and r_{oo} approach the form $\dfrac{1+2n(1-p)}{2+8n(1-p)}$.

Where the distribution of values of p is not confined too closely to the neighbourhood of 0 or 1 and the number of genes is large, the variability is almost wholly determined by the epistatic effects. The correlation between parent and offspring becomes approximately

$$r_{po} = \frac{\Sigma\sigma_\epsilon^2\, r_{\epsilon_p\epsilon_o}}{\Sigma\sigma_\epsilon^2} = \left[\frac{\int_0^1 p^3(1-p)y\,dp}{\int_0^1 p^2(1-p^2)y\,dp}\right]^2.$$

This is exactly the square of the correlation between parent and offspring on the primary scale.

Similarly the correlation between two offspring is approximately:

$$r_{oo} = \frac{\Sigma\sigma_\epsilon^2\, r_{\epsilon_o\epsilon_o'}}{\Sigma\sigma_\epsilon^2} = \left[\frac{\int_0^1 p^2(1-p)(1+3p)y\,dp}{4\int_0^1 p^2(1-p^2)y\,dp}\right]^2.$$

This also is exactly the square of the corresponding correlation on the primary scale.

In the case of no dominance, both the parent-offspring and the fraternal correlation come out $\frac{1}{4}$ under the above conditions. This again is the square of the parent-offspring and the fraternal correlations on the primary scale.

ENVIRONMENTAL COMPLICATIONS.

So far it has been assumed that all variability is due to genetic differences. Assume now that nongenetic factors tend to cause normally distributed variability on the primary scale. The total variance on this scale (σ_S^2) can be analysed into two portions due respectively to heredity, $\sigma^2_{S(H)}$, and to environment $\sigma^2_{S(V)}$. Note that $\sigma^2_{S(H)}$ is the same as the expression $\mu_2(S)$ of equations (13) and (66) (cases of no dominance and complete dominance respectively) in which all variance of the primary scale was assumed to be genetic. We will use C for the secondary character when affected by nongenetic factors. Its variance, σ^2_C, is to be analysed into a portion due to heredity $\sigma^2_{C(H)}$ (equivalent to σ^2_H of the preceding discussion), a portion due to environment, $\sigma^2_{C(V)}$ and a portion due to nonadditive effects of heredity and environment, $\sigma^2_{C(HV)}$.

By (9) and (10) the mean deviation of the secondary character from optimum is

$$\bar{C} = -[\sigma_S^2 + (M-O)^2].$$

This is greater than the value under exclusive genetic variability by the amount $\sigma^2_{S(V)}$. Under our assumption of normality for the nongenetic variability $\mu_3(S)$ of (14) and (67) is unchanged, but $\mu_4(S)$ of (15) and (68) is increased by the term $3\sigma^4_{S(V)} + 6\sigma^2_{S(H)}\sigma^2_{S(V)}$. The variance of the secondary character comes out:

$$\sigma^2_C = \sigma^2_{C(H)} + 4\sigma^2_{S(H)}\sigma^2_{S(V)} + 2\sigma^4_{S(V)} + 4\,(M-O)^2\sigma^2_{S(V)}.$$

The last two terms can be attributed wholly to environment ($\sigma^2_{C(V)}$) while the second term must be considered as due to nonadditive effects of heredity and environment, $\sigma^2_{C(HV)} = 4\sigma^2_{S(H)}\sigma^2_{S(V)}$.

The correlations between relatives are of course reduced by replacing σ_H^2 by σ_C^2 in (36), (37), (63) and (64). The most interesting case is again that in which the mean is at the optimum (giving maximum nonlinearity) and the number of genetic factors is sufficiently large so that $\sigma^2_{C(H)}$ is approximately $2\sigma^4_{S(H)}$ (compare values of σ_H^2 $(=\sigma^2_{C(H)})$ in (17) and (77).

$$\sigma_C^2 = 2\,[\sigma^2_{S(H)} + \sigma^2_{S(V)}]^2 \text{ approximately,}$$

$$\frac{\sigma^2_{C(H)}}{\sigma_C^2} = \left(\frac{\sigma^2_{S(H)}}{\sigma_S^2}\right)^2 \text{ approximately.}$$

The ratio of genetic to total variance on the secondary scale is thus the square of that on the primary scale in this extreme case, and the correlations between relatives are reduced accordingly from the values calculated on the assumption of complete determination by heredity. As it has previously been shown that the parent-offspring and fraternal correlations on the secondary scale are the squares of the values on the primary scale under complete determination by heredity, it now follows that these correlations on the two scales are related in this way irrespective of environmental complications.

SUMMARY.

The formulae for the mean and variance of squared deviations from an optimum are given for the cases of no dominance and of complete dominance, first assuming no environmental complications but later removing this restriction. The variance in each case is analysed into contributions due to (1) additive gene effects, (2) dominance deviations, (3) epistatic deviations, (4) environmental effects and (5) nonadditive joint effects of heredity and environment. In the case of complete dominance, formulae are developed which apply to epistatic relations in general. These lead to formulae for the correlations between parent and offspring and between two offspring.

It appears that in a population in which the mean of some measurable character is at the optimum, the parent-offspring and fraternal correlations in adaptive value are approximately the squares of the corresponding correlations with respect to the character itself, whatever environmental complications there may be. Where the mean is not at the optimum, there is less difference between the correlations in adaptive value and the corresponding ones with respect to the character itself.

REFERENCES.

FISHER, R. A. (1918). "The correlation between relatives on the supposition of Mendelian inheritance." *Trans. roy. Soc. Edinb.* **52**, pt. 2, pp. 399–433.

—— (1930). *The genetical theory of natural selection.* Oxford: at the Clarendon Press. 272 pp.

HALDANE, J. B. S. (1932). *The causes of evolution.* New York and London: Harper and Brothers. 235 pp.

WRIGHT, S. (1929). "The evolution of dominance." *Amer. Nat.* **63**, 556–61

—— (1931). "Evolution in Mendelian populations." *Genetics,* **16**, 97–159.

EVOLUTION IN POPULATIONS IN APPROXIMATE EQUILIBRIUM.

By SEWALL WRIGHT.

(*Department of Zoology, The University of Chicago.*)

CONTENTS.

INTRODUCTION.

THIS paper is concerned with the evolutionary processes in populations in which the selective values of different grades of a character depend on the (squared) deviations of the latter, from an optimum. It is assumed that the effects of different genes on the character combine additively (no epistasis). The cases of no dominance and of complete dominance in these primary effects will be treated in succession. A preceding paper (1934) dealt with the analysis of variability and the correlations between relatives with respect to squared deviations from the optimum. The same symbols will be used here. Equation numbers from the preceding paper will be referred to in square brackets.

CASE OF NO DOMINANCE ON PRIMARY SCALE.

The nature of the evolutionary processes under the conditions described may be visualised by treating the population at a given moment as located at a point in a multidimensional space defined by the set of gene frequencies ($q_1, \dots q_n$) pertaining to the *plus* members ($A_1, \dots A_n$) of gene pairs affecting the character (cf. Haldane, 1931). Ordinates are to be erected measuring the average adaptive value ($\bar{H}$) of the character. The signs of $\bar{H}$ and γ_i (net effect of substitution of A_i for a_i on adaptive value in the population in question) are taken so as to make the high points correspond to optimal values. The effect of substitution of A_i for a_i on the *primary character* is represented by α_i. The position of the optimum on this primary scale is represented by O and that of the mean by M ($=2\Sigma\alpha q$):

$$\bar{H} = -2\Sigma\alpha^2 q\,(1-q) - (M-O)^2, \qquad \text{......(1) } (=[16])$$

$$\gamma_i = \alpha_i\,[\alpha_i(2q_i - 1) - 2(M-O)]. \qquad \text{......(2) } (=[19])$$

The rate at which the mean of a character changes in relation to changes in the frequency (q_i) of one of the genes must equal twice the net effect of a single gene substitution:

$$\frac{\partial H}{\partial q_i} = 2\gamma_i. \qquad \text{......(3) (=[18])}$$

The rate of change of gene frequency per generation in the case of no dominance is well known to be $\Delta q = sq(1-q)$, where s measures the selective disadvantage of one allelomorph (Fisher, 1922). In the general case, the selective disadvantage must be proportional to the momentary net effect of gene replacement on adaptive value:

$$\Delta q_i = s\gamma_i q_i(1-q_i). \qquad \text{......(4)}$$

The rate of change of the mean adaptive level of the population per generation can be written

$$\Delta \bar{H} = \sum_{i=1}^{n} \frac{\partial \bar{H}}{\partial q_i} \Delta q_i = 2s\Sigma\gamma^2 q(1-q). \qquad \text{......(5)}$$

The right-hand member of this equation is proportional to the general formula for the portion of the variance of H which can be attributed to additive gene effects [20].

$$\Delta \bar{H} = s\sigma_G^2. \qquad \text{......(6)}$$

This principle was arrived at in a different way by Fisher (1930). He enunciated it as the "fundamental theorem of natural selection." "The rate of increase in fitness of an organism at any time is equal to its genetic[1] variance at that time."

I have criticised this application of it on the ground that it measures merely the *tendency* toward increase in fitness due to selection. Other evolutionary factors such as recurrent mutation, immigration and the effects of sampling in populations of limited size must also be considered. There must ordinarily be an approximate balancing of these *first* order pressures so that evolutionary change is a *second* order resultant (Wright, 1930).

From the formula as given, $\Delta \bar{H}$ cannot be negative as is of course to be expected. In the absence of other factors, evolutionary change ceases whenever $\Delta \bar{H} = 0$. This occurs at any point at which each q has one of the three values 0, 1 or such a value that $\gamma = 0$.

Consider first the case in which there is complete homozygosis (all q's either 0 or 1). Whether there is stability in the face of low rates of mutation or immigration depends on the gradient. If $q_1 = 0$ and $\frac{\partial \bar{H}}{\partial q_1} > 0$,

[1] Fisher explicitly includes in "genetic variance" only that portion of the variance which can be attributed to additive gene effects.

the homozygote a_1a_1 is unstable, since $\bar{H}$ increases with mutation of a to A. Similarly if $q_1 = 1$ and $\frac{\partial \bar{H}}{\partial q_1} < 0$, A_1A_1 is unstable. These conditions for instability may be written as follows from (3) and (2):

$$(M - O) < -\frac{\alpha_1}{2}, \quad (q_1 = 0) \qquad \ldots\ldots(7)$$

$$(M - O) > \frac{\alpha_1}{2}. \quad (q_1 = 1) \qquad \ldots\ldots(8)$$

Thus if the mean of the character is below the optimum by more than half the effect of a gene which is fixed in the minus phase, this fixation is unstable, or if the mean is above the optimum by more than half the effect of a fixed plus factor, minus mutations will tend to accumulate.

Since the effect of replacing a_1a_1 by A_1A_1 is to increase the mean by $2\alpha_1$, it is possible for both homogenic populations to be unstable. There is stable equilibrium at the point $q_1 = \frac{1}{2} + \frac{M - O}{\alpha_1}$ if all other genes are fixed.

The condition when the population is heterogenic in more than one respect requires further consideration. The condition that the γ's for all unfixed genes be zero is that for each of them

$$q_1 = \tfrac{1}{2} + \frac{M - O}{\alpha_1}. \qquad \ldots\ldots(9)$$

Thus all of the q's (frequencies of plus genes) must be less than $\frac{1}{2}$, all equal to $\frac{1}{2}$ (mean and optimum coincide at the mid-point of the scale where $M = \Sigma\alpha$), or all greater than $\frac{1}{2}$. Those with the same effect (α) must have the same gene frequency at equilibrium. The limiting case for small values of all q's, all α's the same, occurs with the optimum at $\frac{\alpha}{2}$. The other limiting case (all q's close to 1) occurs with the optimum at $2n\alpha - \frac{\alpha}{2}$. Thus no equilibrium of 2 or more unfixed factors is possible unless the optimum is more than half a gene effect within the limits of variation. With unequal gene effects, the possible range of location of the optimum is less.

As to the stability of these equilibria, consider the way in which the γ's vary:

$$\frac{\partial \gamma_i}{\partial q_j} = -4\alpha_i\alpha_j, \qquad \ldots\ldots(10)$$

unless $i = j$, when
$$\frac{\partial \gamma_i}{\partial q_i} = -2\alpha_i^2.$$

$$\partial\gamma_i = \sum_{j=1}^{n} \frac{\partial \gamma_i}{\partial q_j}\delta q_j = 2\alpha_i[\alpha_i \delta q_i - \delta M]. \qquad \ldots\ldots(11)$$

The expression $\delta M(=2\Sigma\alpha\delta q)$ is the deviation of the mean of the character from its value at equilibrium, brought about by the deviations of the q's from their values at equilibrium. If all $\delta\gamma$'s are opposite in sign to the corresponding δq's (as is obviously the case when all δq's are of the same sign) all q's tend to return to equilibrium. If, on the other hand, all $\delta\gamma$'s are of the same sign as the corresponding δq's (as is obviously the case when the signs of the δq's are so balanced that there is no change of mean ($\delta M=0$)), *all* q's tend to depart farther from equilibrium. In intermediate cases, some q's may go toward equilibrium, others go farther away. Clearly there can be no stable equilibrium (under selection alone) when more than one factor is unfixed. The points at which $\Delta H=0$ and more than one gene is unfixed are of the nature of saddles in our multi-dimensional space.

With n factors affecting the character, there may be any number from 1 to $\frac{\underline{|n}}{(\underline{|n/2})^2}$ "peaks," stable to small displacements, relative to adaptive value H. As shown above, there is either fixation of all loci or of all but one, at each peak. The number of such peaks depends on the position of the optimum and the relative magnitudes of the gene effects.

While the number of peaks is in general large (if n is large), the total number of stationary points ($\Delta\bar{H}=0$) is much larger. These are of the two sorts already discussed. All of the 2^n points at which all genes are fixed, minus those which are peaks, are stable only in the absence of mutation or immigration. In addition are the usually numerous 2 to n-dimensional "saddles."

For further discussion it will be convenient to restrict attention to the case in which all genes have the same effect on the character.

This gives the simplification that at stationary points all unfixed genes have the same value of q. Assume that at a stationary point ($\Delta\bar{H}=0$) there are n_1 genes in which the plus phase is fixed ($q=1$), n_0 in which the minus phase is fixed ($q=0$) and n_x which are unfixed ($q=q_x$). Let $O=k\alpha$ be the optimum.

$$q_x=\frac{k-2n_1-\frac{1}{2}}{2n_x-1}, \qquad \text{......(12)}$$

from (9).

There is, of course, a stationary point only if the values of n_1, n_0 and n_x are such that q_x falls between 0 and 1. The number of combinations with the same values of n_1, n_0 and n_x is $\frac{\underline{|n}}{\underline{|n_1}\,\underline{|n_0}\,\underline{|n_x}}$. All of these are stationary points with the same adaptive value, if any one is. The adaptive value $\bar{H}$ reduces to the following expression:

$$\bar{H} = -\left[\frac{n_x\alpha^2}{2} - (M-O)^2(2n_x-1)\right], \qquad \ldots\ldots(13)$$

where $$(M-O) = \left(\frac{k-2n_1-n_x}{2n_x-1}\right)\alpha.$$

The points which are stationary merely because all factors are fixed ($n_x=0$) have, of course, the mean adaptive value

$$\bar{H} = -(M-O)^2 = -(k-2n_1)^2\alpha^2.$$

This may range from 0 to very large negative values. On the other hand, a stationary population with even a large number of unfixed factors is not handicapped very much if its mean is at the optimum. In this case, $\bar{H} = -\frac{n_x}{2}\alpha^2$.

For example, with n factors with unit effects and optimum at the mid-point of the range ($k=n$), the adaptive values of completely fixed combinations may be as low as $-n^2$ while a population in which no factors are fixed, all values of q being $\frac{1}{2}$, would have a value of $\bar{H}$ only $\frac{n}{2}$ points below that of the optimal fixed combination. This is the greatest depression for any "saddle," since the term $(M-O)^2(2n_x-1)$ is necessarily positive for all values of n_x except 0. Thus the population can pass from one peak to any other of the $\frac{\lfloor n}{(\lfloor n/2)^2}$ "peaks" without crossing any but very shallow valleys.

Complete dominance.

It is important to compare the preceding results with those found where there is complete dominance of gene effects on the primary character. Using p_i for the frequency of a recessive gene and α_i for the effect of the corresponding dominant gene (whether plus or minus), the mean adaptive value $\bar{H}$ of the population and γ_i the momentary net effect on adaptive value of replacing a_i by A_i, are as follows:

$$\bar{H} = -\Sigma\alpha^2p^2(1-p^2) - (M-O)^2, \qquad \ldots\ldots(14)\ [=69]$$

where $M = \Sigma\alpha(1-p^2)$,

$$\gamma_i = -\alpha_i p_i[\alpha_i(2p_i^2-1) + 2(M-O)]. \qquad \ldots\ldots(15)\ [=70]$$

As before (equation 5), $\Delta\bar{H}=0$ if each p_i is 0, 1 or such a value that $\gamma_i=0$. In this case if $p_i=0, \gamma_i=0$. There is instability of a fixed dominant

($p_i=0$) in the face of occasional mutation if $\gamma_i<0$, which occurs if the mean exceeds the optimum by more than half the effect of a dominant plus gene or is below the optimum by more than half the effect of a dominant minus gene. The value of γ_i is of course exceedingly small in either case.

If $p_i=1$ (fixation of recessive), dominant mutations will tend to accumulate if $\gamma_i>0$. This occurs if $(M-O)<-\frac{\alpha_i}{2}$, where the mutation has a plus effect, and if $(M-O)>\frac{(-\alpha_i)}{2}$ if α_i is negative.

As the effect of replacing a_ia_i by A_iA_i is to change the mean by α_i, one or the other homogenic population must be stable against rare mutations. It is, however, possible for $\bar{H}$ to remain unchanged for all frequencies of one gene.

The condition that the γ's for all unfixed gene pairs be zero is that for each of them

$$p_i^2=\tfrac{1}{2}-\frac{M-O}{\alpha_i}. \qquad \ldots\ldots(16)$$

But as $M=\Sigma\alpha\,(1-p^2)$, p_i^2 drops out. Thus the value of the character $\bar{H}$ at any of these equilibrium points remains unchanged by change in any one of the gene frequencies. But change of any one at such a point will change the γ's of all others. Thus these points are all of the nature of "saddles." This agrees with conclusions of Fisher (1930) and of Haldane (1932) reached by different methods.

The condition for such a saddle is analogous to that in the case of no dominance. The range of variation of the primary character possible with a given set of equally effective unfixed factors, for n_x of which the plus phase is dominant (effect α) and for m_y of which the minus phase is dominant (effect $-\alpha$), is from $-m_y\alpha$ to $n_x\alpha$. The optimum must fall more than half a gene effect within these limits.

The nature of the "surface" of adaptive values ($\bar{H}$) is similar in many respects to that in the case of no dominance. Again assuming that the effects of all gene differences are the same in magnitude, but assuming that in n pairs the dominant has a plus effect while in m it has a minus effect, we may distinguish the following classes:

Dominant plus genes		Dominant minus genes	
p	No. loci	p	No. loci
0	n_0	0	m_0
1	n_1	1	m_1
$\sqrt{\tfrac{1}{2}-\frac{M-O}{\alpha}}$	n_x	$\sqrt{\tfrac{1}{2}+\frac{M-O}{\alpha}}$	m_y

The number of combinations with the same set of 6 numbers is

$$\frac{\underline{|n} \quad \underline{|m}}{\underline{|n_1}\,\underline{|n_0}\,\underline{|n_x}\,\underline{|m_1}\,\underline{|m_0}\,\underline{|m_y}}. \qquad \text{......(17)}$$

All of these are stationary points if p_x and p_y for unfixed factors fall between 0 and 1. Assuming that the optimum is at $k\alpha$, we have

$$p_x^2 = \frac{n_x + n_0 - m_0 - k - \frac{1}{2}}{n_x + m_y - 1}, \quad p_y^2 = 1 - p_x^2, \qquad \text{......(18)}$$

$$\bar{H} = -\left[\frac{(n_x + m_y)\alpha^2}{4} - (M - O)^2\,(n_x + m_y - 1)\right]. \qquad \text{......(19)}$$

For completely homogenic populations this, of course, reduces to $-(M-O)^2$ which may be a very large negative number. For "saddles" with mean at the optimum it reduces to $-\frac{(n_x + m_y)\alpha^2}{4}$ which is relatively small. Again we find that all peaks are connected by shallow saddles.

Nature of evolutionary process.

If evolution were controlled only by selection, the locus of a population characterised by any given set of gene frequencies would move up the steepest gradient in the field, each gene frequency changing at the rate $\Delta q\ (= s\gamma q\,(1-q))$ and the mean adaptive value rising at the rate $\Delta\bar{H}\ (= s\Sigma\gamma^2 q\,(1-q) = s\sigma_G^2)$ per generation. This process, supplemented by the occurrence of wholly new mutations favourable from the first is that which has been investigated chiefly by Haldane in 1924 and later, and by Fisher (1930).

Having reached a "peak" at the optimum grade of the character in question such evolutionary change must cease until conditions change. Any new mutation must necessarily cause a shift from the optimum and therefore be injurious at its first appearance as Fisher (1930) has pointed out. Of course if the optimum is beyond the current limits of variation, there is the possibility of slow advance through utilisation of new mutations (each with chance of reaching fixation of $2s$ (Fisher, 1930)). But the process ceases with attainment of the optimum grade in all respects.

Indeed it may appear that there is no possibility of further advance by any mechanism. We have seen that with an intermediate optimum there is in general a very large number of separate peaks separated by shallow "saddles." But all of these peaks must be at the same or very nearly the same level, and even if the locus of the population could by some means be moved across a saddle to a new peak it would mean no advance.

The subject presents a somewhat different aspect when we recall that genes in general have multiple effects. The system of peaks relative to one character is not independent of that relative to another. Moreover, it is the harmonious adjustment of all characteristics of the organism as a whole that is the object of selection, not the separate metrical "characters." It is estimated that there are many thousands of genes at least in higher organisms (about 15,000 in *Drosophila melanogaster*, according to Gowen and Gay), and each of these is probably capable of mutating through indefinitely extended series of multiple allelomorphs. No limit can be set to the number of possible combinations, and it seems safe to postulate an inconceivably great number of "peaks," many of them characterised by different harmonious combinations of characters (although many also for the same character). These may be at all levels and of all orders of dominance and subordination in relation to each other. It has seemed to me (1929 *et seq.*) that the central problem of evolution (as of live stock improvement (Wright, 1920, 1922)) is that of a trial and error mechanism by which the locus of a population may be carried across a saddle from one peak to another and perhaps higher one. This view contrasts with the conception of steady progress under natural selection developed in most extreme form by Fisher (1930). Haldane has taken to some extent an intermediate position. He notes (1931) that almost every species is to a first approximation in genetic equilibrium, and after treating mathematically the two-factor case of "metastable equilibrium" he suggests that in many cases "the process of species formation may be a rupture of the metastable equilibrium." The mathematical analysis in the present paper deals with a case in which there may be innumerable separate peaks though all at approximately the same level. It may be looked upon as a simplified model of the complex case in which adaptation of the organism as a whole replaces that of a single metrical character. Consideration of the means by which the locus of a population may be carried across a saddle may be of interest from this standpoint.

The rate of mutation of particular type genes has been found to be of the order of 10^{-5} or 10^{-6} per generation in *Drosophila* (Muller, 1928) and corn (Stadler, 1930). This is enough to prevent complete fixation of any genes in large populations. There may be special cases in which mutation pressure drives the locus of a population from one peak to another against the pressure of selection. In general, however, mutation pressure seems to be so low compared with selection pressure that the population would merely be held at a point a little below a particular peak. In the case of no dominance this point is approximately at the array of values typified by $q_i = 1 - \frac{u_i}{s\alpha_i^2}$ [43] and in that of dominance by $p_i = \sqrt{\frac{u_i}{s\alpha_i^2}}$ (where u_i is the rate of mutation of gene A_i).

If the population is not indefinitely large, the accidents of sampling will cause independent fluctuations about this point of equilibrium in all of the gene frequencies. Each of these distributions is I-shaped if $4Nu$ and $4Ns$ are large (N effective size of breeding population) but J or U-shaped if $4Nu$ and $4Ns$ are less than 1 (Wright, 1931). While the rate of change per generation of gene frequency due to accidents of sampling is low (causing fixation at the rate of $1/2N$ per generation in the absence of selection and mutation) it is possible that in the course of time it may carry the system of gene frequencies across a shallow saddle to a new peak.

The effectiveness of this mechanism is enormously increased if the population is subdivided into many local groups which breed largely within themselves. The distribution of gene frequencies under random sampling is here determined by the relation between effective size of the local group and the cross breeding index. The rate of random drift of each gene frequency may be relatively rapid per generation. The shallow saddles would be crossed so easily that no local group would be expected to stay long at the same peak and no two sufficiently isolated local groups would occupy the same peak at a given time. As far as the metrical character in question is concerned there would be no appreciable changes. The average in all local groups would remain very close to the optimum even when the locus is crossing a saddle. But there would be a kaleidoscopic shifting among other characters, affected by the same genes, and at the time subject only to selection of second order importance. At any time combinations might be reached by chance, in particular local groups, with effects of the first order of importance, leading to expansion of such groups (intergroup selection). This process should be of much greater evolutionary significance if considered with respect to total adaptive value of the organism, instead of approach of a particular character to an optimum.

So far we have assumed that conditions are constant. A drastic change of conditions, resulting in a drastic shift in the position of the optimum, would be followed by steady evolutionary change of the type described by Haldane and Fisher until the lost ground is regained and the mean again coincides with the optimum. But minor changes of conditions, shifting the position of the optimum back and forth by no more than the effect of a single gene, will have evolutionary consequences of a different sort. With such a shift of the optimum, the old peak will be depressed and there will arise in general a large number of new peaks immediately about it. To which of these the population moves will depend on the preceding accidents of sampling. If now the optimum shifts back to its original position, it will be very unlikely that the population will move back to the original peak. Thus trivial oscillations in conditions will be enough to carry the population to any peak in the system with similar consequences for other characters to those suggested above. Again the process will be enormously speeded up if there is local inbreeding and there are slight local differences in conditions.

The combination of the effects of inbreeding and of varying local conditions of selection provide a mechanism for the indefinitely continued process of trial and error among local populations with respect to gene *combinations* which is probably necessary for progressive evolution.

REFERENCES.

FISHER, R. A. (1922). "On the dominance ratio." *Proc. roy. Soc. Edinb.* **42**, 321–41.

—— (1930). *The genetical theory of natural selection.* Oxford: at the Clarendon Press. 272 pp.

GOWEN, J. W. and E. H. GAY (1933). "Gene number, kind, and size in *Drosophila.*" *Genetics,* **18**, 1–31.

HALDANE, J. B. S. (1924). "A mathematical theory of natural and artificial selection. Part I." *Trans. Camb. phil. Soc.* **23**, 19–41.

—— (1931). "A mathematical theory of natural selection. Part VIII. Metastable populations." *Ibid.* **27**, pt. 1, pp. 137–42.

—— (1932). *The causes of Evolution.* London: Longmans Green. 235 pp.

MULLER, H. J. (1928). "The measurement of gene mutation rate in *Drosophila,* its high variability, and its dependence upon temperature." *Genetics,* **13**, 279–357.

—— (1929). "The gene as the basis of life." *Proc. Int. Cong. Plant Sciences,* **1**, 897–921.

STADLER, L. J. (1930). "The frequency of mutation of specific genes in maize." *Anat. Rec.* **47**, 381.

WRIGHT, S. (1920). "Principles of live stock breeding." *Bull. U.S. Dep. Agric.* No. 905, 68 pp.

—— (1922). "The effects of inbreeding and crossbreeding on guinea-pigs. III. Crosses between highly inbred families." *Ibid.* No. 1121, 60 pp.

—— (1929). "Evolution in a Mendelian population." *Anat. Rec.* **44**, 287.

—— (1930). "The genetical theory of natural selection. A review." *J. Hered.* **21**, 349–56.

—— (1931 *a*). "Evolution in Mendelian populations." *Genetics,* **16**, 97–159.

—— (1931 *b*). "Statistical theory of evolution." *Proc. Amer. Stat. Assoc.* **26**, 201–8.

—— (1932). "The roles of mutation, inbreeding, crossbreeding and selection in evolution." *Proc. Sixth Internat. Congress of Genetics,* **1**, 356–66.

—— (1934). "The analysis of variance and the correlations between relatives with respect to deviations from an optimum." *J. Genet.* **30**, 243–256.

17

The Distribution of Gene Frequencies in Populations

Science 85, no. 2212 (1937):504

18

The Distribution of Gene Frequencies in Populations

Proceedings of the National Academy of Science 23, no. 6 (1937):307–20

Introduction

In these two papers, Wright took the dramatic step of incorporating his concept of the mean adaptive value of a population ($\bar{W}$) into his statistical distribution of genes, replacing his former limited selection term e^{4Nsq} with the much richer term $\bar{W}^{2N}$. Since $\bar{W}$ was the surface of selective values, it could incorporate any degree of dominance, any amount of genic or chromosomal interaction, and any number of loci or alleles. Of course the cost of all this added theoretical flexibility was the greater abstractness of the $\bar{W}$ term, which was virtually impossible to calculate in practice.

The first of these papers is an abstract of the second, but the first actually contains the most general form of the statistical distribution of genes, complete with the migration terms. Although Wright considered these two papers to represent a significant advance in his quantitative theory of evolution, he gives little indication of their significance in the papers. I discuss them briefly in *SW&EB*, 305–6. Wright devoted a whole volume of his *E&GP* (vol. 2) to the theory of gene frequencies. This volume should be compared to the large middle section of his 1931 paper, "The Distribution of Gene Frequencies and Its Immediate Consequences," to assess how far Wright was able to realize his aim of bringing his quantitative theory as close as possible to his qualitative theory.

THE DISTRIBUTION OF GENE FREQUENCIES IN POPULATIONS

THE effects of the various evolutionary factors can be reduced to common terms by considering the rates of change which they tend to bring about in the relative frequencies of alleles within a population. In the absence of such factors, there is a constancy of gene frequencies from the symmetry of the Mendelian mechanism. The frequency (q) of a given gene changes at the rate

$$\Delta q = -uq + v(1-q) - m(q-q_t) + \frac{q(1-q)}{2}\frac{\partial}{\partial q}\log \overline{W}$$

where u is the rate of mutation of the gene in question, v is the rate of mutation to it from its alleles, m is the effective amount of exchange between the local population under consideration and the species as a whole (gene frequency q_t), and $\overline{W}$ is the mean selective value of the array of genotypes characteristic of this population. Gene frequency is in equilibrium (stable or otherwise) at any point at which $\Delta q = 0$ except for variation due to the accidents of sampling among the gametes. The sampling variance for one generation is $\frac{q(1-q)}{2N}$ where N is the effective size of the breeding population. The pressure toward a stable equilibrium in value of q, due to mutation, crossbreeding and selection (assuming persistence of the same conditions for a long period), and the divergent tendency due to inbreeding should between them determine a certain probability distribution of values of q for the local population considered. The following formula is reached, assuming that the selective effects of the gene in question are independent of those of other genes.

$$\varphi(q) = \frac{Ce^{4N\int \frac{\Delta q\, dq}{q(1-q)}}}{q(1-q)}$$

$$= C\overline{W}^{2N} q^{4N[mq_t + v]-1} (1-q)^{4N[m(1-q_t)+u]-1}$$

More generally, selective values depend on the interactions of the entire system of genes. It is the harmonious development of all characteristics that determines the success of an organism, not the absolute grades of the separate characters and still less the composition with respect to a single series of alleles. The mean selective values, $\overline{W}$, of populations characterized by different sets of gene frequencies form a multidimensional surface which in general has many peaks. The joint distribution of the gene frequencies is given by the formula

$$\varphi(q_1, q_2 \cdots q_n) = C\overline{W}^{2N} \prod_{i=1}^{n} q_i^{4N[mq_t + v_i]-1} (1-q_i)^{4N[m(1-q_t)+u_i]-1}$$

Sewall Wright

University of Chicago

THE DISTRIBUTION OF GENE FREQUENCIES IN POPULATIONS

By Sewall Wright

Department of Zoölogy, University of Chicago

Read before the Academy, April 26, 1937

The effects of the various evolutionary factors—mutation, cross breeding, selection and inbreeding—can be reduced to common terms by considering the rates of change which they tend to bring about in the relative frequencies of alleles.[1] In the absence of such factors, there is constancy of gene frequencies from the symmetry of the Mendelian mechanism.

The frequency (q) of a given gene changes at the rate $\Delta q = -uq$ per generation under recurrent mutation of the gene to alleles at the rate u. Mutation in the opposite direction at the average rate v per generation changes the gene frequency at the rate, $\Delta q = v(1 - q)$.

If a certain gene has the frequency q in a local population but q_t in the species as a whole, exchange of the proportion m of the local population with an equal number of random individuals from the whole species leads to change of gene frequencies in the former at the rate $\Delta q = -m(q - q_t)$. Cross breeding is, however, most likely to be with neighboring populations which differ but little in value of q. In this case the coefficient m is only a small fraction of the actual amount of exchange. There may be other complications such as selective immigration or emigration, but the above simple form will suffice here to illustrate cross breeding or migration pressure.

The simplest kind of selection is that in which the heterozygote is exactly half way between the two homozygotes in the extent per individual to which it contributes to the next generation. The selective value of zygotes (relative to a certain standard) will be designated w and the mean value for a population, $\bar{w}$.

Zygote	Frequency	w
AA	$(1 - q)^2$	1
AA'	$2q(1 - q)$	$1 - s$
$A'A'$	q^2	$1 - 2s$

$$\bar{w} = 1 - 2sq \qquad \frac{d\bar{w}}{dq} = -2s$$

$$\Delta q = \frac{(1 - s)q(1 - q) + (1 - 2s)q^2}{1 - 2sq} - q =$$

$$\frac{-sq(1 - q)}{1 - 2sq} = \frac{q(1 - q)}{2\bar{w}}\frac{d\bar{w}}{dq}. \qquad (1)$$

In the more general case in which $\bar{w}$ is not related linearly to q, the momentary selective advantage ($-2s$) of replacing A by A' is still given

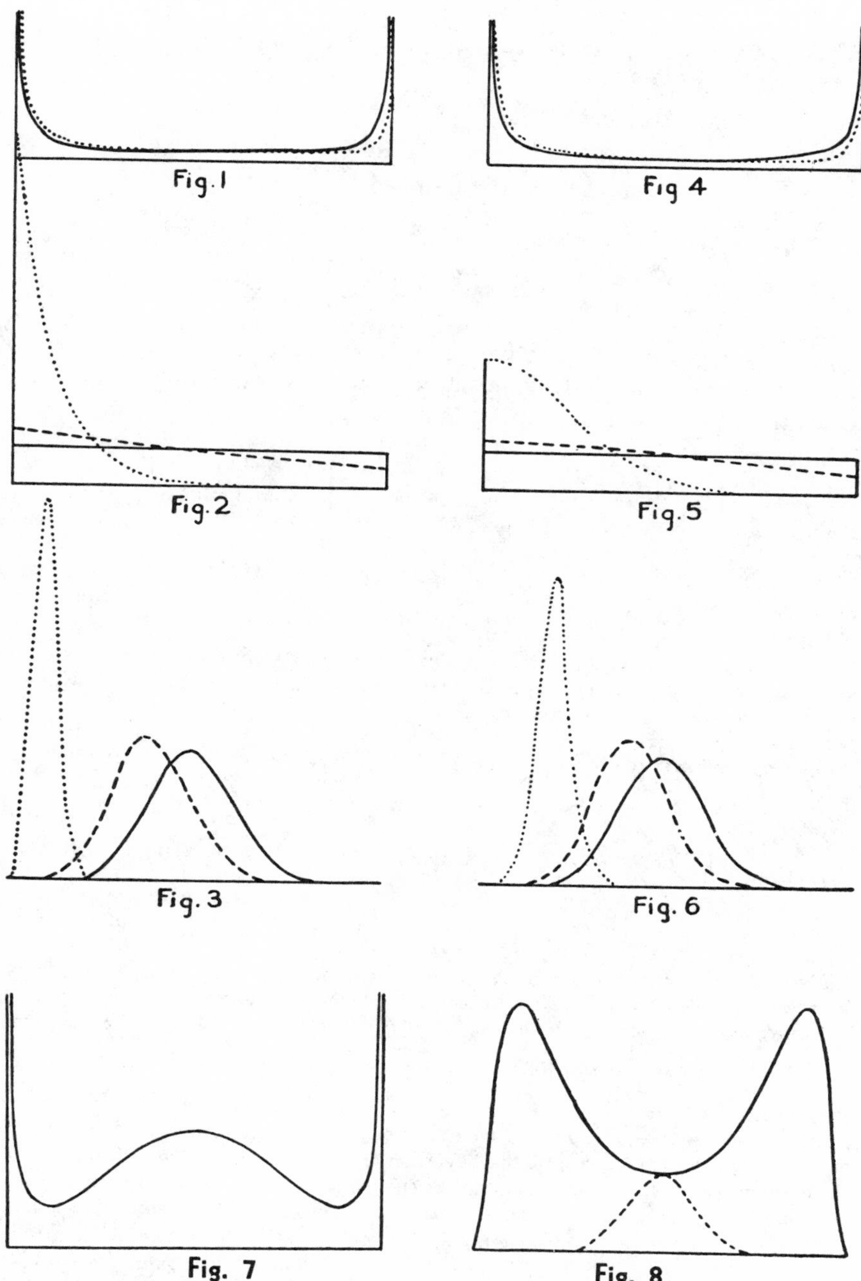

(Captions for figures on next page.)

by $d\bar{w}/dq$ and the value of Δq is given by the same formula as above in terms of w, $d\bar{w}/dq$ and q.

ZYGOTE	FREQUENCY	w	
AA	$(1-q)^2$	1	$\bar{w} = 1 - 2s_1q(1-q) - s_2q^2$
AA'	$2q(1-q)$	$1-s_1$	$\dfrac{d\bar{w}}{dq} = -2s_1 + 2(2s_1 - s_2)q$
$A'A'$	q^2	$1-s_2$	

$$\Delta q = \frac{q(1-q)}{2\bar{w}}\frac{d\bar{w}}{dq}. \tag{2}$$

Still more generally,[2] selective values depend on the interactions of the entire system of genes. It is the harmonious development of all characteristics that determines the success of an organism, not the absolute grades

CAPTIONS FOR FIGURES ON PRECEDING PAGE

Figures 1 to 3. Some of the forms taken by the distribution of gene frequencies in the case of no dominance. ($\varphi(q) = Ce^{4Nsq}q^{4Nv-1}(1-q)^{4Nu-1}$). Mutation rates are assumed constant and equal ($u = v$). Effective size of population is $N = \frac{1}{40v}$, $\frac{10}{40v}$ and $\frac{100}{40v}$ in figures 1, 2 and 3, respectively. In each case the solid line represents the least selection ($s = -v/10$), the broken line selection 10 times as severe (not represented in figure 1 since practically indistinguishable from the preceding) and the dotted line represents selection 100 times as severe.

Figures 4 to 6. Some of the forms taken by the distribution of frequencies of a completely recessive deleterious gene. $Ce^{2Ntq^2}q^{4Nv-1}(1-q)^{4Nu-1}$. Mutation rates are assumed constant and equal ($u = v$). Effective size of population is $N = \frac{1}{40v}$, $\frac{10}{40v}$ and $\frac{100}{40v}$ in figures 4, 5 and 6, respectively. In each case the solid line represents the least selection ($t = -v/5$), the broken line selection 10 times as severe (not represented in figure 4 since practically indistinguishable from the preceding) and the dotted line represents selection 100 times as severe.

Figure 7. One of the forms taken by the distribution of gene frequencies in the case in which there is no adaptive difference between the two homozygous types but the heterozygote is selected over both. $Ce^{4Nsq(1-q)}q^{4Nv-1}(1-q)^{4Nu-1}$. In the case represented, $u = v$, $N = \frac{1}{40v}$, $s = 100v$.

Figure 8. The frequencies along two diagonals of the joint distribution for two series of alleles with equal and additive effects on a character on which adverse selection acts according to the square of the deviation from the mean. The solid line shows the frequencies in populations along the line connecting the two favorable types $A_1A_1a_2a_2$ and $a_1a_1A_2A_2$. The broken line refers to the line connecting the extreme types $a_1a_1a_2a_2$ and $A_1A_1A_2A_2$. $\varphi(q) = C[1 - 2s[q_1(1-q_1) + q_2(1-q_2) + 2(q_1+q_2-1)^2]^{2N}q_1^{4Nv_1-1}(1-q_1)^{4Nu_1-1}q_2^{4Nv_2-1}(1-q_2)^{4Nu_2-1}$. In the case shown, $u_1 = v_1 = u_2 = v_2$; $N = 1/2v_1$, $s = 5v_1$. Along the favorable diagonal $q_1 = (1-q_2)$ the distribution is approximately $Ce^{-20q_1(1-q_1)}q_1^2(1-q_1)^2$. Along the unfavorable diagonal ($q_1 = q_2$) it is approximately $Ce^{-20(1-3q_1(1-q_1))}q_1^2(1-q_1)^2$.

of separate elementary characters, and still less its composition with respect to a single series of alleles. A gene that is favorable in one combination may be deleterious in another. However, if values of w are assigned to each possible combination, the rate of change of the frequency of a particular gene under selection (with specified values of all other gene frequencies) is given by the formula

$$\Delta q_i = \frac{q_i(1 - q_i)}{2\bar{w}} \frac{\partial \bar{w}}{\partial q_i}. \tag{3}$$

If the selection coefficients are small, $\bar{w}$ is close to 1, and the form $^1/_2 q_i(1 - q_i)\partial \bar{w}/\partial q_i$ is sufficiently accurate and is sometimes more convenient.

Two or more of the above factors are usually acting simultaneously. If the rates of change per generation are small, the net rate of change is given sufficiently accurately by the sum.

$$\Delta q = -uq + v(1 - q) - m(q - q_t) + \frac{q(1 - q)}{2} \frac{\partial}{\partial q} \log \bar{w}. \tag{4}$$

Gene frequency is in equilibrium (stable or otherwise) at any point at which $\Delta q = 0$. Opposing mutation pressures, for example, tend to maintain a stable equilibrium at the point $\hat{q} = v/(u + v)$. Mutation opposed by sufficiently strong genic selection $[\Delta q = v(1 - q) - sq(1 - q)]$ gives stable equilibrium at the point $\hat{q} = v/s$. Recessive mutation opposed by sufficiently strong selection $[\Delta q = v(1 - q) - sq^2(1 - q)]$ gives stable equilibrium at the point $\hat{q} = \sqrt{v/s}$. If there is mutation in both directions and sufficiently strong selection against the heterozygote $[\Delta q = -uq + v(1 - q) - s(1 - 2q)q(1 - q)]$ there are two points of stable equilibrium, and one of unstable equilibrium.

Migration pressure, if non-selective, may be written in the same form as mutation pressure, $\Delta q = -m(1 - q_t)q + mq_t(1 - q)$. The theory for migration effects can thus be obtained at any time from that for mutation merely by substituting $m(1 - q_t)$ for u and mq_t for v.

If the population is not indefinitely large, random changes occur in gene frequencies merely as a result of the accidents of sampling among the gametes. Letting N be the effective size of the breeding population, the sample of $2N$ gametes, necessary to replace it, will be distributed according to the expansion of $[(1 - q)A + qA']^{2N}$. The resulting standard deviation of q is $\sqrt{q(1 - q)/2N}$.

The changes in gene frequency due to accidents of sampling are, of course, not correlated in successive generations. Nevertheless, the variance of the probability array for q increases approximately with the number of generations until damped by the approach of q to 0 or 1.

The pressure toward a stable equilibrium point due to mutation, cross breeding and selection, and the divergent tendency due to inbreeding should between them determine a certain *distribution* of values of q which is in equilibrium. The central problem in the genetics of populations is that of finding this distribution under various conditions.

The first attempt at a solution was made by Fisher[3] who used a transformation of scale, $\theta = \cos^{-1}(1 - 2q)$, designed to give a uniform sampling variance, for all values of q. He attempted to express the conditions in a differential equation but reached erroneous conclusions. My first note on the subject was in 1929,[4] the detailed account appearing in 1931.[1] Fisher (1930)[5,6] after inspection of the latter paper in manuscript was able to correct his method so as to yield results in agreement in a number of special cases.

The method followed in the 1931 paper, referred to above, may be summarized as follows. A class of genes with frequency array $[(1 - q)A + qA']$ is distributed in the following generation according to the expansion $[(1 - q - \Delta q)A + (q + \Delta q)A']^{2N}$. The contribution to the class of genes with frequency array $[(1 - q_c)A + q_cA']$ is given by the term in the expansion relating to $2Nq_c A''$s, multiplied by the frequency (f) of the contributing class. The sum of such contributions from all classes of genes should restore the same frequency as in the preceding generation if the form of distribution is in equilibrium. The distinction between gene frequency (q) and frequency (f) of a gene frequency must be kept in mind.

$$f_c = \frac{\lfloor 2N}{\lfloor 2Nq_c \lfloor 2N(1 - q_c)} \sum_{q=0}^{1} [(q + \Delta q)^{2Nq_c}(1 - q - \Delta q)^{2N(1 - q_c)}f]. \quad (5)$$

Replacing summation by integration and letting $f = \varphi(q)/2N$ or $\varphi(q)dq$ according to position in the equation, the equation to be solved for $\varphi(q)$ is as follows

$$\varphi(q_c) = \frac{\Gamma(2N)}{q_c(1 - q_c)\Gamma(2Nq_c)\Gamma(2N(1 - q_c))} \int_0^1 (q + \Delta q)^{2Nq_c}(1 - q - \Delta q)^{2N(1 - q_c)}\varphi(q)dq. \quad (6)$$

If Δq is negligibly small—except for enough mutation from the homallelic classes ($q = 0$ or 1) to make equilibrium possible—it may easily be seen that the solution of this equation is

$$\varphi(q) = \frac{C}{q} + \frac{D}{1 - q}. \quad (7)$$

Only the symmetrical case, however, is in complete equilibrium with the homallelic classes, although the rate of change in other cases is extremely slow.

$$\varphi(q) = \frac{(l)}{2(\log 2N + .577)q(1 - q)} \tag{8}$$

when (l) is the proportion of heterallelic loci. The proportions in the homallelic loci are

$$f(0) = \frac{1}{4Nv} f(1/2N),\ f(1) = \frac{1}{4Nu} f(1 - 1/2N). \tag{9}$$

$$f(0) + l + f(1) = 1$$

If there is no mutation, there is equilibrium of form among the heterallelic classes when all are approximately equally frequent but falling off at the rate $1/2N$ per generation.

$$\varphi(q) = (l_0 e^{-T/2N}) \tag{10}$$

where l_0 is the initial proportion in heterallelic loci and T is the number of subsequent generations. The two preceding results were confirmed by Fisher[5,6] by his method.

If mutation is recurring at appreciable rates such that $\Delta q = -uq + v(1 - q)$, the distribution was shown to take the form

$$\varphi(q) = \frac{\Gamma(4Nu + 4Nv)}{\Gamma(4Nu)\Gamma(4Nv)} q^{4Nv - 1}(1 - q)^{4Nu - 1}. \tag{11}$$

With irreversible mutation, $\Delta q = v(1 - q)$, there is equilibrium of form, with falling off of all class frequencies at the rate v per generation

$$\varphi(q) = (l_0 e^{-vT})4Nvq^{4Nv - 1}. \tag{12}$$

With genic selection but very small mutation rates, $\Delta q = sq(1 - q)$, equilibrium of form was shown for the distribution

$$\varphi(q) = \frac{Ce^{4Nsq} + D}{q(1 - q)}. \tag{13}$$

The case

$$\frac{Ce^{4Nsq}}{q(1 - q)} \tag{14}$$

is that which is in complete equilibrium with the homallelic classes. Another case is the solution

$$\varphi(q) = \frac{C(e^{4Nsq} - 1)}{q(1 - q)} \tag{15}$$

given by Fisher[5,6] for irreversible mutation.

It was shown finally that the formula for mutation and genic selection combined, $\Delta q = -uq + v(1 - q) + sq(1 - q)$, could be written sufficiently accurately

$$\varphi(q) = Ce^{4Nsq}q^{4Nv-1}(1 - q)^{4Nu-1}. \tag{16}$$

The effects of cross breeding could be introduced in place of (or supplementary to) mutation by the substitution previously referred to.

Haldane[7] has criticized the conclusions derived from these formulae

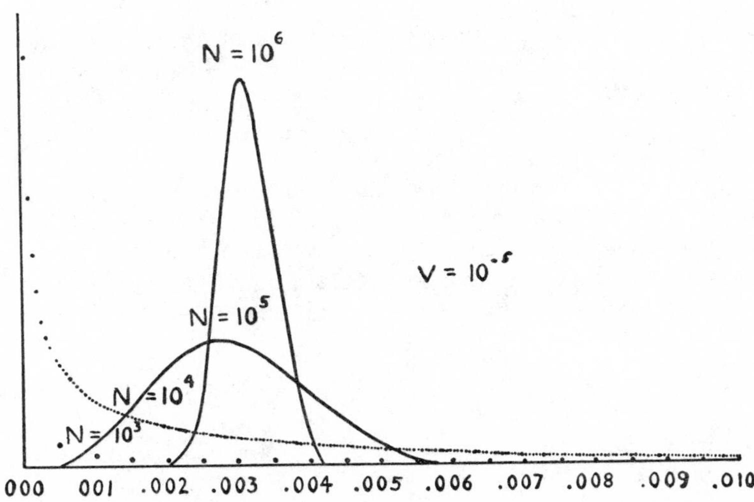

Figure 9. Some of the forms taken by the distribution of frequencies of a recessive lethal gene under different sizes of population. $C(1 - q^2)^{2N}q^{4Nv-1}(1 - q)^{-1}$. The mutation rate ($v$) is taken as 10^{-5}, giving a mean gene frequency of $\sqrt{v} = .0032$ in very large populations. This is approximately realized with $N = 10^6$. With $N = 10^5$, the gene is always present, but $\bar{q}$ is slightly reduced (.0030). In populations with effective size, $N = 10^4$, the gene is absent in about 15%, at any given moment and $q = .0020$. With $N = 10^3$, the gene is absent in 87%, and $\bar{q} = .0008$. The case of $N = 10^2$ is not shown as the gene is absent in about 99% and $\bar{q}$ is only .00026. With $N = 10$, the gene is absent in 99.9%, and $\bar{q} = .00008$. In selfed lines ($N = 1$) the gene is absent in 99.996% giving $\bar{q} = .00002$ ($= 2v$).

on the ground that only one type of selection is considered. It is undoubtedly important to extend these results to more general formulations of the action of selection. This has now been done by the same method as above for selection pressure of the form $\Delta q = (s + tq)q(1 - q)$ which applies to any degree of dominance, assuming s and t to be small. I have not been able to reduce the general solution to a simpler form than an infinite series, but in the case of complete equilibrium with the fixed classes there is reduction to the following

$$\varphi(q) = \frac{Ce^{4Nsq + 2Ntq^2}}{q(1 - q)}. \tag{17}$$

Still more general results can be obtained by another approach (presented as an alternative in the preceding paper in the case of mutation pressure). A frequency distribution, ranging from 0 to 1, in which mean and standard deviation do not change under evolutionary pressure and sampling errors, and whose mode (or modes) can be shown to be correct in the limiting case of large population size, must be a good approximation to the desired type. Let $\varphi(q)$ be any distribution of gene frequencies.

Then

$$\bar{q} = \int_0^1 q\varphi(q)dq \text{ (assuming that } \int_0^1 \varphi(q)dq = 1). \tag{18}$$

Accidents of sampling by themselves can have no effect in changing the mean, the mean of the distribution $[(1 - q)A + qA']^{2N}$ being $2Nq$. The effect of any evolutionary pressure in changing the mean is given by the expression

$$\int_0^1 \Delta q\varphi(q)dq. \tag{19}$$

If Δq is linear in q, and therefore of the form $\Delta q = -K(q - \hat{q})$ where $\hat{q}$ is the equilibrium point, the change of mean is

$$\int_0^1 \Delta q\varphi(q)dq = -K\int_0^1 (q - \hat{q})\varphi(q)dq = -K(\bar{q} - \hat{q}). \tag{20}$$

Thus if the mean of $\varphi(q)$ has reached the equilibrium point $(\bar{q} = \hat{q})$ there is no further change. Mutation pressure, $\Delta q = -uq + v(1 - q) = -(u + v)(q - \hat{q})$, comes under this head. The mean of the distribution of gene frequencies is always $\hat{q} = v/(u + v)$ at equilibrium, irrespective of size of population. Cross breeding also comes under this head, $\Delta q = -m(q - q_t)$. Selection pressure on the other hand, with Δq at least a quadratic function of q should give different values of $\bar{q}$ with change in size of population.

The variance of gene frequencies is also easily found in the case of linear evolutionary pressures. The pressure from both directions toward the equilibrium point must be balanced by the scattering effect of accidents of sampling if there is equilibrium in the form of distribution.

$$\sigma_q^2 = \int_0^1 (q - \bar{q})^2\varphi(q)dq = \int_0^1 (q + \Delta q - \bar{q})^2\varphi(q)dq + \int_0^1 \frac{q(1 - q)}{2N}\varphi(q)dq. \tag{21}$$

Inserting the value $\Delta q = -K(q - \bar{q})$ this reduces to

$$\sigma_q^2 = \frac{\bar{q}(1 - \bar{q})}{1 + 4NK - 2NK^2}. \tag{22}$$

The term in the denominator involving K^2, tracing to the term in (21) involving $(\Delta q)^2$ is negligible for small values of Δq.

Thus if $\Delta q = -K(q - \bar{q})$, the distribution of gene frequency has the mean, $\bar{q}$; variance, $\sigma_q^2 = \frac{\bar{q}(1 - \bar{q})}{1 + 4NK}$; a range limited at $q = 0$ and $q = 1$ and for large values of N condenses about a single mode at the equilibrium point, $q = \bar{q}$. In Pearson's system of frequency distributions, unimodal curves of limited range come under type I. Assuming a range limited at 0 and 1 and a unit area, the formula for type I is $\varphi(q) = \frac{\Gamma(x + y)}{\Gamma(x)\Gamma(y)} q^{x-1}(1 - q)^{y-1}$. Substituting this value of $\varphi(q)$ in the formulae for $\bar{q}$ and σ_q^2 gives $x = 4NK\bar{q}$, $y = 4NK(1 - \bar{q})$. In the case of mutation, $K = u + v$, $\bar{q} = v/(u + v)$.

$$\varphi(q) = \frac{\Gamma(4Nu + 4Nv)}{\Gamma(4Nu)\Gamma(4Nv)} q^{4Nv-1}(1 - q)^{4Nu-1}. \tag{23}$$

In the case of migration, $K = m$, $\bar{q} = q_t$

$$\varphi(q) = \frac{\Gamma(4Nm)}{\Gamma(4Nmq_t)\Gamma[4Nm(1 - q_t)]} q^{4Nmq_t-1}(1 - q)^{4Nm(1-q_t)-1}. \tag{24}$$

These are identical with the formulae derived by the preceding method (cf. 11).

When selection as well as mutation is at work we may write

$$\Delta q = Zq(1 - q) - uq + v(1 - q) \text{ where } Z = {}^1/_2\, d \log \bar{w}/dq. \tag{25}$$

A suggestion for the distribution formula can be obtained from the special case in which N is so large that there is very little spread from the equilibrium point (or points) at $\Delta q = 0$. The modes should approach the equilibrium points as N is increased. We will make the provisional assumption that the selection coefficients appear in a separate factor from those involving the mutation coefficients and that the desired formula is therefore of the type $\varphi(q) = \psi q^{4Nv-1}(1 - q)^{4Nu-1}$ where ψ is the required function of the selection coefficient. Putting $d \log \varphi(q)/dq = 0$ as a condition for any mode:

$$\frac{d}{dq}\log \varphi(q) = \frac{d}{dq}\log \psi + \frac{4Nv - 1}{q} - \frac{4Nu - 1}{1 - q} = 0. \tag{26}$$

Thus

$$\frac{q(1-q)}{4N}\frac{d}{dq}\log\psi + v(1-q) - uq + \frac{2q-1}{4N} = 0. \tag{27}$$

But at the equilibrium point,

$$Zq(1-q) + v(1-q) - uq = 0. \tag{28}$$

Ignoring the term $(2q-1)/4N$ which tends to disappear as N becomes large

$$Z = \frac{1}{4N}\frac{d}{dq}\log\psi. \tag{29}$$

$$\psi = Ce^{4N\int Zdq}. \tag{30}$$

The formula suggested on this basis is thus as follows:

$$\varphi(q) = Ce^{4N\int Zdq}q^{4Nv-1}(1-q)^{4Nu-1}. \tag{31}$$

This can be written as follows by evaluating $\int Zdq$

$$\varphi(q) = C\bar{w}^{2N}\, q^{4Nv-1}(1-q)^{4Nu-1}. \tag{32}$$

We will return to this form later; for the moment it will be convenient to use the following alternative form, easily derived from (25) and (31).

$$\varphi(q) = \frac{Ce^{4N\int \Delta q dq/q(1-q)}}{q(1-q)}. \tag{33}$$

So far this is merely a suggestion derived from a limiting case. It may be tested, however, for equilibrium in the general case. Testing first for shifting of the mean under evolutionary pressure (see 19)

$$\int_0^1 \Delta q\varphi(q)dq = \frac{C}{4N}\int_0^1 \frac{4N\Delta q}{q(1-q)}\, e^{\int 4N\Delta q dq/q(1-q)}dq =$$

$$\frac{C}{4N}\Big|_0^1 e^{4N\int Zdq}q^{4Nv}(1-q)^{4Nu}. \tag{34}$$

Thus there is no shifting of the mean if $4Nv > 0$, $4Nu > 0$.

The test (21) for balancing of the effects on variance of evolutionary pressure and accidents of sampling can be written as follows, omitting the negligible term in $(\Delta q)^2$.

$$2\int_0^1 q\,\Delta q\varphi(q)dq = -\frac{1}{2N}\int_0^1 q(1-q)\varphi(q)dq. \tag{35}$$

The left-hand member can be integrated by parts after substituting the suggested value of $\varphi(q)$ and shown to equal the right-hand member.

$$2C\int_0^1 \frac{q\,\Delta q}{q(1-q)}\, e^{4N\int \Delta q dq/q(1-q)}dq =$$

$$\frac{C}{2N}\int_0^1 qd(e^{4N\int \Delta q dq/q(1-q)}) = \frac{C}{2N}\Big|_0^1 e^{4N\int Z dq}q^{4Nv+1}(1-q)^{4Nu} -$$

$$\frac{C}{2N}\int_0^1 e^{4N\int \Delta q dq/q(1-q)}dq = -\frac{1}{2N}\int_0^1 q(1-q)\varphi(q)dq. \qquad (36)$$

The condition of no change in variance is thus met to the same degree of approximation as in the case of mutation. The special case of mutation and genic selection, $\Delta q = -uq + v(1-q) + sq(1-q)$ gives $\varphi(q) = Ce^{4Nsq}q^{4Nv-1}(1-q)^{4Nu-1}$, identical with that (16) derived by the previous more exhaustive method. This holds even in the limiting case (14) in which $\Delta q = sq(1-q)$ but as already noted equilibrium requires that there be some mutation even though the rates are so low that they may be ignored in Δq. The formula (17), obtained by the more exhaustive method in the case of more or less dominance, $\Delta q = (s + tq)q(1-q)$ is also in agreement with the present result.

The effects of certain differences in severity of selection and in effective size of population on the distribution of gene frequencies, assuming no dominance, are illustrated in figures 1 to 3. Figures 4 to 6 make similar comparisons for the case of a completely recessive unfavorable gene. These figures can also illustrate the joint effects of selection, cross breeding and inbreeding in local populations by replacing u and v by $^1/_2m$.

One of the forms taken by the distribution when there is equal selection against both homozygotes in favor of the heterozygote is illustrated in figure 7. With sufficiently smaller population size or sufficiently less intense selection, the distribution would become U-shaped. With a sufficiently larger population size it would become I-shaped about a mean frequency, $q = 0.5$. Increased severity of selection, without increase in size of population, would also pile up the frequencies about this point.

In a large population, mutation opposed by moderately severe selection tends to hold the deleterious gene at a low frequency, $\hat{q} = v/s$ if semi-dominant, $\hat{q} = (v/s)^{1/2}$ if completely recessive. In a sufficiently small sample from such a population, selection of the same degree of severity becomes ineffective and the mean frequency of the deleterious gene gradually rises to the equilibrium point due to opposing mutation pressures $\hat{q} = v/(u+v)$. If unfavorable mutation is much more frequent than the reverse $(v > u)$ this may mean a shift to approximate fixation of the deleterious gene. The rate of approach to the new mean, after a reduction in size of population is, however, extremely slow, being dependent on mutation pressure.

In the case of a deleterious *recessive* factor, the immediate effect of reduction in size of population is indeed the reverse of that indicated above, though the final effect, a rise in mean frequency of the deleterious gene toward the point $q = v/(u + v)$ occurs at length in this case as well as when dominance is lacking (provided that populations in which the deleterious gene is fixed can persist). The immediate decrease in frequency on reduction of size of population is illustrated in figure 9 in the extreme case of a recessive lethal in which case there can, of course, be no secondary rise in frequency. Taking the dominant as the type, $\bar{w} = 1 - q^2$, $\Delta q = v(1 - q) - q^2/(1 + q)$ giving the approximate equilibrium point $q = v^{1/2}$. The distribution is

$$\varphi(q) = C(1 - q^2)^{2N} q^{4Nv - 1}(1 - q)^{-1}. \qquad (37)$$

The mean, $\bar{q}$, can easily be expressed in Γ functions (on substituting x for q^2). It reduces approximately to $\bar{q} = \dfrac{\Gamma(2Nv + 1/2)}{\sqrt{2N}\,\Gamma(2Nv)}$. For values of $2Nv$ larger than 1, this is close to $v^{1/2}$ i.e., gene frequency varies about the equilibrium point. If on the other hand $2Nv$ is a small fraction, q is approximately $v(2\pi N)^{1/2}$ which means a great reduction in frequency of lethals in populations as the effective size of inbreeding units falls below $1/2v$.

So far we have derived formulae for distributions only where the selection coefficients are constant. But as already noted, it is really the system of gene frequencies that is more or less adaptive, not the isolated genes. No adequate picture of the evolutionary process can be made without taking factor interaction into account.

The momentary selection pressure in cases of factor interaction was given in (3), giving as the distribution of gene frequencies

$$\varphi(q_i) = C\bar{w}^{2N} q_i^{4Nv_i - 1}(1 - q_i)^{4Nu_i - 1}. \qquad (38)$$

Here $\bar{w}$ is the mean selective value with q_i variable, but a specified set of values for the other gene frequencies.

The joint frequency surface must be such that on assigning specified values to all of the q's but one, the distribution for that one is that given by (38). The joint distribution

$$\varphi(q_i, q_2, \ldots q_n) = C\bar{w}^{2N}\Pi_{i=1}^{n} q_i^{4Nv_i - 1}(1 - q_i)^{4Nu_i - 1} \qquad (39)$$

when $\bar{w}$ is the mean selective value in terms of all of the q's as variables, satisfies this condition and is thus the desired form.

As an example consider the case of a character for which the grade of development depends on the additive effects of multiple factors, lacking dominance, but for which the selective value falls off as the square of its deviation from an optimum. The selective value has been shown to be as follows:[2]

$$\bar{w} = 1 - K[2\Sigma\alpha_i^2 q_i(1 - q_i) + (M - O)^2]$$

where α_i is the effect of gene A_i with frequency q_i; $M(= 2\Sigma\alpha_i q_i)$ is the mean and O is the optimal grade.

The multidimensional surface, $\bar{w}$, has in general many peaks, separated by shallow saddles. The shallowest saddles are those between optimal combinations differing only in two pairs of factors, e.g., $A_1A_1a_2a_2$...... and $a_1a_1A_2A_2$.......

The nature of the distribution along and across such a saddle is illustrated in figure 8 in the case of two pairs of factors with equal effects. With smaller N the distribution from $A_1A_1a_2a_2$ to $a_1a_1A_2A_2$ becomes U-shaped. With larger N or weaker selection it becomes I-shaped about $q = 0.5$. Stronger selection pushes the modes toward the favored homallelic types.

The evolutionary implications will not be discussed here in detail. For the most part the present results merely put the conclusions previously reached[1,8,9] on a more definite basis. These conclusions may be summarized briefly as follows.

In large freely interbreeding populations with no secular change in conditions of life for long periods of time, all gene frequencies approach equilibrium at a certain peak $\bar{w}$, not necessarily the highest peak. Under secular change in conditions the surface $\bar{w}$ itself changes and there is evolutionary change in the system of gene frequencies, following the changes in position of the controlling peak. Evolution here may be said to be guided by intragroup selection.

In sufficiently small completely isolated populations, the random divergencies of gene frequencies from their equilibrium values become important, tending to bring about approximate fixation of some random combination of genes which is not likely to be a peak combination. The result is a largely nonadaptive differentiation. In extreme cases there may be the deterioration which characteristically follows excessive inbreeding. Isolation may here be considered the dominating evolutionary factor.

In a large population subdivided into numerous small, partially isolated groups, the combination of directed and random divergencies in gene frequencies, associated with intergroup selection, gives a trial and error mechanism under which the system of gene frequencies may pass from lower to higher peak values of $\bar{w}$ and the species may evolve continuously even without secular changes in conditions (although this process, occurring in all species, itself tends to bring about such secular changes). The combination of partial isolation of subgroups with intergroup selection seems to provide the most favorable conditions for evolutionary advance.

Mutation is always a factor in providing material for evolution but may be said to dominate the course of evolution only in so far as mutants appear which are fertile *inter se* but largely infertile with the parent type, i.e., when mutation is itself an isolating factor.

[1] Wright, S., *Genetics*, **16**, 97–159 (1931).

[2] Wright, S., *Jour. Genetics*, **30**, 243–256 (1934).

[3] Fisher, R. A., *Proc. Roy. Soc. Edinburgh*, **42**, 321–341 (1922).

[4] Wright, S., *Amer. Nat.*, **63**, 556–561 (1929).

[5] Fisher, R. A., *The Genetical Theory of Natural Selection*, Oxford, Clarendon Press, 272 pp. (1930).

[6] Fisher, R. A., *Proc. Roy. Soc. Edinburgh*, **50**, 205–220 (1930).

[7] Haldane, J. B. S., *The Causes of Evolution*, New York and London, Harper Brothers, 235 pp. (1932).

[8] Wright, S., *Proc. Sixth Internat. Congress of Genetics*, **1**, 356–366 (1932).

[9] Wright, S., *Jour. Genetics*, **30**, 257–266 (1935).

19
Size of Population and Breeding Structure in Relation to Evolution

Science 87, no. 2263 (1938):430–31

INTRODUCTION

In this paper Wright introduced for the first time the concept of isolation by distance, while emphasizing in general the importance of population size and breeding structure in relation to the evolutionary process.

Size of population and breeding structure in relation to evolution: SEWALL WRIGHT. Size of population plays an important rôle in evolutionary theory. The effective size (N) of the theory, may, however, differ much from the apparent size, being usually much less. N obviously refers only to the breeding population. If the numbers (N_m, N_f) of mature males and females are different, N depends mainly on the less numerous sex.

$$\left(N = \frac{4\,N_m\,N_f}{N_m + N_f}\right).$$

The surviving offspring are likely not to be derived at random from the parental generation. With N_o parents furnishing varying numbers (k) of gametes to a next generation of equal size ($\overline{K} = 2$),

$$N = \frac{4\,N_o - 2}{2 + \sigma_k^2}.$$

Of greater probable importance in nature are cyclic variations in numbers. In a cycle of not too long a period (in generations) the effective size

$$\left(N = \frac{n}{\sum_{i=1}^{n} \frac{1}{N_i}}\right).$$

is controlled largely by the phase of small numbers. A small N permits random fixation of non-adaptive characters and to some extent control by mutation pressure. In a large species, restrictions on interbreeding may permit differentiation of local populations. The variance of gene frequencies (σ_q^2) takes the value $q_t(1 - q_t)\,f$, where q_t is the mean gene frequency in the species and f is the inbreeding coefficient. In a population distributed continuously over a large area, but with mates always drawn from small groups (size N) the value of f for groups separated by n generations of ancestry (or by $\sqrt{n}$ diameters of the unit area), lies between

$\frac{\Sigma}{2N+\Sigma}$ and $\frac{\Sigma}{2N-\Sigma}$ where $\sum = \sum_{x=1}^{n}(1/X)$. This permits considerable fluctuating local differentiation where N is less than a few hundred but leads to approximate fixation of differences only if N is much smaller. In a species, whose range is essentially one dimensional, Σ has the value $\sum_{x=1}^{n} \sqrt{1/X}$. Differentiation increases much more rapidly with distance than in the preceding case. Another mode of attack is appropriate where the range is subdivided into partially isolated territories. As shown previously σ_q^2 here takes the form $q_t(1-q_t)/4N_m+1)$ where N is the effective size of the local group and m the effective proportion of immigrants from the species as a whole. Both N and m may be much smaller than indicated by actual numbers and amounts of cross breeding with neighboring groups. If small enough, there is random non-adaptive differentiation of local groups. With small m, but not N, there is adaptive differentiation in respects related to differential conditions. These processes may be expected to be supplemented by intergroup selection such that those local groups which happen to acquire combinations of characters of more than local adaptive significance multiply relatively rapidly and supply more than their share of emigrants. The simultaneous action of partial isolation and intergroup selection should result in a more rapid evolutionary process than either isolation alone or intragroup selection alone. Splitting of species requires nearly complete isolation. In some cases (as where translocations become fixed) there is evidence of fixation against very strong selection, likely to occur (in a sexually reproducing species) only if there are numerous outlying territories in which the populations are so isolated and so liable to extinction that the lines of continuity frequently pass through single stray individuals.

20
The Distribution of Gene Frequencies under Irreversible Mutation

Proceedings of the National Academy of Science 24, no. 7 (1938):253–59

21
The Distribution of Gene Frequencies in Populations of Polyploids

Proceedings of the National Academy of Science 24, no. 9 (1938):372–77

22
The Distribution of Self-Sterility Alleles in Populations

Genetics 24 (July 1939):538–52

Introduction

These three papers are attempts by Wright to assess the consequences of various genetic assumptions upon the statistical distribution of genes that he had derived in 1937. The third of these, stimulated by Sterling Emerson's data on distributions of self-sterility alleles in small populations of the desert plant *Oenothera organensis*, led in the 1950s and 1960s to a lively controversy between Wright, Fisher, Ewens, Kimura, and others (see *SW&EB*, 488–91, and Wright's summary in *E&GP* 2:402–16).

THE DISTRIBUTION OF GENE FREQUENCIES UNDER IRREVERSIBLE MUTATION

By Sewall Wright

Department of Zoölogy, The University of Chicago

Communicated June 13, 1938

Under reversible mutation, the frequency (q) of a gene, subject to systematic evolutionary pressure (Δq) and to the accidents of sampling in a limited population (N diploid individuals), varies according to a certain distribution ($\varphi(q)$) discussed in previous papers.[1,2,3] If mutation is irreversible, the distribution curve for such genes should attain constancy of form, but all class frequencies should fall off at a uniform rate (K) as genes drift irreversibly into fixation. The purpose of the present paper is to broaden somewhat the treatment in this latter case.

As previously shown[2] the rate of fixation is approximately half the frequency, $f(1 - 1/2N)$, in the subterminal class. Thus with

$$\int_0^1 \varphi(q)dq = 1$$

$$K = 1/2\, f(1 - 1/2N) = \frac{\varphi(1 - 1/2N)}{4N} \quad \text{approximately.} \qquad (1)$$

The changes in the mean ($\bar{q} = \int_0^1 q\varphi(q)dq$) and the variance ($\sigma_q^2 = \int_0^1 (q - \bar{q})^2\varphi(q)dq$) of gene frequencies, due to fixation in one generation, may be expressed as follows in terms of the systematic evolutionary pressure, Δq, and the variation due to sampling, $\sigma_{\Delta q}^2$

$$\int_0^1 \Delta q\varphi(q)dq = K(1 - \bar{q}). \qquad (2)$$

$$\int_0^1 (q + \Delta q - \bar{q})^2\varphi(q)dq + \int_0^1 \sigma_{\Delta q}^2\varphi(q)dq = (1 - K)\sigma_q^2 + K(1 - \bar{q})^2. \qquad (3)$$

The latter can be reduced to following, ignoring a negligible term in $(\Delta q)^2$

$$2\int_0^1 (q-\bar{q})\,\Delta q\varphi(q)dq + \int_0^1 \sigma^2_{\Delta q}\varphi(q)dq = K[(1-\bar{q})^2 - \sigma^2_q]. \quad (4)$$

If the conditions are such that under equilibrium, with reverse mutation at an indefinitely low rate, there is no important accumulation of genes in the class $q = 1$, complete irreversibility of mutation should make no appreciable difference in the form of distribution. The demonstration of the formula for this case ($K = 0$) can be put in a very simple form.

$$\text{Let} \quad \int \Delta q\varphi(q)dq = \chi(q). \quad (5)$$

Equations (2) and (4) can be written as follows, putting $K = 0$,

$$\chi(1) - \chi(0) = 0. \quad (6)$$

$$\int_0^1 \chi(q)dq - [\bar{q}\chi(0) + (1-\bar{q})\chi(1)] - \frac{1}{2}\int_0^1 \sigma^2_{\Delta q}\varphi(q)dq = 0. \quad (7)$$

Equation (7) is obviously satisfied by the following

$$\chi(q) - [\bar{q}\chi(0) + (1-\bar{q})\chi(1)] - \frac{1}{2}\sigma^2_{\Delta q}\varphi(q) = 0. \quad (8)$$

This means little, until it is shown that (8) also satisfies (6).

As there can be no sampling variance in homallelic populations, $\sigma^2_{\Delta q} = 0$ if $q = 0$ or $q = 1$. Equation (8) reduces to (6) if $q = 0$ or $q = 1$ and both $\varphi(0)$ and $\varphi(1)$ are finite. Equation (8), therefore, satisfies the condition of constancy of the mean as well as that of constancy of the variance.

$$\varphi(q) = \frac{2[\chi(q) - \chi(1)]}{\sigma^2_{\Delta q}}. \quad (9)$$

$[\chi(q) - \chi(1)]$ can be evaluated as follows, using (5) and (9),

$$d \log [\chi(q) - \chi(1)] = \frac{d\chi(q)}{\chi(q) - \chi(1)} = \frac{2\,\Delta q\varphi(q)dq}{\sigma^2_{\Delta q}\varphi(q)}, \quad (10)$$

$$\chi(q) - \chi(1) = \frac{C}{2}e^{2\int\frac{\Delta q dq}{\sigma^2_{\Delta q}}}. \quad (11)$$

$$\varphi(q) = \frac{Ce^{2\int(\Delta q/\sigma^2_{\Delta q})dq}}{\sigma^2_{\Delta q}}. \quad (12)$$

This is the desired expression for the distribution in terms of Δq and $\sigma^2_{\Delta q}$.

Putting $\sigma^2_{\Delta q} = \dfrac{q(1-q)}{2N}$, its value in a population of N diploid individuals, it is the same as the formula given previously[3] (except for the value of the coefficient C).

If K is not zero, this method does not appear to lead to a usable general expression which satisfies (1), (2) and (4). An expression for the rate of decay can, however, be obtained from (2). Let v be the rate of mutation, and let $\overline{W}$ be the mean selective value of genotypes. As shown previously[3]

$$\Delta q = v(1 - q) + q(1 - q)\frac{d \log \overline{W}}{2dq}. \qquad (13)$$

From (2)

$$K = v + \frac{1}{2(1 - \bar{q})}\int_0^1 q(1 - q)\varphi(q) d \log \overline{W}. \qquad (14)$$

If there is no selection ($\overline{W} = 1$)

$$K = v. \qquad (15)$$

The distribution (16) for this case was obtained in a previous paper[2] by substitution of $\Delta q = v(1 - q)$ in what is equation (18) of the present paper. It can easily be verified that (16) also satisfies (1), (2) and (4).

$$\varphi(q) = 4Nvq^{4Nv-1}. \qquad (16)$$

The most important cases are probably those in which the size of population is so small that recurrent mutation has no effect on the form of distribution (v much less than $1/4N$). The selection pressure for any degree of dominance can be written sufficiently accurately

$$\Delta q = (s + tq)q(1 - q). \qquad (17)$$

The condition that the frequency of any class of gene frequencies, q_c, be reconstructed after each generation, except for a uniform decay at rate K can be represented as follows,[2] using $p = 1 - q$ for brevity.

$$(1 - K)\varphi(q_c) = A \int_0^1 (q + \Delta q)^{2Nq_c}(p - \Delta q)^{2Np_c}\varphi(q)dq$$

where

$$A = \frac{\Gamma(2N)}{p_c q_c \Gamma(2Np_c)\Gamma(2Nq_c)}. \qquad (18)$$

If v is so small that practically all genes are fixed in the class $f(0)$, K may be ignored in determining the form of the distribution for unfixed genes. Substituting the value of Δq from (17)

$$\varphi(q_c) = A \int_0^1 q^{2Nq_c}p^{2Np_c}[1 + p(s + tq)]^{2Nq_c}[1 - q(s + tq)]^{2Np_c}\varphi(q)dq. \qquad (19)$$

The following approximations may be used

$$[1 + p(s + tq)]^{2Nq_c} = e^{2Nq_c p(s+tq)}[1 - Nq_c p^2(s + tq)^2]. \qquad (20)$$

$$[1 - q(s + tq)]^{2Np_c} = e^{-2Np_cq(s+tq)}[1 - Np_cq^2(s + tq)^2]. \quad (21)$$

The product of expressions (20) and (21) is approximately

$$e^{2Ns(q_c-q)+Nt(q^2{}_c-q^2)-Nt(q_c-q)^2}\left\{1 - N(s + tq)^2[p_cq_c + (q_c - q)^2]\right\}. \quad (22)$$

Since the random deviations of q have the variance, $\sigma^2_{\Delta q} = \frac{pq}{2N}$, $(q_c - q)^2$ is of the order $1/2N$. The term $N(s + tq)^2(q_c - q)^2$ is thus negligibly small compared with $N(s + tq)^2p_cq_c$ which itself is as small a term as it is necessary to consider. The former may thus be ignored. The constant and variable gene frequencies are separable in the exponential term in (22) except in the term $e^{-Nt(q_c-q)^2}$. The exponent in this case is smaller than $-t$ and the term can be written $[1 - Nt(q_c - q)^2]$ with sufficient accuracy. Equation (19) can now be written as follows:

$$\varphi(q_c) = Ae^{2Nsq_c + Ntq^2{}_c}\int_0^1 q^{2Nq_c}p^{2Np_c}e^{-2Nsq-Ntq^2}\,[1 - N(s + tq)^2p_cq_c - Nt(q_c - q)^2]\varphi(q)dq. \quad (23)$$

Let
$$\varphi(q) = \frac{e^{2Nsq + Ntq^2}}{q\,(1-q)}\,(C_0 + C_1q + C_2q^2 \ldots). \quad (24)$$

This entirely eliminates the exponential terms, leaving (23) in a form which can be solved, using the following sufficiently accurate formula in which terms of the order $\frac{1}{N^2}$ are ignored.

$$\frac{\Gamma(2N)}{\Gamma(2Np_c)\Gamma(2Nq_c)}\int_0^1 q^{2Nq_c-1+x}\,p^{2Np_c-1}\,dq = q_c^x\left[1 - \frac{x(x-1)}{4N}\right] + q_c^{x-1}\left[\frac{x(x-1)}{4N}\right]. \quad (25)$$

The resulting coefficients of the powers of q_c on the right side may be equated to those on the left side leading to the following general expression (in which $C_{-1} = C_{-2} = C_{-3} = 0$).

$$C_m = \frac{m(m+1)}{4N}\,C_{m+1} + C_m - \frac{C_{m-1}}{2}\,(2Ns^2 + t) - C_{m-2}\,(2Nst) - C_{m-3}\,Nt^2. \quad (26)$$

The higher coefficients can all be expressed in terms of C_0 and C_1 and substituted in (24). By letting $C_1 = 2NsC_0 + D$ the terms for which C_0 is the coefficient can be condensed into exponential form. It will be convenient to substitute C for C_0.

$$\varphi(q) = \frac{e^{2Nsq + Ntq^2}}{q(1-q)}\left[Ce^{2Nsq + Ntq^2} + Dq\psi(q)\right], \tag{27}$$

where

$$\psi(q) = 1 + \frac{(2Nsq)^2}{\underline{|3}} + \frac{(2Nsq)^4}{\underline{|5}} + \frac{(2Nsq)^6}{\underline{|7}} \cdots \tag{28}$$

$$+ (2Ntq^2)\left(\frac{1}{\underline{|3}} + \frac{(2Nsq)}{\underline{|3}} + \frac{2(2Nsq)^2}{\underline{|5}} + \frac{2(2Nsq)^3}{\underline{|5}} + \frac{3(2Nsq)^4}{\underline{|7}} \cdots\right)$$

$$+ (2Ntq^2)^2\left(\frac{7}{\underline{|5}} + \frac{2(2Nsq)}{\underline{|5}} + \frac{69(2Nsq)^2}{\underline{|7}} \cdots\right)$$

$$+ (2Ntq^2)^3\left(\frac{27}{\underline{|7}} \cdots\right) + \cdots$$

If $D = 0$

$$\varphi(q) = \frac{Ce^{4Nsq + 2Ntq^2}}{q(1-q)}. \tag{29}$$

This is the case of equilibrium under reversible mutation or irreversible mutation opposed by sufficiently strong selection as can be seen by substituting $\Delta q = (s + tq)q(1-q)$, $\sigma^2_{\Delta q} = \frac{q(1-q)}{2N}$ in (12). They agree except in the coefficient.

The case of irreversible mutation with fixation occurring at a low rate, can be found from (1) and (2), assuming that nearly all genes are in one of the homallelic classes. It should be noted that the formula for $\varphi(q)$ only applies where K is of lower order than $1/2N$, $2Ns^2 + t$ in (26). Mutations to the class $q = 1/2N$ contribute the amount $2Nvf(0) = \frac{1}{2}f(1/2N)$ and these contribute to the change of mean by the amount $f(1/2N)/4N$. The mean, however, must be so low that the term $K(1 - \bar{q})$ may be written K sufficiently accurately. The following relations are all approximate:

$$\int^1 \Delta q\varphi(q)dq + \frac{f(1/2N)}{4N} = K = \frac{f(1 - 1/2N)}{2}. \tag{30}$$

$$f(1/2N) = C\left[1 + \frac{1}{2N} + 2s\right] + \frac{D}{2N}. \tag{31}$$

$$f(1 - 1/2N) = C[e^{4NS + 2Nt}(1 + 1/2N - 2s - 2t)] + D[e^{2Ns + Nt}(1 - s - t)\psi(1 - 1/2N)]. \tag{32}$$

$$\int_0^1 \Delta q \varphi(q) dq = C\left[\frac{e^{4Ns+2Nt}}{4N} - \frac{1}{4N}\right] + \frac{Ds}{2} + \frac{Dt}{3}. \qquad (33)$$

Substituting (31), (32), (33) in (30) and substituting $\psi(1)$ for $\psi(1 - 1/2N)$ (leading terms in difference, $t/3$, $\frac{2Ns^2}{3}$)

$$D = -\frac{Ce^{2Ns+Nt}}{\psi(1)}. \qquad (34)$$

With practically all genes in the class $q = 0, f(0) = 1 = \frac{f(1/2N)}{4Nv} = \frac{C}{4Nv}$

Thus $C = 4Nv$ approximately.

$$\varphi(q) = \frac{4Nve^{4Nsq+2Ntq^2}}{q(1-q)}\left[1 - \frac{e^{2Ns(1-q)+Nt(1-q^2)}q\psi(q)}{\psi(1)}\right]. \qquad (35)$$

For sufficiently small values of Ns and Nt we may take $\psi(q) = 1 + 1/3Ntq^2$ and represent the exponentials by the first two terms of their expansions. The following shows how the hyperbolic distribution $4Nv/q$ is modified by weak selection (s positive for favorable mutation, negative for unfavorable mutation).

$$\varphi(q) = \frac{4Nv}{q}[1 + 2Nsq + 2/3Ntq(2q - 1)]. \qquad (36)$$

The rate of fixation (K) of genes can be found from the left member of (30) (in which inaccuracies in the evaluation of D have less effect than in the right member).

$$K = v\left[e^{4Ns+2Nt} - \frac{(2Ns + 4/3Nt)e^{2Ns+Nt}}{\psi(1)}\right]. \qquad (37)$$

This reduces to $v(1 + 2Ns + 2/3Nt)$ for such small values of Ns and Nt as implied in (36). It is to be noted that irreversible mutation should ultimately lead to fixation of the mutant even when opposed by selection (s negative) but the rate is exceedingly slow unless Ns and Nt are small.

In the special case of genic selection ($t = 0$)

$$\psi(q) = \frac{e^{2Nsq} - e^{-2Nsq}}{4Nsq}. \qquad (38)$$

$$\varphi(q) = \frac{4Nv}{(1 - e^{-4Ns})}\frac{[1 - e^{-4Ns(1-q)}]}{q(1-q)}. \qquad (39)$$

An essentially similar derivation of this formula has been given previously by the author[2] and a different one by Fisher.[4] In this case $K = \frac{4Nvs}{1 - e^{-4Ns}}$ approaching $v(1 + 2Ns)$ as $4Ns$ decreases.

The question of the chance of fixation of an individual mutation must be distinguished from the rate of fixation (K) under recurrent mutation. The chance of fixation is given by the ratio $f(1 - 1/2N)/f(1/2N) = K/2Nv$. In the case of no dominance, this gives $2s/(1 - e^{-4Ns})$ or approximately $2s$ for favorable mutations occurring in a large population, in agreement with Fisher.[4] For indifferent factors it is $1/2N$. Unfavorable mutations have a chance of fixation $2s/(e^{4Ns} - 1)$ but this is small unless $4Ns$ is small.

The results presented here bear on the possibility of a course of evolutionary change determined by mutation pressure, a process which at first sight seems the most obvious implication of modern genetics. The possibility does indeed exist but requires either an almost complete indifference of the mutation with respect to adaptive value or else a very small effective size of population over a long period of time. The most important case in which mutation pressure seems likely to be a major factor is that of extreme degeneration or elimination of organs that have ceased to be useful.[5,6] The degeneration of the eyes and loss of pigment of cave forms is an example of a case in which the conditions make it especially probable that mutation pressure is a real factor. In all of these cases, however, the likelihood that various direct and indirect effects of selection may also play a rôle should not be ignored.[5]

It should be noted that while the average rate of fixation of irreversible mutations is low, the large element of chance with respect to *which* mutations become fixed in each particular case makes this a greater factor in the diversification of small isolated populations than is at first apparent. Indeed there may be much diversification of gene frequencies among such populations under conditions in which there is no appreciable systematic tendency toward fixation of the sort investigated here.

[1] Wright, S., *Amer. Nat.*, **63**, 556–561 (1929).

[2] Wright, S., *Genetics*, **16**, 97–159 (1931).

[3] Wright, S., *Proc. Nat. Acad. Sci.*, **23**, 307–320 (1937).

[4] Fisher, R. A., *Proc. Roy. Soc. Edinburgh*, **50**, 205–220 (1930).

[5] Wright, S., *Amer. Nat.*, **63**, 274–279 (1929).

[6] Haldane, J. B. S., *Ibid.*, **67**, 5–19 (1933).

THE DISTRIBUTION OF GENE FREQUENCIES IN POPULATIONS OF POLYPLOIDS

By Sewall Wright

Department of Zoölogy, The University of Chicago

Communicated August 12, 1938

The theoretical distribution of gene frequencies in populations of diploid organisms, as affected by mutation, selection, migration and inbreeding, has been discussed in previous papers.[1,2] It is desirable to extend the theory to populations of polyploids.

Mutation and Migration.—Assume gene frequencies $[(1 - q)A + q\,a]$ where a is the gene under consideration and A represents its array of alleles (without any implications with respect to dominance). Zygotic frequencies rapidly approach an equilibrium under random mating, the equilibrium distribution in $2k$-ploids being the expansion of $[(1 - q)\,A + q\,a]^{2k}$(Haldane[3]).

Mutation pressure is the same as with diploids. Letting u be the rate of mutation from a, and v the rate of mutation to a, the rate of change of gene frequency per generation is

$$\Delta q = v(1 - q) - uq. \tag{1}$$

Crossbreeding pressure also is the same as with diploids. Letting q_t be the gene frequency of the species as a whole and m the effective amount of exchange between the local population in question and the species as a whole we have:

$$\Delta q = -m(q - q_t). \tag{2}$$

Any results for mutation pressure can be transformed into ones for crossbreeding pressure by substituting $m(1 - q_t)$ for u and mq_t for v. For simplicity, we will develop conclusions for mutation pressure only.

Selection Pressure.—The simplest kind of selection is that in which the selective values (W) of the various zygotes deviate from standard in proportion to the number of replacements of type genes (the case of no dominance).

ZYGOTE	FREQUENCY	W
A^{2k}	$(1 - q)^{2k}$	1
$A^{2k-1}a$	$2kq(1 - q)^{2k-1}$	$1 - s$
......	......	......
a^{2k}	q^{2k}	$1 - 2ks$

$$\bar{W} = 1 - 2ksq. \tag{3}$$

$$\Delta q = -\frac{sq(1-q)}{1-2ksq} = \frac{q(1-q)d\overline{W}}{2k\overline{W}\,dq}. \tag{4}$$

In the more general case in which $\overline{W}$ is not related linearly to q because of more or less dominance, the momentary selective advantage of replacing A by a is still $\frac{d\overline{W}}{dq}$ and the formula for Δq is still given by (4).

Still more generally, if there is interaction between different series of alleles, and a selective value, W, is assigned each combination, the rate of change of the frequency of a particular gene under selection (with specified values of all other gene frequencies) is given by the formula

$$\Delta q = + \frac{q(1-q)}{2k\overline{W}} \cdot \frac{\partial\overline{W}}{\partial q}. \tag{5}$$

Inbreeding.—The effect of restriction in the effective size of population can be worked out by means of path coefficients.[1,4,5] In a $2k$-ploid the zygote may be considered as equally and completely determined by the $2k$ component genes.

Let y be the path coefficient measuring the contribution of gene to zygote and let f be the correlation between genes due to inbreeding. Expressing the complete determination of zygote by genes in an equation:

$$2ky[y + (2k-1)yf] = 1.$$

$$y^2 = \frac{1}{2k[1 + (2k-1)f]}. \tag{6}$$

The path coefficient (x) expressing the contribution of the zygote to a single gene of a gamete must equal the correlation coefficient as there is only one connecting path.

$$x = y + (2k-1)yf$$

$$x^2 = \frac{1 + (2k-1)f}{2k}. \tag{7}$$

$$xy = \frac{1}{2k}. \tag{8}$$

The zygote may also be considered as equally and completely determined by the two uniting gametes and each of these by its component genes. From this point of view, there are two kinds of correlation between component genes: between genes of the same gamete (c) and between genes of different gametes (d). The average correlation (f) is thus:

$$f = \frac{(k-1)c + kd}{2k-1}. \tag{9}$$

Under the simplest theory of segregation in polyploids (Muller[6]) the correlation between genes of the same gamete must be the same as between two *different* genes of the parental zygote. Using primes to designate preceding generations

$$c = f'. \tag{10}$$

There is a possibility (greatest for loci remote from the spindle fibre) that a gamete in a polyploid may contain more than one representative of the same parental gene (Haldane[3]) but the possible modification of the results due to this complication cannot be great and will not be considered here.

The correlation between genes of uniting gametes can be obtained by tracing the connecting paths according to the system of mating.

The proportion of unlike pairs of genes among pairs drawn from zygotes will be represented by p. In diploids this is the percentage of heterozygosis.[1] The relation of p to f is easily found by constructing the correlation table for f. For the present purpose two A's are considered like genes (two non-a's) even though they may actually be different alleles of a.

	A	a	
a	$p/2$	$q - p/2$	q
A	$1 - q - p/2$	$p/2$	$1 - q$
	$1 - q$	q	1

$$f = 1 - \frac{p}{2q(1 - q)}. \tag{11}$$

Self-Fertilization.—In this case $d = x'^2$. From this result and (9), (10) and (7)

$$f = \frac{1 + (4k - 3)f'}{2(2k - 1)}. \tag{12}$$

$$p = \left[\frac{4k - 3}{4k - 2}\right] p'. \tag{13}$$

Following are special cases:

Diploid	$p = {}^1/_2 p'$	Hexaploid	$p = {}^9/_{10} p'$
Tetraploid	$p = {}^5/_6 p'$	Octoploid	$p = {}^{13}/_{14} p'$

Haldane[3] has studied the effects of self-fertilization by a direct consideration of all types of zygotes. He represents the symmetrical case in tetraploids as $p_nA^4 : q_nA^3a : r_nA^2a^2 : q_nAa^3 : p_na^4$ where $2p_n + 2q_n + r_n = 1$, $p_1 = q_1 = 0, r_1 = 1$.

He shows that the following relations hold

$$p_{n+1} = p_n + {}^1/_4 q_n + {}^1/_{36} r_n$$
$$q_{n+1} = {}^1/_2 q_n + {}^2/_9 r_n$$
$$r_{n+1} = {}^1/_2 q_n + {}^1/_2 r_n.$$

The proportion of unlike pairs of genes in zygotes (p of the present paper) is $q + ({}^2/_3)r$. It may easily be seen that $(q_{n+1} + {}^2/_3 r_{n+1}) = {}^5/_6(q_n + {}^2/_3 r_n)$ or $p = {}^5/_6 p'$ in our terminology.

Haldane has similarly derived the consequences of self-fertilization in hexaploids. Our formula, $p = {}^9/_{10} p'$ is in agreement.

GROUPS OF N MONOECIOUS INDIVIDUALS

Random Self-Fertilization.—Extension can easily be made to larger inbreeding groups. In the case of N monoecious individuals with self-fertilization at random the chance of selfing is $\frac{1}{N}$ and of union of gametes from different individuals is $\frac{N-1}{N}$. Giving due weight to these two possibilities

$$d = \frac{1}{N} x'^2 + \frac{N-1}{N} (4k^2 x'^2 y'^2) d'$$
$$= \frac{1}{2Nk} [1 + (2N-1)(2k-1)f' - (2N-2)(k-1)f''] \quad (14)$$

$$p = \frac{1}{2N(2k-1)} [(6Nk - 4N - 2k + 1)p' - (2N-2)(k-1)p'']. \quad (15)$$

This reduces to $p = \left(\frac{4k-3}{4k-2}\right)p'$ if $N = 1$ (self-fertilization) but if N is large it is approximately $p = \left(1 - \frac{1}{2Nk}\right) p'$.

No Self-Fertilization.—With N monoecious individuals and self-fertilization excluded, the chance of matings between siblings is $\frac{2}{N(N-1)}$, between half-siblings is $\frac{4(N-2)}{4(N-1)}$ and of more remote matings

$$\frac{(N-2)(N-3)}{N(N-1)}.$$

Using these weights:

$$d = \frac{1}{N(N-1)} (x'^2 + d') + \frac{N-2}{N(N-1)} (x'^2 + 3d') + \frac{(N-2)(N-3)}{N(N-1)} d'. \quad (16)$$

$$p = \frac{1}{2N(2k-1)}[(6Nk - 4N - 4k + 2)p' - (2Nk - 2N - 4k + 3)p'']. \quad (17)$$

This also reduces approximately to $p = \left(1 - \frac{1}{2Nk}\right) p'$ if N is large.

Mating of Siblings.—If $N = 2$ in the preceding case (mating of siblings)

$$p = p' - \frac{1}{8k-4}(2p' - p''). \quad (18)$$

With constant size of population it is legitimate to put $p/p' = p'/p''$ to find the limiting relation between successive generations.

$$p = \left[\frac{4k - 3 + \sqrt{16k^2 - 16k + 5}}{8k - 4}\right] p' \text{ approximately.} \quad (19)$$

Following are special cases:

	EXACT	LIMITING
Diploid	$p = p'/2 + p''/4$	$p = 1/4(1 + \sqrt{5})p' = 0.80902p'$
Tetraploid	$p = 5/6p' + p''/12$	$p = 1/12(5 + \sqrt{37})p' = 0.92356p'$
Hexaploid	$p = 9/10p' + p''/20$	$p = 1/20(9 + \sqrt{101})p' = 0.95249p'$
Octoploid	$p = 13/14p' + p''/28$	$p = 1/28(13 + \sqrt{197})p' = 0.96556p'$

Bartlett and Haldane[7] gave the limiting result in tetraploids in a form equivalent to $p = 0.92356p'$, from the pertinent solution of an octic equation derived from the iteration equations for all possible types of mating. This agrees with our result.

Sampling Variance.—In a population of N $2k$-ploids with gene frequencies $[(1 - q)A + qa]$ and *random* association of genes in the gametes (i.e., $p = 2q(1 - q)$) the variance of q among progeny populations, resulting from random samples of $2N$ gametes is

$$\sigma^2_{\Delta q} = \frac{q(1-q)}{2Nk}. \quad (20)$$

Unfortunately the association of genes in gametes is not wholly a random one. For example, a tetraploid $AAaa$ produces gametes in the proportions $1/6\,AA : 4/6\,Aa : 1/6\,aa$ (if A is near the spindle fibre) while under random association the proportions would be $1/4\,AA + 2/4\,Aa + 1/4aa$. However, the departure from (20) is slight except in very small populations. In the extreme case of self-fertilization ($N = 1$) it may easily be shown that $\sigma^2_{\Delta q} = \frac{q(1-q)}{4k-2}$.

The Distribution of Gene Frequencies.—A formula for the distribution of gene frequencies in populations subject to evolutionary pressure Δq and sampling variance $\sigma^2_{\Delta q}$ was reached in a previous paper.[8]

$$\varphi(q) = (C/\sigma^2_{\Delta q})e^{2\int(\Delta q/\sigma^2_{\Delta q})dq}. \qquad (21)$$

The approximate formula for the distribution of q in $2k$-ploids is thus as follows, letting $\Delta q = v(1-q) - uq + \dfrac{q(1-q)}{2k\bar{W}}\dfrac{d\bar{W}}{\partial q}$ from combination of (1) and (4) and letting $\sigma^2_{\Delta q} = q(1-q)/2Nk$.

$$\varphi(q) = C\bar{W}^{2n}\, q^{4nkv-1}(1-q)^{4nku-1}. \qquad (22)$$

The joint distribution for a system of multiple genes can be written as follows, letting $\bar{W}$ here be the mean selective value of the population in terms of all gene frequencies as variables.

$$\varphi(q_1 \ldots q_n) = C\bar{W}^{2n}\Pi[q_i^{4nk_iv_i-1}(1-q_i)^{4nk_iu_i-1}]. \qquad (23)$$

It is to be noted that this applies to aneuploids (k variable) as well as to euploids (k constant).

[1] Wright S., *Genetics*, **6,** 111–178 (1921).
[2] Wright S., *Proc. Nat. Acad. Sci.*, **23,** 307–320 (1937).
[3] Haldane, J. B. S., *Jour. Genetics*, **22,** 359–372 (1930).
[4] Wright, S., *Proc. Nat. Acad. Sci.*, **19,** 411–420 (1933).
[5] Wright, S., *Ann. Math. Stat.*, **5,** 161–215 (1934).
[6] Muller, H. J., *Amer. Nat.*, **48,** 508–512 (1914).
[7] Bartlett, M. S., and J. B. S. Haldane, *Jour. Gen.*. **29,** 175–180 (1934).
[8] Wright, S., *Proc. Nat. Acad. Sci.*, **24,** 253–259 (1938).

THE DISTRIBUTION OF SELF-STERILITY ALLELES IN POPULATIONS

SEWALL WRIGHT
The University of Chicago, Chicago, Illinois

Received February 18, 1939

THE phenomenon of self-sterility is known to be very common among higher plants. The genetic mechanism in most cases studied seems to be that discovered by EAST and MANGELSDORF in *Nicotiana alata* and *N. Sanderae*. A single series of alleles, $S_1, S_2 \cdots$, is involved. The growth of a pollen tube is so much inhibited in a style whose cells contain the same allele that it does not reach the embryo sac. It is obvious that all individuals must be heterozygous and that a population cannot persist with less than 3 S-alleles. It is also fairly obvious that selection would tend to increase the frequency of any additional alleles that may appear. DR. STERLING EMERSON has been investigating a series of at least 37 self-sterility alleles of this sort, found in *Oenothera organensis*, a species in which the entire population apparently consists of less than one thousand individuals, scattered in small groups among certain canyons of the Organ Mountains of New Mexico (EMERSON 1938, 1939). I am indebted to him for calling my attention to the interesting questions involved in the effects of selection and inbreeding in this case.

SELECTION PRESSURE

We assume the existence of a series of n self-sterility alleles, $S_1, S_2 \cdots S_n$, with frequencies $q_1, q_2 \cdots q_n$ such that $\Sigma q = 1$ in a population of N diploid individuals. The frequencies of zygotes containing one of these (S_i) must be $2q$ (with the appropriate subscript) since all are heterozygotes. The frequency of functioning S_i female gametes is q, assuming no differential selection. The frequency of functioning S_i pollen grains is not in general the same. S_i pollen has, by hypothesis, no chance of functioning in the styles of zygotes containing S_i, but has a better than average chance in zygotes that lack S_i (frequency $1-2q$) since each zygote of this class inhibits pollen of two of the other kinds. Assume that on non-S_i styles, the ratio of successful S_i pollen grains to successful ones of other types is as $q:R(1-q)$. The total frequency of functioning S_i pollen is then $q(1-2q)/[q+R(1-q)]$. The average frequency of functioning S_i gametes is $q(1-q)(1+R)/2[q+R(1-q)]$ and the change from the previous generations is therefore

$$\Delta q = \frac{q(1-q)(1+R)}{2[q+R(1-q)]} - q = \frac{q}{2}\left[\frac{(1-R)-q(3-R)}{R+q(1-R)}\right].$$

Letting $\hat{q}$ represent the point at which selection has no effect ($\Delta q = 0$),

$$\hat{q} = \frac{1 - R}{3 - R}, \qquad R = \frac{1 - 3\hat{q}}{1 - \hat{q}}, \qquad \Delta q = -\frac{q(q - \hat{q})}{1 - 3\hat{q} + 2\hat{q}q}.$$

If all non-S_i alleles have the same frequency, it is obvious that R will have the value $(n-3)/(n-1)$ since each non-S_i style permits functioning of only $n-3$ of the $n-1$ types of non-S_i pollen grains. This leads to a formula for the frequency of functioning S_i gametes, $q(1-q)(n-2)/(n-3+2q)$ communicated to me by DR. EMERSON for this case. But the frequencies of the various alleles may be expected to drift at random about the equilibrium point as a result of the accidents of sampling. The average value of R will consequently be somewhat less than $(n-3)/(n-1)$ (unless $n=3$, $R=0$) since pollen grains with the more abundant alleles will be inhibited in excess. We will assume that R may be treated with sufficient accuracy as constant in populations that have long been subject to the same conditions (number of individuals, mutation rate in the S series, rate of cross breeding, etc.).

The fluctuating variations in R due to chance variations in the relative frequencies among the non-S_i alleles are not an important qualification but variations correlated with differences in the frequency of S_i itself introduce a systematic error. If there are only four alleles altogether, it is obvious that the three non-S_i alleles may be expected to be somewhat more nearly equal in frequency as a result of selection, if S_i is rare, than if it is common; and that R will consequently be negatively correlated with, q. The upward pressure on rare alleles being approximately $q(1-R)/2R$ the assumption that R is constant slightly overestimates the upward selection pressure on rare alleles carried by a population under given conditions. This effect rapidly decreases in importance as n becomes greater than 4.

MUTATION AND MIGRATION PRESSURES

Mutation from S_i at the rate u per generation, and to it at the rate v, changes gene frequency at the rate $v(1-q)-uq$. Crossbreeding between the population in question and the species as a whole, such that the proportion m of the gametes come from the latter, changes gene frequency at the rate $-m(q-q_t)$ where q_t is the frequency of S_i in the species as a whole. The net rate of change of gene frequency and its sampling variance are as follows:

$$\Delta q = \frac{q(1 - q)(1 + R)}{2[R + q(1 - R)]} - [1 + u + m(1 - q_t)]\, q + (v + mq_t)(1 - q)$$

$$\sigma^2_{\Delta q} = \frac{q(1 - q)}{2N}.$$

THE DISTRIBUTION FORMULA

The general formula for the distribution of gene frequencies determined by any systematic evolutionary pressure Δq, and random variations measured by $\sigma^2_{\Delta q}$, has been shown to be as follows (WRIGHT 1938).

$$\phi(q) = \frac{Ce^{2\int (\Delta q/\sigma^2_{\Delta q})dq}}{\sigma^2_{\Delta q}}.$$

Substituting the values of Δq and $\sigma^2_{\Delta q}$ found for the present case we have

$$\phi(q) = C[R + q(1 - R)]^{2N(1+R)/(1-R)} q^{4N(v+mq_t)-1}(1 - q)^{4N[1+u+m(1-q_t)]-1}.$$

The simplest special case is that of three alleles with negligible rates of mutation and cross breeding. In this case $R=0$ since S_2S_3 can be fertilized only by S_1 pollen. Thus $\hat{q} = \frac{1}{3}$, $\Delta q = (1-3q)/2$.

$$\phi(q) = \frac{\Gamma(6N)}{\Gamma(2N)\Gamma(4N)} q^{2N-1}(1 - q)^{4N-1}.$$

The mean and variance of this distribution are easily found.

$$\bar{q} = \int_0^1 q\phi(q)dq = \frac{1}{3}$$

$$\sigma^2_q = \int_0^1 (q - \bar{q})^2\phi(q)dq = \frac{2}{9(6N + 1)}.$$

The mode is at the point at which

$$\frac{d \log \phi(q)}{dq} = 0$$

$$\text{Mode} = \frac{2N - 1}{6N - 2}.$$

These formulae can easily be generalized to allow for mutation and crossbreeding but not for the presence of a larger number of alleles. However, if values of N, u, v, m, q_t and R are assumed, the ordinates of the distribution may be calculated as multiples of C. The value C required to make the total frequency one can then be found after estimating the sum of all frequencies. The mean ($\bar{q}$) and other statistics of the distribution can also be calculated empirically. Having found $\bar{q}$, the number of alleles (n) present under the conditions is given by the reciprocal ($\bar{q} = \Sigma q/n = 1/n$). The proportion of the alleles lost in each generation by the accidents of sampling can be estimated from the frequencies of the low gene frequencies $1/2N, 2/2N \cdots$, by use of the Poisson series, a question discussed in the next section. At equilibrium, losses must be balanced by accessions of new alleles by mutation or outbreeding. If the values of u, v or m, so calculated, do not agree with those originally assumed and these are large enough to affect appreciably the form of the distribution, it becomes necessary to try other values of R until, by repeated trial and interpolation, agreement is reached.

THE RELATIONS BETWEEN MUTATION RATE, POPULATION SIZE AND NUMBER OF ALLELES

It has been shown previously (WRIGHT 1931) that the rate of loss of alleles is usually about half the frequency of the subterminal class, thus, $K=\frac{1}{2}f(1/2N)=(1/4N)\phi(1/2N)$. In the present case, however, there may be such strong selection pressure tending to increase the frequency of rare alleles that this formula requires modification. It is indeed obvious that it is impossible to lose an allele if only three are present without extinction of the population. If in this case q_1 happens to fall very low at any time (population nearly 100 percent S_2S_3), q_1 automatically rises almost to the opposite extreme ($q_1=.50$) in the next generation. For larger values of n, Δq may be taken as $\hat{q}\, q/(1-3\hat{q})$ where q is small. As $\hat{q}$ is close to $\bar{q}(=1/n)$ except in extreme cases, we may take $\Delta q=q/(n-3)$ as a rough approximation for small q and $n>3$. The chance of loss with a given small value of q is $e^{-2N(q+\Delta q)}$ or $e^{-2Nq(n-2)/(n-3)}$ approximately, making allowance for selection. On testing different extreme types of distribution (such as $\phi(q)=1$, $\phi(q)=C/q$), it may be concluded that the fraction $(n-3)/2(n-1)$ gives with sufficient accuracy for our present purpose, the portion of the subterminal class lost, (0 if $n=3$, 1/6 if $n=4$, $\frac{1}{4}$ if $n=5$, approaching the typical value $\frac{1}{2}$ as n becomes large).

$$K=\frac{(n-3)}{2(n-1)}f(1/2N)\text{ approximately.}$$

Assume now that a limited number (n') of alleles is possible, of which n are present at any moment. The chance that any given allele, absent from the population, will arise in the next generation by mutation is $2Nv$, if v is the rate of mutational origin of the allele in question. Assuming this to be the same for all of the n' alleles, the total number of missing alleles that will appear in a generation is $2N(n'-n)v$. Equating this to the number of losses, Kn, we must have at equilibrium

$$2N(n'-n)v=\frac{n(n-3)}{2(n-1)}f(1/2N).$$

If u is the rate of mutation *from* one allele to all others, $u=(n'-1)v$

$$u=\frac{n(n-3)(n'-1)}{4N(n-1)(n'-n)}f(1/2N).$$

Figure 1 shows the number of alleles (n) maintained in a population of size N by various mutation rates u, assuming that 25 alleles are possible.

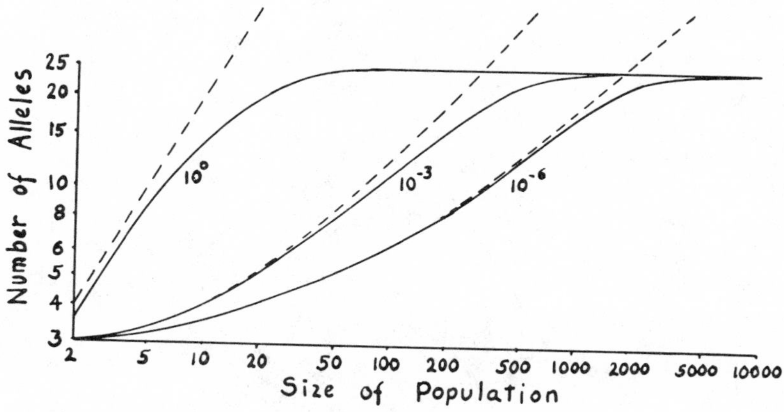

FIGURE 1.—The number of self-sterility alleles maintained in populations of various sizes if 25 different alleles are possible and mutation (or replacement by immigration) is occurring at the rates 10^{-3} or 10^{-6} per generation. The limiting case, complete replacement, is indicated by the curve labelled 10^0. The broken lines show the number of alleles if an indefinitely large number are possible.

If new alleles are being introduced into a population to a significant extent by outbreeding rather than by mutation, corresponding equations can be written to express the balance of losses and gains. It is merely necessary to substitute mq_t for v, $m(1-q_t)$ for u and $n_t(=1/q_t)$, the number of alleles in the species as a whole, for n'.

If the number of possible alleles is indefinitely great it is possible that each mutation will represent a wholly novel allele (v indefinitely small). In this case $(n'-n)/(n'-1)$ approaches 1.

$$u = \frac{n(n-3)}{4N(n-1)} f(1/2N).$$

Figure 2 shows the number of alleles (n) maintained in a population of size N by the occurrence of novel mutations at the rate u per generation. Figures 3, 4 and 5 compare the distributions of frequencies in populations of 50 and 500 individuals, at mutation rates of 10^{-2}, 10^{-4}, and 10^{-6}, respectively. The values of the selection index R, the mean number of representatives of an allele ($2N\bar{q}$), the mean number of alleles (n) and the turnover (K) are given in table 1.

COMPARISON WITH EMERSON'S DATA IN OENOTHERA ORGANENSIS

In EMERSON'S data, 34 alleles were demonstrated in a sample of 135 gametes from a population estimated as less than 1000 plants and very

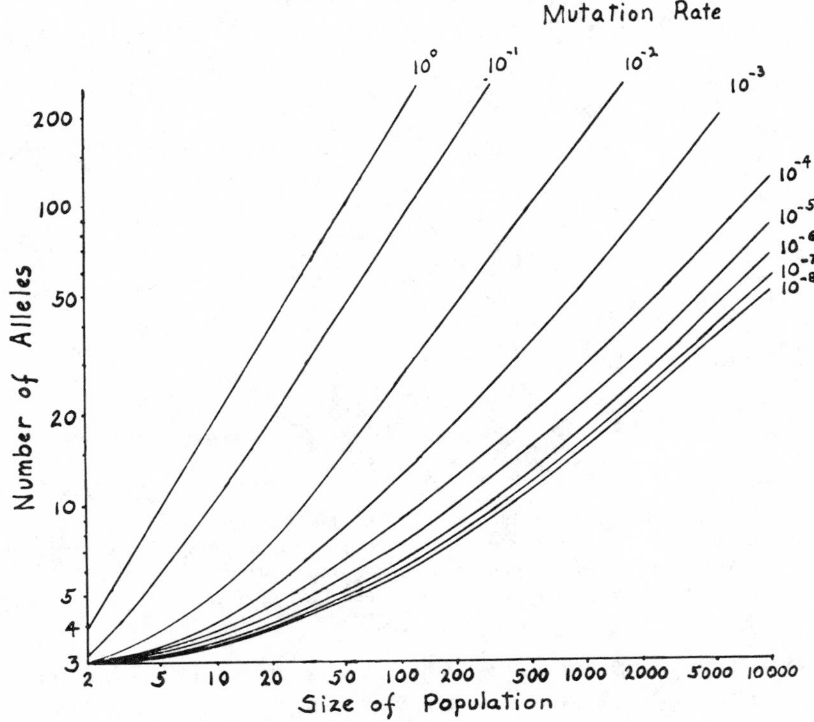

FIGURE 2.—The number of self-sterility alleles maintained in populations of various sizes if the number of possible alleles is indefinitely great and mutation (or replacement by immigration from a very large population) occurs at rates from 10^0 to 10^{-8}.

TABLE I

Comparison of populations of 50 and of 500 individuals at three different rates of mutation (or of immigration from an indefinitely large population).

N NO. OF INDIVIDUALS	u MUTATION RATE	R SELECTION INDEX	$2N\bar{q}$ MEAN NO. OF REPRESENTATIVES OF ALLELE	n NO. OF ALLELES	K FACTOR TURNOVER PERCENT
50	.01	.77	6.8	14.8	7
500	.01	.95	10.4	96.0	10
50	.0001	.59	15.4	6.5	0.15
500	.0001	.87	51.3	19.5	0.51
50	.000001	.49	19.5	5.1	0.002
500	.000001	.82	77.7	12.9	0.008

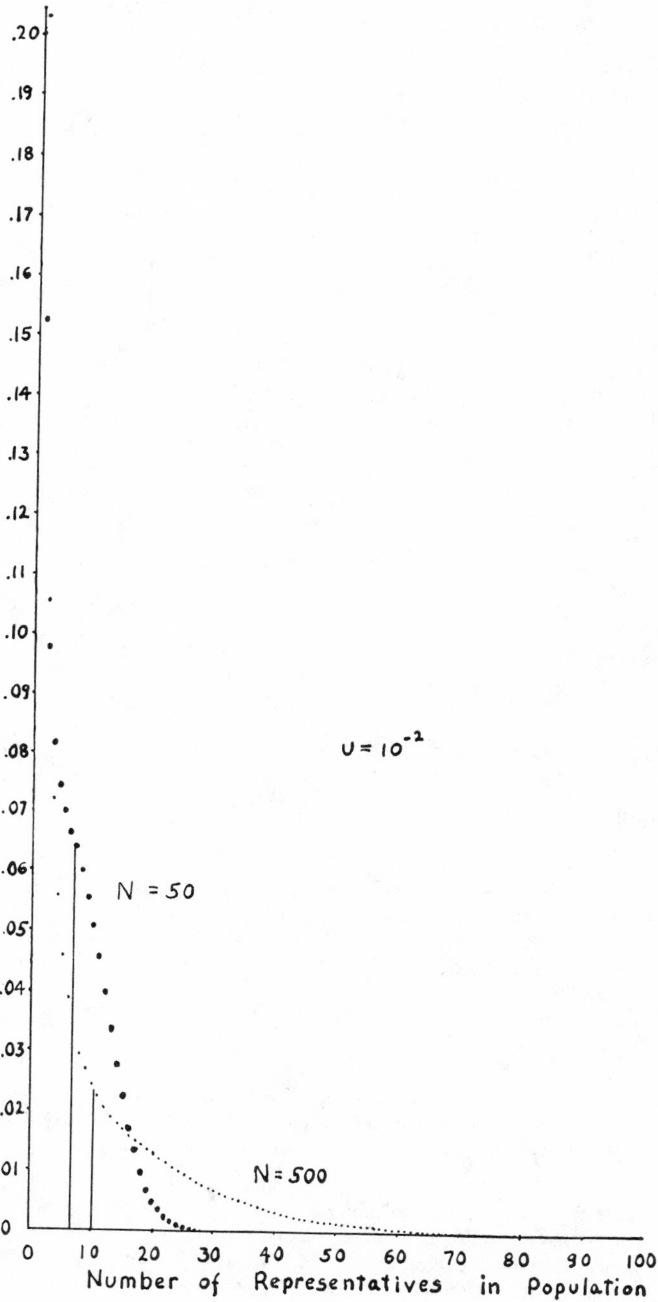

FIGURE 3.—Description under figures 4 and 5.

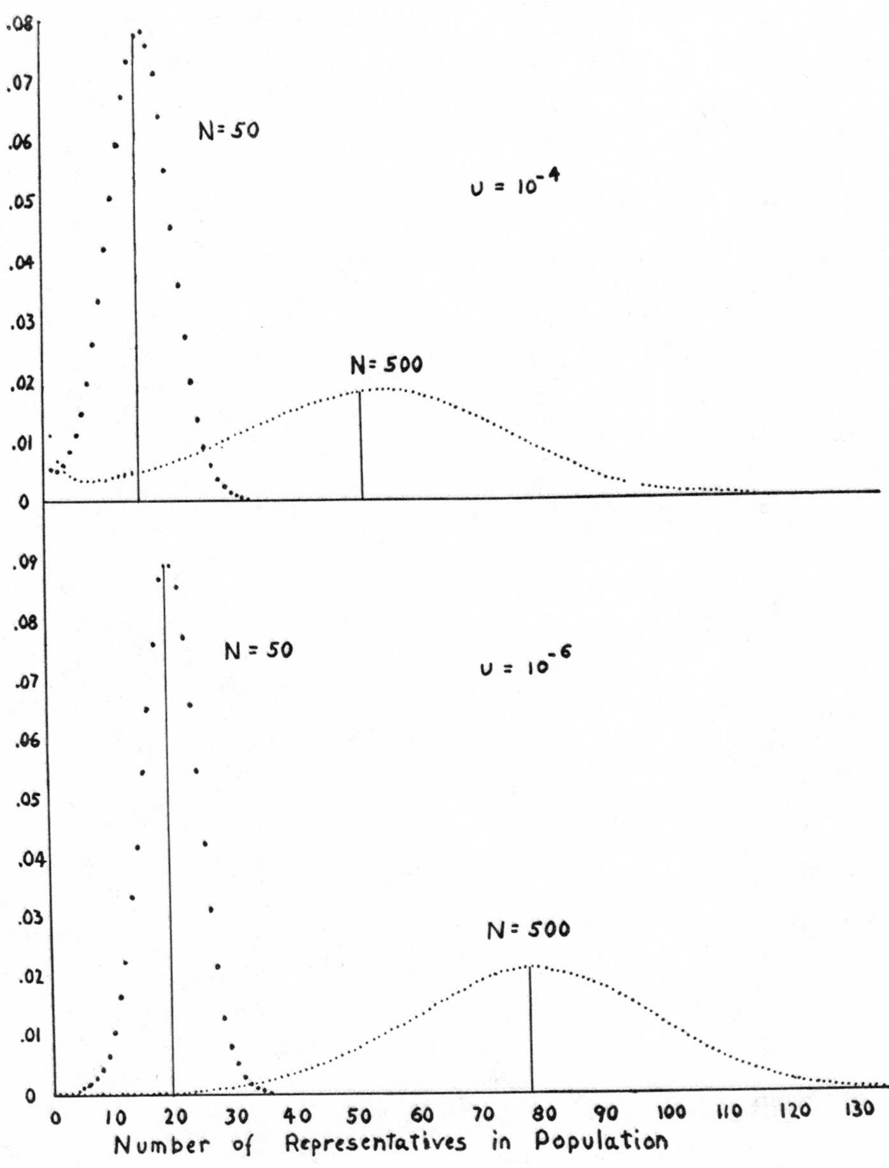

FIGURES 3 to 5.—The probabilities of different numbers of representatives ($2Nq$) of a self-sterility gene (while present at all) compared in populations of 50 and 500 individuals and with mutation (or replacement by immigration from a very large population) at rates of $u=10^{-2}$ (fig. 3), $u=10^{-4}$ (fig. 4, above) or $u=10^{-6}$ (fig. 5, below). An indefinitely large number of alleles is assumed possible. See table 1.

likely less than 500. It is known that several additional alleles are present in the population and doubtless the number would be considerably increased in a sample of size $2N$. From figure 2, it is apparent that it would require a mutation rate of considerably more than 10^{-3} to maintain such a large number of alleles. Mutation rates of 10^{-5} or less would maintain less than 15 alleles. PROFESSOR EMERSON informs me that in preliminary tests no mutations have been found in 45,000 pollen grains, indicative as far as it goes of a mutation rate decidedly too low to account for the observed number of alleles.

One possible explanation would be that the size of the population is much greater than estimated from the data now at hand or, if not greater now, that it has recently been much greater, with loss of alleles at too slow a rate to have reached equilibrium. Inspection of figure 2 indicates that a population of some 4000 to 5000 is required to account for the probable number of alleles. Another possible explanation would be that some alleles are much more mutable than others. An average rate greater than 10^{-3} seems, however, improbable. Finally, there is the possibility that the large number of alleles is a consequence of local inbreeding. This possibility may be considered by extension of the methods used here.

THE EFFECT OF LOCAL INBREEDING

It is obvious from inspection of figure 2 that subdivision into completely isolated groups would greatly increase the number of alleles carried by the species as a whole, although reducing the number found in a single local collection. In the long run each group would come to have an entirely different set of alleles from every other group. As an isolated population of 50 individuals would maintain five or six alleles at mutation rates of 10^{-6} or 10^{-5}, an assemblage of 10 such groups would maintain 50 or 60 alleles instead of the 13 to 15 expected under random mating. With a finer subdivision, the increase would be still greater.

However, it is clear from EMERSON's data that there is nothing approaching complete isolation. Of the 34 alleles found in the 1937 collection, seven were found in all three of the most thoroughly studied canyon populations, 14 were found in two of them, leaving 13 which have so far been found in only one. On the other hand, as he points out, there is not completely random interbreeding. There is almost twice as much chance (114/2699 = .043) that a second gamete from the same canyon as the one chosen first will have the same allele, as that one from another canyon will agree (128/5451 = .024). Even the slightest restriction on random breeding will presumably increase the number of alleles carried by the total population, but whether this increase is an appreciable one where there is as much interbreeding as is indicated in the present case, requires consideration.

We have already considered the distribution of gene frequencies in a local population of size N, receiving the proportion m of its gametes from the species as a whole, with gene frequency q_t. If the species is not indefinitely large, q_t itself is a variable and it is, indeed, the distribution of q_t which we wish ultimately to estimate. However, q_t enters into the formula for $\phi(q)$ in such a way that there is no important error in assuming it constant at the value $\bar{q}_t = 1/n_t$ where n_t is the total number of alleles at which we wish to aim in our calculations. The value of R for the local population may be found as before by trial and interpolation to satisfy the condition of equilibrium.

In considering the species as a whole, it is convenient first to distinguish the proportion $1-2m$ of the fertilizations in which the pollen as well as the ovules come from the local group and the proportion $2m$ in which the pollen is a random sample from the species. In the former, selection pressure depends on the number of alleles in the group and hence on the value of R determined for the group. In the latter it depends on the number of alleles in the whole species. We will consider first the component due to inbreeding.

Assume that there are G subgroups of equal size (N). Let h be the proportion of these that contain the allele S_i. Then q_t/h is the average frequency within these groups. S_i pollen has no chance of success on the $2q_t$ plants that carry S_i. On the $(h-2q_t)$ plants that are of the same subgroups but do not carry S_i, the average frequency of success of S_i pollen to other kinds is in the ratio $q_t/h : R[1-(q_t/h)]$. The frequency of functioning S_i pollen grains is thus

$$\frac{(h - 2q_t)q_t/h}{q_t/h + R[1 - (q_t/h)]}$$

and the frequency of functioning S_i gametes is

$$\frac{q_t}{2}\left[\frac{(h - q_t)(1 + R)}{Rh + q_t(1 - R)}\right].$$

It will not do even as a first approximation, to assume that h is constant. If the allele is represented only once in the whole species ($q_t = 1/2N_t$), h is obviously $1/G$. With increasing representation in the species, h may be expected to increase until a point is reached at which the gene is practically always present in all subgroups ($h = 1$). As a first approximation, we will assume that h varies linearly with q_t between $q_t = 1/2N_t$ and $h = 1$, with such a slope (b) as will yield the required $\bar{q}_t$.

$$h = \frac{1}{G} + b\left(q_t - \frac{1}{2N_t}\right).$$

It will be convenient to write this, $h=a+bq_t$, where $a=(1/G-b/2N_t)$.

In those fertilizations which may be treated as random breeding, the frequency of functioning S_i gametes is

$$\frac{q_t}{2}\frac{(1-q_t)(1+R_t)}{[R_t+q_t(1-R_t]}$$

where R_t depends on the number of alleles in the whole species. If the amount of crossbreeding is small, this term is relatively unimportant and a rough approximation will suffice. Indeed if $m=.01$ or less, it makes little difference whether this term is considered, or whether it is omitted altogether, giving full weight to the inbreeding term. If included, we may use $R_t=(n_t-3)/(n_t-1)$, the value if all alleles are equally frequent, as such an approximation. Giving inbreeding and crossbreeding due weight and introducing mutation pressure, the rate of change of gene frequency may be written

$$\Delta q_t = (1-2m)\frac{(1+R)}{2}q_t\left[\frac{a+q_t(b-1)}{Ra+q_t(Rb+1-R)}\right] + 2m\frac{(1+R_t)}{2}q_t\left[\frac{1-q_t}{R_t+q_t(1-R_t)}\right] - q_t(1+u).$$

The sampling variance of a subgroup is $q(1-q)/2N$ disregarding crossbreeding. That of the whole species is

$$\sigma^2_{\Delta q_t} = \frac{1}{G^2}\Sigma\left(\frac{q(1-q)}{2N}\right),$$

noting that $\Delta q_t=\Sigma\Delta q/G$. A first approximation can be obtained by replacing q by its average value (q_t/h) in those groups (hG in number) that contain it at all.

$$\sigma^2_{\Delta q_t} = \frac{q_t(h-q_t)}{2N_th} = \frac{q_t[a+q_t(b-1)]}{2N_t(a+bq_t)} \text{ approximately.}$$

If q_t is small, this differs little from the value $q_t(1-q_t)/2N_t$ in a random breeding population of the same size. The latter may be used for convenience in dealing with the small crossbreeding term in the ratio $\Delta q_t/\sigma^2_{\Delta q_t}$.

Substitution in the general formula for $\phi(q)$ leads to the following expression.

$$\log_{10}\phi(q_t) = \log_{10}(a+bq_t) - \log_{10}q_t + \left[\frac{4N_ta(1+u)}{(b-1)^2}-1\right]\log_{10}[a+q_t(b-1)] + [\log_{10}e]\left[\frac{2N_tb(1+R)(1-2m)}{1+R(b-1)} - \frac{4Nb(1+u)}{b-1}\right]q_t$$

$$+ (1 - 2m) \frac{2N_t a(1 - R^2)}{[1 + R(b - 1)]^2} \log_{10} \left[Ra + q_t[1 + R(b - 1)] \right]$$

$$+ 4mN_t \frac{(1 + R_t)}{(1 - R_t)} \log_{10} [R_t + q_t(1 - R_t)].$$

Values of b and u may be found by trial that give the required mean ($\bar{q}_t$) and satisfy the balance between loss and mutation.

Four cases have been worked through to obtain an indication of the effect of subdivision into partially isolated populations (table 2). In the first case, a total population of 500 was assumed to be divided into 10 equal

TABLE 2

Comparison of subdivided and random breeding populations of 500 individuals and of the partially isolated groups of the former. The columns N, m and u describe the conditions with respect to size of population, immigration and mutation. Columns R and b give two additional constants of the formulae. Column n gives the number of alleles maintained under the given conditions and column K the percentage turnover of alleles per generation.

	N NO. OF INDIVIDUALS	m IMMIGRATION INDEX	u MUTATION RATE	R SELECTION INDEX	b	n NO. OF ALLELES	K FACTOR TURNOVER PERCENT
{Group	50	.01	.000045	.732	—	11.6	3.6
{Total	500	0	.000045	.732	6.73	20.0	0.23
Random	500	0	.000045	.860	—	17.5	0.25
{Group	50	.01	.0010	.755	—	13.4	5.3
{Total	500	0	.0010	.755	5.29	50.0	2.0
Random	500	0	.0010	.906	—	33.8	3.0
{Group	50	.001	.00020	.658	—	8.2	1.0
{Total	500	0	.00020	.658	2.50	50.0	0.43
Random	500	0	.00020	.880	—	22.1	0.90
{Group	10	.01	.00009	.425	—	5.3	3.3
{Total	500	0	.00009	.425	3.73	50.0	0.18
Random	500	0	.00009	.869	—	19.2	0.45

groups breeding within themselves except for 2 percent foreign pollen. It is required to find the mutation rate that will maintain 20 alleles in the species. By trial it was found that the value of R within groups is .732 and that the mean number of alleles within groups is 11.6 with turnover of 3.6 percent per generation. For the whole population, b was found to be 6.73 and the required mutation rate 4.5×10^{-5}. The rate of turnover of alleles in the species is 0.23 percent. In a completely random breeding population, the same mutation rate would maintain an average of 17.5 alleles with a turnover of 0.25 percent per generation. In this case the

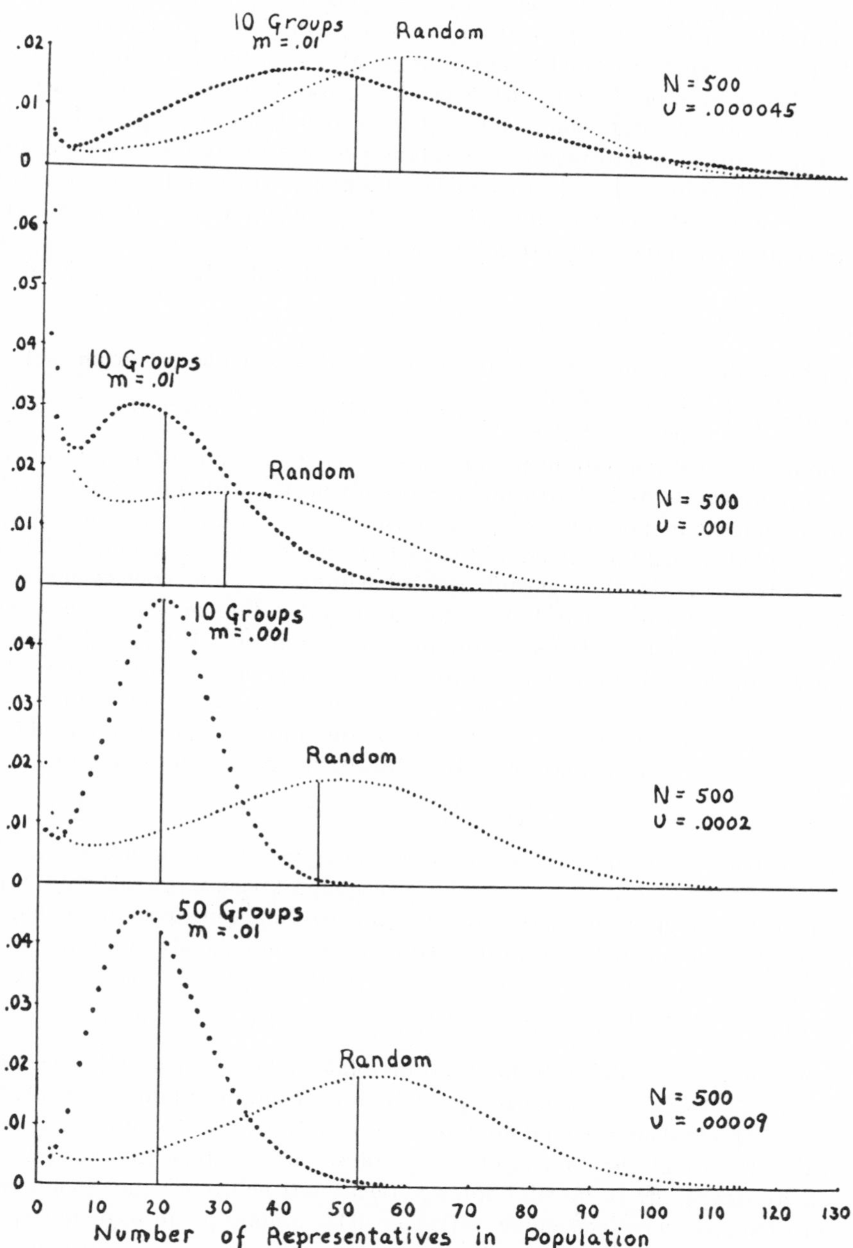

FIGURES 6–9. (See next page for description.)

postulated amount of subdivision increases the number of alleles only about 14 percent. The distribution in the subdivided and random breeding populations are compared in figure 6.

With the same amount of subdivision ($G=10$, $N=50$, $m=.01$) it requires a mutation rate of about 10^{-3} to maintain 50 alleles (turnover 2.0 percent per generation). This mutation rate would maintain only about 34 alleles under random mating. The subgroups carry 13.4 alleles on the average (figure 7).

It seemed of interest next to investigate the effect of more complete isolation. It was again assumed that there are 10 subgroups of 50 individuals each but the crossbreeding index was taken as .001 instead of .01. It only requires a mutation rate of 2×10^{-4} to maintain 50 alleles in the species (turn over 0.4 percent), a mutation rate that would maintain 22 alleles under random mating. The subgroups carry only 8.2 alleles on the average (figure 8).

None of the preceding cases approaches a satisfactory interpretation of the situation described by EMERSON in *Oenothera organensis*. To maintain the probable number of alleles in the species requires a higher mutation rate and more extreme isolation than is probable. The subgroups postulated in all of these cases (50 individuals) are roughly comparable to the populations of separate canyons. More alleles have been observed in the separate canyons (average for the three most studied, 20.7) than is permitted by the theory (about eight in the last case).

The only possible interpretation, accepting 500 as an estimate of the total population, assuming mutation rates less than 10^{-4} and assuming equilibrium, seems to be along the line of a finer subdivision. It was accordingly assumed next that the species is divided into 50 local groups of 10 individuals and that these groups breed within themselves except for 2 percent foreign pollen. With this degree of subdivision, it requires a mutation rate of 9×10^{-5} to maintain 50 alleles in the species (turnover 0.18 percent). This rate would maintain only about 19 alleles under random

---

EXPLANATION OF FIGURES 6–9

FIGURES 6–9.—(from top to bottom). The probabilities of different numbers of representatives of a self-sterility allele (while present at all) in populations of 500 individuals, compared in the cases of random breeding and of subdivision into partially isolated groups.

FIGURE 6.—Ten groups, $m=.01$, 20 alleles in the subdivided population. The implied mutation rate (4.5×10^{-5}) would maintain 17.5 alleles under random mating.

FIGURE 7.—Ten groups, $m=.01$, 50 alleles in the subdivided population. The implied mutation rate (10^{-3}) would maintain 34 alleles under random mating.

FIGURE 8.—Ten groups, $m=.001$, 50 alleles in the subdivided population. The implied mutation rate (2×10^{-4}) would maintain 22 alleles under random mating.

FIGURE 9.—Fifty groups, $m=.01$, 50 alleles in the subdivided population. The implied mutation rate (9×10^{-5}) would maintain 19 alleles under random mating.

mating (figure 9). The groups of 10 plants should carry an average of only 5.3 alleles; but a canyon containing several such groups, would, of course, maintain several times this number.

This last case obviously comes closest to a satisfactory interpretation and even closer inbreeding would be still more satisfactory. It appears that if plants are pollinated in some 98 percent or more of the cases by their immediate neighbors and only 2 percent or less by a random sample of pollen from the species as a whole, it would be possible for a species of only 500 individuals to maintain 40 or 50 alleles by mutation rates of the order of 10^{-5} or 10^{-6} per generation.

LITERATURE CITED

East, E. M., and Mangelsdorf, A. J., 1925 A new interpretation of the hereditary behavior of self-sterile plants. Proc. Nat. Acad. Sci. **11**: 166–171.

Emerson, S., 1938 The genetics of self-incompatibility in *Oenothera organensis*. Genetics **23**: 190–202.

1939 A preliminary survey of the *Oenothera organensis* population. Genetics **24**: 524–537.

Wright, S., 1931 Evolution in Mendelian populations. Genetics **16**: 97–159.

1938 The distribution of gene frequencies under irreversible mutation. Proc. Nat. Acad. Sci. **24**: 253–259.

23
Statistical Genetics in Relation to Evolution

Actualités scientifiques et industrielles, 802: Exposés de Biométrie et de la statistique biologique XIII. Paris: Hermann & Cie, 1939

Introduction

This 62-page monograph was an extensive summary of White's shifting balance theory and his statistical distribution of genes in populations. This monograph is the clearest, most accessible, and comprehensive statement of Wright's views on evolutionary theory before the publication of *E&GP*. Unfortunately, World War II broke out before copies of the monograph could be distributed by the Paris publishers. Wright himself obtained only a handful of copies after the war, and the monograph was never widely read.

If this monograph had been published in the United States and made easily accessible, I suspect it would have become a very influential work during the 1940s, perhaps the best known, instead of one of the least known, of Wright's publications. I discuss aspects of this monograph in *SW&EB*, 307, 311–12.

I

INTRODUCTION

Evolution is a process of cumulative change in the *heredities* characteristic of species. Before 1900 very little was known of heredity beyond the fact of existence. The numerous theories of evolution developed in the 19th century necessarily rested on this rather unsatisfactory basis. Since 1900, genetics, the science of heredity and variation, has been transformed into a subject which challenges any other in the field of biology in the extent to which observations of the most diverse sorts and from the most diverse groups of organisms fall into order from the viewpoint of a simple and definite theory. Yet most present-day discussions of evolution are expressed in terms of the theories of the 19th century, affected only superficially by the development of genetics.

The evolutionary implications of genetics are indeed not wholly obvious. There is often a failure to appreciate that the theory of evolution, having to do, for the most part, with transformations of populations rather than of individual lineages, can make little use of genetic principles until their statistical consequences have been worked out. Fortunately the simplicity and mathematical precision of the Mendelian mechanism make possible a rather detailed elaboration of statistical consequences ; and the wide extent to which simple Mendelian mechanisms have been found to apply in nature gives confidence that the conclusions are of significance in evolutionary theory.

It cannot be pretended, of course, that the present state of genetics furnishes anything approaching a complete basis for a theory of evolution. The ecological factors involved in selection, migration, etc., can be given mathematical treatment only by using highly simplified models. There are important modes of chromosomal change for which there is no sufficiently precise quantitative theory. It is also clear that not all heredity is Mendelian. We know very little about non-Mendelian heredity, too little to make even qualitative verbal deductions that are of much value. Finally in the natural sciences no chain of deductions can be trusted which has not been checked at many points by comparison with observation. Unfortunately our knowledge of the nature of the genetic differences between races and species, not to speak of higher taxonomic categories, lags far behind our knowledge of the genetics of individual differences. Enough is known to indicate that the statistical consequences of the Mendelian mechanism are of first importance in evolutionary theory but how much the theory must be qualified by other considerations is for the future to decide.

II

GENE FREQUENCIES IN HOMOGENEOUS DIPLOID POPULATIONS

MENDELIAN EQUILIBRIUM

It is obvious from the symmetry of the Mendelian mechanism that the relative frequency of alleles in an indefinitely large, closed population remains constant in the absence of any unbalanced pressure due to mutation or selection. If a population is made up with the initial composition $X_m AA + Y_m AA' + Z_m A'A'$ males and $X_f AA + Y_f AA' + Z_f A'A'$ females the *gene frequencies* are $(X_m + 1/2\ Y_m)A + (1/2\ Y_m + Z\ )A'$ in the former, $(X_f + 1/2\ Y_f)A + (1/2\ Y_f + Z_f)A'$ in the latter. In the next and all later generations, the gene frequencies will be the same in males and females, and equal to the unweighted average of these expressions. It is convenient to write these gene frequencies in the form $(1-q)A + qA'$. The zygote frequencies reach equilibrium under random mating according to the expression $(1-q)^2AA + 2q(1-q)AA' + q^2A'A'$ in the first generation after that in which the gene frequencies have become equal in the sexes (Hardy).

The complications due to sex linkage, to polyploidy, and to multiple alleles, and the conditions of equilibrium for two or more series of alleles will be considered later.

MUTATION PRESSURE

With constancy in the system of gene frequencies as the norm, change of gene frequency may be looked upon as the elementary evolutionary process. Haldane was the first to study systematically the changes due to mutation and selection expressing his results in terms of the changes in the ratio of gene frequencies, a quantity equal to $\frac{q}{(1-q)}$ of the present paper. The results of the method of analysis used here are in agreement as far as they overlap, after making the appropriate transformation.

One cause of change in gene frequency is obviously recurrent mutation. The occurrence of mutations of particular genes at the rate u per generation changes the gene frequency at the rate $\Delta q = -uq$. With reverse mutation at the rate v per generation the net rate of change in gene frequency due to mutation is $\Delta q = -uq + v(1-q)$.

SELECTION PRESSURE

Selection whether in mortality, mating or fecundity applies to the organism as a whole and thus to the effects of the whole gene system rather than to single genes. A gene that is more favorable than its allele in one combination may be less favorable in another. It is simplest, however, to consider first the case of a single pair of alleles whose effects on adaptation of the organism are assumed to be independent of all other genes. Discussion of the effects of factor interaction will be deferred to a later section.

The simplest case of all is that in which the heterozygote is exactly half way between the two homozygotes in the extent to which individuals contribute to the next generation. The relative selective value of genotypes will be designated W, and the mean value for a population, $\overline{W}$.

Genotype	W	Frequency
AA	1	$(1-q)^2$
AA′	$1-s$	$2q(1-q)$
A′A′	$1-2s$	q^2

$$\overline{W} = 1 - 2sq$$

$$\frac{d\overline{W}}{dq} = -2s$$

$$\Delta q = \frac{(1-s)q(1-q) + (1-2s)q^2}{1-2sq} - q = -\frac{sq(1-q)}{1-2sq} = \frac{q(1-q)}{2\overline{W}}\frac{d\overline{W}}{dq}.$$

In the more general case in which $\overline{W}$ is not related linearly to q, the *momentary* selective advantage of replacing A by A′ is still given by the slope of the tangent $\frac{d\overline{W}}{dq}$, and the value of Δq is given by the same formula as above in terms of $\overline{W}$, and q, Wright (1937).

Genotype	W	Frequency
AA	1	$(1-q)^2$
AA′	$1-s_1$	$2q(1-q)$
A′A′	$1-s_2$	q^2

$$\overline{W}=1-2s_1 q(1-q)-s_2q^2$$

$$\frac{d\overline{W}}{dq}=-2s_1+2(2s_1-s_2)q$$

$$\Delta q=\frac{q(1-q)}{2\overline{W}}\frac{d\overline{W}}{dq}.$$

If the selection coefficients are small, $\overline{W}$ is close to 1 and the form $\frac{q(1-q)}{2}\frac{d\overline{W}}{dq}$ is sufficiently accurate and sometimes more convenient. Following are certain cases with this simplification.

General	$\Delta q=-[s_1-(2s_1-s_2)q]q(1-q)$
A′ dominant over A, $s_1=s_2$	$\Delta q=-s_2q(1-q)^2$
No dominance, $s_1=1/2\ s_2$	$\Delta q=-\frac{s_2}{2}q(1-q)$
A′ recessive to A, $s_1=0$	$\Delta q=-s_2q^2(1-q)$.

As an example of a case in which $\overline{W}$ in the denominator must be retained, consider that of a recessive lethal.

A′ recessive to A, $s_1=0$, $s_2=1$ $\quad \Delta q=-\frac{q^2}{1+q}$

THE EQUILIBRIUM POINT

Gene frequency is in equilibrium (stable or unstable) at any point at which $\Delta q=0$. Thus opposing mutation pressures, causing net change of gene frequency at the rate $\Delta q=-uq+v(1-q)$ reach equilibrium when $q=\frac{v}{u+v}$. We may write $\Delta q=-(u+v)(q-\widehat{q})$. Equilibrium here is obviously stable.

In the case of selection pressure, $\Delta q=-[s_1-(2s_1-s_2)q]q(1-q)$, there is no change of gene frequency if $q=0$, $q=1$

or $q = \frac{s_1}{2s_1 - s_2} \cdot \widehat{q}$ lies between 0 and 1 if s_1 and $(s_1 - s_2)$ are both positive or both negative. On writing $\Delta q = (2s_1 - s_2)(q - \widehat{q})\, q(1 - q)$ it is obvious that equilibrium is stable if s_1 and $(s_1 - s_2)$ are both negative which is the case in which the heterozygote is better adapted than either homozygote. On the other hand, if s_1 and $(s_1 - s_2)$ are both positive, deviation of q from $\widehat{q}$ is followed by a runaway process. This is the case in which the heterozygote is less well adapted than either homozygote. (Fisher 1922, 1930c).

Ordinarily there is simultaneous pressure on gene frequency from mutation and selection. Unless both rates are high, the net rate of change is given sufficiently accurately by the sum

$$\Delta q = -uq + v(1 - q) + \frac{q(1 - q)}{2\overline{W}} \frac{d\overline{W}}{dq}$$

If selection favors the heterozygote, the equilibrium point lies between that due to selection alone and that due to mutation alone. If the heterozygote is less well adapted than either homozygote, the point of unstable equilibrium due to selection is shifted somewhat by mutation, and two points of stable equilibrium appear at approximately $\widehat{q} = \frac{v}{s_1}$ and $\widehat{q} = 1 - \frac{u}{s_1 - s_2}$.

Following are the approximate points of equilibrium in certain other cases.

Dominant injurious factor, $s_1 = s_2$	$\widehat{q} = \frac{v}{s_2}$
Partially dominant injurious factor, $s_1 = hs_2$	$\widehat{q} = \frac{v}{hs_2}$
Recessive injurious factor, $s_1 = 0$	$\widehat{q} = \sqrt{\frac{v}{s_2}}$

Injurious factors that are completely recessive escape the action of natural selection to a considerable extent and thus should accumulate more in populations than ones which show any degree of dominance in their injurious effect, even though the rates of occurrence of the mutations be the same. Comparison of the above values of $\widehat{q}$ brings this out quantitatively.

Actually, the greater portion of the mutations that are observed to occur are recessive or nearly so. There has been considerable discussion of the physiological and evolutionary basis of this phenomenon which we will not here go into. (Fisher, 1928 '29, '30a, '31, ; Wright, 1929a, b, '34a, ; Haldane, 1930b ; Muller, 1932a ; Plunkett, 1932).

THE DISTRIBUTION OF GENE FREQUENCIES

If the population is not indefinitely large, random changes occur in gene frequencies merely as a result of the accidents of sampling among the gametes. Letting N be the effective size of breeding population, samples of 2N gametes will be distributed according to the expansion of $[(1-q)A + qA']^{2N}$ with a standard deviation of $\sqrt{2Nq(1-q)}$. The standard deviation of gene frequencies in samples is thus $\sigma_{\Delta q} = \sqrt{\dfrac{q(1-q)}{2N}}$.

The changes in gene frequency due to accidents of sampling are of course not correlated in successive generations. Nevertheless the net effect is an increase, the squared standard deviation increasing approximately with the number of generations until damped by the approach of frequencies to 0 or 1. The pressure toward a stable equilibrium point due to mutation and selection, and the tendency to drift away from this point due to accidents of sampling should between them determine a certain probability distribution of values of q. The central problem in the genetics of populations is that of finding this distribution under various conditions (Wright 1929b, 31, 37, 38a).

Let $\varphi(q)$ be the required distribution with total frequency,

$$\int_0^1 \varphi(q)dq = 1 ; \qquad \text{mean,} \qquad \bar{q} = \int_0^1 q\varphi(q)dq.$$

and variance,

$$\sigma^2_q = \int_0^1 (q-\bar{q})^2\varphi(q)dq.$$

The conditions that the mean and variance remain constant in spite of systematic evolutionary pressure (Δq) and accidents of sampling ($\sigma^2_{\Delta q}$) can be written as follows :

$$\int_0^1 \Delta q\varphi(q)dq = 0 \tag{1}$$

$$\int_0^1 (q + \Delta q - \bar{q})^2\varphi(q)dq + \int_0^1 \sigma^2_{\Delta q}\varphi(q)dq = \sigma^2_q \tag{2}$$

The latter can be reduced to the following, ignoring a negligible term in $(\Delta q)^2$

$$2\int_0^1 (q-\bar{q})\Delta q\varphi(q)dq + \int_0^1 \sigma^2_{\Delta q}\varphi(q)dq = 0 \tag{3}$$

$$\text{Let} \quad \int \Delta q\varphi(q)dq = \chi(q)$$

Equations (1) and (3) can be written as follows

$$\chi(1) - \chi(0) = 0 \tag{4}$$

$$\int_0^1 \chi(q)dq - [\bar{q}\chi(0) + (1-\bar{q})\chi(1)] = {}^1/_2 \int_0^1 \sigma^2_{\Delta q}\varphi(q)dq \tag{5}$$

Equation (5) is obviously satisfied by the following

$$\chi(q) - [\bar{q}\chi(0) + (1-\bar{q})\chi(1)] = {}^1/_2\sigma^2_{\Delta q}\varphi(q) \tag{6}$$

This means little until it is shown that (6) also satisfies (4). There can be no sampling variance in homallelic populations, so that $\sigma^2_{\Delta q} = 0$ if $q = 0$ or $q = 1$. Equation (6) reduces to (4) if $q = 0$ or $q = 1$ and both $\varphi(1)$ and $\varphi(0)$ are finite. Equation (6) therefore satisfies the condition of constancy of the mean as well as that of constancy of the variance,

$$\varphi(q) = \frac{2[\chi(q) - \chi(1)]}{\sigma^2_{\Delta q}} \tag{7}$$

$[\chi(q) - \chi(1)]$ can be evaluated as follows, using (7)

$$d \log [\chi(q) - \chi(1)] = \frac{d\chi(q)}{\chi(q) - \chi(1)} = \frac{\Delta q\varphi(q)dq}{\frac{1}{2}\sigma^2_{\Delta q}\varphi(q)}$$

$$[\chi(q) - \chi(1)] = \frac{C}{2} e^{2\int \frac{\Delta q dq}{\sigma^2_{\Delta q}}}$$

$$\varphi(q) = \frac{Ce^{2\int \frac{\Delta q dq}{\sigma^2_{\Delta q}}}}{\sigma^2_{\Delta q}} \tag{8}$$

A check can be obtained from the position of the mode in large populations.

$$\frac{d\varphi(q)}{dq} = \frac{2\Delta q\varphi(q) - \varphi(q)\frac{d}{dq}\sigma^2_{\Delta q}}{\sigma^2_{\Delta q}} = 0$$

$$\Delta q - {}^1/_2 \frac{d}{dq}\sigma^2_{\Delta q} = 0$$

In large populations the sampling variance, $\sigma^2_{\Delta q}$, approaches 0 for all values of q. Thus in such populations, the distribution described by (8) condenses about a point or points at which systematic evolutionary pressures balance each other ($\Delta q = 0$) as must of course be true of the correct formula.

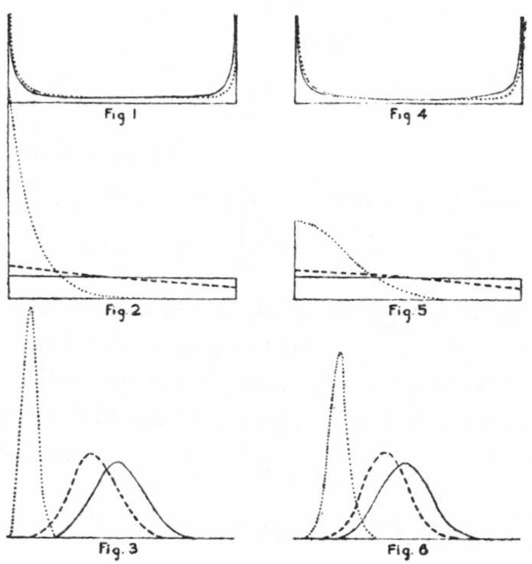

Figures 1 to 3. Some of the forms taken by the distribution of gene frequencies in the case of no dominance. ($\varphi(q) = Ce^{4Nsq}q^{4Nv-1}(1-q)^{4Nu-1}$). Mutation rates are assumed constant and equal ($u = v$). Effective size of population is $N = \frac{1}{40v}, \frac{10}{40v}$ and $\frac{100}{40v}$ in figures 1, 2 and 3, respectively. In each case the solid line represents the least selection ($s = -v/10$), the broken line selection 10 times as severe (not represented in figure 1 since practically indistinguishable from the preceding) and the dotted line represents selection 100 times as severe.

Figures 4 to 6. Some of the forms taken by the distribution of frequencies of a completely recessive deleterious gene. $Ce^{2Ntq^2}q^{4Nv-1}(1-q)^{4Nu-1}$. Mutation rates are assumed constant and equal ($u = v$). Effective size of population is $N = \frac{1}{40v}, \frac{10}{40v}$ and $\frac{100}{40v}$ in figures 4, 5 and 6, respectively. In each case the solid line represents the least selection ($t = -v/5$), the broken line selection 10 times as severe (not represented in figure 4 since practically indistinguishable from the preceding) ant the dotted line represents selection 100 times as severe.

The distribution can be expressed in terms of mutation rates, selection coefficients and size of population by substituting the values of Δq and $\sigma^2_{\Delta q}$,

$$\varphi(q) = C\overline{W}^{2N}q^{4Nv-1}(1-q)^{4Nu-1}$$

This formula gives the frequency distribution for values of q of a single gene during a long period of constant conditions. It can also be interpreted as giving the distribution at any moment for all genes subject to the same conditions. Another interpretation is that of the distribution of a single gene among completely isolated populations subject to the same conditions.

Some of the forms taken by the distribution of gene frequencies are illustrated in figures 1 to 6.

EFFECTS OF CHANGES IN CONDITIONS

The probability array discussed above can be realized only under long continued constancy in the conditions of life. It is important to consider how rapidly a new equilibrium is approached on change of conditions.

The simplest case is that in which the population is so large that accidents of sampling are negligible. The mode of approach of q to its value at equilibrium $(\widehat{q})$, may be obtained approximately by putting Δq equal to $\frac{dq}{dT}$ and solving for q as a function of number of generations (T) (or for T as a function of any required value of q). (Haldane, 1932.)

Loci which have been fixed can leave this state only through mutation in a closed population. As mutation rates appear to be very low as a rule (of the order of 10^{-5} to 10^{-6} from the few studies that have been made) this usually means very slow change. For mutation pressure by itself we have :

$$\frac{dq}{dT} = -(u + v)(q - \widehat{q})$$

$$(q - \widehat{q}) = (q_0 - \widehat{q})e^{-(u + v)T}$$

In this case the frequency of the type gene obviously falls according to the law $(q - \widehat{q}) = (q_0 - \widehat{q})(1 - u - v)^T$ without the approximation involved in substituting $\frac{dq}{dT}$ for Δq. If the initial value of q is not close to 0 or 1 and there is a strong selection pressure, a new equilibrium may be approached relatively rapidly. In the case of genic selection balanced by mutation

$$(\Delta q = -uq + sq(1 - q), \qquad \widehat{q} = 1 - u/s$$

$$\frac{dq}{dT} = -sq(q - \widehat{q})$$

$$q = \frac{q_0\widehat{q}}{q_0 - (q_0 - \widehat{q})e^{-s\widehat{q}T}}$$

Other cases can be worked out similarly.

This type of solution applies either to the effect of change of conditions on a large population or to the effect of great increase in size in a population which had previously been so small that gene frequencies had drifted away from the equilibrium point.

The effect of great reduction in size of population including the case of complete isolation of a small sample, requires consideration of the changes in the distribution of gene frequencies. Starting from a group of loci with the same initial value of q and subject to the same evolutionary pressures, the accidents of sampling cause a scattering of gene frequencies, the standard deviation after one generation being $\sqrt{\frac{q(1-q)}{2N}}$. This scattering continues generation after generation. Eventually certain of the genes drift into fixation. It is to be expected that at length an equilibrium will be reached in the *form* of distribution of the frequencies of the heterallelic loci but with decrease in the *numbers* of these as fixation proceeds at a uniform percentage rate. This rate (K) is easily determined. With equilibrium in a symmetrical form

$$\sigma^2_q = \frac{\Sigma(q - 1/2)^2 f}{\Sigma f}$$

$$\Delta\sigma^2_q = \frac{\Sigma q(1-q)f}{2N\Sigma f} = 1/_{2N}(1/4 - \sigma^2_q)$$

$$\sigma^2_q + \Delta\sigma^2_q = K/4 + (1-K)\sigma^2_q$$

$$(K - 1/_{2N})(\sigma^2_q - 1/4) = 0$$

Thus the rate of decay is 1/2N per generation (Wright 1928a, 31).

In a population in which there are n representatives of a gene the chance of complete loss as a result of the accidents of sampling is e^{-n} by the Poisson law. In the distribution $\varphi(q)$, referring to a population of size N, the frequency of the class with n representatives is $\frac{\varphi(n/2N)}{2N}$. Thus the total amount of loss in one generation from the whole distribution is

$$e^{-1}\left[\frac{\varphi(1/_{2N})}{2N}\right] + e^{-2}\left[\frac{\varphi(2/_{2N})}{2N}\right] + e^{-3}\left[\frac{\varphi 3/_{2N}}{2N}\right]\cdots$$

If all frequencies are the same, the rate of loss is .58 times the class frequency. If $\varphi(q) = \frac{C}{q(1-q)}$, the rate of loss is .46 times the frequency of the subterminal class (that with 1 representative of the gene). Thus over a wide range of types of frequency distribution, the amount of loss is close to one half the subterminal class.

The distribution of heterallelic loci with equilibrium of form but decay at a constant rate due to fixation, not balanced by mutation, should satisfy the following conditions with respect to change of mean and change of variance in one generation.

$$\int_0^1 \Delta q \varphi(q) dq = (1 - \bar{q}) \frac{\varphi(1 - {}^1/_{2N})}{4N} - \bar{q} \frac{\varphi({}^1/_{2N})}{4N}$$

$$\int_0^1 (q + \Delta q - \bar{q})^2 \varphi(q) dq + \int_0^1 \frac{q(1-q)}{2N} \varphi(q) dq$$

$$= \left[1 - \frac{\varphi(1 - {}^1/_{2N})}{4N} - \frac{\varphi({}^1/_{2N})}{4N}\right] \int_0^1 (q - \bar{q})^2 \varphi(q) dq$$

$$+ (1 - \bar{q})^2 \frac{\varphi(1 - {}^1/_{2N})}{4N} + \bar{q}^2 \frac{\varphi({}^1/_{2N})}{4N}.$$

In the case of no selection as well as no mutation, the distribution is symmetrical ($\bar{q} = 1/2$). It may easily be seen that the desired solution is

$\varphi(q) = 1$ or taking cognizance of the decay

$\varphi(q) = l_0 e^{-T/2N}$ where l_0 is the initial number of loci

and T is number of generations. There is a small amount of distortion near the ends, which has been investigated by Fisher (1930) who after previously obtaining a different result (1922), later confirmed the above approximation using a different method of attack.

As loci accumulate in the homallelic classes, mutation to at least one of the subterminal classes should eventually become important. A new equilibrium of form will be approached, with fixation at one end of the distribution balanced by mutation, but with decay due to uncompensated loss at the other end. The formula for irreversible mutation at rate $\Delta q = v(1 - q)$ is easily found.

The conditions with respect to change of mean and change of variance in one generation are as above with omission of the terms involving $\varphi(1/2N)$. It can easily be shown that these are satisfied to a sufficient approximation by the distribution

$$\varphi(q) = 4N\upsilon q^{4N\upsilon - 1}$$

The coefficient $(l_0\, e^{-T\upsilon})$ may be included to take cognizance of decay at the rate υ per generation. A different method of demonstration of these cases has been given in a previous paper (1931).

The introduction of selection introduces mathematical complications that have not been satisfactorily solved in the general case. Simple results have been reached in certain important special cases, however (Wright, 1938a).

First, in large populations in which mutation is opposed by such strong selection that even the equilibrium distribution contains low frequencies in the class homallelic in the mutation $(q = 1)$, there would be no appreciable rate of fixation under irreversible mutation and the form of distribution is the same as in the case of reversible mutation at an indefinitely low rate.

The most important case is probably that of a population so small that mutation rate is negligible and Δq can be written in the form $(s + tq)q(1 - q)$.

It can be shown that for small values of $4Ns$ and $2Nt$ the limiting hyperbolic distribution $\varphi(q) = \frac{4N\upsilon}{q}$ is modified as follows :

$$\varphi(q) = \frac{4N\upsilon}{q}\left[1 + 2Nsq + \frac{2}{3}Ntq(2q - 1)\right]$$

$$K = \upsilon[1 + 2Ns + {}^{2}/_{3}Nt]$$

In the special case of genic selection $(t = 0)$, the formula reduces to a simple form which agrees with a result reached by Fisher (1930a,b) for this case using a different method.

$$\varphi(q) = \frac{4N\upsilon[1 - e^{-4Ns(1-q)}]}{q(1 - q)(1 - e^{-4Ns})}$$

$$K = \frac{4N\upsilon s}{1 - e^{-4Ns}}$$

The chance of fixation of an individual gene is given by $\frac{K}{2N\upsilon}$. In the case of genic selection this gives $\frac{2s}{1 - e^{-4Ns}}$ or approximately $2s$ for favorable mutations occurring in large populations. For indifferent factors it is $\frac{1}{2N}$. Even unfavorable mutations have a chance of fixation $\frac{2s}{e^{4Ns}-1}$ but this is very small unless 4Ns is small.

A change in the size of population does not cause any change in the mean of the distribution of gene frequencies if evolutionary pressures are linear in relation to q. Thus under mutation pressure ($\Delta q = -uq + v(1-q)$), gene frequency is always close to

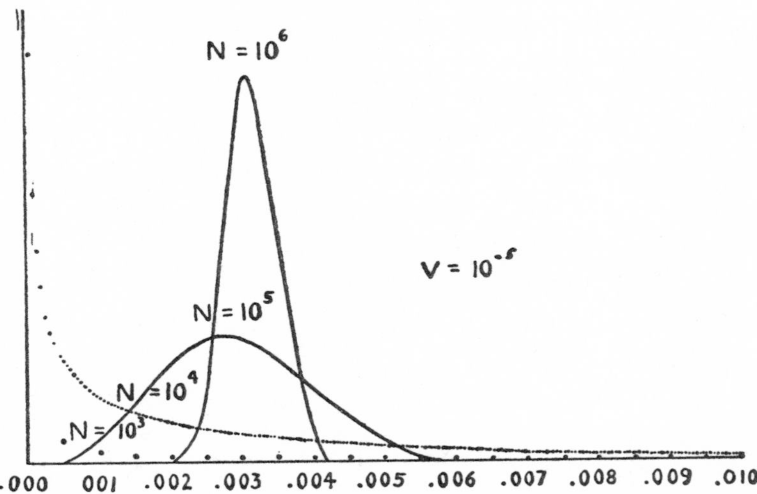

Figure 7. Some of the forms taken by the distribution of frequencies of a recessive lethal gene under different sizes of population.

$$C(1-q^2)^{2N}q^{4Nv-1}(1-q)^{-1}.$$

The mutation rate (v) is taken as 10^{-5}, giving a mean gene frequency of $\sqrt{v} = .0032$ in very large populations. This is approximately realized with $N = 10^6$. With $N = 10^5$, the gene is always present, but $\bar{q}$ is slightly reduced (.0030). In populations with effective size, $N = 10^4$, the gene is absent in about 15 %, at any given moment and $q = .0020$. With $N = 10^3$, the gene is absent in 87 %, and $\bar{q} = .0008$. The case of $N = 10^2$ is not shown as the gene is absent in about 99 % and $\bar{q}$ is only .00026. With $N = 10$, the gene is absent in 99.9 %, and $\bar{q} = .00008$. In selfed lines ($N = 1$) the gene is absent in 99.996 % giving $\bar{q} = .00002$ ($= 2v$).

the point $\bar{q} = \frac{v}{u+v}$ in large populations and while in most small populations there will be fixation of one allele or the other at any moment (U shaped distribution) the mean of the distribution, including the fixed classes is always the same value $\bar{q} = \frac{v}{u+v}$.

The situation is otherwise with non linear evolutionary pressures. A selection pressure may almost eliminate the less favorable allele in a large population. For example $\widehat{q} = \frac{v}{s}$ approximately in the case of a deleterious gene for which $\Delta q = -uq + v(1-q) - sq(1-q)$. The same selection pressure becomes ineffective in small populations, in most of which the gene becomes fixed in one phase or the other with relative frequencies:

$$f(0) = \frac{C}{4Nv} \quad \text{and} \quad f(1) = \frac{Ce^{-4Ns}}{4Nu}.$$

The mean gene frequency of the less favorable gene rises practically to that determined by the opposing mutation pressures $\bar{q} = \frac{v}{u+v}$ which may mean its fixation in most of the population, assuming $v > u$.

With sufficiently severe selection, the mean frequency of a deleterious recessive gene may decrease at least at first (instead of increasing) on subdivision of a large population into small inbreeding populations. This may be illustrated by the extreme case of a recessive lethal (fig. 7) for which, of course, there can be no increase due to fixation. In this case

$$\overline{W} = 1 - q^2, \qquad \Delta q = v(1-q) - \frac{q^2}{1+q}$$

giving equilibrium at approximately $\widehat{q} = \sqrt{v}$ and the frequency distribution

$$\varphi(q) = C(1-q^2)^{2N-1}(1+q)q^{4Nv-1}.$$

The mean gene frequency is approximately $\bar{q} = \frac{\Gamma(2Nv + 1/2)}{\sqrt{2N}\,\Gamma(2Nv)}$.

For values of $2Nv$ larger than 1 this is close to $\sqrt{v}$ but if $2Nv$ is smaller $\bar{q}$ is approximately $v\sqrt{2\pi N}$, a decrease.

III

THE BREEDING STRUCTURE OF POPULATIONS

EFFECTIVE SIZE OF POPULATIONS

The preceding investigation has postulated a population of N diploid individuals, reconstituted in each generation from a *random* sample of 2N gametes. It is important to consider the effects of various deviations from this ideal situation. A precise analysis of the consequences of different systems of mating—e.g. inbreeding of any degree or assortative mating—can be made by a correlation method, the method of path coefficients, as well as by more direct methods in simple cases (Wright 1931, 33). The scope of the present paper, however, permits only a consideration of the approximate sampling variance under certain conditions.

Our ideal system, with sampling variance $\frac{q(1-q)}{2N}$ applies directly to a population of N mature monoecious individuals with random union of gametes (including self fertilization). The exclusion of self fertilization, however, makes little difference in the sampling variance even in very small populations. The mean sampling variance in this case, with the distribution of q found with no mutation or selection, is very nearly $\frac{q(1-q)}{2N+1}$.

In populations consisting of separate sexes, N_m males and N_f females, the sampling variance of the $2N_m$ gametes which produce the males is approximately $\frac{q(1-q)}{2N_m}$ and for those which produce females is approximately $\frac{q(1-q)}{2N_f}$. The effective gene frequency of the population being $q = \frac{1}{2}(q_m + q_f)$, the approximate sampling variance may be written $q(1-q)(1/8N_m + 1/8N_f)$. If there are equal numbers of males and females ($N_m = N_m = N/2$) this reduces to the same formula as for monoecious populations, $\frac{q(1-q)}{2N}$. In other cases, the expression $N_e = \frac{4N_m N_f}{N_m + N_f}$ may be taken as the effective size of population. It is important to note that the effective size of population depends much more on the sex that is more restricted in number than on the other sex. Thus with N_m males but an indefinitely large number of females, N_e is only $4N_m$. The distribution of gene frequencies can be obtained by substituting N_e for N.

The condition of random sampling from the parental generation is likely not to be realized because of correlation in the chances of survival or death among the offspring of the same brood.

Assume that N individuals furnish varying numbers (k) of gametes to the next generation but a total number of 2N. The mean number per individual is thus $\bar{k} = 2$. There is a variance

$$\sigma^2_k = \frac{\sum_1^N (k-2)^2}{N}$$

The proportion of the cases in which two random gametes come from the same parent is

$$\frac{\sum_1^N k(k-1)}{2N(2N-1)} = \frac{2+\sigma^2_k}{4N-2}$$

This should be compared with the corresponding proportion among pairs of gametes drawn from an indefinitely large number to which all of the N parents contribute equally. This is obviously $1/N$. The effective size of population may thus be written $\frac{4N-2}{2+\sigma^2_k}$,

In case of a random sample of size 2N

$$\sigma^2_k = \frac{2(N-1)}{N}$$

The formula for effective size of population reduces to N as expected. If N is large there is here an approach to a Poisson distribution of k, characterized by

$$\sigma^2_k = \bar{k} = 2.$$

If each individual contributes just 2 gametes to the next generation, $\sigma_k^2 = 0$. The effective size of population, $N_e = 2N - 1$, is about twice the apparent size unless this is very small.

The preceding case is highly improbable however, except in a planned experiment. In a natural population it is much more probable that there will be more variability in numbers furnished by parents than expected from random sampling and that the effective size of population will thus be less and often very much less than the apparent size.

A population may vary tremendously in numbers from generation to generation. If there is a regular cycle of a few generations, an approximately equivalent population number can be found. The sampling variance of gene frequency being $\frac{q(1-q)}{2N}$, the average sampling variance for a cycle of of n generations with numbers N_1, N_2 ... N_n is approximately

$$\frac{q(1-q)}{2n}\left[\frac{1}{N_1}+\frac{1}{N_2}\cdots\frac{1}{N_n}\right]=\frac{q(1-q)}{2\left[\frac{n}{\sum\frac{1}{N}}\right]}$$

The effective size of population is thus $\frac{n}{\Sigma\left(\frac{1}{N}\right)}$ which is controlled much more by the smaller than by the larger numbers. Thus, if the breeding population increases in 5 generations in geometric series from 10 to 1,000,000 but returns each year to 10, the effective size of population, 54, is relatively small.

This formula can be applied only if the number of generations is sufficiently small that the amount of approach toward equilibrium under selection and mutation during the phase of large population number does not destroy the cumulative character of the sampling variance at successive periods of small numbers. In cases in which the numbers are greatly reduced on rare occasions all that can be considered from the standpoint of population size is the chance that on one such occasion the system of gene frequencies may become so altered that the systematic evolutionary pressures will thereafter be directed toward a new point of equilibrium.

The actual occurrence of cycles in population and of occasional great reductions has been discussed especially by Elton who has concluded that chance deviations in the character of survivors at times of least numbers may have significant evolutionary effects.

PARTIAL ISOLATION

Even after making allowance for inequalities in numbers of the sexes, for reproductive inequalities among individuals and periodic depletion in numbers, it would appear that the effective size of most species would be so enormous that there could be no important sampling effect. But in large species, there is never even a remote approach to random mating and the number of individuals among which any given individual may find a mate is likely to be rather small. This is obvious in sedentary forms but even in such mobile and migratory forms as birds, recent studies indicate an extraordinary tendency to return always to the same locality for breeding. It is thus important to consider the effects of partial isolation. Actual situations are so complex that it is necessary to deal with highly simplified models for mathematical purposes.

Assume that there is an *effective* exchange of individuals between a local group (gene frequency q) and the species as a whole (gene frequency q_t) to the extent m per generation. The rate of change of gene frequency in a group is obviously

$$\Delta q = -m(q - q_t)$$

This migration or cross breeding pressure can be written

$$-m(1 - q_t)q + mq_t(1 - q)$$

which is in the same form as the expression for mutation pressure, $-uq + v(1 - q)$. Obviously the results for mutation pressure can be transformed into those for migration pressure by substituting $m(1 - q_t)$ for u and mq_t for v. The distribution of gene frequencies for a local group may therefore be written

$$\varphi(q) = C\overline{W}^{2N}q^{4Nmq_t-1}(1 - q)^{4Nm(1-q_t)-1}$$

In this formula N is the effective population number of the local group and $\overline{W}$ the mean adaptive value in it. The mutation terms $4Nv$ and $4Nu$ can also be included in the exponents of q and $(1 - q)$ respectively but are likely to be negligible in comparison with the migration terms.

The variance of the distribution in the absence of selection ($\overline{W} = 1$)

is $\sigma^2_q = \frac{q_t(1 - q_t)}{4Nm + 1}$ Wright 1931, Kolmogorov 1935.

As written it would appear that there can be exchange of only a few individuals per generation if the isolation is to have significant effects. However, as we have seen, N may be much less than the actual size of population and because of the usual similarity of adjacent local populations between which most exchange of individuals takes place, the effective m should usually be much smaller than the actual amount of exchange. The quantity $(4N_m + 1)$ may thus be very much smaller than appears.

However, isolation must be practically perfect to permit the drifting apart of large populations within which there is free interbreeding and the rate of divergence in this case is at best very slow. On the other hand, rapid non-adaptive differentiation is to be expected where there is subdivision on a fine scale even though this is rather imperfect.

LOCAL DIFFERENCES IN SELECTION

Different subgroups of a population are usually subjected to somewhat different conditions of selection. Let s represent here the local selective advantage of the gene, q_t as before the gene frequency in the whole population

$$\Delta q = sq(1 - q) - m(q - q_t) - uq + v(1 - q)$$

If s is much less than m, the local equilibrium frequency $(\widehat{q})$ is approximately as follows (from solution of the quadratic equation, $\Delta q = 0$,

$$\widehat{q} = q_t + \frac{1}{m}[sq_t(1 - q_t) - uq_t + v(1 - q_t)]$$

The values in different local populations are thus clustered closely about q_t. If each local population is large enough so that $\widehat{q}$ may be treated as the mean gene frequency, the following relation should hold in the population as a whole :

$$\bar{s}q_t(1 - q_t) - uq_t + v(1 - q_t) = 0$$

This is independant of m and identical with the value under random mating, a net selection coefficient of $\bar{s}$ and the same mutation rates.

If on the other hand, s is much larger than m, we have approximately

$$\hat{q} = 1 - \frac{1}{s}[m(1 - q_t) + u] \qquad \text{if } s \text{ is positive}$$

$$\hat{q} = \left[\frac{mq_t + v}{-s}\right] \qquad \text{if } s \text{ is negative}$$

If all local populations are subject to the same selection, $\hat{q} = q_t$ and $q_t = 1 - \frac{u}{s}$ or $\frac{v}{-s}$ (approximately) as expected.

If, however, there is strong positive selection in some groups, strong negative in others, the mean gene frequency of the total population has little relation to the values expected under random mating and the same net selection coefficient. Its value may be expected to be closer to .5 than where determined by the opposition of mutation and the net selection coefficient.

IV

COMPLICATIONS FROM DIVERSE MODES OF INHERITANCE

SEX LINKAGE

Sex linkage is sufficiently common to deserve special consideration. In this case, the gene frequencies in the females (assumed here to be XX) must be given twice as much weight as those in the males (assumed here to be XY) in determining the mean gene frequencies of the population. Owing to the asymmetry in relation to sex, the frequencies in males and females, if different, approach equilibrium in an oscillatory fashion. (Jennings, Robbins 1918b.)

The gene frequency of males must equal that of their mothers, while that of the females must be the unweighted average of their parents. Using q_m for gene frequency of males, q_f for that of females in a given generation and letting q'_m and q'_f be the corresponding gene frequencies of the preceding generation we have :

$$q_m = q_f'$$
$$q_f = {}^1/_2[q_m' + q_f']$$

Thus $\bar{q} = {}^1/_3(q_m + 2q_f) = {}^1/_3(q_m' + 2q_f')$ is constant

And $q_m - \bar{q} = -{}^1/_2(q_m' - \bar{q})$

$$q_f - \bar{q} = -{}^1/_2(q_f' - \bar{q})$$

Thus the deviation from equilibrium in either sex is followed in the next generation by half as great a deviation in the opposite direction.

Mutation and migration pressures introduce no new difficulties. The effects of selection, however, may be rather complicated because of the necessity of considering the sexes separately. We will consider here only zygotic selection and assume that the severity is so low that q_m and q_f never differ to an appreciable extent. Using q for this common frequency ($\bar{q}$ above) the rate of change in gene frequency due to selection may be written as follows for females and males respectively

$$\Delta q_f = \frac{q(1-q)}{2} \frac{d}{dq} \log \overline{W}_f$$

$$\Delta q_m = q(1-q) \frac{d}{dq} \log \overline{W}_m$$

The rate of change of the average gene frequency may then be written

$$\Delta q = {}^2/_3 q(1-q) \frac{d \log \overline{W}}{dq} \quad \text{where} \quad \overline{W} = \sqrt{\overline{W}_m \overline{W}_f}$$

or $\Delta q = {}^2/_3 q(1-q) \frac{d\overline{W}}{dq}$ where $\overline{W} = {}^1/_2(\overline{W}_m + \overline{W}_f)$, selection rates low.

Introducing a mutation pressure in opposition to selection, this reduces to the following in the general one factor case

$$\Delta q = v(1-q) - {}^1/_3 q(1-q)[s_m + 2s_1 - 2(2s_1 - s_2)q]$$

giving equilibrium at approximately $\hat{q} = \frac{3v}{s_m + 2s_1}$ (gametes).

The most interesting special case is that of a deleterious recessive ($s_1 = 0$)

$$\Delta q = v(1-q) - {}^1/_3 q(1-q)(s_m + 2s_2 q)$$

$$\hat{q} = \frac{3v}{s_m}$$

The variance of q in samples of N_m X chromosomes in eggs which develop into males is $\frac{q(1-q)}{N_m}$ that for the $2N_f$ X chromosomes in the fertilized eggs which develop into females is $\frac{q(1-q)}{2N_f}$ and for the average

$$(q = {}^2/_3 q_f + {}^1/_3 q_m) \quad \text{we have} \quad \sigma^2_{\Delta q} = \left(\frac{2}{9N_f} + \frac{1}{9N_m}\right) q(1-q)$$

This indicates an effective size of population of $N_e = \frac{9N_m N_f}{4N_m + 2N_f}$ on comparison with the standard case.

The distribution of gene frequencies reduces to the following

$$\varphi(q) = C\overline{W}^{\frac{8N_e}{3}} q^{4N_e v - 1}(1-q)^{4N_e u - 1}$$

If there are an equal number of males and females, $N_e = {}^3/_4 N$

$$\varphi(q) = C\overline{W}^{2N} q^{3Nv - 1}(1-q)^{3Nu - 1}$$

POLYPLOIDY

Zygotic frequencies do not come to equilibrium immediately with polyploids. The rate of approach to an equilibrium distribution, according to the expansion of $[(1-q)A + qA']^{2K}$ in a 2K-ploid is, however, rapid. (Haldane 1930a.)

Mutation and migration pressures apply directly to genes and so introduce no complications.

Selection pressure by an analysis similar to that used in the case of diploids may be expressed as follows (Wright 1938b).

$$\Delta q = \frac{q(1-q)}{2\,K}\frac{d}{dq}\log \overline{W}$$

In a population of N 2K-ploids with random association of genes, the variance of q among progeny populations of the same size is approximately

$$\sigma^2_{\Delta q} = \frac{q(1-q)}{2\,NK}$$

Thus

$$\varphi(q) = \frac{Ce^{4\,NK\int\frac{\Delta q dq}{q(1-q)}}}{q(1-q)}$$

$$= C\overline{W}^{2\,N}q^{4\,NKv-1}(1-q)^{4\,NKu-1}$$

MULTIPLE ALLELES

So far we have dealt only with pairs of alleles. It must be recognized, however, that the large number of multiple allelic series already known make it probable that all loci are capable of many, perhaps an indefinite number of different mutations.

Presumably any particular gene can arise at a single step from only certain of the others and in turn mutate only to certain ones but the latter may be capable of producing mutations which could not have arisen from the former at one step and so on through a branching network of potentially unlimited extent. Since genes as a rule have multiple effects it is probable that in time a gene may come to produce its major effects on wholly different characters than at first. Under these conditions each locus has a history which is not a mere oscillation between approximate fixation of two phases but a real evolutionary process in itself. It is necessary to make allowance for this probability in the statistical treatment.

As with pairs of alleles the relative frequencies of multiple alleles in an indefinitely large closed population remain constant in the absence of mutation or selection. The zygote frequencies (of autosomal genes) reach equilibrium under random mating according to the expansion of $(q_1A_1 + q_2A_2 \cdots + q_nA_n)^2$ in the first generation after attainment of equality of gene frequencies in the sexes.

With recurrent mutation and selection an equilibrium should be reached with a certain set of frequencies. The accidents of sampling should result in a multidimensional distribution of gene frequencies.

The treatment of multiple alleles may, however, be reduced to that of paired alleles by contrasting each gene with the totality of its alleles. In the binomial expression $[(1 - q)A + qA']$, A' may be understood as representing a particular gene (or particular class of alleles) and A as including all others of its series.

Such treatment, however, requires further qualification with regard to the constancy of the various coefficients. It may be assumed that the rate (u) of mutational change of the gene in question (if not a designation of a class) is reasonably constant but its rate (v) of mutational origin from alleles must be expected to change. This may be expected to rise to maximum as genes closely allied to A' in structure become frequent, and to fall to zero as changes accumulate in the locus. Even at its maximum, however, its rate of formation should in general be small compared with its rate of change, simply because it is one and its alternatives many. Thus with n alleles, each mutating to any one of the others $(n-1)$ at the rate of X, the total rate (u) of mutation from the gene is $(n - 1)$X while the rate (v) of mutation to it from the class of other alleles is only X giving the equilibrium frequency for each allele $\hat{q} = \frac{v}{u + v} = \frac{1}{n}$. This disproportion may be even greater in the case of type genes. The commonest type of mutational change (from the standpoint of gene action) is partial or complete inactivation (not necessarily physical loss). Such mutations, however, are probably not the ones most likely to be seized upon and fixed in the course of evolution. The consequence is that the rate of formation of genes of evolutionary significance is likely to be disproportionately small compared with the rate of break down of such genes. The selection coefficient relating to a particular gene becomes a function of the relative frequency of different alleles.

The degree of stability of these constants depends much on the effective size of population. In a sufficiently large, freely interbreeding population, the distribution of gene frequencies should depart but little from the equilibrium point and possibly favorable alleles at two or more removes from those which are most abundant have very little opportunity to occur. In small populations with U shaped distribution of gene frequencies, on the other hand, approximate fixation of A_1 may be followed by approximate fixation of A_2 then by A_3 and so on indefinitely. All of the evolutionary « constants » will exhibit a high degree of lability.

V

THE FREQUENCIES OF INTERACTING FACTORS

EQUILIBRIUM

Combinations of different series of alleles are obviously in equilibrium when there is random combination. Thus a population in which genotypes are distributed according to the expansion of

$$\{[(1 - q_A)a + q_A A]^2[(1 - q_B)b + q_B B]^2 \cdots\}$$

is in equilibrium.

If not present from the first, equilibrium in this respect is approached as a limit (Weinberg). Linkage slows down the rate of approach to equilibrium but has no effect on the ultimate frequencies. (Robbins 1918a).

JOINT FREQUENCY DISTRIBUTION

So far we have derived formulae for distributions of gene frequencies only for single pairs of alleles, with constant selection coefficients. As we have just seen, however, an adequate theory requires that cognizance be taken of factor interactions. The distribution of all genes should be considered simultaneously.

The momentary selection pressure on a particular gene frequency, q_i, associated with a specified set of values of the other gene frequencies is as follows.

$$\Delta q_i = \frac{q_i(1 - q_i)}{2} \frac{\partial}{\partial q_i} \log \overline{W}$$

$$\varphi(q_i) = C\overline{W}^{2N} q_i^{4Nv_i - 1}(1 - q_i)^{4Nu_i - 1}$$

$\overline{W}$ is the mean selective value with q_i variable but the specified set of values of the other q's.

The joint frequency surface must be such that on assigning any set of values to all of the q's but one, the distribution for that one is of the type given above. The formula

$$\varphi(q_1, q_2 \cdots q_n) = C\overline{W}^{2N} \Pi [q_i^{4N\nu_i - 1} (1 - q_i)^{4Nu_i - 1}]$$

where $\overline{W}$ is the mean selective value in terms of all of the q's as variables, satisfies this condition and is thus the desired form. The coefficient in each case is that necessary to make the sum of all frequencies unity.

As a simple illustration consider the interactions of two pairs of genes assuming dominance.

Genotype	Frequency	W
A– B–	$(1-q_a^2)(1-q_b^2)$	1
A– bb	$(1-q_a^2)q_b^2$	$1-s_b$
aa B–	$q_a^2(1-q_b^2)$	$1-s_a$
aa bb	$q_a^2 q_b^2$	$1-s_{ab}$

$$\overline{W}=1-s_a q_a^2(1-q_b^2)-s_b(1-q_a^2)q_b^2-s_{ab}q_a^2q_b^2$$

$$\Delta q_a = -\frac{1}{\overline{W}}\{q_a^2(1-q_a)[s_a+(s_{ab}-s_a-s_b)q_b^2]\}$$

Changes in the frequency of gene a depend on the frequency of b and all of the selection coefficients, becoming independant of b

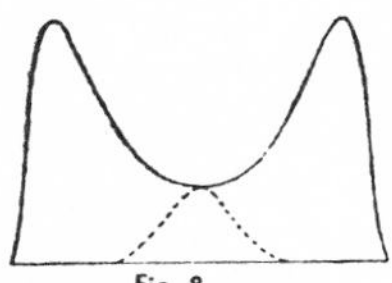

Figure 8. The frequencies along two diagonals of the joint distribution for two series of alleles with equal and additive effects on a character on which adverse selection acts according to the square of the deviation from the mean. The solid line shows the frequencies in populations along the line connectig the two favorable types $A_1A_1a_2a_2$ and $a_1a_1A_2A_2$. The broken line refers to the line connecting the extreme types $a_1a_1a_2a_2$ and $A_1A_1A_2A_2$. $\varphi(q) = C[1-2s[q_1(1-q_1)+q_2(1-q_2)+2(q_1+q_2-1)^2]^{2N} q_1^{4N\nu_1-1}(1-q_1)^{4Nu_1-1} q_2^{4N\nu_2-1}(1-q_2)^{4Nu_2-1}$. In the case shown, $u_1 = \nu_1 = u_2 = \nu_2$; $N = 1/2\nu_1$, $s = 5\nu_1$. Along the favorable diagonal $q_1 = (1-q_2)$ the distribution is approximately $Ce^{-20q_1(1-q_1)}q_1^2(1-q_1)^2$. Along the unfavorable diagonal $(q_1 = q_2)$ it is approximately $Ce^{-20(1-3q_1(1-q_1))}q_1^2(1-q_1)^2$.

only if the two series of genes are cumulative with respect to selection, *i.e.* if $(1-s_{ab}) = (1-s_a)(1-s_b)$.

The distribution may have a saddle shaped form with two distinct peaks. This is the case, for example, if A— bb and aa B—

are both more adaptive than aa bb and A- B-. Another case with two peaks is illustrated in figure 8. With multiple factors, representation of the distribution requires a corresponding number of dimensions for the gene frequency with an additional one for adaptive value, $\overline{W}$. There may be a very large number of peaks. We will return to consideration of this case later.

In populations with equal effective numbers of males and females but with some genes autosomal, some sex-linked, some disomic others polysomic, the formula for the joint distribution can be written

$$\varphi(q_1, q_2 \cdots q_n) = C\overline{W}^{2N}\Pi\,[q_i^{4N_e v_i - 1}(1 - q_i)^{4N_e u_i - 1}]$$

where $\overline{W} = \sqrt{\overline{W}_m \overline{W}_f}$

and $4N_e$ in the exponents of q_i and $(1 - q_i)$ is 4NK in case of a 2K-somic or is 3N in the case of sex-linked genes. The formula breaks down, however, if there are non-additive interactions between sex linked and autosomal factors in a population in which the numbers of the sexes are seriously unequal since in this case the exponent of $\overline{W}$ differs for the two classes of factors.

VI

THE BIOMETRIC PROPERTIES OF POPULATIONS

The distribution of gene frequencies is important in connection with such properties of populations as the mean and variability of characters and the correlations between relatives. The scope of the present paper does not permit extended discussion but certain general considerations in connection with mean and variability may be brought out.

The mean of a quantitative character (due to multiple factors) changes with changes in mean gene frequency as discussed in a previous section. In the case of genes with additive effects (no dominance or factor interactions) the changes are strictly in proportion on giving due weight to the various gene effects. Mean $= 2\Sigma\alpha q$. In order to relate to the preceding discussion we will assume no correlation between gene effect and gene frequency and write, Mean $= 2M\alpha \int_0^1 q\varphi(q)dq$, where M is the number of loci and α is the typical gene effect. With dominance, the contribution of an individual factor is αq^2 (where q is the frequency of the recessive) instead of $2\alpha q$.

The contribution to variability of a gene with no dominance is $2\alpha^2 q(1 - q)$ which is maximum if $q = 1/2$ and disappears if $q = 0$ or 1. In the case of complete dominance, it is $\alpha^2 q^2(1 - q^2)$ which is maximum when $q = \sqrt{1/2}$.

Further discussion will be restricted to the case of no dominance and the same mutation rates and selection coefficients will be assumed for all genes.

In the case of genes which are indifferent to selection (s much less than u and v) the mean remains unchanged through all transformations of the distribution of gene frequencies from U-shaped in small populations to I-shaped in large ones. The variance is small in small populations, rises at first in nearly direct proportion to size of population and reaches half its limiting value at $N = \frac{1}{4(u + v)}$

$$\text{Mean} = \frac{2\,M\,\alpha v}{u + v} \qquad \sigma^2 = \sigma_\infty^2 \left(\frac{4\,N(u + v)}{4\,N(u + v) + 1}\right)$$

$$\text{Where } \sigma_\infty^2 = \frac{2\,M\alpha^2 uv}{(u + v)^2}$$

is the limiting value.

As discussed in a previous section, changes in size of population are not accompanied by instant adjustment of the distribution of gene frequencies. The expected changes in mean and variability also, therefore, show a lag after population has increased or decreased.

To illustrate the case of genes favored by vigorous selection it will be assumed that s is much larger than u and v and that $4Nu$ and $4Nv$ can be treated as negligible in the range of sizes of population with which we are concerned. The distribution of gene frequencies can accordingly be written $\varphi(q) = \frac{Ce^{4Nsq}}{q(1 - q)}$ with fixed classes $f(0) = C/4Nv$ and $f(1) = \frac{Ce^{4Ns}}{4Nu}$.

Since the great majority of loci are homallelic, we can write approximately

$$\text{Mean} = \frac{2\,M\alpha f(1)}{f(0) + f(1)} = \frac{2\,M\alpha ve^{4Ns}}{ve^{4Ns} + u}$$

$$\sigma^2 = 2\,M\alpha^2 C \int_0^1 e^{4Nsq} dq = 2\,M\alpha^2 C \left(\frac{e^{4Ns}-1}{4\,Ns}\right)$$

But treating $f(0)$ and $f(1)$ as including nearly all loci, $C = \frac{4Nuv}{ve^{4Ns} + u}$

$$\sigma^2 = \frac{2\,M\alpha^2 uv}{s}\left(\frac{e^{4Ns}-1}{ve^{4Ns} + u}\right)$$

Thus with increase in size of populations the mean rises very slowly from the value expected from mutation alone $\frac{(2M\alpha v)}{(v + u)}$ which may be very low, to almost the opposite extreme. The variance also rises very slowly after increase in population size and decreases very slowly after a decrease in population size. If $u = v$, half the limiting value is reached where $e^{4Ns} = 3$. If v is much less than u, approximately half the limiting value is reached when $e^{4Ns} = u/v$

$$\text{Mean} = \text{Mean}_\infty \left(\frac{ve^{4Ns}}{ve^{4Ns} + u}\right) \quad \text{where Mean}_\infty = 2M\alpha$$

$$c^2 = \sigma^2_\infty \quad \left(\frac{ve^{4Ns} - v}{ve^{4Ns} + u}\right) \quad \text{where } \sigma^2_\infty = \frac{2M\alpha^2 u}{s}$$

To illustrate the effect of breeding structure, consider first a large population divided into subgroups of effective size N and with m as the effective amount of exchange with the population as a whole (gene frequency q_t). If there are no important differences in conditions of selection among the subgroups so that q_t depends on the mutation rates and net selection pressure for the population as a whole $q_t = \frac{v}{s}$ in the case of no dominance. Within a subgroup, the distribution of values of q about this mean value is similar to that under mutation pressure, replacing v by mq and u by $m(1 - q_t)$. The variance of characters is small in small subgroups, rises with size of subgroup, reaches half its limiting value (that of an indefinitely large random breeding population with gene frequency q) at $N = \frac{1}{4m}$ (assuming no dominance). This would usually be in very much smaller populations than under complete isolation.

The variance of the population as a whole is greater than that of a random breeding population with the same q_t. In the limiting case in which all of the subgroups have U-shaped distribution the contribution of each gene to variance is $4\alpha^2 q(1 - q)$ instead of $2\alpha^2 q(1 - q)$, i.e. is twice as great as under random mating.

If there is much differential selection among subgroups, the mean gene frequency may be expected to be much closer to 1/2 than where dependent on the opposition of mutation rate to net selection rate. The product $q_t(1 - q_t)$ will thus in general be much larger than where there is no differential selection, and variability both of subgroups and of the species as a whole will be correspondingly larger.

VII

THE PROCESS OF EVOLUTION

PROBLEMS AND THEORIES

We are now ready to consider the bearing of statistical genetics on various aspects of the theory of evolution. Several general problems may be distinguished. There is first the problem of evolutionary change in morphological and physiological characters. To be distinguished from this is consideration of the changing organization of the hereditary material itself. We are not here considering the importance of such change as a factor in the evolution of other characters. This may be great but is not necessarily so in the case of balanced rearrangements of loci. But chromosome evolution is of special interest, apart from character evolution, as the most direct indicator of the relationships of living forms (cf. Dobzhansky and Sturtevant, Babcock, etc.). A third general problem is that of speciation in the sense of the branching of phylogenetic lines. The much debated term species is here understood as applying to contemporary populations within which there has been such recent continuity of interbreeding that there is substantial continuity in statistical properties, but between which there has been sufficient discontinuity in interbreeding to have permitted discontinuity in statistical properties. The problem of the origin of cross sterility as the most important factor in causing long-standing discontinuity of interbreeding comes here. Finally is the problem of the origin of categories higher than the species.

It is convenient to begin with a classification of theories of evolution based primarily on the role ascribed to mutation (used here in the broadest sense). In the first category comes evolution under direct pressure of controlled mutation and thus a process of physiological transformations occurring simultaneously along all lineages. Possible mechanisms under this head are differentiated according to the nature of the mutational process.

The second main category includes statistical transformations of populations, the material for which is furnished by recurrent mutations but of which the direction is largely determined either by selection or by the accidents of sampling (isolation) or by a combination of these processes.

The third main category includes evolutionary changes which depend on the occurrence of single major events — conspicuous mutations or hybridizations. Subdivision is made according as the higher categories are the results of cumulative processes or themselves depend on single events.

In a general way, the theories are here arranged according to the demands which they place on chance events.

I. *Evolution by Physiological Transformation.*
   1. Mutation directed by the effects of use and disuse.
   2. Mutation directed by the external environment.
   3. Mutation according to innate constitution (orthogenesis).

II. *Evolution by Statistical Transformation of Population.*
   1. Course determined by natural selection.
   2. Random change under isolation.
   3. Course determined jointly by partial isolation and selection.

III. *Evolution from individual Mutations.*
   1. Elementary species by mutation higher categories cumulative.
   2. Higher categories from individual mutations.

It will be convenient to start with consideration of the possibilities of evolution under the second category as that on which the statistical theory bears most directly.

EVOLUTION BY STATISTICAL TRANSFORMATION

The observed properties of gene mutation — fortuitous in origin, infrequent in occurrence and deleterious when not negligible in effect — seem utterly inadequate as a basis for evolution as long as one considers only pairs of alleles. Under biparental reproduction, however, a limited number of mutations which are not too injurious to be carried by the species in appreciable frequencies, furnish a practically infinite field of possible variations through which the species may work its way under natural selection (East).

Estimates of the total number of genes in the cells of higher organisms range from 1000 up. Hundreds of loci have been reported. as having mutated in *Drosophila melanogaster* and in *Zea mais* during laboratory experience which is extremely limited compared with the history of species in nature. Observations of the chromomeres in the chromosomes of various organisms indicate that the numbers run into the thousands. Presumably, alleles of all type genes are present at all times in any reasonably numerous species. Judging from the frequency of multiple alleles in those organisms which have been studied most, it is reasonably certain that many different alleles of each gene are in existence at all times. With 10 alleles in each of 1000 loci the number of possible combinations is 10^{1000} which is a very large number. It has been estimated that the total number of electrons in the visible universe is much less than 10^{100}.

However, not all of this field is easily available in an interbreeding population. Suppose that each type gene is manifested in 95 % of the individuals and that most of the remaining 5 % have the most favorable of the other alleles, which in general means one with very little differential effect. The average individual will deviate in 50 respects from the type (i. e. 5 % of 1000) and since the standard deviation is about $\sqrt{50}$, practically all individuals will be in the range between 30 and 70 deviations. There is no appreciable chance that any individual will deviate in as many as 100 respects although 1000 are theoretically possible. The population is thus confined to an infinitesimal portion of the field of possible gene combinations, yet this portion includes so many combinations that there is no appreciable chance that any two sexually produced individuals will ever have exactly the same genetic constitution in a species of millions upon millions of individuals persisting over millions of generations. There is no difficulty in accounting for the probable genetic uniqueness of each individual human being or other organism which is the product of biparental reproduction. Monozygotic twins and other clone mates are, of course, here excluded.

Suppose now that every genotype in the field of possible gene combinations is graded with respect to its net selective value (W), under the varying conditions of life of the species. The location of the combinations in proper relation to each other requires $(m - 1)$ dimensions for each set of m alleles, and with n such sets, $n(m - 1)$ dimensions (9000 in the illustration above). An additional dimension is needed for selective value.

As to the nature of this surface, it is possible that a particular combination gives maximum adaptation and that the selective values of the other combinations fall off more or less regularly according to the number of steps by which they are removed from it. A species whose individuals are clustered about some combination other than the highest would move up the steepest gradient toward the peak, having reached which, it would remain unchanged except for the rare occurrence of wholly new favorable mutations.

But even in the two factor case, it is possible that there may be two peaks. This is the case, for example, with semilethals in plants (reciprocal translocations) AB and ab viable, Ab and aB inviable (Belling). Other cases are known involving only slightly deleterious mutations (cf. Gonzalez, Timoféeff-Ressovsky, 1934b). The chance that there may be multiple peaks increases greatly with each additional locus.

In any species that has lived under similar conditions for hundreds of generations, the best adapted individuals are likely to be the ones that are near the average in all measurable morphological characters. In other words, the optimum for each character is near the mean rather than beyond one of the extremes. Any gene with a constant effect on such a character increases fitness in combinations below the optimum but decreases fitness above this point.

As a simple illustration consider a character affected by 4 genes with equal effects. It is convenient and also natural to assume that selective disadvantage varies with the square of the deviation from the optimum. Capital letters are used below to represent the alleles with positive effects on the character and not to indicate dominance (Wright, 1935).

Homozygous. Genotypes . .	aabbccdd	AAbbccdd aaBBccdd aabbCCdd aabbccDD	AABBccdd AAbbCCdd AAbbccDD aaBBCCdd aaBBccDD aabbCCDD	AABBCCdd AABBccDD AAbbCCDD aaBBCCDD	AABBCCDD
Grade of Character	8	9	10	11	12
Selective Value	0.96	0.99	1.00	0.99	0.96

In this case the 6 genotypes of grade 10 are at 6 separate " peaks " of selective value, each separated by at least 2 gene replacements from the other peaks.

The mean selective values ($\overline{W}$) of populations defined by sets of values of q_A, q_B, etc. form a continuous surface. Let C be the grade of the character in the general case, $\overline{C}$ its mean and O its optimum (both grade 10 in the case above).

$$W = 1 - K (C - O)^2 = 1 - K [(C - \overline{C}) + (\overline{C} - O)]^2$$
$$\overline{W} = 1 - K [\sigma_c^2 + (\overline{C} - O)^2]$$

Assume that dominance is lacking in the effects of the genes on C and let α_i measure the effect on C of replacing gene a_i by A_i.

$$\overline{C} = 2\Sigma\alpha q$$
$$\sigma_c^2 = 2\Sigma\alpha^2 q(1 - q)$$
$$\overline{W} = 1 - K [2\Sigma\alpha^2 q(1 - q) + (2\Sigma\alpha q - O)^2]$$
$$\frac{d\overline{W}}{dq_i} = - 2K [\alpha_i^2(1 - 2q_i) + 2\alpha_i(2\Sigma\alpha q - O)]$$

It is easy to show from consideration of the slopes, $\frac{d\overline{W}}{dq_i}$ that there may be a large number of separate peaks (up to $\frac{n!}{[(n/2)!]^2}$ with n pairs of alleles) in which all, or all but one (but never less) of the loci are homallelic. The number of such peaks depends on the position of the optimum and the relative magnitudes of the gene effects. These peaks are connected by very shallow saddles. The situation is essentially similar if there is dominance.

This is the case of a single character with selection directed toward an optimum near its mean. In the case of the selective value of the organism as a whole, the situation should be somewhat similar but more complicated. There are likely to be many different harmonious combinations of characters, adapted often to different external situations. The multidimensional surface of selective values for all possible genotypes and also the surface of mean selective values for all possible populations should be of a very rugged character, with innumerable peaks at different heights, connected by saddles of varying depth.

Returning to the assumption of 1000 sets of 10 alleles each, the chance that a random combination is as adaptive as those charac-

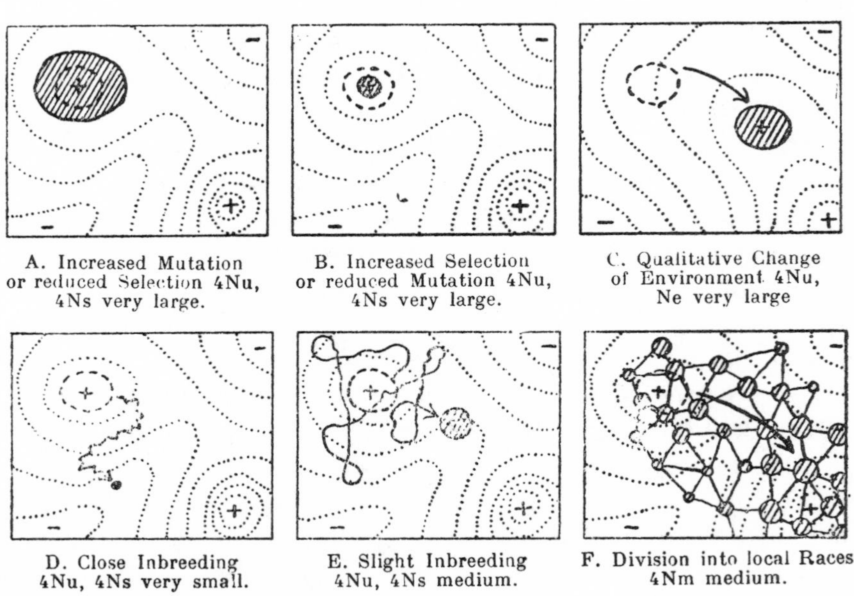

A. Increased Mutation or reduced Selection 4Nu, 4Ns very large.

B. Increased Selection or reduced Mutation 4Nu, 4Ns very large.

C. Qualitative Change of Environment 4Nu, Ne very large

D. Close Inbreeding 4Nu, 4Ns very small.

E. Slight Inbreeding 4Nu, 4Ns medium.

F. Division into local Races 4Nm medium.

Figure 9. Field of gene combinations occupied by a population within the general field of possible combinations. Type of history under specified conditions indicated by relation to initial field (heavy broken contour) and arrow.

teristic of the species may be as low 10^{-100} and still leave room for 10^{800} separate peaks, each surrounded by 10^{100} more or less similar combinations. In such a situation selection will easily guide the species to the nearest peak, but there may be innumerable other peaks which are higher, but separated by valleys. The problem of evolution from this viewpoint is that of a mechanism by which the species may continually find its way from lower to higher peaks in such a system. In order that this may occur there must either be continual change in the conditions or else some trial and error mechanism on a grand scale by which the species may explore the region surrounding the small portion which it occupies at a given time. We will consider the possibilities for evolution under various conditions, from this viewpoint (Wright, 1932).

In an indefinitely large but freely interbreeding species living under conditions which have not changed secularly for a long period of time, each series of alleles should reach a certain equilibrium in frequency. The species comes to occupy a certain field of variation about a peak in the surface (W) of selective values, represented in the parts of figure 9 by a heavy broken contour. The field occupied remains substantially constant although no two individuals are ever identical. Unless conditions change, further evolution can occur only by the appearance of wholly new (instead of recurrent) mutations and ones which happen to be favorable from the first. Such mutations would change the character of the field itself, increasing the elevation of the peak occup ed by the species. But as pointed out in the discussion of multiple alleles there is very little chance of occurence of wholly new alleles in a *large* freely interbreeding population. There is also very little chance that any new mutation will be favorable at its first occurence and even if favorable very little chance that it will attain sufficient frequency to be subject to selection to an appreciable extent. The situation is one in which there would be a high degree of stability of the species in spite of continuing mutation pressure, a continuing field of variability so extensive that no two individuals are ever genetically the same, and continuing selection.

The general rate of mutation may conceivably increase for some reason. The effect (figure 9A) will be a spreading of the field of variability occupied by the species until a new equilibrium is reached. There will be an average lowering of the adaptive level of the species ($\bar{W}$). There is somewhat more possibility of wholly new mutations (alleles at two removes from type). Another possibility of evolutionary advance is that the spreading of the field occupied may go so far as to include another and higher peak, in which case the species will move over and occupy the region about this under pressure of selection. These possibilities appear relatively unimportant from the present viewpoint. We will, however, return to the discussion of mutation pressure in a later section.

The effects of reduced mutation rate (figure 9B) are of course the opposite : a rise in average level ($\bar{W}$) but reduced variability, less chance of novel favorable mutation and less chance of capture of a neighboring peak.

The effect of increased severity of selection (also figure 9B) is of course to increase the average level of adaptation until a new equilibrium is reached. But again this is at the expense of the field of variability of the species and reduces the chance of capture of another peak. Novel favorable mutations, while less frequent would be selected more rapidly, but as selection pressure is very slight at best when gene frequency is small this is of little importance.

Relaxation of selection has the opposite effects and thus effects somewhat like those of increased mutation rate (9A).

However, the environment, living and nonliving, of any species is actually in continual change. In terms of the diagram this means that certain of the high places are being depressed and certain of the low places are becoming higher (fig. 9C). A species occupying a small field under the influence of severe selection is likely to be left in a pit and become extinct, the victim of extreme specialization to conditions that have ceased, but if under sufficiently moderate selection to occupy a wide field, it will merely be kept continually on the move. This may appear to be a mechanism which merely causes change in adaptation without general advance along any line. Under it, however, the species will be shuffled out of low peaks more easily than high ones and thus should gradually work its way to the higher general regions of the field as a whole. Here we undoubtedly have a evolutionary process of major importance. It is essentially that which Darwin and Wallace put first. It is also that to which most attention has been devoted by Haldane and Fisher in their studies of the evolutionary implications of statistical genetics.

It should be added that even slight changes in conditions, of a fluctuating character, may have considerable evolutionary significance. Where the optimum of a quantitatively varying character is near the mean, the surface, of selective values may have a very large number of peaks all at about the same level and separated by shallow saddles. A shifting of the location of the optimum by no more than the effect of a typical gene results in depression of the peak occupied by the population and elevation of many adjacent ones at one remove from it. To which of these the population moves is largely accidental. If now the optimum shifts back to its original position, it is a matter of chance whether the population will move back to the original peak or to one of the many others.

Thus trivial fluctuations in conditions will be enough to allow populations ultimately to reach remote parts of the system. While this has little significance as far as the character in question is concerned, the genes involved may have secondary effects on other characters, which may be important, but not attainable without such a trial and error mechanism.

Fig. 9D illustrates the effect of reduction in size of population below a certain relation to rate of mutation and severity of selection ($4Nu < 1$, $4Ns$ (no dominance) < 1). There is fixation of one or another allele in nearly every locus, largely irrespective of which is most favorable. With multiple alleles, there may be a drift to ones at two or more removes from those fixed at first and ultimately to ones which had not existed at all previously. The character of the whole field is thereby changed opening up new evolutionary possibilities which may be seized upon if there is a greatly increased selection pressure directed toward fixation of some of the new genes. In general, however, the species moves down from its peak in erratic fashion and comes to occupy a much smaller field of variability. In other words there is the deterioration and homogeneity of a closely inbred population. After equilibrium has been reached in variability, change becomes very slow and such as there is, is largely non-adaptive. The most probable end result of extreme inbreeding is extinction of the species.

With an intermediate relation between size of population and mutation rate, gene frequencies drift at random but without reaching the nearly complete fixation in all series of genes of the preceding case (fig. 9E). The species never moves down from the peak but continually wanders in the vicinity. There is some chance that it may encounter a gradient leading to another peak and shift its allegiance to this. Since it will escape more easily from low peaks than from high ones there is here a trial and error mechanism by which in time the species may work its way to the higher parts of the general field. The rate of progress, however, is extremely slow since change of gene frequency is of the order of the reciprocal of effective population size and thus must be of the order of the mutation rate to meet the conditions of this case.

Finally (fig. 9F) consider the case of a large species which is subdivided into small local races each breeding largely within itself but occasionally cross breeding. The field of gene combinations occupied by each of these local races shifts continually in a fashion which is primarily non-adaptive but which is restrained by selection from becoming antiadaptive to an appreciable extent. In addition to chance differentiation of this sort, there will also be primary adaptive differentiation if the conditions of selection vary in different parts of the range. The rate of movement of gene frequencies may be enormously greater than in the preceding case, since the condition for such movement is that the reciprocal of the population number be of the order of the proportion of cross breeding instead of mutation rate. With many local races each spreading over a considerable field of gene combinations and each moving relatively rapidly in the more general field about the controlling peak, there is a good chance that one at least will come under the influence of another peak, or in other words, will turn out to have acquired a preadaptation. If this second peak is a higher one (i. e. corresponds to a superior adaptation) this race will expand in numbers and by cross breeding with other races, as well as by actual displacements of these, will pull the species as a whole toward the new position. Fine subdivision of a species into partially isolated local populations provides a most effective mechanism for trial and error in the field of gene combinations and thus for evolutionary advance by intergroup selection.

Complete isolation of a portion of such a species should result relatively rapidly in differentiation of specific rank, to a large extent non-adaptive but adaptive in so far as there has been differential selection or as primarily non-adaptive changes turn out to be preadaptive. Such isolation is usually geographic in character at the outset, but may be clinched after long continued separation by a gradual chance acumulation of genic and chromosomal differences that in combination bring cross sterility.

The importance of geographic isolation as an evolutionary factor was first effectively presented by M.Wagner, who, however, considered its importance to lie in permitting the unimpeded action of different sets of environmental factors. Gulick was the first to appreciate the possibility of a random drifting apart of isolated races as a statistical consequence of inbreeding, though not of course, in Mendelian terms. He presented abundant evidence that such differentiation actually takes place in his study of land snails (Achatinellidae) of the Hawaiian Islands.

Probably the majority of field biologists in more recent years have stressed the importance of local differentiation as a factor in speciation, however, varied their interpretation (e. g. Jordan, Kellogg, Crampton, Osgood, Osborn, Rensch, Dobzhansky, etc.).

An array of closely allied forms among separate but adjoining territories has come to be accepted as the rule rather than the exception in most groups of organisms. Whether such an array is considered a single species subdivided into subspecies (Rassenkreis) or a complex of separate species (Artenkreis) depends on the amount of intergradation at the boundaries. The distinction is not a sharp one.

The existence of racial differentiation on an even finer scale has been demonstrated in many cases where statistical studies have been made (cf. Schmidt, Sumner, Thompson).

The extent to which the differentiation of neighboring populations is adaptively related to differences in conditions varies greatly.

From a study of the geographical races of the gypsy moth, Lymantria dispar, Goldschmidt concluded that such races depend almost wholly on the occurence of small numbers of mutations which happen to be adaptive to conditions outside the previous range of the species. He concluded that the formation of subspecies and races has no relation whatever to the origin of true species. He held that these can arise only by abrupt and profound reorganization of heredity. A very different situation, however, has been found in other cases. Pickford, in her monograph of the Acanthodriline earthworms of South Africa, gave careful attention to this point but could find almost no indication of adaptive differentiation of subspecies and local races. She states emphatically that the differences are similar in kind to those between species and genera, being merely less in degree. Kinsey, in his study of the Gall wasps of the genus Cynips, found random differentiation predominating in all categories from subspecies to subgenus. He was able to trace approximate continuity from species to species in long branching chains continually returning on themselves in the region between Eastern United States and Central Mexico. In one case such a chain included 86 species grouped in 9 complexes and 2 subgenera (Acraspis, Philonix). In another case (Atrusca) there were 44 species which were grouped in 4 complexes. While the exact points at which the chain was divided into complexes and even subgenera were largely arbitrary, the differences between the extremes were enormous. The species of the same complex occupied different territories (except in certain cases of host isolation on different species of oak) but several connected complexes often occupied the same territory through returns of the chain on itself. The isolation provided by the oak forests of the " island " mountain ranges of Southwestern United States and Mexico seem to have been of such a degree as to provide an exceptionally favorable condition for speciation, but even in Eastern United States there seems to have been enough isolation merely from distance to permit considerable speciation. There was, however, a marked contrast between the small, rather uniform " island " species of the Southwest and the large, highly variable " continental " species of the East.

In other cases, the balance between adaptive and non-adaptive differentiation seems to be more nearly even. Thus in the deer mice (Peromyscus) studied genetically as well as in the field by Sumner and by Dice, most of the color variations (dark, in the forest forms ; nearly white, on the sandy beaches ; pallid, as a rule, in the deserts but dark on *ancient* beds of black lava) are rather clearly adaptive. But other differences described by these authors are distributed in such a way as to make ascription to selection very difficult.

In conclusion in seems clear that both adaptive and non-adaptive differentiations of local races must be recognized as occuring, the relative importance varying widely in different cases. The relative importance may well be in accordance with the degree of differential selection, the degrees of local isolation and the effective local population numbers. In such cases as Lymantria where racial differences are described as almost wholly adaptive while the species differences are essentially non-adaptive, it seems possible that there may have been conditions somewhere at some time under which there could have been a non-adaptive differentiation of local types capable of leading to the observe species differences, even though such conditions are not found today. In cases in which the situation is reversed — predominant, non-adaptive differentiation of local populations, but important adaptive differences between higher categories ; it is a plausible hypothesis that the former have at least played a role in building up differences which later turned out to be adaptive.

There are minor preadaptations, no doubt, which are due to single mutations (Davenport, Goldschmidt) but more important are the cases, emphasized by Cuénot, in wich it seems necessary to postulate the existence of some mechanism by which rather complicated preadaptations may be built up.

EVOLUTION UNDER UNIPARENTAL REPRODUCTION

The statistical situation under exclusive uniparental reproduction, whether vegetative or by self-fertilization, is very different from that in a population in which there is biparental reproduction. A population of the former sort consists of completely isolated lineages. There may be large groups (clones) of individuals, descended relatively recently from a common ancestor, which are of identical genetic constitution and which differ from other such groups by one or more mutations. Under vegetative reproduction these clones are likely to be heterozygous in many respects as there is no tendency toward homozygosis, after occurence of a mutation, except that due to interclone selection. Under self fertilization, on the other hand, a mutation is rapidly fixed or lost ($\text{A}a \times \text{A}a \rightarrow 25\ \%\ \text{AA} + 50\ \%\ \text{A}a \times 25\ \%\ aa$), with the consequence that the clones are always homozygous in most respects unless mutation rates are very high.

In either case, the field of variability presented for selection is very much poorer than under biparental reproduction, because of the absence of recombination. 1000 mutations produce only 1001 types instead of a potential 2^{1000}. Mutations are not selected independently of each other as under biparental reproduction (Muller). The fate of an individual mutation thus depends less on its own contribution to adaptive value than on the character of the clone in which it happens to occur.

These conditions appear to be disadvantageous for evolutionary advance. There is, however, to some extent a counterbalancing advantage in the very fact that there is differential increase or decrease of clones according to the adaptive value of the genotype as a whole instead of mere change of gene frequencies according to the net adaptive values of the separate genes. Genotypes of exceptional adaptive value are not broken up as under biparental reproduction but may rapidly displace the rest of the population.

The combination of prevailing uniparental reproduction with occasional cross breeding gives results with the favorable properties of both systems, especially in cases in which there is the possibility of very rapid multiplication under favorable conditions. The situation is closely similar to that of subdivision of a population into local inbreeding races with occasional intermigration. A rich field of variability is provided even by infrequent cross breeding, while interclone selection provides for the effective selection of types which have adaptive genotypes as wholes. As in the other case, a balance between the two systems provides a more favorable condition for evolutionary advance than either system by itself. Occasional cross fertilization followed by vegetative multiplication or inbreeding and by interclone selection is indeed the method which has been most successful in practical plant breeding.

As between uniparental reproduction and the moderate inbreeding of local biparental races, the advantage depends on the potential rate of reproduction. A local population in which all individuals are of identical genotype may be highly adapted to one set of conditions but very poorly to another. Frequent extinction of local population will not endanger the species if very rapid multiplication of adaptive migrants is possible. In slowly reproducing forms on the other hand the greater genetic flexibility resulting from incomplete fixation of the genotype in biparentally reproducing populations is an advantage.

EVOLUTION BY PHYSIOLOGICAL TRANSFORMATION

To many it has seemed impossible to account for the niceties of adjustment of the parts of any organism to each other or of any organism as a whole to its environment on any evolutionary hypothesis which involves chance. To such persons, evolution, if a natural process at all, can only be conceived of as an essentially physiological one.

In this category, we may put first the theory of evolution as a result of the inheritance of acquired characters, formulated by Lamarck, and still widely held. We will not review here the well known difficulties encountered by this theory in the history of the germ cells and in the epigenetic character of the developmental process. The greatest difficulties with its use in interpreting evolution, at least in multicellular organisms, are, on the one hand, the absence of repeatable experimental evidence for any heritable effect of use or disuse and, on the other, the evidence that the vast body of hereditary variability due to Mendelian mutation seems definitely not to be subject to physiological control. The essence of the gene concept, required by the data, is that of entities whose specific properties (revealed by a repertoire of definite effects each in relation to a set of associated genes and environmentally conditions) are duplicated and transmitted with the greatest precision completely irrespective of the particular mode of manifestation (or lack of manifestation) in the parents. Even the occasional abrupt changes in specificity (gene mutations) that are observed, appear to be of an undirected, accidental character.

It should be said that there is considerably more evidence in one celled organism for hereditary change correlated with physiological adaptation (adaptive Dauermodifikation of Jollos) than in multicellular organisms and it would be dogmatic to assert that such processes can play no role whatever in the evolution of the latter, especially in general cellular characters. But it seems hardly likely that this role can be more than a minor one in comparison with that of the vast body of demonstrable Mendelian variability.

The case is closely similar with the theory of evolution through modifications of heredity of a non-adaptive sort, specifically imposed by the external environment. This theory, suggested by Buffon as an explanation of particular minor differentiations has been widely held in conjunction with the theory of the inheritance of acquired characters. Here again there is considerable evidence for the production of specific non-adaptive « Dauermodifikationen » in one celled forms but no convincing evidence in higher forms. This statement requires the qualification that it has been abundantly demonstrated that particular agents (X-rays, radium, ultra-violet, heat), may greatly increase the rate of occurence of gene mutations. But the bulk of the evidence indicates that this increase is not directed along any particular lines by particular agents. There is merely an increase in the rate of occurence of accidents in the process of gene duplication. Unusual environmental conditions may affect the rate of the evolutionary process by changing mutation rates but the direction seems to be left to statistical processes.

The conception of evolution as an unfolding of innate potentialities, analogous to the process of individual development is one that has attracted a number of biologists (e. g. Nageli, Eimer and more recently Osborn). As a complete explanation, this theory immediately encounters the absurdity of supposing that two systems, the organism and its environment, can continue to fit each other at all, down the ages, if evolving wholly independently. In many of the cases described as orthogenetic nothing more seems to be involved than adaptive advance along the only line open to a form which has, so to speak, irrevocably committed itself by specialized adaptations. There are, howerer, certain minor trends which strongly suggest a sort of organic momentum.

In this group are the cases in which there has been an orderly but seemingly useless or even injurious increase in size or complexity of particular structures (horns of titanotheres and certain deer, tusks of certain proboscidians, spines in various invertebrates, sutures of ammonites, etc.). The suggestion of Julian Huxley, that orthogenetic trends of this sort (where not, after all, adaptive in themselves) are by products of the action of selection on general size, is a plausible one. From the morphological standpoint, the system of genes may be thought of as a system of physiological agents, determining directly or indirectly a system of differential growth rates in relation to the primary pattern of the embryo. Being differential, an increase in extent of growth automatically carries with it changes in proportions (heterogonic growth of Pezard). It is understandable that an alteration of the system of interrelated growth rates might be a matter of considerably greater evolutionary difficulty than change in the extent of growth. Under conditions in which large size is an advantage, subject to natural selection , the resulting increase in average size might well be accompanied by a change in proportions of a neutral or even somewhat injurious character as a byproduct.

As to the possibility of innate trends in gene mutation, it is clear that genes may mutate more easily to certain alleles than to others (Timoféeff-Ressovsky). Thus A_1 may mutate most frequently to A_2, this to A_3 and so on. But as brought out earlier such a process could continue only in very small populations or in the case of genes that are almost wholly indifferent, as otherwise an equilibrium would be reached between mutation and selection.

The type of case in which mutation pressure has been most frequently invoked as a probable evolutionary factor is that of degeneration of useless organs beyond the point to which it would seem possible that they could be carried by selection based on their effects as encumbrances (e. g. hind leg bones of whales, eyes and pigmentation of many cave formes, etc.). Mutations, being of the nature of accidents, are more likely to cause interference with the development of positive characters than enhancement. Inspection of the mutations actually encountered in such a form as Drosophila indicates that mutations are predominantly of this degenerative sort. The degeneration by mutation pressure of organs that have lost their usefulness is indeed merely an expression of the shift in mean gene frequencies discussed in a previous section.

It is probable, however, that selection continues to play an important role in such cases, though of a less obvious sort. Genes have multiple effects and the heredity which has maintained an organ may be deeply involved in the development of other parts. If these are vigorously conserved by selection after the organ under consideration has lost its usefulness, the latter may be maintained for a long time in spite of mutation pressure and weak adverse selection but if there is adaptive modification of the correlated parts under the new conditions, the replacements of genes would tend automatically to bring about degeneration and loss of the useless organ.

One reason for the common belief that some sort of directed hereditary change is necessary for evolution is the belief that the effects of selection must be of obvious life or death value to be effective. But, as we have seen, this is not at all the case under the statistical theory. Small adaptive differences, not only in mortality but also in mating or in fecundity, merely need to give a minute net advantage to a gene (perhaps no more than 10^{-5}) to overbalance ordinary mutation pressures in a large freely interbreeding population. In the case of subdivision into partially isolated races, very slight differences in rates of population growth among the races and hence in capacity to furnish migrants would have rapid evolutionary consequences. If there were a tendency toward inheritance of acquired characters or toward orthogenetic change, too slight to be detected in laboratory experiments, these could have no appreciable evolutionary significance in large populations in the presence of moderate selection pressures.

EVOLUTION FROM MUTANT INDIVIDUALS

At the opposite extreme from those who require a physiologically directed process of evolution are those who are not daunted by contemplating the origin of the whole character complex of a species or even of a higher category by a single chance event. Among the pioneer evolutionists, E. G. St. Hilaire studied monstrosities as a possible source of major evolutionary change. Goldschmidt, in recent years, has held that species and higher categories must originate in " hopeful monsters ".

De Vries first brought the conception of abrupt evolution into the field of experimental science by showing that transmissible mutations, differing from type much as one species differs from another, were actually being produced wholesale in Oenothera lamarckiana. Since then, mutations that might be considered as of this kind have been observed in many diverse organisms.

Many of these major mutations, including most of those observed by de Vries, have been shown to depend on gross aberrations in the transmission of chromosomes. These include doubling or other multiplication of the number of basic sets of chromosomes, duplication or losses of whole chromosomes or of parts, fusions, fragmentations, inversions and translocations. Comparison of the chromosome arrays of related species (e. g. Babcok's studies of Crepis) demonstrate that such changes must have been of frequent occurence in the course of evolution, though of very different frequencies in different groups. The great majority appear to ave been balanced ones (multiplications of the entire set or mere rearrangements of the chromatin within the set) rather than ones involving unbalanced additions or losses of material. The unbalanced type, however, seem to have occured occasionally (Darlington). Whatever the significance of such changes for evolution of characters or for speciation, they are solely responsible for the process of evolution of the chromosome complex considered by itself.

As to character evolution, the balanced rearrangements appear to have no more significance than single gene mutations. Those which have occurred in laboratory experiments either have no visible effect, or effects no greater than single genes (position effects). The numerous inversions described by Sturtevant and Dobzhansky in natural races of Drosophila pseudoobscura have no correlation with recognizable character differences. The numerous polyploid varieties of plants are usually somewhat larger than the diploids but with few or no other differences. The unbalanced chromosome aberrations (e. g. the Datura trisomics described by Blakeslee) produce complex enough effects, but as noted have little chance of fixation in nature.

There is, however, at least one type of chromosomal change which is responsible for the origin of new forms of undoubted specific rank. This is the occurence of fertile true breeding allotetraploids from doubling of the set of chromosomes found in certain sterile hybrids between remote species. Their character differences from both parent species, their fertility inter se and their production of largely sterile hybrids in crosses to the two parent species mark them as new species. The occurence of undoubted allotetraploid species in nature (e. g. Galeopsis tetrahit according to Müntzing), Iris versicolor according to E. Anderson, certain Nicotianas according to Clausen and Goodspeed, etc, demonstrates that there is here a real evolutionary process. The case of Spartina Townsendii, discovered in England in 1870 and almost certainly an allotetraploid from the cross of European S. stricta with American S. alterniflora (Huskins) demonstrates that an important adaptive advance may occur at one step. This plant was not only able to establish itself, but has found a relatively unoccupied ecological niche for itself on muddy foreshores where it has multiplied enormously in recent years. These cases do not prove, however, that allotetraploidy is among the most important of species forming processes among organisms in general. It is a process that must be very rare except in forms in which the primary sterile hybrids can reproduce vegetatively.

Most mutants, chromosomal as well as genic, appear in populations with which they can interbreed. The fact that the effect may seem of specific or higher rank does not remove the statistical problem of accounting for the establishment of the mutation in the population. The same statistical theory applies to all gene mutations, great and small. Most of the types of chromosome aberration can also be treated formally in the same way as gene mutations (with some complications) (e. g. reciprocal interchanges).

Each mutation has a rate of occurence and is subject to selection and to the accidents of sampling.

The chance that any mutation will have a net favorable effect at its first occurence is very slight. The greater the effect, the less the likelihood of being adaptive, a point which Fisher has discussed mathematically. The chance of an immediate adaptive effect is probably greatest in the balanced chromosomal mutations but even these produce progenies, on crossing with the normals of the population, which usually include non-viable types with consequent vigorous selection against the mutation. Thus polyploids produce nearly sterile triploid hybrids and would certainly be eliminated rapidly under predominant biparental reproduction apart from disturbances created in sex determination in many cases (Muller). Translocations produce recombination types with non-viable deficiencies and duplications. In the case of inversions such types are typically produced by crossing over (though exceptions are possible as in Drosophila in which there is no crossing over in males and the unbalanced products of crossing over with inversion go to the polar body (Sturtevant and Beadle). In all of those cases the selection term ($\bar{W}^{2N}$) in the formula for the distribution of frequencies is such that there is no appreciable chance of fixation of even a favorable chromosome rearrangement, under biparental reproduction, unless there is frequent opportunity for the development of populations from isolated individuals. It is interesting in this connection that Dobzhanski and Sturtevant find a multiplicity of different inversions but no translocations among local populations of Drosophila pseudoobscura. A prevalence of partial isolation is indicated. In the case of the closely allied species D. miranda, however, Dobzhansky and Tan find at least 6 translocations (as well as more than 50 inversions) differentiating the arrangement of the bands in the chromosomes of the salivary glands from those of D. pseudoobscura. It seems necessary to postulate numerous origins of populations from few individuals in the period since separation of these species.

Under prevailing uniparental reproduction, the establishment of major mutations obviously presents much less difficulty than under biparental reproduction. This is especially true under vegetative multiplication, under which more or less adaptive heterozygous types can multiply without restraint from the production of inviable segregants. Under exclusive uniparental reproduction, however, the species concept itself loses most of its meaning.

Irregularities in meiosis, due to differences in the chromosome sets derived from the parents, is clearly one cause of hybrid sterility (e. g. in triploids and in those sterile hybrids which produce fertile allotetraploids on doubling of the chromosome complex) (Darlington). Studies of other hybrids, however, have shown that multiple gene differences may be responsible for imperfect development of the gonads in which the germ cells never reach meiosis at all (Dobzhansky). It appears probable now that cross sterility is usually a by-product of the gradual accumulation of genetic differences between geographically isolated forms (Stern, Dobzhansky).

Genetics can make but little direct contribution to the question as to whether the distinguishing characters of categories higher than the species arise by a cumulative process (as held by most biologists) or arise abruptly (as held by Austin Clark and by Goldschmidt). If species may arise from isolation of single mutations, presumably such mutations may occasionally (but much less frequently) produce great enough changes to warrant recognition of a higher category. If, however, species differences are usually built up by gradual accumulation this should be even more the rule for higher categories.

The conclusion that the accumulation of non-adaptive differences, arising under local inbreeding, may play a major role in the building up of the differences between species and genera, does away with the argument that these categories must have arisen at single steps by the occasional occurence of major mutations in those cases in which selection can not be invoked.

VIII

SUMMARY

The various evolutionary factors can be brought under a common viewpoint by considering the rates at which they change gene frequency. Formulae are presented for the rate of change under simultaneous pressure from recurrent mutation, selection and migration and for the variability in this rate resulting from the accidents of sampling. The central problem treated is that of the probability distribution of gene frequencies, tending on the one hand to approach an equilibrium under the pressure of opposing forces and on the other to drift at random under the accidents of sampling. It is brought out that accidents of sampling may be much more important in populations in which interbreeding is restricted than is at first apparent.

The analysis indicates that pressure from recurrent mutations can play only a minor role in directing the course of evolution. The most plausible case is that of orthogenetic degeneration of characters that have become useless. Even here mutation pressure can hardly be important by itself except in the case of small permanently isolated populations.

The considerable importance of the role played by individual major mutations is recognized. The extensive evolution of the chromosome arrangement itself seems to be due largely to major chromosomal aberrations of a balanced nature. Character evolution probably owes much less to major mutations but such cases as allotetraploid species demonstrate the reality of evolution by processes of this sort. It is brought out that there is almost no chance of establishment of major mutations within large freely interbreeding populations. The more neutral aberrations may become established in small isolated populations to spread over a wide territory later. Those aberrations which produce inviable classes in crosses with normals of the population can hardly be established in sexually reproducing forms except in populations derived from a few stray individuals. Under uniparental reproduction, on the other hand, the establishment of a major mutation or aberration is limited only by its adaptive value.

The most favorable conditions for evolution appear to be furnished by a certain balance in the actions of the various evolutionary processes on the material furnished by recurrent minor mutations. Under biparental reproduction low rates of mutation balanced by moderate selection are enough to maintain a practically infinite field of possible gene combinations within the species, among which there should be innumerable peaks representing harmonious combinations separated by less harmonious ones. At any given moment, the portion of the field actually occupied, typically clustered about a single peak, is relatively small, although sufficiently extensive that no two individuals have exactly the same genetic constitution.

In a large freely interbreeding population, with no secular change in the direction of selection over long periods of time, a condition of equilibrium is approached in all gene frequencies which is affected only to a minor extent, by systematic changes in rate of mutation or severity of selection. Under conditions which alter the direction of selection, however, there is movement of the system of gene frequencies toward the neighbouring portion of the field of combinations which has become the most adaptive. The course of evolution is here guided by selection among individuals of the population.

In sufficiently small, completely isolated populations, the random divergencies of gene frequencies from their equilibrium values become important, tending in each case to bring about approximate fixation of some random combination of genes which is not likely to be a peak combination. The result is a largely nonadaptive differentiation. In extreme cases, there may be the deterioration which characteristically follows excessive inbreeding. Isolation may here be considered the dominating evolutionary factor.

In a large population subdivided into numerous small partially isolated groups, the combination of directed and random divergences in gene frequencies, subject to intergroup, as well as intragroup selection, gives a trial and error mechanism under which the species may evolve continuously even without secular changes in conditions (although this process, occurring in all species, itself tends to bring about such secular changes). A certain degree of partial isolation of local populations within a large species seems to provide the most favorable conditions for evolutionary advance.

REFERENCES

ANDERSON, Edgar. 1928. *Annals Missouri Botanical Garden* 15 : 241-332.
BABCOCK, E. B. and M. NAVASHIN. 1930. *Bibliographia Genetica* 6 : 1-90.
BELLING, J. 1914. *Zeitschr. f. ind. Abst. u. Vererb.* 12 : 303-342.
— 1925. *Ibid.* 39 : 286-288.
BLAKESKLEE, A. F. 1932. *Proc. 6th Int. Congress Genetics* 1 : 104-120.
CLAUSEN, R. E. and T. H. GOODSPEED. 1925. *Genetics* 10 : 278-284.
CRAMPTON, H. E. 1916. *Carnegie Inst. Washington Publ.* 228 : 1-311.
— 1932. *Ibid. Publ.* 410 : 1-335.
CUÉNOT, L. 1925. *L'adaptation.* Paris, G. Doin. 420 pp.
DARLINGTON, C. D. 1937. *Recent advances in cytology.* Blakiston's, Philadelphia.
DAVENPORT, C. B. 1903. *The Decennial Publications.* The University of Chicago. 10 : 157-176.
DICE, L. R. and P. M. BLOSSOM. 1937. *Carnegie Inst. Washington. Publ.* 485 : 1-129.
DOBZHANSKY, Th. 1937. *Genetics and the Origin of Species.* Columbia Univ. Press.
DOBZHANSKY, Th. and M. L. QUEAL. 1938. *Genetics* 23 : 239-251.
DOBZHANSKY, Th. and A. H. STURTEVANT. 1938. *Genetics* 23 : 28-64.
DOBZHANSKY, Th. and C. C. TAN. 1936. *Zeitschr. f. ind. Abst. u. Vererb.* 72 : 88-114.
EAST, E. M. 1918. *Amer. Nat.* 52 : 273-289.
ELTON, C. S. 1924. *Brit. J. Exp. Biol.* 2 : 119-163.
FISHER, R. A. 1918. *Trans. Roy. Soc. Edinburgh* 52 : part 2 : 399-433.
— 1922. *Proc. Roy. Soc. Edinburgh* 42 : 321-431, 571-574.
— 1928. *Amer. Nat.* 62 : 115-126.
— 1929. *Ibid.* 63 : 553-556.
— 1930*a*. *The Genetical Theory of Natural Selection.* 272 pp. Oxford. Clarendon Press.
— 1930*b*. *Proc. Roy. Soc. Edinburgh* 50 : 205-220.
— 1930*c*. *Amer. Nat.* 64 : 385-406.
— 1931. *Biol. Rev.* 6 : 345-368.
GOLDSCHMIDT, R. 1933. *Science* 78 : 539-547.
— 1934*a*. *Proc. 6th Congress Genetics.* 1 : 173-184.
— 1934*b*. *Bibliographia Genetica* 11 : 1-186.
GONZALEZ, B. M. 1923. *Amer. Nat.* 57 : 289-325.
GULICK, J. T. 1905. *Carnegie Inst. Washington Publ.* 25 : 1-269.
HALDANE, J. B. S. 1930*a*. *Jour. Genetics* 22 : 361-372.
— 1930*b*. *Amer. Nat.* 64 : 87-90.

— 1932. *The causes of evolution.* Harper & Bros. London. (includes summary of mathematical theory of selection presented in numerous earlier papers).
— 1933. *Amer. Nat.* 67 : 5-19.
— 1935. *Journ. Gen.* 31 : 317-326.
— 1937. *Amer. Nat.* 71 : 337-349.
HALDANE, J. B. S. and C. H. WADDINGTON. 1931. *Genetics.* 16 : 357-374.
HARDY, G. H. 1908. *Science* 28 : 49-50.
HUSKINS, C. L. 1931. *Genetica* 12 : 537-538.
HUXLEY, J. S. 1932. *Problems of relative growth.* 276 pp. New York. Lincoln Mc Veagh.
JENNINGS, H. S. 1916. *Genetics* 1 : 53-89.
JOLLOS, V. 1931. *Verh. Deuts. Zool. Ges.* 252-295.
— 1934. *Genetica* 16 : 476-494.
JORDAN, D. S. 1905. *Science* 22 : 545-562.
KELLOGG, V. L. 1908. *Darwinism Today.* 403 pp. New York. Henry Holt & Co.
KINSEY, A. C. 1929. *Indiana University Studies* N° 84, 85, 86. 577 pp.
— 1936. *Ibid. Science Series* N° 4. 334 pp.
— 1937. *Proc. Nat. Acad. Sci.* 23 : 5-11.
KOLMOGOROV, A. 1935. *Compt. Rend. Acad. Sci. U. R. S. S.* 3 : 129-132.
MULLER, H. J. 1925. *Amer. Nat.* 59 : 346-353.
— 1932*a*. *Proc. 6th Int. Gen. Congress* : 213-255.
— 1932*b*. *Amer. Nat.* 64 : 118-138.
MÜNTZING, A. 1932. *Hereditas* 16 : 105-154.
OSBORN, H. F. 1927. *Amer. Nat.* 49 : 193-239.
— 1933. *Proc. Nat. Acad. Sci.* 19 : 159-163.
OSGOOD, W. H. 1909. *Revision of the mice of the genus, Peromyscus.* North American Fauna 28 : 1-285.
PLUNKETT, C. R. 1932. *Proc. 6th Int. Congress Genetics* 2 : 160-162.
PICKFORD, G. E. 1937. *A monograph of the Acanthodriline Earthworms of South Africa.* 97 pp. Bournemouth Guardian.
RENSCH, B. 1929. *Das Prinzip geographischer Rassenkreise und das Problem der Artbildung.* 206 pp. Berlin. Gebrüder Borntraeger.
ROBBINS, R. B. 1918 *a*. *Genetics* 3 : 73-92.
— 1918 *b*. *Ibid.* 3 : 375-396.
ROBSON, G. C. and O. W. RICHARDS. 1936. *The variation of animals in nature.* London. Longmans Green.
SCHMIDT, J. 1917. *Jour. Gen.* 7 : 105-118.
STERN, C. 1930. *Amer. Nat.* 70 : 123-142.
STURTEVANT, A. H. and G. W. BEADLE. 1936. *Genetics* 21 : 554-604.
STURTEVANT, A. H. and C. C. TAN. 1937. *Jour. Gen.* 34 : 415-439.
SUMNER, F. B. 1932. *Bibl. Genet.* 9 : 1-106.
THOMPSON, D. H. 1931. *Trans. Ill. State Acad. Sci.* 23 : 276-281.
TIMOFÉEFF-RESSOVSKY. 1934*a*. *Biol. Reviews* 9 : 411-457.
— 1934*b*. *Zeitschr. f. ind. Abst. u. Vererb.* 66 : 319-344.
WEINBERG, W. 1909. *Zeitschr. f. ind. Abst. u. Vererb.* 1 : 277-330.
— 1910. *Arch. Rass. u. Ges. Biol.* 7 : 35-49, 169-173.
WRIGHT, S. 1921. *Genetics* 6 : 111-178.
— 1929*a*. *Amer. Nat.* 63 : 274-279.
— 1929*b*. *Amer. Nat.* 63 : 556-561.

— **1931**. *Genetics* 16 : 97-159.
— **1932**. *Proc. 6th Int. Congress Genetics* 1 : 356-366,
— **1933**. *Proc. Nat. Acad. Sci.* 19 : 411-420, 421-433.
— **1934*a***. *Amer. Nat.* 68 : 24-53.
— **1934*b***. *Annals Math. Statistics* 5 : 161-215.
— **1935**. *Journ. Gen.* 30 : 243-256 ; 257-266.
— **1937**. *Proc. Nat. Acad. Sci.* 23 : 307-320.
— **1938*a***. *Proc. Nat. Acad. Sci.* 24 : 253-259.
— **1938*b***. *Proc. Nat. Acad. Sci.* 24 : 372-377.

24

Breeding Structure of Populations in Relation to Speciation

American Naturalist 74 (May–June 1940):232–48

25

The Statistical Consequences of Mendelian Heredity in Relation to Speciation

In *The New Systematics*, ed. by Julian S. Huxley, 161–83. Oxford: Oxford Univ. Press, 1940

Introduction

These are the only two papers that Wright ever devoted entirely to the problem of speciation. He delivered the first paper at a joint symposium entitled Speciation sponsored by the American Society of Zoologists, Genetics Society of America, and the American Association for the Advancement of Science. Dobzhansky organized the symposium, and Wright and Mayr gave the major addresses; their papers were published back to back in the *American Naturalist*. The forum was prestigious. The second paper was commissioned by Julian Huxley for his edited volume *The New Systematics*, also a prestigious forum.

Mayr on several occasions has observed that Wright devoted very little of his efforts in evolutionary biology to the problem of speciation, which to Mayr is the core of the evolutionary process. Mayr's observation is historically accurate, but the influence of these two papers was considerable, even upon Mayr himself.

In Mayr's published paper from the speciation symposium he declared that the definition of species that Wright contributed in his address exhibited "the fewest flaws" of all species definitions (Mayr 1940, 255). Perhaps more important, Wright presented in his paper a clear statement of what Mayr later would term "founder effect," with the difference that Wright considered it most significant when founder effects occurred serially with population flushes after each founder event. Again in this paper, as he had in 1938, Wright suggested the possibility of isolation by distance as a factor in setting the conditions for speciation even in geographically continuous areas. Moreover, especially in the second paper, Wright argued that the process of speciation generally followed a population split that resulted in the daughter populations being located on different peaks on his surface of selective values, or a shift of one population from one adaptive peak to another on his surface of selective values. In Wright's view, such peak shifts required a change of selective values in many genes. Thus Wright's view of speciation in these papers shares many common elements with Mayr's concept of the founder effect (Mayr 1942) and his later concept of genetic revolutions that occasionally followed the founder event (Mayr 1954). Mayr carefully cited and quoted from Wright's papers in both 1942 and 1954 publications. Further, Wright's concept of speciation in these papers shares much with Carson's founder-flush concept of speciation derived primarily from Mayr (1954) and observations and analysis of speciation phenomena in Hawaiian *Drosophila*.

I treat Mayr and Wright on speciation in *SW&EB*, 479–84. Wright readdressed the question of speciation in a paper published in 1982 (Wright 205), which is included in this volume. It is instructive to compare that paper with these two.

BREEDING STRUCTURE OF POPULATIONS IN RELATION TO SPECIATION[1]

PROFESSOR SEWALL WRIGHT
THE UNIVERSITY OF CHICAGO

INTRODUCTION

THE problem of speciation involves both the processes by which populations split into non-interbreeding groups and those by which single populations change their characteristics in time, thus leading to divergence of previously isolated groups.

The first step in applying genetics to the problem is undoubtedly the discovery of the actual nature of the genetic differences among allied subspecies, species and genera in a large number of representative cases. Differences which tend to prevent cross-breeding are obviously especially likely to throw light on the process of speciation, but all differences are important.

Our information here is still very fragmentary. We know enough, however, to be able to say that there is no one rule either with respect to cross-sterility or to other characters. In some cases the most significant differences seem to be in chromosome number and organization. At the other extreme are groups of species among which gross chromosome differences and even major Mendelian differences are lacking, both cross-sterility and character differentiation depending on a multiplicity of minor gene effects. In general, there are differences at all levels (*cf.* Dobzhansky, 1937).

But even if we had a complete account of the genetic differences within a group of allied species, we would not necessarily have much understanding of the process by which the situation had been arrived at. A single mutation is not a new species, except perhaps in the case of polyploidy. The symmetry of the Mendelian mechanism

[1] Read at a joint symposium on "Speciation" of the American Society of Zoologists and the Genetics Society of America, American Association for the Advancement of Science, Columbus, Ohio, December 28, 1939.

is such that any gene or chromosomal type tends to remain at the same frequency in a population except as this frequency is changed either by some steady evolutionary pressure (such as that due to *recurrent* mutation, to various kinds of selection, to immigration and to differential emigration) or by the accidents of sampling, if the number of individuals is small. The elementary evolutionary process, from this view-point, is change of gene frequency.

It is to be expected that the nature of the process will be found to be affected by what I have called the breeding structure of the species, and it is this aspect of the matter that I wish to discuss here. Such a discussion involves at least three steps. First, there is the observational problem of determining what the breeding structures of representative species actually are. Naturalists are only beginning to collect the detailed information which turns out to be necessary, but that which we have indicates situations of great complexity. The second step is that of constructing a mathematical model which represents adequately the essential features of the actual situation while disregarding all unimportant complications. The third step is the determination of the evolutionary implications of a given breeding structure in relation to mutation and selection. As difficult problems of description and mathematical formulation are also involved in the cases of mutation and selection pressures, the whole problem is exceedingly complex. I can only discuss the implications of certain very simple models of breeding structure, chosen partly because they appear to correspond to situations which one might expect to find in nature, but partly also because of mathematical convenience.

Evolution under Panmixia

The simplest situation, under biparental reproduction, is that of a large population, breeding wholly at random (panmixia). If sufficiently large, variability due to accidents of sampling is negligible. Each gene frequency shifts steadily under the pressures of selection and recurrent mutation. Mathematical formulations of these pressures have been given. Letting q be the frequency of a given gene, $(1-q)$ that of its alleles, u and v the mutation rates respectively from and to the gene in question and $\overline{W}$ the mean selective value of all possible genotypes, weighted by their frequencies, the change in gene frequency in a generation is given by the following formula (Wright, 1937):

$$\Delta q = v(1-q) - uq + \frac{q(1-q)\partial\overline{W}}{2\overline{W}\partial q}$$

For a gene which causes the same difference from its allele in all combinations and which lacks dominance, the term for selection pressure reduces approximately to $sq(1-q)$, where s is the selective advantage over the allele.

The numbers of generations necessary for any given shift in gene frequency, under various hypotheses, have been presented by Haldane (1932 and earlier). This sort of process has been taken as typical of evolutionary change by R. A. Fisher (1930), who has compared its unswerving regularity to that of increase in entropy in a physical system.

If, however, conditions are constant, this process comes to an end at an equilibrium point at which opposing pressures balance each other ($\Delta q = 0$). At this point there is stability of the species type in spite of continual occurrence of mutations, an extensive field of variability and continuous action of selection. On the other hand, conditions never are wholly constant. It is possible that evolution, in each series of alleles, may consist of an unswerving pursuit of an equilibrium point, which is itself continually on the move because of changing conditions.

The postulate that variations in gene frequency, due to accidents of sampling, are negligible calls for some comment. The variance in one generation is $\sigma^2_{\Delta q} = \dfrac{q(1-q)}{2N}$ in a diploid population of effective size N. This is cumulative and may cause wide divergence from equilibrium if the population is not too large. The systematic evolutionary pressures directed toward equilibrium and this sampling variance determine between them a certain distribution of values of the gene frequency instead of a single equilibrium point. The general formula can be written as follows (Wright, 1937):

$$\varphi(q) = (C/\sigma^2_{\Delta q})\, e^{2\int(\Delta q/\sigma^2_{\Delta q})dq}$$

For the values of Δq and $\sigma^2_{\Delta q}$ given above this reduces to

$$\varphi(q) = C\overline{W}^{2N}q^{4Nv-1}(1-q)^{4Nu-1}$$

In the special case of no factor interaction and no dominance, the term $\overline{W}^{2N}$ becomes approximately e^{4Nsq}. There is a marked tendency toward chance fixation of one allele or another if $4Ns$, $4Nv$ and $4Nu$ are all less than 1 while such variability is negligible if these quantities are large (*e.g.*, as large as 100).

The possible evolutionary significance of these random variations in gene frequency in a panmictic population has been considered elsewhere (Wright, 1931, 1932) and will not be discussed further here.

Mating never is wholly at random. It is important to determine whether departures from panmixia have significant effects on the evolutionary process and if so whether these consist merely in impeding the pursuit of equilibrium or whether they may not bring about progress of a different sort.

One limitation on the effectiveness of selection in a panmictic population is that it can apply only to the *net effects* in each series of alleles. It is really the organism as a whole that is well or ill adapted. A really effective selection pressure should relate to genotypes not genes. But in a panmictic population, combinations are formed in one generation only to be broken up in the next.

If a selective value (W) is assigned to every one of the practically infinite number of possible combinations of genes of all loci, the array of such values forms a surface in a space of at least as many dimensions as there are loci, more if there are multiple alleles. Because of nonadditive factor interactions, this surface in general has innumerable distinct peaks (*i.e.,* harmonious combinations) each surrounded by numerous closely related but slightly less adaptive combinations and separated from the others by valleys. Selection according to net effect can only carry the species up the gradient to the nearest peak but will not permit it to find its way across a valley to a higher peak. Evolution would have a richer field of possibilities under a breeding system that permitted exploration of neighboring regions in the surface of adaptive values, even at some expense in momentary adaptation.

A somewhat similar situation holds within systems of *multiple alleles* (*cf.* Timoféëff-Ressovsky, 1932). There is presumably a limit to the number of alleles that can arise from a given type gene by a single act of mutation. But each of these mutations presumably can give rise to mutations at two steps removed from the original type gene and so on in an indefinitely extended network. If there is approximate fixation of one allele (to be expected in general under panmixia), only those mutations that are at one or two removes have any appreciable chance of occurrence. There will be continual recurrence of the same mutations without real novelty. A breeding system that tolerates a continually shifting array of multiple alleles in each series in portions of the population, gives the opportunity for a trying out of wholly novel mutations which occasionally may be of great value. The question then is whether there are breeding structures that permit trial and error both within each system of multiple alleles, and within the field of gene combinations, in such a way as to give a richer field of possibilities than under the univalent determinative process in a panmictic population.

EVOLUTION UNDER UNIPARENTAL REPRODUCTION

At the opposite extreme from the system of random mating is that in which there is uniparental reproduction.

Under vegetative multiplication, or under diploid parthenogenesis, each individual produces a clone in which all individuals are of exactly the same genotype, except for occasional mutations. Continued self fertilization also leads to the production of groups of essential identical individuals.

Suppose that a highly variable panmictic population suddenly shifts to uniparental reproduction. Selection then would be between genotypes. Those combinations that are most adaptive would increase, including perhaps rare types that would have been broken up and lost under panmixia. The less adaptive combinations would soon be displaced. Selection would be exceedingly effective until only one clone was left in each ecological niche. But at this point evolution would come to an end, except for the exceedingly rare occurrence of favorable mutations.

It is obvious that a certain combination of the preceding systems should be much more effective than either by itself (*cf.* Wright, 1931). Prevailing uniparental reproduction, with occasional crossing would permit an effective selection by genotypes to operate in a continually restored field of variability. This combination is of course one that has been used most effectively by plant breeders. It is found in many plants and animals in nature and has presumably been an important factor in their evolution.

The demonstration of the evolutionary advantages of an alternation of periods of uniparental reproduction with cross-breeding may seem to prove too much, since it is not usual in those groups that are usually considered to have evolved the most, the higher arthropods and vertebrates. Perhaps, however, there are other systems which also bring about differentiation of types and thus a basis for selection based on type rather than mere net gene effect, and which have more stability than arrays of clones.

Evolution in Subdivided Populations

A breeding structure that happens to be very convenient from the mathematical standpoint is one in which the species is subdivided into numerous small local populations which largely breed within themselves but receive a small proportion of their population in each generation from migrants which can be treated as random samples from the species as a whole. The basis for the partial isolation may be geographical, or ecological or temporal (breeding season). In the latter two cases an adaptive difference is postulated. We are not here considering the origin of this but rather its consequences on other characters.

Whatever the mechanism of isolation, its evolutionary significance can be evaluated in terms of the effective size of population (N) of the isolated group, the effective rate (m) of exchange of individuals between the group (gene frequency q) and the species as a whole (gene frequency q_t) and the local selection coefficient. It will be convenient here to write s for the net selection coefficient and to ignore mutation pressure (Wright, 1931).

$$\Delta q = sq(1-q) - m(q-q_t)$$

If s in a local population is much larger than m, we have approximately

$$\hat{q} = 1 - \frac{1}{s}[m(1-q_t)] \qquad \text{if } s \text{ is positive}$$

$$\hat{q} = \frac{mq_t}{(-s)} \qquad \text{if } s \text{ is negative}$$

If the values of s among local populations show differences greater than m, there will be marked adaptive differentiation of such populations. There is an approach toward fixation of the locally favored gene largely irrespective of the frequency in the species as a whole.

The importance of isolation in evolution seems to have been urged first by M. Wagner as permitting divergent evolution under the control of different environments. Wagner thought of environment as directly guiding the course of evolutionary change, when its effects were not swamped by those of cross-breeding. A similar view has been held by many others since his time who have considered such orderly clines among geographical races as those described by the laws of Bergmann, Gloger, and Allen. While direct control over mutation is not in line with present knowledge of genetics, indirect control through differential selection seems probable enough in these cases (*cf.* Dobzhansky, 1937; Huxley, 1939).

Davenport (1903) and Goldschmidt (1934) have stressed the likelihood of the spreading of the range of species by the diffusion of preadaptive mutations into territories in which they are isolated from the first by the inability of the typical members of the species to live. Goldschmidt has interpreted the major differences among races of *Lymantria dispar* in this way. He finds these differences primarily in such physiological characters as developmental rate, length of diapause, etc. Mathematically, this would be a special case of the foregoing scheme.

Differential selection has been considered so far as a factor making only for divergence of groups within the species and thus tending toward splitting of the latter. There is a possibility, however, that it may be a factor making for progressive evolution of the species as a unit. Particular local populations may, by a tortuous route, arrive at adaptations that turn out to have general, instead of merely local, value and which thus may tend to displace all other local strains by *intergroup* selection (excess emigration). In terms of our multidimensional surface of adaptive values, a particular substrain may be guided from one peak to another by a circuitous route around a valley which would probably not have been found except by such a trial and error mechanism. As different alleles may approach fixation in different populations, mutations at two or more removes from the original type have more opportunity for occurrence than if the population were homogeneous. Thus there may be trial and error within series of alleles as well as between gene combinations.

Let us now turn to the case in which the local selection coefficient is smaller instead of larger than m. The local equilibrium frequency ($\hat{q}$) is approximately as follows.

$$\hat{q} = q_t + \frac{sq_t}{m}(1 - q_t)$$

The values in different local populations in which s is smaller than m are clustered closely about the mean gene frequency, q_t. Selection causes no important differentiation. There may however be variability of each local population due to accidents of sampling if N is small and, consequently, much non-adaptive differentiation among such populations at any given moment.

$$\varphi(q) = Ce^{4Nsq}q^{4Nmq_t-1}(1-q)^{4Nm(1-q_t)-1}$$

Figure 1 shows the form of the distribution for various values of Nm, taking $q_t = \frac{1}{2}$ and assuming no selection ($s = 0$). The variance in this case is as follows.

$$\sigma^2_q = \frac{q_t(1-q_t)}{4Nm+1}$$

The distribution of gene frequencies is U shaped, implying random drifting from fixation in one phase to another if m is less than $\frac{1}{4Nq_t}$ and $\frac{1}{4N(1-q_t)}$. This again would permit trial within each series of alleles, and also between gene combinations.

The latter at least would be important even with larger values of m relative to $\frac{1}{4}N$. With $Nm = 5$, the standard deviation of values of q is 22 per cent. of its limiting value $\sqrt{q(1-q)}$. Such variability tends to become unimportant however if Nm is much larger.

Gulick seems to have been the first to point out the possible significance of isolation in bringing about a non-adaptive differentiation of local races. He has been followed by others, notably recently by Kinsey in his studies of the gall wasps of the genus Cynips (1929, 1936). A study of eleven isolated mountain forests in the Death Valley region by Dobzhansky and Queal (1938) showed a close approach to random mating with no appreciable selection within localities. Between localities on the other hand, frequencies ranged from 51 per cent. to 88 per cent., 2 per cent. to 20 per cent., 8 per cent. to 39 per cent. with standard deviations which can be accounted for by an effective value of Nm of about 5.1. The much greater standard deviation for the range of *D. pseudoobscura* as a whole shows that this differentiation is cumulative with distance.

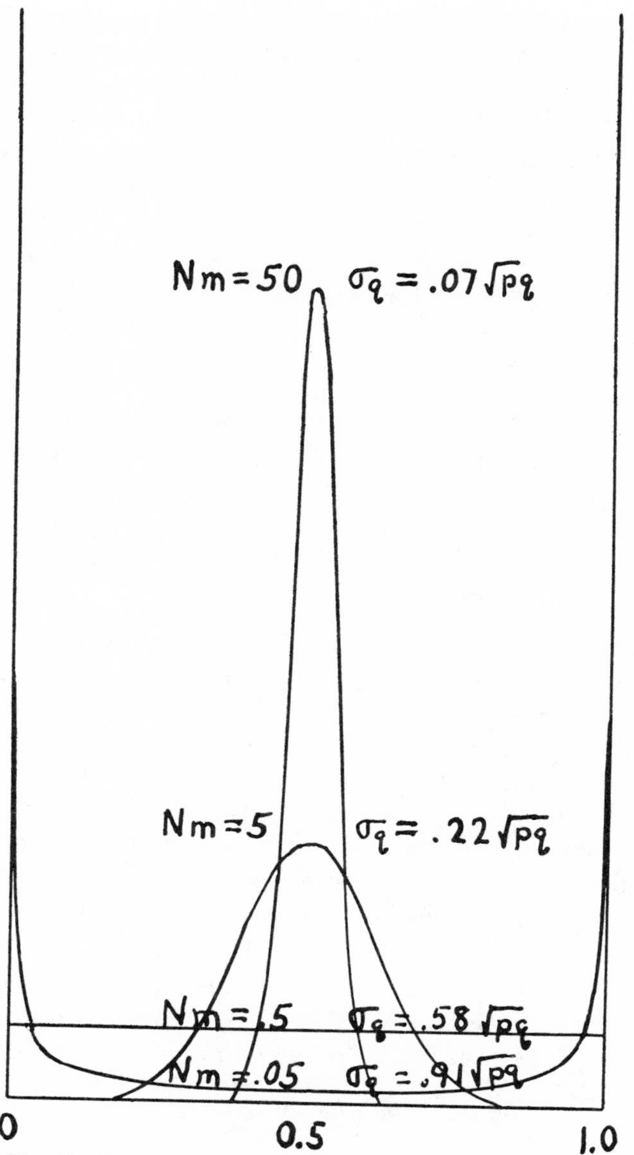

FIG. 1. The distribution of gene frequencies for various values of Nm, taking $q_t = \frac{1}{2}$ and assuming no selection. The symbol p is used for 1-q.

EFFECTIVE SIZE OF POPULATION

There appears to be the difficulty here that the number of individuals in such a form as Drosophila is so enormous that it is difficult to conceive of a limitation in numbers as having any appreciable sampling effects. However, the effective N may be very much smaller than the apparent N (Wright, 1938).

If the number of the two sexes (N_m males, N_f females) is unequal, it can be shown that effective $N = \frac{4N_mN_f}{N_m + N_f}$. With unequal numbers, the effective size of population depends more on the smaller number than on the larger number. Thus with N_m males but an indefinitely large number of females, $N = 4N_m$.

Again, different parents may produce widely different numbers of young. If σ^2_κ is the variance in number of gametes contributed by individuals to the following generation in a population (N_0) that is maintaining the same numbers ($\bar{\kappa} = 2$), $\sigma^2_\kappa = \frac{\Sigma(\kappa - 2)^2}{N_0}$

$$N = \frac{4N_0 - 2}{2 + \sigma^2_\kappa}$$

The effective size of population is twice as great as the apparent in the highly artificial case in which each parent contributes just two gametes. Effective and apparent size of population are the same if the number of gametes contributed by different parents vary at random (Poisson distribution). If, as would often be the case, most of the offspring come from a small percentage of the mature individuals of the parental generation, the effective size would be much less than the apparent size.

A population may vary tremendously in numbers from generation to generation. If there is a regular cycle of a few generations (N_1, N_2 . . . N_n) an approximately equivalent constant population number can be found.

$$N = \frac{n}{\sum_{x=1}^{n} [1/N_x]}$$

This is controlled much more by the smaller than by the larger numbers. Thus if the breeding population in an isolated region increases ten-fold in each of six generations during the summer (N_0 to 10^6N_0) but falls at the end of winter to the same value, N_0, the effective size of population ($N = 6.3N_0$) is relatively small.

In such a cycle, certain individuals in favorable locations are likely to start reproduction earlier than others, perhaps getting a start of a whole generation. In a rapidly breeding form, these few individuals would contribute overwhelmingly more than the average to all later generations. Thus, by a combination of the two preceding principles, the effective size of population may be very small indeed.

The possible evolutionary significance of periodic reduction in the size of natural populations has been discussed by a number of authors. Elton (1934) especially has maintained that chance deviations in the characteristics of survivors at the time of least numbers may have important effects of this sort.

An important case arises where local populations are liable to frequent extinction, with restoration from the progeny of a few stray immigrants. In such regions the line of continuity of large populations may have passed repeatedly through extremely small numbers even though the species has at all times included countless millions of individuals in its range as a whole (*cf.* Fig. 2).

Such mutations as reciprocal translocations that are very strongly selected against until half fixed seem to require some such mechanism to become established. There is an exceedingly deep valley in the surface $\overline{W}$ representing the mean adaptive value in populations with given frequencies of old and new chromosomes, and the term $\overline{W}^{2N}$ in the formula for the joint chromosome frequencies is so small, where N consists of more than some half-dozen individuals that fixation is virtually impossible. Yet translocations have been noted between Drosophila species (*e.g.*, *D. pseudoobscura* and *D. miranda,* Dobzhansky and Tan, 1936) although they are far less common than inversions. The difficulty referred to here does not, of course, apply in species that reproduce vegetatively or by self-fertilization.

We have discussed various considerations that make

effective N much smaller than at first apparent. The effective amount of cross-breeding may also be much less than the actual amount of migration seems to imply. Most of the immigrants are likely to come from neighboring groups, differing less from the receiving population in gene frequency than would a random sample from the

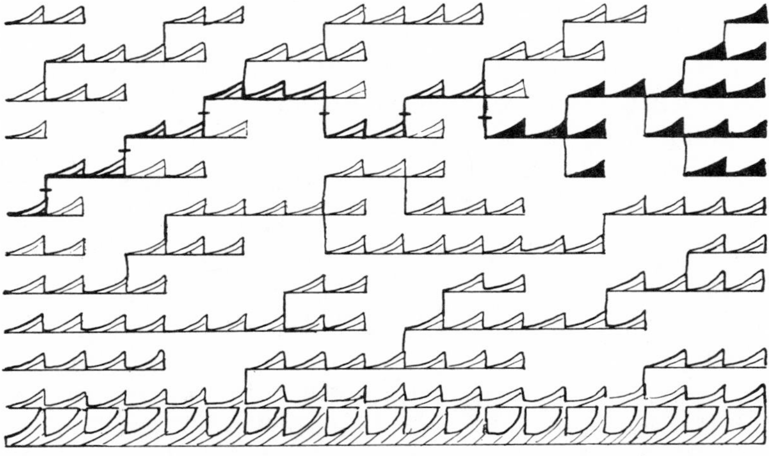

FIG. 2. Diagram of breeding structure in a species in which the populations in certain regions are liable to frequent extinction with reestablishment by rare migrants. Different territories are distinguished vertically. Generations proceed from left to right horizontally. The heavily shaded group represents a large population the entire ancestry of which has passed through small groups of migrants six times in the period shown.

species. If there is a correlation, r, between immigrants and receiving group, the m of the formula must be replaced by $m(1-r)$ if m is to continue to be the actual amount of replacement by immigration.

In the case of *Drosophila pseudoobscura,* it has been noted that Dobzhansky and Queal (1938) found variability in gene frequency among mountain forests of the Death Valley region which implied an effective value of Nm of about 5. For the species as a whole, variability is such that effective Nm must be only about one tenth as large as this (0.5).

ISOLATION BY DISTANCE

This last case leads to another model of breeding structure which may be of considerable importance (Wright, 1938). Suppose that a population is distributed uniformly over a large territory but that the parents of any given individual are drawn from a small surrounding region (average distance D, effective population N). How much local differentiation is possible merely from accidents of sampling? Obviously the grandparents were drawn from a larger territory (average distance $\sqrt{2}\,D$, effective population $2N$). The ancesters of generation n came from an average distance $\sqrt{n}\,D$ and from a population of average size nN. It is assumed that the variance of the ancestral range, either in latitude or in longitude, increases directly with the number of generations of ancestry.

Fig. 3 shows how the standard deviation of gene frequencies for unit territories of various effective sizes increases with distance. If $\sigma_q = .577\sqrt{q_t(1-q_t)}$ and $q_t = 1 - q_t = \frac{1}{2}$ all values of gene frequency are equally numerous ($\phi(q) = 1$). Any larger value implies a tendency toward fixation of one or the other allele in different local populations.

If the parents are drawn from local populations of effective size greater than 1,000, the situation differs little from panmixia even over enormous areas. There is considerable fluctuating local differentiation of unit territories where their effective size is of the order of 100, but not much differentiation of large regions unless effective N is much less.

Kinsey's (1929) description of the gall wasp, *Cynips pezomachoides erinacei,* conforms fairly well to the above model for the case of moderately large N. This subspecies ranges over some 500,000 square miles in northeastern United States. Both the insects and their galls may differ markedly and consistently in collections taken from different trees or small groves at short distances apart, but the same variability is found throughout the range. There is little regional differentiation in this enormous territory, although at still greater distances

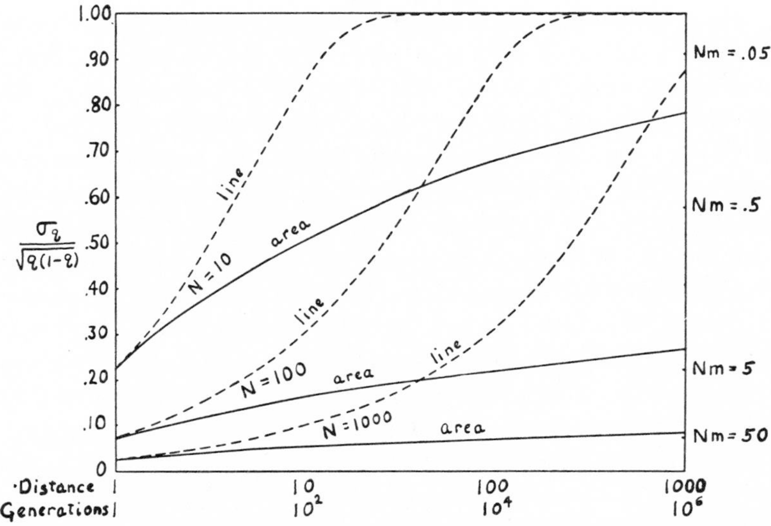

FIG. 3. The standard deviation of the mean gene frequencies of unit random breeding territories (N = 10; N = 100; N = 1000), in relation to mean distance. The case in which the population is distributed uniformly over an area is represented in solid lines, that in which it is distributed along one dimension by broken lines.

the species complex *C. pezomachoides* is subdivided into eight intergrading species.

In a species whose range is essentially one-dimensional (shore line, river, etc.) the ancestors of generation n come from an average distance of $\sqrt{n}\ D$ as before, but the effective size of population is $\sqrt{n}\ N$ instead of nN. Differentiation increases much more rapidly with distance than in the preceding case. This principle was suggested in qualitative terms by Thompson (1931) in his study of differentiation within species of river fish in relation to water distance. In weak swimmers (*e.g.*, Johnny Darters) there was marked increase in average difference in number of fin rays with increased distance in spite of a continuous distribution. The differentiation with distance was not as rapid, however, as that in several species with discontinuous distribution (restricted to the smallest stream). On the other hand the strong swimmers of the rivers showed little or no differentiation throughout their entire ranges.

Such uniformity in breeding structure as implied above is not likely to be closely approximated in nature. Even where there is apparent continuity of a population, it is likely that conditions vary from place to place in such a way that there is excess multiplication at certain centers separated by regions in which the species would be unable to maintain itself permanently were it not for immigration (as in the breeding structure of Figure 2). Moreover, even with complete uniformity of conditions, local differentiation should result in the accidental attainment of more adaptive complexes in some regions than in others. As before, incipient nonadaptive differentiation may lead to a more important adaptive differentiation. The centers in which population is increasing most rapidly will become increasingly isolated from each other by the mere fact that they are centers of emigration.

A process of this sort has been postulated by Sumner (1932) in the case of subspecies of Peromyscus. Within subspecies, he found statistical differentiation of most local populations which may well have been of the type due merely to distance. But at the subspecies boundaries there was typically a zone of relatively rapid change. These boundaries were not necessarily along natural barriers to migration. Sumner compared them with the distributions which would result "if a collection of spherical rubber bags were placed in rigid containers and then strongly but unequally inflated."

The breeding structure of natural populations thus is likely to be intermediate between the model of subdivision into partially isolated territories and that of local inbreeding in a continuous population. In so far as it is continuous, it is likely to be intermediate between area continuity and linear continuity.

Summing up, we have attempted to show that the breeding structure of populations has a number of important consequences with respect to speciation. Partial isolation of local populations, even if merely by distance is important, not only as a possible precursor of splitting of the species, but also as leading to more rapid evolutionary change of the population as a single system and thus more rapid differentiation from other populations from which it is completely isolated. Local differentiation within a species, based either on the nonadaptive inbreeding effect or on local conditions of selection or both, permits trial and error both within series of multiple alleles and between gene combinations and thus a more effective process of selection than possible in a purely panmictic population.

LITERATURE CITED

Davenport, C. B.
1903. *The Decennial Publications,* 10: 157–176. The University of Chicago.

Dobzhansky, Th.
1937. "Genetics and the Origin of Species." New York: Columbia University Press, 364 pp.

Dobzhansky, Th., and M. L. Queal
1938. *Genetics,* 23: 239–251.

Dobzhansky, Th., and C. C. Tan
1936. *Zeit. Ind. Abst. Ver.,* 72: 88–114.

Elton, C. S.
1924. *Brit. Jour. Exp. Biol.,* 3: 119–163.

Fisher, R. A.
1930. "The Genetical Theory of Natural Selection." Oxford: Clarendon Press, 272 pp.

Goldschmidt, R.
1934. "Lymantria," *Bibliographia Genetica,* 11: 1–186.

Haldane, J. B. S.
1932. "The Causes of Evolution." London: Harper and Bros., 235 pp.

Huxley, J. S.
1939. "Bijdragen tot de dierkunde," pp. 491–520. Leiden: E. J. Brill.

Kinsey, A. C.
1929. Studies No. 84, 85, 86. Indiana University Studies, Vol. 16.
1936. Indiana Univ. Publ., Science Series No. 4.

Sumner, F. B.
1932. *Bibliographia Genetica,* 9: 1–106.

Thompson, D. H.
1931. *Trans. Ill. State Acad. Sci.,* 23: 276–281.

Timoféëff-Ressovsky, N. W.
1932. *Proc. 6th Internat. Cong. Genetics,* 1: 308–330.

Wright, S.
1921. *Genetics,* 6: 111–178.
1922. *Bull. no. 1121,* U. S. Dept. of Agr., 59 pp.
1931. *Genetics,* 16: 97–159.
1932. *Proc. 6th Internat. Cong. Genetics,* 1: 356–366.
1937. *Proc. Nat. Acad. Sci.,* 23: 307–320.
1938. *Science,* 87: 430–431.

THE STATISTICAL CONSEQUENCES OF MENDELIAN HEREDITY IN RELATION TO SPECIATION

By SEWALL WRIGHT

THE most obvious starting-point, in attempting to relate laboratory genetics to speciation in nature, is the phenomenon of mutation. The term 'species' primarily means 'kind', and mutations are the elementary observable changes in kind along lines of descent. The abruptness of origin of mutations and their stability after origin appear to give a simple explanation of the traditional distinctiveness and internal homogeneity of species in their essential characters. While it has been obvious that most mutations do not produce changes in kind directly comparable to species-differences, yet from the viewpoint that the term 'species' denotes a certain critical step in the hierarchy of degrees of difference in kind, it has appeared legitimate to identify the origin of species and even of higher categories with mutation, with the qualification that only occasional mutations are of specific rank and still rarer ones of generic, familial, or higher rank. This is a viewpoint taken by certain geneticists, most notably in recent years by Goldschmidt.

Few systematists, however, seem ever to have been at all satisfied with this identification of speciation and mutation (cf. Osborn, 1927). In part this has been because of the apparent dissimilarity in the type of difference referred to above, but more, perhaps, because of a change from the traditional conception of the species, imposed by the findings in those groups which have been studied most exhaustively. It has become necessary to shift the emphasis in the definition of species from the essentially physiological concept, kind, to the ecological one, the interbreeding population. It has come to be recognized that most species must be divided into subspecies, differing rather consistently in kind at their centres of distribution but connected by statistically intergrading populations, and that even within the taxonomically recognizable subspecies, sufficiently careful studies show that no two local populations have exactly the same statistical properties (cf. Schmidt, 1917; Sumner,

1932; Turesson, 1925; anthropologists in general). In the other direction, numerous recognized species living in different regions and supposedly distinct because of marked differences in kind have had to be assembled in species-complexes (or *Rassenkreise*) because of the discovery of intergrading populations between them (cf. Rensch, 1929).

The ideal has been to apply the specific name to groups within which all subdivisions interbreed sufficiently freely to form intergrading populations wherever they come in contact, but between which there is so little interbreeding that such populations are not found. But it is becoming apparent that this ideal is often unrealizable. A chain of intergrading populations may return on itself around a circle, leading to populations which occupy the same territory without interbreeding (e.g. *Peromyscus maniculatus bairdii* (Hoy & Kennicott) and *P. m. gracilis* (Le Conte) in Michigan; Osgood, 1909; Dice, 1931). The most remarkable case of this sort so far reported seems to be that found by Kinsey in the American gall-wasps of the genus *Cynips*. Long branching chains of intergrading 'species' loop back repeatedly to the same region and host, leading ultimately to types so different in kind that they have been considered subgenera. From the standpoint of continuity one of these chains (with 76 to 86 races in one case) must be considered as a single giant *Rassenkreis* (cp. Goldschmidt, 1937), but from the standpoint of sufficient differentiation to permit occupancy of the same territory without interbreeding some 8 or 9 good species must apparently be recognized in the above case, and finally from the standpoint of character-differences there are 86 taxonomically distinguishable 'species' to be grouped arbitrarily in 9 'complexes' and the 2 subgenera, *Acraspis* and *Philonix*.

From the population viewpoint, a single mutation can give rise to a new species only if its descendants form a group not interbreeding in nature with the parent species. But this at once implies coincidence of another sort of evolutionary factor, isolation, unless the mutation itself has isolating effects. Thus such mutations as the trisomics of *Oenothera lamarckiana* Ser. and of *Datura stramonium* L., which can only exist as types segregating from the parent populations, cannot form species in this sense, though the difference in kind may seem fully comparable to that of true species. On the other hand, a tetraploid mutation, fertile by itself, but producing a largely sterile triploid hybrid with the parent species, may give rise to a population which may properly be considered a new species from the population viewpoint even though the character-differences, at least at the time of mutation, may be very slight. Most of the character-differences in species which have had this mode of origin probably trace to other evolutionary processes, the tetraploidy being significant largely as an isolating factor which gives an opportunity for the action of these other processes. Tetraploidy in a species-hybrid, however, produces a type which satisfies the usual criterion for degree of difference in kind from pre-existent species as well as the more essential criterion of isolation. There can be no doubt that both auto- and allotetraploidy have played a role in the multiplication of genuine species, especially in higher plants (cf. Müntzing, 1932, 1936), but also in some groups of animals. Nevertheless, it appears that this role is a subordinate one in the evolutionary process in general. No other type of mutation is known which can bring about the origin of a new species by itself, at least under prevailing biparental reproduction. We will not here consider the meaning of the term 'species' as applied to forms with exclusive or nearly exclusive vegetative reproduction.

Genetic analyses of species-crosses show that in most cases the differences are not such as could have arisen at one step. An array of differences of all grades of coarseness is usually found in each case: gross chromosome-changes such as inversions, translocations, and duplications, gene-differences with major effects, comparable to those of the familiar laboratory mutations, and finally apparent blending differences, due presumably to a multiplicity of individually minute or subliminal gene-effects (cf. Dobzhansky, 1937). Moreover, a certain degree of variability may be an essential characteristic of a species. Kinsey (1929) compares the extraordinary variability of *Cynips erinacei* (Walsh) within all parts of a range of some 500,000 square miles with the approximate uniformity of *C. pezomachoides* (O-S.) in a large adjoining territory. He also (1937) compares the usual high variability of 'continental' species with the relative uniformity of 'insular' ones.

These findings indicate that the occurrence of single mutations can be of little importance in speciation in general. For the significant factors we must look to statistical processes which can carry certain genes to fixation or hold them at certain constant frequencies in the population. We must consider such factors as the pressure of *recurrent* mutation, the pressure of selection, both between individuals and between populations, and finally the effects of isolation, partial or complete.

These constitute a very heterogeneous group of factors. In order to discuss their roles intelligently it is necessary to compare their effects on a common scale. Such a scale is to be found in *gene-frequency*, applicable, it may be noted, to most types of chromosome aberration as well as to single gene-mutations. Because of the symmetry of the Mendelian mechanism, the relative frequencies of two allelic conditions tend to remain unchanged in a large randomly breeding population in the absence of recurrent mutation, selection, or immigration. From this viewpoint, an adequate genetic description of a species would consist of a list of gene-frequencies such as

$$(p_1 a_1 + q_1 A_1)(p_2 a_2 + q_2 A_2)...(p_n a_n + q_n A_n)$$

rather than of a single typical genotype such as

$$A_1 A_1 A_2 A_2 \dots A_n A_n.$$

The elementary evolutionary process becomes change of gene-frequency rather than mutation.

It is easy to see how recurrent mutation tends to change gene-frequency. If q is the frequency of a given gene, $(1-q)$ that of its alleles, and the latter, considered collectively, mutate to the given gene at the rate v per generation, the mutation pressure is obviously $\Delta q = v(1-q)$. Since mutation is known to be a reversible process in many cases (Muller, 1928; Timofeeff-Ressovsky, 1932), reverse mutation should also be represented in the formula. If the rate per generation is u, the net mutation-pressure may be written

$$\Delta q = v(1-q) - uq.$$

A graphic representation is given in Fig. 1.

There is a certain point on the scale at which the opposed pressures balance each other $\left(\Delta q = 0 \text{ at } \hat{q} = \frac{v}{u+v}\right)$. At this point there is stable equilibrium. This situation differs from the persistence of gene-frequencies found in the absence of any evolutionary pressure in the tendency for the population to return always to the same equilibrium point after a change in either direction. A species in which a few important gene-differences or many minor ones are held in equilibrium at frequencies not too close to 0 or 1 is characterized by a constant high degree of variability as well as by constant average values of each character.

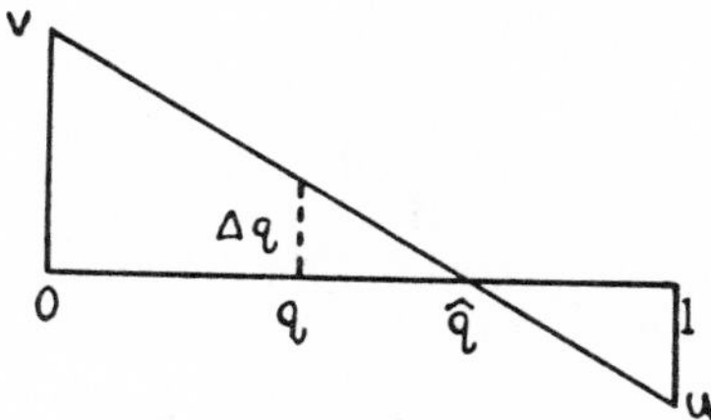

FIG. 1. Mutation-pressure (Δq) on an exaggerated scale in relation to gene-frequency (q). $\Delta q = v(1-q)-uq$. At equilibrium $\Delta q = 0$, $\hat{q} = \frac{v}{u+v}$.

The effect of selection can also be measured on the scale of gene-frequencies. As there can be no selection relative to a gene, if it is either completely fixed or absent ($q = 1$ or $q = 0$), selection-pressure may be expected to contain the term $q(1-q)$. The simplest case is that in which there is a constant difference in adaptive value (W) between the two homozygous conditions, with the heterozygote exactly halfway between

$$(W_{AA} = 1,\ W_{AA'} = 1-s,\ W_{A'A'} = 1-2s).$$

Selection-pressure here takes the simple form $q = sq(1-q)$ (Fisher, 1930). The consequences of many other special cases have been worked out by Haldane, who, however, has expressed his results in terms of the frequency ratio of alleles, equivalent to $q/1-q$ in the symbolism used here.

The assignment of selection coefficients to individual genes does not give as realistic a representation of natural selection as is desirable. It is the organism as a whole that is more or less adaptive in relation to prevailing conditions, not single genes. A gene that produces a more favourable effect than its allele in one combination is likely to be less favourable in others. Thus if a certain proportionality in the development of two parts of an organism is more adaptive than deviations in either direction, genes that increase the relative size of one part will be adaptive in combinations in which this part is unduly small because of other factors, but will be antiadaptive in combinations which make this part too large. It can be shown that if an average adaptive value (W), relative to prevailing conditions, can be assigned each genotype as a whole, the net selection-pressure on a single gene (in a population of diploid organisms) can be expressed by the formula

$$\Delta q = \frac{q(1-q)}{2\overline{W}}\frac{\partial \overline{W}}{\partial q},$$

where $\overline{W}$ is the average of the adaptive values of all possible genotypes, giving due weight to their frequencies.

These frequencies are functions of the frequencies of the component genes. Thus for the array of gene-frequencies given above, the frequencies of all possible genotypes, under random mating, are given by the appropriate terms in the expansion of the expression

$$(p_1 a_1 + q_1 A_1)^2 \ (p_2 a_2 + q_2 A_2)^2 \ \ldots \ (p_n a_n + q_n A_n)^2.$$

This means that in general the selection-pressure on any gene is a function not only of its own frequency but of that of all other genes.

There may be a stable equilibrium due to selection alone under conditions in which a heterozygote is selected over both homozygotes. The extreme case is that of balanced lethals, exhibited in nature in the permanent heterozygosis of many species of *Oenothera*. It is probable that polymorphism in many species rests on a slight advantage of the heterozygote (Fisher, 1930*b*).

In most cases in which one allele has an appreciable selective advantage under the prevailing conditions, equilibrium is probably established between opposed pressures of selection and mutation. The net rate of change of gene-frequency, due to selection and mutation together, may be written as follows by combining preceding formulae:

$$\Delta q = v(1-q) - uq + \frac{q(1-q)}{2\overline{W}}\frac{\partial \overline{W}}{\partial q}.$$

A special case is shown graphically in Fig. 2.

Another possibility of change of gene-frequency is exhibited by the well-known tendency towards fixation of one or another chance combination of genes in closely inbred strains. The effect depends on accidental, rather than selective, factors which determine the individuals which become the parents of the next generation. It requires $2N$ gametes to reconstitute a population of N mature individuals. Random samples of this size, from a population in which a given gene has the frequency q, vary in gene-frequency with a standard deviation,

$$\sigma_{\Delta q} = \sqrt{\frac{q(1-q)}{2N}}.$$

It may seem that this is an insignificant amount of variability if N is reasonably large, especially as the accidental variations

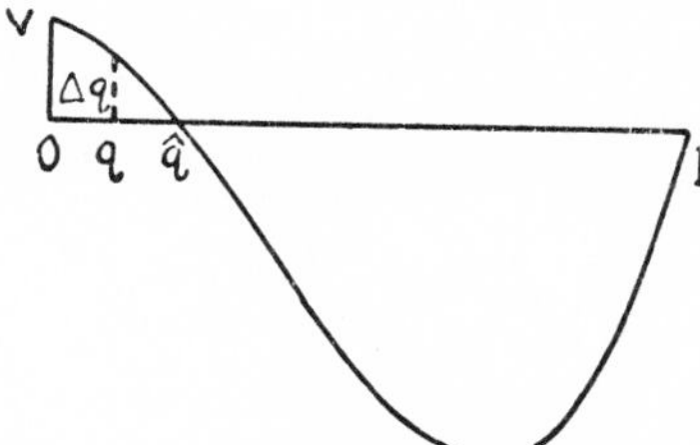

FIG. 2. Opposed pressures of mutation (v) and selection ($t = 25v$) in the case of a deleterious recessive. $\Delta q = v(1-q) - tq^2(1-q)$, $\hat{q} = \sqrt{(v/t)}$.

in the following generations are as likely to be in the reverse direction as in the same direction. However, variance of sampling is cumulative, being nearly doubled in two generations, multiplied threefold in three generations, and so on until damped by approach of q to 0 or 1. In the course of geologic time these chance variations may be expected to bring one allele or the other to fixation, even in rather large populations, in the absence of other factors. The rate of fixation here is $1/2N$ per generation. This process, occurring independently in all loci, may bring about any degree of differentiation of isolated populations in respect to indifferent characters.

We must again, however, consider the simultaneous effects of the various evolutionary factors. As described above, the directed evolutionary pressures, such as selection and mutation, in opposing each other, determine a certain point of equilibrium in gene-frequency. The inbreeding effect is opposed to such factors in a different sense from that in which they oppose each other. It tends to bring about random deviations in either direction from the equilibrium point. The resultant is a probability distribution relative to the scale of gene-frequencies (Fig. 3). The ordinates show how often, in the long run, the gene exhibits each frequency. The curve can also be interpreted as the frequency distribution at a given moment for all genes subject to the same conditions. A third interpretation is the distribution of frequencies for a given gene at a given moment in an array of completely isolated populations.

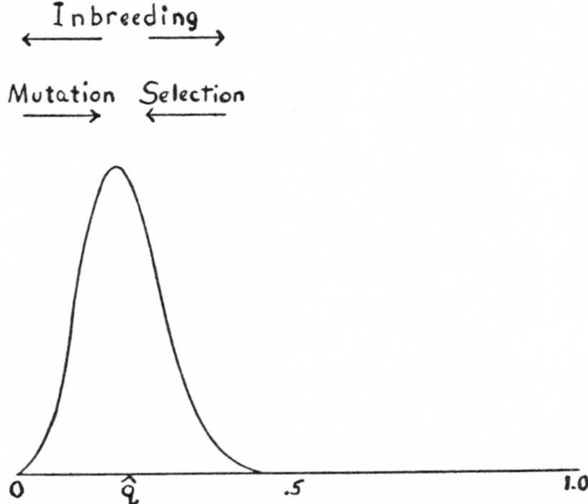

FIG. 3. Distribution ($\phi(q)$) of gene-frequencies for a recessive deleterious gene in a population of limited size (N). Mutation and selection-pressures as in Fig. 2 ($t = 25v$). $N = 1/v$. $\phi(q) = Ce^{-50q^2}q^3(1-q)^{-1}$.

As to the formula for this probability curve, it can be shown that it can be expressed in terms of the rate of directed change in gene-frequency (Δq) and the undirected sampling variance per generation ($\sigma^2_{\Delta q}$).

$$\phi(q) = (C/\sigma^2_{\Delta q})e^{2\int(\Delta q/\sigma^2_{\Delta q})dq}.$$

For the value of Δq given above and for a population of diploid individuals, this resolves into the following:

$$\phi(q) = C\bar{W}^{2N}q^{4Nv-1}(1-q)^{4Nu-1}.$$

Similar formulae can be developed for other cases such as sex-linked genes and polyploids.

The above expression brings together the effects of mutation, selection, and size of population. The ways in which its form varies in relation to severity of selection and size of population are illustrated in Figs. 4–6 for a recessive gene.

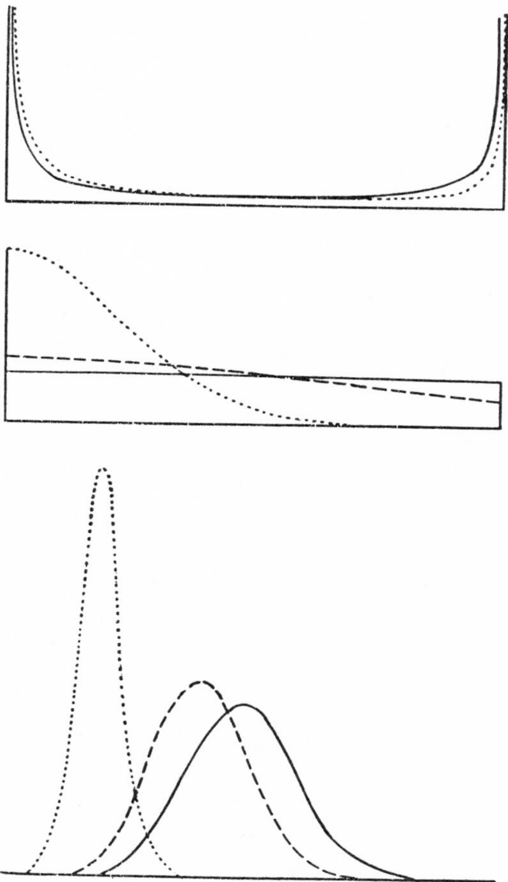

FIG. 4–6. Some of the forms taken by the distribution of frequencies of a recessive deleterious gene, $\phi(q) = Ce^{-2Ntq^2}q^{4Nv-1}(1-q)^{4Nu-1}$. Mutation rates to and from the gene are assumed equal ($u = v$). Effective size of populations is $N = \frac{1}{40v}$, $\frac{10}{40v}$, and $\frac{100}{40v}$ in Figs. 4, 5, and 6 respectively from top to bottom. In each case the solid line represents the least selection ($t = v/5$), the broken line selection 10 times as severe (not represented in Fig. 4, since practically indistinguishable from the preceding), and the dotted line represents selection 100 times as severe.

As already noted, $\overline{W}$ is a function of all gene-frequencies. In the practically infinite field of gene-combinations, possible from differences in only a few thousands, or even hundreds of loci, there are likely to be an enormous number of different harmonious combinations of characters. These would appear as peak values of $\overline{W}$, separated by valleys or saddles in a multi-dimensional surface. The distribution surface for all gene-frequencies, considered simultaneously, will show corresponding peaks.

$$\phi(q_1, q_2, \ldots, q_n) = C\overline{W}^{2N} \prod_{i=1}^{n} [q_i^{4Nv_i-1}(1-q_i)^{4Nu_i-1}].$$

Fig. 7 illustrates a two-factor case with two peaks. It is assumed in this case that the grade of character is determined by two equivalent additive factors, but that the most adaptive combinations are intermediate, as is to be expected in any species long exposed to the same conditions.

Grade of character	*Adaptive value* (W)	*Genotypes*
5	$1-4s$	$A_1A_1A_2A_2$
4	$1-s$	$A_1A_1A_2a_2$, $A_1a_1A_2A_2$
3	1	$A_1A_1a_2a_2$, $A_1a_1A_2a_2$, $a_1a_1A_2A_2$
2	$1-s$	$A_1a_1a_2a_2$, $a_1a_1A_2a_2$
1	$1-4s$	$a_1a_1a_2a_2$

There are two peak combinations that may become fixed in populations ($A_1A_1a_2a_2$, $a_1a_1A_2A_2$) separated from each other in the two-dimensional system of values of q_1 and q_2 by a saddle which is shallow if s is a small fraction. The joint distribution of gene-frequencies shows two peaks determined by selection-pressures directed towards the peak combinations but opposed by mutation-pressures.

The peaks and valleys of the surface of adaptive values ($\overline{W}$) are so much exaggerated in the distribution of gene-frequencies (term $\overline{W}^{2N}$) that we are again confronted with the apparent unlikelihood of any appreciable non-adaptive drift away from the peak occupied by the species at a given time, if the size of population is at all large. The effective size of population may, however, be much smaller than the apparent size for various reasons. It refers, of course, only to sexually mature individuals. If mature individuals of one sex are much less numerous than those of the other, the effective size is largely determined by the former. It cannot be as much as four times as great. If there are wide variations in the numbers of offspring reaching maturity, left by different parents, the effective number may be reduced. If the species goes through a more or less regular cycle of numbers, the effective size is determined largely by the phase of small numbers. The situation is approximately as if the

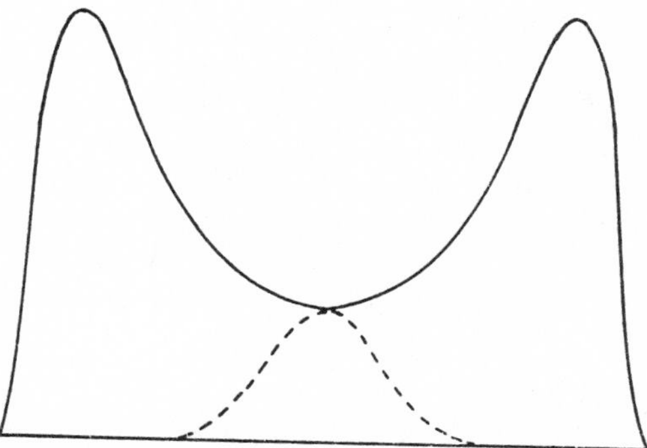

FIG. 7. The frequencies along the two diagonals of the joint distribution for two series of alleles with equal and additive effects on a character on which adverse selection acts according to the square of the deviation from the mean. The solid line shows the frequencies in populations along the line connecting the two favourable types $A_1 A_1 a_2 a_2$ and $a_1 a_1 A_2 A_2$. The broken line refers to the line connecting the extreme types $a_1 a_1 a_2 a_2$ and $A_1 A_1 A_2 A_2$. In the case shown, all mutation rates (u_1, v_1, u_2, v_2) are assumed equal, $N = 1/(2v)$, and $s = 5v$. Along the favourable diagonal $(q_1 = (1-q_2))$ the distribution is approximately $Ce^{-20q_1(1-q_1)}q_1^2(1-q_1)^2$. Along the unfavourable diagonal $(q_1 = q_2)$ it is approximately

$$Ce^{-20(1-3q_1(1-q_1))}q_1^2(1-q_1)^2.$$

population were characterized by the minimum number, but with the interval between generations as long as the length of the cycle. The significance of periodic reductions in size of population in making possible random variations in the character of populations has been stressed by Elton (1924).

In a large species the condition of random mating among all individuals is never realized. There are two limiting cases which can be represented conveniently by simple mathematical models. There may be continuity throughout the range, but restriction of the probable mates of any given individual to a small surrounding territory. The parents, grandparents, great-grandparents, and more remote ancestors trace to ever-widening territories. The standard deviation of gene-frequencies among populations separated by n generations of ancestors (or about $\sqrt{n}$

diameters of the elementary area) is $\sqrt{\{q_t(1-q_t)f\}}$, where q_t is the mean gene-frequency, and where f, the local coefficient of inbreeding, lies between

$$\frac{\sum}{2N+\sum} \text{ and } \frac{\sum}{2N-\sum} \text{ (or 1)}, \sum \text{ being } \sum_{x=1}^{n} (1/x).$$

Under these conditions, the results are almost those of random mating, even for populations separated by thousands of diameters of the elementary territory, if the effective population number of the latter is in the thousands or more. There may be considerable fluctuating local differentiation (such as Anderson, 1936, has described in species of *Iris*) if the effective population number is less than a few hundreds. For an approach to fixation of different alleles in different regions, the elementary population number must be measured in 10's rather than in 100's.

In a species with a range that is essentially one-dimensional (shore-line, river, long chain of mountain ranges) the possibility of non-adaptive differentiation increases much more rapidly with distance than where continuity is over an area (as noted by Thompson, 1931, in a study of differences in number of fin-rays of small fishes in relation to water distance in Illinois). The quantity $\sum$ of the formula has the value $\sum_{x=1}^{n} \sqrt{(1/x)}$ in this case.

A model that fits rather better into the mathematical system developed here is appropriate where the range is subdivided into partially isolated territories. For a territory with effective population number N, replaced each generation to a small extent, m, by migrants derived at random from the species, there is a cross-breeding or migration-pressure $\Delta q = -m(q-q_t)$. This can be thrown into the same form as mutation-pressure, making it possible to introduce it into all of the formulae where mutation-coefficients appear, merely by substituting mq_t for v and $m(1-q_t)$ for u.

$$\phi(q) = C\overline{W}^{2N} q^{4Nmq_t-1}(1-q)^{4Nm(1-q_t)-1}.$$

Taking into account the fact that the effective population number (N) may be much less than the apparent number, and that the effective migration-coefficient (m) may be much less than indicated by the actual amount of immigration (largely from neighbouring, closely related populations), a large amount of non-adaptive differentiation of local populations is easily possible. The standard deviation of q for indifferent loci ($\overline{W} = 1$) is

$$\sigma_q = \sqrt{\left(\frac{q_t(1-q_t)}{4Nm+1}\right)}.$$

The most extreme case is that in which there are many territories in which the populations are so isolated and so liable to extinction that the lines of continuity of populations often pass from territory to territory through single stray individuals. Translocations are violently selected against, so long as rare, because of the production of lethal duplications and deficiencies. There can be little chance of their establishment in a subgroup of a species in which reproduction is exclusively biparental, except by some such process as outlined above. But translocations are among the differentials between closely allied species, e.g. in the case of *Drosophila pseudo-obscura* (Frolown) and *D. miranda* (Dobzh.), studied by Dobzhansky and Tan (1936). The difficulties of fixation of such mutations are obviously much less in species which can multiply vegetatively or under predominant self-fertilization (cf. Blakeslee, 1932).

The cross-breeding coefficient, m, is also important in connexion with the possibility of local differentiation under the influence of differential selection. Taking s as the net selection-coefficient for a local population, the net evolutionary pressure being $\Delta q = sq(1-q)-m(q-q_t)$, it can be shown that there can be no important amount of such differentiation if the variations of s are small compared with m, while extensive adaptive differentiation occurs if the reverse is true.

These statistical deductions from the Mendelian mechanism do not in themselves give a general evaluation of the roles of the various factors in evolution. They bring these factors under a common viewpoint, however, and make it possible to form a judgement as to the conditions under which one or another, or a combination, may dominate the process.

The conditions under which mutation-pressures, at rates like those usually observed in the laboratory, are likely to dominate in the course of evolution appear to be decidedly restricted. Even a very slight selective advantage (e.g. of the order 10^{-4} or even 10^{-5}) would usually be more important. However, under extreme reduction of size of populations ($4Ns$ much less than 1) selection-pressure becomes ineffective, while mutation-pressure is not affected. The one *systematic* effect of mutation seems to be a tendency towards degeneration (as may be seen from a casual survey of the effects of most of the *Drosophila* mutations). Thus a trend towards degeneration of structures of little or no use in small completely isolated populations (e.g. in caves or small oceanic islands) may be due to mutation-pressure. Even here there are possibilities of indirect control by selection which should not be ignored.

Great increases in mutation-rate at certain periods of the earth's history have been postulated by various authors to explain periods of rapid evolutionary advance. The real effect would depend on the prevailing balance with other factors. Such a change in mutation-rate would probably mean merely a degenerative trend unless the effects of all other influences were correspondingly speeded up.

In a large population with sufficient random interbreeding and no secular changes in conditions over a long period of time, all gene-frequencies may be expected to approach equilibrium values largely dominated by selection. Once the population has reached a certain peak in the surface of adaptive values there will be no further significant evolutionary change in spite of continual mutation, persistent variability, and rigorous selection. More rigorous selection will merely concentrate the species as a whole about the peak, raising the mean adaptive value, but reducing the variability on which further evolution must largely depend. The chance of occurrence of a wholly novel mutation, possibly adaptive from the first, is reduced, since the reduction in frequency of non-type alleles reduces the chance of occurrence of untried mutations at two or more removes from the type gene.

A secular change in the conditions to which the species must adapt itself changes the entire set of adaptive values. Peak combinations may be depressed while other genotypes may turn out to be more adaptive under the new conditions. A relatively rapid evolutionary process, controlled by the net selection-pressures on all genes, may be expected to start and to go on until the species reaches equilibrium about a new peak value of $\overline{W}$. Repeated changes in conditions, especially if involving changes in direction, mean a continuing evolutionary process. It can hardly be doubted that this has been one of the most important causes of evolution.

Extreme reduction in numbers, bringing a tendency towards random fixation of one or another combination which is almost certainly less adaptive than the previous type, should result in a degenerative evolution even before there is time for any appreciable effect of mutation-pressure. Extreme inbreeding is a factor that may be expected to lead to extinction of the species rather than to evolutionary advance.

In a large population subdivided into numerous partially isolated groups, both adaptive and non-adaptive differentiation is to be expected. A small value of the cross-breeding coefficient, m, favours both of these processes. On the other hand, the local population-size has opposite effects on these processes. Large numbers do not interfere with differential selection-pressures but, of course, prevent any appreciable random drifting apart, while the small numbers that make the latter possible reduce the effectiveness of selection. The greatest amount of differentiation should occur with a certain intermediate population number ($4Nm$ and $4Ns$ both in the neighbourhood of 1). Under this condition neither does random differentiation proceed to fixation nor adaptive differentiation to equilibrium, but each local population is kept in a state of continual change. A local population that happens to arrive at a genotype that is peculiarly favourable in relation to the general conditions of life of the species, i.e. a higher peak combination than that about which the species had hitherto been centred, will tend to increase in numbers and supply more than its share of migrants to other regions, thus grading them up to the same type by a process that may be described as intergroup selection. There is here a trial-and-error mechanism under which the field of factor combinations surrounding the prevailing peak combination is explored by the various subgroups with occasional crossing of a saddle from a lower to a higher peak. The species as a whole may evolve continuously even without secular change in conditions (although this process, occurring in all species, itself tends to bring about such secular changes and thus periods of rapid change controlled by natural selection). The combination of partial isolation of subgroups with intergroup selection seems to provide the most favourable conditions for evolutionary advance.

Partial isolation of subgroups is here considered as a factor in the evolution of the species as a whole. Splitting of the species depends on more complete isolation. It appears probable that the more or less complete cross-sterility that permanently separates most good species from their nearest allies is usually a by-product of the gradual accumulation of genetic differences in populations isolated at first merely by geography, habitat, &c. A direct origin (except in such cases as speciation by tetraploidy) is very difficult to understand. Selection, however, may play a role in perfecting cross-sterility where two groups which have drifted apart under complete isolation, later come in contact again, or where there is strong differential selection in other respects.

We have considered the possibilities for evolutionary change which may be deduced from the Mendelian mechanism. It is most important that such deductions be compared with conditions in nature.

There is a considerable body of data on the genetics of species and subspecies differences (cp. Dobzhansky, 1937), but more exhaustive investigations, made with the statistical consequences of Mendelian heredity in mind, must be made to establish the connexions on a secure basis.

There is an especially large amount of information on the chromosome differences of related species (cf. Darlington, 1937). Little is known, however, of the significance of these in determining character-differences (other than of the nucleus itself). Many major Mendelian differences have been isolated from crosses, but how these compare in importance with the cumulative effects of multiple minor factors is not clearly brought out in many cases. As it stands, widely different conclusions have been reached in different cases. Thus, Goldschmidt attributes most importance to certain major mutations (and to cytoplasmic differences) in distinguishing the races of *Lymantria dispar* Linn., although recognizing that the sharpness of segregation of the major factors is usually blurred by modifiers. Sumner, on the other hand, found no suggestion of unitary differences capable of analysis, in crosses between subspecies of *Peromyscus maniculatus* (Wagner). There was blending heredity with only a slight average increase of variability in F_2 over that of F_1. The closest approach which he observed to determination by major factors was in the case of a colour difference between subspecies of *P. polionotus* (Wagner) in which F_2 was so variable as to suggest that the difference might be due to as few as three or four factors. The most usual result in the many species and subspecies crosses in plants seems to be a hierarchy of differences ranging from a considerable number of major ones to an indefinitely large number of minor ones.

The genetic basis for cross-sterility is in some cases clearly chromosomal (as where a sterile hybrid produces a fertile allotetraploid), but in other cases is clearly genic (as in species-crosses in *Drosophila*; Dobzhansky, 1933). The relative importance of these mechanisms in different groups and, in the case of genic cross-sterility, the relative importance of single and multiple factors, need determination in more cases.

Data on the amount of hereditary variability within local wild populations are still rather scanty. One aspect is the determination of gene-frequencies where species are frankly polymorphic in respect to major genes. Another is the frequency of lethals and of other major factors that are strongly selected against. Several studies in *Drosophila* have shown that such factors are not uncommon in nature (Dubinin, 1934; Dobzhansky, &c.). Of perhaps greater evolutionary significance are biometric studies of the correlation between parent and offspring in quantitatively varying characters within local populations to determine how generally such populations are heterallelic in minor factors. Many such studies have been made in man, domesticated animals, and cultivated plants, but relatively few have been made for wild species. Sumner's data from *Peromyscus* subspecies are an example.

Estimates of typical mutation-rates are at present based on a very small number of organisms. Stadler found that 7 out of 8 genes of *Zea mays* Linn., chosen merely because of convenient endosperm effects, mutated repeatedly in samples of from 250,000 to 2,500,000 gametes. While the data suggested that rates between 10^{-5} to 10^{-6} per generation were most typical, one gene mutated at as high a rate as 5×10^{-4} and one failed to mutate in $1{\cdot}5\times10^{-6}$ gametes. Estimates of mutation-rates per generation are about the same in *Drosophila*. Haldane has estimated the rate of occurrence of the sex-linked mutation causing haemophilia in man as about 2×10^{-5} per generation. The rate per year is of course enormously less in man than in *Drosophila*. Whether these rates can be taken as typical or whether most loci mutate more rarely we do not know. It may be that all loci mutate much more frequently than is usually suspected, but that most of the mutations are so slight in effect as to make detection difficult (East, 1936). The discovery of genes which systematically affect the mutation-rates of other genes (Demerec, 1937) presents interesting questions on possible evolutionary regulation of mutation-rate itself (cf. Sturtevant, 1937). The extent to which possible mutations at a locus form indefinitely extended branching systems of multiple alleles, each allele capable of giving rise to others which cannot arise from the type gene directly, is a question with very important evolutionary implications, on which only a beginning has been made (cf. Timofeeff-Ressovsky, 1932).

Selection-coefficients have been determined ranging from 100 per cent. (complete lethality) down to perhaps 1 per cent. It is probable that most of the mutations which are important in evolution have much smaller selection-coefficients than it is practicable to demonstrate in the laboratory.

The phase of the theory that is most open to investigation in nature is that of breeding structure. It should be possible in many cases to estimate the effective size of the randomly-breeding units, and the effective amount of cross-breeding, with sufficient accuracy to form some judgement of the role which can be played by partial isolation. The distribution of frequencies for a single approximately neutral gene in a species gives an index of the amount of random differentiation.

That these coefficients may be relatively small even in species with enormous numbers of individuals has been shown in certain cases. Sturtevant and Dobzhansky (1936) have shown that chromosome-inversions in *Drosophila pseudo-obscura* behave as approximately neutral Mendelian units. Dobzhansky and Queal found three different inversions of one chromosome to be present in large numbers in the populations of the isolated mountain forests of the Death Valley region of California. The frequencies among 11 such populations ranged from 51 to 88 per cent. in one case, 2 to 20 per cent. in another, and 8 to 39 per cent. in the third. The standard deviations were such as to indicate an effective value of Nm (i.e. the effective number of migrants per generation) of about 5 (assuming approximate equilibrium). Some seventeen different inversions of the same chromosome have been found within the whole range of the species with frequencies ranging in one case from 0 to 100 per cent. (Dobzhansky and Sturtevant, 1938). For the species as a whole, the effective value of *Nm* is apparently only about one-tenth as great as its value in the restricted Death Valley region. Taking this as an indicator, there is the possibility of a great deal of non-adaptive differentiation in *Drosophila pseudo-obscura*. In the human species, the blood-group alleles are neutral as far as known. The frequencies vary widely from region to region and in such a way as to indicate that the historical factor (i.e. partial isolation) is the determining factor. The frequency distribution indicates a considerable amount of random differentiation even among the largest populations. The greater range of variability among small uncivilized groups (e.g. over 90 per cent. gene *O* in most tribes of American Indians, but about 80 per cent. *A* in Blackfeet and in Blood Indians and about 90 per cent. *B* in Patagonians; Gates, 1935) suggests that, during the major portion of the period in which man was evolving from lower primate ancestry, random local differentiation may have played a much greater role than at present.

Attempts to make general evaluations of the roles of adaptive and random differentiation have led to the most diverse conclusions because of the lack of objective criteria. The publication of Darwin's *Origin of Species* was followed by intensive and ingenious attempts to interpret all sorts of species-differences as adaptive under the belief that natural selection was the sole controlling principle of evolution (Wallace). In recent years Fisher has maintained on theoretical grounds that evolution of organisms is as completely subject to the net selection-pressures on the separate genes as the history of physical systems is subject to the increase of entropy. From the viewpoint of the present paper, however, this does not appear to be a necessary theoretical conclusion. Even from the first, certain authors (e.g. Gulick and Romanes) maintained that it was futile to look for a selective mechanism back of many of the differences between isolated populations living under substantially identical conditions. The majority of systematists have probably been sceptical of the adaptive significance of all taxonomic differences (cf. Robson and Richards, 1936).

The issue is complicated by the fact that the antithesis

'adaptive *v.* non-adaptive' does not necessarily correspond to the antithesis 'selection-pressure *v.* isolation-effect'. On the one hand, seemingly neutral character-differences may be interpreted as superficial by-products of adaptive but not easily observed physiological differences. On the other hand, an obviously adaptive difference does not necessarily imply intragroup selection-pressure as its cause. There may have been random differentiation of subgroups, followed by establishment of different types in different regions as a result of intergroup selection. There may even have been random differentiation accompanied by adoption of appropriately different environments or different ways of life in regions not necessarily different in their general conditions. Davenport has stressed the role of selection of suitable environments by organisms that differ merely by chance.

Goldschmidt, as noted earlier, makes a sharp cleavage between the origin of races and that of higher categories, based primarily on study of the gipsy moth (*Lymantria dispar*). He holds that the racial differences, in this case at least, are due to selection-pressure acting at the time of expansion of the species into regions to which it has been poorly adapted. These differences he finds to be largely quantitative and either directly adaptive (length of diapause, rate of larval development, &c.) or theoretically interpretable as indicators of unknown physiological adaptations (colour, pattern of larvae, &c.). He contrasts these with the qualitative differences between species (adaptive or otherwise) which he believes could not result from any amount of accumulation of the quantitative racial differences. He feels constrained to attribute their origin to mutations of a sort with which we have no experience in the laboratory ('hopeful monsters'). Cuénot is another recent author who has been impressed by the difficulty of accounting for many adaptations of species by a cumulative process and has urged that they must have arisen as pre-adaptations.

One may sympathize with these difficulties, but question whether it is necessary to bring in an unknown factor. Kinsey finds little indication of adaptiveness in the trivial taxonomic differences between adjacent populations of *Cynips*. But these apparently random differences accumulate along the species-chains and lead ultimately to differences apparently as qualitative (in character of galls, for example) as those which distinguish species of *Lymantria*.

There is no theoretical necessity for supposing that evolution has proceeded in the same way in all groups. In some it may proceed largely under direct selection-pressure following change in conditions, in other cases it may be determined by random differentiation of small local populations, with or without inter-group selection. It may even be dominated by mutation-pressure in special cases. It may be a gradual, fine-grained process or at times a coarse-grained process, new species arising directly from hybridization and polyploidy.

It has been pointed out, however, that the most favourable conditions for a continuing evolutionary process are those in which there is, to a first-order, balanced action of all of the statistical evolutionary factors. It is consequently to be expected that in most actual cases indications can be found of simultaneous action of all of them.

REFERENCES

ANDERSON, EDGAR (1936). 'The Species Problem in Iris.' *Annals Missouri Bot. Garden*, **23**, 457–509.

BLAKESLEE, A. F. (1932). 'The Species Problem in Datura.' *Proc. 6th Internat. Congress of Genetics*, I: 104–20.

CUÉNOT, L. (1925). *L'Adaptation.* Paris: G. Doin.

DARLINGTON, C. D. (1937). *Recent Advances in Cytology.* London: J. & A. Churchill.

DAVENPORT, C. B. (1903). 'The Animal Ecology of Cold Spring Sand Spit.' *The Decennial Publications, The University of Chicago*, **10**, 157–76.

DEMEREC, M. (1933). 'What is a Gene?' *J. Hered.* **24**, 369–78.

—— (1937). 'Frequency of Spontaneous Mutations in Certain Stocks of Drosophila melanogaster.' *Genetics*, **22**, 469–78.

DICE, L. C. (1931). 'The Occurrence of Two Subspecies in the Same Area.' *J. Mamm.* **12**, 210–13.

DOBZHANSKY, TH. (1933). 'On the Sterility of the Interracial Hybrids in Drosophila pseudo-obscura.' *Proc. Nat. Acad. Sci.* **19**, 397–403.

—— (1937). *Genetics and the Origin of Species.* New York: Columbia Univ. Press.

—— and QUEAL, M. L. (1938). 'Genetics of Natural Populations. I. Chromosomal Variation in Populations of Drosophila pseudo-obscura inhabiting Isolated Mountain Ranges.' *Genetics*, **23**, 239–51.

—— and STURTEVANT, A. H. (1938). 'Inversions in the Chromosomes of Drosophila pseudo-obscura.' *Genetics*, **23**, 28–64.

—— and TAN, C. C. (1936). 'Studies on Hybrid Sterility. III. A Comparison of the Gene Arrangement in Two Species, Drosophila pseudo-obscura and Drosophila miranda.' *Z. indukt. Abstamm.—u. Vererb Lehre* **72**, 88–114.

DUBININ, N. P., and fourteen collaborators (1934). 'Experimental Study of the Ecogenotypes of Drosophila melanogaster.' *B. Zh.* **3**, 166–206.

EAST, E. M. (1936). 'Genetic Aspects of Certain Problems of Evolution.' *Amer. Nat.* **70**, 143–58.

ELTON, C. S. (1924). 'Periodic Fluctuations in the Number of Animals: Their Causes and Effects.' *Brit. J. Exp. Biol.* **3**, 119–63.

FISHER, R. A. (1930). *The Genetical Theory of Natural Selection.* Oxford: Clarendon Press.

—— (1930). 'The Evolution of Dominance in Certain Polymorphic Species.' *Amer. Nat.* **64**, 385–406.

GATES, R. R. (1935). 'Recent Progress in Blood Group Investigations.' *Genetica,* **18**, 47–65.

GOLDSCHMIDT, R. (1933). 'Certain Aspects of Evolution.' *Science,* **78**, 539–47.

—— (1934). 'Lymantria.' *Bibliographica Genetica,* **11**, 1–186.

—— (1937). 'Cynips and Lymantria.' *Amer. Nat.* **71**, 508–14.

HALDANE, J. B. S. (1932). *The Causes of Evolution.* London: Harper & Bros.

—— (1935). 'The Rate of Spontaneous Mutation of a Human Gene.' *J. Genet.* **31**, 317–26.

KINSEY, A. C. (1929). *The Gall Wasp Genus Cynips.* Studies No. 84, 85, 86. Indiana Univ. Studies, **16**.

—— (1936). *The Origin of the Higher Categories in Cynips.* Indiana Univ. Publications, Science Series, No. 4.

—— (1937). 'An Evolutionary Analysis of Insular and Continental Species.' *Proc. Nat. Acad. Sci.* **23**, 5–11.

MULLER, H. J. (1928). 'The Production of Mutations by X-rays.' *Proc. Nat. Acad. Sci.* **14**, 714–26.

MÜNTZING, A. (1932). 'Cyto-genetic Investigations on Synthetic Galeopsis tetrahit.' *Hereditas,* **16**, 105–54.

—— (1936). 'The Evolutionary Significance of Autopolyploidy.' *Hereditas,* **21**, 263–378.

OSBORN, H. F. (1927). 'The Origin of Species. V. Speciation and Mutation.' *Amer. Nat.* **61**, 5–42.

OSGOOD, W. H. (1909). 'Revision of the Mice of the Genus Peromyscus.' *North American Fauna,* **28**, 1–285.

RENSCH, B. (1929). *Das Prinzip geographischer Rassenkreise und das Problem der Artbildung.* Berlin: Gebrüder Borntraeger.

ROBSON, G. C., and RICHARDS, O. W. (1936). *The Variation of Animals in Nature.* London: Longmans, Green.

SCHMIDT, J. (1917). 'Statistical Investigations with Zoarces viviparus L.' *J. Genet.* **7**, 105–18.

STADLER, L. J. (1930). 'The Frequency of Mutation of Specific Genes in Maize.' Abstract. *Anat. Rec.* **47**, 381. (See also Demerec, 1933.)

STURTEVANT, A. H. (1937). 'Essays on Evolution. I. On the Effects of Selection on Mutation Rate.' *Quart. Rev. Biol.* **12**, 464–7.

—— and DOBZHANSKY, TH. (1936). 'Inversions in the Third Chromosome of Wild Races of Drosophila pseudo-obscura, and Their Use in the Study of the History of the Species.' *Proc. Nat. Acad. Sci.* **22**, 448–50.

SUMNER, F. B. (1932). 'Genetic, Distributional, and Evolutionary Studies of the Subspecies of Deer Mice (Peromyscus).' *Bibl. Genetica*, **9**, 1–106.
THOMPSON, D. H. (1931). 'Variation in Fishes as a Function of Distance.' *Trans. Ill. State Acad. Sci.* **23**, 276–81.
TIMOFEEFF-RESSOVSKY, N. W. (1932). 'Mutations of the Gene in different directions.' *Proc. 6th Internat. Congress Genetics*, **1**, 308–30.
TURESSON, G. (1925). 'The Plant Species in Relation to Habitat and Climate.' *Hereditas*, **6**, 147–236.
WRIGHT, SEWALL (1931). 'Evolution in Mendelian Populations.' *Genetics*, **16**, 97–159.
—— (1932). 'The Roles of Mutation, Inbreeding, Crossbreeding, and Selection in Evolution.' *Proc. 6th Internat. Congress Genetics*, **1**, 356–66.
—— (1935). 'Evolution in Populations in Approximate Equilibrium.' *J. Genet.* **30**, 257–66.
—— (1937). 'The Distribution of Gene Frequencies in Populations.' *Proc. Nat. Acad. Sci.* **23**, 307–20.
—— (1938). 'Size of Population and Breeding Structure in Relation to Evolution.' *Science*, **87**, 430–1.

26
The "Age and Area" Concept Extended

(A review of *The Course of Evolution by Differentiation or Divergent Mutation Rather Than by Selection*, by J. C. Willis)
Ecology 22, no. 3 (1941):345–47

27
The Material Basis of Evolution

(A review of *The Material Basis of Evolution*, by Richard B. Goldschmidt) *Scientific Monthly* 53 (August 1941):165–70

28
Tempo and Mode in Evolution: A Critical Review

(A review of *Tempo and Mode in Evolution*, by George Gaylord Simpson) *Ecology* 26, no. 4 (1945):415–19

Introduction

From 1940 to 1948 Wright published no general papers on his shifting balance theory other than reviews of books by Willis, Goldschmidt, and Simpson. These were no ordinary reviews, but full and critical essays in which Wright examined the evidence presented by these experts, and then, with the explanatory power of his own shifting balance theory, he compared and contrasted their theories for explaining the evidence. These three reviews took so much of Wright's time and effort that for more than fifteen years after his review of Simpson's book, Wright refused all requests (and there were many) to review books.

Wright did not deny the validity of the basic observations that led Willis to his antiselectionist mechanism of evolution or those that led Goldschmidt to his "hopeful monster" mechanism. Wright argued, however, that their mechanisms were doubtful on many grounds. Then in both cases he showed that their observations were fully consistent with his more robust shifting balance theory. Their mechanisms were not needed to explain the phenomena they had observed, and he had a better one.

The situation was different with Simpson's book. Simpson, already much influenced by Wright, had attempted in his book to explain the paleontological record, using what he understood to be Wright's shifting balance theory. Wright of course had nothing but praise for this approach, so his review concentrated upon correcting and extending Simpson's analysis. One of Wright's corrections was aimed at Simpson's argument that intergroup selection could produce nothing new (as opposed to individual natural selection, which could produce new things). Wright responded by working out the first mathematical demonstration of the evolution of an individually disadvantageous character by group selection, a starting point for the later theory of group selection (for reviews see Wilson 1980, 1983). I would emphasize, however, that this example of "group" selection is a logically distinct mechanism from the "intergroup" selection of the shifting balance theory.

I have examined these three reviews in *SW&EB*, 412–17. Wright returned to the questions of speciation and macroevolution particularly in two 1982 papers (Wright 205, 207).

The "Age and Area" Concept Extended [1]

This book is a restatement and further elaboration of the views expressed by Dr. Willis in "Age and Area" published in 1922.

Theories of evolution can usually be classified according to the demands made on chance variation. Those which treat evolution as a preordained developmental process (orthogenesis and allied views) and those which treat it as an extension of individual physiology (*e.g.* Lamarckism) attempt to eliminate chance altogether. In those under which evolution is a statistical process of transformation of populations (*e.g.* Darwinism) chance variations play an important rôle in keeping the species variable although not in determining the course of evolution. Finally are the theories in which single mutations are the crucial events in the origin of new species and even of genera and higher categories, and in which accordingly chance is all that there is.

Willis' theory, at least on first impression, seems to belong in this last group.

"Evolution most probably goes on by definite single mutations, which cause structural alterations, which may, but by no means necessarily must have some functional advantage attached." "The mutations supposed in differentiation would at one step cross the 'sterility line' between species which has always been a great stumbling block to natural selection; and thus at once isolate the new form, preventing its loss by crossing." With respect to mechanism he writes, "Chromosome alterations are probably largely responsible for the mutations that go on."

Willis holds that not only species but higher categories arise abruptly. "Mutation tends to be divergent, especially in the early stages of a family. The family, consisting probably of a one genus and one species is probably first created by a single mutation, whilst later ones are usually less marked than the first and give rise to further genera and species." "Evolution goes on in what one may call the downward direction from family to variety, not in the upward direction required by the theory of natural selection." "Varieties are the last stages in the mutation and are not, as a rule, incipient species."

These quotations seem to imply that the course of evolution is wholly dependent on the chances of mutation. But in the full elaboration of the theory, there is a meeting of the extremes of the above classification of theories of evolution. Thus he writes: "The process of evolution appears not to be a matter of natural selection of chance variations of adaptational value. Rather it is working upon some definite law that we do not yet comprehend. The law probably began its operation with the commencement of life and it is carrying this on according to some definite plan." "Evolution is no longer a matter of chance but of law. It has no need of any support from natural selection." "It thus comes into line with other sciences which have a mathematical basis."

To the reviewer, there appears much less law here, in the scientific sense, than in the statistical theory. There is rather an inscrutable succession of acts of creation.

At other points, the author urges the possible rôle of cosmic rays (to account for the prevalence of endemic species on mountains) and of the continued pressure of flowing water (to account for the extraordinary diversity of the Podostemaceae). This seems to put his views in the second category but these agents are apparently considered merely as releasing preordained tendencies rather than as guiding evolution by physiological means or as inducing chance mutations.

The view to which the author is most systematically opposed is that of evolution by gradual statistical transformation of populations.

"It (selection) comes in principally as an agent to fit into their places in the local economy of the place where they are trying to grow, the forms there furnished to it, whether newly evolved or only newly arrived, killing out those in any way unsuitable." "It has, therefore, not been responsible for the progress that has been made by the actual evolution of new forms, but it has been all important in fitting them into their places in the economy which is always increasing in complexity."

No positive evidence is adduced for the evolutionary process postulated. The book is largely devoted to a critique of natural selection with the implication that if natural selection is inadequate, the only alternative is the view outlined.

Some of the difficulties which the author finds with natural selection have been suggested above. He presents evidence that small endemic species are neither relics of ancient dying species nor adaptations to local conditions as required by natural selection. He considers the only alternative to be that they are species of recent mutational origin. He believes that there has been much less extinction of species than usually held and thus that genera containing numerous species, and families containing numerous genera, are as a rule older than monotypic genera or families respectively and simi-

[1] The Course of Evolution by Differentiation or Divergent Mutation rather than by Selection. By J. C. Willis, Cambridge University Press. 207 pp. $3.00.

larly that widely distributed species or higher categories are older than ones of corresponding rank but restricted geographical range, the principle of "Age and Area."

The author devotes much attention to his discovery of the so-called "hollow curve" of frequencies of genera classified according to number of species. Tabulations show that in almost any higher category there are a large number of monotypic groups, much fewer with 2 or 3 subdivisions and fewer still with any particular larger number, yet some as a rule with very large numbers of subdivisions.

The frequencies have been found to fall off in such a way that a plot of the logarithms of number of genera against the number of contained species, gives a close approach to a straight line. This form of distribution has been shown (by Yule) to agree with expectation on the mutation theory and is held by the author to be wholly at variance with expectation under natural selection.

Again the author presents abundant cases to show that taxonomic differences, including those distinguishing the higher categories, are often of a type that is discontinuous by its very nature. He lists many which he believes can have no adaptive value. He holds that selection is too slow a process and that if it were effective it should produce a few superplants for each ecologic niche or at least a convergence of characters, instead of the rich abundance of utterly diverse types actually found.

The book presents a strong case against the theory that the sole factor in evolution is the action of natural selection among individuals on the basis of minute quantitative variation. Natural selection of this sort is, however, only one of many factors in the modern population theory of evolution. This theory accepts the kinds of mutational change that are actually observed as its basis. These variations may be large or small in effect. A single genic (or chromosomal) change may determine a discontinuous change at one step or may merely cause a slight change in the percentage incidence of such a character. The observed changes, whether large or small, are rarely associated with complete cross sterility with the parent form (and would be still more rarely capable of persisting in nature if there were cross sterility). Thus they determine merely more or less aberrant members of the population and not new species. Even if the character is discontinuous its fixation can only come after a gradual process of increase of gene frequency.

There is no consideration of the kinds of variation actually found in genetic experiments, beyond a few bare statements such as that quoted above in regard to chromosomes, no consideration of the facts of heredity and consequently nothing on the statistical consequences of the Mendelian mechanism. The old criticism of Darwin's views by Fleeming Jenkin (that favorable variations would soon be lost by crossing) is repeatedly mentioned without recognition of the very different aspect in which this matter appears, as shown by Haldane and Fisher, on consideration of the statistical effects of steady selection pressure on a population containing a favorable Mendelian mutation.

In the population theory we must distinguish two processes which are not distinguishable in the mutation theory: the formation of a new species by transformation of an old species as a single group, and the splitting of a species. In the population theory, selection has much more direct importance in the former than in the latter. It is not the only factor here however. The pressure of recurrent mutation (a very different thing from the abrupt origin of new species by mutation) must be considered as a factor, though rarely as a major determining factor. In small isolated populations there is also a random drifting of all gene frequencies which keeps the character of the group from becoming stationary. The most favorable conditions for evolution as a single group seem to be found where a large population is subdivided and resubdivided into small *partially* isolated populations which become differentiated partly under the pressure of selection in relation to different local conditions but partly in a random fashion as a result of accidents of sampling among the parents, generation after generation. The resulting combination of nonadaptive and adaptive differentiation, associated with intergroup selection (through differential population growth and migration) makes for an effective but essentially unpredictable process of change by trial and error of the population as a whole even under static conditions.

If now a portion of the species is completely isolated the two portions will inevitably drift apart even if conditions are essentially the same. Both groups will always have adaptive combinations of traits but different ones because of the combination of non-adaptive with adaptive processes in their evolution. Cross sterility, on this viewpoint, arises essentially as a by-product of differentiation although its later stages may be speeded up by selection against hybridization. In the end, there may be discontinuity between the populations in many respects without there ever having been any extinction of contemporary intermediates. Willis notes briefly the possible significance of isolation in the origin of endemic species but does not recognize that it, not selection, is the real alternative to origin by mutation in the statistical theory of evolution. Under this viewpoint, as well as under the mutation theory, small endemic species may often be young species. It is not necessary to show that one species is more adaptive than its parent. Both are presumably adaptive but different. A multiplicity of diverse types can come about quite as well by isolation as by mutation.

The difference between evolution "downward" from family to species and "upward" from species to family is not as absolute as Willis insists. Under certain conditions the processes considered in the statistical theory may lead to a very rapid, almost explosive, origin of higher categories. One of these is the attainment by a species of an adaptation of *general* importance in contrast with a *specializing* adaptation. Such an attainment leads to relaxation of the selective processes that have previously restrained variation in the species, selection for accessory mechanisms and diversifying selection directed toward occupation of many different ecological niches. Again if a species reaches relatively unoccupied territory a similar very rapid evolution is to be expected. Under either of these conditions, a new higher category may arise very rapidly under statistical processes, and may then evolve "downward" to lower categories in exploiting the new general adaptation or the new territory.

The large amount of attention devoted to the "hollow curve" as evidence for the mutation theory, requires consideration. Such a distribution of generic sizes is undoubtedly to be expected on this theory but it has not been shown that it would not equally be expected on other theories. Let us consider a large family (or higher category) in which the total number of species remains constant, the process of duplication of species (whether by mutation or by isolation aided by selection or by any other mechanism) being balanced by extinction. For a given interval of time there is a certain probability of extinction of a species and certain probabilities that it will have one, two or larger numbers of descendants. By choosing the appropriate interval of time, the variance of this array of probabilities may be made 1 and therefore equal to the mean of the array. Genera containing n species (whose fates are assumed to be independent) should show a distribution of generic sizes after the chosen interval such that the variance is n, again equal to the mean generic size. These conditions insure that the distribution will be at least approximately of the Poisson type. In the long run, the frequencies, $f(x)$, of generic size, x, should reach a certain equilibrium in form. The class of genera with n species is dispersed after the chosen interval of time but receives recruits from each other class (x species) according to the frequency $f(x)$ and to the appropriate terms in the Poisson distribution, viz. $e^{-x}x^n/n!$ The occasional origin of new genera permits equilibrium.

Thus $$\frac{1}{n!}\sum_{1}^{\infty} e^{-x}x^n f(x) = f(n).$$

Replacing summation by integration to obtain an approximate solution

$$\frac{1}{\Gamma(n+1)}\int_0^\infty e^{-x}x^n f(x)dx = f(n).$$

This equation is satisfied if $f(x) = C/x$ where C is any constant.

Thus the form of distribution of generic size that is to be expected if extinction and duplication of species balance, irrespective of the mechanism of origin of species, is a rectangular hyperbola, C/x, a curve which is of the type of Willis' "hollow curve" and which becomes a straight line if plotted on double log paper. No doubt variations from this precise formula are to be expected under various conditions but obviously no conclusions on the mode of origin of species can be drawn merely from the occurrence of a "hollow curve" of generic sizes.

The author's wide acquaintance with plants in nature make his critique of natural selection as the sole factor in evolution of great weight. It does not appear, however, that he has sufficiently considered the alternatives. It does not seem necessary to postulate an unknown sort of mutation. The objections to exclusive determination of the course of evolution by selection can be met, at least to a very large extent, by a fuller consideration of the statistical consequences of known genetic mechanisms working on known classes of mutations.

SEWALL WRIGHT
THE UNIVERSITY OF CHICAGO

THE MATERIAL BASIS OF EVOLUTION[1]

By Dr. SEWALL WRIGHT
PROFESSOR OF ZOOLOGY, UNIVERSITY OF CHICAGO

THEORIES of evolution may be classified according to the demands which they make on chance variation. No demands are made by a group of theories under which evolution is a byproduct of individual physiology, whether directed by environmental influences or by the inheritance of individual adaptations or by an innate developmental process (orthogenesis). Theories of this group have had a wide appeal but have been largely discredited by the results of experiment. Under a second group of theories, evolution is a population matter, with varying emphasis on the pressures of recurrent mutation, of selection among individuals and of the effects of inbreeding, crossbreeding and intergroup selection in the hierarchy of partially isolated subdivisions of the species. Finally come the theories under which unique mutations are the crucial events, with subdivision according to whether such mutations are supposed to give rise merely to species, the higher categories arising by a cumulative process, or whether the higher categories as well as the species are supposed to appear abruptly. Dr. Goldschmidt has been the leader in recent years in advocating this last viewpoint on genetic grounds. He has now assembled pertinent data and his deductions from these, in book form,[1] an elaboration of the Silliman Lectures which he delivered at Yale in December, 1939.

The book is not intended to cover the entire field of evolution. It is an "inquiry into the types of hereditary differences which might possibly be used in evolution to produce the great differences between groups." There are two main parts—"Microevolution" (pp. 8–183) and "Macroevolution" (pp. 184–395)—in accordance with the absolute distinction which Goldschmidt has come to make between evolution within species and the origin of species and higher categories. He reaches the following conclusions with respect to microevolution.

> Microevolution within the species proceeds by the accumulation of micromutations and occupation of available ecological niches by the preadapted mutants. Microevolution, especially geographic variation, adapts the species to the different conditions existing in the available range of distribution. Microevolution does not lead beyond the confines of the species and the typical products of microevolution, the geographical races, are not incipient species. There is no such category as incipient species.

The reasons given for concluding that the processes which determine microevolution have nothing to do with macroevolution may be summarized briefly as follows: the observed discontinuity of species and higher categories can not be explained by these processes without postulating an unreasonable amount of extinction of intermediates; these processes are too slow; they are too limited in scope, since it is inconceivable that mutations of protozoan genes could furnish enough material for the evolution of higher animals, even granting duplication; and finally gene mutations are inadequate in kind. "The germ plasm as a whole controls a definite reaction system controlled as a whole by one agency."

As hereditary cytoplasmic differentiation is ruled out as an important factor, the author concludes that only one possibility is left.

[1] *The Material Basis of Evolution.* R. Goldschmidt. Illustrated. v + 436 pp. $5.00. May, 1940. Yale University Press.

Species and the higher categories originate in single macroevolutionary steps as completely new genetic systems. The genetical process which is involved consists of a repatterning of the chromosomes which results in a new genetic system. The theory of the genes and of the accumulation of micromutants by selection has to be ruled out of this picture. This new genetic system, which may evolve by successive steps of repatterning until a threshold for changed action is reached, produces a change in development which is termed a systemic mutation. Thus selection is at once provided with the material needed for quick macroevolution.

Goldschmidt holds that this theory avoids the insurmountable difficulties which he finds at all points in "neo-Darwinism." As the reviewer is listed among the advocates of neo-Darwinism it may be inferred that he does not find himself in agreement at all points. A detailed critique is not practicable in the scope of this review, but certain general points may be noted.

The antithesis between Goldschmidt's viewpoint and neo-Darwinism is not simply one of preference for systemic chromosomal changes as opposed to micromutations. The neo-Darwinists have not been concerned primarily with the problem which is central in the present book, the types of mutation involved in evolution. Their primary concern is the dynamics of the process, accepting all types of mutation actually observed. As the formal laws of heredity are the same, to a considerable extent, for gene mutations and the simpler chromosomal changes, the same theory applies to both, with minor complications, and has been so applied. Goldschmidt gives no serious discussion of questions of dynamics, and conclusions from statistical genetics are in several cases seriously misinterpreted. Yet the dynamics of the postulated accumulation of subliminal steps in chromosome repatterning and of the establishment of the systemic mutations, once the threshold has been passed, are questions which must be considered, if valid comparisons are to be made with the conclusions reached by neo-Darwinism.

From the latter viewpoint, the species (in sexually reproducing organisms) has a complex reticular structure in time, made up on a fine scale of the conjugating and segregating lineages, and, on coarser scales, of local populations and of subspecies. These groups tend to drift apart under a process that has jointly adaptive and fortuitous aspects (selection in relation to local conditions and effects of inbreeding). They can seldom get very far apart, however, and are continually shifting in character as long as continuity is maintained by crossbreeding. The species as a whole gradually changes its character as a result of intra- and inter-group selection and, under certain conditions, of accidents of sampling. If isolation of any portion of the species becomes sufficiently complete, the continuity of the fabric is broken. The two populations may differ little if any at the time of separation but will drift ever farther apart, each carrying its subspecies with it. The accumulation of genic, chromosomal and cytoplasmic differences tends to lead in the course of ages to intersterility or hybrid sterility, making irrevocable the initial merely geographic or ecologic isolation.

There are certain similarities with results to be expected from Goldschmidt's systemic mutations: subspecies are *not* ordinarily incipient species and there is usually marked discontinuity between species, occurring without any necessary extinction of contemporary intermediates. Under neo-Darwinism, however, these are relative matters, while absolute under Goldschmidt's scheme.

Goldschmidt devotes much space to the "bridgeless gap" between species which is crucial for his theory. To the reviewer, it appears that the data indicate every conceivable intergrade in degrees of morphological and physiological distinction, of chromosomal differentiation and of cross-sterility or hybrid sterility, with the correlation between these criteria far from perfect. Goldschmidt recognizes the lack of perfect correlation. He holds that "the development of intersterility is therefore to be regarded as the decisive step in the isolation of species." This then is the most essential effect of the systemic mutation. He devotes considerable space to the troublesome cases that have been turning up in increasing numbers in which sharply distinct, non-interbreeding populations living in the same region (and thus apparently good species) have been shown to be connected around a circle by a chain of intergrading types (making them merely subspecies). Goldschmidt, taking the latter horn of the dilemma, decides that the criterion for a good species is demonstrated physiologic intersterility or hybrid sterility, in contrast with failure to interbreed which may be due merely to psychologic causes. Thus he unites into one Rassenkreis two species of deer mice, ***Peromyscus leucopus*** and ***P. gossypinus,*** which hybridize rarely if ever in nature, where their ranges overlap, but which have been shown by Dice and Blossom to produce fertile hybrids in the laboratory. He does not, however, refer to the existence of fertile hybrids from crosses between bison and cattle and between pigeons and doves which would relegate accepted genera and families to the level of subspecies by this criterion. Later in the book indeed he recognizes that "good" species (of lepidoptera) may produce fertile hybrids. The question of an absolute difference in kind between the products of microevolution and of macroevolution thus seems to be left in a rather unsatisfactory state. It may be added that he himself to some extent cuts the ground from under the conception of a "bridgeless gap" by suggesting later that genes have no existence except as undefinable regions in the chromosome pattern, that gene mutations are merely highly localized changes in pattern and thus that there is no essential difference in kind between the materials of microevolution and those of macroevolution. Moreover, the hypothesis that repatterning may require many steps before a systemic mutation occurs, tacitly reintroduces "incipient species" though not necessarily in the same sense as the recognized subspecies.

With respect to the alleged slowness of neo-Darwinian evolution it may be said that the theoretical rate depends very much on the conditions. Under reasonably stable conditions, with every ecologic niche occupied, theory indicates equilibrium of all gene frequencies to a first order in spite of continual mutation, considerable variability if the population is large, and severe selection. Nevertheless, second-order processes, in a species with a hierarchy of partially isolated subgroups, permit continuous, if usually rather slow, change. There is always the possibility, moreover, that this may lead to an adaptation of a type that is of general significance, perhaps opening up a new way of life. The result of such an adaptation would be a relaxation of selection, a rise in the frequencies of many mutant genes and great increase in variability and in numbers. With ecologic isolation in the various niches now open to the species, and selection pressures now directed differently in each niche, an explosive evolutionary process is to be expected, an adaptive radiation within a new higher category. The same process may occur from a change of conditions. It would seem inevitable if a species reaches an unoccupied region that offers diverse opportunities. Such events must have been relatively common early in geologic time. A relatively recent example seems to be found in the history of the Drepanids of the Hawaiian Islands (an endemic family of birds with 18 highly diversified genera and 40 species, presumably tracing to a pair or flock of stray migrants of an American species) yet Goldschmidt cites this as unimaginable except by an outburst of systemic mutations of family, generic and specific rank.

It will be well to reiterate that neo-Darwinism is not concerned solely with micromutations. It is quite in accordance with the theory that mutations with conspicuous effects may play an important rôle in the adaptive radiation of populations for which selection is relaxed, provided they do not have deleterious effects on individual physiology or fertility. On the other hand, complex adaptations would be enormously more likely to be reached by a cumulative process, involving trial and error, than by a single step as postulated by Goldschmidt. To the reviewer, the latter's theory seems to make demands on chance that are much too severe, both with respect to the origin of the complex differences between species and their establishment, recalling the postulated abrupt occurrence of intersterility with the parent stock.

The origin of new types by polyploidy is a process in which the individual mutant is the crucial event, which must be recognized by neo-Darwinists as of considerable importance. In these cases, however, there is either little initial change in character (autopolyploidy) or there is a balance of parental character after hybridization (allopolyploidy) favorable to survival. There is a population problem even here, in that the low fertility of triploid hybrids with the parent stock raises severe obstacles to establishment unless vegetative multiplication occurs freely. Goldschmidt himself considers autopolyploidy as falling within the scope of microevolution and does not ascribe primary evolutionary significance to allopolyploidy.

With respect to limitations in the amount of material for evolution provided by gene mutation as compared with chromosome repatterning, it is not clear that the possible combinations of all possible mutations at thousands of loci furnish less material than the possible permutations resulting from repatterning of normal chromosomes. The evidence at present suggests that the material at each locus (which may be thousands of times as much as that in a large protein molecule) may be capable of mutating through an indefinitely extended, branching series of alleles. Every locus may thus have had an exceedingly complex evolutionary history from protozoan to man.

At the root of the differences between Goldschmidt's viewpoint and that of neo-Darwinism seem to be different conceptions of the physiological relations between germ plasm and organism The earlier evolutionists, under the influence of a rigid conception of morphological homology, were seriously troubled by the apparent necessity for assuming independent heredities for each part of each replicated structure. An enormous load was lifted when it was recognized that complex differences in form may trace to simple differences in the developmental process and that the same genetic system that determines, through physiological channels, the form of one structure may be expected to give the basis for more or less similar replications wherever the local conditions are sufficiently similar. Goldschmidt has played a leading rôle in bringing home to geneticists the necessity for a physiological interpretation. This general viewpoint is explicitly accepted by all those whom he lists as neo-Darwinists. Goldschmidt, however, seems to hold that the conception of the organism as an integrated reaction system requires a corresponding *spatial* integration of the germ plasm and that essential change in the reaction system can thus come about only by repatterning of the chromosomes. To others, a *temporal* integration is all that is necessary, or even possible, with the chain reaction as the simplified model. Within the organism as a more or less integrated reaction system, there is a hierarchy of subordinate reaction systems, each with considerable independence, as shown by capacities for self-differentiation. Thus there must be partially isolated reaction systems for each kind of organ and for each kind of cell. It is difficult to see how any spatial pattern in the germ plasm can operate in determining these, but there is no theoretic difficulty with a branching hierarchic system of chain reactions in which genes are brought into effective action whenever presented with the proper substrates, irrespective of their locations in the cells. There is no limit to the number of reaction systems that can be based on the same set of genes, and such systems may obviously evolve more or less independently of each other.

In an introductory chapter, Goldschmidt cites about a score of characters whose evolution he challenges the neo-Darwinist to explain. The actual course of events has usually been so tortuous in cases in which the evidence is reasonably complete that no attempt at reconstruction without direct evidence can be taken very seriously. However, there appear to be no special theoretic difficulties in the cases cited if the neo-Darwinian position is correctly understood. We can only take space for one of these cases but it is typical of several. Goldschmidt holds that alternation of generations could only have arisen by a systemic mutation. Let us consider the case in coelenterates, beginning for sake of argument with a hydra-like polyp, multiplying by budding. Colony formation could arise gradually by the accumulation of micromutations, causing delay in separation of daughter polyps. Differentiation of polyps into nutritive and reproductive zooids involves the same sort of problem as the differentiation of organs within an individual. Alternative reaction systems may become differentiated gradually with expression related to position in the colony. The reaction system of the reproductive zooid may become modified gradually to provide for increasingly regular separation from the colony and for locomotion, until the medusa type is reached. This, incidentally, offers a new way of life and evolution may continue by gradual suppression of the reaction system of the fixed phase and elaboration of that of the free phase.

So far we have compared Goldschmidt's theory with neo-Darwinism largely on the plausibilities with which they explain evolutionary phenomena. But neo-Darwinism takes its premises from the chromosome theory as actually developed by genetics and cytology. Goldschmidt deduces a different type of chromosomal organization, on the basis of the supposed theoretic necessity of his systemic mutations. No data are given that support the conception of a spatial pattern of the germ plasm, correlated with the reaction system of the organism. Against this is the independence of the chromosomes and the apparent absence of any correlation between location of genes and at least the more conspicuous effects of their mutations. Abundant data indicate that a type gene produces its characteristic effect in translocations whether the translocated piece is small or large, or if large whether the locus is near the right end or the left end or the middle. The effect may be weakened or less stable or otherwise modified, but the type alleles of white, of yellow, of *cubitus interruptus*, etc., of *Drosophila melanogaster* still retain their essential specificities. The simplest description of the genetic data is still obtained by attributing specificity, independent of position, to each gene and merely qualifying this by recognition of occasional second order effects of position. The old question whether adjacent genes are completely separate or are merely specialized regions in a continuum is irrelevant in this connection.

Goldschmidt's contention that character differences comparable to those between species can not be brought about without repatterning is refuted by the characters of trisomics in Oenothera, Datura, Nicotiana, etc. In these there is no repatterning of any chromosome but merely quantitative increase in all genes of one chromosome, yet the morphological and physiological effects give the appearance of specific difference. On the other hand, a single moderately long inversion should annihilate the pattern of a chromosome as a whole. Yet inversions either have no effect or effects no greater than single gene mutations. The hypothesis that there is a threshold in repatterning at which the systemic mutation makes its appearance seems to be of a wholly *ad hoc* character.

While the reviewer radically disagrees with the author's central thesis, he wishes to testify to the importance of the book. A great store of well-selected data have been assembled from diverse sources, fairly presented and discussed from viewpoints which must be carefully considered by any one interested in the problem of evolution.

BOOK REVIEWS

Tempo and Mode in Evolution: A Critical Review[1]

A dozen years ago, the dean of American paleontologists wrote: "The attempt to trace the temporal origin of biomechanical adaptations, which paleontology demonstrates are *determinate, orthogenetic, secular, germinal processes involving enormous periods of time,* shows that the mutationists and selectionists are traversing a swamp of useless inquiry led by the will-of-the-wisp of expectation" (Osborn, '33). One of the leading paleontologists of the present day has written a book, an essential part of which he describes as the attempted synthesis of paleontology and genetics. The reviewer, as a geneticist, finds remarkably little difference between Dr. Simpson's interpretation and that which had seemed to him, even before Osborn made the above statement, the natural deduction from genetic principles (Wright, '29, '31, '32). It would certainly be too much to say that paleontologists and geneticists as groups have reached agreement but at least a contact has been established that did not exist a few years ago.

Modern genetics traces its origin to an important extent to the mutation theory of de Vries and the evolutionary speculations of geneticists were long dominated by de Vries' conception of the abrupt origin of species by single mutations. It has indeed been demonstrated that there is one such a process. It is generally conceded, however, that the role of polyploidy in evolution has been a minor one in most groups, certainly including for example, the vertebrates. The most important step on the genetic side toward reconciliation with paleontology is probably the recognition that in general, evolution is a population problem. A mutation, chromosomal or genic, major or minor, must undergo a virtually continuous process of change in frequency before becoming characteristic of a population. The elementary evolutionary process is therefore change of gene (or chromosome) frequency.

Simpson accepts this viewpoint as basic to the genetic interpretation. He discusses the determinants of evolution (Chapter II) under the heads: variability, mutation rate, character of mutation, length of generation, size of population, and selection. For the most part, the presentation is quite acceptable to a geneticist, but discussion of certain points seems desirable.

Data and logical considerations are presented to show that amount of variability is not, and should not be expected to be, at all closely related to rate of evolution. It is held that species carry large stores of potential genetic variability, manifested as actual variability only to slight extents. This is undoubtedly the case, but in the main, I think, for a different reason than that given. Simpson, following Mather ('41), locates this store in the maintenance of systems of genes balanced within chromosomes by means of linkage, and selection against recombination. This mechanism is effective in extreme cases in which linkage is virtually complete (e.g. in inverted sections of chromosomes in *Drosophila*). It is not a very effective mechanism with crossing over at the usual rates and with only slight selection against unbalanced recombinations. Thus in the case of 2 pairs of factors in the same chromosome (with the proportion of crossing over c) and a selective disadvantage s against heterozygous unbalanced types (AB/Ab, AB/aB) in comparison with balanced types (Ab/Ab, aB/aB, Ab/aB, AB/ab) and a selective disadvantage of $4s$ against the homozygous unbalanced types (AB/AB, ab/ab) the proportions among the gametes do not deviate very much at equilibrium from those of random combination unless s is of the same order as c or higher. It can easily be shown that if the array of the 4 kinds of gametes in the symmetrical case is written $(w\,AB + x\,Ab + x\,aB + w\,ab)$, $w = [s + c - \sqrt{c^2 + s^2}]/4s$, $x = [s - c + \sqrt{c^2 + s^2}]/4s$ at equilibrium, or, if s is less than c, about $(\frac{1}{4}) - (s/8c)$ and $(\frac{1}{4}) + (s/8c)$ respectively, instead of $\frac{1}{4}$, expected in both cases in the absence of selection. If a large number of linked factors are involved, selection against the smallest deviation from balance must be of the order of the amount of crossing over between neighboring genes (or higher) to prevent a largely random recombination from taking place.

A species may, however, be expected to carry an enormous store of rather easily available potential variability if it is divided into numerous partially isolated local populations (Wright, '36, '40). Each local population may be expected to approach fixation of a particular balanced combination of the many genes that act on each character. But as there may be a very large number of different balanced genotypes with the same or nearly the same phenotype, each local population may be expected to center about a different genotype. There will then be little apparent variability either within local populations or within the species as a whole, yet an enormous field of potential variability, available by mere increase in the amount of migration between populations. Thus it is probable that any species that inhabits a large territory, with a sufficiently sparse population, or with sufficiently numerous barriers to free interbreeding, carries numerous slightly different alleles at most loci.

[1] **Simpson, George Gaylord.** 1945. Tempo and mode in evolution. N. Y.: Columbia University Press. Pp. i + xviii. + 237. 36 figs. $3.50.

Ten alleles at each of only 100 loci implies an inconceivably great field of potential variability (10^{100} potential homozygous genotypes). I think that Simpson somewhat underestimates the amount of evolution that could occur by mere shifts in gene frequencies without any new mutation and without depletion of variability. The difficulty in evolution is not in the amount of raw material available at any time (in such a species as described above) but in the finding of a path from one harmonious combination to another, through the system of potential combinations, against the enormous and primarily conservative pressure of selection. The situation is such that under exceptionally favorable conditions great evolutionary advance is possible at an explosively rapid rate. It should, of course, be added that while mutation rate is probably not a limiting factor, new alleles may actually be expected to appear following radical changes in gene frequency.

With respect to the character of mutations, Simpson maintains that the paleontological evidence indicates the sort of continuity expected of concurrent change in the frequencies of many minor genetic factors. He finds no substantial support among paleontologists for Goldschmidt's hypothesis of origin of species and higher categories by single systemic mutations (Goldschmidt, '40).

If mutation rate were the limiting factor, length of generation might be expected to show a strong negative correlation with rate of evolution. Simpson finds no close relation of this sort, although he makes some interesting suggestions of a role in particular cases. He notes that eleplants are close to the maximum among mammals in both respects.

The discussion of the effect of size of population is marred by a number of unfortunate typographical errors. On pages 66 and 67 and elsewhere, $\frac{1}{2}N$ is written for $1/(2N)$, $\frac{1}{2}s$ for $1/(2s)$ and $\frac{1}{4}u$ for $1/(4u)$. In a case in which effective size of population (N) is one million the first of these formulae is in error by 10^{12} and the errors may be comparable for selection (coefficient s) and mutation rate (u). The reviewer wishes also to make a correction here to a statement on page 67, similar to ones made by other authors, in which he is credited with the conclusion that the conditions most favorable for rapid evolution are found in populations of a certain intermediate size. The actual statement made in several papers was to the effect that conditions are more favorable in a population of intermediate size than in a very small one or in a very large *random breeding* one (assuming a constant direction of selection). But such a statement has always been followed by the statement that conditions are enormously more favorable in a population which may be large but which is subdivided into many small local populations almost but not quite completely isolated from each other (Wright, '29, '31, '32, '42). This it may be noted is the same sort of population referred to above as that which carries the maximum amount of potential variability.

With respect to selection, Simpson makes the statement that while natural selection acts on variation both within and between groups, "its action on intergroup variation can produce nothing new; it is purely an eliminating, not an originating force." I differ strongly here. A single choice between two types may indeed appear eliminative rather than creative, but a succession of choices has a guiding influence that I think may properly be called creative, irrespective of the level at which the choices are made. I would consider the drastic elimination of families and orders of vertebrates, and the compensatory adaptive radiation of the successful ones, as highly creative with respect to the major course of vertebrate evolution. The main point that I wish to make, however, is that selection between the genetic systems of local populations of a species, operating through differential migration and crossbreeding, has been perhaps the greatest creative factor of all in making possible selection of genetic systems as wholes in place of mere selection according to the net effects of alleles. It is this sort of intergroup selection that makes possible the transition from one adaptive peak to another, which in its most extreme form is Simpson's quantum evolution to be discussed later.

There is another sort of consequence of intergroup selection which has not previously been discussed in terms of mathematical models but which may be considered creative. The following one factor case illustrates the possibility of fixation by this means of a character of value to the population, but disadvantageous at any given moment to the individuals. Assume that A_2 is a mutation from A_1 that at any given time is disadvantageous to individuals to the extent s in heterozygotes and $2s$ in homozygotes; but that increase in its frequency affects favorably (term $(1 + bq)$), the rate of increase of all of the individuals in the population through some product or activity of those that carry it.

Genotype	Frequency (f)	Selective Value (W)
A_1A_1	$(1 - q)^2$	$a(1 + bq)$
A_1A_2	$2q(1 - q)$	$a(1 + bq)(1 - s)$
A_2A_2	q^2	$a(1 + bq)(1 - 2s)$

If the W's are reproductive rates for genotypes, their average, $\bar{W} = \Sigma Wf = a(1 + bq)(1 - 2sq)$ is that for the population as a whole. If b is equal to $2s/(1 - 4s)$ or greater, the population as a whole grows most rapidly if consisting wholly of genotype A_2A_2. The rate of change of gene frequency may be deduced from the general formula for a random breeding population

$$\Delta q = \left[q(1 - q)\, \Sigma W \frac{df}{dq} \right] / 2\bar{W}$$

(Wright 1942). The common factors of the W's cancel, leaving as the rate of change of the frequency of A_2 the expression $\Delta q = -sq(1-q)/(1-2sq)$. This is negative and implies fixation of A_1 in the population (unless there is counteracting mutation pressure $v(1-q)$) to keep A_2 in the population at a low frequency. Thus the socially favorable mutation A_2 tends to be lost or nearly lost in a random breeding population. In a population divided into many small, *completely* isolated groups, selection becomes reduced in efficiency as drift due to accidents of sampling increases. A_2 may occasionally drift into fixation in a local group but cannot spread in the absence of migration. If, however, there is a small amount of migration and crossbreeding, a term $m(q_i - q)$, in which m is the effective amount of replacement by migrants and q_i is the frequency of A_2 among the migrants, must be introduced into the expression for change of gene frequency within a local population.

$$\Delta q = m(q_i - q) - sq(1-q)/(1-2sq)$$

If the local populations are sufficiently small in size (N) and there are not too many immigrants, there will be extensive random variation of the frequency q. The variance has approximately the value $\sigma_q^2 = \bar{q}(1-\bar{q})/(4Nm+1)$, ($\bar{q}$ being the average value of q), (disregarding the effects of selection). Most of the migrants may be expected to come from the centers of rapid population growth which are those in which q has drifted into exceptionally high values, with the consequence that q_i should be considerably greater than q. The term $m(q_i - q)$ should accordingly be greatest with a certain degree of fineness of subdivision (small N) and a certain small proportion of immigration (m). This term may easily be large enough to overbalance the selective disadvantage of A_2 and lead to its fixation. It is indeed difficult to see how socially advantageous but individually disadvantageous mutations can be fixed without some form of intergroup selection. Evolution of such traits may be considered to be as creative a process as that of traits which make for individual survival (*cf.* Allee, '43).

The major portion of the book is, as the title implies, devoted to consideration of the actual rates of evolution indicated by paleontological material and deductions as to the major modes of evolution. A most impressive feature is the documentation of statements by statistical studies.

It is made abundantly clear in Chapter I that there is no general law of rates except that of diversity. Analysis of the evolution of characters of the horse shows that strictly there is no such thing as the rate of evolution along a particular phylum since different characters evolve at widely different rates.

It is, however, of interest to compare the rates in terms of degrees of overall change which paleontologists have recognized as worthy of a given taxonomic rank. We learn that horses and chalicotheres evolved through 8 and 5 genera respectively at about the same rate, 1.8 and 1.7 genera respectively per ten million years while ammonites passed through 8 successive genera at a rate of only half a genus per ten million years. Extensive specific and generic but not familial differentiation has occurred in cricetine rodents, procyonid carnivores and deer in the two million years or so since South America was invaded by North American forms. Well defined subspecific differentiation is attested in rodents in even less than 300 generations.

Especially instructive are tables of first and last appearance of the genera of land carnivores and of pelecypods, and the deductions from them. A rough similarity is found between the survivorship curves of pelecypod genera, carnivore genera and vestigial winged Drosophilas when reduced to a common time scale by equating the respective times of mean survival, 78×10^6 years, 6.5×10^6 years and 14 days, each to 100. We learn that there were 28.2 genera of carnivores on the average (standard deviation 13.1) in each of the 14 stages from Middle Paleocene to Late Pliocene inclusive and that the average turnover in a stage was about 4.

The continuity of the evolutionary process is discussed in Chapter III with especial reference to the origin of the higher categories. Paleontology has relatively little to contribute to the distinction between micro and macroevolution of Dobzhansky's definition in which the dividing line was that between the origins of subspecies and species (Dobzhansky, '37). For the paleontologist, the distinction between small scale and large scale evolution is rather that between the origins of genera and of families. Simpson coins a new term, megaevolution, for large scale evolution of the paleontologist. He strongly supports the idea of continuity from species to species and genus to genus on the basis of the complete continuity of the paleontological record in favorable cases and the indication that discontinuities in the records can be correlated with discontinuities in deposition of strata or (as in the case of horses in Europe) sampling of successive waves of migrants.

He emphasizes, however, that at the level of megaevolution, essentially continuous transitions are virtually absent. Reasons are given for rejection of the idea that the transition forms never existed and that families and higher categories arose by saltation. "In summary, the theory here developed is that megaevolution normally occurs among small populations that become preadaptive and evolve continuously (without saltation but at exceptionally rapid rates) to radically different ecological positions." I have expressed essentially the same idea in the few cases in which

I have ventured into explicit discussion of mega-evolution, a field necessarily rather remote from genetics. "Under certain conditions, the processes considered in the statistical theory may lead to a very rapid, almost explosive, origin of higher categories. One of these is the attainment by a species of an adaptation of *general* importance in contrast with a *specializing* adaptation. Such an attainment leads to a relaxation of the selective processes that have previously restrained variation in the species, to selection for accessory mechanisms, and to diversifying selection directed toward occupation of many different ecological niches. Again if a species reaches relatively unoccupied territory, a similar very rapid evolution is to be expected. Under either of these conditions, a new higher category may arise very rapidly under statistical processes, and may then evolve downward to lower categories in exploiting the new general adaptation or the new territory" (Wright, '41a; *cf.* also Wright, '41b, '42).

With regard to mechanism, Simpson holds that the typical pattern is probably that "a large population is fragmented into numerous small isolated lines of descent. Within these, inadaptive differentiation and random fixation of mutations occur. Among many such inadaptive lines, one or a few are preadaptive, i.e. some of their characters tend to fit them for available ecological stations quite different from those occupied by their immediate ancestors. Such groups are subjected to strong selection pressure and evolve rapidly in the further direction of adaptation to the new status. The very few lines that successfully achieve this perfected adaptation then become abundant and expand widely at the same time becoming differentiated and specialized on lower levels within the broad new ecological zone." Later he states that the "conditions most propitious to such fixation is that the populations be small and completely isolated." This agrees in part with the views which I have expressed but I would insert the words "almost but not quite" before "completely isolated" in the last quotation. This may appear a very trivial distinction but from the standpoint of genetic mechanism it is all important. The reasons may be put as follows:

Let N be the effective size of the typical local population, u the mutation rate per generation at a locus and s the selective advantage of one allele over the others, considered collectively. It can be shown that the conditions for evolutionary change with jointly adaptive and nonadaptive aspects is that $4Nu$ be of the order 1 to 10 and $4Ns$ not much larger. If these expressions are much less than 1, the inbreeding effect: fixation of some random and therefore deleterious combination of genes dominates the situation. The end result is extinction. If $4Ns$ is very much larger than 10, selection pressure in the current direction dominates the situation. At intermediate values, there is the sort of trial and error process which makes possible the transition from one adaptive peak to another. Unfortunately there may be no value of N which meets both conditions and if there is, the process is at best exceedingly slow. This is because mutation rates rarely exceed 10^{-5}. To meet the conditions effective N should be of the order 100,000 to 1,000,000 which implies an exceedingly slow non-adaptive drift and s should not be very much greater than mutation rate, which constitutes decidedly weak selection. But if there is a small amount of migration and crossbreeding, the immigration coefficient m takes the place of u, mutation rate in the formulae. The difficulty with this scheme is the opposite of that with complete isolation. Unless m, the proportion of replacement by immigrants is decidedly small, the result is virtually equivalent to panmixia over the entire range of the species. If the species occupies its range sparsely, however, m may be small enough to provide the favorable conditions. If m is of the order .01 to .001, s of the same order or somewhat greater, and effective N (of the local population) only of the order of 100, nonadaptive drift is a rapid process in terms of geologic time and selection pressure on the whole genetic system is also strong. A very rapid process of trial and error among populations and of intergroup selection through dispersion of migrants from the more successful centers, is the consequence. There is a reasonable chance that at one center a genetic system may be arrived at that marks the transition from the currently dominating adaptive peak to a widely different and higher one, which may become the starting point of a new higher category. The reason that mutation rate can largely be ignored here is essentially because of the enormous store of potential, easily available but not actually manifested, variability which such a species possesses. Mutation rate is not the limiting factor that it is under the scheme of moderately small completely isolated populations (Wright, '31, '32 and later).

Returning to Simpson's discussion of rates of evolution, he finds different distributions of rates in different groups of animals (e.g. mammals and pelecypods) (Chapter IV). Certain lines in most groups fall outside the standard distribution in both directions. He suggests the term horotelic for rates within the normal range, tachytelic for the rare cases of extraordinarily rapid evolution which we have discussed, and bradytelic for the moderately common cases in which cumulative change seems virtually to have ceased. Among living groups he lists the lingulids (since the Ordovician), limulids (since the Triassic), coelocanths (since the Devonian), sphenodonts (since the Triassic), crocodiles (since early Cretaceous) and opossums (since late Cretaceous) as examples of bradytely. These are interpreted essentially as groups that have reached the limits of evolution that are organically probable along certain lines. They were progressive when they first appeared and in most cases have existed in continuously large breeding populations. They may occasionally give rise to horotelic or tachytelic branches but the bradytelic stem is likely to persist.

Chapter V deals with "Inertia, trend and momentum." Most paleontologists have stressed the tendency of phyla to continue to evolve in much the same direction for considerable periods of time. Mystical notions of organic inertia, of unfolding of innate tendencies, and of working toward the best, independently of selection or other environmental control, have been common. Simpson accepts rectilinear evolution as an approximately correct description of a common evolutionary phenomenon but as noted treats it as merely a temporary incident in the total history. It should be said that even Osborn, who, as indicated in the quotation at the beginning of this review, treated "aristogenesis" as the primary evolutionary problem, also recognized the importance of adaptive radiation which of course implies a tortuous course of evolution in the l..ng run. Simpson rejects all mystical notions on the subject. He treats orthogenesis as merely the course followed by a continuously large population which has attained such specialization along a certain line that only farther enhancement of that one line of specialization is open to it. He rejects control by directed mutation (except in the trivial sense that the nature of possible mutational effects is limited at any stage by the organization already attained) and accepts selection as the essential controlling factor, conclusions in which the reviewer thoroughly concurs. There is a destructive analysis of supposed cases of organic momentum. "There is no good evidence that a trend has ever continued by momentum beyond a point of advantageous or selectively neutral modification or has ever been a direct cause of extinction."

In Chapter VI on Organism and Environment, the author develops a classification of environments into discontinuous adaptive zones and subzones, corresponding to ways of life. There is for example the adaptive zone occupied by carnivorous mammals and subzones corresponding to the ways of life of a canid and of a felid. This concept is obviously related to the reviewer's concept of peaks in the array of adaptive values of the genetic combinations momentarily possible in a species. There is a difference in point of view, however, and also the difference that there may be numerous adaptive peaks in the genotypic field of a species, relating to the same adaptive subzone as well as ones relating to different subzones and zones.

The final chapter introduces terms designed to focus attention or three fundamental modes of evolution, related to but not identical with the three fold classification of evolutionary tempos. The old term speciation refers to the mere multiplication of subspecies and species—partially nonadaptive, and adaptive only in relation to minor subzones—that occurs in bradytelic groups as well as in ones that are evolving directionally.

Simpson points out that this is the only mode open to investigation by experimental biology, neozoology and genetics but is that which is least open to the paleontologist. Phyletic evolution is the sustained directional shift in average direction to which, according to the author, 90 per cent of the data of paleontology pertains. It includes strictly orthogenetic trends but is defined as not necessarily rectilinear. The rates are in general those defined as horotelic. The third mode is quantum evolution, the rapid (tachytelic) shift from one major adaptive zone to another through a position of instability. It is the attempted establishment of the existence and characteristics of this mode of evolution which the author considers to be the most important but admittedly controversial outcome of his investigations.

This is a long review of a relatively short book. So much thought is condensed into its compass, however, that it has not been practicable here to do more than discuss a few of the many suggestive ideas that are presented. If considerable space has been devoted to critical discussion of certain points this should not obscure the enthusiasm which the reviewer feels for the compact marshalling of pertinent paleontological data and the stimulating attempt at interpretation by genetic principles.

Sewall Wright

Department of Zoölogy,
The University of Chicago

References Cited

Allee, W. C. 1943. Science **97**: 517–525.

Dobzhansky, Th. 1937. Genetics and the origin of species. 1st edition. New York: Columbia University Press.

Goldschmidt, R. 1940. The material basis of evolution. New Haven: Yale Univ. Press.

Mather, K. 1941. Jour. Genetics **41**: 159–163.

Osborn, H. F. 1935. Proc. Nat. Acad. Sci. **19**: 159–163.

Wright, S. 1929. Anat. Rec. **44**: 287.

——. 1931. Genetics **16**: 97–159.

——. 1932. Proc. 6th International Congress of Genetics **1**: 356–366.

——. 1936. Jour. Genetics **30**: 257–266.

——. 1940. The American Naturalist **74**: 232–248.

——. 1941a. Ecology **22**: 345–347.

——. 1941b. The Scientific Monthly **53**: 165–170.

——. 1942. Bull. Amer. Math. Soc. **48**: 223–246.

29
Isolation by Distance

Genetics 28 (March 1943):114–38

30
Analysis of Local Variability of Flower Color in *Linanthus parryae*

Genetics 28 (March 1943):139–56

31
Isolation by Distance under Diverse Systems of Mating

Genetics 31 (January 1946):39–59

Introduction

The quantitative theory of isolation by distance is one of Wright's most significant contributions to modern evolutionary biology. On two previous occasions, he had briefly discussed the possibility of isolation by distance as a means of genetic differentiation of demes within a large and apparently continuous population (Wright 94, 430–31; 100, 244–48), but had not really developed the theory mathematically. Stimulated by the wonderful set of data on *Linanthus parryae* that Carl Epling and his associates had gathered in the spring of 1941 and later published by Dobzhansky and Epling (1942), Wright wrote the first two of the papers in this section, the first presenting his theory of isolation by distance and the second applying the theory to the *Linanthus* data. Wright's further development of *F*-statistics (measuring degrees of inbreeding) is a notable aspect of his theory of isolation by distance.

The story of the interaction between Wright, Dobzhansky, and Epling over *Linanthus parryae* and the theory of isolation by distance is complex and yields much insight into the relationship between theoretical models and field research at this time (see *SW&EB*, 370–81, 484–88; *E&GP* 2:295–335).

In 1946 Wright extended his model of isolation by distance to include various systems of mating within the continuous population.

ISOLATION BY DISTANCE*

SEWALL WRIGHT
The University of Chicago[1]

Received November 9, 1942

STUDY of statistical differences among local populations is an important line of attack on the evolutionary problem. While such differences can only rarely represent first steps toward speciation in the sense of the splitting of the species, they are important for the evolution of the species as a whole. They provide a possible basis for intergroup selection of genetic systems, a process that provides a more effective mechanism for adaptive advance of the species as a whole than does the mass selection which is all that can occur under panmixia.

RANDOM DIFFERENTIATION UNDER THE ISLAND MODEL

Mathematical consideration requires the use of simple models of population structure. The simplest model is that in which the total population is assumed to be divided into subgroups, each breeding at random within itself, except for a certain proportion of migrants drawn at random from the whole. Since this situation is likely to be approximated in a group of islands, we shall refer to it as the island model.

The gene frequency (q) of a subgroup tends to vary about a certain equilibrium point ($\hat{q}$) in a distribution curve ($\phi(q)$) determined by the net systematic pressure (measured by Δq, the net rate of change of gene frequency per generation from recurrent mutation, immigration, and selection) in conjunction with the cumulative effects of accidents of sampling (random deviation δq, variance per generation $\sigma^2_{\delta q}$) (WRIGHT 1929, 1931, 1942).

$$\phi(q) = (C/\sigma^2_{\delta q}) \exp\left[2\int(\Delta q/\sigma^2_{\delta q})dq\right]. \tag{1}$$

Let N be the effective size of the subgroup, m the effective proportion of its population replaced in each generation by migrants, and q_t the gene frequency in the total population. The rate of change of gene frequency per generation in a subgroup, taking account only of immigration pressure, is $\Delta q = -m(q-q_t)$. In a random breeding population $\sigma^2_{\delta q} = q(1-q)/2N$. Substitution in (1) gives the following, choosing C so that $\int_0^1 \phi(q)dq = 1$ (WRIGHT 1931, 1942).

$$\phi(q) = \frac{\Gamma(4Nm)}{\Gamma(4Nmq_t)\Gamma[4Nm(1-q_t)]} q^{4Nmq_t-1}(1-q)^{4Nm(1-q_t)-1} \tag{2}$$

$$\bar{q} = \int_0^1 q\phi(q)dq = q_t \tag{3}$$

* A portion of the cost of composing the mathematical formulae is borne by the Galton and Mendel Memorial Fund.

[1] Acknowledgment is made to the DR. WALLACE C. and CLARA A. ABBOTT MEMORIAL FUND of the UNIVERSITY OF CHICAGO for assistance in connection with the calculations.

$$\sigma_q^2 = \int_0^1 (q - \bar{q})^2\phi(q)dq = q_t(1 - q_t)/(4Nm + 1). \tag{4}$$

In the derivation of (1), it was assumed that Δq is sufficiently small that terms involving $(\Delta q)^2$ might be ignored. A more accurate value of σ_q^2 may be obtained directly. The deviation of a local gene frequency from the average, $(q-q_t)$, tends to be reduced to $(1-m)(q-q_t)$ in the next generation. The mean sampling variance of $(q+\Delta q)$ is

$$\frac{1}{2N}\int_0^1 [q - m(q - q_t)][1 - q + m(q - q_t)]\phi(q)dq = [q_t(1 - q_t) - (1 - m)^2\sigma_q^2]/2N. \tag{5}$$

Thus with a steady balance between the effects of immigration and of the accidents of sampling

$$\sigma_q^2 = (1 - m)^2\sigma_q^2 + [q_t(1 - q_t) - (1 - m)^2\sigma_q^2]/2N \tag{6}$$

$$\sigma_q^2 = q_t(1 - q_t)/[2N - (2N - 1)(1 - m)^2]. \tag{7}$$

This is approximately the same as (4) for small values of m but becomes $q_t(1-q_t)/2N$, the sampling variance, in the limiting case of no isolation whatever $(m=1)$. This is about twice as great as given by (4) in this extreme case.

The variance, excluding the immediate sampling variance may be obtained by multiplying (7) by $(1-m)^2$ as indicated in (6). Formula (4) lies between the values with and without the immediate sampling variance.

Under exclusive uniparental reproduction, whether vegetative or by self-fertilization, the distribution of alternative genotypes may be treated by the same theory except for replacement of 2N by N. Immigration pressure is the same but the sampling variance is $q(1-q)/N$.

THE INBREEDING COEFFICIENT

Departures from panmixia may be expressed in terms of the average inbreeding coefficient of individuals, relative to the total population under consideration. This coefficient has been defined as the correlation between uniting gametes with respect to the gene complex as an additive system. It has been shown that its value can be found for any pedigree by finding all paths by which one may trace back from the egg to a common ancestor (A) and thence forward to the sperm along a wholly different path. According to the theory of path coefficients, the correlation between uniting gametes is the sum of contributions from all such paths (WRIGHT 1921, 1922b).

$$F = \sum [(1/2)^{n_S+n_D+1}(1 + F_A)] \tag{8}$$

where F and F_A are the inbreeding coefficients of the individual and of a common ancestor of sire and dam, respectively, and n_S and n_D are the numbers of generations from sire and dam, respectively, to this common ancestor. In a population in which the average inbreeding coefficient is F, the frequencies of genotypes (one pair of alleles) are as follows (WRIGHT 1921, 1922a).

	Genotype	Frequency
(9)	AA	$x_t = q_t^2(1 - F) + q_tF$
	Aa	$y_t = 2q_t(1 - q_t)(1 - F)$
	aa	$z_t = (1 - q_t)^2(1 - F) + (1 - q_t)F$

The inbreeding, measured by F, may be of either of two extreme sorts: sporadic mating of close relatives with no tendency to break the population into subgroups, and division into partially isolated subgroups, within each of which there is random mating. The latter is the case in which we are primarily interested here. Assume that there are K subgroups each of size N. The proportion of heterozygotes within a subgroup is $2q'(1-q')$ where q' is the gene frequency in the parental generation, including immigrants.

(10) $$y_t = 2\sum_1^K q'(1 - q')/K = 2q_t - 2(\sum q'^2)/K.$$

The variance of the gene frequencies of the subgroups, not allowing for accidents of sampling in the last generation, is

(11) $$\sigma^2_{q'} = \sum_1^K (q' - q_t)^2/K = (\sum q'^2)/K - q_t^2$$

(12) $$y_t = 2q_t(1 - q_t) - 2\sigma^2_{q'} \quad \text{from (10) and (11)}$$

(13) $$\sigma^2_{q'} = q_t(1 - q_t)F \quad \text{from (9) and (12).}$$

This formula does not allow for the contribution to variance due to accidents of sampling in the last generation. Thus it gives $\sigma^2_{q'} = 0$ instead of $\sigma^2_q = q_t(1-q_t)/2N$ for $F = 0$. To compare with (7) it must be divided by $(1-m)^2$.

(14) $$\sigma^2_q = q_t(1 - q_t)F/(1 - m)^2$$

(15) $$F = (1 - m)^2/[2N - (2N - 1)(1 - m)^2] \quad \text{from (14) and (7)}$$

(16) $$m = 1 - \sqrt{2NF/[(2N - 1)F + 1]}$$

(17) $$\sigma^2_q = q_t(1 - q_t)[(2N - 1)F + 1]/2N.$$

The formula $F = 1/[4Nm+1]$ given in a preceding paper (Dobzhansky and Wright 1941) is a satisfactory approximation if m is small.

This island model is not likely to be exactly realized in nature. In most cases, the actual immigrants to a population come from immediately surrounding localities in excess and thus are not a random sample of the species. This can be remedied to some extent by multiplying the proportion of replacement by an appropriate factor to obtain the effective immigration index. If q_m is the gene frequency in the actual immigrants (varying from group to group) the appropriate factor would be $(q-q_m)/(q-q_t)$. Unfortunately the values for effective m for different loci may be very different.

LOCAL INBREEDING IN A CONTINUOUS AREA

At the opposite extreme from the island model is that in which there is complete continuity of distribution, but interbreeding is restricted to small distances by the occurrence of only short range means of dispersal. Remote populations may become differentiated merely from *isolation by distance* (WRIGHT 1938, 1940).

Each individual has its origin at a particular place. Assume that its parents originated at distances from this place with a certain variance both in longitude

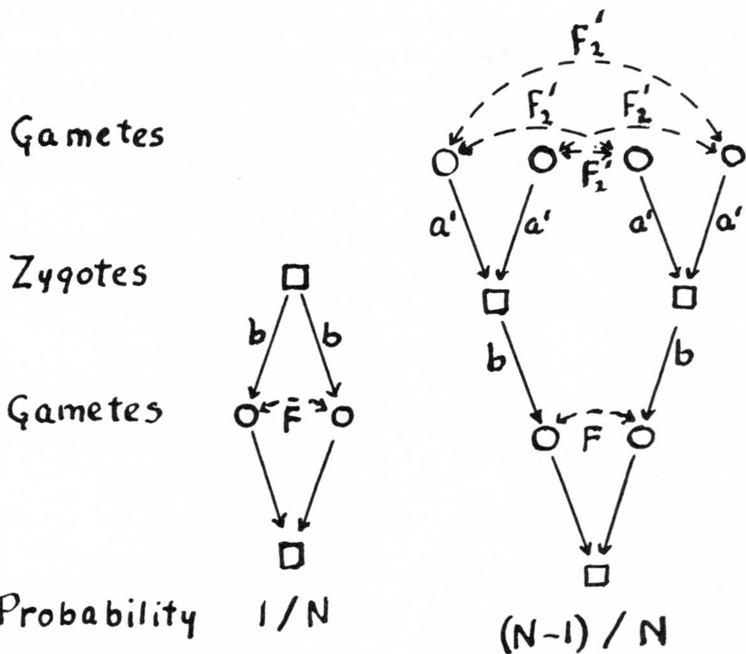

and in latitude. If the same condition held in preceding generations, the grandparents originated at distances with twice this variance in longitude and in latitude and the ancestors of generation K originated at distances with K times this variance in both directions. The parents may be considered as if drawn at random from a territory with a certain radius R and effective population size N. The ancestors of generation K may then be considered as drawn similarly from a territory of radius $\sqrt{K}$ R and effective population size KN.

We shall use the term parental group for the population (effective size N) from which the parents of an individual may be considered to be drawn; the term random breeding or panmictic unit will be used for any local population of the same effective size as the parental group.

The assumption of random union of gametes, including self fertilization (probability 1/N) can be made with sufficient accuracy even though there is actually no self fertilization. It has been shown that such unions in a population of constant size N lead to fixation at the rate 1/2N in comparison with the rate $[(N+1)-\sqrt{(N^2+1)}]/2N$ either in a population of size N equally divided between males and females or in a population of N monoecious individuals in which self fertilization does not occur. As the latter formula may be written $[1-(1/2N)\cdot\cdot\cdot]/2N$ the difference is ordinarily negligible (WRIGHT 1931).

The inbreeding coefficient of individuals in such a population can be calculated from its definition as the correlation between uniting gametes. Let F_x be the correlation between random gametes drawn from a population of size xN and use primes to indicate preceding generations as in the text figure (p. 117). The inbreeding coefficient itself would be F_1 in this terminology. The values of these coefficients can be expressed in terms of coefficients for preceding generations by tracing all connecting paths and noting that the path coefficient b, relating gamete to parental zygote, has the value $\sqrt{(1+F')/2}$ and that the path coefficient, a, relating offspring zygote to one of the gametes that produced it, has the value $\sqrt{1/[2(1+F)]}$. The compound coefficient $ba'=\frac{1}{2}$ (WRIGHT 1921). It may easily be seen that (8) can be deduced at once from these considerations.

In the case of continuity

$$(18)\quad \begin{cases} F = \frac{1}{N} b_2 + \frac{N-1}{N} 4b^2a'^2F_2' = \frac{1}{N}\left(\frac{1+F'}{2}\right) + \frac{N-1}{N} F_2' \\ F_2' = \frac{1}{2N}\left(\frac{1+F''}{2}\right) + \frac{2N-1}{2N} F_3'' \\ F_3'' = \frac{1}{3N}\left(\frac{1+F'''}{2}\right) + \frac{3N-1}{3N} F_4''' \text{ etc.} \end{cases}$$

(19) Thus

$$F = \frac{1+F'}{2N} + \frac{N-1}{N}\left\{\frac{1}{2N}\left(\frac{1+F'}{2}\right) + \frac{2N-1}{2N}\left[\frac{1}{3N}\left(\frac{1+F'''}{2}\right) + \cdots\right]\right\}.$$

If the same population structure has continued indefinitely, primes may be dropped.

$$(20)\quad F = \left(\frac{1+F}{2N}\right)\left[1 + \frac{1}{2}\left(\frac{N-1}{N}\right) + \frac{1}{3}\left(\frac{N-1}{N}\right)\left(\frac{2N-1}{2N}\right) + \frac{1}{4}\left(\frac{N-1}{N}\right)\left(\frac{2N-1}{2N}\right)\left(\frac{3N-1}{3N}\right)\cdots\right].$$

This is an infinite series, but in practice the value of F that is of interest is that relative to some finite population. The correlation between random gametes in a population of size KN is F_K which may be taken as zero, thereby stopping the series at $(K-1)$ terms. Let t_x be the xth term in the series in brackets and $\sum_1^{K-1}t$ the sum of first $(K-1)$ such terms

(21) $$F = \sum_{1}^{K-1} t \Big/ \left[2N - \sum_{1}^{K-1} t\right]$$

(22) $$t_x = \frac{(x-1)N - 1}{xN} t_{(x-1)}.$$

Let $t_{(x-0.5)} = (t_x + t_{(x-1)})/2$ and $\Delta t_{(x-0.5)} = t_x - t_{(x-1)}$

(23) $$\frac{\Delta t_{(x-0.5)}}{t_{(x-0.5)}} = -\frac{2(N+1)}{N(2x-1)-1}.$$

If the values of t are treated as ordinates of a curve with abscissas x, we may write t and x in place of $t_{(x-0.5)}$ and $(x-0.5)$, respectively. The following then hold approximately

(24) $$\frac{dt}{tdx} = -\frac{2(N+1)}{2Nx-1}$$

(25) $$t = C\left(x - \frac{1}{2N}\right)^{-(N+1)/N}$$

(26) $$\sum_{K_1}^{K_2-1} t = \int_{K_1-0.5}^{K_2-0.5} tdx \quad \text{approximately}$$

(27) $$\sum_{K_1}^{K_2-1} t = CN\left[\left(K_1 - \frac{1}{2} - \frac{1}{2N}\right)^{-1/N} - \left(K_2 - \frac{1}{2} - \frac{1}{2N}\right)^{-1/N}\right].$$

The value of the constant C can be obtained by equating actual and estimated values of t. Estimates for all but the first few terms in the series are in close agreement. Thus if N = 10

Actual series $[1+.45+.285+.206625+\cdots]$
Estimated series $C[1.05805+.47969+.30423+.22067+\cdots]$

The estimated value of C from the first term is .9451, from the second term .9381, from the third term .9363. The limiting value is .935774. The value of C approaches 1 as N increases. Thus for N = 100, C = .994157.

Estimates of $\sum_1^{K-1} t$ directly from (27) are not good approximations, but most of the error is in the first few terms. Good estimates can be made by using the actual values from (22) for these terms and the estimates from (27) for the later terms. For N = 10

	Actual (22)	Estimate (27)	Error of Estimate
$\sum_{1}^{3} t$	1.73500	1.86782	+.13282
$\sum_{4}^{9} t$	.79002	.79250	+.00248
$\sum_{10}^{39} t$	.99511	.99541	+.00030
$\sum_{40}^{99} t$	.57228	.57228	+.00000

A priori, one would expect F to approach 1 as a limit as the size of population is increased without limit. This requires that $\sum_1^\infty t$ approach N. Trial for values of N from 10 to 10,000 indicates that this is actually the case and thus gives a good check on the theory. Following are examples:

	N = 10	N = 20	N = 50		N = 100
$\sum_1^{39} t$ from (22)	3.52013	3.86519	4.09266	$\sum_1^{9} t$	2.797
$\sum_{40}^{\infty} t$ from (27)	6.47987	16.13481	45.90734	$\sum_{10}^{\infty} t$	97.203
	10.00000	20.00000	50.00000		100.000

LOCAL INBREEDING ALONG A LINEAR RANGE

In a species with an essentially one dimensional range (parents drawn from the whole width) the extent along the range from which the ancestors of generation K are drawn is proportional to $\sqrt{K}$ as with area continuity, but the effective size of the corresponding population is $\sqrt{K}\,N$ instead of KN. By analogous reasoning

$$(28) \qquad F = \sum t/(2N - \sum t)$$

where

$$\sum t = \left[1 + \frac{1}{\sqrt{2}}\left(\frac{N-1}{N}\right) + \frac{1}{\sqrt{3}}\left(\frac{N-1}{N}\right)\left(\frac{\sqrt{2}N-1}{\sqrt{2}N}\right)\cdots\right]$$

$$(29) \qquad t_x = \frac{N\sqrt{(x-1)} - 1}{N\sqrt{x}}\, t_{(x-1)}$$

$$(30) \qquad \frac{\Delta t_{(x-0.5)}}{t_{(x-0.5)}} = \frac{2N(\sqrt{x-1} - \sqrt{x}) - 2}{N(\sqrt{x-1} + \sqrt{x}) - 1}\,.$$

Treating this expression as the slope at the mid-interval and replacing (x−0.5) by x

$$(31) \qquad \begin{aligned} \frac{dt}{tdx} &= \frac{2N(\sqrt{x-0.5} - \sqrt{x+0.5}) - 2}{N(\sqrt{x-0.5} + \sqrt{x+0.5}) - 1} \\ &= -\frac{N[1 + 1/(32x^2) + \cdots] + 2\sqrt{x}}{2Nx[1 - 1/(32x^2) + \cdots] - \sqrt{x}}\,. \end{aligned}$$

Ignoring $1/(32x^2)$ and smaller terms in the brackets, this yields

$$(32) \qquad t = Ce^{-2\sqrt{x}/N}[\sqrt{x} - (1/2N)]^{-[1+(1/N)^2]}.$$

This seems to be as accurate an approximation as is warranted after replacement of $\Delta t/t$ by dt/tdx.

Comparisons of actual and calculated values of t indicate that estimates of C approach stability after a few terms. For N = 10, C = 1.1529 (from 30th to

40th terms). For $N = 100$, $C = 1.01465$ (from 9th and 10th terms). For larger values of N, especially if x is 10 or more, it may be sufficiently accurate to take dt/tdx as $-(N+2\sqrt{x})/2Nx$, $C = 1$

$$t = e^{-2\sqrt{x}/N}/\sqrt{x} \quad \text{approximately.} \tag{33}$$

In this case

$$\sum_{K_1}^{K_2-1} t = N(e^{-2\sqrt{K_1}/N} - e^{-2\sqrt{K_2}/N}) \tag{34}$$

The value of $\sum_1^{K-1}t$ can be approximated by finding actual $\sum_1^9 t$ from (29), estimating $\sum_{10}^{K-1}$ from (34) and multiplying the latter by the mean ratio of t from (32) to that from (33). Calculation of $\sum_1^\infty t$, $N = 10$, by this method (by steps) gave 10.008 (instead of theoretical 10) and for $N = 100$ gave 100.07 instead of theoretical 100. These theoretical values are on the assumption that the limiting value of F is 1 which again is seen to be verified.

CORRELATION BETWEEN ADJACENT INDIVIDUALS UNDER UNIPARENTAL REPRODUCTION

The effect of isolation by distance on the frequencies of two alternative types in a population with exclusive uniparental reproduction can be treated similarly, again assuming that there are no complications from other factors. The treatment, however, cannot be in terms of the inbreeding coefficient. Let E be the correlation between adjacent individuals, and assume that there is short range dispersion in each generation such that individuals are derived from a parental group of effective size N. With area continuity, the ancestors of the Kth generation are drawn from a population of effective size KN. The correlation between adjacent individuals can be analyzed into two components, that due to the chance, $1/N$, of derivation from the same parent and that due to the chance $(N-1)/N$, of derivation from different individuals of the group, the correlation between which may be represented by E_2' in analogy with F_2' in the case of biparental reproduction. This in turn can be analyzed into a component due to the chance $1/2N$ of derivation from the same individual of the second preceding generation and that due to the chance $(2N-1)/2N$ of derivation from different individuals of this group, the correlation between which may be represented by E_3''.

$$\left\{\begin{aligned} E &= \frac{1}{N} + \frac{N-1}{N}E_2' \\ E_2' &= \frac{1}{2N} + \frac{2N-1}{2N}E_3'' \\ E_3'' &= \frac{1}{3N} + \frac{3N-1}{3N}E_4''' \text{ etc.} \end{aligned}\right. \tag{35}$$

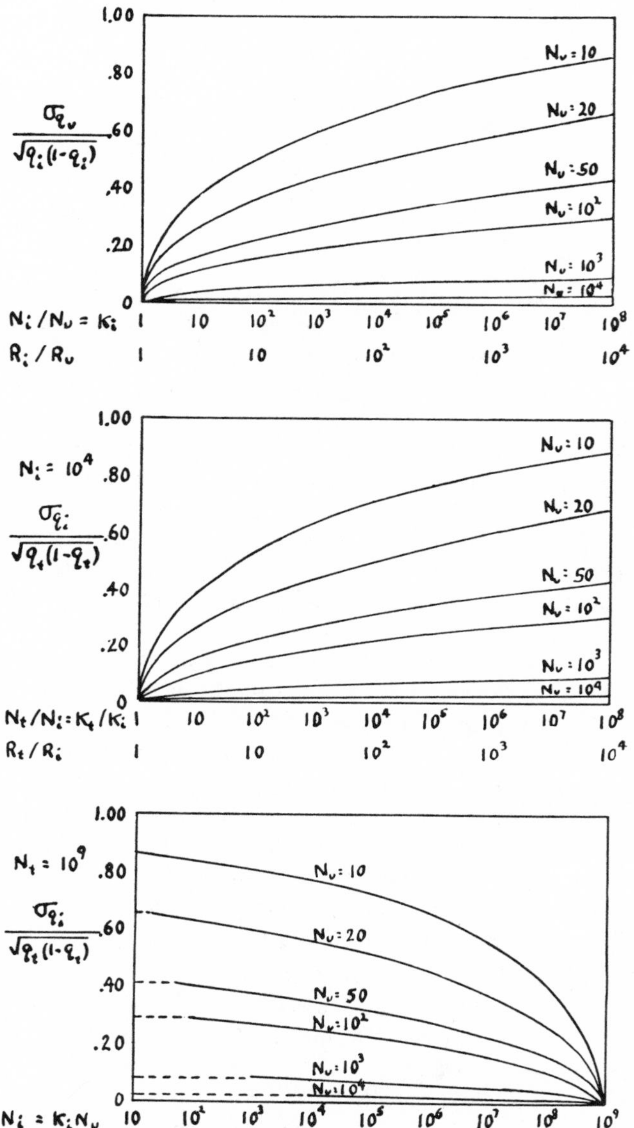

Figures 1 to 3. Variability of gene frequencies of local populations within a continuously inhabited area that extends indefinitely in all directions. It is assumed that there is no appreciable long range dispersal or mutation. Each curve applies to a particular size (N_u) of random breeding unit and thus to a certain amount of short range dispersal. Variability is measured by $\sigma_x/\sqrt{q_y(1-q_y)}$ where q_x represents the gene frequencies of the subgroup in question and q_y that of the comprehensive population.

FIGURE 1 (top).—The variability of gene frequencies (q_u) of the random breeding units themselves, within areas up to 10^4 times their radius (R_i/R_u) or 10^8 times their population size (N_i/N_u).

(Legend continued on next page.)

Again we may drop primes if the same population structure has continued for a large number of generations.

$$(36)\quad E = \frac{1}{N}\left[1 + \frac{1}{2}\left(\frac{N-1}{N}\right) + \frac{1}{3}\left(\frac{N-1}{N}\right)\left(\frac{2N-1}{2N}\right) + \frac{1}{4}\left(\frac{N-1}{N}\right)\left(\frac{2N-1}{2N}\right)\left(\frac{3N-1}{3N}\right)\cdots\right].$$

$$(37)\quad E = \sum t/N.$$

The series $\sum t$ is the same as encountered in the case of biparental reproduction, but the formula for E differs from that for F. It resembles it in approaching 1 as a limit, as is to be expected *a priori*, but for a given N, E is about twice as great as F for small values of $\sum t$, and the difference from the limit is only about half as great if $\sum t$ is close to 1. These relations are illustrated in figures 7 and 1 dealing with uniparental and biparental reproduction, respectively.

In the case of linear continuity and derivation of individuals from a parental population of N, the effective size of the population of the Kth ancestral generation is $\sqrt{K}\,N$, again as under biparental reproduction. By analogous reasoning $E = \sum t/N$ where $\sum t$ is the same series as in the biparental case. The relation of E to F for the same N is similar to that described above in the case of area continuity.

RANDOM DIFFERENTIATION OF PANMICTIC UNITS IN A CONTINUUM

Returning to biparental reproduction, the situation in a random breeding unit imbedded in a continuous population of defined size may be compared in some respects with that in an "island" whose population is replaced to such an extent in each generation by migrants representative of the whole that the inbreeding coefficient of individuals is the same. There is the important difference that adjacent groups should be closely similar in the former but uncorrelated in the latter. Nevertheless the amount of differentiation among groups taken at *random* from the whole should be the same in both cases since equations (9) to (17) apply in both. It is most convenient to use $\sqrt{F} = \sigma_q/\sqrt{q_t(1-q_t)}$ (from (13)) to measure this differentiation. It should be noted that this excludes the variability due to the immediate effect of sampling.

The theoretical variabilities of random breeding units of various sizes (10 to 10,000) within populations up to 10^8 times the size of the units (or 10^4 times the radius), continuous in all directions, are compared in figure 1. In interpreting this variability, it may be noted that if $q_t = \frac{1}{2}$, a value of $\sqrt{F}$ (ordinate)

K_i is the average number of generations of separate ancestry of random individuals of the population N_i.

FIGURE 2 (middle).—The variability of gene frequencies (q_i) of populations of a given size, $N_i = 10^4$, within areas up to 10^4 times their radius (R_t/R_i) or 10^8 times their population size (N_t/N_i). Note the similarity to Figure 1.

FIGURE 3 (bottom).—The variability of gene frequencies (q_i) of populations of any size, N_i, within a region with a population of a given size, $N_t = 10^9$.

greater than .577 means a U-shaped distribution of gene frequencies and thus very great differentiation. The situation is similar to that found where Nm is less than 0.5 in the island model. There is important differentiation down to at least $\sqrt{F} = .22$ (equivalent to Nm = 5). There is only slight differentiation if $\sqrt{F}$ is less than .07 (equivalent to Nm = 50) (*cf.* fig. 1, WRIGHT 1940).

It is apparent from figure 1 (this paper) that there is a great deal of local differentiation if the random breeding unit is as small as 10, even within a territory the diameter of which is only ten times that of the unit. If the unit has an effective size of 100, differentiation becomes important only at much greater relative distances. If the effective size is 1000, there is only slight differentiation at enormous distances. If it is as large as 10,000 the situation is substantially the same as if there were panmixia throughout any conceivable range.

The situation is very different as may be seen from figure 4 in a species whose range is essentially one dimensional (for example, a shore line). Different alleles may approach fixation in different parts of a range only 100 times the length of the random breeding unit if the effective size of the latter is less than 100. The range must be about 1000 times the length of the unit if the latter has a size of 1000 and about 10,000 times its length if the size of the unit is 10,000 to give this result. This difference between area and linear continuity has been suggested on *a priori* grounds by THOMPSON (1931) in connection with a study of the correlation between water distance and amount of differentiation within species of fish.

RANDOM DIFFERENTIATION IN A HIERARCHY OF SUBDIVISIONS

The attempt to apply these conclusions to actual cases is hampered by the difficulty of determining what are the random breeding units and their effective sizes. To obviate this, we should find how groups of any arbitrary size vary within a more comprehensive population.

Consider a total population, size N_t, subdivided into H groups of intermediate size N_i and these in turn subdivided into K random breeding groups of size N_u. The inbreeding coefficient of individuals is zero relative to the unit groups, F_i relative to the intermediate groups and F_t relative to the total. Both H and K, in contrast with N_u, will be treated as large numbers.

The variance of the gene frequency (q_u) of unit groups within the intermediate groups is given by (17) using the proper subscripts. The average value of this variance will be represented by $\sigma^2_{u \cdot i}$. The variance of the mean gene frequencies of the intermediate groups in the total will be represented by $\sigma^2_{i \cdot t}$ and that of q_u in the total by $\sigma^2_{u \cdot t}$.

(38) from (17) $$\sigma^2_{u \cdot i} = \sum_1^H [q_i(1-q_i)][(2N_u-1)F_i+1]/2HN_u$$

(39) $$\sigma^2_{u \cdot i} = [q_t(1-q_t)-\sigma^2_{i \cdot t}][(2N_u-1)F_i+1]/2N_u$$

(40) from (17) $$\sigma^2_{u \cdot t} = q_t(1-q_t)[(2N_u-1)F_t+1]/2N_u$$

(41) $$\sigma^2_{u \cdot t} = \sigma^2_{u \cdot i} + \sigma^2_{i \cdot t}$$

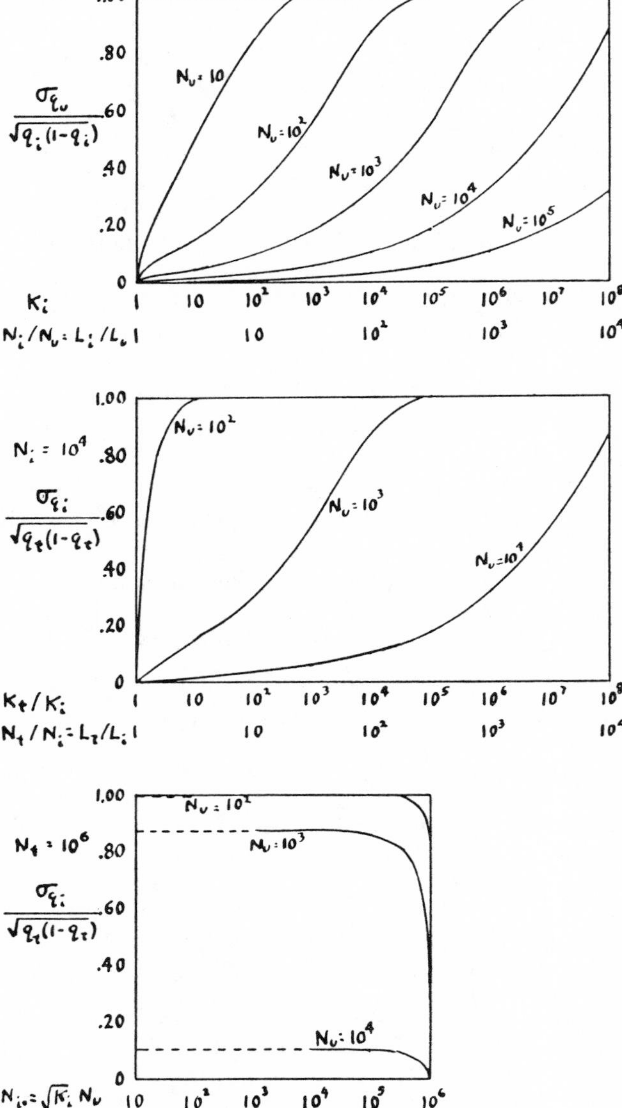

Figures 4 to 6. Similar to figures 1 to 3, respectively, except that a linear range (such as a shore line) is postulated.

FIGURE 4 (top).—The variability of gene frequencies (q_u) of the random breeding units themselves, within ranges up to 10^4 times their length (L_i/L_u) or population (N_i/N_u).

FIGURE 5 (middle).—The variability of gene frequencies (q_i) of populations of a given size, $N_i = 10^4$, within ranges up to 10^4 times their length (L_t/L_i) or population (N_t/N_i). Note the dissimilarity to figure 4 in contrast with the similarity of figures 1 and 2.

FIGURE 6 (bottom).—The variability of gene frequencies (q_i) of populations of any size (N_i) within a range with a population of a given size, $N_t = 10^6$.

(42) from (39) and (41) $\sigma^2_{u\cdot t} = [q_t(1-q_t) - \sigma^2_{i\cdot t}][(2N_u - 1)F_i + 1]/2N_u + \sigma^2_{i\cdot t}$.

(43) Equating (40) and (42) $\sigma^2_{i\cdot t} = q_t(1-q_t)[F_t - F_i]/[1-F_i]$.

This demonstration involves the assumption that there is inbreeding relative to the intermediate groups because these are subdivided. It may be noted that the same value of $\sigma^2_{i\cdot t}$ may be derived as follows without this assumption.

(44) From (9) $y_t = 2q_t(1 - q_t)(1 - F_t)$.

But y_t is also the average heterozygosis of the intermediate groups

(45)
$$y_t = \sum_1^H [2q_i(1 - q_i)(1 - F_i)]/H = 2(1 - F_i)\left(q_t - \left(\sum_1^H q_i^2\right)/H\right)$$

(46)
$$\sigma^2_{i\cdot t} = \sum_1^H (q_i - q_t)^2/H = \left(\sum_1^H q_i^2\right)/H - q_t^2.$$

(47) From (45) and (46) $y_t = 2[1 - F_i][q_t(1 - q_t) - \sigma^2_{i\cdot t}]$.

(48) From (44) and (47) $\sigma^2_{i\cdot t} = q_t(1 - q_t)[F_t - F_i]/[1 - F_i]$.

In neither demonstration is there any assumption as to the geographic distribution of the values of the mean gene frequencies, q_i, within the total. They may be distributed at random as implied in the island model or there may be gradients as expected with continuity.

The quantity

$$\sqrt{\frac{(F_t - F_i)}{(1 - F_i)}} = \frac{\sigma_{i\cdot t}}{\sqrt{q_t(1 - q_t)}}$$

may be used as an index of the amount of differentiation among populations of any size N_i within a more comprehensive population (N_t). The variabilities of populations of effective size $N_i = 10{,}000$ are considered in figure 2 (area continuity) and figure 5 (linear continuity). In the case of area continuity the curves are somewhat similar to those shown in figure 1 for unit groups. It appears that populations of 10,000 (or any other size) exhibit about the same amount of differentiation within a whole whose population is a certain multiple of their own as the unit groups exhibit in a population that is the same multiple of their size. Whatever the size of the subpopulations considered the variability depends on the size of the inbreeding unit. There is an important amount of differentiation among large regions if the unit group is as small as 10, appreciable differentiation if the unit group is as large as 100 but little if it is as large as 1000. It should be said that there are important qualifications if there are other factors (mutation, rare long range dispersal or selection) which will be considered later.

The situation differs considerably in the case of linear continuity. Groups of size $N_i = 10{,}000$ approach the limiting amount of differentiation within populations only three times their length of range if $N_u = 100$ or less. There must be virtually complete fixation of one allele or the other over long distances with only short regions of transition. If $N_u = 1000$ there is relatively little differentiation within 10-fold lengths (that are heterallelic at all) but an approach to 100 percent differentiation in 100-fold lengths. Thus transition regions are of the order of 10 lengths. If $N_u = 10{,}000$, the transition regions are of the order of 10^3 lengths, and such groups approach 100 percent differentiation within 10^4 lengths.

The interpretation of figures 1, 2, 4, and 5 is somewhat complicated by the fact that these do not measure variability on a constant scale. The denominators of the ordinates (namely, $\sqrt{q_t(1-q_t)}$ in 2 and 5) increase with the abscissas. The tendency toward fixation of large populations means that at the lower abscissas the average value of q_t must be close to 0 or 1, making $\sqrt{q_t(1-q_t)}$ small. The structure of a population is exhibited in perhaps the most easily interpreted form by considering a constant comprehensive population N_t and showing how much differentiation there is among subdivisions of all sizes from the random breeding units up to major subdivisions ($\sigma_{q_i}/\sqrt{q_t(1-q_t)}$ plotted against N_i). Here the denominator is constant so that variability is always on the same scale.

Figure 3 shows that with area continuity, the amount of differentiation falls off slowly with the size of the subdivision considered. If $N_u = 10$ and N_t is 10^9 (or any other size in the absence of other factors) there is marked differentiation among populations that are 10 percent of the total, although much less than among subdivisions of smaller sizes. If $N_u = 100$, there is only moderate differentiation among the smaller subdivisions and very little among ones that are as large as 10 per cent of the total. In the case of linear continuity (fig. 6) there is virtually complete fixation of all subdivisions up to 10 percent of the total if N_u is 100 or less. If, however, N_u is 1000 there is a considerable proportion of these unit groups that are not fixed ($\sigma_{q_u}/\sqrt{q_t(1-q_t)} = .87$). The differentiation among larger populations up to $N_i = 0.1\ N_t$ is not appreciably less than among the unit groups. If $N_u = 10{,}000$, $\sigma_{q_u}/\sqrt{q_t(1-q_t)}$ is only .10, but this index is practically as great among larger populations up to 10 percent of the total. Thus with linear continuity most of the differentiation is that among large subdivisions of the total (of the order of 10 percent of its size). With area continuity, differentiation is more uniformly distributed at all levels.

Area and linear continuity as well as the island model are ideal cases. There may be all grades of intermediacy between area and linear continuity as exhibited in branching and reticular distributions. Even with rather complete area continuity there are almost certain to be variations in density of population. The ancestry of individuals in the centers of high density would spread out less rapidly than under the ideal theory with the consequence that there would in general be more differentiation among such centers than indicated, unless this is interfered with by other factors, which must now be considered.

COMPLICATING EFFECTS OF MUTATION AND LONG RANGE DISPERSAL

The foregoing theory indicates the possibility of an approach to fixation of different alleles in large areas of the same continuous population without the help of any differential action of selection. It is obvious, however, that this very slow process would be greatly affected by other factors that change gene frequency. The very fact of persistence of more than one allele over a long period of time tends to indicate that such factors are present in some sort of balance. Thus there may be reversible mutation, selection opposed by mutation or selection against both of two homozygotes. Moreover, the short range means of dispersal that have been postulated are likely to be supplemented by occasional long range dispersal. All of these tend to prevent fixation of one type even locally. On the other hand, selection may favor one allele in some places and others in other places. This would tend to increase local differentiation. It is necessary to consider how such processes affect the situation.

It will be well to review first the joint effects of recurrent mutation and long range dispersal in the case of the island model (WRIGHT 1931). The rate of change of gene frequency under recurrent reversible mutation varies linearly with the gene frequency: $\Delta q = v(1-q) - uq = -(u+v)(q-\hat{q})$ where v is the mutation rate to the allele in question, u is the rate of mutation from it and $\hat{q}(=v/(u+v))$ is the value of q at equilibrium, which is the same in this case as $\bar{q}$ the mean value of q. This is similar in form to the expression for the effects of long range dispersal: $\Delta q = -m(q-q_t)$.

If both processes are occurring, the expressions merely need to be added:

$$(49) \qquad \Delta q = v(1-q) - uq - m(q-q_t) = -(m+u+v)(q-\hat{q})$$

where

$$\hat{q} = \bar{q} = (mq_t + v)/(m+u+v)$$

for a local population in which $\bar{q}$ is not necessarily the same as gene frequency for the whole species (q_t), since other factors may be at work in other localities. The long time distribution for such a population is approximately

$$(50) \qquad \phi(q) = Cq^{4N(mq_t+v)-1}(1-q)^{4N[m(1-q_t)+u]-1}$$

$$(51) \qquad \sigma_q^2 = \bar{q}(1-\bar{q})/[4N(m+u+v)+1].$$

If conditions are the same in all islands, $\bar{q} = q_t = v/(u+v)$ and the variance $q_t(1-q_t)/[4N(m+u+v)+1]$ is not only the long time variance for a single island but also the variance of q, at any time, among the islands.

The variance of subpopulations (inbreeding coefficient F_i) in a total relative to which the inbreeding coefficient is F_t has been given (43, 48) as $q_t(1-q_t)[F_t-F_i]/[1-F_i]$ applicable to any case, including both the island model and that of a continuous population with only short range dispersal. The effective value of the immigration index in the latter may be obtained by equating with the expression for σ_q^2 given in (51).

$$(52) \qquad m = [1-F_t]/[4N(F_t-F_i)].$$

At first sight it might appear that the rate of change of gene frequency in cases in which there is long range dispersal and reversible mutation (joint coefficient m_1) in addition to predominant short range dispersal (coefficient m_2 from (52)) might be obtained by simply adding the contributions from these sources as calculated from their effects by themselves. This, however, overlooks the likelihood of an important interaction effect. It is necessary to go back to the formula for the correlation between uniting gametes (18) and determine how it is affected by mutation and long range dispersal.

Assume that the proportion m_1 of the gametes represent a random sample from the whole species. The identity of the theories of long range dispersal and mutation make it possible to let m_1 here represent $(m+u+v)$ of preceding formulae. Cases in which one or both of the uniting gametes are included in this proportion make no contribution to the correlation between uniting gametes. The proportion which makes a contribution is $(1-m_1)^2$. Equations (18) are accordingly to be modified as follows:

$$(53)\quad \begin{cases} F = (1 - m_1)^2\left[\dfrac{1}{N}\left(\dfrac{1+F}{2}\right) + \left(\dfrac{N-1}{N}\right)F_2'\right] \\ F_2' = (1 - m_1)^2\left[\dfrac{1}{2N}\left(\dfrac{1+F''}{2}\right) + \dfrac{2N-1}{2N}F_3''\right] \text{ etc.} \end{cases}$$

Again primes may be dropped, if the same situation has held for a long time.

$$(54)\quad F = \left[\frac{1+F}{2N}\right]\left[(1-m_1)^2 + \frac{(1-m_1)^4}{2}\left(\frac{N-1}{N}\right) + \frac{(1-m_1)^6}{3}\left(\frac{N-1}{N}\right)\left(\frac{2N-1}{2N}\right)\cdots\right]$$

$$(55)\quad t = C(1-m_1)^{2x}[x-(1/2N)]^{-(N+1)/N}$$

$$(56)\quad \sum_{K_1}^{K_t-1} t = C\int_{K_1-0.5}^{K_2-0.5}(1-m_1)^{2x}[x-(1/2N)]^{-(N+1)/N}dx \text{ approximately.}$$

This is a less convenient expression than obtained where $m_1=0$, but approximate values can be obtained by taking values of K at short enough intervals, finding

$$\left[\int_{K_1-0.5}^{K_2-0.5}(1-m_1)^{2x}dx\right]\left[C\int_{K_1-0.5}^{K_2-0.5}[x-(1/2N)]^{-(N+1)/N}dx\right]$$

and correcting according to the percentage error where both factors are of the form $[\int_0^1 e^{Kx}dx]$ with the K's chosen so as to give the same ratios of terminal ordinates.

$$(57)\quad F = \sum_1^{K-1} t \Big/ \left[2N - \sum_1^{K-1} t\right].$$

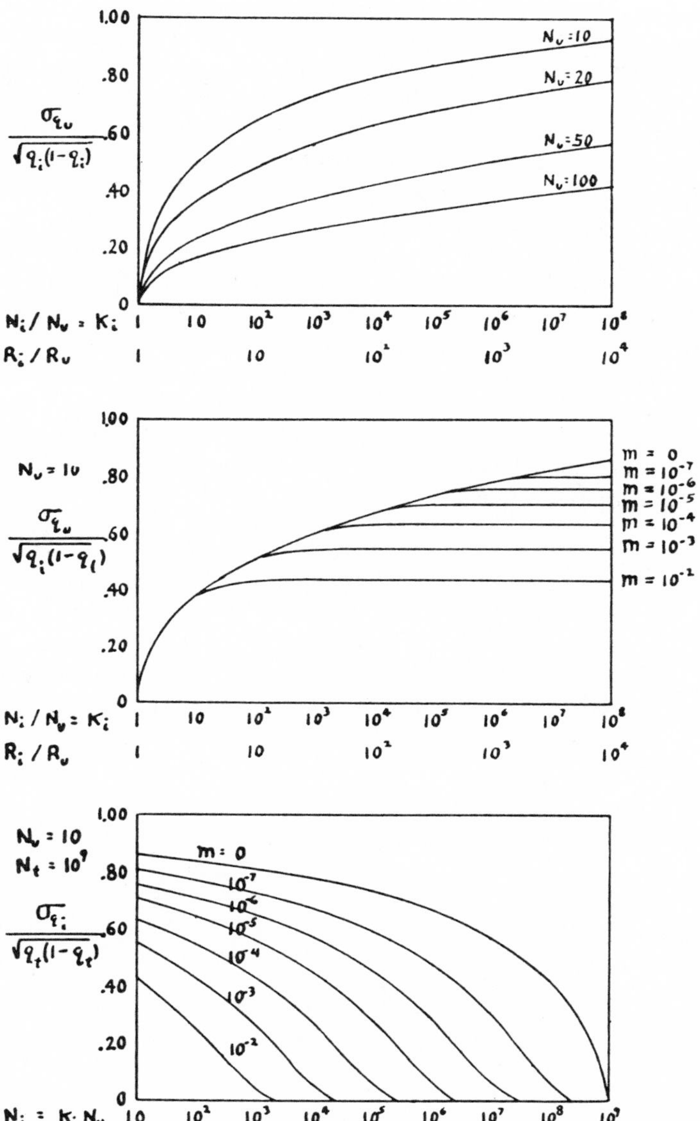

FIGURE 7 (top).—Similar to figure 1 except that exclusive uniparental reproduction is assumed. N_u is the population size of the group from which the parents of adjacent individuals are drawn at random and thus measures the extent of dispersal. The curves show the variability of gene frequencies (q_u) of such unit groups within areas up to 10^4 times their radius (R_i/R_u) or 10^8 times their population (N_i/N_u). Each curve applies to a particular extent of short range dispersal.

FIGURE 8 (middle).—The effect of occasional long range dispersal or mutation (rates up to $m = 10^{-2}$) on the variability of gene frequency of random breeding units of size $N_u = 10$ within areas

Figure 8 shows how $\sigma_{q_u}/\sqrt{q_i(1-q_i)}=\sqrt{F_i}$, for parental populations of size $N_u=10$, rises with the size of the population ($N_i=KN_u$) in the presence of random replacement (m_1) in the proportions 10^{-7} to 10^{-2}. The variability of the unit population is substantially the same as if there were no mutation or long range dispersal if N_i is less than $1/m_1$, but rather abruptly approaches a limit in larger populations. Instead of approaching 1 as when $m_1=0$, $\sqrt{F_i}$ approaches 0.81 if $m_1=10^{-7}$, 0.76 if $m_1=10^{-6}$, 0.70 if $m_1=10^{-5}$, 0.63 if $m_1=10^{-4}$, 0.55 if $m_1=10^{-3}$, and 0.44 if $m_1=10^{-2}$.

Figure 9 shows how the variability of subpopulations of any size N_i within a total population of size 10^9 is affected by the value of m_1. It is again assumed that short range dispersal is such as to give $N_u=10$. There is very little differentiation in this case of subpopulations larger than $30/m_1$. It is clear that it requires only a small amount of long range dispersal or mutation to prevent the differentiation of large populations.

The amount of differentiation of populations, that are a given multiple (K_i) of the unit population, falls off rapidly with increase of N_u. But the multiple beyond which differentiation virtually ceases is largely controlled by the factor $(1-m_1)^2$ and is thus nearly the same for all values of N_u under which there is any appreciable differentiation at any level. The value of $(1-m_1)^{2x}$ is reduced to approximately 10 per cent of its value each $\log_e 10/2m_1$ generations (assuming m_1 to be small).

Among populations of a given absolute size (N_i) there is, therefore, a certain range of dispersal (determining N_u) that is most favorable to differentiation in a continuous population. On the one hand, if the range of dispersal is such that N_u is larger than 1000, there is very little differentiation, but on the other hand, if N_u is so small that there are more than $3/m_1$ random breeding units in the population under consideration, there is also virtually no differentiation.

Linear continuity may be treated similarly, by multiplying the terms of (32) by $(1-m_1)^{2x}$.

Under exclusive uniparental reproduction, the chance that an individual is derived from the parental population without mutation is $(1-m_1)$, instead of $(1-m_1)^2$. Each term in the series $\sum t$ is accordingly to be multiplied by $(1-m_1)^x$.

The formula for the distribution of gene frequencies among subpopulations of a given size, N_i, in the total N_t, may be written approximately as follows:

$$(58) \qquad \phi(q) = Cq^{[(1-F_t)/(F_t-F_i)]q_t-1}(1-q)^{[(1-F_t)/(F_t-F_j)](1-q_t)-1}.$$

Here the F's incorporate the effects of mutation and long range dispersal as

---

up to 10^4 times their radius or 10^8 times their population size. The highest curve ($m=0$) is the same as the highest curve in figure 1.

FIGURE 9 (bottom).—The effect of occasional long range dispersal or mutation (rates up to 10^{-2}) on the variability of gene frequencies of populations of any size, N_i, within a region with a population of a given size, $N_t=10^9$. The random breeding unit is assumed to be $N_u=10$. The highest curve ($m=0$) is the same as that in figure 3.

well as of short range dispersal. This distribution has the mean q_t and the variance $q_t(1-q_t)[F_t-F_i]/[1-F_t]$ derived above. It differs considerably from the distribution

$$\phi(q) = Cq^{4N(m_1+m_2)q_t-1}(1-q)^{4N(m_1+m_2)(1-q_t)-1} \tag{59}$$

if m_2 is the estimate of effective m from (52) based on the value of F_i and F_t under short range dispersal in the absence of other factors. It is legitimate, however, if m_1 is known, to write $\phi(q)$ in the form of (59) with the understanding that m_2 measures the effect of short range dispersal in the presence of the other factors measured by m_1 with full allowance for the interaction effect. Indeed this seems to be the only practicable method to use in analyzing data from actual populations in view of the fact that no ideal model such as area or linear continuity is likely to be exactly realized.

THE EFFECTIVE SIZE OF INBRED POPULATIONS

The effect of inbreeding on the effective size of populations is a matter that requires some consideration. Size of population enters into the formulae for the distribution of gene frequencies principally through the sampling variance which is $q(1-q)/2N$ in a random breeding diploid population. Assume that individuals have an inbreeding coefficient F_i relative to an island population. It makes a difference in the sampling variance whether this is due to mating of relatives, not resulting in any territorial subdivision, or whether it is due to partial isolation of subdivisions that breed at random within themselves. In the former case, the increased frequency of homozygotes causes an increased sampling variance of the whole island. If there were nothing but homozygotes, $(q_iAA+(1-q_i)aa)$, as under long continued self-fertilization, the sampling variance would be $q_i(1-q_i)/N_i$, twice that under random mating. With random bred and inbred components in the array of equations (9) in the proportions $(1-F_i)$ to F_i, the sampling variance would be the weighted average.

$$\begin{aligned} \sigma^2_{\delta q_i} &= (1-F_i)q_i(1-q_i)/2N_i + F_iq_i(1-q_i)/N_i \\ &= q_i(1-q_i)(1+F_i)/2N_i. \end{aligned} \tag{60}$$

If on the other hand, the island population is subdivided into partially isolated groups that breed at random within themselves and if each group tends to maintain its numbers (that is, there is no intergroup selection) the sampling variance of the total island population is *less* than if there were random mating throughout. In each subgroup, the sampling variance is $q_u'(1-q_u')/2N_u$, average $\sigma^2_{\delta q_u}=\sum_1^K q_u'(1-q_u')/2N_uK$. The sampling variance for the mean gene frequency of the island would be $\sigma^2_{\delta q_i}=\sigma^2_{\delta q_u}/K=\sum_1^K q_u'(1-q_u')/2N_iK$ if $N_i = KN_u$. But from (10) $y_i=2\sum_1^K q_u'(1-q_u')/K$. Thus $\sigma^2_{\delta q_i}=y_i/4N_i$. From (9) $y_i = 2q_i(1-q_i)(1-F_i)$ giving

$$\sigma^2_{\delta q_i} = q_i(1-q_i)(1-F_i)/2N_i. \tag{61}$$

The situation in an arbitrarily delimited region in a continuum resembles the second. Effective N in such a formula as (59) is thus $KN_u/(1-F_i)$.

COMPLICATING EFFECTS OF SELECTION

Consider next the complications introduced by selection. The effects of various kinds of selection on gene frequency (contributions to Δq) and the form taken by $\phi(q)$ on substitution in (1) have been discussed in previous papers (WRIGHT 1931, 1942). These are applicable directly to the island model. The case of arbitrarily delimited portions of a continuum can be treated in the same way, but if so, m_2 of formula (59) includes the interaction effect of selection as well as of mutation (and of long range dispersal if this can be distinguished from the short range dispersal). The index m_2 is to be interpreted as the effective amount of replacement of the subpopulations in question by representatives of the species as a whole under the conditions of mutation and selection that actually hold. As noted in connection with the complications introduced by mutation and long range dispersal, this seems to be the most practicable method of dealing with concrete data. It is important, however, to determine the theoretical relations between the values of m among subdivisions of different sizes under various ideal population structures.

For such theoretical consideration of the interaction of selection with the effects of short range dispersal, it is necessary to return to the derivation of F by path coefficients (18) in analogy with the treatment of the complications due to mutation and long range dispersal (53). But in attempting to carry out the analogy we encounter a serious difficulty.

Long range dispersal (by definition) and mutation may be treated as introducing a random admixture into the local population in constant proportion m_1. Selection may also be treated as introducing a certain random admixture, but it is not in constant proportion. The amount of such admixture in the case of mutation and long range dispersal may be represented as

$$[-\Delta q/(q-\hat{q})] = (m+u+v) = m_1. \tag{62}$$

This formula may be applied where Δq also involves selection pressure. Consider the case of a balance between opposing pressures of mutation and selection in the simplest case, that of no dominance, and assume that the same situation holds throughout the species.

$$\Delta q = v(1-q) - sq(1-q) = -s(1-q)(q-\hat{q}) \quad \text{where} \quad \hat{q} = v/s \tag{63}$$

$$m_1 = [-\Delta q/(q-\hat{q}] = s(1-q). \tag{64}$$

The joint effect of mutation and selection in this case is equivalent to immigration of a random sample, but to an extent that is a function of the local gene frequency. A rough idea of the effect may be obtained by substituting $\hat{q}$ for q and treating $m_1 = s(1-\hat{q}) = s-v$ as a constant. If s is much larger than v we may indeed simply take $m_1 = s$ and use $(1-s)^2$ in place of $(1-m_1)^2$ in the theory developed for mutation and long range dispersal. Inspection of figures 8 and 9 shows how selection of this sort interferes with the differentiation that would occur within the continuous population under the specified conditions if there were no complication of this sort.

As another example consider the case of selection against both of two homozygotes. Representing the relative selective values of AA, Aa and aa by $1-s_{AA}$, 1 and $1-s_{aa}$ respectively

(65) $$\Delta q = -(s_{AA} + s_{aa})q(1-q)(q-\hat{q})$$

where $$\hat{q} = s_{aa}/(s_{AA} + s_{aa})$$

(66) $$m_1 = (s_{AA} + s_{aa})q(1-q).$$

While selection does nothing to local populations that have become fixed and the equivalent immigration index m_1 is accordingly 0 if q is either 0 or 1, the average value may well be such as to severely restrict differentiation of even rather small subdivisions of a continuous population. Again a rough idea of the effect may be obtained by substituting $\hat{q}$ for q. It should be noted that if there are numerous alleles and selection for heterosis is general, selection tends to increase differentiation.

In a recent paper (WRIGHT, DOBZHANSKY and HOVANITZ 1942) an attempt was made to interpret the frequencies of lethals in a continuous population of *Drosophila pseudoobscura* on Mt. San Jacinto. The following formula was arrived at for the rate of change of the frequency of a typical lethal gene.

(67) $$\Delta q = \bar{v}(1-q) - m(q-\bar{q}) - q(\bar{s}+F) - q^2(1-3\bar{s}-2F)$$

where $\bar{v}$ is the mean mutation rate per generation, $\bar{s}$ the mean selective disadvantage of heterozygotes, $\bar{q}$ the mean gene frequency, F the inbreeding coefficient, and m the effective immigration coefficient of the territory under consideration. It was shown that approximately the same variance of gene frequencies was reached by replacing the above expression by one in which the component of Δq, measuring the tendency toward increase of gene frequency—namely, $(\bar{v}+m\bar{q})(1-q)$, is balanced by the linear expression that gives the same mean as the correct expression namely, $-(\bar{v}+m\bar{q})(1-\bar{q})q/\bar{q}$

(68) $$\Delta q = -(m + \bar{v}/\bar{q})(q-\bar{q}) \text{ approximately}$$

(69) $$m_1 = (m + \bar{v}/\bar{q}) \text{ approximately.}$$

DIFFERENTIATION OF SUBDIVISIONS BY SELECTION

If selection acts differently in different regions, it is obvious that none of the formulae given here apply to the distribution of values q among these regions, but only to the long term distribution within single ones. As a basis for discussion consider the following simple case, which refers to rate of change of gene frequencies in an island as affected by the local conditions of selection measured by s (assuming no dominance) and the amount of immigration measured by m (WRIGHT 1931, 1940).

(70) $$\Delta q = sq(1-q) - m(q-q_t).$$

In a local population in which s (whether plus or minus) is smaller in absolute value than m, gene frequency can depart only slightly from the average of the species ($\hat{q}=q_t+(s/m)q_t(1-q_t)$) approximately. Crossbreeding here swamps the tendency toward selective differentiation. On the other hand, local gene frequency tends to be dominated by the local conditions of selection in populations in which s is larger than m in absolute value $\hat{q}=1-(m/s)(1-q_t)$ or $\hat{q}=(-m/s)q_t$ approximately, depending on whether s is positive or negative.

The effectiveness of selection here is not related directly to the size of the island population. However, there is likely to be indirect relationship. This may be illustrated by considering three situations.

First, consider islands with various populations but the same absolute amount of immigration (as might well be the case if the areas are the same but population densities differ). Among such islands, Nm is constant. All have the same amount of nonadaptive differentiation (measured by $1/(4Nm+1)$ but a given selection pressure is more effective on the islands with larger population (and hence smaller m) than among those with smaller populations.

A second situation is that in which size of population is proportional to area and the number of immigrants is proportional to the extent of boundary ($Nm \propto \sqrt{N}$). Here there is more nonadaptive differentiation on the smaller islands and more adaptive differentiation of the larger ones, although the latter effect is less marked than in the preceding case.

Finally, if both size of population and amount of immigration are proportional to the area (m constant), there is markedly more nonadaptive differentiation on the smaller islands but no relationship between adaptive differentiation and size of population.

Summing up, any sort of differentiation is favored by small m, but the large populations tend on the whole to exhibit predominant adaptive differentiation, while the smaller ones exhibit predominantly nonadaptive differentiation.

The situation in a continuous population is similar in that nonadaptive differentiation should be most conspicuous locally and adaptive differentiation among larger subdivisions. The most significant thing, however, given a certain amount of differential action of selection, is the size of the random breeding unit. If this is large—for example, over 1000, very little nonadaptive differentiation is to be expected and only rather strong differences in the action of selection avoid swamping. If on the other hand, there is only short range dispersal—for example, $N_u=10$, large regions tend to become adaptively differentiated under the influence of slight differences in selection, and superimposed on this should be a large amount of nonadaptive differentiation of small regions. The maximum amount of nonadaptive differentiation among populations of a given size however, is not found with the smallest N_u, but at a certain optimum value.

If a population spreads over a large territory in which the environmental conditions are substantially uniform, there would primarily be only nonadaptive differentiation, the amount depending on the value of m or of N_u depending on the model that is most appropriate. With such differentiation occurring simultaneously but more or less independently in all series of alleles, each locality would have a slightly different genetic system from every other locality. These systems may be expected to differ in their success in meeting the environmental conditions. Among those which are relatively successful, adaptation is likely to have a slightly different basis in each case. The populations with such systems tend to become denser and to send out more than their share of migrants and thus enlarge in extent. Each would tend to perfect the line of adaptation on which it had started. Thus permanent differential action of selection would soon be brought into play in spite of the postulated uniformity of the conditions.

The expansion of centers of population characterized by certain genetic systems and contraction of those characterized by other systems is the process of intergroup selection referred to in the opening paragraph. The genetic system, including its state of heterogeneity as well as its central type, is the basis of selection instead of merely the net favorable or unfavorable effect of each single gene, which is the only basis for selection under panmixia; or the single genotype, which is the most probable basis under self-fertilization or vegetative multiplication. The present analysis indicates that this most favorable basis for evolutionary advance of the species as a whole may be present under certain conditions in a continuous population as well as in one consisting of partially isolated groups.

SUMMARY

Formulae are derived relating the variance of the gene frequencies of subgroups (σ_q^2) to the effective population number of these (N), the effective proportion of replacement per generation by immigrants (m), the inbreeding coefficient of individuals relative to the total population (F), and the mean gene frequency in the latter (q_t). Thus $\sigma_q^2 = q_t(1-q_t)/[2N-(2N-1)(1-m^2)] = q_t(1-q_t)F/(1-m)^2$ including the immediate sampling variance, but $\sigma_q^2 = q_t(1-q_t)F$ excluding this.

The effect of isolation by distance in a continuous population in which there is only short range dispersal in each generation is worked out on the hypothesis that the parents of any individual may be treated as if they were taken at random from a group of a certain size (N). It is shown that the inbreeding coefficient of individuals in such a population relative to a population of size KN can be expressed in the form $F = \sum_1^{K-1} t/[2N - \sum_1^{K-1} t]$ where $\sum t$ is the sum of a series of terms in which $t_1 = 1$ and $t_x = t_{(x-1)}[(x-1)N-1]/xN$ or approximately $C[x-(1/2N)]^{-(N+1)/N}$ where C is a constant close to 1. The value of $\sum_1^{K-1} t$ can be obtained sufficiently accurately by actual calculation of the first few terms, supplemented by the approximate formula

$$\sum_{K_1}^{K_2-1} t = CN\left[\left(K_1 - \frac{1}{2} - \frac{1}{2N}\right)^{-1/N} - \left(K_2 - \frac{1}{2} - \frac{1}{2N}\right)^{-1/N}\right]$$

for later terms. The limiting value $\sum_1^{\infty} t$ is N. Thus F approaches 1 in an indefinitely large continuous population.

The preceding results apply to area continuity. With continuity in a linear range (for example, shore line), $F=\sum t/[2N-\sum t]$ as above, $t_1=1$ but $t_x=t_{(x-1)}[N\sqrt{x-1}-1]/N\sqrt{x}$ or approximately $Ce^{-2\sqrt{x}/N}[\sqrt{x}-(1/2N)]^{-(N^2+1)/N^2}$.

In a continuous population with exclusive uniparental reproduction, the correlation between adjacent individuals is of the form $E=\sum t/N$ where $\sum t$ is the same as above for area or for linear continuity as the case may be.

The variance of gene frequencies in subdivisions of any size, N_i, within a more comprehensive population N_t is given by the formula $\sigma^2_{i.t}=q_t(1-q_t)[F_t-F_i]/[1-F_i]$ where F_i and F_t are the inbreeding coefficients relative to the populations of size N_i and N_t, respectively.

It is shown that in the absence of disturbing factors, short range dispersal (N less than 100 in the case of area continuity) leads to considerable differentiation not only among small subdivisions but also of large ones. Values of N greater than 10,000 give results substantially equivalent to panmixia throughout a range of any conceivable size. With linear continuity, there is enormously more differentiation than with area continuity. There is somewhat more differentiation under uniparental than under biparental reproduction.

Recurrent mutation, long range dispersal and selection are factors that restrict greatly the amount of random differentiation of large (but not small) subdivisions of a continuous population. A term $(1-m_1)^{2x}$ under biparental, $(1-m_1)^x$ under uniparental, reproduction is introduced into the expressions for t referred to above. In this $m_1=[-\Delta q/(q-q_t)]$ where Δq is the rate of change of gene frequency (q) which such factors tend to bring about.

The effective size of a population characterized by the inbreeding coefficient F depends on whether F is due to a tendency toward mating of relatives not associated with territorial subdivision, or to such subdivision. In the former case the sampling variance is $\sigma^2_{\delta q}=q(1-q)(1+F)/2N$, in the latter, $q(1-q)(1-F)/2N$, in contrast with $q(1-q)/2N$ in a random bred population.

If different regions are subject to different conditions of selection, the amounts of both adaptive and nonadaptive differentiation depend on the smallness of m (if subdivision into partially isolated "islands") or of N, size of the random breeding unit (if a continuous distribution). If these are sufficiently large there is no appreciable differentiation of either sort; if sufficiently small there is predominantly adaptive differentiation of the larger subdivisions with predominantly nonadaptive differentiation of smaller subdivisions superimposed on this. Even under uniform environmental conditions, random differentiation tends to create different adaptive trends in different regions and a process of intergroup selection, based on gene systems as wholes, that presents the most favorable conditions for adaptive advance of the species.

LITERATURE CITED

DOBZHANSKY, TH., and S. WRIGHT, 1941 Genetics of natural populations. V. Relations between mutation rate and accumulation of lethals in populations of *Drosophila pseudoobscura*. Genetics **26**: 23–51.

THOMPSON, D. H., 1931 Variation in fishes as a function of distance. Trans. Illinois Acad. Sci. **23**: 276–281.

WRIGHT, S., 1921 Systems of mating. Genetics **6**: 111–178.

1922a The effects of inbreeding and crossbreeding on guinea pigs. III. Crosses between highly inbred families. Bull. U. S. Dept. Agric. No. **1121**.

1922b Coefficients of inbreeding and relationship. Amer. Nat. **56**: 330–338.

1929 The evolution of dominance. Amer. Nat. **58**: 1–5.

1931 Evolution in mendelian populations. Genetics **16**: 97–159.

1938 Size of population and breeding structure in relation to evolution. Science **87**: 430–431.

1940 Breeding structure of populations in relation to speciation. Amer. Nat. **74**: 232–248.

1942 Statistical genetics and evolution. Bull. Amer. Math. Soc. **48**: 223–246.

WRIGHT, S., TH. DOBZHANSKY, and W. HOVANITZ, 1942 Genetics of natural populations. VII. The allelism of lethals in the third chromosome of *Drosophila pseudoobscura*. Genetics **27**: 363–394.

AN ANALYSIS OF LOCAL VARIABILITY OF FLOWER COLOR IN LINANTHUS PARRYAE*

SEWALL WRIGHT

The University of Chicago[1]

Received November 9, 1942

THE DISTRIBUTION OF LINANTHUS PARRYAE

EPLING and DOBZHANSKY (1942) have recently published a very detailed account of the distribution of two alternative characteristics, blue and white flowers, of a diminutive annual plant, *Linanthus Parryae,* in a portion of its range in the Mojave desert. It seemed of interest to make an analysis of their data from the standpoint of the theory of isolation by distance discussed in the preceding paper (WRIGHT 1943).

The region in question is about 80 miles long and averages about 10.5 miles wide. It stretches in an east-west direction along the piedmont north of the San Gabriel and San Bernardino Mountains and is largely isolated from other populations of the species. Data were obtained at stations every half mile along the principal roads, including two or three parallel roads at most places. At each station if the plants were present, counts were made of four samples of 100 plants each, spaced at intervals of approximately 250 feet at right angles to the road.

The vegetation in this piedmont belt is stated to be homogeneous. A number of species of shrubs are listed as characteristic. "The spacing of these shrubs is wide, and it is doubtful if they cover as much as 60 percent of the ground. Around the base of most of the bushes, the soil has accumulated so as to form a slight mound. *Linanthus Parryae* occupies the ground between the mounds, forming a widespread reticulum which is interrupted only by the stream beds or depositions alluded to above." The authors find nothing to suggest any selective differential.

Some of the conclusions reached by the authors are given as follows. "The apparent complexity of the distribution pattern of white and blue flower color in *Linanthus Parryae* can be reduced to a relatively simple scheme. The blue was found principally in three or four 'variable areas.' Outside these areas the blue was encountered sporadically, as would be expected if it were introduced there only on rare occasions through mutation or through occasional transport of 'blue' pollen or seed. Within the variable areas, the white and blue occurred side by side, and the population was differentiated into an extremely fine mosaic of microgeographic races. Pure white and pure blue colonies occurred at distances as small as 500 feet. Nevertheless, populations found one mile or less apart, resemble each other more than do populations taken at random in

* A portion of the cost of composing the mathematical formulae is borne by the Galton and Mendel Memorial Fund.

[1] Acknowledgment is made to the DR. WALLACE C. and CLARA A. ABBOTT MEMORIAL FUND of the UNIVERSITY OF CHICAGO for assistance in connection with the calculations.

the variable areas." This is followed by comparison with the U-shaped distribution of gene frequencies which the present author has deduced as characteristic in effectively small populations.

THE HIERARCHY OF SUBDIVISIONS

For more detailed mathematical analysis, it is convenient to define a hierarchy of subdivisions. Six compact, approximately equal, primary subdivisions are recognized along the length of the range. Each of these includes five secondary subdivisions. Each of these in turn includes four tertiary subdivisions. The tertiary subdivisions were chosen so as to include three stations as far as possible (two stations in six cases, four stations in two cases). The stations, as noted above, typically include four samples, but many of them contain less.

SMALLER SUBDIVISIONS	PRIMARY SUBDIVISIONS (EAST TO WEST) I	II	III	IV	V	VI	TOTAL
Secondary	5	5	5	5	5	5	30
Tertiary	20	20	20	20	20	20	120
Stations	57	59	60	60	61	59	356
Samples	198	211	214	214	218	203	1258

The population density in 1941 was found to vary from 1 to 26 per square foot (average 9.7) in the variable areas and from 1 to 48 per square foot (average 7.4) in the predominantly white areas. The area occupied by an average sample of 100 of the plants may thus be taken as about 12 square feet. It is stated, however, that in unfavorable years the species may be found only in sparse concentration or abundant only locally. Since the effective size of a population depends much more on the number of productive individuals in unfavorable than in favorable years, it is probable that the typical density is very much less than 100 per 12 square feet.

The average distance between samples at a typical station (four samples spaced at 250 foot intervals in a line) is 417 feet. Thus a station may be considered as representative of a circle of about this radius and hence of an area of about 5.4×10^5 square feet ($417^2\pi$) or 0.020 square mile. A station would contain about 45,000 sample areas if these were closely packed. However, since it is stated that about 60 percent of the ground is occupied by other vegetation, this estimate is to be reduced accordingly. Using round numbers, this indicates that there were about 2×10^4 sample areas in the population represented by a station in 1941. But if unfavorable years are taken into account, this number would probably have to be reduced enormously.

The average distance between stations in a group of three at half mile intervals is two-thirds of a mile. A tertiary subdivision may thus be considered to be representative of an area of about 1.4 square miles ($=(\frac{2}{3})^2\pi$) and thus would contain about 70 station equivalents. No doubt there should be some reduction to allow for interruptions. We shall use 50 as a round number for the station equivalents in a tertiary subdivision.

The secondary subdivisions typically include 12 stations spaced at half mile intervals, often along a straight road, but more often involving roads at right angles to each other or two close parallel roads. If along a straight line, the average distance between included stations is $2\frac{1}{6}$ miles. We shall consider a secondary subdivision to represent an area of about 14 square miles.

The area occupied by the entire population is about 840 square miles. The average area of one of the six primary subdivisions is thus about 140 square miles. If each of these were completely filled by its five recorded secondary subdivisions, the average area represented by each of the latter would be 28 square miles, just twice that estimated above. There were, however, large territories far from the roads which were not sampled. The smaller estimate accordingly seems preferable. These estimates are summarized below on the basis of the numbers in 1941 and on an arbitrary hypothesis.

TABLE 1

The hierarchy of subdivisions.

	AREA	ESTIMATED PLANTS IN 1941	NUMBER OF UNITS (A) 1941	(B) ARBITRARY
Total population	840 sq. mi.	6×10^{10}	6×10^{8}	6×10^{6}
Primary subdivisions	140 sq. mi.	10^{10}	10^{8}	10^{6}
Secondary subdivisions	14 sq. mi.	10^{9}	10^{7}	10^{5}
Tertiary subdivisions	1.4 sq. mi.	10^{8}	10^{6}	10^{4}
Stations	0.02 sq. mi.	2×10^{6}	2×10^{4}	200
Samples	12 sq. ft. (A) 1200 sq. ft. (B)	100	1	1

It would appear that the total number of plants of *Linanthus Parryae* in this region was between 10^{10} and 10^{11} in 1941. The average effective size of breeding population over a period of years is probably much less. For the sake of comparison, two widely different hypotheses will be used for the area from which the parents of individuals are drawn: (A) the area occupied by 100 plants in 1941 (12 square feet on the average), (B) an area 100 times as large to allow for years in which the population is sparse. Estimate (A) would be practically the minimum possible even if 1941 were a typical year. Estimate (B) is quite arbitrary.

THE SIGNIFICANCE OF DIFFERENTIATION WITH STATIONS

The first statistical question that requires consideration is whether or not there are greater differences among samples from the same station than are expected from the accidents of sampling. Let p be the actual but unknown frequency of blue in a homogeneous local population. Let p_0 be the observed frequency in a random sample of N individuals. Let $\delta p = p_0 - p$. Then $\sigma_{\delta p}^2 = p(1-p)/N$. This may be estimated from the observed frequency by applying the usual Gaussian correction after substituting p_0 for p. Thus for L samples from the same homogeneous population

$$\frac{1}{L}\sum_{1}^{L} p_0(1-p_0) = \frac{1}{L}\sum_{1}^{L}(p+\delta p)(1-p-\delta p) \tag{1}$$

$$= p(1-p) - \sigma^2_{\delta p} \quad \text{if} \quad L \doteq \infty$$

$$= (N-1)\sigma^2_{\delta p}$$

$$\sigma^2_{\delta p} = p_0(1-p_0)/(N-1) \tag{2}$$

taking $p_0(1-p_0)$ as the best estimate of the theoretic mean, obtainable from a single sample.

In a group of K samples of N each, not necessarily drawn from a homogeneous population, and with observed variance $\sigma_{p_0}{}^2 = \sum_1^K (p_0 - \bar{p}_0)^2/K$

$$\overline{\sigma^2_{\delta p}} = \frac{1}{K}\sum_{1}^{K}[p_0(1-p_0)/(N-1)] \tag{3}$$

$$\overline{\sigma^2_{\delta p}} = [\bar{p}_0(1-\bar{p}_0) - \sigma^2_{p_0}]/(N-1). \tag{4}$$

The deviation of the frequencies, p_0, of samples about the unknown actual frequency, $\bar{p}$, of the whole heterogeneous population may be analyzed into two independent components, the deviation from the observed mean frequency $\bar{p}_0$ of the K samples and the deviation of this from $\bar{p}$

$$(p_0 - \bar{p}) = (p_0 - \bar{p}_0) + (\bar{p}_0 - \bar{p}) \tag{5}$$

$$\sigma^2_{(p_0-\bar{p})} = \frac{1}{L}\sum_{1}^{L}\sum_{1}^{K}(p_0-\bar{p}_0)^2/K + \sigma^2_{(p_0-\bar{p})}/K \tag{6}$$

$$\frac{K-1}{K}\sigma^2_{(p_0-\bar{p})} = \frac{1}{L}\sum_{1}^{L}\sigma^2_{p_0}. \tag{7}$$

If the observed variance $\sigma_{p_0}{}^2$ be treated as representative of the theoretical average $\sum\sigma_{p_0}{}^2/L$, we obtain the formula with ordinary Gaussian correction for the uncertainty of the mean. This, however, will not do in this case, because $\sigma_{p_0}{}^2$ is not independent of $\bar{p}_0$. If, for example, the frequencies of blue in four samples of 100 plants are 100, 100, 0, 0, respectively, the observed variance, $\sigma_{p_0}{}^2 = \frac{1}{4}$, is the theoretical maximum, and application of the Gaussian correction gives the impossible estimate $\sigma^2_{(p_0-\bar{p})} = \frac{1}{3}$. Since $\sigma_{p_0}{}^2$ necessarily approaches 0 as $\bar{p}_0$ approaches either 0 or 1, assume as a first approximation that $\sigma_{p_0}{}^2 = C\bar{p}_0(1-\bar{p}_0)$ where C is a constant for a given group of samples. Then

$$\frac{K-1}{K}\sigma^2_{(p_0-\bar{p})} = \frac{C}{L}\sum_{1}^{L}\bar{p}_0(1-\bar{p}_0) \tag{8}$$

$$= C\bar{p} - \frac{C}{L}\sum_{1}^{L}\bar{p}_0{}^2 \tag{9}$$

(10) But $$\sigma^2_{(\bar{p}_0-\bar{p})} = \frac{1}{L}\sum_{1}^{L}\bar{p}_0^2 - \bar{p}^2 \quad \text{and also} \quad \sigma^2_{(p_0-\bar{p})}/K.$$

(11) Thus $$\frac{K-1}{K}\sigma^2_{(p_0-\bar{p})} = C\bar{p}(1-\bar{p}) - \frac{C}{K}\sigma^2_{(p_0-\bar{p})}$$

(12) $$\sigma^2_{(p_0-\bar{p})} = KC\bar{p}(1-\bar{p})/(K+C-1).$$

The best estimate of $\bar{p}(1-\bar{p})$ is $\bar{p}_0(1-\bar{p}_0)$. An approximation for the variance of sample frequencies, corrected for uncertainty of the mean is thus

(13) $$\sigma^2_{(p_0-\bar{p})} = K\sigma^2_{p_0}/(K+C-1) \quad \text{where} \quad C = \sigma^2_{p_0}/\bar{p}_0(1-\bar{p}_0).$$

It may be noted that if C is small there is an approach to the ordinary Gaussian correction, but if $C=1$ (as in the extreme case cited above) there is no Gaussian correction at all, and impossible estimates are avoided.

This variance includes the sampling errors in the determination of the local frequencies as well as the variance due to real differentiation of local populations. To obtain an estimate of the latter, the mean sampling variance, $\overline{\sigma^2_{\delta p}}$ as given by (4) must be subtracted.

(14) $$\sigma_p^2 = \frac{K\sigma_{p_0}^2}{(K+C-1)} - \frac{\bar{p}_0(1-\bar{p}_0) - \sigma_{p_0}^2}{(N-1)}.$$

While this seems to be as good an estimate as it is practicable to obtain from a single group of samples, it has serious limitations if $\bar{p}_0$ is close to 0 or 1. There is only one type of distribution in a group of four samples of 100 plants each for which $\bar{p}_0=.0025$—namely, samples with the frequencies 0, 0, 0, 1. Formula (14) gives very nearly $\sigma_p^2=0$ (exactly if C is treated as 0), but obviously no information is given (or can be given) on differentiation among samples from stations for which $\bar{p}$ (as opposed to $\bar{p}_0$) is .0025. Again, in such a station as 103 with frequencies 0, 0, 0, 5 the estimate of σ_p^2 is the maximum possible from a group of four samples with mean $\bar{p}_0=.0125$ but is much less than might occur among stations for which $\bar{p}=.0125$. The estimate from $\bar{p}_0$ and $\sigma_{p_0}^2$ is satisfactory for values of $\bar{p}_0$ less than .25 or greater than .75 (if four samples) only if the extreme type of distribution (0, 0, 0, n) is rare. In the present data, distributions of this extreme type are abundant below $\bar{p}_0=.05$ and above $\bar{p}_0=.95$ (24 in 40 stations, excluding those in which all plants or all but one were alike). There were 61 stations in which there were from 5 to 95 percent blues. Among these, only two showed the most extreme possible differentiation for their average (namely, stations 137 with 0, 0, 30, 0 blues and station 371 with 0, 100, 100 blues). The following estimates are based on these 61 stations.

NO. OF SAMPLES PER STATION	NO. OF STATIONS	ESTIMATED σ_p^2	ESTIMATED $\sigma^2_{\delta p}$
2	8	.0394	.0012
3	16	.0704	.0013
4	37	.0269	.0017
Total	61	.0400	.0015

The estimates of σ_p^2 varied enormously. In one case (station 292) the estimated variance was less (by an insignificant amount) than that expected from the accidents of sampling. At the opposite extreme were stations 371 (0, 100, 100) and 27 (0, 99, 100), which were largely responsible for the high mean estimate of σ_p^2 in stations with three samples. On the average, the total estimated variance is about 28 times that expected from accidents of sampling. There is thus no doubt of the reality of the differentiation among samples from the same station.

SIZE OF THE PARENTAL POPULATION ON FOUR HYPOTHESES

Of greater interest for our present purpose is the variability of gene frequencies. Unfortunately this depends on the answers to a number of questions which could be obtained only by experiments which have not yet been made.

It is conceivable that the difference between blue and white is not genetic at all, but this is highly improbable. Assuming that blue and white differ genetically, it makes a difference whether there is exclusive cross pollination, exclusive self fertilization, or some intermediate condition. While self pollination appears improbable, we shall consider this possibility as well as that of cross pollination. Under exclusive self pollination, the mode of inheritance makes no difference in the distribution of the blue and white clones. If, however, there is cross pollination, the estimates of gene frequencies from observed phenotypic frequencies depend on the mode of inheritance. We shall consider three extreme hypotheses—namely, that blue is recessive, that it is dominant, and that it depends on multiple factors and a threshold.

The primary purpose of the analysis will be to find the effective size of the population from which parents (of adjacent individuals in the case of exclusive self fertilization) must be drawn to account for the observed distribution as a cumulative consequence of sampling, according to the theory of isolation by distance recently presented (WRIGHT 1943). The possibility that the distribution may be affected by differential selection must also be examined. We shall consider first the differentiation demonstrated above to occur within stations and after this the differentiation among larger populations.

If there is exclusive self fertilization, we are concerned with the quantity $E=\sigma_p^2/\bar{p}(1-\bar{p})$, which measures the correlation between adjacent plants relative to the population of the station (WRIGHT 1943). In this formula, $\bar{p}$ is the mean, σ_p^2 the variance of the frequency of blue among the unit populations from which adjacent plants are drawn. However, since these populations are unknown, we can only calculate E from the samples of 100 plants. This has been done separately for each of the 61 stations in which $\bar{p}_0$ was between .05 and .95, using formula (14) to estimate σ_p^2 in each case. The values of E ranged from 0.00 to 1.00 in a very asymmetrical distribution. The distribution of $\sqrt{E}$ was more nearly normal.

If the hypothesis of exclusive self fertilization is correct, the variation of $\sqrt{E}$ should be independent of $\bar{p}_0$. This was tested by calculating the correlation and regression coefficients. The statistical constants are given in table 2. It turns out that the regression coefficient of $\sqrt{E}$ on $\bar{p}_0$, $.22\pm.10$, is slightly more than twice its standard error. This is not in good agreement with the hypothesis, although it cannot be considered to eliminate it.

The effective size of the parental population may be estimated from the average value of E. In these 61 cases the average value was .210. (It may be noted that this is considerably larger than the square of the average value of $\sqrt{E}$ given in table 2 because of the great variability among stations. In fact $\bar{E}=(\overline{\sqrt{E}})^2+\sigma^2_{\sqrt{E}}=.155+.055=.210$.)

TABLE 2

Statistical constants under four different genetic hypotheses. $\bar{p}_0$, $\bar{q}$ and $\bar{m}$ are the stations means, and $\bar{\bar{p}}_0$, $\bar{\bar{q}}$ and $\bar{\bar{m}}$ are the means of these means for the group of stations considered. In the first column, b is the regression of $\sqrt{E}$ on $\bar{p}_0$ and r is the correlation between these variables. Similarly b and r refer to the corresponding variables in the other columns.

(1) SELF FERTILIZATION		CROSS FERTILIZATION SINGLE GENE DIFFERENCE			MULTIPLE ADDITIVE FACTORS AND THRESHOLD	
			(2) BLUE DOMINANT	(3) BLUE RECESSIVE		
RANGE ($\bar{p}_0$)	.05 to .95	RANGE (q)	.10 to .90	.10 to .90	RANGE ($\bar{m}$)	−1.85 to +1.85
NO.	61	NO.	47	63	NO.	57
$\bar{\bar{p}}_0$	.440	$\bar{\bar{q}}$	.380	.517	$\bar{\bar{m}}$	−.212
$\sigma_{\bar{p}}$	.254	$\sigma_{\bar{q}}$	.202	.245	$\sigma_{\bar{m}}$	.932
$\overline{\sqrt{E}}$	.394	$\overline{\sqrt{F}}$	.380	.373	$\overline{\sqrt{F}}$	.399
$\sigma_{\sqrt{E}}$	.234	$\sigma_{\sqrt{F}}$	.256	.229	$\sigma_{\sqrt{F}}$	.232
r	$+.24\pm.10$	r	$+.56\pm.10$	$-.21\pm.12$	r	$+.16\pm.13$
b	$+.22\pm.10$	b	$+.70+.16$	$-.20+.12$	b	$+.040+.033$
$\bar{E}$	.210	$\bar{F}$	.210	.192	$\bar{F}$	.213
N (A)	45	N (A)	25	27	N (A)	25
N (B)	25	N (B)	14	15	N (B)	14

If for the moment we assume that the samples correspond in size to the groups from which the parents of any individual are drawn (hypothesis (A)) an estimate can be made of the effective population number of these groups from the theory of area continuity (WRIGHT 1943, fig. 7). Under this theory it requires a parental population of about 45 to give a value of E of .210 in a total (station) including 2×10^4 of these unit populations. This can agree with the actual number of plants in a sample (100) on taking account of the fact that these varied enormously in productivity. However, as noted, 1941 was a year of exceptional abundance. On the average it might require a considerably larger area than indicated in this year to provide an effective population number of 45. But in this case there would be less than 2×10^4 such groups in the area represented by a "station," which in turn would require a smaller estimate of the population number of the parental group. If, for example, a station includes only 200 parental groups (hypothesis (B)) instead of 20,000, the effective population number of these under the theory would be about 25.

Consider next estimates under the hypothesis of prevailing cross pollination with blue dependent on a single differential gene. If blue is dominant, its gene frequency in a sample is given by $q = 1 - \sqrt{1 - p_0}$. A variation, δp in phenotypic frequency, implies the variation $\delta q = \delta p / 2\sqrt{1-p}$ in gene frequency. Thus the sampling variance of gene frequencies is given by the formula

$$\overline{\sigma_{\delta q}^2} = \frac{1}{K}\sum_1^K\left[\frac{\sigma_{\delta p}^2}{4(1-p_0)}\right] \tag{15}$$
$$= \frac{1}{K}\sum_1^K\left[\frac{p_0(1-p_0)}{4(N-1)(1-p_0)}\right] = \frac{\bar{p}_0}{4(N-1)}.$$

The mean gene frequency, $\bar{q}$, and the variance, σ_q^2, were estimated for each station by the following formulae in which N was 100 and K usually 4.

$$\bar{q} = \frac{1}{K}\sum_1^K q \tag{16}$$

$$\sigma_q^2 = \frac{K\sigma_{q_0}^2}{(K + C - 1)} - \frac{\bar{p}_0}{4(N-1)} \quad \text{where} \quad \sigma_{q_0}^2 = \frac{1}{K}\sum_1^K q^2 - \bar{q}\sum q \tag{17}$$
$$\text{and} \qquad C = \sigma_{q_0}^2/\bar{q}(1-\bar{q}).$$

The quantity $F = \sigma_q^2/\bar{q}(1-\bar{q})$ (slightly different from the provisional quantity C) should be independent of $\bar{q}$ on a valid hypothesis. This was tested by calculating the regression of $\sqrt{F}$ on $\bar{q}$ for all cases in which $\bar{q}$ was between .10 and .90 (47 in number). There turns out to be a highly significant positive regression $+.70 \pm .16$. This means that there is a great deal too much variability at stations in which blue is common as compared with that in stations in which it is rare to be compatible with this hypothesis.

The average value of F on this hypothesis is .210 which would imply a parental population of about 25 if it is assumed that there are 2×10^4 or more of these in the area represented by a station (hypothesis (A)). However, if there are fewer per station, the estimate must be decreased. If, for example, there are only 200 parental populations per station area (hypothesis (B)) the estimate of effective size is about 14.

If blue is assumed to be recessive, its gene frequency in a sample is given by $q = \sqrt{p_0}$. Phenotypic variation δp implies $\delta q = \delta p / 2\sqrt{p}$ and mean sampling variance $\sigma_{\delta q}^2 = (1 - \bar{p}_0)/4(N-1)$. There were 63 stations for which $\bar{q}$ was between .10 and .90 under this hypothesis. The values of F, $\sqrt{F}$, and σ_q^2 in each of these were estimated by formulae analogous to those used under the hypothesis of dominance.

It turned out that the regression of $\sqrt{F}$ on $\bar{q}$ is negative, $-.20 \pm .12$. There is not enough variability within stations in which blue is common to satisfy this hypothesis in contrast with the situation if blue is assumed dominant. The regression is less than twice its standard error, however, so that the data cannot be considered to be incompatible with recessiveness of blue.

The average value of F on the hypothesis of recessiveness is .192 and thus only slightly different from that on the hypothesis of dominance. The estimates of effective size of parental group are accordingly practically the same (27 under hypothesis (A), 15 under hypothesis (B)).

The third hypothesis with respect to the mode of inheritance, assuming cross fertilization, is that the alternative characters blue and white depend on whether the cumulative effects of multiple factors (with no dominance or interaction) exceed a certain threshold. The common type of polydactyly in guinea pigs is an example of a character in which there is a superficial simulation of simple Mendelian heredity but in which data from F_3 and later generations indicate the above mechanism (WRIGHT 1934). The methods used in the analysis of polydactyly may be applied except that the correction for the sampling error was given incorrectly. (Fortunately the correction was so small that this had no appreciable effect on the results.)

Assume that the multiple factors determine a substantially normal distribution on a primary scale but that the phenotype depends on whether the value on this scale is above or below a threshold. It is convenient to take the threshold as the zero point and take as the unit of measurement the standard deviation (on this primary scale) characteristic of a sample in the station under consideration. The proportion above the threshold in a given sample with mean at m on the scale (and thus threshold at $-m$ relative to the mean) is as follows in terms of probability integral,[2]

$$\text{pri } x_1 = \frac{1}{\sqrt{2\pi}} \int_{-\infty}^{x_1} e^{-x^2/2} dx$$

$$p = 1 - \text{pri } (-m) = \text{pri } m \tag{18}$$

$$m = \text{pri}^{-1}\, p. \tag{19}$$

A phenotypic variation δp implies $\delta m = \delta p/y$, where y is the ordinate of the unit normal curve at the threshold. Thus the sampling variance for m is $\sigma_{\delta m}{}^2 = p_0(1-p_0)/y^2(N-1)$. This was found for each sample and averaged to find the correction to be applied in calculating $\sigma_m{}^2$.

$$\sigma_m{}^2 = \left[\sum_1^K m^2 - \bar{m}\sum_1^K m\right]/(K-1) - \sum_1^K [p_0(1-p_0)/y^2]/K(N-1). \tag{20}$$

To estimate the inbreeding coefficient F from $\sigma_m{}^2$ it may be noted first that with multiple factors and no dominance or factor interaction, F measures the proportional *decrease* in the variance of characters within random breeding

[2] In a number of papers (WRIGHT 1926, 1934, etc.) prf x_1 has been used for the area of the probability curve between the mean and the deviation x_1, as a multiple of the standard deviation. This has the disadvantage that the inverse function is two-valued unless areas below the mean are treated as negative. The form used above avoids this difficulty. The recognized form erf $x_1 = (2/\sqrt{\pi})\int_0^{x_1} e^{-x^2}dx$ is inconvenient for several reasons.

subgroups (WRIGHT 1921). The contribution of a pair of alleles A, a to the variance of a subgroup under the above conditions is $2q(1-q)(A)^2$ where (A) is the effect of replacing a by A. The average contribution in an array of such subgroups is $2[\bar{q}(1-\bar{q})-\sigma_q^2](A)^2$. But $\sigma_q^2=\bar{q}(1-\bar{q})F$, where F is the inbreeding coefficient of individuals relative to the total. Thus the average contribution of A, a to intragroup variance may be written $2(1-F)\bar{q}(1-\bar{q})(A)^2$. The total intragroup variance is merely the sum of such contributions under the conditions.

(21) $$\sigma_g^2 = 2(1 - F)\sum [\bar{q}(1 - \bar{q})(A)^2].$$

Next it may be noted that F measures the proportional *increase* in the variance of the *total* population resulting from subdivision into inbred lines. This total variance (σ_t^2) is compounded of the intragroup variance (σ_g^2) given above and the intergroup variance (σ_m^2). The contribution of A, a to the mean of each subgroup is of the form $2q(A)$. The variance of these is $4\sigma_q^2(A)^2 = 4F\bar{q}(1-\bar{q})(A)^2$.

(22) $$\sigma_m^2 = 4F\sum [\bar{q}(1 - \bar{q})(A)^2]$$

(23) $$\sigma_t^2 = \sigma_g^2 + \sigma_m^2 = 2(1 + F)\sum [\bar{q}(1 - \bar{q})(A)^2].$$

(24) Thus $$F = \sigma_m^2/(2\sigma_g^2 + \sigma_m^2).$$

In the present case, $\sigma_g^2=1$ by hypothesis.

(25) $$F = \sigma_m^2/(2 + \sigma_m^2).$$

Estimates of F have been made separately by this formula for each of 57 stations in which $\bar{m}$ was between -1.85 and $+1.85$. Again $\sqrt{F}$ should be independent of $\bar{m}$ if the hypothesis is valid. The regression of $\sqrt{F}$ on $\bar{m}$ turned out to be positive but not significant ($+.040\pm.033$). From this standpoint this hypothesis is satisfactory, more satisfactory in fact than any of the others considered.

The mean value of F was .213, substantially the same as under the hypothesis that blue depends on a single dominant (F=.210) or a single recessive (F=.192). The estimate of the effective size of the parental group is accordingly approximately the same under all of these hypotheses: 25 to 27 if there are 2×10^4 or more such groups in the area represented by a station but less if the number of random breeding units in a station is less than 2×10^4 (about 14 or 15 if the number of groups is as small as 200).

The effects of partial determination of the character by environmental factors are obvious under the last hypothesis. If there are environmental effects on individuals, not related to their location (for example, change in color with aging), the genetic component of intragroup variance would be less than 1, but σ_m^2 would not be affected. In this case, F is larger than estimated above, requiring N (size of random breeding unit) to be smaller to account for the observed differentiation within stations. In the case of *Linanthus Parryae*, the observations of EPLING and DOBZHANSKY tend to rule this out.

If, on the other hand, there are environmental influences (such as character of soil) that make a difference between samples but rarely between individuals of the same sample, it is σ_m^2 that must be reduced if it is to represent genetic differentiation and the estimate of F becomes smaller than calculated above. In this case, a larger value of N is implied. If intersample variance is almost entirely environmental, even though genetic segregation occurs within samples, F is almost zero, and it is implied that N is so large (for example, over 1000) that there is virtual panmixia throughout the station. There is little likelihood, however, that flower color is due to such environmental effects in *Linanthus Parryae.*

ANALYSIS OF VARIABILITY WITHIN THE RANGE

It is next of interest to analyze the variability in the region as a whole. The mean gene frequencies in the primary and secondary subdivisions were as follows, assuming blue to be recessive.

TABLE 3

Gene frequencies in the primary and secondary subdivisions assuming blue to be recessive.

PRIMARY SUBDIVISION	SECONDARY SUBDIVISION A	B	C	D	E	TOTAL
I	.573	.504	.717	.657	.302	.551
II	.339	.032	.007	.005	.008	.078
III	.009	.000	.000	.000	.000	.002
IV	.010	.005	.000	.000	.068	.017
V	.126	.004	.002	.000	.000	.027
VI	.000	.106	.411	.224	.014	.151
Total						.137

The distribution on the map is shown in figure 1. This brings out the three separate centers of high frequency, one in I overlapping II, a second in VI and a third in the southern parts of IV and V, to which EPLING and DOBZHANSKY called attention.

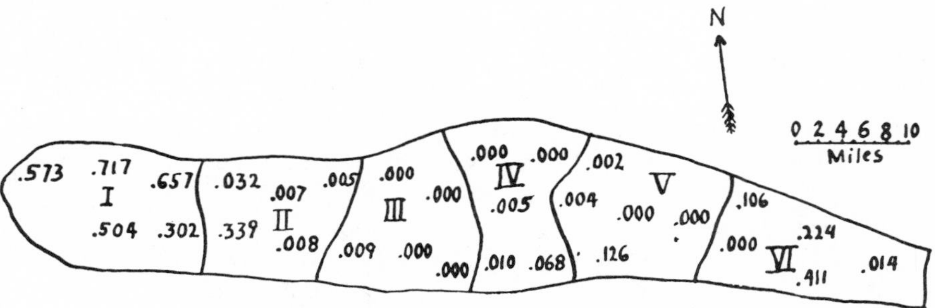

FIGURE 1.—The geographical distribution of the frequency of the gene for blue flowers (if recessive) in the portion of the range of *Linanthus Parryae* investigated by EPLING and DOBZHANSKY. The six primary subdivisions are indicated by Roman numerals.

It will be convenient to symbolize the various levels in the hierarchy by subscripts: 1 for a primary subdivision, 2 for a secondary, 3 for a tertiary, 4 for a station, and 5 for a sample. Thus q_4 represents the mean gene frequency in a station, K_4, the number of recorded stations in a tertiary population, etc., $\sigma_{4.3(G)}^2$ is used for the gross variance of q_4 (stations) within the next higher category (with the modified Gaussian correction for uncertainty of the mean, but without correction for the sampling variance of the next lower level); $\sigma_{4.3}^2$ is the final estimate including this last correction. Finally, $\sigma_{4.t}^2 (= \sigma_{4.3}^2 + \sigma_{3.2}^2 + \sigma_{2.1}^2 + \sigma_{1.t}^2)$ is the variance of q_4 within the total population.

The variance of samples within stations was obtained for all of the 356 stations.

$$\sigma^2_{5\cdot4(G)} = \frac{\sum_1^{356} [\sum q_5^2 - q_4 \sum q_5]}{\sum_1^{356} [K_5 - 1 + \overline{C}_4]} = \frac{7.2523}{959.3} = .0076$$

$$\sigma^2_{5\cdot4} = .957\sigma^2_{5\cdot4(G)} = .0073.$$

The summations in the bracket refer to the K_5 (usually four) samples of a station. $\overline{C}_4 = .161$ is the average value of $C_4 (= \sigma_{q_0}^2 / \bar{q}(1 - \bar{q}))$ for the 63 stations in which q_4 was between .10 and .90. The mean sampling variance of q_5 in these 63 stations was such that $\sigma_{5.4}^2 / \sigma_{5.4(G)}^2$ was .957. This ratio is carried over to all stations.

The variance of stations within the 120 tertiary subdivisions was as follows.

$$\sigma^2_{4\cdot3(G)} = \frac{\sum_1^{120} [\sum q_4^2 - q_3 \sum q_4]}{\sum_1^{120} [K_4 - 1 + \overline{C}_3]} = \frac{4.7112}{267.9} = .0176$$

$$\sigma^2_{4\cdot3} = \sigma^2_{4\cdot3(G)} - 356\sigma^2_{5\cdot4(G)}/1258 = .0176 - .0021 = .0154.$$

The number of stations within tertiary subdivisions (K_4) was usually 3. There were 30 tertiary subdivisions in which q_3 was between .10 and .90.

$$\overline{C}_3 = \frac{1}{30} \sum_1^{30} \left[\frac{\sum q_4^2 - q_3 \sum q_4}{K_4 q_3 (1 - q_3)} \right] = .266.$$

Similarly the variance of tertiary subdivisions within the 30 secondary subdivisions was as follows.

$$\sigma^2_{3\cdot2(G)} = \frac{\sum_1^{30} [\sum q_3^2 - q_2 \sum q_3]}{\sum_1^{30} [K_3 - 1 + \overline{C}_2]} = \frac{2.1386}{97.1} = .0220$$

$$\sigma^2_{3\cdot2} = \sigma^2_{3\cdot2(G)} - 120\sigma^2_{4\cdot3(G)}/356 = .0220 - .0059 = .0161.$$

Here K_3 was regularly 4. The average value of $\overline{C}_2$, calculated from ten secondary subdivisions for which q_2 was between .10 and .90, was .237.

discrepancy is greater, since the variability within stations indicates a random breeding unit of about 27 on this hypothesis, while the figures for the higher categories are only slightly modified.

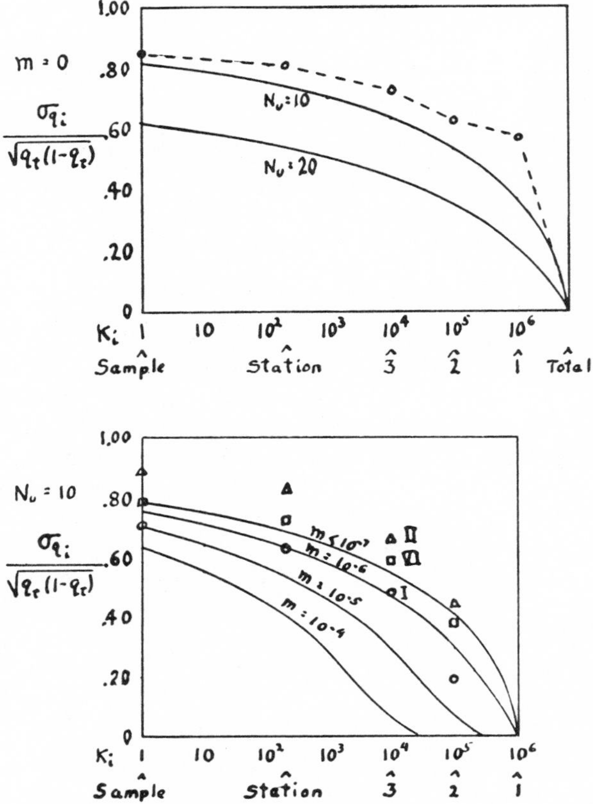

FIGURE 2.—The small circles connected by broken lines indicate the calculated variability of the frequency of the gene for blue (if recessive) among subgroups (samples, stations, tertiary, secondary and primary subdivisions of the range studied). Comparisons are made with the theoretical variability on the hypotheses that the effective number of individuals in a random breeding unit is 10 or 20, and that long range dispersal or mutation is negligible.

FIGURE 3.—The triangles indicate the calculated variability of the gene frequency of blue (if recessive) among subgroups of the primary subdivision II. The squares and circles do the same for primary subdivisions VI and I, respectively. These are compared with the theoretical amounts of variability on the hypothesis that $N = 10$ and that there is long range dispersal or mutation at rates up to $m = 10^{-4}$.

The most serious discrepancy is in the great variability of the primary subdivisions. In this case, however, the comparison with theory is hardly a fair one. The elongated range along the piedmont would favor a greater amount of differentiation than indicated by the theory of area continuity. There is some approach to the conditions of linear continuity.

The variance of secondary subdivisions within the six primary subdivisions was as follows.

$$\sigma^2_{2\cdot1(G)} = \frac{\sum_1^6 [\sum q_2^2 - q_1 \sum q_2]}{\sum_1^6 [K_2 - 1 + \bar{C}_1]} = \frac{.3218}{25.0} = .0129$$

$$\sigma^2_{2\cdot1} = \sigma^2_{2\cdot1(G)} - \sigma^2_{3\cdot2(G)}/4 = .0129 - .0055 = .0074.$$

Here K_2 was regularly 5. There were three primary subdivisions with q_1 considered large enough to warrant estimation of C_1—namely, I, II, and VI with values of C_1 of .084, .237 and .182 and $\bar{C}_1 = .168$.

Finally the variance of the six primary subdivisions within the total was as follows.

$$\sigma^2_{1\cdot t(G)} = \frac{\sum_1^6 q_1^2 - q_t \sum_1^6 q_1}{(K_1 - 1 + C_t)} = \frac{.2197}{5.3} = .0414$$

$$\sigma^2_{1\cdot t} = \sigma^2_{1\cdot t(G)} - \sigma^2_{2\cdot1(G)}/5 = .0414 - .0026 = .0388.$$

Here

$$C_t = \frac{\sum q_1^2 - q_t \sum q_1}{6q_t(1-q_t)} = .309.$$

There is little doubt of the reality of differentiation at all of these levels except for the case of secondary within primary subdivisions in which $\sigma_{2.1(G)}^2$ is only 2.3 times the variance expected from accidents of sampling at the next lower level. The extreme departures from normality in the distribution of q, however, must be kept in mind. They make the applicability of FISHER'S (1938) z test somewhat dubious.

The variance of the groups at each level within the total may be obtained by adding the variances down to the level in question. Thus $\sigma_{1\cdot t}^2 = .0388$, $\sigma_{2\cdot t}^2 = .0462$, $\sigma_{3\cdot t}^2 = .0623$, $\sigma_{4\cdot t}^2 = .0777$ and $\sigma_{5\cdot t}^2 = .0850$. By dividing these expressions by $q_t(1-q_t)$ where $q_t = .137$ we obtain values of $\sigma_{1\cdot t}^2/q_t(1-q_t)$ etc. This expression has been shown to be equal to $(F_t - F_i)/(1-F_i)$ under area continuity, in the absence of disturbing factors (WRIGHT 1943). It is .717 for samples, .656 for stations, .525 for tertiary groups, .390 for secondary groups, and .327 for primary groups. In figure 2 the square roots of these figures are plotted against number of random breeding units and compared with the theoretical curves for $N = 10$ and $N = 20$ deduced in the paper referred to above. The number of random breeding units is as given in table 1 using the arbitrary assumption (B) that there are 200 of these to a station.

From inspection of this figure it appears that there is considerably more variability of the higher categories than expected from that within stations. It would require a random breeding unit of considerably less than ten to account for this variability in contrast with about 15 indicated by the variability within stations on the assumption made above. If the samples are considered to be the random breeding units (2×10^4 per station equivalent) the

TABLE 4

Analysis of variability in six primary subdivisions (I to VI). The numbers in column 1 designate the level in the hierarchy. Column 2 gives the number of groups at each level included in the primary subdivision. Column 3 gives the sum of squared deviations of q from the mean of the next higher category. This is divided by the entries in column 4, $\sum(K-1+C)$ to give the entries in column 5, which are gross intragroup variances not corrected for the accidents of sampling. The sampling variances are given in column 6. Subtraction of these from the entries in column 5 give the net intragroup variances of column 7. The running sum of these (column 8) gives the variance of the groups at each level within the primary subdivisions. In column 9 these are divided by $q_1(1-q_1)$ where q_1 is the mean gene frequency for the primary subdivision (table 3) to give the estimate of the quantity $(F_t-F_i)/(1-F_i)$. The square roots of these quantities are the ordinates in figure 3. Column 10 gives the ratio of the gross variance (column 5) to that expected from sampling (column 6).

	(1)	(2)	(3)	(4)	(5)	(6)	(7)	(8)	(9)	(10
I	2	5	.1037	4.2	.0249	.0172	.c077	.0077	.031	1.4
	3	20	1.1140	16.2	.0688	.0161	.0527	.0604	.244	4.3
	4	57	1.9389	42.3	.0458	.0072	.0386	.0990	.400	6.3
	5	198	3.7814	150.2	.0252	.0009	.0243	.1232	.498	—
II	2	5	.0854	4.2	.0205	.0062	.0143	.0143	.199	3.3
	3	20	.3990	16.2	.0246	.0069	.0178	.0321	.446	3.6
	4	59	.8993	44.3	.0203	.0017	.0186	.0507	.704	11.9
	5	211	.9825	161.5	.0061	.0002	.0059	.0565	.786	—
III	2	5	.0001	4.2	.0000	.0000	.0000	.0000	.006	—
	3	20	.0003	16.2	.0000	.0000	.0000	.0000	0	—
	4	60	.0037	45.3	.0001	.0001	.0000	.0000	0	—
	5	214	.0675	163.7	.0004	.0000	.0004	.0004	.213	—
IV	2	5	.0034	4.2	.0008	.0006	.0002	.0c02	.014	1.3
	3	20	.0382	16.2	.0024	.0025	(−.0001)	.0001	.004	0.9
	4	60	.3440	45.3	.0076	.0007	.0069	.0070	.428	10.8
	5	214	.4094	163.7	.0025	.0001	.0024	.0094	.576	—
V	2	5	.0125	4.2	.0030	.0005	.0025	.0025	.096	5.8
	3	20	.0339	16.2	.0021	.0025	(−.0004)	.0021	.081	0.8
	4	61	.3484	46.3	.0075	.0011	.0065	.0085	.331	7.0
	5	218	.6391	166.8	.0038	.0001	.0037	.0122	.473	—
VI	2	5	.1167	4.2	.0280	.0085	.0195	.0195	.152	3.3
	3	20	.5533	16.2	.0342	.0090	.0252	.0446	.348	3.8
	4	59	1.1768	44.3	.0266	.0026	.0240	.0686	.535	10.2
	5	203	1.3724	153.5	.0089	.0003	.0086	.0772	.603	—

ANALYSIS OF VARIABILITY WITHIN THE PRIMARY SUBDIVISIONS

Because of the above consideration it is desirable to analyze the variability within the primary subdivisions. This has been done, with the results presented in table 4. Judging from the ratio of the gross variance to that expected from sampling (column 10), there appears to be significant differentiation at all levels within the primary subdivisions in which blue was not rare—namely, I($\bar{q}$=.551), II($\bar{q}$=.078), and VI ($\bar{q}$=.151) with one exception. The exception is the absence of significant differentiation of secondary subdivisions in I. In addition, there was significant differentiation among secondary subdivisions in V and of stations in both IV and V in spite of the rarity of blues in these areas ($\bar{q}=.027$ in V, $\bar{q}=.017$ in IV). The data from III, which was almost uniformly white flowered ($\bar{q}=.002$), are wholly inadequate for any estimates. The significance of differentiation among samples within stations has not been determined separately for the primary subdivisions. As noted, the variance in stations in which $\bar{q}$ was between .10 and .90 was about 24 times that expected from the accidents of sampling and undoubtedly significant.

In figure 3 the values of $\sigma_{x.t}/\sqrt{q_t(1-q_t)}$ as estimates of $\sqrt{(F_t-F_i)/(1-F_i)}$ under the theory of area continuity are plotted against the estimates of the number of random breeding units included in the category in question, within primary subdivisions I, II and VI. Again it is assumed arbitrarily that there are 200 random breeding units in a station.

The curves for the three primary subdivisions agree with each other as well as can be expected. They are somewhat more nearly parallel to the theoretical curve for $N=10$ than when differentiation was considered relative to the entire range. There is still, however, more variability of the higher categories at least within II and VI than expected from the theory of area continuity. Again the discrepancy would be much more serious if the random breeding units are identified with the 1941 samples, 2×10^4 to a station.

There are other factors, however, that must be considered. Even if there are only 200 random breeding units per station, the number in one of the primary subdivisions is of the order 10^6 (and in the whole range considered here, 6×10^6). This would be the typical number of generations to common ancestors of individuals that are far apart if the theory of area continuity applies strictly. In this case it would require something like a million years for a local colony to spread over this area, which is certainly highly improbable. However, means of dispersal to great distances so rare that only a minute fraction of the population of any occupied region has such an origin would enable the species to spread over a large suitable range in a few years. On the other hand, the effects of even a minute amount of replacement by a random sample of the species are not negligible.

It was shown in the preceding paper that such replacement in the proportion m (whether due to long range dispersal or mutation or, as accurately as possible, of uniform selection) removes nearly all random differentiation of populations more than $1/m$ times the random breeding unit and considerably reduces such variability in populations one tenth of this size. Thus reversible mutation between blue and white at rates of the order of 10^{-6} per generation should practically eliminate random differentiation of primary subdivisions and somewhat reduce that of secondary subdivisions, even under hypothesis B (in which these are 10^6 and 10^5 times the random breeding unit respectively). Admixture of a random sample of the species into all populations at the rate 10^{-4} per generation would practically eliminate all random variability of tertiary and larger subdivisions under the same hypothesis. Under hypothesis A this would eliminate random variability even among stations. The variability of samples within stations, however, would not be affected under hypothesis B and not very much under hypothesis A. The expected variability at each level in the hierarchy is shown in figure 3 for $N=10$ and $m=10^{-4}$, 10^{-5}, 10^{-6} and $m<10^{-7}$.

The large amount of differentiation actually found at all levels up to the highest indicates that some other factor than mere accumulation of sampling differences has been at work.

One possibility is that there are irregularities in the distribution, including differences in density. This would cause a greater amount of random differentiation of the larger categories than expected under a uniform distribution.

The most obvious possible factor that could counterbalance the effects of long range dispersal and mutation, however, is differential selection. Mr. W. HOVANITZ, in a personal communication, suggests that the climatic conditions may differ sufficiently near the ends of the region studied (I and VI, in which blue was relatively common) from those in the middle (where blue was rare) to make such an interpretation plausible. It is less plausible for the differences among secondary and tertiary subdivisions of the same primary subdivision.

There is a possibility, however, of selective differentiation even in the absence of any environmental differentiation. As noted in previous papers, the random differences in gene frequency, occurring in all series of alleles up to a certain level in the hierarchy, create a unique genetic system in each locality. Slightly different adaptive systems may be arrived at in different localities. If the gene or genes which distinguish blue and white play a role in any such systems, this would give a basis for locally different selection pressures.

The distribution of blue and white can be accounted for most easily by supposing that most of the differentiation of the smaller categories is random in character and due to the accumulation of sampling accidents in random breeding groups of one or two dozen productive individuals per year but that at the higher levels, processes which tend to pull down random differentiation such as mutation and especially occasional long range dispersal are counterbalanced by selective differentials between local genetic systems.

SUMMARY

The detailed account of the distribution of blue and white flowers of the annual plant *Linanthus Parryae* in a region of the Mojave desert by EPLING and DOBZHANSKY provides interesting material for comparison with the theoretical amount of random differentiation in a population that is continuous but in which dispersal is severely restricted.

For this purpose the 840 square miles studied is broken up into a hierarchy of subdivisions. There proves to be highly significant differentiation of samples (of 100 plants) within stations (representative of about 0.02 square miles). There is significant differentiation of stations within tertiary subdivisions (about 1.4 square miles), of these within secondary subdivisions (about 14 square miles), of these within primary subdivisions (about 140 square miles), and very marked differentiation of the primary subdivisions along the somewhat narrow piedmont zone.

Assuming that the difference is a genetic one, four hypotheses are considered in connection with the variability of samples within stations. It is improbable that reproduction is by self fertilization, but if it is, it would require that the population from which adjacent individuals are derived be about 45 to account for the observed variability, accepting the density of population found in 1941 as typical. This, however, was an unusually favorable year. Arbitrarily assuming 200 units per station instead of 20,000 as indicated in 1941, an effective population number of about 25 per unit is indicated.

If there is predominant cross fertilization, it makes no appreciable difference whether blue depends on a single dominant or a single recessive gene or on multiple factors and a threshold. Under any of these hypotheses, the effective population number of the random breeding unit comes out 25 to 27 if the 1941 estimate of numbers is accepted and 14 or 15 if the area occupied by a parental unit is assumed to be 100 times as large.

The amount of differentiation of the higher categories is somewhat greater than expected as a random consequence of that of the lower categories, under continuity of area and with negligible rates of long range dispersal and mutation (rates less than 10^{-7} per generation) and no differential or other selection. This is especially true of the primary subdivisions, but here the theory is unsatisfactory because of the elongated character of the range, which favors excessive differentiation.

However, there is somewhat too much variability of the higher categories even within the compact primary subdivisions to be accounted for as wholly random under the assumption above. As it is highly probable that the theoretical values should be substantially reduced because of long range dispersal at rates greater than 10^{-6} per generation, there is probably a counterbalancing influence of differential selection. This would not necessarily depend on differential environmental conditions. It could be a by product of the development of different genetic systems by the process of random differentiation.

LITERATURE CITED

Epling, C., and Th. Dobzhansky, 1942 Genetics of natural populations. VI. Microgeographical races in *Linanthus Parryae*. Genetics **27**: 317–332.

Fisher, R. A., 1938 Statistical methods for research workers. 7th ed. 356 pp. Edinburgh: Oliver and Boyd.

Wright, S., 1921 Systems of mating. Genetics **6**: 111–178.

1926 A frequency curve adapted to variation in percentage occurrence. J. Amer. Statist. Ass. **21**: 162–178.

1934 An analysis of variability in number of digits in an inbred strain of guinea pigs. Genetics **19**: 506–536.

1943 Isolation by distance. Genetics **28**: 114–138.

ISOLATION BY DISTANCE UNDER DIVERSE SYSTEMS OF MATING

SEWALL WRIGHT
The University of Chicago[1]

Received August 3, 1945

INTRODUCTION

THE effects of restricted dispersion on the genetic properties of a continuous population have been treated mathematically in a previous paper (WRIGHT 1943a). The conclusions have been applied to the interpretation of observed local variability in a population of a plant, *Linanthus Parryi* (WRIGHT 1943b) and in one of an animal, *Drosophila pseudoobscura* (DOBZHANSKY and WRIGHT 1943). The mathematical treatment was based on the assumption of completely random union of gametes within each neighborhood and thus would rarely be strictly applicable to actual cases. The purposes of the present paper are to compare the effects of various systems of mating within continuous populations, and to present a more accurate method than before for estimating from data an important theoretical quantity, N, the effective size of population of a "neighborhood" in the sense discussed below.

It was postulated in the previous paper that a population of uniform density occupies either an indefinitely large area (area continuity) or a strip of indefinitely great length but of such narrow width that dispersion occurs across it within a single generation (linear continuity). Both uniparental and biparental reproduction were considered. In the former case, it was postulated that the locations of parents at some phase of the life cycle are distributed, relative to the corresponding locations of their progeny, according to a normal probability curve with a standard deviation σ, if there is linear continuity, or in a bivariate normal distribution with standard deviation σ, for both x and y coordinates of the parental locations (relative to their progeny) if there is area continuity. In the case of biparental reproduction the locations of the parents were assumed to be uncorrelated with each other. This is assumed to hold in the present paper except as qualified by self fertilization or brother-sister mating in excess of random.

A term is needed to designate the local population of which the parents may be considered as representative. Various terms were used in the preceding paper. "Panmictic unit" applies only if there is completely random mating locally and is thus not sufficiently general. "Parental group" is better but is somewhat awkward. An essential property of the population in question is that the individuals are neighbors in the sense that their gametes may come together. The term "neighborhood" is thus an appropriate one for this important unit.

[1] Acknowledgment is made to the DR. WALLACE C. and CLARA A. ABBOTT MEMORIAL FUND of the UNIVERSITY OF CHICAGO for assistance in connection with the calculations.

THE POPULATION NUMBERS OF NEIGHBORHOODS

To obtain numerical results, it was assumed in the previous paper that the mating system is equivalent to the random union of gametes produced by a population of N monoecious individuals, thus involving the proportion 1/N of self fertilization. This raises the question of the relation of N to the density of the population of mature individuals and the standard deviation of the parental distribution. Consider first the situation in a population of uniform density along a linear range. Assume that the location points of parents (x) relative to offspring may be represented sufficiently accurately by the normal distribution $y=(1/\sigma\sqrt{2\pi})\exp(-x^2/2\sigma^2)$. Let n be the number of potential parents in a strip of length 2σ. The density per unit distance is then $d=n/2\sigma$. The average length of the territory occupied by each individual may be written as follows

$$1/d = 2\sigma/n = \int_{x_i-\sigma/n}^{x_i+\sigma/n} dx \tag{1}$$

The chance that a particular gamete comes from a particular member of the parental generation at distance x_i is y_i/d. The chance that two uniting gametes came from the same individual is thus $\Sigma(y_i/d)^2$ where the summation applies to all individuals in the parental distribution. This is the expression to be equated to 1/N. Making use of (1):

$$\frac{1}{N} = \sum_{x_i=-\infty}^{+\infty}\left[\frac{2\sigma}{n} y_i^2 \int_{x_i-\sigma/n}^{x_i+\sigma/n} dx\right] = \frac{2\sigma}{n}\int_{-\infty}^{+\infty} y^2 dx \text{ approximately}$$

$$= \frac{2\sigma}{n}\int_{-\infty}^{+\infty} [e^{-x^2/\sigma^2}/2\pi\sigma^2]dx$$

$$= \frac{1}{n\sqrt{\pi}}\int_{-\infty}^{+\infty} [e^{-x^2/2(\sigma/\sqrt{2})^2}/\sqrt{2\pi}(\sigma/\sqrt{2})]dx$$

$$= 1/n\sqrt{\pi} \tag{2}$$

$$N = \sqrt{\pi}\, n = 2\sqrt{\pi}\,\sigma d = 3.545\sigma d \tag{3}$$

Effective N is thus equivalent to the number of reproducing individuals along a strip 3.5 σ long. About 92.4 per cent of the actual parents of individuals should fall within the range $\pm\sqrt{\pi}\,\sigma d$.

This method of relating N to σ and d can be extended to populations that are continuous over an area. Let $y=[1/2\pi\sigma^2]\exp[-(x_1^2+x_2^2)/2\sigma^2]$ be the distribution of the locations of parents relative to those of their offspring. Let n be the number of reproducing individuals in a square, 2σ on a side. The density per unit area is $d=n/4\sigma^2$. The average area occupied by an individual (coordinates x_1, x_2) may be written as follows:

$$\frac{1}{d} = \frac{4\sigma^2}{n} = \int_{x_2-\sigma/n}^{x_2+\sigma/n}\int_{x_1-\sigma/n}^{x_1+\sigma/n} dx_1 dx_2 \tag{4}$$

The chance that two uniting gametes come from the same individual may be written as follows, using (4):

$$\frac{1}{N} = \frac{4\sigma^2}{n}\int_{-\infty}^{+\infty}\int_{-\infty}^{+\infty} y^2 dx_1 dx_2 = \frac{1}{n\pi} \tag{5}$$

$$N = \pi n = 4\pi\sigma^2 d = 12.566\sigma^2 d \tag{6}$$

Effective N is equivalent to the number of reproducing individuals in a circle of radius 2σ. Such a circle would include 86.5 per cent of the parents of individuals at the center.

In an analysis of data from natural populations of *Drosophila pseudoobscura* (DOBZHANSKY and WRIGHT 1943), effective N was taken as equivalent to the breeding population in a circle of such radius ($\sqrt{2}\sigma$) that the volume of a cylinder erected upon it would equal that of the bivariate normal distribution with $\sigma_{x_1} = \sigma_{x_2} = \sigma$ and central ordinate equal to the height of the cylinder. This does not do justice to the dispersive effect of the more extreme parent-offspring distances. The estimate of effective N for a given parental standard deviation should be just twice that arrived at in this way. As the estimate in this application could at best be considered as giving only the order of magnitude, the effect of this correction is not very important.

The distribution of location points of grandparents relative to those of their grandchildren is compounded of their distributions about the location points of the parents and the similar distribution of these about the individuals in question. Thus the variance of the grandparental distribution is twice that of the parental distribution and its standard deviation is $\sqrt{2}\sigma$. The standard deviation for ancestors of generation K is similarly $\sqrt{K}\sigma$. The effective size of the population from which ancestors of generation K are taken is $\sqrt{K}N$ if the range is linear but KN if it extends in all directions.

CASE I. MONOECIOUS POPULATIONS WITH EQUAL DISPERSION FROM MALE AND FEMALE PARENTS

This case was considered in the previous paper, subject to the postulate that union of gametes from the neighborhood is completely random (apart from the differential weighting of probabilities by distances of parental from offspring location). This implies self fertilization at the rate 1/N, a highly arbitrary postulate. It is desirable to determine the consequence under any specified percentage of self fertilization.

Let N be the effective population number in neighborhoods.

Let q be the gene frequency in a neighborhood.

Let h be the proportion of self fertilization.

The following quantities are relative to the population from which ancestors of generation K were drawn.

Let N_K be the effective population number.

Let q_K be the average gene frequency.

Let S be the correlation between gametes from the same individual.

Let D be the correlation between gametes from different individuals from the neighborhood, which unite or contribute to adjacent individuals. Let D_X be that between gametes from different individuals from the Xth ancestral generation—weighted according to their likelihood of contributing to the $(X-1)$st ancestral generation.

Let E be the correlation between gametes which contribute to adjacent individuals. Let E_X be the correlation between gametes from the Xth generation which contribute to the $(X-1)$st ancestral generation.

Let F be the correlation between uniting gametes. This is the inbreeding coefficient relative to the population of size N_K. It must be distinguished from the inbreeding coefficient relative to the neighborhood, which is $[h-(1/N)]/[2-h-(1/N)]$ from (16) below ($K=1$, $\Sigma t_A=0$) or from (16) and (18) by the formula $(F-E)/(1-E)$, Σt_A with $(K-1)$ terms.

Let $a(=\sqrt{1/2(1+F)}$, WRIGHT 1921) be the path coefficient relating zygote to one of the gametes which united in its production.

Let $b(=\sqrt{(1+F')/2}$, where F' is the value of F in the preceding generation, WRIGHT 1921) be the path coefficient relating gamete to the zygote which produced it. The compound path coefficient relating a gamete to one of the two from which it traces a generation earlier is $ba'=1/2$ where a' is the value of a in the preceding generation.

It will be assumed that the same population structure has continued indefinitely. Under this assumption primes may be dropped.

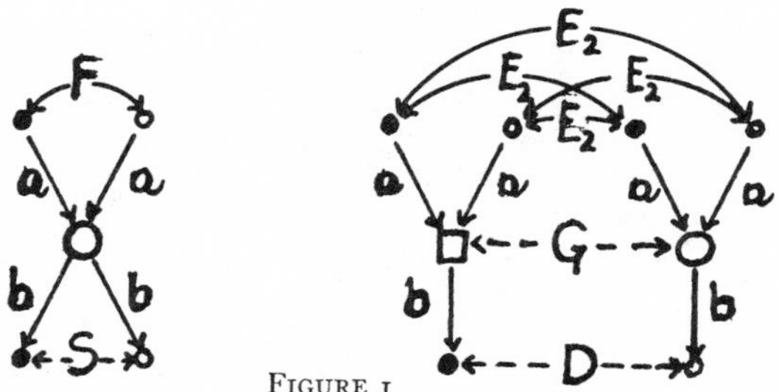

FIGURE 1

The following equations may be written:

$$E = \frac{1}{N}S + \frac{N-1}{N}D \quad \text{by definition} \tag{7}$$

$$F = hS + (1-h)D \quad \text{by definition} \tag{8}$$

$$S = b^2 = (1+F)/2 \quad \text{from fig. 1} \tag{9}$$

$$D = 4b^2a^2E_2 = E_2 \quad \text{from fig. 1} \tag{10}$$

Similarly

$$D_X = E_{X+1} \quad \text{by analogy with (10)} \tag{11}$$

Consider first the case of area continuity in which $N_K = KN$

$$E_X = \frac{1}{XN}S + \left[1 - \frac{1}{XN}\right]E_{X+1} \tag{12}$$

$$\text{Thus } E = \frac{1}{N} S + \frac{N-1}{N} E_2$$

$$E_2 = \frac{1}{2N} S + \frac{2N-1}{2N} E_3$$

$$E_3 = \frac{1}{3N} S + \frac{3N-1}{3N} E_4$$

$$E_K = 0$$

$$E = \frac{S}{N}\left[1 + \frac{1}{2}\left(\frac{N-1}{N}\right) + \frac{1}{3}\left(\frac{N-1}{N}\right)\left(\frac{2N-1}{2N}\right)\cdots\right] \qquad (13)$$

The series in brackets, with K-1 terms, will be designated Σt_A. Formulae for its approximate evaluation were given in the preceding paper (formulae 22 and 27). Its limiting value for $K=\infty$ is N.

$$E = S\sum t_A/N \qquad \text{from (13)} \qquad (14)$$

$$D = S(\sum t_A - 1)/(N-1) \qquad \text{from (7) and (14)} \qquad (15)$$

$$F = [h(N-1) + (1-h)(\sum t_A - 1)]/[(N-1) + (1-h)(N - \sum t_A)] \qquad \text{from (8), (9), (15)} \qquad (16)$$

$$S = (N-1)/[(N-1) + (1-h)(N - \sum t_A)] \qquad \text{from (9) and (16)} \qquad (17)$$

$$E = (N-1)\sum t_A/N[(N-1) + (1-h)(N - \sum t_A)] \qquad \text{from (14) and (17)} \qquad (18)$$

$$D = (\sum t_A - 1)/[(N-1) + (1-h)(N - \sum t_A)] \qquad \text{from (15) and (17)} \qquad (19)$$

The amount of differentiation among neighborhoods within the comprehensive population N_K can be found from a formula given in WRIGHT 1943a. It was shown there that if there is random mating within each of a number of populations, the variance of the values of q of these populations is given by the formula $\sigma_q^2 = q_K(1-q_K)F$ using the symbols defined here. In the present paper, random mating within neighborhoods is not assumed. However, E is defined in such a way that it would be the coefficient of inbreeding (F) if there were random mating in the last generation.

$$\sigma_q^2 = q_K(1 - q_K)E \qquad (20)$$

This formula may be derived in another way. The allele present in a gamete taken at random from a neighborhood may be considered to be determined jointly by the gene frequency (q) of the latter and an uncorrelated deviation. The degree of determination is the square of the correlation between gametic value and q. It gives the portion of the variance of gametes, $q_K(1-q_K)$, for which differentiation of neighborhoods, measured by σ_q^2, is responsible. But the correlation (E) between pairs of gametes taken at random from neighborhoods is also given by this same squared correlation. Thus $E=\sigma_q^2/q_K(1-q_K)$ in agreement with (20).

If self fertilization occurs wholly at random at all times ($h=1/N$), we have the case treated in the preceding paper (formula 21, WRIGHT 1943a).

$$E = F = \sum t_A/(2N - \sum t_A) \qquad \text{from (18), } h = 1/N \quad (21)$$

If there is no self fertilization ($h=0$), the result differs little, unless N is very small.

$$E = (N - 1)\sum t_A/N(2N - \sum t_A - 1) \qquad \text{from (18), } h = 0 \quad (22)$$

Under exclusive self fertilization ($h=1$) the amount of differentiation of neighborhoods is the same as that relative to two alternatives under any form of uniparental reproduction. The result agrees with formula (37) of the preceding paper.

$$E = \sum t_A/N \qquad \text{from (18), } h = 1 \quad (23)$$

The correlation between genotypes from a neighborhood (assuming additive effects) is independent of h in Case 1. The quantity, to be called G, is equal to D/b^2 as may be seen from figure 1.

$$G = D/b^2 = (\sum t_A - 1)/(N - 1) \quad (24)$$

Analogous results can be obtained for the case of linear continuity. As brought out in the previous paper, it is necessary merely to substitute the following series Σt_L for Σt_A. The approach to the limiting value N as the number of terms is increased is much more rapid.

$$\sum t_L = \left[1 + \frac{1}{\sqrt{2}}\left(\frac{N-1}{N}\right) + \frac{1}{\sqrt{3}}\left(\frac{N-1}{N}\right)\left(\frac{\sqrt{2}N-1}{\sqrt{2}N}\right)\cdots\right] \quad (25)$$

The conclusions reached on the relation of effective N to density of population (d) and standard deviation of parental distances (σ), assuming random mating within neighborhoods, may seem to require reconsideration in this case in which a specified amount (h) of self fertilization is assumed. The quantity $1/N$, however, may be defined as the amount of self fertilization that there would be if there were random union of gametes in the neighborhood in the last generation and is independent of the amount of self fertilization that there actually is or has been. The chance that two gametes that enter into the production of two adjacent individuals (instead of two uniting gametes) come from the same individual may be written $1/N=\Sigma(y/d)^2$, leading to formulae (3) and (6).

CASE 2. PERMANENT PAIRS

The simplest system of mating with separate sexes is that in which reproduction is wholly by permanent pairs and the amount of dispersion of males and females is the same. The analysis is somewhat similar to that in the preceding case if the pair is treated as the unit and located at the point at which pairing first occurs. The treatment of individuals as located at the points at which they began their careers as fertilized eggs (or at any later phase) would be complicated by a correlation between the parental localizations relative to those of offspring, beyond that which may be due to a tendency to brother-sister mating. As before, correlations are relative to a population of specified size (KN_P).

Let $N_P(=N/2)$ be the number of pairs in a neighborhood.

Let h here be the proportion of brother-sister mating.

Let C be the correlation between gametes produced by siblings.

Let D be the correlation between gametes of mated or adjacent non siblings, and D_X that between gametes of non siblings of ancestral generation $(X-1)$ relative to pairs.

Let E be the correlation between gametes of adjacent individuals and E_X be that between gametes of individuals belonging to ancestral generation $(X-1)$ relative to pairs.

Let F be the correlation between uniting gametes (the inbreeding coefficient).

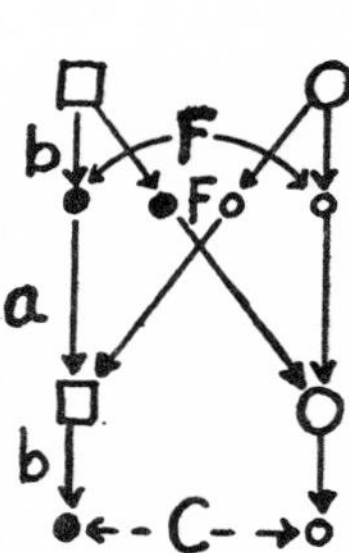

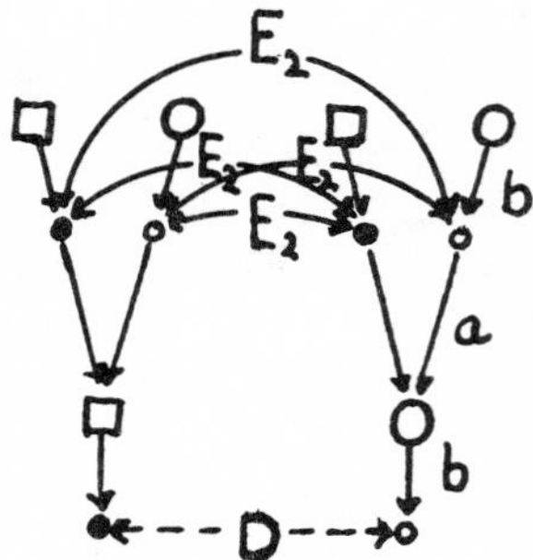

FIGURE 2

From inspection of figure 2

$$C = b^2a^2(2b^2 + 2F) = (1 + 3F)/4 \tag{26}$$

$$D = 4b^2a^2E_2 = E_2 \tag{27}$$

$$D_X = 4b^2a^2E_{x+1} = E_{x+1} \tag{28}$$

$$E = C/N_p + E_2(N_p - 1)/N_p \qquad \text{by definition and (27)} \tag{29}$$

$$E_X = C/XN_p + E_{x+1}(XN_p - 1)/N_p \qquad \text{by analogy} \tag{30}$$

$$E_K = 0 \qquad \text{by definition of } KN_p \text{ as the population of reference} \tag{31}$$

$$E = C\sum t_A/N_p \qquad \textit{cf.}\ (13), (14) \tag{32}$$

$$D = (N_pE - C)/(N_p - 1) = C(\sum t_A - 1)/(N_p - 1) \qquad \text{from (29), (32)} \tag{33}$$

$$F = C[h(N_p - 1) + (1 - h)(\sum t_A - 1)]/(N_p - 1)$$

by definition and (33) (34)

$$F = [h(N_p - 1) + (1 - h)(\sum t_A - 1)] / [(N_p - 1) + 3(1 - h)(N_p - \sum t_A)] \quad \text{from (26), (34)} \quad (35)$$

$$E = (N_p - 1)\sum t_A / \{N_p[(N_p - 1) + 3(1 - h)(N_p - \sum t_A)]\} \quad \text{from 32, 26, 35} \quad (36)$$

The amount of variation of gene frequency among neighborhoods is the same as that given in (20), $\sigma_q^2 = q_K(1 - q_K)E$.

If brother-sister mating occurs at random at all times within neighborhoods, $h = 1/N_P$

$$E = \sum t_A / [4N_p - 3\sum t_A] \quad \text{from (36), } h = 1/N_P \quad (37)$$

A case which approximates that in a continuous human population of low mobility is that in which $h = 0$ (no brother-sister mating).

$$E = (N_p - 1)\sum t_A / [N_p(4N_p - 3\sum t_A - 1)] \quad \text{from (36), } h = 0 \quad (38)$$

With exclusive brother-sister mating ($h = 1$) all lines become homallelic, and the differentiation among neighborhoods becomes the same as under uniparental reproduction, except that pairs, instead of individuals, must be taken as the units.

$$E = \sum t_A / N_p \quad \text{from (36), } h = 1 \quad (39)$$

The correlation between random zygotes from the same neighborhood, assuming no dominance or factor interaction, is given in this case by E/b^2.

$$G = (N_p - 1)\sum t_A / \{N_p[(N_p - 1) + (1 - h)(N_p - \sum t_A)]\}. \quad (40)$$

In the case of a population with a linear range, it is necessary merely to substitute Σt_L for Σt_A in the preceding formulae.

The relation between effective number of pairs, standard deviation (σ_p) of distances between parent-offspring mating sites, and density of the pairs (d_p) can be found as in case 1.

$$N_p = 2\sqrt{\pi}\,\sigma_p d_p \quad \text{with linear continuity} \quad (41)$$

$$N_p = 4\pi\sigma_p^2 d_p \quad \text{with area continuity} \quad (42)$$

Since $N = 2N_p$ and $d = 2d_p$ and $\sigma = \sigma_p$ (if individuals are located at the site of first mating), formulae (41) and (42) apply to individuals by dropping all subscripts. They become identical with (3) and (6).

CASE 3. SEPARATE SEXES AND RANDOM MATING

Assume that males and females are both distributed uniformly but not necessarily with the same density over either an extensive linear range or an extensive area, that each offspring is produced by a separate random mating, and that there is equal dispersion of sons and daughters. Let N_m and N_f be respectively the effective numbers of mature males and females per neighborhood. The diagrams below represent three of four possible relations between gametes of mated or adjacent individuals.

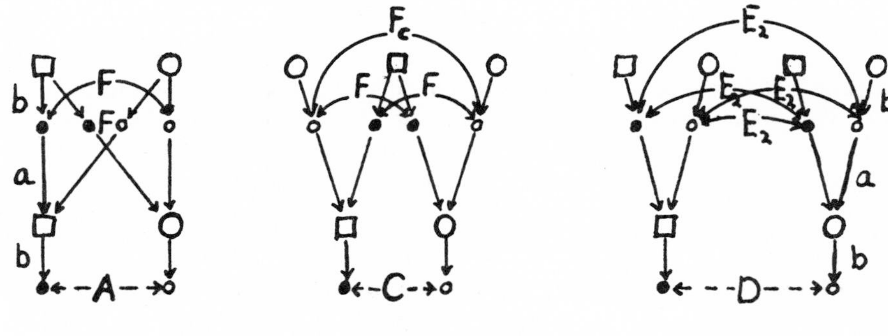

FIGURE 3

All correlations are assumed to be relative to a population of specified size $K(N_m+N_f)$.

Let A be the correlation between gametes produced by full brother and sister.

Let B and C be the correlations between gametes produced by half brothers and sisters with common mother and father, respectively.

Let D be the correlation between gametes of mated or adjacent individuals without a common parent, and D_X that between gametes of non-siblings of ancestral generation $(x-1)$ relative to matings.

Let $E=F$ be the correlation between gametes of mated or adjacent individuals and E_X that between gametes of individuals of ancestral generation $(x-1)$ relative to matings.

Let F_C be the correlation between gametes of individuals that mate at some time with the same individual.

Assuming random mating, the probabilities of the four cases and their contributions to E are readily determined.

$$E = [A + (N_m - 1)B + (N_f - 1)C + (N_m - 1)(N_f - 1)D]/N_mN_f \quad (43)$$

$$A = b^2a^2(2b^2 + 2F) = (1 + 3F)/4 \quad (44)$$

$$B = C = b^2a^2(b^2 + 2F + F_C) = (1 + 5F + 2F_C)/8 \quad (45)$$

$$D = 4b^2a^2E_2 = E_2 \quad (46)$$

$$D_X = 4b^2a^2E_{X+1} = E_{X+1}. \quad (47)$$

Collecting terms and multiplying N_m and N_f by X,

$$E_X = \left(\frac{N_m + N_f}{XN_mN_f}\right)\left(\frac{1 + 5F + 2F_C}{8}\right) + \left[1 - \left(\frac{N_m + N_f}{XN_mN_f}\right)\right]E_{X+1} - \left(\frac{F + F_C - 2E_{X+1}}{2X^2N_mN_f}\right) \qquad (48)$$

In the last term the numerator does not include population numbers and is twice the difference between two quantities that are both less than one and nearly the same if X is small; the denominator involves X^2. This term must always be much smaller than any of the others. It will be assumed that it can be ignored.

It will be convenient to write N′ for $N_mN_f/(N_m+N_f)$.

$$E = [(1 + 5F + 2F_C)/8N'] + (N' - 1)E_2/N' \quad \text{approximately}$$
$$E_2 = [(1 + 5F + 2F_C)/16N'] + (2N' - 1)E_3/2N' \quad \text{approximately}$$
$$E_X = [(1 + 5F + 2F_C)/8XN'] + (XN' - 1)E_{X+1}/XN'$$
$$E_K = 0$$

$$E = [(1 + 5F + 2F_C)/8N']\sum t_A. \qquad (50)$$

The correlation F_C cannot differ appreciably from F (and E) although theoretically slightly smaller. Substituting E for both F and F_C, the following approximate solution can be obtained.

$$E = \sum t_A/(8N' - 7\sum t_A). \qquad (51)$$

Special cases of interest are that in which $N_m=N_f=N/2$, $N'=N/4$, and that in which N_f is indefinitely large in comparison with N_m with the consequence that $N'=N_m$.

The amount of differentiation among neighborhoods is given as in other cases by formula 20, $\sigma_q^2=q_K(1-q_K)E$.

Again it is necessary merely to substitute Σt_L for Σt_A to obtain the formula applicable to a linear range.

To relate N_m and N_f to the densities d_m and d_f of males and females respectively and to the standard deviation of parent-offspring distances $\sigma_m=\sigma_f=\sigma$, let $y=(1/\sigma\sqrt{2\pi})\exp(-x^2/2\sigma^2)$ be the distribution of parents to offspring in a linear range. Let n_m and n_f be the numbers of males and females in a strip 2σ long. The mean space occupied by males is $1/d_m=2\sigma/n_m$ and by females is $1/d_f=2\sigma/n_f$. The chance that adjacent individuals have a common father is

$$\frac{1}{N_m} = \sum_{x=-\infty}^{+\infty}(y/d_m)^2 = \sum_{x=-\infty}^{+\infty}\frac{1}{d_m}y^2\int_{x-\sigma/n_m}^{x+\sigma/n_m}dx$$
$$= 1/(2\sqrt{\pi}\,\sigma d_m) \qquad (52)$$

Thus $N_m = 2\sqrt{\pi}\,\sigma d_m = 3.5\sigma d_m$ (53)

Similarly $N_f = 2\sqrt{\pi}\,\sigma d_f = 3.5\sigma d_f$.

The result is the same per individual as in cases 1 and 2. This is also true for area continuity.

$$N_m = 4\pi\sigma^2 d_m = 12.6\sigma^2 d_m \qquad (54)$$
$$N_f = 4\pi\sigma^2 d_f = 12.6\sigma^2 d_f.$$

CASE 4. DISPERSION BY GAMETES OF ONLY ONE SEX

The assumption that the distribution of parent-offspring distances is the same for both parents often does not apply. Thus among plants, dispersal of seed may be negligible in comparison with that of pollen. The assumption of different variances for male and female parents leads unfortunately to great algebraic complexity. It must suffice here to consider the limiting case in which all dispersal is through one parent (taken as the male). For simplicity, we shall assume all individuals to be hermaphrodites but with self fertilization occurring to any specified extent. The population will be assumed to be of uniform density, either along a strip of indefinitely great length, negligible width, or over an area of indefinite extent in all directions. In the former case, the distribution of pollen parent about progeny (and ovule parent) will be assumed to be univariate normal with standard deviation σ, apart from excess (or defect) of the middle class due to excess (or defect) in amount of self fertilization relative to that under random fertilization. Letting n be the number of individuals in a strip 2σ long, the average density is $d=n/2\sigma$ and the average interval between individuals is $2\sigma/n$. Let N be the effective number of individuals in the neighborhood that function as pollen parents, without taking cognizance of excess self fertilization. Then the chance of self fertilizat on under random union is $(1/N)=(2\sigma/n)y_0$ where y_0 is the midordinate of the normal distribution $y=(1/\sigma\sqrt{2\pi})\exp(-x^2/2\sigma^2)$. It is assumed that N is sufficiently large that the middle class of the distribution can be represented reasonably well by the product of midordinate into class range. Thus

$$1/N = 2\sigma/n\sigma\sqrt{2\pi} = (1/n)\sqrt{2/\pi}$$
$$N = n\sqrt{\pi/2} = \sqrt{2\pi}\,\sigma d = 2.50\sigma d. \qquad (55)$$

In the case of area continuity, we assume that the distribution of pollen parents is bivariate normal relative to ovule parents and progeny, $y=(1/2\pi\sigma^2)\exp[-(x_1^2+x_2^2)/2\sigma^2]$. Letting n be the number of individuals in a square 2σ on a side, the average density is $d=n/4\sigma^2$, and the average space per individual is $4\sigma^2/n$. Defining N as above, the chance of self fertilization under random mating of gametes in the neighborhood is $4\sigma^2 y_0/n$.

$$1/N = 4\sigma^2/(n(2\pi\sigma^2)) = 2/\pi n$$
$$N = \pi n/2 = 2\pi\sigma^2 d = 6.28\sigma^2 d. \qquad (56)$$

N is thus equivalent to the number of individuals in a circle of radius $\sqrt{2}\sigma$ in this case. It is half as great as in the case of an equally dense population of hermaphrodites with the same amount of dispersion of both male and female gametes as of male gametes in this case. If the variance of pollen dispersal is twice that of the preceding cases but d is the same, formulae (55) and (56) become the same as (3) and (6), respectively.

Let r be the proportion of the pollinations that may be considered as at random from the neighborhood and $(1-r)$ the excess self fertilization. Assuming that N is at least 2, the variance of pollen parents in any direction is $r\sigma^2$ (area continuity), since self fertilization contributes nothing. The total proportion of self fertilization is $h=(r/N)+(1-r)$.

The correlation (E) between ovules and pollen grains that contribute to adjacent zygotes and that (F) between ones that unite may be analyzed as indicated in figure 4.

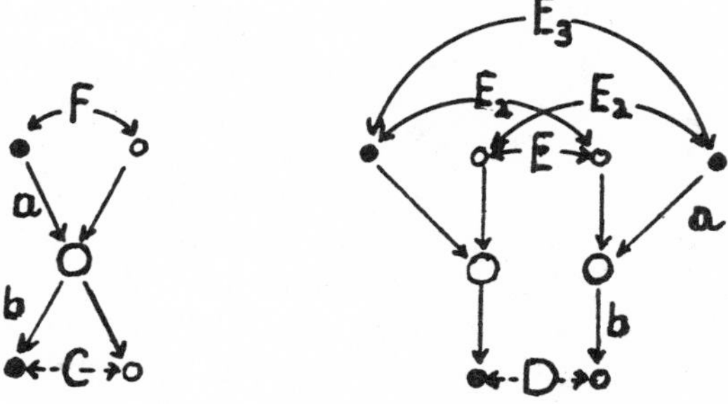

FIGURE 4

$$E = (C/N) + D(N - 1)/N \tag{57}$$

$$F = hC + (1 - h)D. \tag{58}$$

In cases of cross pollination, the variance of father's mother relative to mother's mother is the same as that of father relative to mother (σ^2) because of the postulated absence of dispersion in the female line. The correlation between ovules from which fathers and mothers (where different) were derived is thus E. The variance of father's father about mother's mother (father and mother different) is $(1+r)\sigma^2$, corresponding to an effective population of $(1+r)N$ if area continuity or $\sqrt{1+r}\,N$ if linear continuity, assuming that N is at least 2. The same is true for father's mother relative to mother's father. The correlation between pollen and ovule produced by these grandparents will be called E_2. The variance of father's father relative to mother's father (father and mother different) is $(1+2r)\sigma^2$, corresponding to an effective population of $(1+2r)N$ if area continuity and $\sqrt{1+2r}\,N$ if linear continuity. The correlation between pollen grains produced by these grandparents will be called E_3.

The spores produced by the Xth ancestral generation, where the parents are different, are drawn from populations with variances $[1+(X-2)r]\sigma^2$, $[1+(X-1)^2r]\sigma^2$ and $[1+Xr]\sigma^2$ according as ovule-ovule, ovule-pollen or pollen-pollen. The effective population numbers are proportional,

$[1+(X-2)r]N$, etc., if area continuity and proportional to the square roots, $\sqrt{1+(X-2)r}\ N$, etc., if linear continuity. Thus if D_X is the correlation between spores from different individuals of this generation,

$$D_X = (E_X + 2E_{X+1} + E_{X+2})/4 \tag{59}$$

$$C = b^2 = (1 + F)/2. \tag{60}$$

With area continuity,

$$\left.\begin{aligned} E &= \frac{C}{N} + \frac{N-1}{4N}(E + 2E_2 + E_3) \\ E_2 &= \frac{C}{(1+r)N} + \frac{(1+r)N - 1}{4(1+r)N}(E_2 + 2E_3 + E_4) \\ E_X &= \frac{C}{[1+(X-1)r]N} + \frac{[1+(X-1)r]N - 1}{4[1+(X-1)r]N}(E_X + 2E_{X+1} + E_{X+2}) \\ E_K &= 0. \end{aligned}\right\} \tag{61}$$

The equations above (except the last) are obviously satisfied if all E's equal 1, which, as in the preceding cases, is the limiting value approached as the area of reference is made indefinitely large. The attempt to solve for E with finite N_K leads, as in case 3, to an unmanageable series. An approximate solution may be obtained as follows:

Let $$(E_X + 2E_{X+1} + E_{X+2})/4 = E_{X+1} + \Delta_{X+1} \tag{62}$$

$$\begin{aligned} E = \frac{C}{N}\Bigg\{1 &+ \left(\frac{1}{1+r}\right)\left(\frac{N-1}{N}\right) \\ &+ \left(\frac{1}{1+2r}\right)\left(\frac{N-1}{N}\right)\left[\frac{(1+r)N-1}{(1+r)N}\right]\cdots \\ &+ \Delta_2\left(\frac{N-1}{N}\right) + \Delta_3\left(\frac{N-1}{N}\right)\left[\frac{(1+r)N-1}{(1+r)N}\right]\cdots\Bigg\} \end{aligned} \tag{63}$$

As $\Delta_{X+1}=\frac{1}{4}[(E_X-E_{X+1})-(E_{X+1}-E_{X+2})]$, the Δ's are equal to fourths of the second differences of the E's. They can easily be shown to be negligibly small, and a first approximation may be obtained by ignoring them.

The series that is left in brackets differs from those considered before and may be called $\Sigma t_A'$.

$$E = C\sum t_A'/N \quad \text{approximately} \tag{64}$$

It can easily be found from the equations given above, including the approximation $D=E_2=(NE-C)/(N-1)$ from (57), the first equation of the set (61), (58) and (60) that

$$C = (1 + rE)/(1 + r) = (1 + F)/2 \tag{65}$$

$$F = [1 + r\,(2E - 1)]/(1 + r) \tag{66}$$

$$\text{Thus}\quad E = \left(\frac{1 + rE}{1 + r}\right)\frac{\sum t_A'}{N} = \frac{\sum t_A'}{N + r(N - \sum t_A')}\quad \text{approximately} \tag{67}$$

The series $\Sigma t_A'$ can be evaluated in the same way as Σt_A (WRIGHT 1943a). If self fertilization occurs only at random in the neighborhood ($r=1$), $\Sigma t_A'$ becomes in fact Σt_A and

$$E = \sum t_A/[2N - \sum t_A]\quad \text{approximately} \tag{68}$$

This is the same formula as that obtained earlier for hermaphrodites with equal dispersion of both parents relative to offspring. However, there is greater differentiation for a given density of population and given dispersion of pollen, since effective N is only half as great if there is no dispersion of ovules as with dispersion of ovules equal to that of pollen.

If $1/r$ is an integer (H), the series $\Sigma t_A'$ can be expressed in terms of Σt_A. Let $N'=rN$ and assume, as before, that this is at least 2. Σt_A below is defined in terms of N' instead of N.

$$\sum t_A' = H\left[\frac{1}{H} + \left(\frac{1}{H+1}\right)\left(1 - \frac{1}{HN'}\right) + \left(\frac{1}{H+2}\right)\left(1 - \frac{1}{HN'}\right)\left[1 - \frac{1}{(H+1)N'}\right] \cdots \tag{69}$$

$$\text{If}\quad r = 1/2,\ H = 2,\ \sum t_A' = HN'(\sum t_A - 1)/(N' - 1) \tag{70}$$

$$E = 2(\sum t_A - 1)/[3N' - \sum t_A - 2] \tag{71}$$

COMPARISON OF CASES

The effects of different systems of mating on populations with uniform distributions over large areas are compared in tables 1 to 4. Consider first the variance (σ_q^2) of gene frequencies of neighborhoods with populations of effective size 200 (table 1) (excluding in this and all other cases the variance, $(q(1-q)/2N)$, due immediately to accidents of sampling). In an indefinitely extended population of hermaphrodites in which the gametes combine at random in the neighborhood and are dispersed equally along both parental lines, σ_q^2 reaches only 2.45 per cent of its maximum within an area that includes two million individuals and only 4.75 per cent of the maximum if 20 billion are included. Complete exclusion of self fertilization makes no appreciable difference. There is also no appreciable difference if the neighborhoods consist of separate sexes that mate at random to form 100 permanent pairs, assuming equal dispersion of males and females (σ_q^2 for neighborhoods rises to 2.51 per cent of the maximum within an area including one million pairs and to 4.97 per cent of the maximum within an area including ten billion pairs). If each offspring is produced by a separate random mating but otherwise conditions are the same, differentiation of neighborhoods is slightly greater ($\sigma_q^2=2.64$ per

TABLE I

The correlation (E) between gametes of adjacent individuals belonging to neighborhoods of size $N(=200)$, relative to populations of size $N_K(=KN)$, which in turn are portions of a population distributed with uniform density (d) over an indefinitely large area. The effective population of a neighborhood is determined by the density and the variance (σ^2) of parent-offspring distances. $N=4\pi\sigma^2 d$. E also measures the variance, σ_q^2, of gene frequencies of neighborhoods relative to the limiting value $q_K(1-q_K)$ of the population of size N_K. $E=\sigma_q^2/q_K(1-q_K)$.

Case 1 is that of a population of hermaphrodites with various amounts (h) of self fertilization. Case 2 is that of permanent pairs with various amounts (h) of brother-sister mating. Case 3 is that of a separate random mating for each offspring. The results would be the same for $N_m=50$, $N_f=\infty$ as for $N_m=N_f=100$. Equal dispersion in the male and female lines is postulated in cases 1, 2 and 3. Case 4 is that of hermaphrodites with no dispersion in the female line, dispersion of double variance in the male line.

K	N_K	HERMAPHRODITES, N=200: CASE 1 EQUAL DISPERSION OF ♀ AND ♂ GAMETES $\sigma_f^2=\sigma_m^2=\sigma^2$				CASE 4 $\sigma_f^2=0$ $\sigma_m^2=2\sigma^2$	
		h=0	h=.005	h=.500	h=1	h=.005	h=.5025
1	200	0	0	0	0	0	0
10	2×10^3	.0071	.0071	.0094	.0141	.0071	.0128
10^2	2×10^4	.0129	.0130	.0172	.0256	.0130	.0277
10^3	2×10^5	.0187	.0187	.0248	.0368	.0187	.0428
10^4	2×10^6	.0244	.0245	.0323	.0478	.0245	.0577
10^5	2×10^7	.0302	.0302	.0398	.0587	.0302	.0725
10^6	2×10^8	.0359	.0360	.0473	.0695	.0360	.0872
10^7	2×10^9	.0416	.0417	.0548	.0801	.0417	.1018
10^8	2×10^{10}	.0474	.0475	.0622	.0906	.0475	.1162

K	N_K	SEPARATE SEXES $N_m=100$; $N_f=100$: CASE 2 PERMANENT PAIRS $\sigma_f^2=\sigma_m^2=\sigma^2$				CASE 3 RANDOM MATING $\sigma_f^2=\sigma_m^2=\sigma^2$
		h=0	h=.01	h=.50	h=1	
1	200	0	0	0	0	0
10	2×10^3	.0071	.0071	.0113	.0280	.0073
10^2	2×10^4	.0130	.0131	.0207	.0505	.0135
10^3	2×10^5	.0189	.0191	.0300	.0722	.0198
10^4	2×10^6	.0249	.0251	.0393	.0933	.0264
10^5	2×10^7	.0309	.0312	.0486	.1140	.0331
10^6	2×10^8	.0370	.0373	.0580	.1341	.0401
10^7	2×10^9	.0432	.0435	.0674	.1538	.0473
10^8	2×10^{10}	.0494	.0497	.0768	.1731	.0547

cent of the maximum in areas including one million of each sex and 5.47 per cent in areas including ten billion of each). If, in a population of monoecious plants, there is no appreciable dispersion of seed but dispersion of pollen of

TABLE 2

The standard deviations $\sigma_{q(i)}$ of gene frequencies of populations of size N_i relative to that of size 2×10^{10} where the effective size of neighborhood is 200. Cases as in table 1

$$\sigma_{q(i)}/\sigma_{q(K)} = \sqrt{(E_K - E_i)/(1 - E_i)}$$

K	N_i	HERMAPHRODITES N = 200					
		CASE 1 EQUAL DISPERSION OF ♀ AND ♂ GAMETES $\sigma_f^2=\sigma_m^2=\sigma^2$				CASE 4 $\sigma_f^2=0$ $\sigma_m^2=2\sigma^2$	
		h = 0	h = .005	h = .500	h = 1	h = .005	h = .5025
1	200	.2176	.2179	.2494	.3011	.2179	.3409
10	2×10^3	.2014	.2017	.2309	.2787	.2017	.3236
10^2	2×10^4	.1868	.1870	.2141	.2584	.1870	.3017
10^3	2×10^5	.1710	.1712	.1959	.2365	.1712	.2770
10^4	2×10^6	.1533	.1535	.1758	.2121	.1535	.2491
10^5	2×10^7	.1332	.1333	.1526	.1843	.1333	.2170
10^6	2×10^8	.1091	.1092	.1250	.1509	.1092	.1782
10^7	2×10^9	.0773	.0774	.0886	.1070	.0774	.1267
10^8	2×10^{10}	0	0	0	0	0	0

K	N_i	SEPARATE SEXES N_f = 100, N_m = 100				
		CASE 2 PERMANENT PAIRS $\sigma_f^2=\sigma_m^2=\sigma^2$				CASE 3 RANDOM MATING $\sigma_f^2=\sigma_m^2=\sigma^2$
		h = 0	h = .01	h = .50	h = 1	
1	200	.2222	.2230	.2772	.4160	.2338
10	2×10^3	.2064	.2071	.2574	.3864	.2185
10^2	2×10^4	.1919	.1926	.2394	.3593	.2043
10^3	2×10^5	.1762	.1768	.2197	.3298	.1885
10^4	2×10^6	.1584	.1590	.1976	.2966	.1705
10^5	2×10^7	.1380	.1385	.1721	.2584	.1493
10^6	2×10^8	.1133	.1137	.1413	.2171	.1233
10^7	2×10^9	.0806	.0809	.1005	.1509	.0882
10^8	2×10^{10}	0	0	0	0	0

twice the variance postulated in the preceding cases (in populations of the same density), the amount of differentiation is practically the same as in the case of hermaphrodites with equal dispersion along both parental lines. This rule applies in random breeding populations, irrespective of density, and applies under linear as well as area continuity. It may be surmised that in such populations the effects with unequal dispersion are substantially the same as if there were equal dispersion to the extent of the average variance along the two parental lines. Returning to the case of neighborhoods of size 200 in a population continuous over an area, it may be added that with dispersion of pollen equal to that in case 1 and no dispersion of seed, σ_q^2 is 5.24 per cent instead of

TABLE 3

The correlation (E) between gametes of adjacent individuals, belonging to neighborhoods of size N=20, relative to populations of size N_K, which in turn are portions of a population distributed with uniform density over an indefinitely large area. The cases are as in table 1. In case 3 the results would be the same for neighborhoods containing five males and an indefinite number of females as with ten of both.

K	N_K	HERMAPHRODITES N=20					
		CASE 1 EQUAL DISPERSION OF ♀ AND ♂ GAMETES $\sigma_f^2=\sigma_m^2=\sigma^2$				CASE 4 $\sigma_f^2=0$ $\sigma_m^2=2\sigma^2$	
		h=0	h=.05	h=.50	h=1	h=.05	h=.525
1	20	0	0	0	0	0	0
10	2×10^2	.0699	.0716	.0918	.1336	.0716	.1260
10^2	2×10^3	.1269	.1298	.1634	.2297	.1298	.2593
10^3	2×10^4	.1821	.1860	.2304	.3136	.1860	.3799
10^4	2×10^5	.2362	.2409	.2937	.3883	.2409	.4856
10^5	2×10^6	.2890	.2943	.3534	.4548	.2943	.5764
10^6	2×10^7	.3401	.3460	.4094	.5141	.3460	6535
10^7	2×10^8	.3894	.3956	.4617	.5669	.3956	.7181
10^8	2×10^9	.4366	.4430	.5104	.6140	.4430	.7716

K	N_K	SEPARATE SEXES $N_f=10$, $N_m=10$				
		CASE 2 PERMANENT PAIRS $\sigma_f^2=\sigma_m^2=\sigma^2$				CASE 3 RANDOM MATING $\sigma_f^2=\sigma_m^2=\sigma^2$
		h=0	h=.10	h=.50	n=1	
1	20	0	0	0	0	0
10	2×10^2	.0723	.0779	.1124	.2525	.0932
10^2	2×10^3	.1378	.1476	.2062	.4092	.1936
10^3	2×10^4	.2071	.2206	.2980	.5310	.3124
10^4	2×10^5	.2799	.2963	.3871	.6275	.4423
10^5	2×10^6	.3545	.3730	.4715	.7041	.5708
10^6	2×10^7	.4289	.4486	.5496	.7649	.6856
10^7	2×10^8	.5013	.5213	.6203	.8133	.7792
10^8	2×10^9	.5699	.5894	.6829	.8517	.8500

2.45 per cent of the maximum in areas including two million individuals and is 9.82 per cent instead of 4.75 per cent in areas including twenty billion.

A tendency toward self fertilization or toward brother-sister mating somewhat increases the amount of differentiation of neighborhoods. But even 50 per cent self fertilization in the case of hermaphrodites and 50 per cent brother-sister mating in the case of permanent pairs raise σ_q^2 as a percentage of the maximum within a population of 20 billion only from 4.75 per cent to 6.22 per cent in the former and from 4.97 per cent to 7.68 per cent in the latter. The ef-

TABLE 4

The standard deviations of gene frequencies of populations of size N_i relative to that of a population of size 2×10^9, where the effective size of neighborhood is 20. Cases as in table 1.

K	N_i	HERMAPHRODITES N=20, CASE 1, EQUAL DISPERSION OF ♂ AND ♀ GAMETES $\sigma_f^2=\sigma_m^2=\sigma^2$				CASE 4 $\sigma_f^2=0$ $\sigma_m^2=2\sigma^2$	
		h=0	h=.05	h=.50	h=1	h=.05	h=.525
1	20	.6585	.6657	.7144	.7836	.6657	.8784
10	2×10^2	.6279	.6325	.6789	.7447	.6325	.8595
10^2	2×10^3	.5956	.6000	.6440	.7064	.6000	.8317
10^3	2×10^4	.5579	.5620	.6031	.6616	.5620	.7948
10^4	2×10^5	.5123	.5160	.5538	.6075	.5160	.7457
10^5	2×10^6	.4557	.4590	.4927	.5404	.4590	.6789
10^6	2×10^7	.3824	.3852	.4135	.4535	.3852	.5839
10^7	2×10^8	.2781	.2801	.3007	.3298	.2801	.4359
10^8	2×10^9	0	0	0	0	0	0

K	N_i	SEPARATE SEXES $N_f=10$ $N_m=10$, CASE 3, PERMANENT PAIRS $\sigma_f^2=\sigma_m^2=\sigma^2$				CASE 4 RANDOM MATING $\sigma_f^2=\sigma_m^2=\sigma^2$
		h=0	h=.10	h=.50	h=1	
1	20	.7549	.7677	.8264	.9229	.9220
10	2×10^2	.7324	.7448	.8017	.8953	.9136
10^2	2×10^3	.7079	.7200	.7749	.8654	.9022
10^3	2×10^4	.6764	.6879	.7404	.8269	.8842
10^4	2×10^5	.6346	.6454	.6947	.7758	.8550
10^5	2×10^6	.5777	.5875	.6324	.7063	.8065
10^6	2×10^7	.4969	.5054	.5440	.6075	.7231
10^7	2×10^8	.3710	.3773	.4061	.4535	.5662
10^8	2×10^9	0	0	0	0	0

fect of self fertilization is naturally much greater in the case of no dispersion of seed but dispersion of pollen of two fold variance. With about 50 per sent self fertilization, σ_q^2 within populations of 20 billion is raised from 4.75 per cent to 11.62 per cent of the maximum.

The differentiation of neighborhoods with 200 individuals that reproduce by exclusive self fertilization is measured by a variance 9.06 per cent of maximum within a population of 20 billion. The case of neighborhoods consisting of 100 brother-sister pairs is the same as for ones of 100 individuals that reproduce by self fertilization. The variance of q reaches 17.31 per cent of maximum within a population of 20 billion. All individuals of course become homozygous in these cases, but neighborhoods become homallelic only to a slight extent, assuming as throughout, no differential selection. Exclusive vegetative multiplication differs in not leading to homozygosis. The differentiation of neighborhoods is the same as in the above cases, taking q as the frequency of one of two alternative genotypes.

Even the largest of the values of σ_q^2 in table 1 may appear rather small, suggesting that, if the effective size of neighborhoods is as large as 200, there is no important differentiation due to the cumulative effects of accidents of sampling even within enormous areas. However, the square roots of these figures give perhaps a fairer idea of amount of differentiation. Thus values of σ_q of 22 per cent to 24 per cent of the possible maximum, occurring independently at all heterallelic loci, are large enough to give the basis for great diversity among the gene systems of neighborhoods and thus a basis for intergroup selection. But groups of only 200 individuals are rather small, and it is desirable to consider the amount of differentiation among larger groups. Table 2 shows the standard deviation $\sigma_{q(i)}$ for populations of various effective sizes N_i within a comprehensive population of 20 billion, in relation to the maximum standard deviation. For populations consisting of 10 per cent of the total, mean gene frequency varies about one-third as much as it does for neighborhoods. It varies about half as much among groups that include 1 per cent of the total and about two-thirds as much among ones that include 0.1 per cent of the total (that is, populations of 20 million within one of 20 billion). There is therefore significant, if not very great, chance differentiation of large groups, if the effective size of the neighborhood is as large as 200. As noted in the previous paper, there is virtual equivalence to panmixia with regard to chance differentiation where neighborhoods are larger than 1000 (WRIGHT 1943a). All of this discussion, it should be noted, applies only to area continuity. Under linear continuity there is enormously more differentiation.

Tables 3 and 4 present the same cases for $N=20$ that are presented in tables 1 and 2 for $N=200$. They show that there is very much greater differentitaion where the neighborhood is small. There is also much more difference in the results of diverse systems of mating. Thus for neighborhoods of 20 random breeding hermaphrodites the variance of q within areas including the equivalent of 100 million neighborhoods is 44.3 per cent of its maximum, for neighborhoods of ten permanent pairs within the same total it is 58.9 per cent of maximum, and for neighborhoods of ten males and ten females, mating at random to produce each offspring, it is 85.0 per cent of maximum, assuming equal dispersion along both lines in all these cases. Table 4 gives the standard deviation of q among populations of various sizes N_i within a constant total of two billion individuals. We note that populations including 1 per cent of this total are strongly differentiated where neighborhoods are as small as 20. With neighborhoods of 20 hermaphrodites with equal dispersion along both lines, the standard deviation of mean gene frequency of populations of 20 million within 2 billion is 38.5 per cent of maximum, for neighborhoods of ten permanent pairs the corresponding figure is 50.5 per cent, and for neighborhoods in which each offspring is the product of a separate random mating among ten males and ten females, the figure is 72.3 per cent.

It was noted in the preceding paper (WRIGHT 1943a) that either reversible mutation or occasional long range dispersion (rate m_1) puts a sharp upper limit on the amount of differentiation among neighborhoods and prevents differentiation of groups larger than about $3/m_1$ neighborhoods. A correction should be made to a statement on p. 131, referred to again in the summary. "Under exclusive uniparental reproduction, the chance that an individual is derived from the parental population without mutation is $(1-m_1)$ instead of $(1-m_1)^2$. Each term in the series is accordingly to be multiplied by $(1-m_1)^x$ (where X is the rank of the term in question)." This is correct if q refers to the frequency of one of only two alternative genotypes, which are multiplying vegetatively. In reference to gene frequencies of a diploid organism the factor to be applied to terms in Σt is $(1-m_1)^{2x}$, irrespective of system of mating.

It should be added that the occurrence of mutations in an indefinitely extended series of multiple alleles, far from limiting the amount of local differentiation, would greatly increase it. The slow process of diffusion of new alleles from small neighborhoods may not keep up with further mutation and thus each allele may have only a local distribution.

Uniform selection pressure limits differentiation, while local differences in the conditions of selection may enormously increase it, provided that dispersion is sufficiently slow to prevent swamping by panmixia. But even with no differences in the external conditions in different localities, mere chance differentiation at all loci of the sort discussed here may be expected to bring about differences in the direction of selection pressure on individual loci in regions in which different gene systems have been arrived at. The differences in selective trends may be expected to be cumulative, with intergroup selection by means of differential rates of dispersal as a further consequence.

The variability of frequencies of two alleles, due to the cumulative effects of accidents of sampling is thus merely the foundation for much more significant evolutionary processes.

SUMMARY

The properties of large, uniformly distributed populations depend on the system of mating and the effective population number (N) of the random breeding "neighborhoods." It is shown that with density d, of breeding individuals and with standard deviation σ, of the coordinates of location of parents at some phase of their life cycles relative to the corresponding phase for offspring, $N=2\sqrt{\pi}\sigma d$ if the population has an essentially one dimensional distribution (for example, shore line, river), and $N=4\pi\sigma^2 d$ if there is continuity over an extensive area. Random mating and equal parental dispersions are assumed here. If there is no dispersion along one parental line (for example, no dispersion of seed) but dispersion measured by variance $2\sigma^2$ along the other line (for example, of pollen), the formulae are the same.

The differentiation of mean gene frequencies among neighborhoods or larger areas, within still larger areas, is investigated in four cases. (1) Populations of hermaphrodites, derived from equally dispersed gametes derived at random from the neighborhood except for a specified tendency toward self fertilization. (2) Populations consisting of permanent pairs, derived by random mating from the neighborhood, except for a specified tendency toward brother-sister mating, with equal dispersion of the sexes. (3) Populations in which each individual is produced by a separate random mating from the neighborhood. The densities of the populations of the two sexes may differ in this case, but dispersion is assumed to be the same. (4) Populations of hermaphrodites, derived from union of ♀ gametes that are not dispersed to any appreciable extent with ♂ gametes derived at random from the neighborhood, except for a specified tendency toward self fertilization.

If the effective breeding population of neighborhoods is 200, there is a moderate amount of differentiation among large subgroups as well as among neighborhoods, within still larger groups inhabiting an indefinitely large area. There are only slight differences among the four cases. Even 100 per cent self fertilization in case 1 or 100 per cent brother-sister mating in case 2 increases differentiation only to a rather slight extent.

If the effective breeding population of neighborhoods is only 20, there is great differentiation among large subgroups as well as among neighborhoods. The amount of differentiation is considerably greater in case (2) than in case (1) or (4) and considerably greater in case (3) than in case (2).

It is noted that this differentiation, due to cumulative effects of accidents of sampling, may be expected in actual cases to be complicated by the effects of occasional long range dispersal, mutation, and selection, but that in combination with these it gives the foundation for much more significant evolutionary processes than these factors can provide separately.

LITERATURE CITED

DOBZHANSKY, TH., and S. WRIGHT, 1943 Genetics of natural populations. X. Dispersion rates in *Drosophila pseudoobscura*. Genetics **28**: 304–340.

WRIGHT, S., 1921 Systems of mating. Genetics **6**: 111–178.

1943a Isolation by distance. Genetics **28**: 114–138.

1943b An analysis of local variability of flower color in *Linanthus Parryae*. Genetics **28**: 139–156.

32
Statistical Genetics and Evolution

Bulletin of the American Mathematical Society 48, no. 4 (1942):223–46

33
The Differential Equation of the Distribution of Gene Frequencies

Proceedings of the National Academy of Science 31, no. 12 (1945):383–89

Introduction

Wright wrote two papers during the 1940s elaborating his statistical distribution of genes. The first, delivered as the Josiah Willard Gibbs Lecture of the American Mathematical Society in 1941, was basically a summary of Wright's publications during the 1930s. Wright's approach up to and including this paper was primarily intended to elaborate an integral equation of gene frequencies as contrasted with Fisher's approach of primarily working with differential equations of gene frequencies. Both Wright and Fisher had concentrated upon the analysis of the distribution of gene frequencies when in statistical equilibrium.

But Wright was keenly interested in treating the distribution of gene frequencies as a system in dynamic change. Shortly after the end of the war, the Russian mathematician A. Kolmogorov sent Wright his initial attempt to construct a differential equation for modeling a population having its gene frequencies in a state of continuous flux, that is, treating the continuous change of gene frequencies as a stochastic process (Kolmogorov 1935, cited by Wright in the second paper above). Wright immediately became interested in applying the Kolmogorov approach in his own way to his distribution of gene frequencies. This Wright did in the second paper of the series. At the time Wright wrote this paper, he was unaware that physicists had earlier applied the mathematical equivalent of the Kolmogorov approach to continuous diffusion effects, calling it the Fokker-Planck partial differential equation. Although Wright was not primarily a mathematician, this paper shows again how quickly he could adapt mathematical approaches for his own purposes. Of particular interest in this paper is the careful way Wright related the new approach to his earlier results and to those of Fisher as well.

For a summary of Wright's later extensions of his work utilizing the Fokker-Planck equation and his treatment of the later important extensions of Kimura, see Wright (*E&GP*, vol. 2, chap. 13), and Crow and Kimura (1970, chap. 8).

STATISTICAL GENETICS AND EVOLUTION

SEWALL WRIGHT

Introduction. When Darwin developed the theory of evolution by natural selection, practically nothing was known of hereditary differences beyond their existence. Since 1900, a body of knowledge on the mechanism of heredity and on mutation has been built up by experiment that challenges any field in the biological sciences in the extent and precision of its results. The implications for evolution are not, however, immediately obvious. It is necessary to work out the statistical consequences.

Studies in the field of statistical genetics began shortly after the rediscovery of Mendelian heredity in 1900. Those of J. B. S. Haldane [**7**] and R. A. Fisher [**4**] have been especially important with respect to the application to evolution. My own approach to the subject came through experimental studies conducted in the U. S. Bureau of Animal Industry on the effects of inbreeding, crossbreeding and selection on populations of guinea pigs [**21, 22, 23, 37**] and through the attempt to formulate principles applicable to livestock breeding [**19, 20, 24, 25, 13, 34**]. On moving into the more academic atmosphere of the University of Chicago, I have become more directly concerned with the problem of evolution.

I should note that the deductive approach, to which I shall confine myself here, involves many questions that can only be settled by observation and experimental work on natural populations and that a remarkable resurgence of interest in such work is in progress [**2, 9**].

Postulates. It will be desirable to begin with a brief review of the more important factors of which account must be taken.

The basic fact of modern genetics is that heredity can be analyzed into separable units, "genes," whose most essential property is that of duplicating themselves with extraordinary precision, irrespective of the characteristics of the organism in whose cells they are carried. We shall restrict consideration to changes in the system of genes and aggregates of genes (chromosomes). There are relatively rare and obscure hereditary changes which must be attributed to other cell components but our knowledge of these does not warrant the elaboration of a statistical theory.

Fortunately the same theory applies to a large extent to gene muta-

The sixteenth Josiah Willard Gibbs Lecture, delivered at Chicago, Illinois, September 3, 1941, under the auspices of the American Mathematical Society; received by the editors October 20, 1941.

tions and to most classes of grosser chromosomal changes (duplications, deficiencies, inversions, translocations, and so on). It will be assumed here that a given kind of mutation occurs at a constant rate per generation. Observed rates in organisms as remote as corn plants, vinegar flies and man are of the order of 10^{-5} or less per generation. Reverse mutation may occur at measurable rates.

It is simplest to deal with mere pairs of alternative conditions (alleles) but the theory remains seriously inadequate unless capable of extension to multiple alleles.

In general I shall assume that the reproductive cells are haploid (that is, contain just one representative from each set of alleles) and that their union results in diploid individuals (with two such representatives in all cells, until reduction occurs in the formation of the germ cells). This is the usual case but there are species in which other situations prevail (tetraploids, hexaploids, aneuploids, and so on). The group of sex linked genes constitutes an important special case in many otherwise completely diploid organisms (including man). I shall not go far into the extension to these cases.

It is simplest to assume that the members of different series of alleles are distributed at random in the reduction division by which the reproductive cells receive a half sample of the genes of the individuals producing them (that, for example, individual *AaBb* produces germ cells *AB*, *Ab*, *aB* and *ab* in equal numbers). The phenomenon of partial linkage, exhibited by genes carried in the same chromosome should, however, be taken into account. These are the principal postulates as far as the mechanism of heredity is concerned though others are required in special cases.

The relations of genes to observed characteristics are important. In general, any measurable character is affected by genes at many loci and a single gene often has multiple apparently unrelated effects. The effects of genes in combinations are often roughly cumulative but marked exceptions are also very common. Account must be taken of noncumulative effects within series of alleles (dominance) and between series (gene interaction).

The breeding structure of the population is important. The situation in nature is so complex that models must be chosen that are compromises between mathematical simplicity and biological adequacy [**35**]. I shall introduce only the simplest models in the course of the present discussion.

Natural selection is an exceedingly complex affair. Selection may occur at various biological levels—between members of the same brood, between individuals of the same local population, between such populations (as through differential increase and migration) and finally between different species, a subject that carries us outside the field of genetics and which has been discussed mathematically by Lotka [**12**], Volterra [**17**] and Nicholson and Bailie [**14**]. Selection among individuals may relate to the mating activities of one or both sexes, to differences in rate of attainment of maturity, to differential fecundity and to differential mortality at all ages. Selection may act steadily or may vary both in intensity and direction in different regions and at different times. Again I can only deal here with the simplest models.

Gene frequency. In such a complex situation, verbal discussion tends toward a championing of one or another factor. We need a means of considering all factors at once in a quantitative fashion. For this we need a common measure for such diverse factors as mutation, crossbreeding, natural selection and isolation. At first sight these seem to be incommensurables but if we fix attention on their effects on populations, rather than on their own natures, the situation is simplified. Such a measure may be found in the effects on *gene frequency* in each series of alleles.

Because of the complete symmetry of the Mendelian mechanism, gene frequency has no tendency to change in an indefinitely large closed population not subject to mutation or selection. Each homozygote (for example, A_1A_1, A_2A_2 or A_3A_3) produces only one kind of germ cell. Each heterozygote (for example, A_1A_2, A_1A_3, A_2A_3) produces two kinds in *equal* numbers. In a population in which the array of gene frequencies is $(q_1A_1+q_2A_2+\cdots+q_mA_m)$ (letting the q's represent the frequencies, and the A's the genes) the frequencies of genotypes come to equilibrium according to the terms in the expansion of $(q_1A_1+q_2A_2+\cdots+q_mA_m)^2$ in the first generation of random mating after attainment of equality of gene frequencies in the sexes [**8**]. Under sex linkage [**10, 15**] and in polyploids [**6**] equilibrium is not reached at once but is rapidly approached. Inbreeding and assortative mating change the relative frequencies of homozygotes and heterozygotes but not the gene frequencies.

One immediate consequence of this persistence of gene frequencies is that variability tends to persist. But the slightest continuing unbalanced pressure on gene frequency tends to cause cumulative change. It is obvious that recurrent mutation, immigration, selection, and the accidents of sampling in an isolated population of small size are all factors that can bring about such change.

The frequencies of combinations of different series of alleles (for example, $A, a; B, b$) reach equilibrium in a random breeding population in which gene frequencies are constant only when the genes are combined at random (terms of $[(1-q_A)a+q_AA]^2[(1-q_B)b+q_BB]^2$). Equilibrium is not reached immediately, however. The departure from equilibrium is halved in each generation of random mating in the case of two pairs of alleles in different chromosomes. In general, the departure is reduced by the proportion $\bar{c}$, where $\bar{c}$ is the mean chance of recombination [**18**, **11**, **16**].

Systematic changes of gene frequencies. The rate at which gene frequency changes under recurrent mutation is obvious [**27**]. Let q be the frequency of the gene and u the rate at which it mutates to its alleles as a group and Δq the rate of change of q per generation $\Delta q = -uq$.

If reverse mutation occurs at the rate v per generation, the net rate of change of q is

$$\Delta q = v(1 - q) - uq. \tag{1}$$

In the case of multiple alleles, v is the weighted average for the various alleles of the gene in question and is thus a function of their relative frequencies. It is, however, independent of q.

The effect of crossbreeding is similar if we adopt the simplest model [**27**]. If a population with gene frequency q exchanges the proportion m each generation with a random sample of immigrants from the whole species (gene frequency q_t) the rate of change in gene frequency is

$$\Delta q = -m(q - q_t). \tag{2}$$

In actual cases the immigrants are not likely to be a random sample from the whole species but to come largely from neighboring populations. Effective m is thus, in general, smaller than the apparent amount of immigration and is not necessarily the same for all loci. There may also be selective migration. The simplest model must suffice here. It permits identification of the theories of mutation and immigration by substituting mq_t for v, and $m(1-q_t)$ for u.

The effects of selection have been considered extensively by Haldane [**7**] in terms of the frequency ratio $(q/1-q)$ and by Fisher [**4**]. As there can be no selection pressure without at least two alternatives, any expression for it, applicable to all values of q, must include the factor $q(1-q)$, excluding certain limiting cases. Thus the form $aq(1-q)$ has been used by Fisher as the basis for general discussion. For the present purpose somewhat less general forms are more useful. Consider first the case of a random breeding population of diploid individuals in which the combinations of paired alleles may be assigned constant relative selective values [**30**].

Genotype	*Frequency (f)*	*Value (W)*
AA	q^2	W_{AA}
AA'	$2q(1-q)$	$W_{AA'}$
$A'A'$	$(1-q)^2$	$W_{A'A'}$

For the frequency of A after a generation

$$q_1 = [W_{AA}q^2 + W_{AA'}q(1-q)]/\overline{W}$$

where

$$\overline{W} = \sum fW = W_{AA}q^2 + 2W_{AA'}q(1-q) + W_{A'A'}(1-q)^2,$$

$$(3)\ \Delta q = q_1 - q = q(1-q)[W_{AA}q + W_{AA'}(1-2q) - W_{A'A'}(1-q)]/\overline{W},$$

$$\Delta q = \frac{q(1-q)}{2\overline{W}}\,\frac{d\overline{W}}{dq}\,.$$

Selection, however, really applies to the organism as a whole not to single series of alleles. If the population is heterallelic in n pairs of pertinent alleles, the number of possible combinations is 3^n. Each of these has a certain frequency and a certain relative selective value, the latter of which we here assume to be constant. If the three phases in the A series of alleles are combined at random with the combinations of the other series, the average selective values of AA, AA' and $A'A'$ are independent of q_A although functions of the other q's. Thus

$$(4) \qquad \Delta q_A = \frac{q_A(1-q_A)}{2\overline{W}}\,\frac{\partial\overline{W}}{\partial q_A}\,.$$

We have assumed only pairs of alleles, but as any group of alleles may be treated formally as one, this formula may be applied to multiple allelic series. The selective values W_{AA}, $W_{AA'}$ and $W_{A'A'}$ are then functions of the relative frequencies within the group of alleles of the gene under consideration, but not of q_A.

In previous general discussions (for example [**35**, **36**]) I have restricted myself to this convenient model of selection pressure. As this has given rise to misapprehension [**5**], it should be emphasized that it applies only under the conditions implied in its derivation.

If there are selective differences between the sexes, as is very likely to be the case, there are departures from random combination within series of alleles. These are, however, unimportant for most purposes unless there is rather strong selection.

Selection itself tends to bring about departures from random combination among different series of alleles. Again the effects are unimportant in most cases, especially if all relative selective differences are slight.

The formula must be written in a more generalized form to include polyploidy [**32**] and sex linkage. For small selective differences in a $2k$-ploid

$$\Delta q_A = \frac{q_A(1-q_A)}{2k\overline{W}}\frac{\partial \overline{W}}{\partial q_A}. \tag{5}$$

This applies (approximately) under sex linkage if $k=3/4$ and $\overline{W}=(\overline{W}_m\overline{W}_f)^{1/2}$ where $\overline{W}_m$ and $\overline{W}_f$ are the mean selective values in males and females, respectively.

There may be departures from random mating because of a constant tendency toward mating of relatives, giving the following genotypic frequencies within a series of alleles [**23**, **3**].

Genotype	*Frequency*
AA	$(1-F)q^2+Fq$
AA'	$2(1-F)q(1-q)$
$A'A'$	$(1-F)(1-q)^2+F(1-q)$

Random combination between series of alleles is not disturbed appreciably if the selective differences are small or if the inbreeding coefficient F is small, giving the following formula in which $\overline{W}_R$ and $\overline{W}_I$ are mean selective values of the random bred and inbred components of the frequencies relative to the A series [**3**]

$$\begin{aligned}\Delta q_A &= \frac{q_A(1-q_A)}{\overline{W}}\left[\frac{(1-F)}{2}\frac{\partial \overline{W}_R}{\partial q_A} + F\frac{\partial \overline{W}_I}{\partial q_A}\right]\\ &= \frac{q_A(1-q_A)}{2\overline{W}}\left[\frac{\partial \overline{W}}{\partial q_A} + F\frac{\partial \overline{W}_I}{\partial q_A}\right].\end{aligned} \tag{6}$$

Inbreeding that leads to subdivision into partially isolated groups is best dealt with by a different mathematical model.

Under assortative mating based on similarity in characteristics there are very great departures from random combination of different series of alleles [**20**]. Again we may best consider such a mating system as one leading to subdivision of the population into partially isolated groups.

Returning to consideration of random breeding populations, it may easily be seen that if the W's are functions of q_A

$$\Delta q_A = \frac{q_A(1 - q_A)}{2\overline{W}}\left[\frac{\partial \overline{W}}{\partial q_A} - \sum\left(f\frac{\partial W}{\partial q_A}\right)\right]. \tag{7}$$

Intra-brood selection is an example of a case in which $\partial\overline{W}/\partial q_A = 0$ but gene frequency nevertheless changes. As in our model case, the process does not necessarily lead to fixation of the most favorable of the genotypes possible from the genes present in the population.

Finally, if the genes under consideration affect the system of mating itself, as is the case of the self-sterility alleles of many plants [**33**] or of genes that determine self fertilization [**5**], the changes in gene frequency can only be found from the composition of the population in successive generations.

We may note here that while, in principle, selection must be considered to apply to the organism as a whole, one may analyze the organism into character complexes which evolve largely independently through changes in largely independent systems of genes, the components of which are distributed at random among the chromosomes. The case which we have chosen as a model of selection pressure (4) and its generalization (5) should apply sufficiently well to most reaction systems in freely interbreeding populations and gives an insight into certain aspects of the effects of selection which cannot be obtained as easily from the more complex special cases.

If a character complex is affected by n loci and m_i alleles at a particular locus, it requires

$$\sum_{i=1}^{n} (m_i - 1)$$

dimensions to represent the system of gene frequencies and

$$\prod_{i=1}^{n} [m_i(m_i + 1)/2]$$

kinds of genotypes are possible. Assuming that the relative selective values of these genotypes are independent of their frequencies, the mean selective values ($\overline{W}$) of possible random breeding populations form a surface relative to this multi-dimensional system of gene frequencies, the gradient of which determines the way in which the population tends to change under the influence of selection. The number of loci that may affect even the simplest characters are known in certain cases to be great and many, if not all such loci are probably represented at all times by multiple alleles. Thus the number $\prod_{i=1}^{n} m_i$ of homozygous types possible from genes actually present in a species and affecting a particular character complex may often run into astronomical figures. Under these conditions it is to be expected that in general the surface $\overline{W}$ for any character complex will have numerous peaks, corresponding not only to different combinations of genes that give the same character [**29**] but also to different harmonious combinations of elementary characters that permit the organism to overcome the same conditions in different ways.

Mutation, immigration and selection may all be occurring simultaneously. The net rate of change of gene frequency may be obtained by simply adding the contributions of these factors (1), (2), (4), if these are small. In our ideal case of a random breeding population of diploid individuals subject to reversible mutation, immigration and constant selective differences between genotypes [**27**, **30**]

$$\Delta q = v(1 - q) - uq - m(q - q_\iota) + \frac{q(1 - q)}{2\overline{W}} \frac{\partial \overline{W}}{\partial q}. \tag{8}$$

There is equilibrium, stable or unstable, if $\Delta q = 0$. With reversible mutation, there must be at least one gene frequency other than 0 or 1 that is in stable equilibrium. There may also be stable equilibrium as a result of opposing selection pressures alone.

Accidents of sampling. There is another possibility of change of gene frequency to be considered. In a population that is not indefinitely large, gene frequency may be expected to change from generation to generation merely from the accidents of sampling. The composition of a population of N diploid individuals depends on that of $2N$ gametes produced by the preceding generation. If these are a random sample from the array $[(1-q)A' + qA]$, the probability array for values of q in the next generation is $[(1-q)A' + qA]^{2N}$ with the standard deviation $(q(1-q)/2N)^{1/2}$. We will call a random deviation of q of this sort δq in contrast with the systematic deviation Δq produced by mutation, migration or selection

$$\sigma^2_{\delta q} = \frac{q(1 - q)}{2N}. \tag{9}$$

It might seem that these random deviations would be negligible in any reasonably large population but in the absence of any systematic pressure toward equilibrium, the squared standard deviation for later generations increases approximately with the number of generations until there is an approach to the limiting value $q(1-q)$ of complete fixation, $[(1-q)A'A' + qAA]$. The exact value for the nth generation is $q(1-q)[1-(1-1/2N)^n]$. The rate of fixation of heterallelic genes approaches $1/2N$ per generation.

The effective value of N should often be much smaller than its apparent value [**27**, **35**]. It obviously applies only to individuals that reach maturity. If there is cyclic variation in population size, N is more closely related to the minimum than to the maximum number. It is also reduced if there is excessive variability in the number of mature offspring from different parents.

In a $2k$-ploid population [**32**],

$$\sigma^2_{\delta q} = \frac{q(1-q)}{2kN} \text{ approximately.} \tag{10}$$

For sex linked genes it is approximately $q(1-q)[2/9N_f+1/9N_m]$ where N_f and N_m are the effective numbers of females and males, respectively, and thus is $2q(1-q)/3N$ if these are equal.

The distribution of gene frequencies in the case of equilibrium. The tendency toward a stable equilibrium in the value of q, found when there are opposing systematic pressures, and the tendency to drift away from this point, due to the sampling variance, should result in a probability distribution which one might expect to find realized by the values taken by the frequency of a particular gene over a long period of generations in the ideal case of a population in which all conditions remain constant. It would also be the distribution of values of q taken by this gene at a given time in an array of completely isolated populations, all of which are subject to the same conditions. Finally, all genes subject to systematic pressures of the same magnitude should exhibit such a distribution at one time in a single population. While these are ideal cases, not likely to be approached in actual cases, it is of primary importance in the genetics of populations to be able to reach conclusions on the nature of such distributions.

The distribution of gene frequencies in the case of equilibrium must satisfy the conditions of stability of the mean

$$(\overline{q + \Delta q + \delta q} = \bar{q})$$

and stability of the variance $(\sigma^2_{(q+\Delta q+\delta q)} = \sigma^2_q)$. The possible values of q must range from 0 to 1. It is convenient to use integration for summation in expanding these expressions. Let $\phi(q)$ be the ordinate of the required distribution. The formula of the distribution may be derived as follows [**30**, **31**]

$$\int_0^1 (q + \Delta q + \delta q)\phi(q)dq = \int_0^1 q\phi(q)dq,$$

$$\int_0^1 (q + \Delta q + \delta q - \bar{q})^2\phi(q)dq = \int_0^1 (q - \bar{q})^2\phi(q)dq.$$

Since the mean value of δq is 0 and since δq is not correlated with $(q+\Delta q)$, these reduce to the following, omitting a term involving $(\Delta q)^2$, negligible if Δq is small

$$\int_0^1 \Delta q\phi(q)dq = 0,$$

$$2\int_0^1 (q - \bar{q})\Delta q\phi(q)dq + \int_0^1 \sigma_{\delta q}^2\phi(q)dq = 0.$$

Let $\int \Delta q\phi(q)dq = \chi(q)$ and integrate the first term of the preceding equation by parts

$$\chi(1) - \chi(0) = 0,$$

$$\int_0^1 \chi(q)dq - [\bar{q}\chi(0) + (1 - \bar{q})\chi(1)] - (1/2)\int_0^1 \sigma_{\delta q}^2\phi(q)dq = 0.$$

It may be found by trial that both of these conditions are satisfied by the following equation if $\phi(0)$ and $\phi(1)$ are finite. Note that $\sigma_{\delta q}^2 = 0$ if $q = 0$ or if $q = 1$, since there can be no sampling variance unless there are alternatives

$$\chi(q) - \chi(1) = (1/2)\sigma_{\delta q}^2\phi(q),$$

$$d \log [\chi(q) - \chi(1)] = \frac{d\chi(q)}{\chi(q) - \chi(1)} = \frac{2\Delta q dq}{\sigma_{\delta q}^2},$$

$$[\chi(q) - \chi(1)] = \frac{C}{2} e^{2\int (\Delta q/\sigma_{\delta q}^2)\, dq},$$

$$\phi(q) = (C/\sigma_{\delta q}^2)e^{2\int (\Delta q/\sigma_{\delta q}^2)\, dq}, \tag{11}$$

where C is a constant such that $\int_0^1 \phi(q)dq = 1$.

The frequency of a particular value of q is approximately $f(q) = \phi(q)/2N$. The amount of exchange between the subterminal and terminal classes is approximately half the frequency of the former from consideration of the Poisson distributions of the classes that are close to fixation [27].

For the model case in which $\Delta q = v(1 - q) - uq - m(q - q_t) + (q(1-q)/2\bar{W})(\partial \bar{W}/\partial q)$ and $\sigma_{\delta q}^2 = q(1-q)/2N$ equation (11) reduces to the following

$$\phi(q) = C\overline{W}^{2N} q^{4N(mq_t+v)-1}(1-q)^{4N[m(1-q_t)+u]-1}, \tag{12}$$

$$\begin{aligned} f(0) &= f(1/2N)/4N[mq_t + v], \\ f(1) &= f(1-1/2N)/4N[m(1-q_t)+u]. \end{aligned} \tag{13}$$

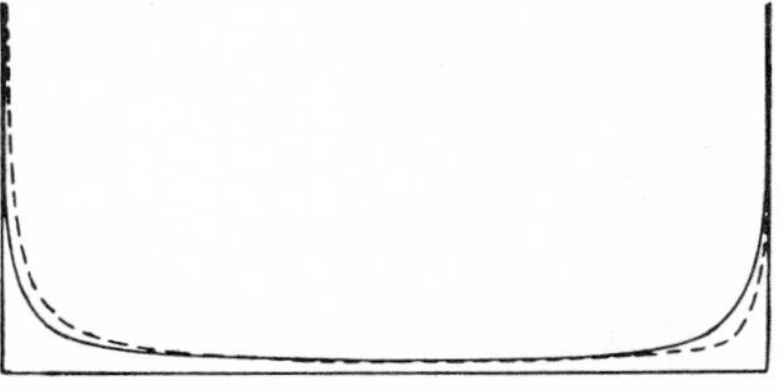

FIG. 1

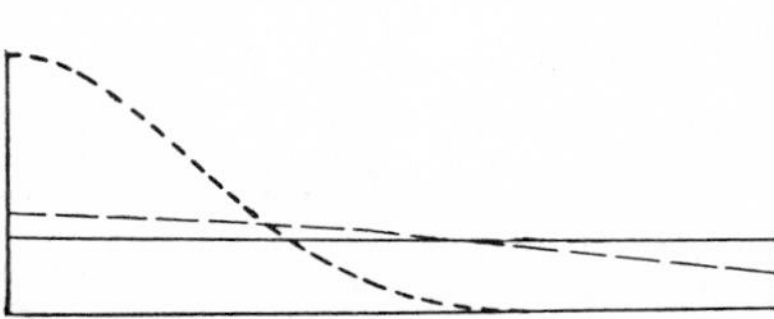

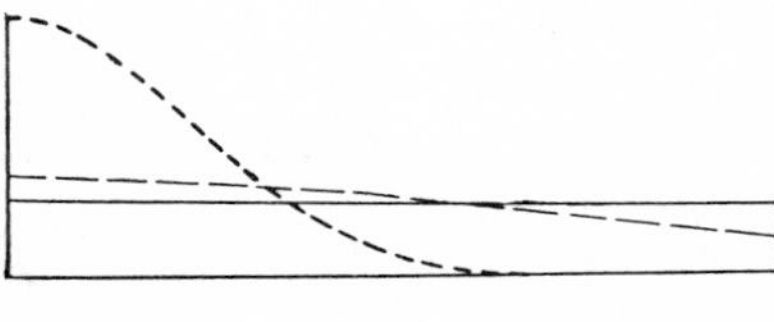

FIG. 2

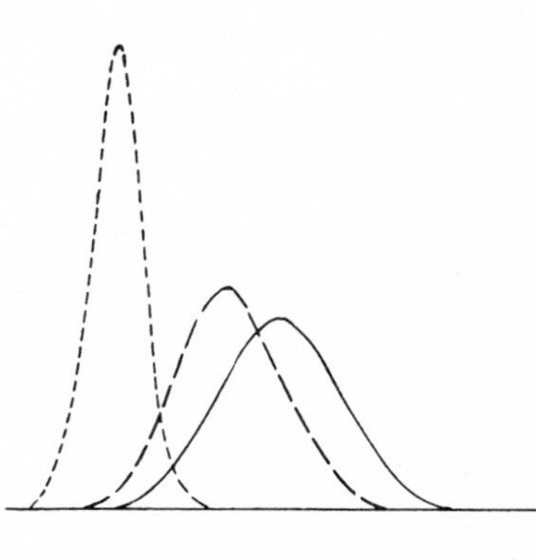

FIG. 3

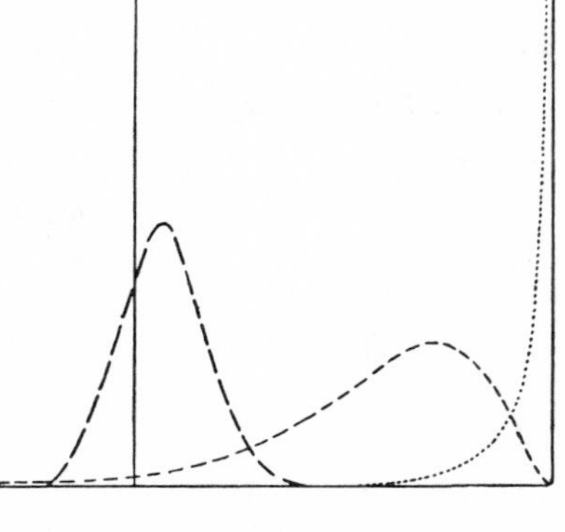

FIG. 4

Figures 1 to 3. Some of the forms taken by the distribution of frequencies of a completely recessive gene. Mutation rates are assumed equal in both directions ($u=v$). $N=1/40v$ in Figure 1, $10/40v$ in Figure 2 and $100/40v$ in Figure 3. In each case the solid line represents the least selection ($t=-v/5$), the long dashes represent selection 10 times as severe (omitted in Figure 1 since practically indistinguishable from the preceding) and the short dashes represent selection 100 times as severe $\phi(q)=Ce^{2Ntq^2}q^{4Nv-1}(1-q)^{4Nv-1}$.

Figure 4. Frequency distribution of a semidominant gene in subgroups of a large population in which the varying conditions of selection among subgroups has lead to a mean gene frequency, $q_t=.25$. The subgroups represented are assumed to be of the same size ($N=1000$) and subject to the same selection pressure ($W_{AA}=1$, $W_{AA'}=.9975$, $W_{A'A'}=.995$) but to different degrees of isolation (long dashes: $m=.01$, short dashes: $m=.001$, dots $m=.0001$), $\phi(q)=Ce^{10q}q^{1000m-1}(1-q)^{3000m-1}$.

This brings the effects of reversible mutation, crossbreeding, selection and size of population into a single formula. Figures 1 to 4 show the forms taken by this distribution in certain special cases. The U-shaped distributions in small populations (Figure 1) may be compared with the I-shaped ones in large populations (Figure 3). The figures bring out the relatively slight effects of selection in small populations.

The joint frequency distribution for multiple pairs of alleles may be written as follows for the model case that we have been considering [**30**]

$$(14)\quad \phi(q_1, q_2, \cdots, q_n) = C\overline{W}^{2N} \prod_{i=1}^{n} q_i^{4N(m_i q_{ti}+v_i)-1}(1 - q_i)^{4N[m_i(1-q_{ti})+u_i]-1}.$$

This applies to $2k$-ploids if $4Nk$ is substituted for $4N$ in the exponents of q_i and $(1-q_i)$. In the case of sex linkage and equal numbers of the sexes, $3N$ is to be substituted for $4N$ in these exponents. As the exponent of $\overline{W}$ is not affected in these cases, the formula applies to joint distributions including different degrees of ploidy (aneuploids) and both autosomal and sex-linked genes.

Figure 5 illustrates the frequencies along two diagonals of the joint distribution for two pairs of alleles which act cumulatively on the

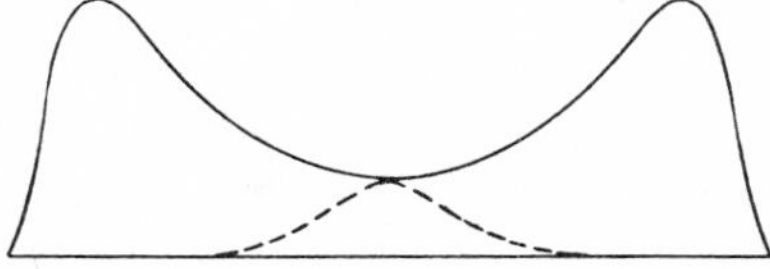

FIG. 5

The frequencies along the diagonals of the joint distribution for two series of alleles with equal and additive effects on a character on which adverse selection acts according to the square of the deviation from the mean ($W_{AAbb}=W_{AaBb}=W_{aaBB}=1$, $W_{AABb}=W_{AaBB}=W_{Aabb}=W_{aaBb}=1-s$, $W_{AABB}=W_{aabb}=1-4s$). The solid line shows the frequencies along the line connecting the two favorable types, $AAbb$ and $aaBB$. The dashes show the frequencies along the line connecting the extreme types $AABB$ and $aabb$. In the case shown, $u_a=v_a=u_b=v_b$, $N=1/2v_a$, $s=5v_a$.

same character of which the midgrade is optimum. There are two peak frequencies corresponding to approximate fixation of two different genotypes that give the midgrade of the character. In cases involving large numbers of genes there may be an indefinitely large number of peak frequencies.

The distribution of gene frequencies under irreversible mutation. It is also important to determine the form taken by the distribution of gene frequencies when fixation of one of the alleles is an irreversible process. The distribution curve should reach constancy of form, but all class frequencies (except that in which fixation is occurring) should fall off at a uniform rate K. The conditions may be expressed as follows [**31**]

$$\overline{q + \Delta q + \delta q} = (1 - K)\bar{q} + K,$$
$$\sigma^2_{(q+\Delta q+\delta q)} = (1 - K)\sigma^2_q + K(1 - \bar{q})^2,$$
$$K = (1/2)f(1 - 1/2N).$$

It can be found by trial that with mutation at rate v, no selection, and $\sigma^2_{\delta q} = q(1-q)/2N$ those conditions are satisfied by the following equation, with decay at rate $K = v$ per generation [**31**]

$$f(q) = 2vq^{4Nv-1}. \tag{15}$$

I have not been able to obtain a general solution comparable to equation (11) but formulae have been obtained for an important class of cases by another method [**27**, **30**, **31**]. Random breeding diploid populations with frequency array $[(1-q)A'+qA]$ are distributed in the following generation according to the expression $[(1-q-\Delta q)A'+(q+\Delta q)A]^{2N}$. Letting $p = 1-q$, the contribution to the frequency $f(q_c)$ of populations characterized by gene frequency q_f, is thus $[(2N)!/(2Np_c)!(2Nq_c)!]\,(p-\Delta q)^{2Np_c}(q+\Delta q)^{2Nq_c}f(q)$. The condition that this frequency be reconstructed after a generation except for a reduction by the amount K can be represented sufficiently accurately as follows:

$$(1 - K)\phi(q_c) = \frac{\Gamma(2N)}{p_c q_c \Gamma(2Np_c)\Gamma(2Nq_c)} \int_0^1 (p - \Delta q)^{2Np_c}(q + \Delta q)^{2Nq_c}\phi(q)dq.$$

If $K = 0$, $\Delta q = 0$, this equation is satisfied by $\phi(q) = Cq^{-1}+D(1-q)^{-1}$ for any values of C and D. For irreversible mutation at rate v, it yields equation (15) and for migration and reversible mutation, but no selection, the same result as obtained from equation (12) [**27**, **31**].

Selection pressure gives more difficulty. An important case that can be solved is that for very rare mutations ($4Nv$ negligibly small) subject to selection pressure of the fairly general form $\Delta q = (s+tq)q(1-q)$. Here K may be taken as 0

$$2Nf(q_c) = A \int_0^1 p^{2Np_c}q^{2Nq_c}[1 - q(s + tq)]^{2Np_c}[1 + p(s + tq)]^{2Nq_c}\phi(q)dq,$$

where

$$A = \frac{\Gamma(2N)}{p_c q_c \Gamma(2Np_c)\Gamma(2Nq_c)}.$$

The following approximate relations may be used

$$[(1-q(s+tq)]^{2Np_c} = e^{-2Np_cq(s+tq)}[1 - Np_cq^2(s+tq)^2],$$
$$[(1+p(s+tq)]^{2Nq_c} = e^{2Nq_cp(s+tq)}[1 - Nq_cp^2(s+tq)^2].$$

Let $\phi(q) = e^{2Nsq+Ntq^2}(C_0 + C_1q + C_2q^2 + \cdots)/q(1-q)$ and use the approximate relation

$$\frac{\Gamma(2N)}{\Gamma(2Np_c)\Gamma(2Nq_c)}\int_0^1 p^{2Np_c-1}q^{2Nq_c-1+x}dq$$
$$= q_c^x + [q_c^{x-1} - q_c^x][x(x-1)/4N].$$

It may be found that $C_m = [(4N^2s^2 + 2Nt)C_{m-2} + 8N^2stC_{m-3} + 4N^2t^2C_{m-4}]/m(m+1)$ ignoring terms in which the exponent of N is less than the sum of those of s and t. After further reduction

$$(16)\quad f(q) = [Ce^{4Nsq+2Ntq^2} + Dqe^{2Nsq+Ntq^2}\psi(2Nsq, 2Ntq^2)]/q(1-q)$$

where C and D are any constants and $\psi(a, b)$ is as follows

$$\psi(a,b) = 1 + \frac{a^2}{3!} + \frac{a^4}{5!} + \frac{a^6}{7!} + \frac{a^8}{9!} + \cdots$$
$$+ b(1+a)\left[\frac{1}{3!} + \frac{2a^2}{5!} + \frac{3a^4}{7!} + \frac{4a^6}{9!} + \cdots\right]$$
$$+ b^2\left[\frac{7}{5!} + \frac{2a}{5!} + \frac{69a^2}{7!} + \frac{6a^3}{7!} + \frac{282a^4}{9!} + \frac{12a^5}{9!} + \cdots\right]$$
$$+ b^3\left[\frac{27}{7!} + \frac{27a}{7!} + \frac{348a^2}{9!} + \frac{204a^3}{9!} + \cdots\right]$$
$$+ b^4\left[\frac{321}{9!} + \frac{132a}{9!} + \cdots\right]$$
$$+ b^5\left[\frac{2265}{11!} + \cdots\right] + \cdots.$$

The special cases of most interest are those of irreversible mutation in one direction or the other and of equilibrium. Consider a population in which $2Ns$ and $2Nt$ have given values but N is indefinitely large and hence s and t are indefinitely small. If mutation is occurring from the class $q=0$ at an exceedingly minute rate v with irreversible fixation in the class $q=1$, the frequency of the subterminal class, $q=1/2N$, must be approximately $4Nvf(0)$ while that of the other subterminal class ($q=1-1/2N$) must be so much smaller as to be negligible. The following are sufficiently accurate, letting $f(0)=1$

$$f(1/2N) = 2N\left[C + \frac{D}{2N}\right] = 4Nv,$$

$$f(1 - 1/2N) = 2N\left[Ce^{4Ns+2Nt} + De^{2Ns+Nt}\psi(2Ns,\ 2Nt)\right] = 0.$$

As $\psi(2Ns, 2Nt)$ is of the order of e^{2Ns+Nt}, $D/2N$ is negligible compared with C

$$C = 2v,$$

$$D = -\frac{2ve^{2Ns+Nt}}{\psi(2Ns,\ 2Nt)},$$

$$(17)\quad f(q) = \frac{2v}{q(1-q)}\left[e^{4Nsq+2Ntq^2} - qe^{2Ns(1+q)+Nt(1+q^2)}\,\frac{\psi(2Nsq,\ 2Ntq^2)}{\psi(2Ns,\ 2Nt)}\right].$$

In the subterminal regions selection is practically inoperative, $\Delta q = s/2N$, approximately for $q = 1/2N$; and $\Delta q = (s+t)/2N$ for $q = (1-1/2N)$. The formula of the curve in the neighborhood of $q = 1/2N$ is therefore approximately $2v/q$. Thus with given v but different values of N, $f(1/2N)(=4Nv)$ always falls very nearly on the same smooth curve. The relation between the selection pressure and the sampling variance (which measures the additive effect of sampling) is constant for given values of $2Ns$, $2Nt$ and q, $(\Delta q/\sigma^2_{\delta q}) = (2Ns+2Ntq)$. Thus the smoothed probability curve, $f(q)$, should be the same throughout with given $2Ns$ and $2Nt$ irrespective of the values of N and of s, t separately. Thus equation (17) is the general formula for the case of irreversible mutation from the class at $q = 0$.

The determination of $\psi(a, b)$ in specific cases is a rather formidable task since it involves two variables and converges slowly. For the case of semidominance however, $t = 0$ and $\psi(2Nsq, 0)$ reduces to $(e^{2Nsq} - e^{-2Nsq})/4Nsq$

$$(18)\qquad f(q) = \frac{2v}{q(1-q)}\,\frac{(1 - e^{-4Ns(1-q)})}{(1 - e^{-4Ns})}\,.$$

This agrees with a result obtained by Fisher [**4**] by a different method, involving a transformation of scale $(\theta = \cos^{-1}(1-2q))$ designed to make sampling variance uniform, and expression of the conditions in the form of a differential equation. The chance of fixation of a mutation in this case is given by the ratio of the subterminal classes $(f(1-1/2N)/f(1/2N))$ and is thus $2s/(1-e^{-4Ns})$ where s is the selection favoring the heterozygote. This is practically constant $(2s)$ in large populations. There is a small chance of fixation of even unfavorable mutations $(2s/(e^{4Ns}-1)$. Figure 6 illustrates the distribution curve for various intensities of selection (but with the selection favoring heterozygotes represented by $(1/2)s$ instead of s).

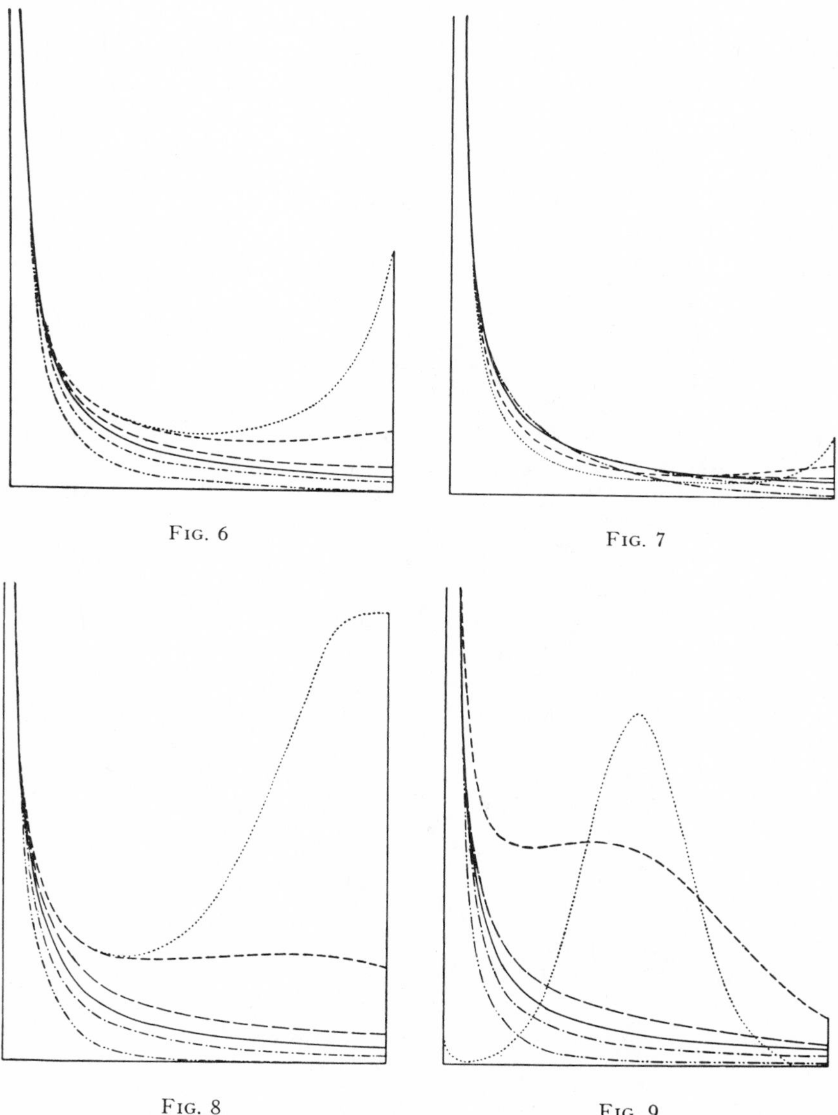

FIG. 6

FIG. 7

FIG. 8

FIG. 9

Figures 6 to 9. Distributions of gene frequencies under irreversible mutation at indefinitely low rates (v). Six cases are shown in each figure. Dash and two dots, $s=-4/2N$; dash and dot, $s=-1/2N$; solid line, $s=0$ ($f(q)=2v/q$ in all figures); long

In the case of recessive mutations, $s=0$ and $\Delta q = tq^2(1-q)$. The law of the series for $\psi(0, b)$ is that $E_m = (E_{m-1}+E_{m-2})/2m(2m+1)$ where E_m is the coefficient of b^m. The distribution curve for a number of values of $2Nt$ are shown in Figure 7 (but with s instead of t as the selection coefficient). The chance of fixation was determined empirically from the ratio of subterminal frequencies up to $2Nt=12$. For larger values (up to $2Nt=64$) it was more convenient to calculate it from the ratio of the flux $(2N\,\Delta q\,f(q))$ in the region of maximum Δq, $(q=2/3)$ to the proportion of mutations $(2Nv)$ (usable only for large $\Delta q/\sigma^2_{\delta q}$). For values of t ranging from $4/2N$ to $64/2N$ the average chance of fixation came out $1.1(t/2N)^{1/2}$, apparently approaching $(t/2N)^{1/2}$ and thus a function of N even in large populations contrary to the case of semidominance.

In the case of dominant mutations, $\Delta q = (s-sq)q(1-q)$ if mutations are taken as occurring from the class $q=0$. It is more convenient to assume that they are occurring from the class $q=1$ $(\Delta q = -sq^2(1-q))$ since this merely requires evaluation of $\psi(0, \;-2Nsq^2)$ instead of the two-dimensional series $\psi(2Nsq, \;-2Nsq^2)$. From considerations analogous to those discussed above, it appears that for $\Delta q = (s+tq)q(1-q)$ but irreversible mutation from the class at $q=1$, $C=0$, $D=2ve^{-2Ns-Nt}/\psi(2Ns, 2Nt)$ to a sufficient approximation,

$$(19) \qquad f(q) = \frac{2v}{q(1-q)}\left[qe^{-2Ns(1-q)-Nt(1-q^2)}\frac{\psi(2Nsq,\ 2Ntq^2)}{\psi(2Ns,\ 2Nt)}\right].$$

dashes, $s=1/2N$; short dashes, $s=4/2N$; dots, $s=16/2N$. The ordinate at $q=1/2N$ is the same for all curves $(f(q)=4Nv)$ and far above the range of the figures (except in the case $s=16/2N$, Figure 9, in which all ordinates are greatly reduced).

Figure 6. The case of a semidominant mutation A' ($W_{AA}=1$, $W_{AA'}=1+s/2$, $W_{A'A'}=1+s$). The probabilities (P) of fixation of a mutation are as follows: For $s=-4/2N$, $P=0.075/2N$; for $s=-1/2N$, $P=0.58/2N$; for $s=0$, $P=1/2N$; for $s=1/2N$, $P=1.6/2N$; for $s=4/2N$, $P=4.1/2N$; for $s=16/2N$, $P=16/2N$; for large s, $P=s$.

Figure 7. The case of a recessive mutation a, ($W_{AA}=W_{Aa}=1$, $W_{aa}=1+s$). For $s=-4/2N$, $P=0.12/2N$; for $s=-1/2N$, $P=0.70/2N$; for $s=0$, $P=1/2N$; for $s=1/2N$, $P=1.3/2N$; for $s=4/2N$, $P=2.3/2N$; for $s=16/2N$, $P=4.3/2N$; for large s, $P=(s/2N)^{1/2}$.

Figure 8. The case of a dominant mutation A ($W_{aa}=1$, $W_{Aa}=W_{AA}=1+s$). For $s=-4/2N$, $P=0.042/2N$; for $s=-1/2N$, $P=0.49/2N$; for $s=0$, $P=1/2N$; for $s=1/2N$, $P=1.9/2N$; for $s=4/2N$, $P=6.6/2N$; for $s=16/2N$, $P=31/2N$; for large s, $P=2s$.

Figure 9. The case of a mutation A', selected only in heterozygotes ($W_{AA}=W_{A'A'}=1$, $W_{AA'}=1+s$). For $s=-1/4N$, $P=0.23/2N$; for $s=-1/2N$, $P=0.71/2N$; for $s=0$, $P=1/2N$; for $s=1/2N$, $P=1.4/2N$; for $s=4/2N$, $P=3.2/2N$; for large s, P approaches 0 and all loci tend to become heterallelic.

Figure 8 shows the form taken by the distribution in special cases (with s as the selection favoring the dominant mutation and mutation taken as occurring from the class $q = 0$). The chance of fixation of favorable dominant mutation with large N is approximately $2s$ which is the same as for favorable semidominant mutations provided that s is the selection favoring the heterozygotes in both cases. This is to be expected since homozygotes are relatively rare until mutation has passed through the point of maximum selection pressure.

Figure 9 shows the distribution curve in the case in which there is no selective difference between the homozygotes but selection favors or opposes the heterozygotes. In this case in which $\Delta q = s(1 - 2q)q(1 - q)$, formula (17) was used for values of $2Ns$ up to 4. For large values of $2Ns$, there is so little fixation that the distribution under equilibrium may be used.

By combining the formulae for irreversible mutation in each direction in such a ratio that the amounts of fixation are equal, we obtain a distribution identical with that of reversible mutation occurring at rates at which $4Nv$ and $4Nu$ are negligible. The result agrees with the limiting value obtained from equation (12)

$$f(q) = \frac{C}{q(1-q)} \left[e^{4Nsq + 2Ntq^2}\right]. \tag{20}$$

The evolutionary process. I can go only briefly into the implications for evolution. We must distinguish two processes (a) the transformation of a single population until it has become so different that a new species or higher category must be recognized and (b) the cleavage of species.

Consider first the possibilities of transformation in a very large, closed, freely interbreeding population, living under conditions that are the same on the average for thousands of generations [**28**]. In such a population, random changes in gene frequency are negligible. Gene frequencies can change only according to the systematic pressures of mutation and selection, a process which must stop when all Δq's become zero, unless there is a flow of untried mutations that are favorable from the first. We have a theory of the stability of species in spite of variability due to continually occurring mutations and in spite of continuous action of selection.

A stable state of this sort may be far from being the most adaptive of the systems possible from the genes actually present. Consider here the situation with respect to characters to which our model selection pressure applies. As noted, many distinct peak values are to be expected in the surface of selective values, $\overline{W}$, relative to the multidimensional system of gene frequencies. In a species located at a particular point in the system, each gene frequency will change until all Δq's are zero. These changes will be such that the mean selective value of the populations changes approximately by the amount $\Delta \overline{W} = \Sigma(\Delta q \partial \overline{W}/\partial q)$, the species moving up the steepest gradient in the surface $\overline{W}$ except as affected by mutation pressures [**29**]. It stops when $\Delta \overline{W} = 0$, a point in the neighborhood of the peak toward which selection has been directed, but not at the highest point because of the mutation terms in the Δq's. In general there will be other peaks on the surface $\overline{W}$ that are higher but the species cannot reach them.

Perhaps, however, we have been too hasty in assuming that all Δq's would ever reach zero simultaneously. It is probable that the potential alleles at each locus form an indefinitely extended series in which any one allele can give rise to certain others, these to ones at two removes from the first, and so on. A continual flow of untried favorable mutations may keep the population in a state of flux. In general, however, it would seem probable that the rate of the movement toward the equilibrium point indicated by the existing genes would be of a higher order of magnitude than the rate of elevation of this peak by the occurrence of mutations of this very unusual sort.

In a population in approximate equilibrium, the variability due to the balance between mutation and selection is not likely to be great. If, for example, $\Delta q = sq(1-q) - uq$, $q = 1 - u/s$ at equilibrium. As mutation rates are typically of the order 10^{-5} or less, q is close to 1 if the gene in question has an appreciable advantage. Loci in which there are opposing selection pressures would contribute more to variability. This may occur where a heterozygote combines favorable effects of two genes (Δq of form $q(1-q)[s_2-(s_1+s_2)q]$ with stable equilibrium at $q = s_2/(s_1+s_2)$ if both s_1 and s_2 are positive). It may also occur where different homozygotes have advantages in different ecological niches occupied by the species in such a way that the net selective values are related inversely to the frequencies (a case to which our model applies only approximately). With only one or the same few alleles maintained at high frequencies by the population the chance for the occurrence of fixation of untried favorable mutations at several removes from those present is small. The extreme improbability of such mutations justifies use of the formulae for irreversible mutation at very low rates ($4Nv$ much less than 1) in spite of the fact that we are considering large populations (17), (19) (Figures 6 to 9). There is the possibility of an indefinitely continuing evolutionary process but the rate is restricted by the low probability of the necessary mutations and the incomplete utilization of the potentialities for adaptation provided by the genes actually present (cf. however Fisher [**4**] and critique [**26**]).

Conditions are, however, continually changing. Selection of increased severity but unchanged direction merely carries the location of the system of gene frequencies closer to the peak and increases somewhat the chance for a novel favorable mutation to reach fixation. With secular changes in the *direction* of selection, on the other hand, peaks in the surface $\overline{W}$ may become depressed and low places elevated. In species which are sufficiently labile to avoid extinction, the system of gene frequencies is kept continually on the move. It is likely to be shuffled into regions of $\overline{W}$ that are in general the higher ones. This process is undoubtedly of great importance for evolutionary change.

In sufficiently small populations, the random divergences of gene frequencies from their equilibrium values become important. In very small populations, these tend to bring about approximate fixation of some random (and therefore, in general, non-adaptive) combination of genes (Figure 1). Moreover, while selection pressure is less effective in small populations than in large ones, mutation pressure remains the same. Random mutations are more likely to be degenerative than adaptive. Long continued reduction in the size of a population is thus likely to lead to extinction. On the other hand, the less extreme random variations found in populations of intermediate size ($4Nv$ or $4Ns$ of the order 1 to 10) (cf. Figures 2, 3, 5) act somewhat like changes in the direction of the selection. The system of gene frequencies is kept continually on the move and this gives a trial and error process which at times may lead to adaptive combinations which would not have been reached by direct selection. The rate of change of this sort is slow under the required conditions.

Consider next a large population that is divided into many small partially isolated races. These may differ in size and degree of isolation and in the direction and severity of the selection to which they are subjected. The conditions are present for an extensive testing by trial and error of a relatively large number of alleles at each locus and of different combinations of these.

Local differences in direction of selection are effective if the selection coefficient s (writing selection pressure in the form $\Delta q = sq(1-q)$) exceeds the crossbreeding coefficient m (cf. Figure 4). While these differences are primarily of merely local significance, there is the possibility of acquirement of an adaptation that turns out to be of general value which then may spread throughout the species. Moreover, if different alleles come to be characteristic of different races, the store of variability of the species as a whole is increased. This is also true of each local race as a consequence of crossbreeding.

In races in which $4Nm$ is of the order 1 to 10 there is considerable random differentiation without the approach to fixation in equally small completely isolated groups. Such changes may occur with considerable rapidity in this case and while non-adaptive are not necessarily anti-adaptive to an appreciable extent. It is merely that the location of the system of gene frequencies on the surface of selective values in our model case is not constrained to move up the *steepest* gradient but may move up gradients that are not the steepest and occasionally even down hill. Among the many local races exploring the neighborhood of a peak in the surface of selective values, one or more may reach a gradient leading to a higher peak (cf. Figure 10).

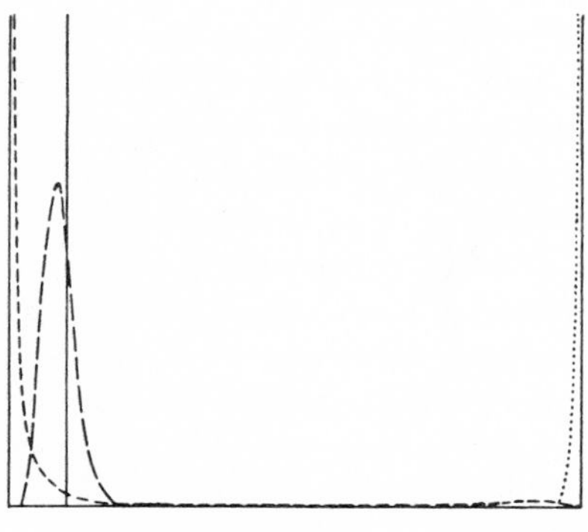

FIG. 10

The frequencies along the diagonal (0, 0) to (1, 1) of the joint distribution of frequencies of recessive genes, a and b in subgroups of the same size ($N=1000$) but different degrees of isolation (long dashes: $m=.01$; short dashes: $m=.001$; dots: $m=.0001$) in a large population in which the conditions are everywhere the same (each recessive with selective disadvantage of .001 relative to the type $A-B-$ but the double recessive with an advantage .01 over type). With mutation rates $u_a=v_a=u_b=v_b=10^{-5}$ there would be two positions of stable equilibrium, one at about (0.1, 0.1) which is assumed to hold for the major portion of the species for historical reasons, and one at about (.999, .999) which is more adaptive

$$\phi(q_a, q_b) = C[1 - .002(q_a^2 + q_b^2) + .012q_a^2q_b^2]^{2000}(q_aq_b)^{400m-1}[(1-q_a)(1-q_b)]^{3600m-1}.$$

Local populations for which $m=.01$ vary only slightly from the frequencies characteristic of the species. Most of those for which $m=.001$ show a close approach to fixation of the type genes ($AABB$) but occasionally there is an approach to fixation of the more adaptive double recessive. Those with $m=.0001$ are largely homallelic in *aabb*. In the long run such superior subgroups might be expected to pull the whole species to this position by intergroup selection. The scales are not the same in this figure.

By expansion of numbers and excess migration such races tend to bring the species as a whole under control of this peak. Intergroup selection of this sort, with respect to racial differentiation that has jointly adaptive and non-adaptive aspects, seems to provide the most effective mechanism for testing many alleles at each locus and many combinations of these and is thus the most effective mechanism for a continuing evolutionary process [**26**, **27**, **28**, **29**, **35**, **36**].

It should be emphasized that we are not concerned here with local races as incipient species. As long as isolation is incomplete the races are bound together by crossbreeding and thus are carried along by the evolution of the species as a whole although subject to the minor kaleidoscopic changes in character which according to this theory play a major role in the evolution of the whole.

The cleavage of species depends on virtually complete isolation of portions of the species from each other. Even if there are no significant character differences at the time of separation and even if conditions remain substantially the same for the two portions, the process described above will insure that they drift apart. Each continues to be adapted to the conditions but in somewhat different ways. They may be expected to move to increasingly more remote peaks on the surface $\overline{W}$. In the course of time genic and chromosomal differences may be expected to accumulate that prevent crossing and so clinch the specific distinction. Before this point is reached, the occasional occurrence of hybridization may transfer blocks of genes from one species to the other or lead to the origin of a completely hybrid species, presenting a mechanism of reticulate evolution, analogous to that described above but on a coarser scale (cf. [**1**]).

Under certain conditions the multiplication and diversification of species may be a very rapid process. These include a relaxation of the general selection pressure on the species permitting great increase in numbers and great variability; the opportunity for the occupation of widely distinct ecological niches associated with almost complete isolation of the groups seizing these opportunities and with subdivision of these groups into partially isolated local populations. A species that is the first of its general kind to reach unoccupied territory finds most at least of these conditions realized. This is also the case with a species that by any means acquires an adaptation of first rate general significance which gives its subgroups an advantage over species already established in various ecological niches, that more than compensates for the initial lack of special adaptations for these niches.

The most general conclusion that can be drawn from the attempt to develop a mathematical theory of the simultaneous effects of all statistical processes that affect the genetic composition of populations is that in general the most favorable conditions for evolutionary advance are found when these are balanced against each other in certain ways, rather than where any one completely dominates the situation. Finally it may be said that the more detailed knowledge of heredity and mutation that is now available confirms Darwin's general contention that evolution is a process of statistical transformation of populations.

LITERATURE CITED

1. E. Anderson, *The species problem in iris*, Annals of the Missouri Botanical Garden, vol. 23 (1936), pp. 457–509.

2. T. Dobzhansky, *Genetics and the Origin of Species*, Columbia University Press, New York, 1941, 445 pp.

3. T. Dobzhansky and S. Wright, *Genetics of natural populations*. V. *Relations between mutation rate and accumulation of lethals in populations of Drosophila pseudoobscura*, Genetics, vol. 26 (1941), pp. 23–51.

4. R. A. Fisher, *The Genetical Theory of Natural Selection*, Clarendon Press, Oxford, 1930, 272 pp. (Much earlier work is here summarized.)

5. ———, *Average excess and average effect of a gene substitution*, Annals of Eugenics, vol. 9 (1941), pp. 53–63.

6. J. B. S. Haldane, *Theoretical genetics of autopolyploids*, Journal of Genetics, vol. 22 (1930), pp. 359–372.

7. ———, *The Causes of Evolution*, Harper, London, 1932, 235 pp. (Much earlier work is here summarized.)

8. G. H. Hardy, *Mendelian proportions in a mixed population*, Science, vol. 28 (1908), pp. 49–50.

9. J. S. Huxley (editor), *The New Systematics*, Clarendon Press, Oxford, 1940, 583 pp.

10. H. S. Jennings, *The numerical results of diverse systems of breeding*, Genetics, vol. 1 (1916), pp. 53–89.

11. ———, *The numerical results of diverse systems of breeding with respect to two pairs of characters, linked or independent, with special relation to the effects of linkage*, Genetics, vol. 2 (1917), pp. 97–154.

12. A. J. Lotka, *Elements of Physical Biology*, Williams and Wilkins, Baltimore, 1925, 460 pp.

13. H. C. McPhee and S. Wright, *Mendelian analysis of the pure breeds of live stock*. III. *The shorthorns*, Journal of Heredity, vol. 16 (1925), pp. 205–215.

14. A. J. Nicholson and V. A. Bailie, *The balance of animal populations*, Proceedings of the Zoological Society of London, 1935, pp. 551–598.

15. R. B. Robbins, *Application of mathematics to breeding problems*. II, Genetics, vol. 3 (1918), pp. 73–92.

16. ———, *Some applications of mathematics to breeding problems*. III, Genetics, vol. 3 (1918), pp. 375–389.

17. V. Volterra, *Leçons sur la théorie mathématique de la lutte pour la vie*, Paris, 1931.

18. W. Weinberg, *Über Vererbungsgesetze beim Menschen*, Zeitschrift für induktive Abstammungs und Vererbungslehre, vol. 1 (1909), pp. 277–330.

19. S. Wright, *Principles of livestock breeding*, Bulletin no. 905, U. S. Department of Agriculture, 1920, 68 pp.

20. ———, *Systems of mating*, Genetics, vol. 6 (1921), pp. 111–178.

21. ———, *The effects of inbreeding and crossbreeding on guinea pigs*. I. *Decline in vigor*, Bulletin no. 1090, U. S. Department of Agriculture, 1922, pp. 1–36.

22. ———, *The effects of inbreeding and crossbreeding on guinea pigs*. II. *Differentiation among inbred families*, ibid., pp. 37–63.

23. ———, *The effects of inbreeding and crossbreeding on guinea pigs*. III. *Crosses among highly inbred families*, Bulletin no. 1121, U. S. Department of Agriculture, 1922, 59 pp.

24. ———, *Mendelian analysis of the pure breeds of livestock*. I. *The measurement of inbreeding and relationship*, Journal of Heredity, vol. 14 (1923), pp. 339–348.

25. ———, *Mendelian analysis of the pure breeds of livestock*. II. *The Duchess family of shorthorns as bred by Thomas Bates*, ibid., pp. 405–422.

26. ———, *The genetical theory of natural selection* (review of book by R. A. Fisher), Journal of Heredity, vol. 21 (1930), pp. 349–356.

27. ———, *Evolution in Mendelian populations*, Genetics, vol. 16 (1931), pp. 97–159

28. ———, *The roles of mutation, inbreeding, crossbreeding and selection in evolution*, Proceedings of the 6th International Congress of Genetics, vol. 1 (1932), pp. 356–366.

29. ———, *Evolution in populations in approximate equilibrium*, Journal of Genetics, vol. 30 (1935), pp. 257–266.

30 ———, *The distribution of gene frequencies in populations*, Proceedings of the National Academy of Sciences, vol. 23 (1937), pp. 307–320.

31. ———, *The distribution of gene frequencies under irreversible mutation*, Proceedings of the National Academy of Sciences, vol. 24 (1938), pp. 253–259.

32. ———, *The distribution of gene frequencies in populations of polyploids*, ibid. pp. 372–377.

33. ———, *The distribution of self-sterility alleles in populations*, Genetics, vol. 24 (1939), pp. 538–552.

34. ———, *Genetic principles governing the rate of progress of livestock breeding*, Proceedings of the American Society of Animal Production, 1939, pp. 18–26.

35. ———, *Breeding structure of populations in relation to speciation*, American Naturalist, vol. 74 (1940), pp. 232–248.

36. ———, *The statistical consequences of Mendelian heredity in relation to speciation* (chapter in *The New Systematics*, edited by J. S. Huxley), Clarendon Press, Oxford, 1940.

37. S. Wright and O. N. Eaton, *The persistence of differentiation among inbred families of guinea pigs*, Technical Bulletin no. 103, U. S. Department of Agriculture, 1929, 45 pp.

UNIVERSITY OF CHICAGO

THE DIFFERENTIAL EQUATION OF THE DISTRIBUTION OF GENE FREQUENCIES

By Sewall Wright

Department of Zoölogy, The University of Chicago

Communicated October 15, 1945

The first attempt to determine the mathematical form of the distribution of gene frequencies in populations was based on the setting up of differential equations for certain special cases (Fisher, 1922,[1] 1930[2]). A correction and extension of these results came from expression of the conditions in an integral equation (Wright, 1929,[3] 1931[4]). A general solution has since been obtained for fully stationary distributions by a third method (Wright, 1937,[5] 1938[6]). The case of uniform flux has been treated less generally (Wright, 1938,[6] 1942[7]). Dr. A. Kolmogorov[8] has recently been kind enough to send me a reprint of an important paper on this subject which was published in 1935 but which had not previously come to my attention. While the application is restricted to a particular stationary distribution, the method of approach points to a more systematic formulation than before.

The situation discussed by Kolmogorov is that of a large population, consisting of many subgroups of size n, each of which receives a certain number (k) of immigrants from the general population but otherwise breeds within itself. The average rate of change of the gene frequency of subgroups, in which p is the frequency of a given gene, is represented by $A = \Sigma(\Delta p) = (k/n)(\bar{p} - p)$ where $\bar{p}$ is the mean value of p in the whole population. The variance of p, due to accidents of sampling in one generation, is represented by $B = \Sigma(\Delta p)^2 = pq/2n$, $q = 1 - p$. It is stated, without demonstration, that the distribution $u(p)$ of gene frequencies among subgroups after a stationary state has been reached, answers to the differential equation

$$\frac{1}{2}\frac{\partial^2}{\partial p^2}(Bu) - \frac{\partial}{\partial p}(Au) = 0. \tag{1}$$

The pertinent solution is given as

$$u(p) = p^{4k\bar{p}-1}q^{4k\bar{q}-1}/B(4k\bar{p}, 4k\bar{q}). \tag{2}$$

The effect of selection in this situation is discussed briefly without, however, modifying $u(p)$ by introduction of the selection term, $\alpha p^2 q$, into A.

It is noted that the same formula (2) had previously been derived by the present author[4] by a different method. Equation (1) has, however, broader implications if valid for the general case $A = \Sigma(\Delta p)$, $B = \Sigma(\Delta p)^2$ and not merely for the particular case $A = (k/n)(\bar{p} - p)$, $B = pq/2n$.

The Immediate Factors of Evolutionary Change.—The immediate factors that tend to cause systematic changes (Δq) in gene frequency (q) may be listed exhaustively[4, 5] as (*a*) mutation pressure, $\Delta q = v(1 - q) - uq$, where v and u are the rates of mutation to and from the gene in question, (*b*) immigration pressure, $\Delta q = m(q_i - q)$ (Kolmogorov's A) where m is the proportion of replacement by immigrants and q_i is the gene frequency in these, and (*c*) selection pressure, which may take widely diverse forms but in the important case of constant relative selective values (W) for each multiple factor genotype in a random breeding population takes the form $\Delta q = q(1 - q) \dfrac{\partial \bar{W}}{\partial q}/r\bar{W}$, where r is 1 in haploids, 2 in diploids, the usual case, 1.5 for sex linked genes (if equal numbers of males and females), 4 in tetraploids, etc.[7, 9] In addition to these systematic pressures are (*d*) the random variations, δq, due to accidents of sampling, the variance of which is $\sigma^2_{\delta q} = q(1 - q)/rN$ in a population of effective size N (Kolmogorov's B). The diploid case ($r = 2$) will be assumed in what follows.

The Stationary Distribution of Gene Frequencies.—Systematic pressure toward the gene frequency, at which $\Delta q = 0$ and the cumulative effects of accidents of sampling determine a probability curve $\varphi(q)$ describing the frequencies which would be exhibited in the long run by the value of q for a particular gene in a population subject to constant conditions. This distribution may also be interpreted as that exhibited at one time by the values of q in a group of populations that are all subject to the same conditions (as in the case of Kolmogorov's $u(p)$). The deviations from the binomial square formula for genotypic frequencies in the total population, depend on the variance of $\varphi(q)$ under this interpretation.[4, 8, 10] In other cases $\varphi(q)$ may be used as the distribution at any time within either a class of non-allelic genes or an extensive series of multiple alleles,[11] all subject to the same conditions.

That equation (1) is, in fact, completely general for the stationary form of distribution may be shown by a slight modification of a method[5, 6] that has been used for derivation of $\varphi(q)$.

The conditions for stability of the distribution (including the terminal classes $q = 0$, $q = 1$) may be represented by two equations expressing the persistence of the mean and variance, respectively

$$\int_0^1 (q + \delta q + \Delta q)\varphi(q)dq = \int_0^1 q\varphi(q)dq. \qquad (3)$$

$$\int_0^1 (q - \bar{q} + \delta q + \Delta q)^2 \varphi(q)dq = \int_0^1 (q - \bar{q})^2 \varphi(q)dq. \qquad (4)$$

Noting that the mean value of δq is zero, and that δq is not correlated with q or Δq, these equations reduce to the following if the term in $(\Delta q)^2$ in (4) may be ignored. It may be noted in this connection that this term is negligible if Δq is of the same order as $\sigma^2_{\delta q}$ or less, while if of higher order, systematic pressure dominates the results so completely that the distribution formula itself becomes unimportant.

$$\int_0^1 \Delta q \varphi(q) dq = 0. \tag{3a}$$

$$2 \int_0^1 (q - \bar{q}) \Delta q \varphi(q) dq + \int_0^1 \sigma_{\delta q}^2 \varphi(q) dq = 0. \tag{4a}$$

Putting $\Delta q \varphi(q) dq = d\chi(q)$ these conditions become

$$\chi(1) - \chi(0) = 0. \tag{3b}$$

$$2\int_0^1 \chi(q) dq - 2[\chi(1) + \bar{q}(\chi(1) - \chi(0))] - \int_0^1 \sigma_{\delta q}^2 \varphi(q) dq = 0. \tag{4b}$$

Substituting (3*b*) in (4*b*) the latter becomes

$$\int_0^1 [2\chi(q) - 2\chi(1) - \sigma_{\delta q}^2 \varphi(q)] dq = 0. \tag{4c}$$

A solution is obtained by removing the integral sign since the resulting equation not only satisfies (4*c*) but also (3*b*) (noting that $\sigma_{\delta q}^2 = 0$ if $q = 0$ or if $q = 1$, there being no sampling variance without alternatives in the sample).

$$\chi(q) - \chi(1) = {}^1/_2 \sigma_{\delta q}^2 \varphi(q). \tag{5}$$

This can be solved for $\varphi(q)$ by differentiating the logarithm of the left-hand number and making the appropriate substitutions.[6]

$$\varphi(q) = (C/\sigma_{\delta q}^2) e^{2\int (\Delta q/\sigma_{\delta q}^2) dq} \tag{6}$$

where C is a constant such that $\int_0^1 \varphi(q) dq = 1$.

Since q increases by steps of $1/2N$ in a population of size N, the frequency of a given value of q is $f(q) = \varphi(q)/2N$. From a study[4] of simple cases ($N = 2$ or 3) in which the frequencies in the stationary state can be determined algebraically and from a more elaborate investigation by R. A. Fisher[2] of the subterminal region in certain cases, it appears that the frequencies are given with considerable accuracy by the formula except for the terminal classes, $q = 0$, $q = 1$. Consideration of the exchanges which occur between the terminal and neighboring classes leads[4] to the following approximate estimate for the terminal class, $q = 0$. That for $q = 1$ is analogous.

$$f(0) = f(1/2N)/4N[mq_i + v]. \tag{7}$$

The differential equation for the completely stationary case is given by differentiation of (5). It comes under equation (1).

$$\frac{1}{2} \frac{d}{dq} (\sigma_{\delta q}^2 \varphi(q)) - \Delta q \varphi(q) = 0. \tag{8}$$

Since $\Delta q \varphi(q)$ is the proportion of the distribution which tends to be carried past a specified value of q by the systematic pressure Δq, the other term must represent the net proportion which tends to be carried in the opposite direction by accidents of sampling in each generation.

The Case of Steady Flux.—There may be a practically stationary state of the proportions in all intermediate values of q in spite of steadily increasing frequency of one terminal class at the expense of the other, provided that the proportion lost by the donor terminal class is negligible. This cannot be the case if either mutation rate or immigration rate is appreciable, but may hold in the presence of strong selection pressure since selection pressure is nil in populations in which $q = 0$ or $q = 1$.

The differential equation for the case of steady flux must differ from (8) by a constant term (D), the net proportion of the total (excluding the recipient class) that is carried past each value of q in each generation.

$$\frac{1}{2}\frac{d}{dq}(\sigma^2_{\delta q}\varphi(q)) - \Delta q\varphi(q) + D = 0. \tag{9}$$

This is the general form given by one integration of (1) which is therefore the general differential equation for a steady state of the intermediate classes. It may be reduced to a linear equation of the first order by making the substitution, $y = \sigma^2_{\delta q}\varphi(q)$.

$$\frac{dy}{dq} - 2\left(\frac{\Delta q}{\sigma^2{}_q}\right)y + 2D = 0. \tag{10}$$

The solution for $\varphi(q)$ is as follows:

$$\varphi(q) = [e^{2\int(\Delta q/\sigma^2_{\delta q})dq}/\sigma^2_{\delta q}]\,[C - 2D\textstyle\int e^{-2\int(\Delta q/\sigma^2_{\delta q})dq}dq]. \tag{11}$$

The simplest special case is that in which Δq may be treated as zero (although there could be no flux if it were absolutely zero).

$$f(q) = \frac{C}{q(1-q)} - \frac{2D}{1-q}. \tag{12}$$

The case under (12) that is most important genetically is that of irreversible mutation at a rate so low that the donor class ($q = 0$, or $q = 1$) is not appreciably depleted. According to direction of mutation,

$$f(q) = 2v/q, \text{ or } f(q) = 2v/(1-q). \tag{13}$$

The ratio of the subterminal classes ($1/2N$ in this case) gives the probability that a single neutral mutation may reach fixation instead of elimination.

Returning to (12) the case in which $D = 0$ yields the corresponding simplest solution for a completely stationary state

$$\varphi(q) = 1/[2(0.577 + \log 2N)q(1-q)] \text{ (terminal classes excluded)}. \tag{14}$$

The case in which there are constant relative selection coefficients for all genotypes ($\Delta q = q(1 - q)\frac{\partial \bar{W}}{\partial q}/r\bar{W}$) gives an apparently simple but in general rather refractory form (assuming a given set of frequencies of other genes)

$$\varphi(q) = [\bar{W}^{2N}/\sigma^2_{\delta q}]\,[C - 2D\int \bar{W}^{-2N}dq]. \qquad (15)$$

It will be convenient for later reference to cite the less general case $\Delta q = q(1 - q)(s + tq)$, $\sigma^2{}_q = q(1 - q)/2N$ which allows for any degree of dominance, provided s and t are both small.

$$f(q) = [e^{4Nsq+2Ntq^2}/q(1 - q)]\,[C - 2D\int e^{-(4Nsq+2Ntq^2)}dq]. \qquad (16)$$

Non-stationary States.—The general case, in which the proportion at each value of q is a function of time as well as of q itself, is given by the following, of which equation (1) is the case in which the left-hand member is zero. Time (T) is measured in generations.

$$\frac{\partial \varphi(q, T)}{\partial T} = \frac{1}{2}\frac{\partial^2}{\partial q^2}\,[\sigma^2_{\delta q}\varphi(q, T)] - \frac{\partial}{\partial q}\,[\Delta q\varphi(q, T)]. \qquad (17)$$

This can be reduced to an ordinary differential equation in the case in which the distribution has reached stability of form, with all classes (except the terminal ones) falling off at the same rate. Let $K = -\frac{1}{\varphi(q, T)}\frac{\partial \varphi(q, T)}{\partial T}$ be the rate of decay per generation.

$$\frac{1}{2}\frac{d^2}{dq^2}\,(\sigma^2_{\delta q}\varphi(q)) - \frac{d}{dq}\,(\Delta q\varphi(q)) + K\varphi(q) = 0. \qquad (18)$$

It may easily be verified that for the case in which fixation is occurring under the uncomplicated effect of inbreeding ($\Delta q = 0$, $K = 1/2N$) the only solution that does not involve negative frequencies is

$$\varphi(q) = 1, \qquad \text{or } f(q, T) = C_0e^{-T/2N}. \qquad (19)$$

In the case of irreversible mutation at an appreciable rate, $\Delta q = v(1 - q)$ the rate of decay is easily shown to be $K = v$. Equation (18) is satisfied by the following value, originally derived by a different method.

$$f(q) = 2vq^{4Nv-1}. \qquad (20)$$

An analogous solution applies to the effect of swamping by immigration from a population in which the gene in question is fixed ($\Delta q = m(1 - q)$, $K = m$)

$$f(q) = 2mq^{4Nm-1}. \qquad (21)$$

Comparison with Results by Other Methods.—The first attempt at determining the distribution of gene frequencies was made by R. A. Fisher[1] who arrived at differential equations for certain special cases, in terms, however, of a different variable than gene frequency, $\vartheta = \cos^{-1}(1 - 2q)$, used in order to make the sampling variance constant. A discrepancy between the rate of decay ($K = 1/4N$), derived by him for the case in which $\Delta q = 0$, and the value, $1/2N$, given by a general method[12, 13] for determining the rate of fixation of genes under any system of mating, led the present author[3, 4] to a different approach. The condition for a stationary state of the intermediate classes except for possible decay at rate K, was represented by the following equation in which q and x are recipient and donor classes, respectively, in the exchanges which occur from one generation to the next.

$$(1 - K)\frac{\varphi(q)}{2N} = \frac{(2N)!}{(2Nq)![2N(1-q)]!}\int_0^1 (x + \Delta x)^{2Nq}(1 - x - \Delta x)^{2N(1-q)}\varphi(x)dx. \quad (22)$$

It could easily be seen that if $\Delta x = 0$, the equation is satisfied by $\varphi(q) = \varphi(x) = 1$, $K = 1/(2N + 1)$, the latter at least a close approximation to the rate of decay expected in this case. For the simplest stationary state, $K = 0$, $\Delta q \doteq 0$, the expression $\varphi(q) = Aq^{-1} + B(1 - q)^{-1}$ is indicated (cf. 12). Approximate solutions could also readily be obtained for the linear pressures of mutation and migration. Selection presented more difficulty.

On inspection of these results in manuscript, Fisher[2] was able to correct and extend his equations to obtain the following:

I. Case of uniform decay ($\Delta q = 0$)

$$\frac{\partial y}{\partial T} = \frac{1}{4n}\left[\frac{\partial^2 y}{\partial \vartheta^2} + \frac{\partial}{\partial \vartheta}(y \cot \vartheta)\right]. \quad (23)$$

$$y = A_0 e^{-T/2n} \sin \vartheta \qquad \text{(cf. (19))}. \quad (24)$$

II. Stationary state, no selection ($\Delta q \doteq 0$)

$$\frac{dy}{d\vartheta} + y \cot \vartheta = -4nB. \quad (25)$$

$$y = A \operatorname{cosec} \vartheta + 4nB \cot \vartheta \qquad \text{(general, cf. (12))}. \quad (26)$$

$$y = A \operatorname{cosec} \vartheta \qquad \text{(symmetrical case, cf. (14))}. \quad (27)$$

$$y = 4nB(\operatorname{cosec} \vartheta + \cot \vartheta) \qquad \text{(unidirectional mutation, cf. (13))}. \quad (28)$$

III. Stationary state, selection, no dominance, $\Delta q = aq(1 - q)$

$$\frac{dy}{d\vartheta} - (2an \sin \vartheta - \cot \vartheta)y = -4anA. \quad (29)$$

$$y = \operatorname{cosec} \vartheta(2A + Be^{-2an \cos \vartheta}) \quad \text{(general, cf. (16), } t = 0). \quad (30)$$

$$y = 4 \operatorname{cosec} \vartheta \frac{(1 - e^{-2an(1+\cos \vartheta)})}{(1 - e^{-4an})} \quad \text{(unidirectional mutation)}. \quad (31)$$

In cases I and II, the results agreed with those obtained from the integral equation (22), as may be seen by making the substitutions $\cos \vartheta = 1 - 2q$, $yd\vartheta = \varphi(q)dq$, $d\vartheta/dq = 1/\sqrt{q(1-q)}$. In case III it was the author's turn to make a correction in the selection term (published first[3] as e^{2Nsq}), by taking cognizance of a series of small terms erroneously thought to be negligible but which actually doubled the exponent. With this correction, there was agreement.[4]

The most general result[5] obtained for the completely stationary state by solution of (22) took into account all of the factors of change in the form $\Delta q = v(1 - q) - uq - m(q - q_i) + q(1 - q)(s + tq)$, $\sigma^2_{\delta q} = q(1 - q)/2N$.

$$\varphi(q) = Ce^{4Nsq+2Ntq^2}q^{4N(mq_i+v)-1}(1 - q)^{4N[m(1-q_i)+u]-1}. \quad (32)$$

This agrees with that obtained by substituting these values of Δq and of $\sigma^2_{\delta q}$ in (6).

The most general result[5,6] obtained by this method for the case of steady flux was for $\Delta q = q(1 - q)(s + tq)$.

$$f(q) = [e^{4Nsq+2Ntq^2}/q(1 - q)]\,[C - 2Dqe^{-(2Nsq+Ntq^2)}\psi(2Nsq, 2Ntq^2)]. \quad (33)$$

where

$$\psi(a, 0) = 1 + \frac{a^2}{3!} + \frac{a^4}{5!} + \frac{a^6}{7!} \cdots = (e^a - e^{-a})/2a$$

$$\psi(0, b) = 1 + \frac{b}{3!} + \frac{7b^2}{5!} + \frac{27b^3}{7!} \ldots E_m b^m$$

$$E_m = (E_{m-1} + E_{m-2})/2m(m + 1).$$

No recurrence formula was recognized for the joint terms, $\psi(2Nsq, 2Ntq^2)$ but the coefficients were calculated[7] up to those pertaining to q^9.

The probability of fixation of a single mutation ($C = 2v$, $D = ve^{2Ns+Nt}/\psi(2Ns, 2Nt)$ for irreversible mutations from class $q = 0$, or $C = 0$, $D = -ve^{-(2Ns+Nt)}/\psi(2Ns, 2Nt)$ for irreversible mutations from class $q = 1$), could be calculated from the ratios of the subterminal classes, (Prob. $= \sqrt{s/2N}$ for a recessive mutation with selective advantage s, Prob. $= 2s$ for a dominant mutation with selective advantage s, or for a semidominant with selective advantage s in the heterozygote). The last agrees with Fisher's conclusion.[2] Equation (33), with $t = 0$, is indeed equivalent to (31).

Comparison of (33) with (16) shows that if the former is correct, the following must hold:

$$\psi(2Nsq,\ 2Ntq^2) = (e^{2Nsq+Ntq^2}/q)\int e^{-(4Nsq+2Ntq^2)}dq. \qquad (34)$$

This was tested by expanding the two exponentials in (34), integrating each term of the second one and combining. The coefficients were in all cases identical with those published[7] for $\psi(2Nsq, 2Ntq^2)$.

Equation (22) also gave the solution (20) for the case of uniform decay under an appreciable mutation rate.[4]

The integral equation (22) and the differential equation (18) are clearly equivalent to a close approximation. They are not exact mathematical equivalents, however, as may be seen from the fact that K must be put $1/(2N + 1)$ in (22) if $\Delta x = 0$ to give the solution $\varphi(q) = 1$, while it takes its true value $1/2N$ in (18) to give the same result. In the other cases (except (12)) second order terms have been omitted in the series, obtained as solutions of the integral equation, which do not appear in the solutions of the differential equation. Neither equation, of course, represents the natural conditions exactly since integration is substituted for summation and differentials for minimal steps $(1/2N)$ in gene frequency.

1 Fisher, R. A., *Proc. Roy. Soc. Edinburgh*, **42**, 321–341 (1922).
2 Fisher, R. A., *Ibid.*, **50**, 205–220 (1930).
3 Wright, S., *Amer. Naturalist*, **63**, 556–561 (1929).
4 Wright, S., *Genetics*, **16**, 97–159 (1931).
5 Wright, S., these PROCEEDINGS, **23**, 307–320 (1937).
6 Wright, S., *Ibid.*, **24**, 253–259 (1938).
7 Wright, S., *Bull. Amer. Math. Soc.*, **48**, 223–246 (1942).
8 Kolmogorov, A., *C. R. de l'Acad. des Sciences de l' U.R.S.S.*, **3** (7), 129–132 (1935).
9 Wright, S., these PROCEEDINGS, **24**, 372–377 (1938).
10 Wahlund, S., *Hereditas*, **11**, 65–106 (1928).
11 Wright, S., *Genetics*, **24**, 538–552 (1939).
12 Wright, S., *Ibid.*, **6**, 111–178 (1921).
13 Wright, S., *Amer. Naturalist*, **61**, 330–338 (1922).

34
On the Roles of Directed and Random Changes in Gene Frequency in the Genetics of Populations

Evolution 2, no. 4 (1948):279–94

35
Fisher and Ford on "the Sewall Wright Effect"

American Scientist 39, no. 3 (1951):452–58, 479

Introduction

The controversy between Fisher and Ford on the one side and Wright on the other, over evolution in the moth *Panaxia dominula*, is one of the most colorful and influential battles of modern evolutionary biology. I have devoted a long section in *SW&EB* (pp. 420–37) to the controversy and its effects and will not attempt to summarize here, except to state the basic chronology.

Fisher and Ford published the first paper in the controversy in 1947 in volume 1 of the new journal *Heredity*. They argued that random drift did not play any significant role in evolution in nature. Wright answered them in 1948 with the first paper in this section. Fisher and Ford wrote a rebuttal to Wright and submitted it to Mayr, the editor of *Evolution*, but he refused to publish it unless they toned down their sharply worded criticisms of Wright. They then published their paper in *Heredity* instead (Fisher and Ford 1950). Wright replied again with the second paper in this section. It was published in *American Scientist*, which boasted a readership of over fifty thousand.

The repercussions from the intense debate over evolution in *Panaxia dominula* have continued to the present. For recent reviews of the status of the debate, see Wright (*E&GP* 4:171–77) and Ford (1975, chap. 7).

ON THE ROLES OF DIRECTED AND RANDOM CHANGES IN GENE FREQUENCY IN THE GENETICS OF POPULATIONS [1]

SEWALL WRIGHT
The University of Chicago, Chicago 37, Illinois

Received July 16, 1948

INTRODUCTION

Science has largely advanced by the analytic procedure of isolating the effects of single factors in carefully controlled experiments. The task of science is not complete, however, without synthesis: the attempt to interpret natural phenomena in which numerous factors are varying simultaneously. Studies of the genetics of populations, including their evolution, present problems of this sort of the greatest complexity. Many writers on evolution have been inclined to ignore this and discuss the subject as if it were merely a matter of choosing between single factors. My own studies on population genetics have been guided primarily by the belief that a mathematical model must be sought which permits simultaneous consideration of all possible factors. Such a model must be sufficiently simple to permit a rough grasp of the system of interactions as a whole and sufficiently flexible to permit elaboration of aspects of which a more complete account is desired.

On attempting to make such a formulation (Wright, 1931) it was at once apparent that any one of the factors might play the dominating role, at least for a time, under specifiable conditions, but it was concluded that in the long run "evolution as a process of cumulative change depends on a proper balance of the conditions which at each level of organization—gene, chromosome, cell, individual, local race—make for genetic homogeneity or genetic heterogeneity of the species."

The purpose of the present paper is to reiterate this point of view in connection with certain misapprehensions which have arisen and in particular to analyze certain data which have been presented recently by R. A. Fisher and E. B. Ford (1947) as invalidating what they consider my point of view. This discussion leads to mathematical comparisons between the amount of random drifting of gene frequency expected in small populations merely from accidents of sampling and that expected in large ones from variations in degree and direction of selection or in amount and character of immigration.

SMALLNESS OF POPULATION SIZE AS A FACTOR

In spite of his repeated emphasis on dynamic equilibrium among all factors as the most favorable condition for

[1] This investigation was aided by a grant from the Wallace C. and Clara A. Abbott Memorial Fund of the University of Chicago.

evolution, the author has often been credited with advocating the all importance of a single factor, viz. sampling effects in very small populations. Thus in Goldschmidt's stimulating book "The Material Basis of Evolution" (1940) he states: "The adherents of such a view" (NeoDarwinism) "derive much comfort from the results of population mathematics, especially Wright's calculation (1931) showing that small isolated groups have the greatest chance of accumulating mutants even without favorable selection. . . . It is the contention that small isolated populations have the greatest chances from the standpoint of population mathematics." What I actually stated on this matter in the summary of the paper referred to was as follows: "In too small a population there is nearly complete fixation, little variation, little effect of selection and thus a static condition, modified occasionally by chance fixation of rare mutations, leading inevitably to degeneration and extinction." The same conclusion has been reiterated in all more recent general discussions.

Population Structure as a Factor

This misapprehension may have arisen from the emphasis put on population structure later in the same summary: "Finally in a large population, divided and subdivided into partially isolated local races of small size, there is a continually shifting differentiation among the latter, intensified by local differences in selection, but occurring under uniform and static conditions, which inevitably brings about an indefinitely continuing, irreversible, adaptive, and much more rapid evolution of the species" (than in a comparably large, random breeding population). It should be noted that a favorable population structure, under this view, may be prevented by excessive density of population, as well as by too small a total number (cf. Wright, 1943). The implied relation of size of population to rate of evolution is not a simple one. I suspect that "inevitably" is too strong a word as used above, but otherwise this quotation still represents my position on the importance of population structure in evolution. The ways in which it is important were, of course, brought out more completely elsewhere in this paper and have been developed further in later papers.

This leads to consideration of a recent inaccurate statement of my views by R. A. Fisher and E. B. Ford. Their first reference is in the main an acceptable statement of the role which I have attributed to random shifts in gene frequencies, provided that the word "partially" is inserted before "isolated" and it is clearly understood that the primary significance of the process is as one of a number of adjuncts to intergroup selection. "Great evolutionary importance has been attached by Sewall Wright (1931, 1932, 1935, 1940) to the fact that small shifts in the gene ratios of all segregating factors will occur from generation to generation owing to the errors of random sampling in the process by which the gametes available in any one generation are chosen to constitute the next. Such chance deviations will, of course, be greater the smaller the isolated population concerned. Wright believes that such nonadaptive changes in gene ratio may serve an evolutionary purpose by permitting the occurrence of genotypes harmoniously adapted to their environment in ways not yet explored and so of opening up new evolutionary possibilities."

The next sentence indicates, however, that the authors have wholly missed the major point stressed in all of the cited papers. "Consequently he claims that subdivision into isolated groups of small size is favorable to evolutionary progress not as others have held through the variety of environmental conditions to which such colonies are exposed but even if the environment were the same for all, through nonadaptive and casual changes favored by small population size." Actually the point stressed most in these papers was the simultaneous treatment of all factors by the inclusion of coefficients measuring the effects of all of them on gene frequency in a single formula. Thus in the 1931 paper the formula for distribution of gene frequencies in a partially isolated local population included the coefficient N_1 for effective size of the local population, s_1 for selection due to local conditions, m_1 for rate of immigration and q_m for gene frequency among the immigrants (mutation pressure was here assumed to be the same throughout the species). This simultaneous treatment made it possible to specify the conditions under which one or another process would dominate with respect to a particular gene. Sampling fluctuations were treated as only one of a number of processes which lead to trial and error among local populations (Wright, 1931, p. 151). This has been reiterated in all later discussions. Thus in the 1940 paper (p. 175) it was noted that "in a large population, subdivided into numerous partially isolated groups, both adaptive and nonadaptive differentiation is to be expected" and it was concluded that conditions are most favorable for evolution when both processes are occurring. The 1935 paper dealt with quantitative variability, due to multiple factors, with the optimum near the mean. The principal conclusion was that slight oscillations in the position of the optimum in local populations provide an important mechanism by which all gene combinations with approximately the same effect in respect to the character under consideration come to be tried out with respect to secondary effects. It was recognized that such oscillations must be of considerable period to be important. Some significance in causing random changes in gene frequency had, however, been attributed to short time oscillations in severity of selection (Wright, 1931) but this effect was not then analyzed mathematically. This question will be taken up later in this paper.

Fisher and Ford continue with an analysis of annual fluctuations of the frequency of a certain gene in an isolated population of the moth *Panaxia dominula*. They decide that the fluctuations are too great to have been due to accidents of sampling and hence conclude that they must have been due to fluctuations in the action of selection. They arrive at the following generalization.

"The conclusion that natural populations in general, like that to which this study is devoted, are affected by selective action, varying from time to time in direction and intensity and of sufficient magnitude to cause fluctuating variations in all gene frequencies is in good accordance with other studies of observable frequencies in wild populations. We do not think, however, that it has been sufficiently emphasized that this fact is fatal to the theory which ascribes particular evolutionary importance to such fluctuations in gene ratios as may occur by chance in very small isolated populations. . . . Thus our analysis, the first in which the relative parts played by random survival and selection in a wild population can be tested, does not support the view that chance fluctuations can be of any significance in evolution."

Thus Fisher and Ford insist on an either-or antithesis according to which one must either hold that the fluctuations of *all* gene frequencies that are of any evolutionary significance are due to accidents of sampling (attributed to me) or that they are *all* due to differences in selection, which they adopt. As already noted, I have consistently rejected this antithesis and have consistently accepted both sorts as playing important, complementary, and interacting roles. According to the criteria developed in the 1931 paper and later, the genes in a population may be put into 3 classes with respect to the roles of selection and random sampling. One class of segretion must therefore have been responsible for them.

In making these calculations, two years are included in which there are no estimates of population size. If the collections for the 6 years, 1941–1946, for which such estimates were made, are tested merely for heterogeneity, we find that χ^2 is 11.8 which, with 5 degrees of freedom, indicates a probability of about .04 that accidents of sampling, based merely on the limited size of the collections, could have given rise to as large deviations from the average. Thus there is no very compelling reason from the fluctuations themselves for assuming that there were any real fluctuations at all, during this period. It is necessary to include the two earlier years to get convincing evidence for real fluctuations. If this is done, χ^2 rises to 35.9, which, with 7 degrees of freedom, is far beyond even the 0.1% value, 24.3, leaving no doubt of the reality of a shift in gene frequency. The big shift is that between these two earlier years as a group (av. 9.8%) and the following 6 years (5.2%) for which the difference is 4.9 times its standard error, with a probability of about 10^{-6} of arising from accidents of sampling. The question thus largely resolves into what happened between summer 1940 and the next summer to cause such a marked drop in gene frequency and what happened at some undetermined time or times between 1928 when gene frequency was only 1.2% in collections and 1939 to cause a rise to 9.2%.

These changes may well have been due to shifts in the conditions of selection as supposed by the authors but there is nothing in the data as presented to rule out the alternative possibility of reduction to an exceptionally small effective population number on two (or more) occasions, an interpretation for such fluctuations that has long been urged by Elton. Nothing is stated of population size in the period 1928–1939 and all that is stated of the year 1940 is that the moths were not as common in this or later years as in 1939. The number actually collected in 1940 (117) was much less than in any other year.

The mean square of the 7 apparent shifts in gene frequency from 1939 to 1946 is .000488. The mean variance to be expected merely from the sizes of the samples collected is .000269. The difference, .000219, is an estimate of the variance of real shifts in gene frequency in the population.

The authors give no estimate of the amount of variation in selective value necessary to account for such fluctuations. Assume for mathematical simplicity that the heterozygote is intermediate between the two homozygotes. It may be noted that it makes no appreciable difference for this purpose what assumption is made about the rare homozygote bimacula. Assume that selective value is independent of gene frequency and has no trend but varies according to nonsecular fluctuations in conditions from year to year.

If medionigra has a selective advantage of s over type in a particular year, bimacula of $2s$, gene frequency tends to shift by the amount $\delta q = sq(1 - q)$.[2] The estimate for the variance of such shifts is thus $\sigma^2_{\delta q} = \sigma_s^2 \Sigma q^2(1 - q)^2/n$, using σ_s^2 for the variance of fluctuations in the value of the selection coefficient s and n for the number of years used in calculating an average. For the 7 observed shifts in gene frequency, $\sigma^2_{\delta q} = .00453\sigma_s^2$. On equating this to the estimate of the real fluctuations (.000219) we find $\sigma_s^2 = .0483$, $\sigma_s = .22$. It is to be noted that this means an absolute standard deviation of 22%, not a mere 22% of the value of s (which in fact is here assumed to have an average value of zero). A standard deviation of .22 means that in the course of half a century the selective value of medionigra heterozygotes would vary between semilethality (or semisterility) ($s = -.50$) to a 50% advantage over type ($s = +.50$).

[2] See Appendix.

gating genes in any population may be expected to be almost wholly dominated by selection in one way or another, another class almost wholly by accidents of sampling, while an intermediate class, to which special importance was attributed, will show important joint effects.

Fluctuations in Gene Frequency in Panaxia dominula

To which of these classes the particular gene studied by Fisher and Ford belongs makes no difference from this standpoint. The use of criteria may, however, be illustrated by further analysis of the data.

A highly isolated population of the moth *Panaxia dominula* near Oxford, England, was studied. A rather conspicuous color variety, *medionigra*, has been present in this colony for at least 20 years but has not been found elsewhere. A more extreme variety, *bimacula*, which occurs in this colony much less frequently has been shown to be the homozygote. The frequency of the medionigra gene is indicated by study of all available collections to have been less than 1.2% before 1929. The frequency had risen to 9.2% by 1939 when careful study was begun and reached its peak 11.1% in 1940. It dropped to 6.8% next year and since then has varied between 4.3% and 6.5%, without any well defined trend. From 1941 to 1946, the total number of imagines that emerge each season has been estimated from the proportions of captured and marked moths, recaptured after release. The most important data and estimates are shown in table I.

The authors make a statistical analysis of the data for the years 1939 to 1946 on the assumption that there are 1000 parents of each generation. They find that the chance that such great fluctuations could arise from accidents of sampling on this basis is less than 1% ($\chi^2 = 20.806$, 7 degrees of freedom, 1% value 18.475). They conclude that fluctuations in the action of natural selec-

Table I

	Estimates of Population Size N	Types of moth collected				q_m Frequency of m	δq_m Shift in q_m
		$+/+$	$+/m$	m/m	Total moths		
Up to 1928	—	164	4	—	168	.012	(+.080)
1939	—	184	37	2	223	.092	+.019
1940	—	92	24	1	117	.111	−.043
1941	2000–2500	400	59	2	461	.068	−.014
1942	1200–2000	183	22	—	205	.054	+.002
1943	1000	239	30	—	269	.056	−.011
1944	5000–6000	452	43	1	496	.045	+.020
1945	4000	326	44	2	372	.065	−.022
1946	6000–8000	905	78	3	986	.043	—

Estimated population size (N), numbers of each genotype and total moths captured, frequency (q_m) of medionigra gene (m) and shift in frequency (δq_m) to next generation (at least 11 generations in first case).

The homozygotes, on the hypothesis adopted, would range from complete lethality (or sterility) to a selective value twice that of type. This last aspect, however, is quite certainly not tenable since bimacula appeared as regularly when gene frequency dropped as when it rose. Its frequency in the total data was, as the authors note, very close to what would be expected in the absence of selection (11 observed, 10.2 expected).

It seems very probable that the heterozygotes actually have a slight net advantage over both homozygotes, as suggested by the authors. The major shifts in frequency are however so much greater than the possible systematic shifts that the latter can be ignored for the present purpose.

It is of interest to see how small the effective size of population (N) would have to have been to account in full for the estimated real variance of fluctuations. The variance for one generation is $q(1 - q)/2N$. For the period of years in question, we take $\overline{q(1 - q)}/2N = .0648/2N$ as the mean variance to be expected from this cause. On equating to the estimated variance of real fluctuations, .000219, we find $N = 150$.

The effective size of population over a succession of generations is the harmonic mean of the effective sizes for the separate ones and thus may be very much smaller than the arithmetic mean. Its value is largely dominated by occasional very small values in particular years, if these occur. We come back to the conclusion that if the effective number of parents of the 1941 moths was very small (some 100 or less) it is possible to account for all fluctuations as those of small populations, even though the effective number from 1941 to 1946 were as much as 1000.

However, 117 moths were actually captured in the summer of 1940. The question of the relation between effective and apparent population size requires further consideration.

The harmonic mean of the authors' estimates of total number of imagines per year from 1941 is about 2000. The females are stated to lay some 200–300 eggs each. Thus the total number of eggs is typically more than 200,000. The females are described as not flying after emergence until they have laid a considerable proportion of their eggs, probably within the first 24 hours after fertilization. Thus the broods are largely concentrated in single spots, subject to the same environmental vicissitudes. The larvae are stated to winter in the 3rd instar. It is estimated that there may be over 50,000 well grown larvae in the spring to be reduced typically to a thousand or a few thousands by the time of emergence of the adults. The greatest menace to the late larvae is considered to be virus disease and to the pupae, mice.

The authors tacitly assume that this elimination of more than 99% of the individuals from egg to imago is random with respect to broods, unless some allowance is intended by using 1000 instead of 2000 as the population number in their calculations. The possibility of a much greater discrepancy between apparent and effective population number is a matter that would seem to require investigation. Thus analyses of some dozen pure breeds of live stock, including cattle, horses, sheep and swine by means of the inbreeding coefficient of hypothetical progeny from random parents (McPhee and Wright, 1925; Lush, 1943), have shown that in each the effective population number is of the order of 100 or at most a few hundred, a very small proportion of the total numbers registered per generation (tens of thousands in most cases). The special reasons responsible for the extreme smallness of N in these cases would not hold in nature. It is not, however, safe to assume without investigation that there are not other reasons for a considerable reduction of N. If, in the present case, there are any conditions of weather or disease that tend to destroy whole broods at any time in the annual cycle, the effective population number would be cut down accordingly. Again recent work by Dobzhansky and associates (1942, 1944) has revealed an extraordinary prevalence of segregating factors in *Drosophila pseudoobscura* in nature with high selection coefficients under some or all environmental conditions. At first sight this would seem to strengthen the view that selective differences are so important that fluctuations due to limitation in the number of parents would be negligible. This is certainly true for the genes or chromosome conditions in question but very heavy selective elimination in such respects would bring about a correlation between brood mates in fate and so automatically lower the effective population number for other more neutral segregating factors. In a sense, the ensuing shifts in the frequencies of these could be attributed to selection, but if the shifts in these neutral genes are equally likely to occur in either direction they must be credited to sampling variability. The situation is similar, except for the element of intent, to one that is familiar to livestock breeders. With very intensive selection for particular characters, others must be allowed to vary at random if numbers are to be maintained (cf. Wriedt, 1930). It would seem possible, a priori, that the causes of coincident elimination of broodmates (common environment, and common highly selective heredity) may be as important as causes of random elimination. If the former should produce an effect equivalent to elimination of 90% of the broods while the latter are causing random elimination of 90% within the favored broods, leaving the observed 1% to carry on, the effective population number would be only 10% of the apparent number and easily capable of accounting for the observed annual fluctuations.

The alternative is the hypothesis that variety medionigra had an average selective advantage of about 20% from 1928 to 1940 whatever fluctuations there may have been, and then shifted to approximate semisterility or semilethality in that year, followed by fluctuations without trend thereafter.

The present author is certainly in no position to decide between these hypotheses (or some combination of them). In a recent study of the frequency of lethals in nature in one of the chromosomes of *Drosophila pseudoobscura*, it was found that these were only about one fourth as numerous as they should have been on the basis of the adequately determined mutation rate and the necessary rate of elimination as recessive lethals in a large random breeding population (Wright, Dobzhansky and Hovanitz, 1942). On introducing all factors, that might shift gene frequency, into equations describing the observed results it appeared that the discrepancy could be accounted for either by a slight selection against heterozygotes (coefficient s) or by a slight excess tendency toward brother-sister mating (average inbreeding coefficient F) or a combination. In fact the coefficients entered the equation together (to the first order) as the sum ($s + F = .018$) so that it was impossible to separate them. Direct tests for selection (not included in the above statement) indicated no significant selection but the required amount was so slight that more extensive tests would have been necessary. This indeterminacy was unfortunate but it seems likely that more progress will be made in analyzing the difficult problems of the genetics of natural populations if all possibilities are treated symmetrically and any indeterminacy is brought clearly into view rather than concealed by an approach from the point of view of advocacy of any single factor, even one of such undoubted importance as selection.

Comparison of Diverse Type of Fluctuations

It is noted above that the observed annual fluctuations in the frequency of medionigra could be due to fluctuations in selective value above and below zero with a standard deviation of .22. Even if this is the correct interpretation for this gene, it does not follow that all segregating genes show comparable fluctuations. Genes with an average frequency of .05 in an indefinitely large population, and with no net selective advantage or disadvantage ($\bar{s} = 0$) but with fluctuations in selective value measured by a standard deviation $\sigma_s = .10$, would vary in gene frequency to about the same extent as would genes with the same average frequency but with no fluctuations in degree of selection in a population of effective size 1000. In the former case, $\sigma_{\delta q} = \sigma_s q(1-q)$ approximately, by the formula given earlier. With the values given above, $\sigma_{\delta q} = .10 \times .05 \times .95 = .0048$. In the case of neutral genes in a population of 1000, $\sigma_{\delta q} = \sqrt{q(1-q)/2N} = \sqrt{.05 \times .95/2000} = .0049$.

If this sort of calculation is applied to genes with frequency .50, it would appear that fluctuating selective values with standard deviation of only .04 would be equivalent to the effect of sampling in a population of 1000. This, however, does not allow for the fact that the new values of gene frequency with q larger or smaller than .50, that result from the fluctuations, are subject to greater subsequent fluctuations from sampling than from selection, because of the relatively slow falling off of the term $\sqrt{q(1-q)}$ in comparison with $q(1-q)$ as q deviates from .50. A more thorough analysis of the effects of the two kinds of fluctuation is necessary.

In the long run, random fluctuations in gene frequency, whatever their cause, tend to result in a certain distribution of gene frequencies ($\varphi(q)$) about whatever equilibrium value ($\hat{q}$) is determined by the systematic pressures (Δq). The formula of these distribution curves may be found (see appendix). Figures 1 to 7 show such curves (formulae in appendix) under various postulated conditions. The abscissas are the possible gene frequencies from 0 to 1. The ordinates are the frequencies with which

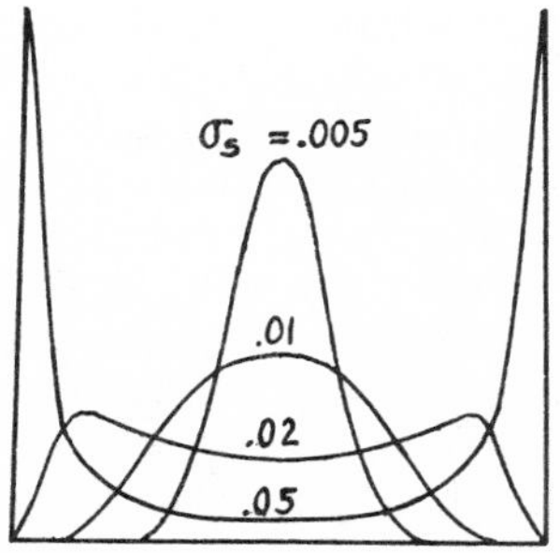

FIG. 1. The distribution of frequencies for genes with different amounts of random fluctuation of selection pressure ($\sigma_s = .005, .01, .02, .05$) in a population in which the only systematic pressure is that due to immigration ($\bar{s} = 0$, $m = .0001$) directed to a mean $\hat{q} = .50$. Size of population is assumed to be so great ($N > 10^6$) that fluctuations of gene frequency due to sampling are negligible.

these values may be expected to occur in the long run if the conditions remain the same. In figures 1 and 2 it is assumed that the frequency of a gene in a local population always tends to move toward .50 because of immigration from the rest of the species, in which this is the average gene frequency. The rate is assumed to be $m = .0001$ (one out of every 10,000 individuals in each generation has become a member of the population by immigration). In figure 1 the population is assumed to be very large, but gene frequency fluctuates about .50

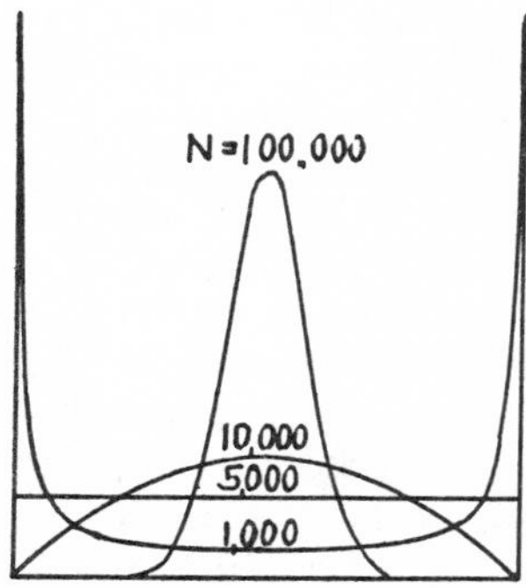

FIG. 2. The distribution of gene frequencies in small populations ($N = 1000, 5000, 10{,}000, 100{,}000$) in which the only systematic pressure is the same as in figure 1 ($\bar{s} = 0$, $m = .0001$, $q_I = .50$) but the only random changes in gene frequency are those due to sampling.

because of variations in the direction and severity of selection (mean selective differential zero ($\bar{s} = 0$), but standard deviation, σ_s, taking values .005, .01, .02 and .05 in the cases illustrated). The value $\sigma_s = .005$ means that the selective value of the heterozygote relative to type ranges from about 99% to 101% of type (deviations of about $2\sigma_s$ in each direction from 100%) and that the homozygotes have twice this range. In this case, the immigration pressure toward gene frequency .50 practically restricts gene frequency to the range .25 to .75. In the last case, however, the cumulative effects of the relatively violent fluctuations in selective value ($\sigma_s = .05$) keep the gene close to fixation or loss most of the time in spite of the continual return of both alleles by immigration. These curves may be compared with those in figure 2 in which no selective differentials are assumed at any time but the effective size of population is assumed to be so small (N = 100,000, 10,000, 5,000 or 1,000 in the 4 cases illustrated) that the cumulative effects of accidents of sampling cause random drift about gene frequency .50, in spite of continual pressure toward this value from the immigration.

It may be seen that a standard deviation of .005 in selective value of heterozygotes (twice this in homozygotes) in a population of millions is only slightly more effective than mere accidents of sampling in a population of 100,000. A standard deviation of .01 in a population of millions is less effective than sampling in a population of 10,000. A standard deviation of .05 is roughly equivalent to sampling in a population of 1000. This is considerably greater than the value .04 reached by comparisons at $q = .50$.

If the immigration pressure were 10 times as great ($m = .001$) the same distribution curves would be obtained with values of σ_s that are 3.2 times as great and with values of N one-tenth as great. In the case of fluctuations in selection the form of the curve depends merely on the value of the parameter m/σ_s^2 and in the case of sampling drift on the value of Nm.

It may be noted that even if m is as great as .01, the variations of gene frequency with $\sigma_s = .05$ in a very large population are only slightly greater than those due to sampling in a population of effective size 1000 (same distributions as those shown for $\sigma_s = .005$ in figure 1 and N = 100,000 in figure 2 with m assumed to be .0001 in both cases). Thus Fisher and Ford's contention that variations in the direction and intensity of selection cause much greater fluctuations in the frequencies of *all* genes than does sampling amounts to the contention that all segregating genes in a population such as that studied are subject to fluctuations in selective value with standard deviations much greater than .05.

Figures 3, 4 and 5 describe the situation where selection tends to increase gene frequency ($\bar{s} = .0002$) while this is opposed by immigration (rate $\bar{m} = .0001$) from populations which lack the gene. In figure 3 it is supposed that fluctuations are due wholly to ones in selective value. It requires a standard deviation

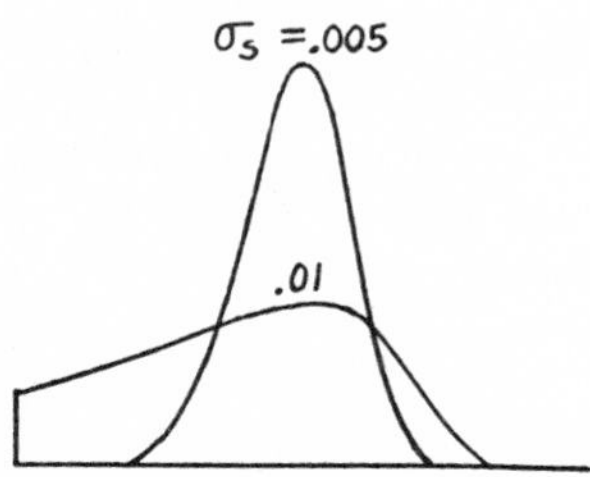

FIG. 3. The distribution of frequencies for genes with different amounts of random fluctuation of selection pressure (σ_s = .005, .01) where there is pressure of favorable selection $\bar{s} = .0002$ in heterozygotes, twice this in homozygotes, opposed by pressure of immigration ($m = .0001$) from populations in which the gene is absent ($q_I = 0$). Size of population is assumed to be so great ($N > 10^6$) that fluctuations of gene frequency due to sampling are negligible. Mutational origin at a very low rate is required to prevent permanent loss of the gene if $\sigma_s = .01$.

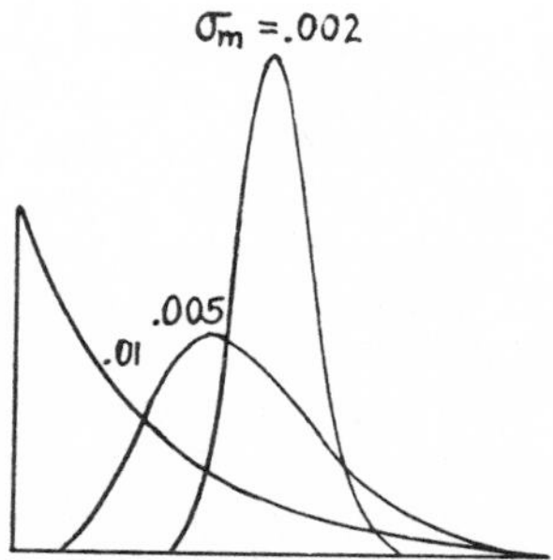

FIG. 4. The distribution of frequencies for genes where the systematic pressures are the same as in figure 3 ($\bar{s}$ = .0002, no dominance, $\bar{m}$ = .0001, q_I = 0) but the amount of immigration fluctuates to different extents (σ_m = .002, .005, .01) and all other random fluctuations are negligible. Mutational origin at a very low rate is required to prevent permanent loss of the gene if σ_m = .01.

approaching σ_s = .005 to cause much departure from equilibrium ($\hat{q}$ = .50). With σ_s = .01 there is wide departure in the direction of low frequencies and with still larger σ_s, the fluctuations in selective values tend to bring about complete and permanent loss of the gene under the postulated conditions. Figure 4 shows distribution of gene frequencies due to fluctuation in immigration rate (a factor which has no effect of this sort where immigration is responsible for the only systematic pressure as in figures 1 and 2). The standard deviation must again be relatively enormous (σ_m = .002, where $\bar{m}$ = .0001) to cause much cumulative effect. There is a much greater effect and a lowering of mean gene frequency if σ_m = .005 and loss of the gene becomes inevitable if σ_m exceeds .01. It may be shown that fluctuations in the gene frequency of immigrants are even less effective. Figure 5 permits comparison with the effects of small population size, under like conditions, except that in this case it is assumed that complete loss is prevented by mutation at the rate of 10^{-6} per generation (s = .0002, m = .0001, v = .000001). It may be seen that the effect of sampling in a population of 100,000 is roughly comparable to that of fluctuation of selective value σ_s = .005 or of fluctuation in the amount of immigration σ_m = .002. In a population of 10,000 the gene would inevitably be lost unless restored by mutation. Even under the assumed conditions, the gene would be completely absent more than 50% of the time in this case.

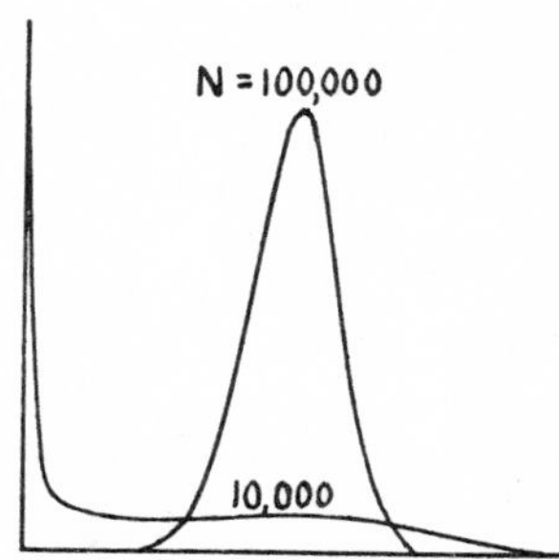

FIG. 5. The distribution of gene frequencies in populations of 10,000 and 100,000 where the systematic pressures are the same as in figures 3 and 4 (s = .0002, no dominance, m = .0001, q_I = 0) except that mutation at the rate $v = 10^{-6}$ is assumed to prevent permanent loss. Random fluctuations of frequency are assumed to be due wholly to sampling.

It seems probable that there are many genes, especially those involved in quantitative variability, that have low net selective values ($\bar{s}$ less than .001 and even than .0001) and correspondingly small fluctuations in value. Fisher indeed has urged that an important phenomenon, the prevalence of dominance of type over deleterious major mutations, is due to genes that act solely as modifiers of dominance, the *maximum* selection pressure on which according to his calculation (Fisher, 1928, 1930) is of the same order as the mutation rate of the mutant gene in question, and thus of the order 10^{-6} per generation. While I have had qualms about accepting the existence of a large class of such nearly neutral genes (Wright, 1929a, b, 1934) [3]

[3] The efficacy of selection in modifying dominance where the heterozygote is the direct object of selection (a subject to which Fisher has devoted much attention (Fisher, 1935; Fisher and Holt, 1944)) does not require the existence of a class of genes with maximum selective differentials anything like as small as 10^{-6} and was explicitly accepted in my first discussion of this matter.

I have nevertheless been inclined to adopt the view that most evolutionary change is due to changes in frequencies of genes of the sorts of which Harland wrote "the modifiers really constitute the species." The "isoalleles" of Stern and Schaeffer (1943) belong in this category.

While annual fluctuations in selective value can have little significance with respect to such factors, variations in selective value of longer period (such as the writer had in mind in his 1935 paper) are, of course, more important. Still more important are the relations between net pressures on genes over very long periods and the effects of sampling. According to the criteria given in the 1931 paper, sampling variance is significant only when the net pressures are small. This is illustrated by changing the interpretations of figures 2 and 5 along the lines indicated earlier. On multiplication of the net pressures due to selection, immigration and mutation by any factor, the distribution curve for gene frequencies remains the same if the hypothetical size of population is divided by the same factor. Immigration pressure is not necessarily the same for all genes but is the same on the simplest model. Selection pressure (and mutation pressure) may, however, differ enormously among different genes and thus may put these in different classes with respect to the importance of sampling. Figure 6 deals with the case of genes which tend to persist at 50% frequency because of equal selection against both homozygotes and which are prevented from ever being permanently lost by mutation rates of 10^{-6} per generation in both directions. Effective population size is taken as 10,000. If the selective advantage of the heterozygote is 1% ($s = .01$), or more, there is very little variation of gene frequency about the equilibrium frequency, .50. If, however, the advantage is only 0.1% there is rather wide variation. Genes for which $\bar{s} = .0003$ vary in frequency from one extreme to the other and each homozygote is fixed about 18% of the time. For genes for which $s = .0001$, each homozygote is fixed more than 30% of the time and frequencies between 10% and 90% are rare in spite of the pressure toward 50% due to both selection and mutation. Without the latter, indeed, one or the other phase would soon become permanently fixed even with genes for which $s = .0003$. It has sometimes been held that nonadaptive differentiation due to sampling can only occur in populations of a few hundred (cf. Lack, 1947) but given enough time and sufficiently neutral segregating genes, this can occur in completely isolated populations whenever $4Nv$ is less than 1, v being the mutation rate from the allele most likely to be fixed under the conditions. Thus with mutation rates of 10^{-6}, there can be random fixation up to a population size of 250,000.

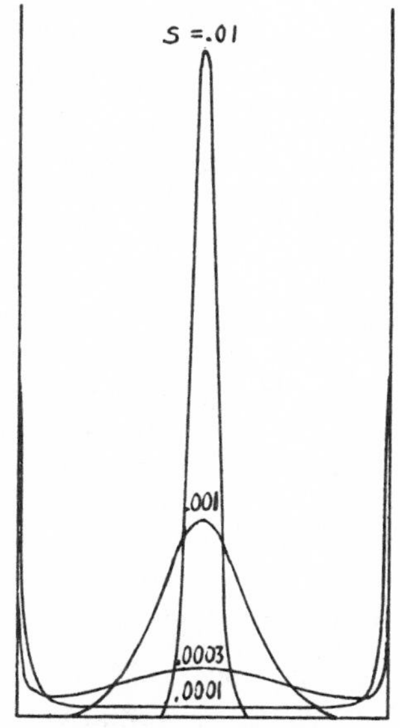

FIG. 6. The distributions of frequencies of 4 classes of genes in a population of effective size 10,000. Selection is assumed to favor the heterozygotes equally over both homozygotes but to different extents (s = .01, .001, .0003, .0001) in the 4 classes. Low rates of mutation equal in both directions ($v = u = 10^{-6}$) are assumed to prevent complete fixation of either homozygote.

Figure 7 deals with a more complex situation in a population of 10,000. Immigration at rate $m = .0001$, tending to establish a gene frequency of .10, is here supposed to be supplemented by selection in which the heterozygous mutant has advantages ranging from .00001 to .001 and homozygotes twice as much in each case. Those genes for which $s = .001$ are almost fixed by

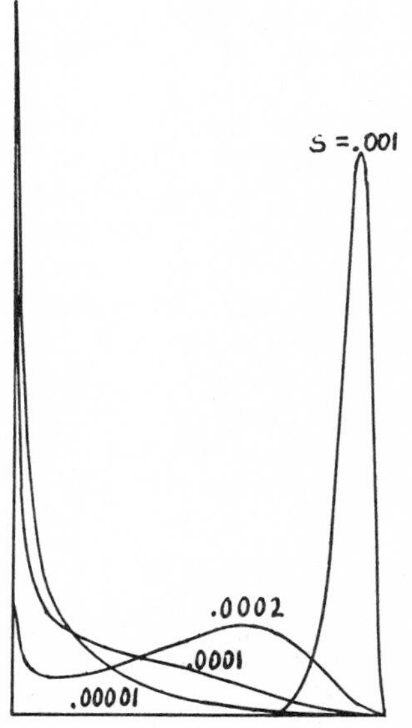

FIG. 7. The distributions of frequencies of 4 classes of genes in a population of effective size 10,000 as in figure 6. In this case, selection is assumed to favor the genes to different extents in the 4 classes ($s = .001, .0002, .0001, .00001$) with no dominance, but with opposition by immigration ($m = .0001$) from population in which gene frequency q_I is .10.

selection in spite of the adverse immigration pressure. Genes for which $s = .0001$ are, on the other hand, largely dominated by the immigration pressure but their frequencies vary over a wide range. At certain intermediate values (e.g. $s = .0002$) there are peak frequencies both above and below 50%. Genes for which $s = 10^{-5}$ are so much dominated by the sampling effect that their frequencies are below 0.5% more than 20% of the time and below 5% about half the time in spite of the pressure of immigration toward a gene frequency of 10% and of selection toward complete fixation.

The importance of random fluctuations of gene frequency, whatever their cause, cannot be adequately evaluated by considering single pairs of alleles in a single population. What is important is simultaneous variation of gene frequencies at many loci, each with many alleles which differ only slightly in selective value, in each of many local populations. The significance is in contributing to the material for the selection of genetic systems as wholes, which may be expected to take place through the welling up of population growth and emigration from those centers in which at the moment the most adaptive systems happen to have been arrived at and the modification by immigration of those centers in which population growth has become relatively depressed because of less successful general adaptation.

It must be emphasized again that this contribution of random differentiation of local population with respect to the more neutral sets of alleles is only one aspect of the whole trial and error process. The great importance of the contribution of selective differentiation among such populations with respect to less neutral alleles is obvious. Moreover, the nonadaptive differentiation is obviously significant only as it ultimately creates adaptive differences.

SUMMARY

Evolution is a process in which many diverse factors are acting simultaneously. Mathematical treatment requires simultaneous treatment of all determinate factors, including both the systematic pressures (recurrent mutation, immigration and selection), which are wholly determinate in principle, and the cumulative effects of random fluctuations of gene frequencies, of which only the variance is determinate. Any one of the factors may play the dominating role for a time under specifiable conditions.

Conditions are held to be most favorable for a continuing process where there are certain states of dynamic equilibrium at many loci. This situation is found to the greatest degree in a population in which there is sufficient isolation of many local centers of population growth and emigration to provide the condition for continual trial and error. The "errors" are the relatively indeterminate elements in the situation: novel mutations at the gene level, and the effects of accidents of sampling, and of variations in the local conditions of selection and migration at the level of the local population. The "trials" are the orderly reactions of the species to changes of gene frequencies, due to more determinate factors, of which selection dominates in the long run. Where there are important secular changes in external relations, selection pressure dominates the situation almost completely until a new state of approximate adjustment is reached and trial and error processes again take over.

The reiteration of the above viewpoint has been stimulated by certain misinterpretations which have arisen. Several authors have attributed to the author the view that accidents of sampling constitute the most important evolutionary process and that conditions are more favorable for evolution the smaller the total population. This is a view that has been emphatically rejected from the first.

Especial attention is devoted to a somewhat different misinterpretation in a recent paper by R. A. Fisher and E. B. Ford. They hold that fluctuations of gene frequencies of evolutionary significance must be supposed to be due wholly either to variations in selection (which they accept) or to accidents of sampling. This antithesis is to be rejected. The fluctuations of some genes are undoubtedly governed largely by violently shifting conditions of selection but for others in the same populations, accidents of sampling should be much more important and for still others both may play significant roles. It is a question of the relative values of certain coefficients.

An analysis of fluctuations in the frequency of a certain gene in a population of the moth *Panaxia dominula*, observed by Fisher and Ford and used by them as the basis for their generalizations, raises considerable doubt as to which category this particular gene belongs.

A comparison is made between the distribution of gene frequencies due to the cumulative effects of various types of random fluctuation in the presence of various pressures. Such comparisons support the view that random drift due to sampling in small local populations is an important member of this category and thus a factor of which account must be taken where there is a favorable type of population structure.

Appendix

The distribution of gene frequencies under diverse causes of fluctuation

The factors of evolution may best be brought under a common viewpoint by measuring each by the change which it tends to bring about in each generation in the gene (or chromosome) frequency (q) under consideration. They may be classified conveniently according to the degree of determinacy in their effects.

The term "pressure" has been used figuratively (Wright, 1929, 1931) for all evolutionary factors with recurrent directed effects capable at least in principle of precise mathematical formulation. The symbol Δq is used. The pressures of recurrent mutation and immigration have linear effects on gene frequency. They can introduce a gene into populations in which it has been absent. Selection pressure can be defined sufficiently broadly to include all processes (such as differential mortality, differential fecundity, differential emigration) which tend to change gene frequency systematically without either change of the hereditary material itself (mutation) or introduction from without (immigration). Selection pressure is necessarily zero either if $q = 0$ or $q = 1$ in contrast with mutation and immigration pressures.

Random fluctuations in gene frequency may be due to accidents of sampling or to fluctuations in the values of the coefficients involved in the evolutionary pressures. The symbol δq has been used for such a change. Direction is indeterminate but the variance ($\sigma^2{}_{\delta q}$) can be specified at least in principle.

In addition it seems necessary to recognize a category that is indeterminate both in direction and variance. This includes events that are unique or nearly so in the history of the species: nonrecurrent mutations, unique hybridizations, nonrecurrent selective events, unique extreme reduction in numbers, etc. The distinction between recurrent events and ones so rarely recurrent as to require treatment as indeterminate is, of course, an arbitrary one.

The systematic pressures (Δq) may lead to fixation or loss of a gene. If not, they tend to hold its frequency at a certain equilibrium point (or points): any value of q at which $\Delta q = 0$, provided that this is stable (Δq opposite in sign to $(q - \hat{q})$ where $\hat{q}$ is the equilibrium value). The random fluctuations (δq) tend to cause random drift from an equilibrium point. The resultant is a probability distribution $\varphi(q)$ which exhibits the frequencies with which the frequency of the gene in question takes all values between 0 and 1 in the long run. It is also the distribution of frequencies of this gene at any moment among a number of populations all subject to the same conditions. The formula for $\varphi(q)$ is a relatively simple function of Δq and $\sigma^2{}_{\delta q}$ (Wright, 1938).

$$\varphi(q) = (C/\sigma^2{}_{\delta q})e^{2\int(\Delta q/\sigma^2{}_{\delta q})dq},$$

where C is a constant such that

$$\int_0^1 \varphi(q)dq = 1.$$

Figures 1 to 7 are obtained by substituting various values of Δq and $\sigma^2{}_{\delta q}$ in this formula. Consider first, however, a somewhat more general case in which heterozygous mutants have a selective advantage s, homozygous mutants an advantage of $2s$ and immigration displaces the proportion m in each generation by immigrants with gene frequency q_I. We must distinguish the mean values of these coefficients ($\bar{s}$, $\bar{m}$, $\bar{q}_I$) and their standard deviation σ_s, σ_m and σ_{qI}, if they are subject to fluctuation. For small values of $\bar{s}$:

$$\Delta q = \bar{s}q(1 - q) - \bar{m}(q - \bar{q}_I).$$

Mutation pressure may be introduced by taking advantage of the equivalence of the theories of immigration pressure and mutation pressure. Letting $\bar{v}$ be the mean rate of mutation to the gene and $\bar{u}$ that from it per generation

$$\begin{aligned}\Delta q &= \bar{s}q(1-q)+(\bar{v}+\bar{m}\bar{q}_I)(1-q) \\ &\qquad -[\bar{u}+\bar{m}(1-\bar{q}_I)]q \\ &= \bar{s}q(1-q)-(\bar{m}+\bar{u}+\bar{v}) \\ &\qquad \times[q-(\bar{v}+\bar{m}\bar{q}_I)/(\bar{m}+\bar{u}+\bar{v})].\end{aligned}$$

For simplicity in most of what follows, $\bar{m}$ will be used in place of $(\bar{m} + \bar{u} + \bar{v})$ and $\bar{q}_I$ in place of $(\bar{v} + \bar{m}\bar{q}_I)/(\bar{m} + \bar{u} + \bar{v})$.

As indicated above random fluctuations in gene frequency may be due to fluctuations in selection (σ_s), in amount of immigration (σ_m), or in gene frequency of immigrants (σ_{qI}) as well as to accidents of sampling, and in general are due to some combination of all of these. We wish, however, to compare their separate effects.

(a) Fluctuations only in selection

$$\begin{aligned}\delta q &= (s-\bar{s})q(1-q) \\ \sigma^2{}_{\delta q} &= \sigma_s{}^2q^2(1-q)^2 \\ \varphi(q) &= C[q/(1-q)]^{2[\bar{s}-m(1-2q_I)]/\sigma_s{}^2} \\ &\quad \times q^{-2}(1-q)^{-2}e^{-2m[q_I(1-q)+(1-q_I)q]/\sigma_s{}^2q(1-q)}\end{aligned}$$

(b) Fluctuations only in amount of immigration

$$\delta q = -(m - \bar{m})(q - q_I)$$
$$\sigma^2_{\delta q} = \sigma_m^2 (q - q_I)^2$$
$$\varphi(q) = C(q - q_I)^{2[s(1-2q_I) - \bar{m} - \sigma_m^2]/\sigma_m^2} \times e^{-2s[(q-q_I)^2 + q_I(1-q_I)]/\sigma_m^2(q-q_I)}$$

(c) Fluctuations only in gene frequency of immigrants

$$\delta q = m[q_I - \bar{q}_I]$$
$$\sigma^2_{\delta q} = m^2 \sigma^2_{q_I}$$
$$\varphi(q) = Ce^{[6m\bar{q}_I q - 3q^2(m-s) - 2sq^3]/3m^2\sigma^2_{q_I}}$$

(d) Fluctuations only from sampling (Wright, 1931)

$$\sigma^2_{\delta q} = q(1-q)/2N$$
$$\varphi(q) = Ce^{4Nsq} q^{4Nmq_I - 1}(1-q)^{4Nm(1-q_I)-1}$$

Figures (1) and (2) deal with the special case of genes for which the only pressure is that due to immigration (m = .0001) and gene frequency in the immigrants is .50.

Fig. 1

$$\Delta q = -m(q - .5)$$
$$\sigma^2_{\delta q} = \sigma_s^2 q^2 (1-q)^2$$
$$\varphi(q) = Cq^{-2}(1-q)^{-2} e^{-m/\sigma_s^2 q(1-q)}$$

Fig. 2

$$\sigma^2_{\delta q} = q(1-q)/2N$$
$$\varphi(q) = C[q(1-q)]^{2Nm-1}$$

Figures 3, 4 and 5 relate to cases in which the pressure of favorable selection ($\bar{s}$ = .0002) with no dominance is opposed by the pressure of immigration (m = .0001, q_I = 0). Mutational origin of the gene at a low rate is required to prevent its loss in extreme cases. In figure 5 this rate is taken as $v = 10^{-6}$.

Fig. 3

$$\Delta q = \bar{s}q(1-q) - mq$$
$$\sigma^2_{\delta q} = \sigma_s^2 q^2 (1-q)^2$$
$$\varphi(q) = C[q/(1-q)]^{2(\bar{s}-m)/\sigma_s^2} \times q^{-2}(1-q)^{-2} e^{-2m/\sigma_s^2(1-q)}$$

Fig. 4

$$\sigma^2_{\delta q} = \sigma_m^2 q^2$$
$$\varphi(q) = Cq^{2(s - \bar{m} - \sigma_m^2)/\sigma_m^2} e^{-2sq/\sigma_m^2}$$

Fig. 5

$$\Delta q = sq(1-q) - mq + v(1-q)$$
$$\sigma^2_{\delta q} = q(1-q)/2N$$
$$\varphi(q) = Ce^{4Nsq} q^{4Nv-1}(1-q)^{4Nm-1}$$

Figures 6 and 7 make comparisons between genes subject to different degrees of selection in a population of limited size (N = 10,000). In figure 6 it is assumed that selection favors heterozygotes equally over both homozygotes but varies in degree (s = .01, .001, .0003 or .0001).

Low rates of mutation, equal in both directions ($u = v = 10^{-6}$), are assumed to prevent permanent fixation.

Fig. 6

$$\Delta q = -[2v + 2sq(1-q)][q - .5]$$
$$\sigma^2_{\delta q} = q(1-q)/2N$$
$$\varphi(q) = Ce^{4Nsq(1-q)} q^{4Nv-1}(1-q)^{4Nv-1} = Ce^{40{,}000sq(1-q)}[q(1-q)]^{-.96}.$$

In figure 7, it is assumed that immigration at rate m = .0001 tends to maintain gene frequency at q_I = .10 but that this is disturbed to varying extents by favorable selection (no dominance). $s = 10^{-5}$, 10^{-4}, 2×10^{-4} and 10^{-3} respectively in a population of 10,000.

Fig. 7

$$\Delta q = sq(1-q) - m(q - q_I)$$
$$\sigma^2_{\delta q} = q(1-q)/2N$$
$$\varphi(q) = Ce^{4Nsq} q^{4Nmq_I - 1}(1-q)^{4Nm(1-q_I)-1} = Ce^{40{,}000sq} q^{-.6}(1-q)^{2.6}$$

The frequencies in fixed classes in any of these cases are estimated where necessary from the subterminal classes by formulae (Wright, 1931)

$$f(0) = [f(1/2N)]/4N(mq_I + v),$$
$$f(1) = \{f[(2N-1)/2N]\} / 4N[m(1-q_I) + u],$$

where

$$f(q) = \varphi(q)/2N.$$

Literature Cited

Dobzhansky, Th., A. M. Holz, and B. Spassky. 1942. Genetics of natural populations. VIII. Concealed variability in the second and fourth chromosomes of *Drosophila pseudoobscura* and its bearing on the problem of heterosis. Genetics, **27**: 463–490.

Dobzhansky, Th. and B. Spassky. 1944. Genetics of natural populations. XI. Manifestation of genetic variants in *Drosophila pseudoobscura* in different environments. Genetics, **29**: 270–290.

Elton, C. S. 1924. Periodic fluctuations in the number of animals: their causes and effects. Brit. J. Exp. Biol. **3**: 119–163.

Fisher, R. A. 1928. The possible modification of the response of the wild type to recurrent mutations. Amer. Nat., **62**: 115–126.

——. 1930. The genetical theory of natural selection. Oxford, Clarendon Press. 272 pp.

——. 1935. Dominance in poultry. Phil. Trans. Roy. Soc. B, **225**: 197–226.

Fisher, R. A., and S. B. Holt. 1944. The experimental modification of dominance in Danforth's short tailed mutant mice. Ann. Eug., **12**: 102–120.

Fisher, R. A., and E. B. Ford. 1947. The spread of a gene in natural conditions in a colony of the moth *Panaxia dominula* L. Heredity, **1**: 143–174.

Goldschmidt, R. 1940. The material basis of evolution. Yale Univ. Press, New Haven. 436 pp.

Harland, S. C. 1936. The genetical conception of the species. Biol. Rev., **11**: 83–112.

Lack, David. 1947. Darwin's finches. Cambridge at the University Press. 208 p.

Lush, J. L. 1943. Animal breeding plans. The Iowa State College Press, Ames, Iowa. 437 pp.

McPhee, H. C., and S. Wright. 1925. Mendelian analysis of the pure breeds of live stock. III. The shorthorns. Jour. Hered., **16**: 205–215.

Stern, Curt, and Elizabeth W. Schaeffer. 1943. On wild type iso-alleles in *Drosophila melanogaster*. Proc. Nat. Acad. Sci., **29**: 361–367.

Wriedt, C. 1930. Heredity in live stock. Macmillan & Co., London. 179 pp.

Wright, S. 1929. Fisher's theory of dominance. Amer. Nat., **63**: 274–279.

——. 1929. The evolution of dominance. Amer. Nat., **63**: 556–561.

——. 1931. Evolution in Mendelian populations. Genetics, **16**: 97–159.

——. 1932. The roles of mutation, inbreeding, crossbreeding and selection in evolution. Proc. 6th Internat. Congress of Genetics, **1**: 356–366.

——. 1934. Physiological and evolutionary theories of dominance. Amer. Nat., **68**: 25–53.

——. 1935. Evolution in populations in approximate equilibrium. Jour. Gen., **30**: 257–266.

——. 1938. The distribution of gene frequencies under irreversible mutation. Proc. Nat. Acad. Sci., **24**: 253–259.

——. 1940. The statistical consequences of Mendelian heredity in relation to speciation. In "The New Systematics," pp. 161–183 (edited by Julian Huxley). Clarendon Press, Oxford.

Wright, S., Th. Dobzhansky, and W. Hovanitz. 1942. Genetics of natural populations. VII. The allelism of lethals in the third chromosome of *Drosophila pseudoobscura*. Genetics, **27**: 363–394.

FISHER AND FORD ON "THE SEWALL WRIGHT EFFECT"

By SEWALL WRIGHT

Department of Zoology, University of Chicago

IN A PAPER in 1947, R. A. Fisher and E. B. Ford published data on the fluctuations in frequency of a certain color factor in a colony of a moth, *Panaxia dominula* [8]. They claimed to prove that these fluctuations were too great to have been due to accidents of sampling from generation to generation, and that the fluctuations must be supposed, by elimination, to have been due largely to variability in the direction and severity of selection. The authors held that their demonstration was fatal to the theory of evolution which they ascribe to me: that there is "a special evolutionary advantage to small isolated communities."

In a reply [22] I pointed out that their statement of the theory omitted its most essential features. As to the fatal effect of their demonstration, it was noted that a great many gene frequencies (possibly including all relating to genes with major differential effects) might fluctuate largely from selective variability without affecting significantly the role actually ascribed to accidents of sampling in partially isolated communities of a large species. While not relevant to this main point, it was also noted that the supposed demonstration of these investigators was highly questionable in its biological premises and in the statistical handling of the data. It seemed at least as probable from the published data that the observed fluctuations were actually due merely to the small and highly variable numbers of parents as to variations in selection from year to year.

In a reply to this paper [9] Fisher and Ford do not discuss these points but reiterate their criticism of the theory. It may be noted here that the use of my name for the evolutionary effect of inbreeding is hardly appropriate. The first author to suggest that random differentiation among small isolated populations was something that must be taken into account seems to have been Gulick, in 1872 [11]. I have ascribed only minor evolutionary importance to this process *by itself*. It is even more apparent than before that we are writing almost wholly at cross purposes with respect to what my theory actually is.

It is accordingly desirable to try to define the issue by listing certain points on which there is probably agreement: (1) Evolutionary transformation consists almost wholly in changes in frequencies in the system of chromosome patterns and Mendelian alleles of populations. (2) The course of such transformation is controlled largely by selection. (Fisher and Ford do not seem to recognize the agreement here.) (3) Mutations merely furnish random raw material for evolution and rarely, if ever, determine the course of the process. (Polyploidy may in a sense be considered an exception.) (4) Mutations with slight effects are more likely to be fixed by natural selection than are those with large effects.

(5) Fluctuations in gene frequencies in small completely isolated communities rarely if ever contribute to evolutionary advance, but merely to trivial differentiation, or in extreme cases to degeneration and extinction. (Fisher and Ford accept at least the first part of this statement but do not seem to recognize my agreement as stated in 1931 and in many later papers.)

Real disagreement seems to center on whether or not control by selection is so direct and complete as to preclude any significant role of random processes above the level of mutation. Fisher's belief in the virtual completeness of control by direct action of selection is illustrated in his theory of the prevailing dominance of type over deleterious mutations, which he advanced in 1928 and has recently reiterated [5, 7]. He pointed out that genetic factors modifying the appearance of a rare intermediate heterozygote, in the direction of type, tend to increase at a rate which he showed to be of the order of the mutation rate of the deleterious gene in question, or less, if multiple modifiers are required to make the latter completely recessive. As there are several hundred loci at which deleterious recessives occur in such a form as *Drosophila melanogaster*, the hypothesis postulates the occurrence of alleles at several hundred loci (or even thousands, if multiple modifiers are typical) at which the selective differential due to the modifying effect is only of the order of 10^{-6} or less per generation, and at which there is no differential effect in the absence of the mutation that takes precedence over this [15, 19, 12, 14].

In the present paper [9], the above belief seems to be involved in the contention that a demonstration that year-to-year fluctuations in the frequency of a particular gene are predominantly selective is fatal to the view that random fluctuations due to inbreeding may have a significant function for any genes.

I shall not attempt a full exposition of my views here, but refer to a recent nonmathematical summary [24]. The mathematical aspects, apart from the details of the theory of population structure, have also been summarized in a recent paper [23], and the latter aspect of the problem in another [25]. The central concept is that certain states of labile balance among the factors (mutation rate, inbreeding, cross breeding, and selection) give a more effective evolutionary mechanism than the operation of any single factor, even of selection, in excess.

The divergence in views probably traces ultimately to the view which I have held, and Fisher and Ford reject, that the effectiveness of selection is restricted in a random breeding population by the prevalence of pleiotropic effects of alleles and of nonadditive effects in combination. If such effects are unimportant in relation to adaptive value as a character, mass selection in a random breeding population guides the genetic system undeviatingly toward the single, most favorable complex possible from the genes that are present. If, however, such effects are important, there tend to be many distinct harmonious combinations (adaptive peaks). Mass selection tends merely to guide the genetic system toward the one that happens to have been most nearly approached for historical reasons, and this is not likely to be the highest.

Fisher's theory requires that mutations occur at an appreciable rate to provide raw material for the continuing operation of selection, but he does not ascribe special evolutionary advantage to the typical mutation.

Similarly, I have held that random changes in the genetic complexes of small local populations within a species are unlikely to be of evolutionary advantage to the populations individually, and that this is also true of the much larger deviations that accumulate from the interplay of such changes and mass selection. Nevertheless, it is held that in the whole array of such differentiated local populations there is the raw material for a different sort of selection in which the genetic complexes as wholes, rather than the mere net differential effects of the individual genes, are the units. This process (intergroup selection) depends on a certain balance between local isolation and cross breeding. There must be sufficient isolation to permit local differentiation, but there are no consequences of importance to the species as a whole, unless there is also sufficient possibility of excess emigration from centers in which superior genetic complexes have appeared and have been improved by local selection, to permit grading up of neighboring regions to the point at which selective improvement along the same line becomes autonomous in them. Once started such a process may spread through the species step by step. There may be spread from several centers coincidentally. Interaction and differentiation along the way insure continual shifting of the centers. The course of evolution of the species as a whole is, of course, guided predominantly by the genetic complexes of the population *sources* rather than of the population *sinks*. Haldane has supported this concept, with especial reference to the human species during the thousands of centuries of the Old Stone Age [13]. The idea that the variation in the frequencies of the blood groups is a relic of such a prehistoric condition of mankind [16, 17, 18] has been supported as at least one of the factors, by Boyd [1].

While the author has stressed especially the role of accidents of sampling in causing random differentiation among neighborhoods, other random processes have also been considered important. Among these are effects of long-time fluctuations in the positions of the optima of quantitatively varying characters among local populations [20]. Other categories are discussed in a later paper [22]. In addition to these, selection related to differences in local conditions has been treated at length [16]. This, however, is not a random process. Its primary significance lies in current adaptability of the species, but it also has significance on the one hand in maintaining a store of potential variability within the species and, on the other, in leading occasionally to cleavage. It has, however, been noted as possible that an adaptation to special local conditions may occasionally prove to confer a general advantage, and thus contribute to intergroup selection in a way similar to random differentiation.

The crucial questions with respect to the place of the inbreeding effect in evolution are: (1) how far genetic complexes may have selective significance beyond the sum of the effects ascribable to the component genes, (2) to what extent favorable population structures exist in nature, (3) whether the selective differentials of alleles at many loci are sufficiently slight, and (4) whether the consequences brought about by intergroup selection on the basis of complexes of such genes may be important in evolution.

I do not pretend that it is easy to answer these questions from observations. The theory was developed from the attempt to decide on the most favorable conditions for evolution indicated by the mathematical theory on taking simultaneous account of all of the factors. These are not necessarily the conditions most likely to hold in a typical species at a given time. Other processes were indicated to be effective, but less than this.

With respect to (1), it is a commonplace that major genes nearly always have pleiotropic effects and that the effects of combinations are rarely if ever wholly predictable from the separate effects and are often utterly unpredictable. We do not, however, know much about pleiotropic and combination effects of minor factors. With respect to (2), the mathematical analysis of the consequences of restricted dispersal in continuous populations indicates that the conditions for a sufficiently fine-scaled structure, while severe in a species with uniform density, are by no means impossible and are probably often realized where density is not uniform [21, 25]. Realization is especially likely where there are numerous largely isolated colonies or "demes" [3]. With respect to (3), it is not necessary to assume that many alleles have such exceedingly slight selective differentials as required by Fisher's theory of dominance. Differentials of the order of 10^{-3} or 10^{-4} are small enough and can easily occur in multifactorial systems, and presumably between isoalleles. As to (4), the theory is undoubtedly concerned most directly with the quantitative variability of the species and very little with conspicuous polymorphism.

It is probable that conspicuous polymorphism is usually a device for adaptation to diverse conditions encountered by the species. It may relate to adaptation to different seasonal conditions as shown by Dobzhansky in the case of chromosome patterns in *Drosophila pseudoobscura* [4, 26]. In other instances it may give a basis for adaptation to diverse microenvironments as discussed in the last reference. The observations of Cain and Sheppard on the frequencies of different color patterns in the land snail *Cepaea nemoralis* under different ecological conditions suggest this interpretation [2].

Polymorphism in which the advantage to the species comes from the mere fact of diversity, rather than in the special properties of alleles, does not come under this head. The occurrence of multiple self-incompatibility alleles of many plants is an example. The prevalance of antigenic polymorphism in animals (e.g., blood groups of man) suggests that diversity may be of advantage in this case also. In such examples the differences among local populations may be due to stochastic processes either wholly or in part, even though the maintenance of the multiple alleles depends on selection. If wholly stochastic, the differences have, of course, only trivial importance in evolution. Let me emphasize that it is only as stochastic differentiation gives a basis for selection in the form of differential population growth and migration that it contributes significantly to evolution.

Judgments of the importance of quantitative variability in evolution vary. Goldschmidt, for example, gives it little weight [10]. I shall not attempt to go into the reasons for which I (and I think also Fisher and Ford) ascribe major importance to it.

It may be stated, however, that the processes of random differentiation and intergroup selection, operating on complexes of modifiers may be of great importance indirectly, and are perhaps necessary, for the establishment of just such major mutations as Goldschmidt would consider significant. The primary objection to the hypothesis of evolution by major mutations is the negligible chance of harmonious adjustment to the rest of the organism. Suppose, however, that one of these mutations offers the possibility of a major advance if only such an adjustment can come about. If the mutation recurs with ordinary frequency and is of such low penetrance as to withstand rapid elimination, it may be expected to appear in all local populations of a large species from time to time. Random drift can never carry the mutation beyond very low frequencies, however. There seems no appreciable chance of improving its adjustment and bringing it to fixation by selection of modifiers, because of its rarity and the consequent failure of selection of modifiers to rise above the threshold imposed by their effects in the absence of the mutation. If, however, the frequencies of numerous almost neutral modifiers are drifting largely at random in many local populations, there is a good chance that somewhere, at some time, a complex will be acquired which carries the adjustment of the major mutation, if it appears locally, beyond the threshold for successful adaptation, and it (and later the complex of modifiers) comes under direct favorable action of selection. The stage is then set for the spread of the whole complex through the species by the process previously described. Observations at any given time would be interpreted as indicating direct control by selection, yet the processes of random drift and intergroup selection have been of primary importance in the process.

The criticism stressed by Fisher and Ford as completely "fatal" to the author's theory, is that "it is not only small isolated populations, but also large populations that experience fluctuations in gene ratio," and this because of fluctuations due to variable selection. They correctly note that "it is only the random sampling fluctuation which is accentuated by the small size of an isolated population; other causes like selective survival varying from year to year will influence large populations equally. . . . This central criticism seems to have escaped Wright's attention, so that in a recent article in *Evolution* he has attributed to us opinions entirely contrary to those which we hold and clearly express in our paper. Thus on p. 291 he says: 'They hold that fluctuations of gene frequencies, of evolutionary significance, must be supposed to be due wholly to variations in selection (which they accept) or to accidents of sampling. This antithesis is to be rejected.' This passage constitutes a direct misstatement of our published views. There is nothing in our article even to suggest the antithesis which Wright ascribes to us. Not only do we presume throughout that accidents of sampling produce their calculable effects in causing fluctuation in gene ratios, but we take some care to evaluate them. An earlier and slightly different statement by Wright to the same effect occurs on p. 281. 'Thus Fisher and Ford insist on an either-or antithesis according to which one must either hold that the fluctuations of *all* gene frequencies, that are of any evolutionary significance, are due to accidents of sampling (attributed to me) or that they are *all* due to differences in selection, which they adopt.' "

I certainly had no intention of stating that Fisher and Ford had ignored *trivial* fluctuations due to accidents of sampling. I was considering only fluctuations of *evolutionary significance* and introduced this phrase into both sentences to guard against such misinterpretation. The first quotation is perhaps somewhat ambiguous out of context. The next two sentences were as follows: "The fluctuations of some genes are undoubtedly governed largely by violently shifting conditions of selection, but for others in the same population, accidents of sampling should be much more important, and for still others both may play significant roles. It is a question of the relative values of certain coefficients."

The antithesis which Fisher and Ford claim not to have suggested inheres in their use of the word "fatal" in the connection indicated above. A demonstration that the fluctuations of a particular gene frequency are largely selective could not be held by them to be fatal to my hypothesis unless they ascribed to me the view that the significant fluctuations of *all* gene frequencies are due to accidents of sampling, and they themselves adopted the antithetic view that *all* changes of possible evolutionary significance are selective. If, however, they will accept my phrase "of evolutionary significance" in place of their substitute "calculable," and rewrite one of their sentences quoted above as follows, "Not only do we presume throughout that accidents of sampling produce effects of evolutionary significance in causing fluctuations in gene ratios but we take some care to evaluate them," I will be glad to acknowledge that I misinterpreted them.

The authors go on to accuse me of misinterpreting my own earlier published views: "Nothing could be further from our actual criticism of the particular contribution to evolutionary theory which is due to Sewall Wright. He tells us that he now attaches importance of accidents of gene sampling only as one of many factors, and (p. 281) that he has always done so. This latter statement is hard to reconcile with his earlier writings. Thus in the *Statistical Theory of Evolution* he says of 'nonadaptive radiation' (p. 208): 'In short, this seems from statistical considerations to be the only mechanism which offers an adequate basis for a continuous and progressive evolutionary process.' He ends the same paper with the sentence: 'In particular, a state of subdivision of a sexually reproducing population into small, incompletely isolated groups provides the most favorable condition, not merely for branching of the species, but also for its evolution as a single group.' "

The sentences quoted above are again taken out of context. The two sentences in my paper that immediately preceded the former of these were as follows: "This process of *intergroup selection* may be very rapid as compared with mass selection of individuals among whom favorable combinations are broken up by the reduction-fertilization mechanism in the next generation after formation. With partial isolation and differentiation accompanying expansion of the successful subgroups, the process may go on indefinitely." The antecedent of "this" in the quoted sentence was thus "intergroup selection" instead of "nonadaptive radiation" as stated by Fisher and Ford. Intergroup selection depends on the balanced action of multiple processes as suggested in the rest of the quotation. In the case of the last sentence, the context also shows that the reference was to the process of intergroup selection, not to random drift *per se*.

I do not, of course, wish to maintain that my views on evolution have stood entirely still since 1931 when the above statements were written. The former of these is undoubtedly too extreme. Qualifications and additions have been made, beginning in 1932, and have continued up to the present time. The interplay of directed and random processes in populations of suitable structure has, however, continued to be the central theme.

REFERENCES

1. Boyd, W. C. *Genetics and the races of man.* Little, Brown & Co., Boston 1950. 453 pp.
2. Cain, A. J., and Sheppard, P. M. Selection in the polymorphic land snail Cepaea nemoralis. *Heredity, 4*, 275–294, 1950.
3. Carter, G. S. *Animal evolution.* Sidgwick and Jackson, Limited, London, 1951. 368 pp.
4. Dobzhansky, Th. Genetics of natural populations. IX. Temporal changes in the composition of populations of Drosophila pseudoobscura. *Genetics, 28*, 162–186, 1943.
5. Fisher, R. A. The possible modification of the response of the wild type to recurrent mutations. *Amer. Nat., 62*, 115–126, 1928.
6. ——. *The genetical theory of natural selection.* Oxford, Clarendon Press, 1930. 272 pp.
7. ——. *The theory of inbreeding.* Oliver & Boyd, Edinburgh, 1949. 120 pp.
8. Fisher, R. A., and Ford, E. B. The spread of a gene in natural conditions in a colony of the moth Panaxia dominula L. *Heredity, 1*, 143–174, 1947.
9. ——. The "Sewall Wright Effect." *Heredity, 4*, 117–119, 1950.
10. Goldschmidt, R. *The material basis of evolution.* Yale University Press, 1940. 436 pp.
11. Gulick, J. T. On the diversity of evolution under one set of external conditions. *Linnean Soc. Jour. Zool., 11*, 496–505, 1872.
12. Haldane, J. B. S. A note on Fisher's theory of the origin of dominance; and on a correlation between dominance and linkage. *Amer. Nat., 64*, 87–90, 1930.
13. ——. Human evolution: past and future. Chapter 7, pp. 405–418, in: *Genetics, paleontology and evolution,* edited by G. L. Jepsen, G. G. Simpson, and E. Mayr. Princeton University Press, 1949.
14. Lush, J. L. The theory of inbreeding (by Ronald A. Fisher). *Amer. Jour. Human Genetics, 2*, 97–100, 1950.
15. Wright, S. Fisher's theory of dominance. *Amer. Nat., 63*, 274–279, 1929.
16. ——. Evolution in Mendelian populations. *Genetics, 16*, 97–159, 1931.

17. WRIGHT, S. Statistical theory of evolution. *Amer. Stat. Jour.* (March supplement), 201–208, 1931.
18. ——. The roles of mutation, inbreeding, crossbreeding and selection in evolution. *Proc. 6th Internat. Congress of Genetics*, *1*, 356–366, 1932.
19. ——. Physiological and evolutionary theories of dominance. *Amer. Nat.*, *68*, 25–53, 1934.
20. ——. Evolution in populations in approximate equilibrium. *Jour. Genetics*, *30*, 257–266, 1935.
21. ——. Isolation by distance. *Genetics*, *28*, 114–138, 1943.
22. ——. On the roles of directed and random changes in gene frequency in the genetics of populations. *Evolution*, *2*, 279–294, 1948.
23. ——. Adaptation and selection. Chapter 20, pp. 365–389, in: *Genetics, paleontology and evolution*, edited by G. L. Jepsen, G. G. Simpson, and E. Mayr. Princeton University Press, 1949.
24. —— Population structure in evolution. *Proc. Amer. Phil. Soc.*, *93*, 471–478, 1949.
25. ——. Genetical structure of populations. *Annals of Eugenics*, 1951 (in press).
26. WRIGHT, S., and DOBZHANSKY, TH. Genetics of natural populations. XII. Experimental reproduction of some of the changes caused by natural selection in certain populations of Drosophila pseudoobscura. *Genetics*, *31*, 125–156, 1946.

36
Evolution, Organic

Encyclopaedia Britannica, 14th ed. revised (1948), 8:915–29

37
Genetics of Populations

Encyclopaedia Britannica, 14th ed. revised (1948), 10:111–15

Introduction

Wright continued in the finest tradition of the *Encyclopaedia Britannica* by contributing two extremely thoughtful and clear articles on evolution and the genetics of populations. He spent a great deal of time writing these articles. The shifting balance theory was submerged in comparison to his usual presentation of the evolutionary process, but even so was not far below the surface.

Crow has pointed out to me that in "Genetics of Populations" Wright presented a new model that assumed mutant alleles were introduced serially into the population rather than preexisting there. Fifteen years later, Kimura and Crow developed this approach as the "infinite allele model" (Kimura and Crow 1964), which has had considerable influence among mathematical population geneticists. In a letter of July 17, 1985, Crow added: "We developed this idea completely independently. I only read Wright's article many years after it was written and after we had published our paper. It never occurred to me to read an encyclopedia article for a totally new idea. Although Wright referred to our paper in his publications and in conversation, he never asserted his priority for the idea."

experiment along such lines is now being conducted. This seems to bear more directly upon the weight to be accorded evidence than upon admissibility of evidence under technical legal rules. Yet many of these rules seek justification in the idea that the evidence or witnesses excluded by them fall into classes of low general credibility; and supposed conditions of special credibility are invoked to justify exceptions to such rules. Hence law teachers and psychologists are working in teams at Yale and Columbia universities to ascertain whether these fundamental classifications and sub-classifications should be accepted as sound. This work has barely begun, but such application of scientific method may ultimately lead to important results.

The topic just touched upon indicates a very real difficulty in bettering rules of evidence. Psychology sounds academic and unlawyer-like. Practical criticism by Wigmore, a law teacher, broke down Münsterberg's early proposals. The law teacher in turn is deemed an unworldly adviser by the active American lawyer. Justification for such distrust must be removed, and the feeling itself allayed, if harmonious progress is to be made. The American Law Institute shrewdly provides for participation by every branch of the profession in the composition of its restatements. When the Legal Research Committee of the Commonwealth Fund put a special committee under the chairmanship of Prof. E. M. Morgan to work on problems of evidence, this special committee spent nearly five years upon detailed enquiry into opinions held by hundreds of lawyers and judges. In the end it unanimously recommended five important revisions of evidential rules. Significantly and encouragingly, the States of New York and Rhode Island have already adopted one of the recommendations. On the whole, there is reason to hope that wise persistence will cause steady, if slow, rationalization and improvement of judicial proof in the United States.

BIBLIOGRAPHY.—Text-books: Simon Greenleaf, *Treatise on the Law of Evidence* (16th ed. by J. H. Wigmore and others, 1899); J. B. Thayer, *Preliminary Treatise on Evidence at the Common Law* (1898); J. H. Wigmore, *Treatise on the Anglo-American System of Evidence in Trials at Common Law* (2d ed., 1923); C. F. Chamberlayne, *Treatise on the Modern Law of Evidence* (1911–16); *Jones' Commentaries on the Law of Evidence in Civil Cases* (based upon the work of B. W. Jones, 2d ed., by J. M. Henderson, 1926); H. C. Underhill, *Treatise on the Law of Criminal Evidence* (3rd ed., 1923). Essays, etc.: J. B. Thayer, *Legal Essays*, pp. 207 *et seq.* on "Bedingfield's Case," "Law and Logic," and "A Chapter of Legal History in Massachusetts" (1908); Z. Chafee, Jr., "Progress of the Law, 1919–1922: Evidence," 35 *Harv. L. Rev.*, pp. 302, 428, 673 (1922); V. H. Lane, "Right of the Jury to Pass upon the Admissibility of Dying Declarations," 1 *Mich. L. Rev.* 624 (1903); J. M. Maguire and C. S. S. Epstein, "Preliminary Questions of Fact in Determining the Admissibility of Evidence," 40 *Harv. L. Rev.* 392 (1927); Eustace Seligman, "An Exception to the Hearsay Rule," 26 *Harv. L. Rev.* 146 (1912); J. M. Maguire, "The Hillmon Case—Thirty-three Years After," 38 *Harv. L. Rev.* 709 (1925); E. M. Morgan, "A Suggested Classification of Utterances Admissible as Res Gestae," 31 *Yale L. J.* 229 (1922); F. A. Ross, "Applicability of Common Law Rules of Evidence in Proceedings before Workmen's Compensation Commissions," 36 *Harv. L. Rev.* 263 (1923); E. M. Morgan and others, "The Law of Evidence—Some Proposals for its Reform" (report for Legal Research Committee of Commonwealth Fund, 1927). Psychological studies: C. T. McCormick, "Deception-Tests and the Law of Evidence," 15 *Calif. L. Rev.* 484 (1927); Robert M. Hutchins and Donald Slesinger, "Some Observations on the Law of Evidence," *Col. L. Rev.* 432 (1928); second article, 41 *Harv. L. Rev.* 860 (1928); and *see* J. H. Wigmore, *Principles of Judicial Proof* (1913), a valuable collection of actual problems in weighing evidence. (J. M. MAG.)

EVIL EYE. The belief that certain persons can injure and even kill with a glance, has been and is still very widely spread. The power was often supposed to be involuntary (*cf.* Deuteronomy xxviii. 54); and a story is told of a Slav who, afflicted with the evil eye, at last blinded himself in order that he might not be the means of injuring his children (Woyciki, *Polish Folklore*, trans. by Lewenstein, p. 25). In Rome the "evil eye" was well recognized and special laws were enacted against injury to crops by incantation, excantation or fascination. The power was styled *βασκανία* by the Greeks and *fascinatio* by the Latins. Children and young animals of all kinds were thought to be specially susceptible. Charms were worn against the evil eye both by man and beast, and were of three classes: (1) those the intention of which was to attract the malignant glance on to themselves; (2) charms hidden in the bosom of the dress; (3) written words from sacred writings. Of these three types the first was most numerous. They were often of a grossly obscene nature. They were also made in the form of frogs, beetles and so on. Spitting was among the Greeks and Romans a most common antidote to the poison of the evil eye. Gestures, too, often intentionally obscene, were regarded as prophylactics on meeting the dreaded individual. The evil eye was believed to have its impulse in envy, and thus it was unlucky to have any of your possessions praised. It was, therefore necessary to use some protective or prophylactic phrase such as—Unberufen, absit omen, As God will or God bless it.

The powers of the evil eye seem indeed to have been most feared by the prosperous. Its powers are often quoted as almost limitless. The modern Turks and Arabs think that their horses and camels are subject to the evil eye. Among the Neapolitans the *jettatore*, as the owner of the evil eye is called, is so feared that at his approach a street will clear: everybody will rush into doorways or up alleys to avoid the dreaded glance. The evil eye is still much feared for horses in India, China, Turkey, Greece and almost everywhere where horses are found. In India the belief is universal. Modern Egyptian mothers thus account for the sickly appearance of their babies. In Turkey passages from the Koran are painted on the outside of houses to save the inmates, and texts as amulets are worn upon the person, or hung upon camels and horses by Arabs, Abyssinians and other peoples. One of the most striking facts about superstitions in the New World is that the evil eye seems to be foreign to the whole hemisphere. Its absence is often cited as a good example of the occurrence of unexpected human divergences.

See Johannes Christian Frommann, *Tractatus de fascinatione novus et singularis*, etc., etc. (Nuremberg, 1675); F. T. Elworthy, *Evil Eye* (1895); R. C. Maclagan, *Evil Eye in the Western Highlands* (1902); E. Thurston, *Omens and Superstitions of S. India* (1912); Herklotz, *Islam in India* (ed. W. Crooke) (1921); R. E. Enthoven, *Folklore of Bombay* (1924); W. Crooke, *Religion and Folklore of Northern India* (1926).

EVIRATO: *see* CASTRATO.

EVOLUTION, in arithmetic is the operation of finding a root of a number. The Greeks found the square root of a number by a method similar to the one given in elementary arithmetic and algebra books, and this passed on to or was independently developed by the Hindus. The latter gave rules for the cube root. Higher roots attracted the attention of various writers of the 16th century, the method being based upon the expansion of a binomial. (*See* BINOMIAL FORMULA.) The word "root," the Latin *radix*, is due to mediaeval translations from the Arabic, whence came the idea of "extracting the root," and so Recorde (*fl.* 1542) speaks of "pulling out" a root. The symbol for root is $\sqrt{\ }$, as in $\sqrt{2}$. For roots beyond the square root the *index* is written, as in $\sqrt[3]{5}$ and $\sqrt[4]{9}$. The symbol denotes the *principal root*. Every number has *n* *n*th roots, only one of which is indicated by the symbol $\sqrt[n]{\ }$. For example, the three cube roots of 1 are 1, $-\frac{1}{2}+\frac{1}{2}\sqrt{-3}$, and $-\frac{1}{2}-\frac{1}{2}\sqrt{-3}$, but the principal cube root is 1.

EVOLUTION, ORGANIC. The speculations of ancient Greece on the origin of man and animals foreshadow dimly the modern theory of organic evolution. In detail, however, these speculations seem almost as fantastic as the abrupt metamorphoses of men and animals that occupy an important place in the folklore of all parts of the world. Evolution, as we understand it, is a process of cumulative change. The aspect of relative persistence is quite as important as that of change. In the development of a sound concept of the origin of forms of life, the first step probably had to be a firmly grounded belief in the autonomous reproduction and permanence of species.

Among the Greeks, Aristotle (384–322 B.C.) and Theophrastus (born 372 B.C.) began the patient accumulation of data on animal and plant life on which sound inductions might ultimately be based. But even Aristotle held that certain highly organized animals were generated spontaneously. Zoological writings of the middle ages were full of recipes for the generation of such forms as flies, bees and even mice from nonliving constituents. The experimental demonstration by F. Redi (1626–1698) that maggots do not appear spontaneously in spoiling meat, but come only from eggs of flies like those into which they themselves develop, was a landmark in biological thought. It was not until the 19th century, however, that carefully controlled experiments of L. Pasteur (1822–1895) put an end to the belief in the spontaneous generation of observable species of micro-organisms.

The modern conception of species began with John Ray (1627–1705) and reached a first culmination in the system of classification of C. Linnaeus (1707–1778). Linnaeus began with the belief that he was cataloguing creation in naming and defining the species of plants and animals and grouping them in genera. Later he questioned whether the genera might not be the separately created forms, an admission of some evolutionary change, but to most of his immediate followers the absoluteness of species was a basic dogma.

As noted, this was probably a necessary step in biological thought. There were, however, a few of the more philosophical of the biologists of the 18th and early 19th centuries who pointed to the implication of a continuity of descent in the "natural" system of classification that was being arrived at, and speculated on possible mechanisms of transformation. G. L. Buffon (1707–1788) suggested a direct moulding influence of the environment, perpetuated by heredity, and Erasmus Darwin (1731–1802) and J. B. Lamarck (1744–1829) both advocated evolution through the inheritance of acquired adaptations.

The great weight of Georges Cuvier (1769–1832), the founder of vertebrate palaeontology, was thrown against such speculations. He discovered numerous bizarre prehistoric forms of life, but these did not suggest stages in the evolution of modern forms. They seemed rather to justify the concept of a succession of cataclysms, each followed by creation of a new fauna and flora. Geoffroy Saint-Hilaire (1772–1844), indeed, tried to combine the ideas of continuity of descent with discontinuity of form by the hypothesis that new species and higher categories start from the occasional appearance of monsters capable of flourishing in an appropriate environment.

It was reserved for the naturalist, Charles Darwin (1809–1882), grandson of Erasmus, to win general acceptance of evolution, not only on the part of scientists but also the educated public, by showing by a most thorough and objective analysis of the data from all fields of biology how all of these fell into order under the organizing power of this concept. Moreover, Darwin, simultaneously with A. R. Wallace (1823–1913), put forward a simple logical possible explanation of the process, which was of much wider applicability and greater plausibility than those previously advanced. They pointed out the inevitability of evolutionary change in a world in which each species tends to produce more individuals than can find a place, provided that there is sufficient variability, at least part of which is transmissible to the next generation, to give material for a natural selection within and between the species, comparable with the artificial selection practised by animal and plant breeders.

A great variety of evolutionary theories was formulated during the latter part of the 19th century. The importance of geographical isolation in the splitting of species was advanced from different viewpoints by M. Wagner (1813–1887) and J. T. Gulick (1832–1923). The former laid most emphasis on the effects of different environments on isolated forms, while the latter pointed out the inevitability of divergence even without difference in environment. These factors may be considered as supplementary to Darwin's theory. Other authors proposed alternative theories. Among these was Karl Naegeli (1817–1891) who postulated that an inner directive force guides the course of evolution largely independently of the environment. Th. Eimer (1843–1898) also held that evolution tends to proceed along direct lines not necessarily adaptive, but attributed this "orthogenesis" to the direct moulding influence of the environment. E. D. Cope (1840–1897) led a revival of Lamarckian views based largely on palaeontological observations. Finally may be mentioned the revival of the hypothesis of the abrupt origin of species by mutations by Hugo de Vries (1848–1935) based on experiments which will be referred to later.

The weakest link in all of these theories, including Darwin's, was the lack of detailed knowledge of the laws of heredity and variation. Thus the complaint was often made that the theory of natural selection was essentially negative: it accounted for the extinction of some forms, persistence of others by the "survival of the fittest," but threw no light on the "arrival of the fittest." Unknown to Darwin, the key which was to transform the subjects of heredity and variation from the most mysterious in biology to the most deeply understood was in process of discovery in the half dozen years after the publication of his *Origin of Species* in 1859. The results of the beautiful series of experiments with the garden pea conducted by Gregor Mendel (1822–84) in the gardens of the monastery at Brünn in Moravia were published somewhat obscurely and did not come to the general attention of biologists until 1900 when the Mendelian principles of heredity were rediscovered practically simultaneously by De Vries, Carl Correns and Erich Tschermak. The rapid development of the science of genetics after that date, at first under the leadership especially of William Bateson and T. H. Morgan, made it possible to replace the speculations of the 19th century by deductions from securely established knowledge of heredity and variation. The earlier interpretations were largely along the lines of De Vries' mutation theory but more recently the analysis of the statistical consequences of the Mendelian mechanism has led to a return to views more nearly like those of Darwin.

The Species Concept.—Under the hypothesis of separate creation, the distinctions between true species should be absolute: members of the same species must all trace to the same originally created individuals, while members of different species can have no common ancestry whatever. Under the hypothesis of evolution no such absoluteness is to be expected. The definition and characteristics of species are thus crucial questions in arriving at conclusions on the reality of evolution and on its mode, if it is occurring.

The most obvious criterion for a species is a consistent difference in kind from all other species, but this immediately encounters difficulties. Offspring from the same parents may differ conspicuously from each other, not only among human beings, domestic animals, and cultivated plants, but also in wild species; and these differences can often be shown to be hereditary, and fixable by inbreeding.

It can not be said that the difference in kind that marks a species is necessarily greater than that which marks a mere variant within a species. The difference in such characters as colour of body, colour and size of eyes, size of wings, size and distribution of bristles, number of tarsal segments, etc., which distinguish laboratory strains of the much studied fruit fly, *Drosophila melanogaster,* are in many cases much more conspicuous than the criteria used in distinguishing recognized species of *Drosophila.* It is rather typical, to be sure, for similar species to differ slightly in every character studied with sufficient care, but for hereditary variants to be indistinguishable except in one conspicuous respect. But some mutations affect many parts of the body and, in any case, by combining mutations the number of differences between strains can be multiplied almost without limit. Moreover, subspecies of the same species differ in the same way as (though usually less than) species of the same genus.

Other possible criteria are given by physiological characters. Thus injection of blood cells, or of blood serum, from individuals of one species into another leads to the development of antibodies by the latter which tend to react specifically with the cells or serum of the donor species. But this is not an absolute criterion. Marked similarity may be shown in certain reactions between forms such as man and the anthropoids, which are admitted by all to belong to different species, while conspicuous differences may occur in other respects even between offspring of the same parents (as in the case of the human blood groups).

Since the discovery that heredity is carried by certain minute threadlike structures, the chromosomes, present in the nuclei of cells in characteristic number and size in each species, the possibility has arisen that these characteristics might be used to define the species. It has turned out, however, that conspicuous differences in the chromosome set can occur between strains of the same species and, indeed, between offspring of the same parents, while in other cases the set differs little if at all in two clearly defined species.

In bacteria, blue-green algae, and sporadic forms among higher organisms, in which reproduction seems to be wholly by fission or parthenogenesis, there is no other criterion for recognition of species than degree of hereditary difference in characters of the sort referred to above. As persistent differences are continually arising among the descendants of single individuals, the naming of species in these cases is wholly a subjective matter. In forms with biparental reproduction, however, there is another criterion which must be considered, the ability to cross. Perhaps these forms can be put in groups within which crossing is possible but between which crossing is impossible; and these groups may be the true species. Crosses have been made by H. H. Newman between the killifish, *Fundulus heteroclitus,* and the mackerel, *Scomber scombrus,* forms that are put in different orders and are as different as any of the teleost fishes. The hybrids die at an early stage but develop far enough to show characters of both parents. There is every gradation from such cases in which inviable, monstrous hybrids are formed to ones in which the hybrids are vigorous but sterile. The mules and zebroids from crosses among horses, donkeys and zebras are examples of the latter. There are a few apparently authentic reports of fertile mules, and examples can be given to illustrate every gradation from rare to complete fertility of hybrids. The attempt to use any grade to delimit the true species is arbitrary and frustrated by inconsistencies. It is common to find that the sexes of hybrids differ in viability or fertility. It is also common to obtain utterly different results from reciprocal crosses.

Especially disconcerting for the attempt to delimit the true species by such means are cases in which two forms which are clearly of different species by such a criterion would each have to be merged with the same third form by the same criterion. Thus several recognized species of *Datura* (*D. stramonium,* the Jimson weed, *D. ferox* and *D. quercifolia*) produce fertile hybrids *inter se,* suggesting, perhaps, that they should be considered mere subspecies of the same true species. Other recognized species, *D. innoxia, D. metaloides* and *D. metel,* form a similar interfertile group, none of which produce viable embryos on crossing with the first group. But *D. Leichardtii* is a form which (as ovule parent) produces viable fertile hybrids with members of both groups (A. F. Blakeslee). Many other examples have been described among plant hybrids. Among animals all of these complications have been found in intensive study of hybrids of *Drosophila* by J. T. Patterson and associates, and examples of some of these can be drawn from work on other groups.

Another complication appears in many cases in which forms which produce fertile hybrids in the laboratory live in the same territory in nature without apparent hybridization. This is, for example, the case with the wood mouse, *Peromyscus leucopus,* and the cotton mouse, *P. gossypinus,* in the lower Mississippi valley (L. R. Dice). In other somewhat similar cases (*e.g.,* the blue-winged and golden-winged warblers, *Vermivora pinus* and *V. chrysoptera* respectively) breeding hybrids are found in nature but are so rare that the sharpness of the specific distinction is not materially blurred in the region in which the ranges overlap. In other cases, however, an intergrading but highly variable hybrid population occurs in the region of contact.

There is only a rough correlation between the degree of difference in characters and the degree of incompatibility in crosses. There are pairs of species such as *Drosophila melanogaster* and *D. simulans* which differ so little in any character that they were not distinguished until it was discovered that no fertile hybrids (and few hybrids of any sort) could be obtained from crosses between individuals from certain populations. On the other hand *Drosophila virilis* and *D. americana* differ in many conspicuous respects and in chromosome patterns but produce fertile hybrids. Similarly such forms as the bison, yak and domestic cattle, which are so different that they are put in different genera, nevertheless produce fertile hybrids. Such hybrids have also been obtained from matings of pigeon and dove, birds that are put in different genera.

In spite of these difficulties, those who are working with those groups of higher plants and animals in which knowledge of the forms in nature is approaching completeness are usually in substantial agreement as to the species. This has required, however, a shift in concept.

A classification into distinct species of the forms in one locality usually offers little serious difficulty. If, however, collections of one of the species are made at increasing distances from this locality, it is usual to find that the averages of such characters as colour, size and proportions of parts of the body change more or less gradually. No line can be found at which one kind of organism stops and another begins; yet, if the range is sufficiently large and dispersion sufficiently slow, it is likely to be found that remote collections differ so consistently that there would have been no hesitation in recognizing two species if they had been found together. Such observations led in the latter part of the 19th century to the recognition of the subspecies or geographical race as a necessary category in classification.

The subspecies concept was at first used rather sparingly, but as knowledge of such forms as the land vertebrates approached maturity it has come to be recognized that an enormous number of supposed species described from different regions show intergradation in intermediate regions and must be relegated to subspecific status. Thus at one time 35 species of otter (genus *Lutra*) were recognized. More complete knowledge has reduced these to four, each with many subspecies (B. Rensch).

According to this concept, the species names among contemporary biparentally reproducing organisms refer to populations or groups of populations within which there is continuity of interbreeding in some sense, rather than to the kinds of organisms in the qualitative sense. As the lives of species are to be measured in terms of hundreds of thousands or even millions of years, it is not desirable to give different specific names to two populations merely because they happen to be completely isolated from each other at the moment. Ideally, the species should consist of populations, or groups of populations, within which there is no genetically determined barrier to the establishment of a chain of intergrading populations between any two components, but between which and all other populations there is such a barrier. Unfortunately it is not usually practicable to determine whether or not such a barrier exists between forms that are separated by a geographical gap.

Even with geographical continuity there are borderline cases in which it is a matter of judgment whether two species should be recognized or merely two subspecies. If fertile hybrids are found in the region of overlap, it must be decided whether they are sufficiently numerous to make the inhabitants of this region a single, if rather variable, population; or whether they are so rare that they do not obscure the existence of two intermingling distinct populations.

If two distinguishable populations of forms with exclusive biparental reproduction occupy the same territory with little or no hybridization, this is ordinarily considered enough to justify the recognition of two species. There are, however, a rapidly increasing number of cases in which two such apparent species have been found to be connected around a circuit by a chain of intergrading subspecies. Thus two sharply distinct kinds of herring gull, *Larus argentatus,* and a form, *graellsii,* are found in the British Isles, and each is represented by a slightly different form to the northeast (Murman coast). Collections across northern Europe, Asia and North America have established continuity of intergradation from *graellsii* to *argentatus.* Thus only one species should be recognized (according to B. Stogmann).

Other cases are known in birds, mammals and reptiles. Perhaps the most remarkable, however, is one in the gall wasps (*Cynips*), in which according to A. C. Kinsey an apparently continuous chain of 76 forms returns on itself repeatedly with the consequence that there are 2 or 3 apparently distinct species in some localities.

Where there is geographical isolation, some indication is given by breeding tests, but these are not in general conclusive. If fertile hybrids are produced by crosses between the two forms, it by no means follows, as already noted, that contact in nature would be followed by the production of an intergrading population. The failure to produce fertile hybrids in extensive tests is relatively conclusive evidence of specific difference but only after all possible connecting chains through other forms have been tested. In most cases it is, of course, hardly practicable to make the necessary extensive breeding tests, or any breeding tests at all, and the systematist must use other criteria. Two geographically separated forms are usually considered to be merely subspecies if there is so much overlapping of characters that some adult individuals of both sexes can not be classified without reference to place of collection, while if all of at least one of the sexes can be so classified they are considered to be different species. There may be overlapping of all directly measurable characters, but if some function of the measurements, or some difference in habitus, recognizable by a trained observer, makes it possible to diagnose all individuals, two species are usually recognized.

It is obvious that the recognized species can not be expected to correspond perfectly with the ideal species as defined above. There are undoubtedly cases in which good species (in the ideal sense), living in the same territory, overlap in characters, or are even practically indistinguishable except by breeding tests (*e.g., Drosophila pseudoobscura* and *D. persimilis*). This may well be the case for many supposed subspecies that are geographically isolated. On the other hand, the absence of overlapping in local populations that are recognized to be merely subspecies because of intergradation through a chain of connecting forms, makes it likely that many geographically isolated supposed species would merge if brought together.

The shift to the population concept of species as the ideal has lagged among those working with invertebrates but there is reason to believe that it is widely applicable from special cases that have been studied intensively. The situation in plants is complicated by an abundance of forms within species (ecotypes) each of which occurs more or less consistently wherever the same ecological niche is found. Such differences have been found to depend in part on direct environmental influences which complicate the expression of underlying genetic differentiation more than is usually found in animals.

In some groups of plants (*e.g., Rosa, Crepis, Rubus,* etc.) predominant uniparental reproduction, whether vegetative or by apomixis, occurs among types with various numbers of sets of chromosomes; and occasional hybridization, followed by segregation of new types, has given rise to complexes of intermingled clones that present an almost insuperable problem to the classifier.

While sexual species can be defined more objectively than any other category in the system of classification it must be concluded that the difficulties of definition, whether by gross morphology, by chromosome sets, by physiology, by incompatibilities in crossing, or even by failure of intergradation with other species, are quite out of harmony with the absoluteness expected under separate creation. They all point toward an evolutionary interpretation of some sort.

An important question with respect to possible modes of evolution is the adaptiveness or lack of it in differences between subspecies of the same species, or species of the same genus. It has long been recognized that species of birds and mammals with a long north-south range tend to exhibit certain rather uniform trends. In the northern hemisphere, the northern forms tend to be lighter coloured (Gloger's rule), larger (Bergmann's rule) and to have relatively smaller extremities—ears, feet, tail (Allen's rule) than those to the south. These are probably to be interpreted as selective adaptations to systematic differences in exposure to the environment.

In some cases such trends are so uniform that any separation into subspecies is wholly arbitrary. There is what is called a cline (J. S. Huxley). Usually, however, widely ranging species, whether or not they show a north-south trend, are divided into subspecies that are separated by rather narrow zones of rapid transition. At least some of the differences between such subspecies are rather clearly of the nature of minor specializations in relation to different ecological conditions. Among animals, differences in concealing coloration, in relation to the prevailing colour of the soil or other environmental features, are common. Thus desert forms and forms living on white sand beaches or dunes are characteristically pale in comparison with those of humid woodlands. The observation that rodents living in the desert on black lava are exceptionally dark is the kind of exception that proves the rule (L. R. Dice and P. M. Blossom). Physiological adaptations to different ecological conditions are no doubt more important than adaptive coloration but more obscure.

The boundary between forms exhibiting what seem to be clearly adaptive differences may not coincide with any natural boundary, suggesting that differential population pressures may take precedence over close adaptation. In many cases indeed, differences in minor morphological characters seem to have no relation to adaptation. This may be due in part merely to lack of knowledge. Conspicuous but apparently neutral differences may possibly be physiological correlates of intangible but adaptively important characters. It is difficult, however, to avoid the impression that a large portion of the differences merely happen to have arisen and have no adaptive significance.

Local collections within the central range of recognized subspecies usually show differences in the means of variable characters that are statistically significant but so slight that individuals could not possibly be classified by their own characters. Here especially the differences have the appearance of being almost wholly random in character.

The differences between species of the same genus are for the most part similar to those between subspecies, though usually greater in degree. Again there are differences that are obviously adaptive in relation to diverse ecological niches. For example, a termite, *Amitermes excellens,* of the tropical rain forest of British Guiana constructs nests on the trunks of trees that have remarkable rain shedding devices (A. E. Emerson). Another species, *Amitermes meridionalis,* in northern Australia, constructs a vertical bladelike nest with north-south orientation, probably as an adaptation to control of temperature and humidity. The morphological differences between the workers that construct the nests are slight. It is indeed difficult to cite strikingly adaptive morphological differences at the species level though not at the level of genus or family. Differences in protective coloration are, however, very common. Among the most remarkable are those found in butterflies, concerned with mimicry of noxious species.

There is a large class of species differences involving colour, form and behaviour which may be interpreted as recognition marks in which mere difference in otherwise rather similar species of the same region seems to be of value to the species in maintaining their integrity.

Another class of rather striking differences are those in which there has been degeneration of an organ in one of the species. The cave fish, *Anopthicthys jordani,* shows almost complete loss of the eyes, with correlated changes in the skull and a compensating notable increase in the number of taste buds on the head and body as compared with the otherwise very similar normal river fish, *Astyonix mexicana,* as noted by C. M. Breder, Jr. These are classified in different genera, but the existence of a cave population with intermediate conditions indicates that the relationship is very close, perhaps merely that of two subspecies.

The overwhelming majority of the morphological differences by which species of the same genus can be distinguished are ones to which it is difficult to assign an adaptive significance. The general conclusion that seems to be indicated is that the evolutionary process at the levels of the subspecies and the species has both adaptive and accidental aspects.

Evidence of Evolution of Higher Categories.—Even before the acceptance of evolution, it was generally recognized by biologists that there is a single natural system of classification of organisms to be distinguished from the practically infinite number of possible arbitrary systems. The basis for this system was clarified by Sir Richard Owen (1804–1892), an opponent of evolution, who distinguished between two kinds of similarity: analogy and homology. *Analogous* parts of two organisms serve a similar function and show only such similarities as are explicable by functional adaptation, supplemented perhaps by coincidence. The wings, legs and eyes of a fly are, for example, analogues of the structurally very different parts of a bird, that are given the same names. *Homologues* show a similarity in structure, mechanism, mode of development and relationship to other parts that goes beyond that explicable by functional adaptation and coincidence. The corresponding legs, wings and eyes of two kinds of insect meet all of the requirements for a judgment of homology. Corresponding statements can be made in regard to the legs, wings and eyes of two species of bird.

Comparisons of the wing of a bird, the foreleg of a frog, the foreleg of a horse, the arm of a man and the flipper of a whale, parts which have the same relation to the rest of the body in these animals, brings out a similarity in the pattern of bones, muscles and other parts that is remarkable in view of the dissimilar functions. These must all be judged to be homologues.

Such similarities are especially noteworthy where the part in question is a functionless vestige in one of the forms compared. Whales have no hind limbs that protrude from the body but in some of them there are groups of small bones in the appropriate position within the body which are obvious homologues of the hind leg bones of land mammals. The vestigial wing of flightless birds, the vestigial eyes of blind cave fishes, the vestigial appendix of man are homologous to functional organs of other forms.

While homology is usually treated as an all or none matter, it is obvious that the implied similarity is a matter of degree. There are always considerable differences between homologues where function is markedly different. Homology between parts of two kinds of organism may only be traceable by comparison of both with another or through a chain of other organisms. Thus there is little direct basis for considering the incus and malleus of mammals, two minute bones that aid in conveying sound vibrations from the ear drum across the middle ear, as homologues of the powerful upper and lower jaws of sharks. Yet the mammalian ear ossicles are clearly homologous to the quadrate and articulare, in the fossilized skulls of cynodont reptiles. These small bones, by which the jaws articulate, are obvious homologues of the larger articulating bones of other reptiles and amphibians and teleost fishes. A comparision of the teleost jaw with that of a shark indicates that, in the former, the jaw is a composite structure in which the articulating parts are homologues of the whole jaw of the latter, while the tooth-bearing elements of the former are comparable with dermal scutes. The chain is further lengthened on recognizing homology of the jaws of the shark with anterior gill-supporting structures of the jawless cyclostomes.

Even in cases in which the functions of homologues are the same, there may be great differences in the degrees of similarity of structure. The wings of birds and bats are constructed on such utterly different plans for carrying out their common function, that they can be considered as merely analogous as wings. Yet the basic patterns equally clearly show an underlying homology as forelimbs. Similarly the homologies of the single-hoofed legs of the horse and the cloven-hoofed legs of the antelope can be traced best by comparing both with mammalian legs that are not specialized for running.

The statement that homologous organs show similar relationships to the rest of the body carries the implication that it is possible to determine what are corresponding parts. This can be done without hesitation with two species of bird, with more uncertainty for certain parts of bird and fish, only in the roughest sense, if at all, in the case of bird and insect, and finally not at all in that of bird and plant (except for intracellular structures). Thus it is usual to find that if two organisms show the type of similarity in one part that indicates homology they do in others and relations among these parts are the same. Apparent exceptions tend to disappear on analysis. Thus the eye of the octopus is incomparably closer in plan to the eye of bird than is the eye of an insect, although homologies in other organs are no more traceable in one case than the other. But this can be accounted for by coincidence. There are only two basic optical mechanisms for an image-forming eye. It happens that bird and octopus have eyes of the camera type while the insect has a system of optically isolated radiating tubes. The detailed carrying out of the plan of the camera eye in the octopus is so different from that in the bird (and all other vertebrates) that the eyes must be considered as merely analogues by this criterion by itself.

In cases in which homology can be traced throughout two organisms, particular homologous organs may show a degree of similarity that far exceeds that of other organs. Thus the horses and an extinct group of South American hoofed mammals show an extraordinary similarity in the structure of their legs and their general conformation, although not notably similar in other respects. It may be supposed that coincidence in the mode of adaptation to a similar function has here supplemented the basic similarity due to homology to bring about convergence.

In some cases particular organs of two forms appear to be homologous in structure but are located in different places in bodies that are otherwise homologous throughout. The locations of the kidney elements in cyclostomes, fishes and reptiles present such a problem. Such cases bring up the relation between the phylogenetic homology which is under consideration and a different sort of homology, that of similar structures in the same organism, which will be discussed later.

Because of the prevailing parallelism in the degrees of similarity of homologous parts of organisms, any serious attempt at classification by morphology leads to a certain branching order, the natural system, referred to above. Species can be grouped into genera within which homologies can be traced in minute details. Within the higher categories, families, orders and classes, homologies can be traced for most organs but the similarities become less. The phyla are bound together as natural groups by only a few broad homologies; in the chordates, for example, largely a dorsal neural tube, an underlying supporting structure of a peculiar type (the notochord), and pharyngeal gill slits, each present, perhaps, in only a phase of the life cycle.

Difficulties of classification arise from departures from parallelism in the similarities of homologues, which may often be related to divergent or convergent adaptation, as in cases discussed above. The parts which are least subject to change of adaptation are generally recognized to be those on which most emphasis should be placed in classification.

The definition of homology, as a similarity greater than can be accounted for by common function or by coincidence, raises the question of its explanation. Only two explanations have been proposed. The first is that homology is an aspect of the design of creation. The alternative is that it is a consequence of relatively unmodified common heredity. The branching order of the natural system of classification is expected under evolution. It is perhaps unwarranted to speculate on what sort of order would be expected under separate creation. It may be pointed out, however, that creative activity as we know it in man is not restricted to a branching order in its products. Ideas are transferable from one field to others, with the consequence that human creations are related in a multidimensional network. At best, the concept of design is one that is more plausible as an *a priori* hypothesis than as an interpretation of concrete cases such as the sporadic occurrence of vestiges, homologous to functional organs of other forms.

If common heredity is the explanation of homology, this sort of similarity should be traceable at all stages of development. It is indeed often the case that homologies are more easily recognized in early stages than in the adults. For example, the mammals have a heart which consists of two muscular tubes that do not communicate directly. The right side receives the venous blood from the body and pumps it to the lungs, whence after aeration, it is carried to the left side of the heart to be pumped to the tissues through branches of a single great artery, the aorta. There is little apparent similarity with the circulatory system of a fish, in which the heart is a single muscular tube which pumps the venous blood through several paired aortae to gills in the walls of a series of slits in the pharynx through which water is kept moving. If, however, embryos are compared there is no difficulty in establishing homologies. In both there is a simple tubular heart, paired aortic arches and gill slits. The rudiment of the mammalian lung is seen to be homologous to an organ which in some cases functions as an accessory respiratory organ, the swim bladder of the fish. The adult condition of the fish is arrived at rather directly, while that in the mammal is reached only by suppression of all of the gill slits except the first (the ear opening), atrophy of most of the aortic arches and the growth of a partition through the heart in such a way that all of the blood must pass through the lungs at each circuit. The mammal seems to pass through a fish stage. As there are other reasons for believing that all land vertebrates are descended from aquatic forms, this seems to justify the aphorism "ontogeny recapitulates phylogeny." This expression is, however, seriously misleading.

There is an appearance of recapitulation wherever a progressive form that has elaborated the later developmental stages is compared with a conservative form that develops relatively directly. It happens, however, that, in the case of mammal and fish, the progressive form has also complicated the earliest stages by producing a membrane about the embryo, the amnion, which can not be interpreted as a recapitulation. Again evolution may also proceed by dropping later stages. The tunicates are saclike marine animals which spend their adult lives attached to rocks, filtering food particles from a current of water drawn through the mouth and passed out through slits in the wall of the pharynx. There are no characters of the adult tunicate and the adult frog to suggest homology. Yet the tunicate develops from a tadpole with much more than a superficial resemblance to that of a frog. The recapitulation theory suggests a tadpole-like common ancestor, but as W. Garstang has pointed out there is considerable reason for a different interpretation: an originally simple but free larval phase of the sessile tunicate may have been elaborated to give more efficient dispersal. The new way of life thus evolved may then have taken precedence over the fixed life. The vertebrates may thus have had their origin in the development of sexual maturity in an elaborated tunicate tadpole, and the suppression of the later fixed stages. (*See* Embryology; Embryology, Human; Invertebrate Embryology; Vertebrate Embryology.)

Among invertebrates there are many examples which seem to require elaboration of early stages or suppression of either early or late stages as well as ones which indicate the addition to late stages, postulated by the recapitulation theory. The homologies between adult annelids and arthropods leave little doubt of the derivation of the latter from the former. Yet nothing could be more different than the ciliated trochophore larvae of the marine annelids and the characteristic nauplius larvae of the marine crustacea with its jointed appendages and absence of cilia. There is, moreover, little resemblance between the modes of development of the terrestrial and marine members of either phylum. On the other hand it seems hardly likely that the detailed similarity between marine annelids and most molluscs in the spiral cleavage of the egg, in the fate of the various blastomeres and in the similar trochophore larvae can be wholly coincidence or common adaptation. Ultimate derivation of these otherwise very different phyla from the same type of unsegmented worm is suggested. The similarity of the nauplius larvae of the barnacles to those of the lower crustacea leave no doubt of the ancestry of barnacles. Again *Sacculina*, a formless parasite of the crab, has nothing in its adult phase to suggest arthropod affinities but its nauplius larva leaves no doubt of its relation to the barnacles.

The allocation of forms to their proper position in the natural system of classification requires consideration of possible homologies of all parts at all stages of development with due weight to the possibilities of adaptive elaboration or simplification at any stage.

Classification by morphological homologies tends to carry with it a grouping by physiological similarity. It has already been noted that the proteins, or compound proteins, of species show a high degree of specificity, as revealed by serological reaction, but that this is not absolute. Attempts have been made on a large scale in both animals and plants to make classifications based on the similarities indicated by such reactions. These agree remarkably closely with those arrived at on morphological grounds. Thus antihuman serum gives a very striking precipitin reaction with serum from anthropoid apes (Simiidae), a somewhat weaker reaction with serum from old world monkeys (Cercopithecidae), still weaker with serum from South American monkeys (Cebidae, Hapalidae) and no reaction (at the titre considered) with serum from lemurs (Lemuridae), in agreement with relationships deduced on morphological grounds.

The epidermal cells of vertebrates characteristically produce keratin, while chitin is the characteristic product in arthropods. Arginine phosphate plays a role in the energy transformation of crustacean muscles analogous to that played by creatine phosphate in vertebrates as far as known. Oxygen transport in vertebrates is always based on the properties of haemoglobin. This is not the case among most invertebrates, in some of which a copper containing protein, haemocyanin, plays a similar role. Haemoglobin does however occur in some molluscs, annelids and arthropods.

In general, however, classification primarily by physiology gives less consistent results than by morphology except probably in the case of protein specificity. The reason seems to be that the basic physiological processes are characteristic of all living matter and furnish perhaps the strongest evidence that all life as we know it now has a common origin. The erratic appearance of haemoglobin as an oxygen carrier throughout the animal kingdom becomes more intelligible when it is learned that its peculiar prosthetic group, based on porphyrin, is found in oxidizing enzymes of almost universal occurrence (cytochrome, cytochrome oxidase, catalase) and also in the chlorophyll of green plants.

The most direct evidence for evolution comes from the actual remains of past organisms buried in sediments, the relative ages of which can be determined. As noted earlier, Cuvier, the founder of vertebrate palaeontology, was a vigorous opponent of evolution. At that time, the estimates of past time allowed only a few thousand years since creation. It was obvious from old inscriptions that man and many animals had not changed appreciably in 3,000 or 4,000 years and it was ridiculous to suppose that they could have evolved from common ancestors in a few more thousand years. It required a drastic revision of the earth's time-scale before evolution could seem a reasonable hypothesis. But such a revision was forced on the attention of science by the studies of rock formations. The examination of thousands of feet of sandstone, shale and limestone strata revealed characteristics that could only be interpreted as those of slowly accumulating deposition, interrupted from time to time by erosion. Estimates based on probable rates of deposition were in terms of many millions of years.

Up to the end of the 19th century, the minimum estimates of geologists were far in excess of the maximum which the physicists would allow for the age of the solar system on the basis of known sources of the energy radiated by the sun. When the enormously greater energy from conversion of mass became known, there was no difficulty in reconciling the estimates. It became possible indeed to determine the age of certain deposits from knowledge of the rate of transformation of uranium into lead with much more confidence than from estimates of rates of deposition. With the Cenozoic era estimated to have begun 70,000,000 years ago, the Mesozoic 190,000,000 years ago and the Palaeozoic 550,000,000 years ago, the lack of apparent evolutionary change in the last 6,000 years is seen to be of no significance.

The sequence in which types appear in the strata is quite in accordance with the deductions from morphology and other lines of evidence. The first vertebrates, for example, are jawless forms in the Ordovician period, jawed fishes appear in Silurian, amphibia in the Devonian, reptiles in the Pennsylvanian, mammals in the Jurassic and man only in the Pleistocene after 99.9% of this history had passed. (*See* PALAEONTOLOGY.)

The most complete evidence comes from the transformations of some of the more abundant marine invertebrates observed in thick continuous deposits (*e.g.*, the ammonite *Kosmoceras*, the brachiopod *Spirifer* and the echinoderm *Micraster*). At a given level, such a species shows a wide range of variability as in a living species. In passing up through deposits that represent hundreds of thousands or even millions of years of almost continuous deposition, a gradual shifting of the mean can be observed until at length the array no longer overlaps that in the lowest stratum. Forms far apart in such series would unquestionably be considered as different species if living contemporaneously.

In an earlier section, the delimitation of the contemporary species was based in part on the absence of intergrading populations. This criterion breaks down in successive species, in which there is the same sort of continuity as in a geographical cline. But while arbitrary stages in a geographical cline need merely be considered as subspecies, a temporal cline must be divided arbitrarily into species.

Significant events in evolution would be expected to be localized at particular places as well as at particular times if dispersion over the earth has been as severely restricted as it is today for most organisms. The geographical distribution of species interpreted as more or less closely related should provide evidence on evolution. It seems indeed to have been evidence of this sort which impressed Darwin and Wallace more than any other. Darwin noted characteristic differences between mammalian faunas of South America and the old world for which climate could offer no explanation. South America contained many species and genera, living and extinct, of armadillos, sloths and anteaters. The whole order to which these belong is not found elsewhere (except for a few migrants into southern North America). If species (and higher categories) of armadillos were separately created, why created only for South America? Darwin saw that evolution from an ancestral armadillo under conditions of isolation from the rest of the world offered the simplest explanation.

The results of palaeontological as well as neological study after Darwin's time fall into a consistent pattern on the assumption that South America was isolated from North America near the beginning of the Cenozoic and until relatively recently (Pliocene) received only a few stray forms from other continents in addition to an initial very primitive mammal population. The effects of easier exchange of North and South American forms since the Pliocene can be traced in the fossils and living forms of both continents. The even more peculiar mammalian fauna of Australia is readily interpreted by a longer and more complete isolation. The fossil record of North America and Eurasia on the other hand indicates an alternation of long periods of isolation with ones of relatively easy communication. (*See* DISTRIBUTION OF ANIMALS; ZOOLOGICAL GEOGRAPHY.)

Darwin was especially impressed by the peculiar situation which he found in the Galapagos Islands. There were numerous species of ground finches found nowhere else in the world. Separate creation gave no explanation of the fact that these species were more like each other than like anything elsewhere but also more like forms from the West Indies than from the old world. Separate creation could also give no explanation of the highly unbalanced character of the fauna which suggested instead derivation from a few accidental immigrants belonging to types which might conceivably have been carried across a large body of salt water under exceptionally favourable conditions. Darwin was irresistibly drawn to the evolutionary interpretation. Intensive study of the faunas and floras of other oceanic islands has given abundant confirmation.

The Course of Evolution.—Accepting evolution as a fact, it is of interest to examine the conclusions that have been reached on the course of evolution in various groups to see what generalizations are indicated on the nature of the process. It appears immediately that there has been the utmost diversity in every respect, including rate, directness of trend, progress in complexity and amount and nature of branching of phyletic lines.

There are existing genera for which evolution seems virtually to have ceased for hundreds of millions of years. The longest record for stability is probably that of the brachiopod genus *Lingula* which has persisted from the Ordovician to the present, and with only slight change from *Lingulella* of the Lower Cambrian some 500,000,000 years ago. There are other brachiopods (*Crania, Discrania*), bryozoa (*Stomatopora, Berenicea*) and a pelecypod (*Nucula*) which have also lived since the Ordovician. More than a dozen other pelecypods, a gastropod (*Capulus*), several genera of foraminifera and a few of the lower crustacea (*Estheria, Apus, Cytheria*) trace to the Palaeozoic. Some 20 more pelecypods, a few gasteropods, a cephalopod (*Nautilus*), an arthropod (*Limulus*), a crinoid (*Isocrinus*) are among those that have come down to the present from the Triassic. The oldest living genera of vertebrates are several of the sharks (*Hexanchus, Heterodontus, Galeus* and *Rhina*) and a skate (*Rhinobatus*), all from the Jurassic. There are, however, the lung fish, *Ceratodus* and the recently discovered coelocanth, *Latimeria,* which have changed little from their ancestral genera of the Devonian and thus far antedate the great diversification of teleost fishes in the late Mesozoic. A reptile (*Sphenodon*) has changed little since the Lower Triassic, a period within which the dinosaurs had their beginning, their spectacular diversification and their extinction. The opossum has seen the origin and differentiation of the orders of placental mammals from the Cretaceous. Among plants the horsetails (*Equisetum*) and club mosses (*Lycopodium*) are relics of the dominant groups of land plants of the mid-Palaeozoic and certain gymnosperms, notably *Araucaria* and *Ginkgo,* represent types of flowering plants of the late Palaeozoic and early Mesozoic that far antedate the appearance in the Cretaceous of the present dominant flowering plants, the angiosperms.

Leaving out of account such cases of extreme stability, there are characteristic differences between the rates found in different groups. Thus G. G. Simpson estimates the average life of the extinct genera of pelecypods to have been 78,000,000 years in comparison with 6,500,000 years for extinct genera of carnivores.

Rapid as has been the evolution of mammals during the Cenozoic, it seems to have been enormously more rapid in all lines during a brief period just preceding this. The only placental mammals known in the Upper Cretaceous were insectivores. Well differentiated carnivores, primates, edentates, rodents, and lagomorphs, as well as several extinct orders of ungulates, appeared in the Palaeocene and most of the remaining orders including such divergent forms as bats and whales were present in the Eocene. Similar periods of extremely rapid evolution are indicated for fishes in the Devonian, amphibians in the Mississippian, reptiles in the Permian. The turtles, for example, seem to have evolved from generalized reptiles at some time in the Permian. By the middle of the Triassic their peculiar structure was essentially as it is today. It is characteristic of these periods of rapid transition that fossils are exceedingly rare indicating small or sparse populations.

Much of the palaeontological record consists of gradual transformations along direct courses. The transformation of Lower Eocene *Eohippus* into the modern horse through a succession of nine genera is perhaps the most complete history of this sort.

In the long run, however, evolution has been far from a direct process. The tetrapod vertebrates had hardly more than emancipated themselves from the water as primitive reptiles than some of them, the mesosaurs, ichthyosaurs and plesiosaurs, proceeded to reverse this direction by readaptations to marine life. Similar reversals occurred in the whales, sirenians and seals among the mammals. The resulting forms were, however, very different from fishes. The peculiarities of the starfishes and other mobile echinoderms would hardly have been arrived at except by a succession of abrupt changes in direction of evolution. The same is true of the salps, larvaceae and perhaps the ostracoderms as free derivatives of the fixed tunicates. The history of the horse from ostracoderm or even tunicate to *Eohippus* was certainly a most tortuous process.

Directedness and rate are related. Typically a lineage exhibits a zigzag course in which periods of slow advance of constant direction are separated by periods of very rapid remoulding toward a new direction.

A comparison of early Palaeozoic and recent forms of either animals or plants no doubt indicates increased complexity of organization as a characteristic evolutionary trend. But here also there is no absolute rule. Under certain conditions, especially parasitism, the trend is toward simplification.

Other characteristics in which the evolutionary process differs in different cases are the amount and nature of branching. The very limited amount of branching in the history of the Equidae since the Eocene contrasts with the rich branching of the Bovidae in the shorter period since the Miocene. The Muridae are another family of mammals notable for multiplication of genera and families without much significant change of type. This is also the case with passerine birds and many groups of teleost fishes, many families of insects, large groups of angiosperms.

In further contrast is the more extreme type of adaptive radiation which leads to widely diverse types, each adapted to an utterly different way of life. This is illustrated by the rapid evolution of many types of ostracoderms in the Silurian, tracing presumably from an unarmoured lower chordate, of fishes in the Devonian, from an ostracoderm; of amphibians in the Mississippian, from a crossopterygian fish; of reptiles in the Permian, from a labyrinthodont amphibian; of archosaurs in the Triassic, from a thecodont reptile; of birds in the Upper Cretaceous, tracing through reptilelike birds to a Jurassic archosaur; of placental mammals in the Palaeocene, tracing to Triassic therapsid reptiles.

The best examples of abrupt change of direction and also of extremely rapid evolution are drawn from such cases in which many changes of direction occur simultaneously. The basis is probably in all cases the achievement of an adaptation or complex of adaptations by the parent form that gives an advantage of a general sort over the forms already occupying many ecological niches. Adaptive radiation in the cases cited above seems to have been based on armour in the ostracoderms, jaws in the fishes, the leg and pentadactyl foot in the amphibians, the amnion and other devices giving emancipation from the water in the reptiles, effective locomotion on land in the archosaurs; wing, feathers and temperature regulation in the birds; effective limbs, hair, temperature regulation and mammary glands in mammals. The primary adaptive complex may or may not have been a rapid achievement itself.

A corollary of adaptive radiation is the extinction of outmoded types. The character of the fauna and flora of the world has been revolutionized by the extinction or near extinction of such once dominant groups as the trilobites, eurypterids, ammonites, ostracoderms, placoderms, palaeoniscids, labyrinthodonts and archosaurs among animals; equisetales, lycopodales, pteridosperms, among plants, and the expansion of the dominant groups of today from apparently insignificant beginnings. Only rarely does a once dominant group that has been supplanted by another, as the Permian and early Triassic therapsids were by the archosaurs, leave a specialized line that later supplants its supplanters as the mammals did after an interval of 100,000,000 years.

Summing up, the characteristic evolutionary process may be described as the *emergence* of a complex of adaptations of general significance, the rapid exploitation of this in diverse ways of life by *adaptive radiations* at successively lower levels, leading ultimately to gradual *orthogenetic* advance along each line, accompanied in some cases by extensive diversification of genera and species with jointly nonadaptive and minor adaptive aspects. On rare occasions a new relatively general adaptive complex may emerge at any stage in the process initiating a new cycle. The broad course of evolution has the appearance of being guided by *selective* expansion and elimination among the higher categories.

In spite, however, of the relative abruptness of changes in direction and the rapidity of evolution in certain cases, the nature of the changes in morphology indicates essential continuity.

The possibility of adaptations to widely different ways of life at different stages of the life cycle (tunicate, amphibian), or in an alternation of generations (alga, bryophyte, hydromedusa) provides a basis for the accomplishment of violent changes in the direction of evolution in a continuous manner by a mere shift in emphasis.

The reorganization of the circulatory system of the fishes, required in land vertebrates by the shift from gills to swim bladder as the primary respiratory organ did not occur by an abrupt change in plan. It required hundreds of millions of years to reach the degree of adaptation to land life comparable with that with which existed for marine life in fishes. All the changes in the heart and blood vessels were of the nature of differential growth or suppression, actually repeated in the individual careers of higher vertebrates. Similarly a comparison of the reptilian jaws, articulated by a chain of bônes and the mammalian jaw with direct articulation at a different point, at first sight seems to require change within a single generation. Yet the conditions found in the jaws of the therapsids and ictidosaurs show that a continuous series of intermediate conditions actually occurred.

New organs characteristically make their appearance as simple outpocketings or outgrowths (*e.g.,* swim bladder, amnion, horns, etc.) or as readaptations of old organs (*e.g.,* teeth, lungs, legs, feathers, wing, etc.). It is very common for an indefinite number of simple undifferentiated elements to become transformed into a definite number of complex individualized structures. The gill arches appear to be very simple and variable in number in lower chordates (tunicates, amphioxus, Agnatha). The number seven becomes standardized in the fishes. The first arch becomes differentiated as a jaw and later, as already noted, becomes the articulation for the jaw and finally (in mammals) reduced in size and changed in function to become a chain of ear ossicles. The second (hyoid) arch also becomes somewhat specialized, while the posterior arches largely disappear with the loss of the gill apparatus. The dermal bony armour of the Agnatha gives rise not only to the bony scales of higher fish but to the highly individualized dermal bones of the skull and to the teeth. The latter are simple, undifferentiated and highly variable in number and location in lower vertebrates. Beginning in therapsids and carried on by the mammals there is individualization of incisors, canines, premolars and molars and a tendency toward fixation of number. The acanthodians among the earliest jawed fishes show various numbers of paired ventral fins (up to seven) as well as unpaired dorsal and anal fins. Two pairs of appendages become standardized in the higher vertebrates. The five-toed foot of the amphibian emerges from a lobe fin with an unstandardized number of branching and radiating lines of small bones. The same principles can readily be illustrated from invertebrates. The indefinite number of similar segments of the annelid, apparently arising by transformation of a method of asexual reproduction into one for producing a compound single organism, becomes in the arthropods a more or less definite number of individualized segments, with more or less individualized appendages.

Occasionally the reverse process occurs. The dental pattern of three incisors, one canine, four premolars and three molars seems to have become standardized in the ancestral placental mammals and thereafter evolved largely by differentiation and loss of elements. In the toothed whales, however, dedifferentiation has occurred and a return to an indefinite number of simple undifferentiated teeth.

Structures of wholly different origin, but which happen to be close together, may become merged in a new complex. There is little similarity between the cartilaginous cranium which protects the brain and sense organs of the lowest fishes and the bony scutes in the skin. In the higher vertebrates, the former become ossified and the latter become associated with them to form the compact cranium in which cartilage and dermal bones can be distinguished only by their modes of development. The complex mammalian ear integrates a number of relatively unrelated structures of lower vertebrates. Evolution involves a continual remoulding of structure.

The Genetic Basis for Evolution.—As evolution implies change of the hereditary characteristics of a species, an adequate explanation of an evolutionary process must be expressed in terms of mechanisms of heredity. The basis for existing views on this subject was laid by the discovery at about the beginning of the 19th century that all higher plants and animals are composed of essentially similar entities, the cells, each of which has the essential properties of an organism on its own account. Cells were found to arise only by division of pre-existent cells and to carry on the essential metabolic processes for their own maintenance and growth. The conception of the organism as an integrated aggregation of cells led to the cell theory of heredity. Under this, heredity is the persistence of the essentials of cell organization through cell division and through the cell fusion that occurs at fertilization. The development of the individual need not proceed with extraordinary precision since it is outside the stream of heredity which consists merely of the line of germ cells. Development is a step by step elaboration of pattern, as cells, derived by division of the fertilized egg, and all of the same essential genetic constitution, react differentially in relation to the diverse external and internal relations that arise in the process.

Experimental evidence of the most direct sort has made it clear that the essential persisting organization of cells (at least in all forms above the bacteria and Cyanophyceae) resides in the threadlike chromosomes of the cell nucleus. Before each ordinary division (mitosis) each of the chromosomes duplicates. At division the duplicants are drawn to opposite poles with the consequence that each daughter cell receives a set of chromosomes identical with that in the parent cell. As the egg and sperm both contain the set characteristic of the species, the fertilized egg has a double set. This is maintained by mitosis in all cells of the body (animals, plant sporophytes). In the formation of reproductive cells in animals, spores in plants, there are typically two peculiar divisions constituting the process of meiosis. Homologous paternal and maternal chromosomes pair and after one duplication and possible exchange of blocks between homologous strands, the four resulting strands are segregated into four cells each of which accordingly receives only a single set of chromosomes (*see* CYTOLOGY). These cells are the germ cells in animals. In plants there is more or less multiplication in this phase (gametophyte generation) before the germ cells are formed by mitosis. The union of two germ cells restores the double (diploid) set and completes the cycle. The exchanges between homologous chromosomes, and their segregation which occurs independently among the chromosomes, make possible a detailed analysis of the composition of chromosomes with respect to differences that have differential effects on characters of the organism. It turns out that the chromosomes are differentiated along their lengths into hundreds or even thousands of elementary blocks, each corresponding to a Mendelian unit of heredity or gene (*q.v.*).

While the mechanisms by which the set of chromosomes is maintained intact through cell division and cell fusion are extraordinarily precise, irregularities occur from time to time resulting in genetic change and giving a possible basis for evolution. If nuclear division is not followed by cell division, the entire array of chromosomes is doubled. The four fold (tetraploid) set is maintained thereafter with almost the regularity of the usual diploid set. There is rarely much effect on the characteristics of the organism except for a usual increase in size. If, however, the process occurs in a hybrid that is sterile because the parents are so different that their chromosomes do not pair, causing irregularities in meiosis, the resulting amphidiploid, with pairing restored, may be of full fertility and breed true to the hybrid type. Many such hybrids have been produced.

In other cases an irregularity in mitosis may result in increasing or decreasing the normal set by one or more chromosomes. The disturbance of the normal balance tends to affect every part of the organism. The condition can rarely be fixed, however, since the union of two germ cells with the same loss or duplication of whole chromosomes gives types that are usually unable to live.

The chromosomes may be broken temporarily by various agencies, including certain chemicals, ultra-violet and ionizing radiations. When two breaks are present simultaneously, the reunion of broken ends may lead to duplication or deficiency of a block of genes, to reciprocal translocation between chromosomes or to inversion of the order within a block. Deficiencies even of small regions are usually lethal if received from both parents. Duplications, especially if small, may however be capable of being fixed. The effect on characters is likely to be more extensive than that of gene mutations. The balanced changes (translocations and inversions) also often may be fixed. As a rule there is either no effect on characters or an effect comparable with that of a gene mutation. Thus to a first order, the actions of the genes appear to be independent of their arrangement in the chromosomes.

Changes within the genes (gene mutations) have been observed to occur at measurable rates. There is a class of highly mutable genes responsible for certain types of variegation, as in variegated corn. More typical are rates of recurrence of the order of 10^{-5} to 10^{-6} per generation, observed for example in ***Drosophila***, maize and man. There are probably many more genes, however, with lower rates. (*See* GENE.)

There are cases in which more than 40 alternative conditions (alleles) of the same gene have been demonstrated in natural species by breeding experiments. There is probably no limit to the number of possible changes at a locus.

These are the principal causes of genetic variability in a form in which there is exclusive uniparental reproduction. Under biparental reproduction, the variability due to these causes is enormously amplified by recombination. Each new mutation adds merely one new type under uniparental reproduction. Under biparental reproduction it may be combined with all existing combinations, thereby doubling the number of types. Thus with 10 alleles at each of only 100 loci, the number of possible true breeding combinations is the inconceivably great number 10^{100} under biparental reproduction.

The relation between gene and character is important in considering the mechanism of evolution. The concept of homology seems to imply a separate heredity for each part of the anatomy which can be homologized in related species. Thus one of the objections formerly raised against the theory of evolution by natural selection was the difficulty in conceiving of adequate simultaneous selection of all parts of the body. It was thought for example that the right and left eyes of vertebrates must have been perfected independently. The complicated adaptive architecture of each bone was thought to have its own separate evolutionary history. This implies the old doctrine of preformation.

The evidence from genetics indicates, on the contrary, that every cell in the body has the same array of genes, but that it is an array that gives the cells a wide repertoire of possible responses in relation to different local conditions.

The closest approach to a one-to-one relation between gene and character is in the determination of elementary metabolic processes of cells. In the mould, *Neurospora*, for example, each of seven successive steps in the synthesis of the amino acid arginine is under the control of a separate gene. There is indeed some reason to suppose that the only direct action of genes apart from self duplication (probably as nucleoprotein molecules), is in regulating the pattern of other protein molecules, some of which as enzymes are the primary regulators of metabolism.

The relation of genes to gross morphological characters with which the evolutionist is largely concerned must be very indirect. The localization of gene action must depend not on localization of the genes but of the conditions under which their actions are invoked. Action may be restricted to a particular tissue, *e.g.*, epidermis, pigment cell, blood cell. In some cases, it may be invoked wherever a given tissue occurs but more often only in parts of the body where some condition in the tissue in question is above (or below) a certain threshold. Localized direct effects may have secondary consequences that ramify from one tissue to others producing extensive changes throughout an organ or the whole body.

No distinction in kind can be made between the indirect effects of a single gene replacement and those of an altered environmental factor on the pattern of a developing embryo. The normal pattern as a whole, however, must be considered to be the epigenetic consequence of differential action of the system of genes in a mass of cells in which there are from the first physiological gradients.

The action of any gene usually depends on the prior action of other genes as illustrated in the simple case of the synthesis of arginine. The genes may accordingly be grouped into reaction systems which, however, may overlap. Experiments indicate that these systems are almost wholly independent of the spatial relations of the genes in the chromosomes. The calling into action of one whole system or of another may depend on which first step is invoked by local conditions. In cases of homoeosis in which a single gene may, as in the fruit fly, *Drosophila melanogaster,* replace antennae by legs, certain mouth parts by legs, balancers by wings, etc., the gene is to be looked upon not as a germinal representative of the whole complex structure but as a switch which alters conditions so as to set going a long established reaction system in a strange location.

The term homology is used not only for corresponding structures in different species but for similar structures in different parts of the same organism. There are bilateral homologues such as the eyes, serial homologues such as the hands and feet, and other replicative homologues such as hairs, feathers, etc. The basis of such ontogenetic homology is clearly the calling into action of the same or similar reaction systems by conditions that recur in development. The interpretation to be put on phylogenetic homology is closely related. Both depend on the calling into play of more or less similar reaction systems by more or less similar local conditions. In ontogenetic homology the total array of genes is identical, the conditions may differ, while in phylogenetic homology every gene of the ancestor may be replaced by a somewhat differently acting allele without altering the pattern of development enough to prevent closely similar conditions and responses from appearing in corresponding places. In both cases homology is properly a matter of degree rather than an absolute.

The action of the system of genes responsible for the general structure of bone, including the property of deposition of bone in adaptive relation to prevailing stresses is doubtless common to all bones and this gives a certain degree of homology to all. The action of the system of genes responsible for the form of a particular bone is more related to the local conditions. There may be almost perfect homology as between the bones of right and left leg or less perfect homology as between bones of arm and leg or merely the low degree of homology of all bones referred to above.

Turning to phylogeny, the frequent transformation of an indefinite number of undifferentiated replicated homologues into a definite number of differentiated partial homologues, or the reverse, to which earlier reference has been made, is readily understandable. Elements that develop under virtually identical reaction systems in the ancestor may hardly be homologues at all in the descendant from the standpoint of common gene action. Conversely, structures such as chondrocranium and dermal scutes which depend initially on almost wholly different reaction systems may, in this sense, acquire a degree of ontogenetic homology (cartilage and dermal bones). A reaction system that is called into play only in certain anterior somites in a primitive vertebrate, to produce pronephric tubules, may be called into play in modified form in posterior somites of descendants, to produce partially homologous mesonephric or metanephric tubules.

THEORIES OF EVOLUTION

Theories of evolution may be classified according to the roles assigned to chance variation and to influence of the environment. Both are denied in the Aristotelian conception of evolution as analogous to individual development, the realization of an innate potentiality, irrespective of environment (except for provision of the necessary conditions for life). In modern times, this idea appears in the orthogenesis of Naegeli and as the core of H. F. Osborn's aristogenesis. According to another group of theories in which chance plays no role, evolution is a continuation of individual physiology through the inheritance of characters acquired either by a moulding influence of the environment (Buffon, Eimer) or by active adaptation to the latter (Lamarck). Under the mutation theory, on the contrary, a single chance mutation is the crucial event in the origin of species. In the more extreme forms of this theory, it is held that the higher categories even up to the phyla arise only from mutations of appropriate magnitude, almost a denial of evolution as a cumulative process (Saint Hilaire, J. C. Willis, Richard Goldschmidt). In the mutation theory of De Vries, species arise by single mutations but the higher categories are cumulative products. Finally come those theories under which evolution is a statistical process of transformation of populations. Random variability furnishes the raw material, but the course, at least in the long run, is indirectly determined by the relation to the environment through natural selection (Darwin, J. B. S. Haldane, R. A. Fisher, S. Wright). (*See* GENETICS OF POPULATIONS.)

Orthogenesis leaves unexplained the branching of the tree of life into millions of species, each marvelously adapted to some niche in nature. In modern times, indeed, it has usually been applied only to those cases in which there has seemed to be a uniform linear trend for a considerable period in a direction difficult to account for by adaptation (*e.g.,* elaboration of the sutures of ammonites, the increase in size of horns of titanotheres and of the Irish elk to unwieldy proportions, etc.). Most authors, however, believe that these cases are open to alternative interpretations in terms more in harmony with our present knowledge of heredity. A certain element of truth in this principle would, of course, be generally recognized in the principle that possible directions of "random" genetic change are restricted, sometimes severely, by the organization already attained.

The inheritance of acquired characters seems at first sight to be the simplest interpretation of certain kinds of adaptations. Thus the evolution of complicated patterns of instinctive behaviour suggest heredity to be a sort of racial memory that is continually being modified by individual experience (R. Semon). Unfortunately for this view the most complicated instincts are probably those exhibited by the sterile workers, but not by the reproductive castes, among the social insects. The universality of such a caste system and also of nest building among the termites indicates that both of these characteristics trace back to the origin of the order. The differences in the nests even among species of the same genus, as noted earlier in the cases of *Amitermes excellens* and *A. meridionalis,* indicate a continuing evolution in the behaviour of the workers which can not be due to direct transmission of changes in heredity based on their individual experiences.

The whole theory of the transmission of changes in anatomical characters or in behaviour, acquired by parents under special conditions, to descendants that have not been subject to these conditions, runs counter to the principles of heredity and development established by experiment. The basic property of genes is the persistence of their properties, irrespective of the characters which develop from interactions with other genes or environment. The occasional mutational changes which occur are due to direct modifications of an individual gene or chromosome on absorption of thermal, chemical or radiant energy and are unrelated to the characters of the organism as a whole. The theory moreover implies a one-to-one relation between such characters and elements of heredity. Such a relation seems indeed to hold for protein specificity, but for anatomical characters as well as for behaviour any such preformation in the germ cells is quite out of harmony with the observed epigenetic course of development.

It may be added that there have been no repeatable experiments that indicate that such a character acquired by the parent under special conditions tends to develop in the offspring in the absence of these conditions. This class of theories is therefore unacceptable.

The mutation theory has the merit that the only sort of case in which the origins of recognized species have been duplicated in the laboratory is by a process that may be considered under this head. A mutant resulting from doubling of the entire set of chromosomes may be fully fertile by itself while producing a sterile triploid hybrid on crossing back to either parent species. In the case of autotetraploids the character change is so slight that a specific difference is likely not to be recognized unless by chromosome count. If of sufficiently ancient origin, however, the reproductive isolation from the parent species will probably have led to a drifting apart from the latter by processes discussed later. An amphidiploid hybrid on the other hand has all of the essential properties of a new species from the first. There is no doubt that many species of flowering plants have arisen in this way. Thus plants which are virtually indistinguishable from a natural species, *Galeopsis tetrahit* with 32 chromosomes, have been produced by Arne Müntzing from hybrids of *G. pubescens* × *G. speciosa,* each with 16 chromosomes. A considerable number of other cases are indicated.

The origin of a new species in nature has been observed in the case of *Spartina townsendii,* a plant which was first found in 1870 in largely unoccupied mud flats in southern England. After 1900 it multiplied enormously. Its characters and its chromosome count (126) indicate that it is an amphidiploid hybrid between European *S. stricta* with 56 chromosomes and *S. alterniflora,* an American species with 70 chromosomes, seed of which was probably carried to Europe accidentally (C. L. Huskins). Its pre-adaptation to an almost empty niche in nature is noteworthy.

There are many cases in which forms which have all the characteristics of new species have been produced in the laboratory, among which one of the most remarkable is the generic amphidiploid Raphanobrassica (36 chromosomes) produced by G. D. Karpetchenko from the radish *Raphanus sativus* with 18 chromosomes and the cabbage *Brassica oleracea* also with 18 chromosomes. Raphanobrassica is a luxuriant plant which is very different from either parent.

There are probably cases of autotetraploid and amphidiploid species among hermaphroditic and parthenogenetic animals, judging from chromosome numbers. The relative prevalence of exclusive biparental reproduction in higher animals is probably the reason for the much lower frequency of all forms of polyploidy among them than among the higher plants as a means of multiplication of species. There is little chance of establishment in nature unless there is the possibility of extensive multiplication by uniparental reproduction.

Types of chromosome change (other than polyploidy) and gene mutations, whatever the magnitude of their effects on the organism, merely produce segregating variations within the populations in which they appear. This was the case for example with a number of forms of evening primrose (*Oenothera*) which De Vries found growing wild in Holland. These differed from the predominant form, *Oenothera lamarckiana,* in every respect, suggesting that they were different species. Since *Oenothera lamarckiana* was an American form, escaped from cultivation, it seemed to De Vries that he was observing here the abrupt origin of new species. His mutation theory had its origin primarily in these forms. It has turned out that most of these have the chromosomal complement of *Oenothera lamarckiana* (14 chromosomes), plus an extra chromosome. The latter is transmitted only in ovules, since pollen containing it is abortive. These types thus can exist only as segregants of a *lamarckiana* population and can not be considered as new species. In other cases, including especially inversions, translocations and small duplications, a new chromosomal type can be produced in true breeding form, but only by statistical processes within the population or an isolated portion of it.

Comparisons of the chromosomes of closely related species often show that an accumulation of many changes must have occurred in one or both after their separation. Crosses between species, however, also indicate multiple gene differences. Occasionally it is possible to demonstrate the segregation of alleles with major effects on one or more characters. Most of the differences between characters, however, show the type of segregation that indicates differences in a large number of genes with individually slight effects.

Genetic analysis of the variability actually found within local populations reveals differences of the same nature as those found to distinguish species. Differences in chromosome pattern are occasionally found. Especially common are differences in order within chromosomes. Thus in *Drosophila pseudoobscura* 15 different orders are found within the third chromosome alone (A. H. Sturtevant and T. Dobzhansky). These are distributed so that three or four are usually found segregating in each locality. Extensive breeding tests of wild populations always reveal the presence in low frequencies of alleles of many genes. There are many species including man, species of mammals, birds, fish, grouse locusts, snails, butterflies, etc., in which there is conspicuous polymorphism because of the presence of two or more alleles at high frequencies. In addition to alleles with such conspicuous effects that genetic analysis can readily be made, there is always considerable genetic variability of all measurable characters—including general size, relative proportions of all parts, intensity of colour, etc.

It is clear that the genetic differences between closely related species are not such as could have arisen at a single step and that the raw material for the accumulation of the differences is present in the genetic variability existent in all species. It is thus necessary to consider the processes which may bring about a gradual transformation in the genetic composition of a population.

The Statistical Transformation of Populations.—Under the mutation theory of evolution, a species is treated as if characterized by a single typical array of genes, organized in a single particular pattern in the chromosomes. Under the statistical theory, the genetic constitution of a species can be adequately represented only in terms of sets of *gene and chromosome frequencies.* It is recognized that observed mutation rates are such that any reasonably large population may be expected to carry more than one allele, usually many, at many if not all loci. Because of the symmetry of the Mendelian mechanism, the frequency of each allele in the population tends to remain constant in the absence of disturbing factors. The proper representation of the genetic constitution of a random breeding population of diploids is of the type $(p_1A_1+p_2A_2 \ldots +p_mA_m)^2(q_1B_1+q_2B_2 \ldots +q_nB_n)^2 \ldots$ Here the A's represent one set of alleles and the p's their relative frequencies; the B's represent another set, the q's their frequencies and so on. The arrays of each set are squared to take account of the postulated random unions of gametes. The arrays for the different sets are represented as multiplied in accordance with the fact that the combinations within each set may be expected to be combined at random with those in other sets, in the long run, including even those in the same chromosome.

This mode of representation may appear rather complicated. Really, however, there is an enormous simplification in attending to the stock of genes carried by the population instead of to the array of genotypes of the actual individuals. Assume 4 alleles (10 combinations) in each of 1,000 sets. The formula for the species would require symbols for 4,000 genes, each with a symbol for its frequency. This is a rather large number but it is incomparably less than the potential number of genetically different kinds of individuals, 10^{1000}.

The system of gene frequencies may be expected to remain substantially constant for many generations, but the array of actual individual genotypes can not be expected to have a single element in common in any two generations. Consideration of the gene frequencies shows how certain statistical properties of the population such as the average and amount of variability in each respect may be expected to persist. It is to be noted that variability itself is a property of major importance in adapting a population to its environmental vicissitudes.

The elementary evolutionary process from this viewpoint is simply change of frequency of a gene or of a chromosome pattern. The immediate processes may be grouped according to the degree of determinacy. First are three types of systematic change: (a) *mutation pressure* due to recurrent change of a given sort in the hereditary material; (b) *immigration pressure,* due to introduction of different heredity from without; (c) *selection pressure,* due to any systematic cause by which the gene or chromosomal arrangement in question tends to increase or decrease in frequency without either mutation or immigration. Differential mortality, differential rate of attainment of maturity, differential mating, differential fecundity and differential emigration are such causes. Next are random changes in gene frequency due to *accidents of sampling,* indeterminate in direction but determinate in variance. Finally, ***nonrecurrent mutation*** is a wholly indeterminate factor (*see* GENETICS OF POPULATIONS).

Recurrent mutation, immigration and selection all tend to have systematic effects. The alleles that are present in a population are, in general, ones for which these systematic pressures tend to oppose each other in one way or another. Thus the tendency to elimination of a given gene by selection may be balanced at a certain frequency by its continued reintroduction by mutation or immigration. Selection favouring a heterozygote (A_1A_2) over both of the corresponding homozygotes (A_1A_1 A_2A_2) may maintain both alleles at certain frequencies. Presumably all mutations that are likely to arise at one or two steps from the more abundant genes present in the population have been tried out by natural selection and found wanting, and thus are found at negligibly low frequencies if at all. There may be very valuable mutations which could only arise through a succession of unfavourable ones but these will have very little chance of occurring. The result is a system of equilibrium frequencies, any departure from which tends to be followed by restoration. Here we have a theory of the stability of subspecies and of species in spite of continuing mutation, continual interbreeding at a low rate with neighbouring populations and continuing selection.

Selection relates to the organism as a whole or to the social group, not to single genes except as a net resultant. Genes that are favourable in one combination may be unfavourable in another. This should indeed be the rule for the class of genes carried at intermediate frequencies in the population. Suppose, for example, that there are many sets of alleles that affect the relative lengths of legs and body in an animal that depends on speed. Maximum speed requires a certain proportionality. Genes that increase the relative length of the legs would be favoured or opposed by selection according as they happen to be associated with an array of other genes that tend to make the legs too short or too long. If we consider merely the combination of two pairs of alleles there may be two different "peak" combinations with respect to adaptive value, *e.g.*, AAbb and aaBB in contrast with both aabb and AABB. A species may be held to approximate fixation of one of these even though the other is somewhat more favourable because the passage through populations of lower average adaptive value, *e.g.* $[(\frac{1}{2})A+(\frac{1}{2})a)]^2[(\frac{1}{2})B+(\frac{1}{2})b]^2$, may be too improbable. In the system of adaptive values of the combinations of all sets of alleles, thousands in number, there may be a virtually infinite number of distinct peaks, each corresponding to a different harmonious combination of traits. The direct pressure of selection tends to carry the system of gene frequencies to the neighbourhood of the peak on the slope of which it already lies, but this is not likely to be the highest peak. The genotypes present in the population when equilibrium is reached about this peak will all be at relatively few removes from the peak genotype, although no two of them may be expected to be exactly the same.

Increase in the severity of selection will bring about a higher level of adaptation but only by forcing a denser clustering of genotypes about the peak type. Variability tends to be reduced.

Increase in the general rate of mutation has the opposite effect, a wider dispersal of genotypes from the peak type, giving a lower average of adaptation but more variability. There is an increased chance of production of individuals highly adapted in a different way, *i.e.*, with a genotype in the neighbourhood of another and perhaps higher peak combination; but, as the genotype is broken up in the formation of germ cells, this does not create an evolutionary trend in the new direction. The problem of evolution from this viewpoint is to find a process by which a species may pass from control of one peak system to control by a higher one. The most obvious possibility here is the occurrence of a novel mutation. There are, however, other processes that are probably more frequent. So far we have assumed constancy of the conditions of selection. But conditions are constantly changing as a result of changes in the abundance of other species, the evolution of these and of secular changes in the nonliving environment. The adaptive values of peak genotypes may be depressed while those of other genotypes may be elevated. Selection under these conditions keeps the species continually on the move in the direction, whatever it may be, in which the pressure is momentarily greatest. Here we have a very important evolutionary process. In the long run the species tends to move into the predominantly higher portions of its general field of potentialities.

We have assumed random interbreeding within the range of the species. If it occupies a large range in which the conditions of selection are not the same in all regions, the local selection pressures may prevail over the tendency toward homogeneity due to migration, provided that dispersal is sufficiently restricted by geographic or ecologic barriers or if it is slight in comparison with the range of the species. The result is the splitting up of the species into subspecies characterized by adaptive differences. The species as a whole has a more extensive field of genetic potentialities than under random mating through increase in the number of alleles carried at each locus. This is also true of the separate subspecies because of interbreeding with each other at boundaries. There is a greater chance than in a homogeneous species that one of the subspecies, in increasing its adaptation to the local conditions, may hit upon an adaptation of general rather than merely local significance, which thereupon may spread through the whole species.

Under uniparental reproduction each mutation gives rise to a clone of individuals that are genetically identical until split into diverse types by a new mutation. Because of this, selection would be enormously more effective than under biparental reproduction if the number of genotypes on which it could operate were comparable. As noted earlier, however, the field of potential genotypes is actually incomparably less than under biparental reproduction. The advantages of the two systems are combined if there is predominant uniparental reproduction but sufficient crossing to provide an unlimited variety of potential kinds of clones. Such a balance between the methods of reproduction provides a trial and error method by which the species may pass from control by one peak combination to another. It is especially effective in organisms whose numbers increase enormously under favourable conditions but are forced to contract on return of unfavourable conditions. The maintenance of a relatively stable population requires more labile local adaptations than provided by this system.

No consideration has yet been given to accidents of sampling as a cause of change of gene frequency. The resultant of the continual tendency of gene frequencies to move toward certain equilibrium values under the systematic pressures that have been discussed, and the tendency to drift at random due to accidents of sampling, is a certain distribution of probabilities which should be exemplified in the course of time. With given systematic pressures the mean deviation from equilibrium due to this cause is slight in large populations, but large in small ones. In very small populations, in fact, the cumulative effect of accidents of sampling dominates the situation so completely that one allele or another tends to drift into fixation at each locus. Being random, the resulting genotype is virtually certain to be poorly adaptive. The population suffers the well-known degenerative effects of too close inbreeding, with extinction as the probable end result.

With a certain balance between the effects of accidents of sampling and the pressures of mutation and selection, loss of variability and average lowering of fitness may be slight, although all gene frequencies are drifting rather widely about their equilibrium values. The constitution of the population keeps moving in a largely random fashion in the field of possible gene frequencies but under loose control of an adaptive peak. It may, however, wander across a "valley" to come under control of another peak, corresponding to the emergence of a new adaptive type. There is a trial and error process here but one that is too slow to be of much evolutionary significance.

A similar process may, however, be of great evolutionary significance when occurring in each of a large number of small, largely but not completely, isolated local races. In this case, the constitution of each race drifts separately under loose control of the peak type. The rate, based here on the balance between the effects of accidents of sampling and immigration pressure may be enormously more rapid than in the preceding case (in which mutation rate is a limiting factor).

With continually shifting, only partially adaptive, differentiation among numerous local populations there is an excellent chance that the constitution of one will cross a valley to come under control of a higher peak and become the principal source of immigrants to other populations, with the consequence that these will be drawn toward it in composition. In the end the whole species may come under the control of this higher peak. This trial and error process may continue indefinitely.

This process may be described as one of intergroup selection. The selection depends on the genetic system of each local race as a whole, instead of merely on the net effects of single genes, all that is possible in a homogeneous population. Characters may be fixed that are favourable to the group even though disadvantageous in individual competition. The consequences of this system are obviously somewhat similar to those of the balance between uniparental and biparental reproduction discussed earlier. The properties of the local race are, however, more favourable to the maintenance of reasonably stable population numbers than are those of the clone.

It is to be noted that the stock of potential but largely unexpressed variability carried in the species as a whole is so great, if there is extensive subdivision into partially isolated local races, that evolution is not likely to be limited by the rate of occurrence of mutations. The limitation is in the difficulty of passage from domination by one adaptive type to domination by another, against the conservative pressure of selection.

The Splitting of Species.—As frequently noted, a mutation, however great its effect, can not be considered as a new species if it persists only as a segregant in the population in which it appeared. The splitting of a species implies the formation of two populations which do not exchange genes. A daughter species may arise from individuals that by a rare accident reach a geographically isolated region. The mere isolation is, of course, not enough by itself. Few biologists would be willing to accept the conclusion that individuals born as members of one species, become members of another, merely by isolation, or that their undifferentiated descendants may be considered as of a different species. In the long run, however, the processes that lead to transformation of single species would usually insure that two isolated groups would become sufficiently different, even without any differences in the conditions of selection, to become recognizable species by the usual criteria.

They would not be separate species by the ideal definition, however, until the appearance of differences that would prevent interbreeding on contact. Among the conditions that can clinch more or less permanently a species difference are psychic or morphologic differences that interfere with cross mating; physiological inhibition of fertilization; abnormal development of hybrids; hybrid sterility, whether due to imperfect development of gonads or to failure of gametogenesis from accumulation of chromosomal differences; and if all else fails, prevailing abnormal development in F_2 segregants.

For the most part, such differences can obviously come about only by the accumulation of steps each one of which has no appreciable isolating effect from that immediately preceding.

These three processes, primary isolation, differentiation in characters and the permanent clinching by genetic barriers to interbreeding, may occur in various relations to each other. The sequence—geographical isolation followed by a secondary drifting apart in morphology and in conditions that prevent possible fusion—has probably been of primary importance. On the other hand, two populations that retain an intergrading connection may become as different as typical species and then graduate from subspecific to specific status by the practical definition, without further change, merely by extinction of the connecting population. They may even be unable to interbreed on restoration of contact and thus be distinct species by the ideal definition as shown by the species rings discussed earlier.

All aspects of speciation may proceed coincidentally in two adjacent subspecies which are becoming adapted to different ecological conditions. Here an intergrading population may be progressively reduced in number and ultimately eliminated by selective favouring of bars to the production of hybrids poorly adapted to both conditions.

The splitting of a species within a single territory is probably very unusual in biparentally reproducing forms (E. Mayr), but geographical isolation need not be as great, as otherwise to permit speciation, if correlated with marked ecological differences. Moreover in forms with very weak means of dispersal, species may arise in relatively small, adjacent regions, especially if these are ecologically diverse. In special cases a single mutation may permit speciation within the territory of the parent form by its isolating action. The cases of autopolyploidy and amphidiploidy have already been discussed. A mutation may bring about a change in the time of flowering in a plant so great that no interbreeding with the parent form is possible.

Directly isolating mutations are, however, unlikely to become established within a population that reproduces wholly by means of separate sexes. Prevailing uniparental reproduction enormously increases the chance of success of such mutations but, also as already noted, reduces the definiteness of the species concept itself.

GENETIC INTERPRETATION OF MAJOR EVOLUTIONARY PROCESSES

The prevalence of apparently directed evolutionary trends has given palaeontologists a strong bias toward the conception of an innate perfecting principle or of an organic momentum. There seems no overwhelming difficulty, however, of interpretation by statistical genetics. A form that has been fitted for life in one ecological niche, in a region in which all niches are occupied, is likely to be in a better position to modify its organization toward more perfect adaptation to its niche, or toward minor changes that occur progressively in this niche, than any other species, but is correspondingly in a poorer position than others to make over its organization for adaptation to other niches. General control by natural selection seems an adequate explanation of such cases as the somewhat orthogenetic sequence of genera of the horses from the Eocene to the present. This is not incompatible with considerable divergence of species and subspecies of jointly adaptive and nonadaptive character. Trial and error processes at these levels may indeed furnish the basis for apparent orthogenetic advance at the higher level. This was essentially W. D. Mathew's interpretation of the palaeontological record of the horses. In terms of the system of adaptive values of gene combination such forms may be said to be ones for which there is only a single available line of increasing peak values.

More difficulty has perhaps been felt in cases in which there is an apparent trend toward the development of an injurious character leading to ultimate extinction. A trend preceding extinction may merely mean that the form has become committed by its past history to a mode of adaptation that is inherently less effective than that of a new rival form, forcing it into a line of ever narrower specialization which eventually becomes a *cul-de-sac*. In other cases, as J. S. Huxley has emphasized, selection between individuals of the species in respects that contribute to success in mating may direct evolution along a course opposed to the welfare of the species. Horns, tusks and total size may become unwieldy through such a process. Again, as Huxley has also pointed out, strong selection in one respect, especially toward increased size, may alter the proportions of parts in a way that suggests organic momentum because of the correlated effects of a system of differential growth rates of the parts.

One sort of trend that requires special consideration is that toward degeneration of parts that have ceased to be useful. A certain amount of reduction may of course be accounted for by direct selection against useless encumbrances, but evolution often carries degeneration beyond the point at which this explanation seems plausible. Another possibility is direct mutation pressure. The most general characteristic of mutations, due to their accidental nature, is a degenerative effect, as is well illustrated by a random array of *Drosophila* mutations in which defective eyes, wings, bristles and pigmentation are the rule. Mutation alone may be adequate especially in small isolated populations. There are, however, indirect effects of selection that may be more important. Genes may have effects on seemingly unrelated characters. In a homogeneous population, alleles are selected on the basis of their net effects. Certain genes, maintained at high frequencies because of roles in the development of an important organ, may be less favourable than certain of their alleles in other respects. If this organ ceases to be useful, selection in the other respects takes precedence, with consequent continued degeneration of the useless organ. On the other hand, a useless organ may be maintained in spite of a slight encumbering effect if the genes that are involved in its development have effects that are tightly interwoven into the general developmental system. In general, these indirect effects of selection may be expected to speed up the elimination of useless organs in a species that is being rapidly made over, but are likely to bring about their retention as vestiges in a conservative form.

The long continued cessation of evolution in some forms is another problem. The apparently obvious explanation that mutation rate is exceptionally low is probably not the primary one. It is more probable that such a form has reached the limit of adaptation along its line of specialization. In terms of the system of adaptive values of gene combinations, they are held to the neighbourhood of the only peak immediately available. However, as mutation rate is itself known to be subject to gene action, it is possible that mutation rate may be secondarily lowered by selection in such a case.

In the opposite direction is the problem of the apparently abrupt emergence of a new higher category and its rapid branching by adaptive radiation. Here the most obvious explanations are either exceptionally rapid mutation or the occurrence of individual mutations with exceptionally great morphological effects. But rapid mutation merely increases the number of freaks, and the greater the morphological effect the less the chance of a harmonious, adaptive result. The explanation is to be sought rather in the magnitude of an opportunity. The field of gene combination is so extensive that, even with complete cessation of mutation, explosively rapid evolution is possible for a considerable period under favourable conditions, particularly ones in which the conservative pressure of selection is relaxed.

Favourable conditions fall into two major categories. First is introduction into unoccupied territory. A large portion of the land birds of the Hawaiian Islands belong to one family, the Drepanididae, found nowhere else in the world. There are 18 highly diversified genera and some 40 species. Presumably this family traces to the immigration of a few stray individuals of an American species. This species, whatever it was, had no such extraordinary outburst of evolutionary activity in its original home in which all ecological niches were occupied. The system of adaptive values of gene combinations in Hawaii (in contrast with that in America) doubtless included a very large number of peaks, easily available by trial and error processes. Species with different adaptations might become established on different islands, even though there were no important differences in conditions on these. Stray migrants between islands after speciation had been clinched would therefore find unoccupied ecologic niches until all niches were occupied on all islands.

The rapid adaptive radiation of the few primitive types of mammals that reached South America between the Palaeocene and Pliocene, and presumably that of the marsupials in Australia, are similar examples on larger stages.

A comparable opportunity is presented to a species that attains by any means a general adaptation that opens up a new way of life or an advantage in many ways of life. There may be nothing especially unusual or rapid in the first steps before a critical threshold is passed in the attainment of the adaptation. Thus while the amphibian leg and foot seem to have emerged very rapidly from the crossopterygian fin, the complex of adaptations which made the mammals a successful class required most of the Mesozoic for its development before the threshold for widespread adaptive radiation was reached at the beginning of the Cenozoic.

The Origin and Nature of Life.—Evolution has been treated here as based on the properties of the nucleated cell. The nucleated cell is itself, however, a most remarkable product of nature.

In speculating on the origin of life, attention is naturally directed toward those constituents of cells, the genes, which themselves have the most essential property of an organism, that of reproduction of their own kind, independently of the specific natures of their associates. The similarity of genes to the entities responsible for the virus diseases in size, chemical constitution (nucleoprotein) and capacity for reproduction, including reproduction as of the new sort after undergoing mutation, indicates that they belong in the same general category. Several of the viruses have been shown to have the molecular mode of organization by their capacity for crystallization. It is thus probable that genes also are to be looked upon as molecular patterns rather than as systems of colloidal phases comparable with cells, a viewpoint which carries us closer to possible derivation from nonliving materials than if the cell is considered as the starting point.

The proteins consist of long chains of amino acids of which the score or more kinds may be arranged in diverse sequences. Nucleic acid consists of long chains of nucleotides in exactly the same spacing, 3.34 A, as the amino acids of the protein, suggesting a one-to-one linkage in the nucleoprotein molecule (W. Astbury). Each nucleotide consists of a N-base (purine or pyrimidine) comparable with the amino acids in size, a five-carbon sugar and phosphoric acid.

It has often been suggested that life originated at a time when there was no free oxygen in the atmosphere and ammonia and reduced compounds of carbon (as well as CO_2) were present in abundance. The crucial event was the combination of these to form molecules with the self-duplicating property. Once formed, such molecules would multiply until the accumulated supply of one of the immediate components was exhausted. A premium would then be put, as pointed out by N. H. Horowitz, on a variant able to catalyze the formation of this component from a precursor, still present in abundance, a beginning of evolution by mutation and natural selection. Selection would also favour molecules capable of breaking down others and reorganizing their constituents according to their own patterns, the beginning of predation.

Symbiotic associations of nucleoproteins, each capable of carrying out one of the steps in the synthesis of necessary components from available substances, would also be favoured. The next step might be the evolution of more or less orderly mechanisms for the maintenance of the composition of such systems on fission, after attainment of a certain size. A symbiotic system of this sort might be expected to include not only diverse kinds of self-duplicating molecules but products of their activity, the beginning of a distinction between genes and cytoplasm. The bacteria and blue-green algae seem to be at about this level.

The continual exhaustion in nature of the necessary immediate constituents of the nucleoprotein molecules by such systems would prevent the origin of new nonsymbiotic self-duplicating molecules. Predatory nonsymbionts might persist and evolve in adaptation to their hosts of which they would constitute virus diseases.

The greatest step in the evolution of life, after its origin, would seem to have been the organization of the genes into linear strands, the chromosomes, associated with mechanisms for the precise apportionment of duplicants to daughter cells, for the union of two cells, and for the compensatory reduction division. These make available the enormous field of genetic variability due to recombination within a population and, in short, make the species, instead of the individual lineage, the evolutionary entity. On the physical side, the cohesion of cells to form multicellular organisms, in which large size and an extensive division of labour is possible, presents an adequate stage for the display of the almost unlimited potentialities of the evolutionary mechanism in a population of biparentally reproducing individuals.

It is a far cry from an autocatalytic nucleoprotein molecule to man. There is, however, at least the possibility of tracing continuity from molecule to man as a complex physicochemical system, consisting of ordinary atoms which are continually entering, taking a place in the system and departing. But the most important aspect of man is his mind: the perceptions of his changing environment, the correlation of impressions, the construction of systems of thought and the capacity for choice among possible courses of action. Some evolutionists (*e.g.*, A. R. Wallace), impressed by the gap in mental accomplishment between man and the morphologically most similar animals, abandon evolution at this point and postulate the incursion of complete novelty. To most, however, the application of the criteria which lead each of us to assume minds in other human beings, although we can not enter into their streams of consciousness, leads to the ascribing of minds to at least the higher vertebrates and to the hypothesis of an evolution of mind. The gap in the genetic basis of mind between animal and man may indeed have been considerably less than appears from the differences in mental accomplishment. Increase in intelligence seems to have been a well-defined trend among the higher primates, perhaps associated with the capacity for manipulating objects given by a type of hand, originally evolved for arboreal locomotion. The special premium on this trend, following the adoption of erect posture and complete freeing of the hand for its new use may have made it possible for man's immediate precursor to rise in intelligence above a threshold that permitted the invention of symbolic speech with the consequent emergence of a new and disproportionately rapid sort of cumulative process, the evolution of ideas, in which transmission is from speaker to listener instead of through the germ line. The transition from animal to man may well have been almost an abrupt one in terms of geologic time.

If mind be granted to the higher vertebrates, the argument from the observed continuity in mode of behaviour, indicates that it must be granted the lower vertebrates and indeed to all animals including the protozoa. No distinction, moreover, can be made between the protozoa and the protophyta.

No line can be drawn between free one-celled organisms and the cells of which higher animals and plants are composed. As the multicellular individual represents an integration of cells both in structure and functioning, so apparently must it be with respect to the minds of the former. The complex structural pattern and mind of the adult develops from that of single cells in each generation. The organization of the egg cell results from the union of two cells from different individuals which demonstrably contribute equally on the average both to heredity of structure and heredity of mind. The potentiality for both the structure and mind of the adult must somehow reside in the array of genes, which, as noted, probably means an array of nucleoprotein molecules.

Certain authors, *e.g.*, C. Lloyd Morgan, have postulated an emergence of mind from inert matter as a process comparable with the emergence of an organ from one of different use. There is a qualitative difference between these cases that makes this unacceptable. A thoroughgoing carrying out of the evolutionary viewpoint seems to lead to the conclusion that mind is an irreducible aspect of all reality; that the world consists primarily of a multiplicity of minds, largely but not completely isolated from each other. Each may construct an external world from the incursions of others into it. Each therefore has two aspects, both mental: as it is to itself (mind) and as it seems, as an incursion into the mind of another (matter). On this view a physical organization (atom, small molecule, gene, cell, multicellular individual) is the external aspect of an internal integration of minds.

The apparently deterministic laws of nature may be interpreted, as pointed out by Karl Pearson, as mere statistical descriptions of more or less consistent behaviour in a world in which every event has unique aspects. The task of science is to describe the statistical regularities in the world as they affect human minds, not to attempt the impossible task of interpreting phenomena as they seem from within. The scientist, as such, must restrict himself rigorously to the objective, mechanistic type of interpretation lest he present duplicate interpretations as if they involved independent factors. In regard to the autonomous behaviour of organisms, he can merely note the hierarchy of trigger mechanisms which make possible freedom (within limits) of the whole without appreciable divergence from the customary behaviour of the parts.

Evolution, above the level of the bacteria and the blue-green algae, is largely concerned with a loosely knit entity, the species. The species may be looked upon as an organism with literal protoplasmic continuity in the space-time of the physicist, through the continual union of lines of germ cells. In many animals this integration is supplemented by that based on social behaviour. It has been noted that the most persistent aspect of the species is a certain stock of genes, consisting probably of a few thousand kinds of nucleoprotein molecules. The individuals are the ever changing front with which the species confronts its environment. But while above the individual from the standpoint of composition, the species is an incomparably less integrated organism.

In most of the discussion, each species has been treated as if it evolved independently of the rest of the world, except for a rather mechanical-seeming connection through the concept of selection pressure. Actually the evolution of each species is merely an aspect in the evolving pattern of life as a whole and indeed of the world as a whole. Natural selection is an abstraction of the complicated reciprocal process in which the pressure of the species to find a larger place for itself in the world, and that of the world to keep the species in its place, results through devious ways, in a general trend toward progressive elaboration of the patterns of organization of both.

BIBLIOGRAPHY.— C. Darwin, *The Origin of Species* (1859), *The Descent of Man* (1872); V. L. Kellogg, *Darwinism Today* (1908); H. Woods, *Palaeontology, Invertebrate* (1937); Th. Dobzhansky, *Genetics and the Origin of Species* (2nd ed. 1941); E. Mayr, *Systematics and the Origin of Species* (1942); J. S. Huxley, *Evolution, the Modern Synthesis* (1943); G. G. Simpson, *Tempo and Mode in Evolution* (1944); A. S. Romer, *Vertebrate Palaeontology* (2nd ed. 1945).

(S. WT.)

where it could be said that Yahweh had recorded his name and would bless his worshippers (Ex. xx, 24). They were abhorrent to the advanced ethical teaching of prophets and of those imbued with the spirit of Deuteronomy (*cf.* ii Kings xviii, 4 with *v.* 22), and it is patent from Jeremiah, Ezekiel and Is. lvi–lxvi that even at a late date opinion varied as to how Yahweh was to be served. It is significant, therefore, that the narratives in Genesis (apart from P) reflect a certain tolerant attitude; there is much that is contrary to prophetical thought, but even the latest compilers have not obliterated all features that, from a strict standpoint, could appear distasteful. Although the priestly source shows how the lore could be reshaped, and Jubilees represents later efforts along similar lines, it is evident that for ordinary readers the patriarchal traditions could not be presented in an entirely new form, and that to achieve their aims the writers could not be at direct variance with current thought.

Southern Interests.—There is relatively little tradition from north Israel; Beersheba, Beer-lahai-roi and Hebron are more prominent than even Bethel or Shechem, and there are no stories of Gilgal, Shiloh or Dan. Yet in the nature of the case there must have been a great store of local tradition accessible to some writers and at some periods. Interest is taken not in Phoenicia, Damascus or the northern tribes, but in the east and south, in Gilead, Ammon, Moab and Ishmael. Particular attention is paid to Edom and Jacob, and there is good evidence for a close relationship between Edomite and allied names and those of south Palestine (including Simeon and Judah). Especially significant, too, is the interest in traditions which affected the south of Palestine, that district which is of importance for the history of Israel in the wilderness and of the Levites. It is noteworthy therefore, that while different peoples had their own theories of their earliest history, the first-born of the first human pair is Cain, the eponym of the Kenites, and the ancestor of the beginnings of civilization (iv, 17, 20–22). This "Kenite" version had its own view of the institution of the worship of Yahweh (iv, 26); it appears to have ignored the Deluge, and it implies the existence of a fuller corpus of written tradition. Elsewhere, in the records of the Exodus, there are traces of specific traditions associated with Kadesh, Kenites, Caleb and Jerahmeel, and with a movement into Judah, all originally independent of their present context. Like the prominence of the traditions of Hebron and its hero Abraham, these features are not fortuitous, though the problems they bring cannot be discussed here (see *Camb. Anc. History,* ii, 359 *sqq.,* iii, 472 *sqq.,* vi, 185 *seq.*).

Bibliography.—S. R. Driver's commentary (*Westminster Series*) deals thoroughly with all preliminary problems of criticism, and is the best for the ordinary reader; Dillmann (6th ed., E. trans.) is technical, Ryle (*Cambridge Bible*) and Bennett (*Century Bible*) more popular. Spurrell, *Notes on the Text of Genesis,* and Ball (in Haupt's *Sacred Books of the O. T.*) appeal to Hebrew students. Addis, *Documents of the Hexateuch,* Carpenter and Harford-Battersby, *The Hexateuch* and C. F. Kent, *Beginnings of Hebrew History,* are important for the literary analysis. J. Wellhausen's sketch in his *Proleg. to Hist. of Israel* (E. trans., pp. 259–342) is admirable, as also is the general intro. (trans. by W. H. Carruth, 1907) to Gunkel's valuable commentary. Fuller bibl. information will be found in the works already mentioned, in the articles in the *Ency. Bib.* (G. F. Moore), and Hastings's *Dict.* (G. A. Smith), and in the fine volume by J. Skinner in the *International Critical Series.* (S. A. C.)

GENET, EDMOND CHARLES (1763–1834), French minister plenipotentiary to the United States in 1793, was born on Jan. 8, 1763, at Versailles. He was for a time attached to the embassy at Berlin and later to the embassy at Vienna; and at the age of 18, following his father's death, succeeded him as secretary interpreter at the ministry of foreign affairs. In 1787 he was sent to the embassy at St. Petersburg where he remained until July 1792, when his liberal views made him *persona non grata.* After a brief stay in Paris, where he came more fully under the influence of the Revolution, "citizen" Genet was sent as French minister to the congress of the United States. He was assuming a position which would require much tact, but his impetuous nature combined with the ovations accorded him by the Democratic-Republicans, led him into misjudging public opinion regarding American neutrality. His activities in instigating military operations against the Spanish possessions of Florida and Louisiana and against Canada, the fitting out of privateers in American ports, his acrimonious debates with the federal government and his caustic attacks on the president, demonstrate conclusively his lack of diplomacy. Genet's threat to override the executive by appealing to the people caused Washington to ask the French republic to recall its representative. His successor, "citizen" Fauchet, brought orders to arrest him and send him back to France for trial, but Washington refused to permit the extradition. He subsequently became a naturalized American citizen. In 1794 he married Cornelia Tappen Clinton, daughter of the governor of New York and in 1814, four years after the death of his first wife, married Martha Brandon Osgood, daughter of the first postmaster general. He died on July 14, 1834.

Bibliography.—American Historical Assn., *Report 1896, 1897;* G. Bathe, *Citizen Genet, Diplomat and Observer* (1946); M. Minnigerode, *Jefferson, Friend of France: The Career of Edmond Charles Genet* (1928); *Miss. Valley Historical Review,* vol. 6 (1919); F. J. Turner, "Genet's Attack on La. and the Floridas," *American Historical Review,* 3:650–71 (July 1898).

GENET, a catlike mammal belonging to the family Viverridae (*see* Carnivora), found chiefly in Africa. The common genet (*Genetta genetta*) occurs throughout Africa and also in southern Europe and Palestine; it is about the size of a cat, but more slender and longer of leg. The fur is dark gray, thickly spotted with black, and having a dark streak along the back, while the tail, which is nearly as long as the body, is ringed with black and white. It frequents the banks of streams and feeds on small mammals and birds. It differs from the true civets in that the anal pouch is a mere depression and contains only a faint trace of the characteristic odour of the former. (J. E. Hl.)

Genet (Genetta servalina) of the Congo, Africa

GENETICS, a term coined by W. Bateson to designate that portion of biology concerned with heredity, variation, development and evolution. It is the science which seeks to account for the resemblances and the differences which are exhibited among organisms related by descent. Its problems are those of the cause, the material basis, and the method of maintenance of the specificity of germinal substance; in other words, "how the characters of parents and offspring are related, how those of the adult lie latent in the egg, and how they become patent as development proceeds." Its methods are those of observation, experimental breeding, cytology and experimental morphology. Its prosecution demands a knowledge of general physiology and of mathematics. It has both scientific and practical application: its principles impinge upon all doctrines of evolution and upon agricultural, animal and plant breeding practices. Its possible applications to human affairs have created the need for and the development of the applied science of eugenics (*q.v.*).

Out of the accumulated facts of genetical experimentation, there has been developed the theory of the gene (*q.v.*) intended to accommodate these facts. It states (1) that the hereditary characters of the individual are referable to paired elements (the genes) in the germinal material (the chromosomes, *q.v.*) which are held together in a definite number of linkage groups; (2) that the members of each pair of genes separate when the germ cells mature in accordance with Mendel's first law, and that in consequence each ripe germ cell comes to contain one set only; (3) that the members of different linkage groups assort independently in accordance with Mendel's second law; (4) that an orderly interchange—crossing-over—also takes place, between the elements in corresponding linkage groups; and (5) that the frequency of crossing-over furnishes evidence of the linear order of the genes in each linkage group and of the relative position of the genes with respect to each other.

The gene, a conception as reasonable and as real as the atom, is to be looked upon as a particular state of organization of the chromatin at a particular point in the length of a particular chromosome. (*See* Animal Breeding, Heredity, Cytology, Mendelism and Plant Breeding.) (F. A. E. C.)

Bibliography.—E. Altenburg, *Genetics* (1945); B. Bateson, *William Bateson, Naturalist* (1928); E. C. Colin, *Elements of Genetics* (1946); L. T. Hogben, *Introduction to Mathematical Genetics* (1946); T. H. Morgan, *The Theory of the Gene* (1928); E. W. Sinnott and L. C. Dunn, *Principles of Genetics* (1939).

GENETICS OF POPULATIONS. Many of the pioneer geneticists to whom is due the extraordinary development of knowledge of heredity after 1900 entered the field because of interest in the theory of evolution or in practical problems of animal and plant breeding. It seemed to them that the simple rules that were found for the heredity of many conspicuous variations, and the parallel behaviour of the thread-like chromosomes, visible in the nuclei of cells, provided keys that would rapidly reveal the solutions of their problems. To some extent their expectations were directly realized. But the harmonious combination of many subtle differences by which one species usually differs from its closest allies, and which similarly characterizes a successful strain of livestock or crop plant, continued to present difficult problems. It has become necessary to consider the statistical properties of populations in which many genetic differences are present simultaneously as the cumulative result of past mutation and recombination.

The reader must be referred to the articles on heredity (*q.v.*) and the gene (*q.v.*) for presentation of the principles of genetics (*q.v.*), without which the following discussion will not be intelligible.

Gene Frequency.—The basic concept of the genetics of populations is that of gene frequency. Let A_1 represent a particular autosomal gene and A_2 its array of alleles. Let $(x_m A_1A_1 + y_m A_1A_2 + z_m A_2A_2)$ represent the array of genotypes and their relative frequencies $(x_m + y_m + z_m = 1)$ in mature males and assume a similar array with subscript f in mature females. The frequencies of gene A_1 in gametes of males and females respectively may be written: $q_m = [x_m + (\frac{1}{2})y_m]$, $q_f = [x_f + (\frac{1}{2})y_f]$. Since there is equal inheritance of autosomal genes from both parents, the gene frequencies become the same in males and females in the next generation. In symbols, $q_m = q_f = (\frac{1}{2})(q'_m + q'_f)$, using primes to designate the preceding generation. In general we may write $q = (\frac{1}{2})(q_m + q_f)$ as the effective gene frequency of a population, irrespective of the absolute numbers of males and females. It is obvious from the symmetry of the Mendelian mechanism, and the absence of any contaminating influence of alleles on each other in heterozygotes, that gene frequencies tend to remain constant in populations $(q = q')$.

In the important special case of sex-linked genes, one sex (to be taken as male in what follows) has only one chromosome (transmitted to half the gametes) that carries the locus in question, while the other sex has the usual pair. As males inherit such genes only from their mothers, while females inherit equally from both parents, $q_m = q'_f$, $q_f = (\frac{1}{2})(q'_m + q'_f)$. The effective gene frequency of the population is here $q = (\frac{1}{3})(q_m + 2q_f) = (\frac{1}{3})(q'_m + 2q'_f) = q'$. While the frequencies of sex-linked genes tend to remain the same generation after generation in the population as a whole, the frequencies in the males and females separately do not come to immediate equilibrium. There are rapidly damped oscillations about the mean, $(q_m - q) = (-\frac{1}{2})(q'_m - q')$. In polyploids, also, the frequencies of the various genotypes (*e.g.*, $A_1A_1A_1A_1$, $A_1A_1A_1A_2$, $A_1A_1A_2A_2$, $A_1A_2A_2A_2$ and $A_2A_2A_2A_2$ in a tetraploid) approach equilibrium only gradually, while gene frequency tends to remain constant.

Departures from random mating do not affect gene frequencies, provided that all genotypes contribute proportionately to the next generation. The frequencies of genotypes are, however, affected. Thus the random union of the array of sperms $[qA_1 + (1-q)A_2]$ with a similar array of eggs produces a population with the genotypic array $q^2A_1A_1 + 2q(1-q)A_1A_2 + (1-q)^2A_2A_2$, which is stable in large populations as long as mating continues at random. But if the system of mating is changed to universal self-fertilization, the homozygotes, A_1A_1 and A_2A_2, produce only their own kind while the heterozygotes, A_1A_2, produce 25% of each of the homozygous classes and only 50% of their own kind. With a 50% reduction in the frequency of heterozygotes in each generation it requires only a small number of generations to convert the population practically wholly into a mixture of homozygous lines $[qA_1A_1 + (1-q)A_2A_2]$. Gene frequency has not changed but the aspect of the population has changed enormously. Recessives that are manifested in only 0.01% of the individuals of the random bred population become fixed in 1% of the inbred lines. Random mating would, however, immediately restore the initial genotypic composition (assuming, of course, that there has been no selection).

The tendency toward indefinite persistence of variability under random mating and its rapid elimination within each closely inbred line, shown above to be immediate consequences of the Mendelian mechanism, are not expected under blending heredity, under which variability should fall off 50% in every generation of random mating unless replenished by mutation at this enormous rate. Comparative data from random bred and inbred lines, and from crosses between inbred lines, indicate the practical universality of Mendelian heredity.

The extension of these principles to systems of multiple alleles is obvious. Under random mating, the frequencies of genotypes are given by the appropriate terms in the expansion of the square of the gene array, $(\Sigma q_i A_i)^2$.

Recombination.—We must next consider the frequencies of combinations of genes at different loci. Suppose that a population produces gametes with combinations of alleles at the A and B loci in the array $[wA_1B_1 + xA_1B_2 + yA_2B_1 + zA_2B_2]$. Let c_m and c_f be the frequencies of recombination in the reduction division in eggs and sperms respectively, with c the unweighted average. The quantity $D(= wz - xy)$ is a measure of the departure from random combination. This falls off at the rate c per generation under random mating. Thus if A and B are in different chromosomes, D is halved in every generation. With 1% crossing over it is halved every 69 generations. Since randomness of combination is approached with respect to any two loci, it is approached with respect to all loci simultaneously. It is sometimes suggested that a correlation between two characters in a population may be due to genetic linkage. This may be true in a population that has arisen recently from a mixture of two strains that differ in both respects but there can be no appreciable correlation of this sort in a population that has bred within itself at random for many generations unless linkage is complete or the approach to random combination is hampered by severe selection.

The most probable genotypic frequencies under long continued random mating are given by the appropriate terms in the expansion of $\prod_{i=1}^{n}\left[\sum_{j=1}^{k} q_{ij} A_{ij}\right]^2$ where A_{ij} is used for the j'th of k alleles at the i'th of n loci and q_{ij} is used for its frequency. Actually all potential genotypes can not be realized in a finite population. With only 4 alleles (10 combinations) at each of only 100 loci, the number of potential genotypes is the inconceivably great number 10^{100}. A very limited number of mutations thus provide a virtually infinite field of potential variability. It is probable that no two individuals in a species with exclusive sexual reproduction ever have exactly the same genetic constitution.

Systematic Changes in Gene Frequency.—The immediate causes of change of gene frequency may be classified exhaustively into four categories: (1) Changes may occur in the hereditary material itself, *mutation* in the broad sense. (2) Genes may be introduced into the population in novel proportions by ***immigration***. (3) The genotypes of a given generation may contribute disproportionately to the following generation through any of the diverse forms of *selection*. (4) The offspring constitute a finite sample and may be expected to deviate slightly from the parents by mere *accidents of sampling*. Back of these factors are, of course, all of those that determine changes in their incidence.

Recurrent mutation, immigration and selection exert directed pressures on each gene frequency. The rate of change per generation due to these will be symbolized by Δq. The accidents of sampling give rise to undirected changes in gene frequencies to be symbolized by δq. There may be a question whether gene mutations are ever more than superficially recurrent. In the case of structural mutations, which usually involve two chromosome breaks, exact recurrence must certainly be a very rare event. Nonrecurrent mutation is undirected, but in a different sense from the accidents of sampling. The former extends the number of alleles at the expense of a very slight reduction in the frequency of those already present, while the latter produces random changes within the group present.

Recurrent mutation from A_1 to A_2 at the rate u per generation with reversal at the rate v per generation obviously tends to change the frequency of A_1 at the net rate $\Delta q = v(1-q)-uq$. There is equilibrium, in the absence of other factors of change, at the value $\hat{q}=v/(u+v)$. If, however, A_2 represents an array of multiple alleles of A_1, v is the weighted average of the rates of mutation from these to A_1 and is thus a function of their varying frequencies. If all mutations are unique, $v=0$.

The effect of immigration depends on the extent (m) to which the immigrants displace the native population in each generation and on the difference between the gene frequency (q_I) of the immigrants and that of the natives (q). $\Delta q = m(q_I - q)$. The theory of immigration pressure may be identified with that of mutation pressure in the case of two alleles by substituting mq_I for v and $m(1-q_I)$ for u.

The effect of selection on gene frequency depends on the frequencies (f) of the genotypes and their selective values (w), defined as the net rates of reproduction per generation on taking account of mortality, emigration, rates of attainment of maturity, duration of reproductive period, success in mating and fecundity. Letting $\bar{w}(=\Sigma wf)$ be the average selective value for the whole population and $\bar{w}_1$ that for gene A_1 (giving all combinations which involve A_1A_1 full weight and those involving heterozygosis of A_1 half weight), the frequency q_1 becomes $q_1\bar{w}_1/\bar{w}$ in the next generation and we may write $\Delta q_1 = q_1(\bar{w}_1 - \bar{w})/\bar{w}$ as the general expression for selection pressure.

Some of the special cases are instructive. We shall assume random mating and consider a pair of alleles with mean selective values w_{11} for A_1A_1, w_{12} for A_1A_2, and w_{22} for A_2A_2, and let q be the frequency of A_1. It may easily be verified that under these conditions $\Delta q = q(1-q)[w_{11}q + w_{12}(1-2q) - w_{22}(1-q)]/\bar{w}$, which may be written in the form $\Delta q = q(1-q)\Sigma w \frac{df}{dq}/2\bar{w}$. If the w's are constants this reduces to $\Delta q = q(1-q)\frac{d\bar{w}}{dq}/2\bar{w}$.

Assume that A_1A_1 has the selective disadvantage s relative to the heterozygote, and A_2A_2 the selective advantage t over the latter, $(w_{11}=1-s, w_{12}=1, w_{22}=1+t)$. Then $\Delta q = q(1-q)[q(t-s)-t]/\bar{w}$. An unfavorable gene may be maintained in the population by recurrent mutation. If A_1 is semidominant ($s=t$), the net pressure may be represented approximately by $\Delta q = v(1-q) - sq(1-q)$, with equilibrium at $\hat{q}=v/s$. If recessive ($t=0$), we have $\Delta q = v(1-q) - sq^2(1-q)$, approximately, with equilibrium at $\hat{q}=\sqrt{v/s}$, much higher than for a more or less dominant gene with the same values of v and s (*see* fig. 1).

The net effects of selection and immigration can be investigated similarly. In a population in which s (case of semidominance) is smaller than m, local differences in conditions of selection are largely swamped by crossbreeding ($\hat{q}=q_I-(s/m)q_I(1-q_I)$, approximately). Local selection dominates the situation, however, if s is larger than m, $\hat{q}=mq_I/s$, approximately, in localities in which selection is adverse.

Two alleles may both be kept at high frequencies if the heterozygote has an advantage over both homozygotes. Letting $w_{11}=1-s_1$, $w_{12}=1$, $w_{22}=1-s_2$ $\Delta q = -q(1-q)(q-\hat{q})$, approximately, where $\hat{q}=s_2/(s_1+s_2)$. If there are multiple alleles, and all heterozygotes have the same selective value, $w_{ij}=1$, while all homozygotes are selected against, $w_{ii}=1-s_i$, $\Delta q_i = q_i[1-s_iq_i-\bar{w}]/\bar{w}$. Thus $s_1\hat{q}_1 = s_2\hat{q}_2$ etc., $\hat{q}_i=(1/s_i)/\Sigma(1/s)$. Each allele is maintained at a frequency inversely proportional to the selective disadvantage of its homozygote relative to the heterozygotes. There is reason to believe that complementary effects of alleles, somewhat of this sort, are not uncommon.

This is not, however, the only mechanism by which two or more alleles may all be maintained at high frequencies within a random breeding population. The territory occupied by the population may contain diverse ecological niches for each of which one allele is superior to the others. Each allele is favoured when rare, but selected against when abundant, assuming that the individuals that carry it can exercise some choice of abode. Suppose that $w_{11}=1+s-tq$, $w_{12}=1$, $w_{22}=1-s+tq$. The heterozygote is always exactly intermediate but $\Delta q = q(1-q)(s-tq)/\bar{w}$, giving stable equilibrium at $\hat{q}=s/t$. Multiple alleles may, of course, more easily be kept at high frequencies in a heterogeneous territory in which the effects of different local conditions of selection are not swamped by too much crossbreeding.

If selective values are constant, selection tends to drive the gene frequencies to a set of values at which the reproductive rate of the population as a whole is at a peak. This is not the case where the selective values are, as in the preceding case, functions of the frequency of the gene in question. As a more striking example, consider a case in which at all times A_2A_2 has a selective advantage s over A_1A_2 and $2s$ over A_1A_1. But assume on the other hand that increase in the frequency (q) of A_1 has a favourable effect (term $a+bq$) on the rate of increase of all individuals in the population through some product or activity of those that carry it. $w_{11}=(a+bq)(1-s)$, $w_{12}=(a+bq)$, $w_{22}=(a+bq)(1+s)$. In the formula for Δq in a random breeding population any factor such as $(a+bq)$ that is common to all w's, cancels leaving the result the same as if there were no such term. Thus a gene with effects that are of advantage to the population but disadvantageous to individuals can not be fixed by selection within the group. It may be fixed, however, by intergroup selection in a population divided into small partially isolated groups.

So far we have disregarded the effects of other series of genes that may interact with the one under consideration. But selection applies to the organism as a whole (or to a group of organisms), not to separate genes. Most characters are affected by many genes and since the best-adapted grade may be expected to be near the mid-grade in a population that has lived long under about the same conditions, a gene that has a favourable effect in combinations below the mid-point should have unfavourable effects in combinations above the mid-point. Moreover if the mean selective values ($\bar{w}$) of populations are imagined to be plotted against locations in the many-dimensioned system of gene frequencies, it is to be expected that these will form a surface with a large number of distinct peaks, each centring about a different harmonious combination of genes. In the case cited above all of these give the same character, the optimum, but there may also be harmonious combinations that represent different ways of life open to the species in question and these peaks may differ greatly in height.

If random combination is assumed, the rate of change of a given gene A is given by the formula $\Delta q_A = q_A(1-q_A)\left(\Sigma w \frac{\partial f}{\partial q_A}\right)/2\bar{w}$, where the w's are the selective values of the combinations of all pertinent pairs of alleles, the f's the corresponding compound frequencies, and the summation has 3^n terms if n loci are involved. The w's may be functions of any or all of the gene frequencies. If constant, however, the rate formulae reduce to the type $\Delta q_A = q_A(1-q_A)\frac{\partial \bar{w}}{\partial q_A}/2\bar{w}$. Under this assumption, the mean selective value of the population tends to move toward its peak value in the surface $\bar{w}$ at the rate $\Delta\bar{w} = \Sigma\left(\frac{\Delta q \partial \bar{w}}{\partial q}\right)$, the summation applying to all pairs of alleles. The peak value toward which it moves is not necessarily the highest. On the other hand, as noted in the one-factor case, selection may carry the population to lower values of $\bar{w}$, if the w's are functions of the gene frequencies.

It should be pointed out that the assumption that all series of alleles are combined at random can only be an approximation, since selection itself disturbs the randomness of combination in the parental generation and the return among the offspring is not complete. It can be shown, however, that the effect is slight if the selective disadvantage of genotypes at one step from a mid-grade optimum is less than the proportion of crossing over.

While models based on pairs or finite numbers of alleles are convenient mathematically, the probability that all loci may, in the long run, evolve through an indefinitely extended branching series of alleles must not be forgotten. In this situation, all selection coefficients are subject to undefinable changes.

Random Changes in Gene Frequency.—The fourth and last mode of change of gene frequency is by accidents of sampling. The composition of a population of N diploid individuals depends on $2N$ gametes produced by the preceding generation. If these are a random sample from the array $[qA_1+(1-q)A_2]$, the array of probabilities for the next generation is $[qA_1+(1-q)A_2]^{2N}$. The variance of q is $\sigma^2_{\delta q}=q(1-q)/2N$. In a k-ploid population $\sigma^2_{\delta q}=q(1-q)/kN$, approximately. For sex-linked genes, it is approximately $2q(1-q)/3N$ if there are equal numbers of males and females.

It might seem that these random deviations would be negligible in any reasonably large population; but the variance of the array of probabilities for later generations increases approximately linearly with the number of generations, until damped by approach to the limiting value, $q(1-q)$, of complete fixation one way or the other. Moreover, the effective value of N should often be much smaller than its apparent value. It obviously applies only to individuals that reach maturity. If there is cyclic variation in population size, effective N is the harmonic mean of the numbers in the generations of the cycle, which is largely dominated by the population minimum. The sampling effect is, however, much more important in connection with a population structure in which there are numerous small partially isolated local strains than in a homogeneous random breeding population.

The tendency toward a stable equilibrium in gene frequency due to opposing systematic pressures and the tendency to drift away from the point of equilibrium due to accidents of sampling result in a distribution of probabilities which one might expect to find realized by the values taken by the frequency of the gene in question over a long period of generations in the ideal case in which all conditions remain constant. Even if never actually realized, the distribution formula is important in reaching an appreciation of the degree of control by the systematic pressures.

The formula for this distribution must satisfy the conditions of persistence of the mean, $\Sigma(q+\Delta q+\delta q)f=\Sigma qf$, and of all higher moments $\Sigma(q+\Delta q+\delta q-\bar{q})^n f=\Sigma\ (q-\bar{q})^n f$. It can be shown that these conditions are satisfied to a first approximation by the differential equation $\frac{d}{dq}\ \sigma^2_{\delta q}\varphi(q)-2\,\Delta q\varphi(q)=0$, the solution of which is the probability curve $\varphi(q)=(C/\sigma^2_{\delta q})\ \exp\ [2\int(\Delta q/\sigma^2_{\delta q})\ dq]$, $\int_0^1\varphi(q)dq=1$. Since there are $(2N+1)$ possible values of q among $2N$ gametes, the frequency of a given class is approximately $f(q)=\varphi(q)/2N$. The formula gives good approximations even for the subterminal classes $f(1/2N)$ and $f[(2N-1)/2N]$ but the frequencies of fixation must in general be determined from the subterminal classes by consideration of the balance between loss of the gene in question or its allele by accidents of sampling (approximately half the subterminal class per generation) and reintroduction by mutation or immigration. Thus, $f(0)=f(1/2N)/(4N[mq_1+v])$.

As an illustration, consider a case in which there is systematic pressure from reversible mutation, immigration and selection measured by $\Delta q=v(1-q)-uq-m(q-q_1)+q(1-q)\frac{d\bar{w}}{dq}/2\bar{w}$ and accidents of sampling measured by $\sigma^2_{\delta q}=q(1-q)/2N$. Substitution yields $\varphi(q)=C\bar{w}^{2N}q^{4N(mq_1+v)-1}(1-q)^{4N[m(1-q_1)+u]-1}$. Examples are given in figs. 1 and 2.

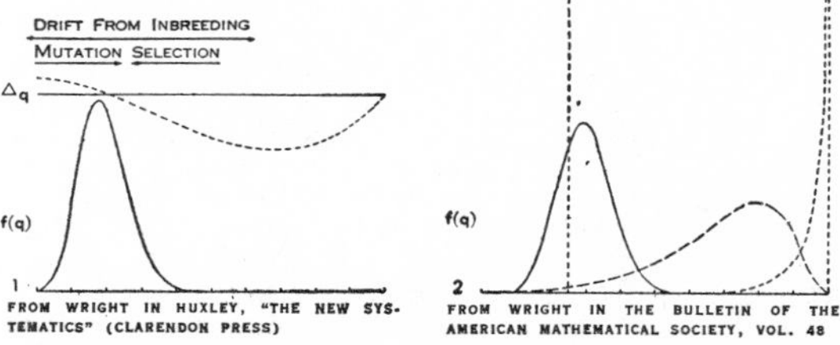

FIG. 1.—ABOVE: SYSTEMATIC RATE OF CHANGE (Δ q) OF THE FREQUENCY (q) OF A DELETERIOUS RECESSIVE GENE FOR WHICH S=25V, AND REVERSE MUTATION (u) IS NEGLIGIBLY LOW. THE SCALE OF Δq IS GREATLY EXAGGERATED. BELOW: DISTRIBUTION OF RELATIVE FREQUENCIES, F (q), OF THE VALUES OF q OVER A LONG PERIOD OF TIME IF V=1/N. FIG. 2.—DISTRIBUTION OF GENE FREQUENCIES q FOR GENE A IN THREE SUBGROUPS OF A POPULATION OF MEAN GENE FREQUENCY .25. THESE SUBGROUPS ARE ASSUMED TO BE OF THE SAME EFFECTIVE SIZE (N=1,000) AND SUBJECT TO THE SAME SELECTION PRESSURE (W_{AA}=1, W_{Aa}=.9975, W_{aa}=.995) BUT TO DIFFERENT DEGREES OF ISOLATION (SHORT DASHES AT .25: NO ISOLATION; SOLID LINE: m=.01; LONG DASHES: m=.001; SHORT DASHES TO RIGHT: m=.001)

This formula can be extended to the case of a finite number of multiple alleles, assuming that mating is at random and that the rate of mutation to each allele from each of the others, is the same (a severe restriction), $\varphi(q_1, q_2 \ldots q_n)=C\bar{w}^{2N}\Pi q^{4N(mq_1+v)-1}$, where the product includes terms for all alleles. This is also the frequency distribution for all series of alleles considered simultaneously, if the product term includes all loci. This formula describes a frequency surface in a space of $\sum^n (k_i-1)$ dimensions where k_i is the number of alleles per locus and n is the number of loci. There are in general a large number of peak frequencies, corresponding roughly to the peaks in the surface ($\bar{w}$) of selective values. Under certain conditions there is an appreciable chance of drifting from control by one peak value to control by a higher one.

The number of alleles that may be carried by a population under various conditions is important. Consider first a series in which each mutation yields a new allele and assume that alleles that are completely neutral, as regards selection, are arising at a constant rate u. The distribution for such alleles is $f(q)=Cq^{-1}(1-q)^{4Nu-1}$ with $f(0)$ infinite. The number of new alleles of the above sort per generation is $2Nu$. The number lost by accidents of sampling is approximately $(n/2)f(1/2N)$ where n is the number present. Putting $\Sigma f(q)=1$ in this case, there is equilibrium between mutations and loss if $n=2u\Sigma q^{-1}\ (1-q)^{4Nu-1}$ with summation including values of q from $1/2N$ to 1. With $u=10^{-6}$, a population of 250,000 may be expected to carry an average of 13.7 alleles. In larger populations, there should be a somewhat less than proportional increase, *e.g.*, 132 alleles if N is increased tenfold. These results are not changed appreciably for mutations that are slightly selected against.

Most of the alleles in these cases are present at very low frequencies. The chance that a single completely neutral mutation will reach approximate fixation is $1/2N$. Even if selected against, there is specifiable chance of displacing the type allele, $2s/(e^{4Ns}-1)$, in case the selective disadvantage of heterozygotes is s, of homozygotes $2s$. With a corresponding selective advantage, the chance is approximately $2s$. It is approximately $\sqrt{s/2N}$ for a completely recessive mutation with an advantage s when homozygous.

Population Structure.—Departures from random mating have important consequences which it is now desirable to consider. A population has a structure in time that consists of a complicated network of paths of descent. An adaptation of the theory of multiple correlation, the method of path coefficients, is convenient in dealing with such networks.

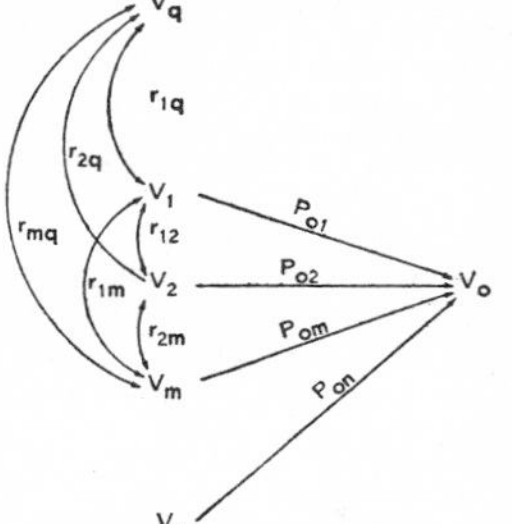

FIG. 3.—A SYSTEM OF INTERRELATED VARIABLES

The basis for application of this method is a diagram which represents a point of view as to which variables in a system of linear relationships are to be treated as immediate functions of which others. Fig. 3 is a simple diagram of this sort. A single-headed arrow indicates the direct effect of one variable upon another while double-headed arrows are used for residual relations tracing to unrepresented factors which may be considered as located at the mid-points of such "arrows." Assume that x_o, x_1, etc., are variables measured in standard form, $x_o=(V_o-\bar{V}_o)/\sigma_o$, where V_o is the original variable, $\bar{V}_o$ its mean and σ_o its standard deviation. Assume that x_o is a linear function of known variables x_1 *to* x_m and that the array of factors is made formally complete by addition of a hypothetical variable x_n, independent of the others. In the expression $x_0=p_{01}\ x_1+p_{02}\ x_2+\ldots p_{om}\ x_m+p_{on}\ x_n$, the coefficients are path coefficients. The correlation (r) between two variables is defined as their average product when expressed in standard form. Thus $r_{oq}=\sum_{i=1}^{n} p_{oi}\ r_{iq}$. The special case of self-correlation $r_{oo}=1$ yields the equation $\sum_{i=1}^{n} p_{oi}\ r_{oi}=1$. This may be analyzed into two components, $p_{on}\ r_{on}(=r^2_{on})$ which measures the portion of the variance (σ^2_o) determined by unknown factors and $\sum_{i=1}^{m} p_{oi}\ r_{oi}$ expressing that due to known factors. The latter is the squared coefficient of multiple correlation.

Factors x_1, x_2, x_q, etc., may in turn be represented as linear functions of variables a step farther back in this network, making possible analysis of the correlation terms, in the above equations. However extensive the network, the following principle holds. Any correlation between variables in a network of directed linear relations can be analyzed into contributions from each of the paths by which the two variables are connected through a common factor. The value of each contribution is the product of the coefficients (of which only one may be a correlation coefficient) pertaining to the elementary paths.

Fig. 4 represents the relations between parents and offspring in the network of descent in the case of an autosomal factor. Variations in the characters of individuals are represented as determined by additive effects of environmental and genotypic variations (path coefficients e and h respectively, $e^2 + h^2 = 1$. The path coefficient a relates genotype to one of the gametes that produced it. Assume that each allele at the locus under consideration is assigned a value and that the value of a genotype is the sum of the values of the two gametes that produced it. As there is complete determination $2a(a + aF) = 1$, where F represents the correlation between uniting gametes. Thus, $a = \sqrt{(1/2)/(1+F)}$. The path coefficient b relates gamete to the genotype that produced it. This is also the correlation coefficient since there is only one connecting path. This correlation must be the same as

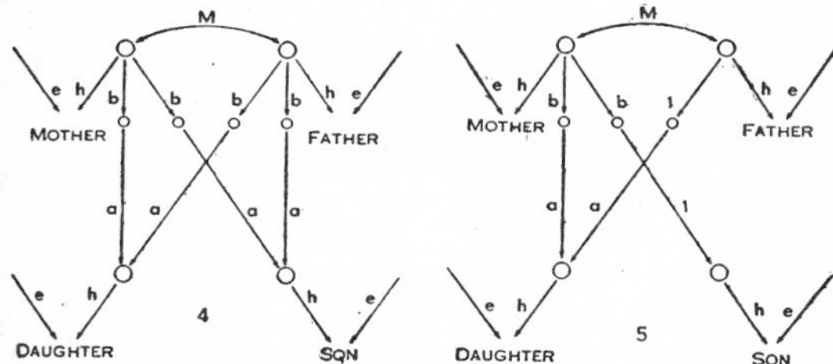

FIG. 4.—RELATIONS AMONG GENOTYPES AND CHARACTERS OF PARENTS AND OFFSPRING, AS DETERMINED BY AUTOSOMAL GENES. FIG. 5.—RELATIONS AMONG GENOTYPES AND CHARACTERS OF PARENTS AND OFFSPRING AS DETERMINED BY SEX-LINKED GENES

that between parental genotype and one of the gametes that produced it, with respect to the series of alleles in question. Thus, $b = a' + a'F' = \sqrt{(1+F')/2}$ where primes are used to indicate the preceding generation. The compound path coefficient relating a gamete to one of those back of it a generation earlier is $ba' = 1/2$.

It may readily be seen by tracing the connecting paths that the correlation between parent and offspring is $h^2 ab\,(1+M)$ relative to a locus and that between brothers is $2h^2a^2b^2(1 + M)$ where $M (= F/b^2)$ is a correlation coefficient that sums up all possible connections between the parental genotypes. These results apply to the combined effects of multiple loci, provided that these are associated at random and that effects are throughout additive. In a random breeding population ($M = F = O$) both of the above correlations reduce to $.50h^2$.

Sex-linked genes can be treated similarly, noting that the coefficient relating male genotype to egg must equal 1 and that relating X-bearing sperm to genotype producing it must also be 1 since there is complete determination in both cases (fig. 5). Coefficients a and b in other cases have the same values as with autosomes. The correlation between any relatives can be found from a diagram that shows all of the ancestral connections.

In actual cases, nonadditive relations are likely to be involved in the relations between genotype and character (as a consequence of dominance or factor interactions) and there may be nonadditive relations between genotype and environment. These introduce complications which can not be discussed in the scope of this article.

The structure of a population is perhaps best expressed in terms of F, which, as the correlation between uniting gametes, is the most appropriate coefficient of inbreeding. The array of genotypes relative to a pair of alleles may be expressed as follows in terms of the frequency, q, of one of the alleles (A_1), and the proportion of heterozygotes p:

$$[(q - \tfrac{1}{2}p)A_1A_1 + pA_1A_2 + (1 - q - \tfrac{1}{2}p)A_2A_2].$$

It may easily be verified that the correlation between the gametes whose union is here indicated is $F = (p_o - p)/p_o$ where p_o $(= 2q\,(1-q))$ is the proportion of heterozygosis under random mating. It may be shown that with multiple alleles as well as with mere pairs the frequencies of homozygotes are of the type $[(1-F)q^2_1 + Fq_1]$ for A_1A_1 and of heterozygotes of the type $[2(1-F)q_1q_2]$ for A_1A_2. The correlation between uniting gametes comes out $F = (p_o - p)/p_o$ where p is the total frequency of all heterozygous classes, irrespective of the values assigned the various alleles.

The value of F can be obtained from inspection of the appropriate diagram in simple cases. Thus for the progeny of an isolated brother-sister mating it is one-fourth and for progeny of a first-cousin mating one-sixteenth. Recurrence formulæ can be obtained in cases of regular systems of mating. Thus if a population is broken up into lines of exclusive brother-sister mating, $F = b^2a'^2(2F' + 2b'^2) = \tfrac{1}{4}(1 + 2F' + F'')$; $p = (\tfrac{1}{2})p' + (\tfrac{1}{4})p''$. Putting $p/p' = p'/p''$, $p/p' = (\tfrac{1}{4})(1 + \sqrt{5}) = .809$, indicating that heterozygosis decreases about 19.1% per generation under continued brother-sister mating. In populations of size N with completely random union of gametes $F = \frac{1}{N}b^2 + \frac{N-1}{N}F' = \frac{1}{2N} + \left(\frac{2N-1}{2N}\right)F'$, $p = \frac{(2N-1)}{2N}p'$, indicating fixation at the rate of $1/2N$ per generation in this case. With N_m males and N_f females similar analysis gives $p = p' - \left(\frac{N_m + N_f}{8N_mN_f}\right)(2p' - p'')$ with fixation at rate $[(1/8N_m) + (1/8N_f)]$ if the N's are large.

A case of great importance is that in which a population has a uniform distribution over a large area but dispersion is limited to small neighbourhoods. The ancestors of any given individual are drawn from ever widening circles as one goes back in time. With random union of gametes within neighbourhoods of size N, the value of F relative to a population of size KN (that from which the Kth ancestral generation was drawn) is $[\Sigma t/(2N - \Sigma t)]$ where Σt is the sum of K-1 terms of the series $\left[1 + \tfrac{1}{2}\left(\frac{N-1}{N}\right) + \tfrac{1}{3}\left(\frac{N-1}{N}\right)\left(\frac{2N-1}{2N}\right)\ldots\right]$ which can be evaluated. Its limiting value is N and that of F is thus 1. If the effective value of N is less than 100 there is considerable differentiation not only of neighbourhoods but also of large territories. If on the other hand, N is more than 1,000, the situation differs little from universal random mating. If there is continuity along only a narrow strip such as a shore line or a river instead of over an extensive area, there is, however, enormously more differentiation with a given size of neighbourhood.

It can readily be seen that the general formula for the inbreeding coefficient, F, applicable to irregular pedigrees is $\Sigma[(\tfrac{1}{2})^{n+n'+1}(1 + F_A)$ where n and n' are the numbers of generations from sire to dam respectively to a common ancestor with inbreeding coefficient F_A, and the summation applies to all connecting paths which include only one common ancestor. By taking random lines of ancestry back of the sires and dams of animals chosen at random and noting the proportion (T) of cases in which those include the same animal, it is possible to estimate $F (= (\tfrac{1}{2})T(1 + F_A))$ for a whole breed. This has been done for several breeds of horses, cattle, sheep and hogs with rather similar results. The inbreeding coefficient rises at rates between 0.2% and 1.2% per generation, indicating that concentration on the sons and grandsons of noted animals has given a system equivalent to the use of only 10 to 60 sires in the whole breed. (Rate $= 1/8N_m$, assuming N_m males and indefinitely many females.)

The inbreeding coefficient measures inbreeding relative to a certain foundation stock. Thus the coefficient 26.0% in Shorthorn cattle of 1920 implies that heterozygosis was 26.0% less than expected from random mating within the cattle population of northern England of about 1780, the period to which the pedigrees trace. The correlation between random pedigree lines of 1920 was 24.6%, indicating that the departure from random mating within the breed in 1920 was not great. The extensive loss of heterozygosis indicated by the value 24.6% applies to the breed as a whole. If we let $F_{I \cdot T}$ be the inbreeding coefficient of individuals relative to some comprehensive population, $F_{I \cdot G}$ that of individuals relative to groups within the latter and $F_{G \cdot T}$ the correlation between random gametes within the groups, $(1 - F_{I \cdot T}) = (1 - F_{I \cdot G})(1 - F_{G \cdot T})$. The structure of a population may be analyzed by application of this formula.

The subdivision of a population into numerous partially isolated groups leads to differentiation of these in gene frequencies, measured by $\sigma^2_q = q_T(1 - q_T)F_{G \cdot T}$ where q_T is the gene frequency of the comprehensive population. The differentiation with respect to characters is, of course, closely related. The variance of a character due to semidominant genes in a population breeding at random is $\sigma^2_R = 2\Sigma[q.(1-q)(\alpha^2)]$ where α is the effect of replacement of a gene by its allele and the summation relates to all pertinent series of alleles. If the population is divided into partially isolated groups, breeding at random within themselves ($F_{I \cdot G} = 0$, $F_{G \cdot T} = F_{I \cdot T} = F$), the variance of group means is $2F\sigma^2_R$, the variance within groups is $(1-F)\sigma^2_R$ and the variance of individuals in the total is $(1 + F)\sigma^2_R$.

The primary consequence of subdivision of a population into more or less isolated groups is, of course, that differences in the conditions of selection may bring about direct differentiation. But even if the conditions of selection are everywhere the same, such groups may be expected to drift apart as a result of the accidents of sampling. This would not be likely to go far by itself but may initiate trends toward diverse ways of meeting the conditions and thus create differences in direction of selection. In terms of the mathematical model, the systems of gene frequencies of different groups may come under the control of different peak adaptive values. If isolation is not complete, the stage is set for selection at the group level. The localities, in which the most successful genetic systems have been arrived at, become the principal sources of migrants to other regions. In a random breeding population selection operates merely according to the net effects of the genes in all combinations. In a population of vegetatively reproducing clones, the genotype as a whole is the object of selection. This process may be very effective, especially in a form capable of rapid multiplication under favourable conditions if associated with occasional crossbreeding to provide new clones for trial more rapidly than by the slow process of mutation. In a large population divided into numerous partially isolated groups the object of selection is the entire genetic system and the situation is that most favourable for an indefinitely continuing evolutionary process. (*See* EVOLUTION, ORGANIC.)

BIBLIOGRAPHY.—R. A. Fisher, *The Genetical Theory of Natural Selection* (1930); J. B. S. Haldane, *The Causes of Evolution* (1932); S. Wright, "Evolution in Mendelian Populations," *Genetics*, vol. 16, pp. 97–159 (1931), "Isolation by Distance," *Genetics*, vol. 28, pp. 114–138 (1943) and vol. 31, pp. 39–55 (1946) and "Statistical Genetics and Evolution," *Bull. American Mathematical Society*, vol. 48, pp. 223–246 (1942). (S. WT.)

GENEVA, a city and canton of Switzerland, situated at the extreme southwest corner both of the country and of the Lake of Geneva or Lac Léman. The canton is, save Zug, the smallest in the Swiss confederation, while the city, long the most populous in the land, is now surpassed by Zürich, Basle and Berne.

The Canton.—The canton has an area of 108.9 sq.mi., of which 11½ are lake. It is entirely surrounded by French territory (the department of Haute Savoie to the south, and that of the Ain west and north), save for about 3½ mi. on the extreme north, where it borders on the Swiss canton of Vaud. The Rhône flows through it from east to west, and then along its southwest edge. The turbid Arve is its largest tributary and flows from the snows of the chain of Mont Blanc, the only other affluent of any size being the Allondon. Market gardens, orchards and vineyards occupy a large proportion of the soil, the apparent fertility of which is largely due to the unremitting industry of the inhabitants. In 1941 there were 11,094 cattle, 1,979 horses, 3,802 swine, 1,157 goats and 5,642 sheep. Besides building materials, such as sandstone, slate, etc., the only mineral to be found within the canton is bituminous shale, the products of which can be used for petroleum and asphalt. The canton is served by broad gauge railways and electric tramways. It was admitted into the Swiss confederation in 1815 and ranks as the junior of the 22 cantons. In 1815–16 it was increased by adding to the old territory belonging to the city (just around it, with the outlying districts of Jussy, Genthod, Satigny and Cartigny) 16 communes (to the south and east, including Carouge and Chêne) ceded by Savoy, and 6 communes (to the north, including Versoix), cut off from the French district of Gex.

In 1941 the canton had 174,619, the city 124,442, inhabitants. This population was divided as follows in point of religion (1930 figures for the city are within brackets): Roman Catholic 72,073 (49,631), Protestants 88,979 (66,016) and Jews 2,345 (2,224).

In point of language 131,753 (93,058) were French-speaking, 24,213 (18,717) German-speaking, 10,099 (7,762) Italian-speaking, while there were also 215 (186) Romansch-speaking. Nationality was as follows: 57,604 (38,546) were Genevese citizens, and 72,874 (92,693) Swiss citizens of other cantons. In 1919 the canton contained 62,611 (51,740) foreigners, but by 1930 the number had fallen to 51,721 (42,599) in consequence of the emigration during and after World War I.

As a result of World War II the number of foreigners in Geneva fell to 40,888 (31,428), of whom 17,577 were French, 12,704 Italians, 3,938 subjects of Germany and 6,669 citizens of various other countries. The League of Nations had ceased to be a centre of attraction for visitors and delegates.

HISTORY

In prehistoric times a great lake city, built upon piles which may still be seen, existed where the waters from the Alpine lakes spread out over the plain before narrowing into the channel of the Rhône. This city was the prehistoric Geneva. After the end of the period of lake dwellings the inhabitants established themselves on the hill on the left bank of the lake and the river.

Caesar states that Geneva was a town (*oppidum*) situated in the extreme north of the country of the Allobroges; the Rhône separated it from the territory of the Helvetii, whose invasion Caesar repelled. The community (*vicus*) of Geneva was one of those dependent on the city of Vienne. It was of some size, and had temples, aqueducts, ports and ships. It was built on the usual plan of intersecting roads meeting in a central forum. One road ran from the south to the lake ports, and the other from the east to the bridge over the Rhône. When the district of Vienne was made into a province, Geneva became a Roman city (*civitas*) with part of what is now Savoy dependent on it. When the empire became Christian, a bishop was appointed at Geneva. After the Barbarian invasions the city shrank to half its former size. It was now concentrated on the high ground; at the foot of the hill the forum constituted a separate township, the *Bourg de Four*. The pagan temples were converted into Christian churches. At the top of the hill rose St. Peter's, while St. Victor's was built in the detached part of the town.

Order had been restored by the Burgundian kings in the 5th century, but Gundibald was defeated by Clovis and his sons were dethroned by the Franks (534). Geneva owed its importance to its bridge over the Rhône. In 563 the bridge was carried away by a flood caused by a landslide at the other end of the lake; it was, however, immediately rebuilt. Geneva lay on the path of the armies marching to the conquest of Italy. Charlemagne held an assembly there in 773. After the break-up of his empire, a new kingdom was set up in Burgundy, that of the Rudolphians. During the feudal period the Burgundian kings had more to fear from the hereditary counts of Geneva than from the elected bishops. Rudolph III conferred estates on the bishops and favoured them at the expense of the counts. On his death in 1032 the emperors of the Holy Roman empire inherited his lands. Frederick Barbarossa confirmed the temporal powers of the bishop of Geneva, who became a prince of the empire, and made the church independent of the nobles of the district. The count of Geneva had a residence in the town, the old royal château, but had to do homage to the bishop for the château and for other fiefs.

The sole direct ruler of Geneva was the prince bishop. But the Genevese were always characterized by their passion for independence, and imitating the example of the Italian towns, with which they traded, they attempted towards the end of the 13th century to create a municipal organization for themselves. They were able to play off against one another the rival rulers of the district.

Savoy.—In Maurienne, a remote district of the country, there presently arose a count, who came to be known as the count of Savoy and was on bad terms with both the count of Geneva and the bishop. Peter of Savoy, who was well received in England by the queen, his niece, acquired the rights of the elder branch of the counts of Geneva, succeeded in depriving the younger branch of the county of Vaud, and entered into relations with the city of Geneva. His nephew Amadeus the Great declared himself the protector of the citizens, who had formed themselves into a municipality with syndics and other officers. The count of Geneva was reduced to a mere vassal of his cousin of Savoy, while the bishop was compelled to yield to the latter his palace, together with the *vidomnat*, the office empowering him to administer summary justice in the city. Finally the bishop recognized the municipality, after the citizens, posted on the towers of St. Peter's, had withstood bombardment by the count of Geneva from his castle. This castle was dismantled in 1320. In the meantime the citizens had defeated the count's army near the lake (June 6, 1307), a victory comparable with that of the Swiss over the duke of Austria at the other end of Switzerland (Morgarten, 1315). But by calling in the count of Savoy the Genevese had fallen out of the frying pan into the fire. They had been able to free themselves from the count of Geneva and to defy the bishop, but they discovered that their protector, not content with the office of *vidomne*, intended to make himself "prince" of the city. He still retained some partisans, however, although some of the bishops did more to

38
Adaptation and Selection

In *Genetics, Paleontology, and Evolution*, ed. by G. L. Jepson, G. G. Simpson, and E. Mayr, 365–89.
Princeton: Princeton University Press, 1949

Introduction

This paper was Wright's contribution to the famous Princeton Conference held under the auspices of the National Research Council at Princeton, New Jersey, on January 2–4, 1947. Mayr has recently argued that this conference "constitutes the most convincing documentation that a synthesis [in evolutionary biology] had occurred during the preceding decade." He further argued that the participants had generally agreed that the evolutionary process was gradual (at the level of microevolution) and dominated by natural selection (Mayr and Provine 1980, 42–43).

Wright's contribution was basically a summary of his shifting balance theory with a discussion of its consequences and significance for the evolutionary process as a whole. This paper is therefore comparable to Wright's 1932 paper delivered at the Sixth International Congress of Genetics (Wright 70), which was designed for a similar role. In the 1932 paper, Wright had emphasized the role of isolation and random genetic drift in producing nonadaptive differences between subspecies, species, and even genera. In the paper for the Princeton Conference, however, Wright did not even mention such a role for isolation and random genetic drift. The difference between these two papers on this issue represents a change of emphasis in Wright's shifting balance theory of evolution. By the late 1940s, he retained the firm conclusion that random genetic drift was very important, but only as a mechanism for generating new interaction systems upon which natural selection acted. The resulting evolutionary process was already adaptive at the lowest taxonomic levels.

This change of emphasis within the shifting balance theory exemplifies what Stephen Jay Gould has termed the "hardening of the synthesis" toward a more thoroughly selectionist view (Gould 1980, 1982, 1983). Wright denies that he became significantly more selectionist during the 1940s (see particularly his statement in Wright 207, 11–13). For my views on this issue, see *SW&EB*, chapter 12.

· 20 ·

ADAPTATION AND SELECTION

BY SEWALL WRIGHT[1]

THE process of adaptive evolution is one of central interest from both the philosophical and scientific points of view. As with other natural phenomena there is probably both an internal and an external aspect. To avoid possible duplication of the same factor under different aspects, an attempt at scientific analysis must be restricted to the latter. In the present discussion, no consideration will be given to hypotheses which attribute evolution to inscrutable creative forces whether these be supposed to bring about gradual orthogenetic advance (Nägeli, Osborn), abrupt emergence (Lloyd Morgan) or are assigned to individual organisms (Lamarck). The discussion will also be restricted to factors for which there is a substantial basis in genetics. This consideration excludes the hypothesis that evolution is an extension of individual physiology (Lamarck, Eimer, Cope, etc.). Mutations of chromosomes and genes, and recombinations of these, are with minor qualifications the only known sources of hereditary change. The available evidence indicates that these occur independently of physiological adaptations of the individual. We are left with the hypothesis that phylogenetic adaptation is ultimately preadaptive. The variations come first, the organisms do the best they can with them, and natural selection is the arbiter (Darwin). Under any hypothesis, the results of natural selection may be considered as the measure of adaptation, and some degree of guidance of the course of evolution is virtually inevitable. It appears now that natural selection is the only verifiable factor making for cumulative adaptive change. This statement, however, by no means exhausts the possibilities of analysis.

EVOLUTION FROM SINGLE MUTATIONS

The earliest hypothesis along Mendelian lines was that significant evolutionary change is due to the occasional occurrence of mutations of such a nature as to give rise to new species (or higher categories) at once (deVries, Goldschmidt).

The abrupt origin of species by mutation requires that the most essential feature of the mutation be an isolating effect. This hypothesis has in its support the evidence that polyploidy has been important in the multi-

[1] Ernest D. Burton Distinguished Service Professor of Zoology, The University of Chicago.

plication of species, especially in cases in which it has occurred in a sterile hybrid, converting the latter into a fertile amphidiploid.

It is, however, impossible to base all evolution on the addition of chromosome sets. There is no other class of mutation which combines the properties of full fertility by itself, the production of sterile hybrids with the parent form or forms, a balanced organization favorable to adaptation, and (in amphidiploids) striking novelty. The other balanced types of chromosome mutation (translocation, inversion) result, when viable, in minor variants within the species of origin, often recognizable only by examination of the chromosomes themselves. Simple aneuploidy may bring complex character changes, suggestive of those distinguishing species, but not isolation. Most such changes are indeed incapable of existence except as heterozygous segregating varieties. Compound aneuploidy may, however, be a rare species-forming process (Darlington). Small duplications have no doubt been very important in evolution in increasing the number of genes, but hardly in the abrupt origin of new species. Such minor chromosomal changes can be treated statistically for the most part as gene mutations.

Evolution by Gradual Transformation

The most obvious alternative to abrupt origin of species by mutation is the conception of gradual transformation through the occurrence, orderly increase in frequency, and ultimate fixation of mutations that give a selective advantage to the individuals that carry them. Haldane has developed mathematically the nature of this process in a great variety of cases (1924 and later). Fisher (1937) has investigated the properties of the wave of diffusion from a portion of a population in which a mutation has become fixed.

Acceptance of any hypothesis of evolution by statistical transformation of populations permits differences of opinion as to whether it is the fixation of rare major adaptive mutations or the accumulation of a large number of minor ones that is usually responsible for a new species, whether by transformation of the entire parent species or of an isolated portion of the latter. This issue is somewhat confused by the ambiguity of the term bigness as applied to a mutation. The most literal sort of bigness is in the degree of physical change in the chromosomes. There is, however, no consistent correlation between this sort of bigness and conspicuousness or complexity of the effects on characters. There may be extensive rearrangements of the material of the chromosomes without conspicuous effect, and conversely single gene mutations may produce abnormalities of the most extreme sort. Again mere conspicuousness (as of many color mutations) may have little relation to morphological or physiological complexity. Finally bigness in any of these ways is far from being positively correlated with ecological potentiality.

Analyses of the actual genetic differences between related species have demonstrated all of the viable types that have arisen in the laboratory (cf. Dobzhansky, 1941). Changes in chromosome number and rearrangements are common but appear to have only minor importance, where any, with respect to distinguishing characters other than of fertility of hybrids. Conspicuous unitary differences are sometimes found especially with respect to color. The mode of segregation for most characters, however, indicates a multiplicity of genetic changes with effects that are either individually slight, or (if the character itself is of an all-or-none sort) of low penetrance.

The view that these species differences are largely due to mutations that were not only adaptive from the first but were carried directly to fixation by favorable selection is undoubtedly too simple, as indeed has probably been recognized in some degree by all who accept the general hypothesis of transformation. It makes the occurrence of adaptive mutations the limiting factor in evolution to an extent that severely restricts the possible rate of evolutionary change. There is evidence that evolution can proceed with great rapidity, given an adequate ecological opportunity.

It is necessary to examine the genetic properties of populations to find whether these may not provide a basis for rapid exploitation of opportunities without waiting for the accidental occurrence of the right mutation.

Gene Frequency

The basic concept of statistical genetics is gene frequency. Because of the symmetry of the Mendelian mechanism, gene frequencies tend to persist unchanged from generation to generation. The state of the species with respect to a particular locus can be described in such a form as

$$\Sigma_{i=1}^{k}(q_i A_i)$$

where the A's designate k alleles and the q's are their frequencies.

The random unions of eggs and sperms, each with the array of alleles characteristic of the species, give a stable genotypic array at each locus of the type

$$[\Sigma_{i=1}^{k}(q_i A_i)]^2$$

(Hardy, 1908; Weinberg, 1908). The frequencies of combinations among loci are in the long run those of random combination, in a random breeding population, irrespective of linkage, unless the latter is complete (Weinberg, 1909, 1910; Robbins, 1918; Geiringer, 1944, 1948). The array may be represented by

$$\Pi_{L=1}^{n}[\Sigma_{i=1}^{k}(q_{Li} A_{Li})]^2$$

where L is any one of n different loci. Thus 4 alleles (10 combinations) at each of 100 loci provide the potentiality for 10^{100} different genotypes.

A genetic description of a population can obviously be given much more economically in terms of gene frequencies than of genotypic frequencies. Moreover the tendency toward persistence of gene frequencies insures relative persistence of the statistical properties of the population, if the system of mating remains the same, even though no genotype is ever duplicated.

A change in the system of mating changes the array of genotypic frequencies and consequently the statistical properties of the population, but does not in itself tend to change the arrays of gene frequencies. Thus a tendency toward mating of relatives, whether sporadic or from a breaking up of the population into more or less isolated groups, gives a genotypic array at each locus of the type

$$(1-F)[\Sigma(q_iA_i)]^2 + F\Sigma(q_iA_iA_i)$$

where F is the inbreeding coefficient, defined as the correlation between uniting gametes with respect to additive gene effects (Wright, 1921, 1922). The increased homozygosis, measured by F, gives rise to the well-known effects of inbreeding, but the q's are not changed, and a return to the original system of mating is followed by return to the original genetic situation. Similarly, pure assortative mating with respect to any character causes departures from randomness of combination among the loci and consequent changes in the statistical properties of the population, but again the process does not in itself affect the gene frequencies and random combination is gradually restored on resumption of random mating (Wright, 1921).

The Elementary Factors of Evolution

The elementary evolutionary process in a reasonably large homogeneous population may be considered to be change of gene frequency. It is convenient for the present purpose to distinguish three primary modes of change according to the degree of determinacy in the changes which they bring about. First are modes of systematic change, termed the evolutionary pressures (Wright, 1929, 1931). These are capable at least in principle of precise mathematical formulation which would make possible prediction of evolutionary trend if there were no processes of an indeterminate nature. Second are the random fluctuations in gene frequency, of which only the variance is determinate. Finally, it is convenient, although somewhat arbitrary, to distinguish from both of these, events that are unique or nearly so in the life of the species.

Modes of Immediate Change of Gene Frequency

1. *Systematic change* ($\triangle q$ determinate in principle).
   A. Pressure of recurrent mutation.
   B. Pressure of immigration and crossbreeding.
   C. Pressure of intragroup selection.
2. *Random fluctuations* (δq indeterminate in direction but determinate in variance).
   A. From accidents of sampling.
   B. From fluctuations in the systematic pressures.
3. *Nonrecurrent change* (indeterminate for each locus).
   A. Nonrecurrent mutation.
   B. Nonrecurrent hybridization.
   C. Nonrecurrent selective incidents.
   D. From nonrecurrent extreme reduction in numbers.

Back of these processes are all factors that cause secular changes in mutation rates, conditions of selection, size and structure of the population and the possibilities of ingression from other populations.

Mutation Pressure

Mutation pressure refers to the tendency of known genes to mutate to known alleles at definite rates. In the two-allele case, the net rate per generation obviously may be written

$$\triangle q = \frac{dq}{dt} = v(1-q)-uq$$

where u is the rate of mutation from the gene in question (frequency q) to its allele, v is the rate of reverse mutation, and time is measured in average generations. Haldane (1926) has shown that the effects of overlapping of generations are of only secondary importance in the transformation of populations. $\triangle q$ may be treated as dq/dt in order to obtain the approximate amount of change in a given number of generations (t) by integration.

The extents of systems of multiple alleles, known at many loci in many organisms, make it probable that each locus is capable of evolutionary change through a branching system of indefinitely great extent, and that many alleles normally coexist at many loci in any large population. Taking cognizance of this, the description of mutation pressure requires a system of simultaneous equations of the type

$$\triangle q_c = \frac{\partial q_c}{\partial t} = \Sigma(u_{ci} q_i) - (\Sigma u_{ic}) q_c$$

where u_{ci} is the rate of mutation to allele A_c from one of its alleles A_i, u_{ic} is the rate of reverse mutation and summation applies to all alleles of A_c.

It is obvious that the distinction between recurrent mutation and non-recurrent mutation must be arbitrary.

Immigration Pressure

Immigration pressure refers to the effects of recurring invasion of the territory under consideration by individuals, capable of producing cross-breds that enter into the population. Here

$$\Delta q_c = -m(q_c - q_{c(I)})$$

where $q_{c(I)}$ is the frequency of the gene among the immigrants and m is a coefficient measuring the amount of replacement of population by the immigrants per generation. The theory is closely similar to that of mutation pressure because of the linear nature of its effect. As with mutation pressure there is no sharp line between recurrent immigration and a type of hybridization so infrequent in the history of the species that each occurrence is best treated as a unique event.

Selection Pressure

Immigration pressure may be considered as a process of intergroup selection, less drastic than expulsion or extermination. We shall, however, use the term selection pressure, where not qualified, as relating only to intragroup selection. Selection pressure in this sense may be defined so as to include exhaustively all systematic modes of change of gene frequency which do not involve physical transformation of the hereditary material (mutation) or introduction from without (immigration). It includes the effects of differences in mating rate, fecundity, mortality rate, rate of attainment of maturity, and emigration rate. It differs mathematically from the pressures of mutation and immigration in being nil for either $q = 0$ or $q = 1$. There can be no selection in the absence of alternatives.

The general formula for selection pressure is

$$\Delta q_c = \partial q_c / \partial t = q_c\,(W_c - \overline{W})/\overline{W}$$

where W_c is the momentary selective value of the gene in question, giving due weight to the frequencies and selective values of all types of zygote into which it enters, and

$$\overline{W}\ (= \Sigma_{i=1}^{k} W_i q_i)$$

is the mean selective value for the population as a whole.

The selective value of a given type of zygote (fertilized egg) is assumed to be measured by its average contribution under the prevailing conditions to the array of zygotes produced a generation later in such a way that $\overline{W}$ is the ratio of the effective size of the population in the following generation to that in the one under consideration.

$$\overline{W} = \Sigma W f$$

in terms of the selective values (W) and frequencies (f) of genotypes.

It is important to distinguish two kinds of selection: that in which the selective values of genotypes are constant under standard conditions, and that in which they are functions of the relative frequencies of genotypes. Relative constancy is to be expected where selection depends directly on a constant environment external to the species. Secular changes in both absolute and relative values of the selective values may indeed be expected to occur with changes in the density of the population but these are independent of the relative frequencies of genotypes in this sort of selection. If, however, there is selection based directly on the relations of different kinds of individuals of the species to each other, the selective values are necessarily functions of the frequencies of these genotypes. In the following discussion, random mating is assumed unless otherwise specified.

CONSTANT SELECTIVE VALUES. SINGLE LOCI

It is convenient to take up first the case in which the selection coefficients for genotypes of the locus in question are independent of frequencies at other loci and are constant and alike in the sexes. In this case the rate of change in frequency of gene A_c is as follows (Wright, 1937, 1942).

$$\Delta q_c = \partial q_c/\partial t = q_c(1-q_c)(\partial\overline{W}/\partial q_c)/2\overline{W}$$

It is assumed that the frequencies of all alleles of the gene, A_c, are expressed in the form

$$q_i = r_{ic}(1 - q_c)$$

in $\overline{W}$ before differentiating. Here r_{ic} is the frequency of A_i among the alleles of A_c. Thus

$$\partial q_i/\partial q_c = -r_{ic} = -q_i/(1 - q_c).$$

The composition of the population relative to a series of alleles may be represented by a point in a $(k-1)$ dimensional space. Thus in the case of three alleles, the gene frequencies may be represented by the distances from the sides of an equilateral triangle of unit height. The mean selective values ($\overline{W}$) for populations with given sets of gene frequencies, corresponding to points within this, may be represented by ordinates from the plane of this figure, defining a surface. Similarly in the case of four alleles the gene frequencies may be represented by distances from the four faces of a regular tetrahedron of unit height. The mean selective values of populations require an additional dimension $\overline{W}$ which may be indicated by shells of equal selective value, the contours of a "surface" in four-dimensional space.

This "surface" of selective values may be of various sorts. There may be a continuously upward gradient leading from each point toward a single peak value. If this is at one of the corners, it means that selection leads inevitably to fixation of the corresponding homozygote, except in so far as mutation or immigration maintains other alleles. If it is on an edge, or face, or in the interior of the figure, it is implied that selection by itself leads to a condition of stable equilibrium among two or more alleles. A depression from which gradients lead up in all directions implies a position of unstable equilibrium. There may also be saddle points, toward which selection drives the set of gene frequencies from certain directions but from which selection tends to cause ever-increasing departure, after a slight initial departure in certain other directions.

All alleles are maintained in equilibrium, if all heterozygotes (A_iA_j) are alike in selective value and are superior to all of the homozygotes (A_iA_i, A_jA_j etc.). Let

$$W_{ij} = a,\ W_{ii} = a(1 - s_i),\ W_{jj} = a(1 - s_j):$$
$$\overline{W} = a[1 - \Sigma_{i=1}^{k} s_i q_i^2]$$
$$\Delta q_c = a q_c (\Sigma_{i=1}^{k} s_i q_i^2 - s_c q_c)/\overline{W}$$
$$\hat{q}_c = (1/s_c)/\Sigma_{i=1}^{k}(1/s_i)$$

where $\hat{q}_c$ is the value of q_c at equilibrium all Δq's equal 0.

If the heterozygotes differ in selective value, the conditions for maintenance of all alleles are somewhat restricted. Thus if all homozygotes (k alleles) have the same selective value, a, and all of the heterozygotes except one have the value $a\ (1 + s)$ with s positive, and that one (A_1A_2) has the value $a\ (1 + t)$

$$\overline{W} = a[1 + s(1 - \Sigma_{i=1}^{k} q_i^2) + 2(t - s) q_1 q_2]$$
$$\Delta q_1 = a q_1 [s(\Sigma_{i=1}^{k} q^2 - q_1) - (t - s)(2q_1 q_2 - q_2)]/\overline{W}$$
$$\Delta q_3 = a q_3 [s(\Sigma_{i=1}^{k} q^2 - q_3) - (t - s) 2 q_1 q_2]/\overline{W}$$
$$\hat{q}_1 = \hat{q}_2 = s/[ks - (k - 2)(t - s)] \text{ if } 0 < t < 2s$$
$$\hat{q}_3 = \hat{q}_k = (2s - t)/[ks - (k - 2)(t - s)] \text{ if } 0 < t < 2s$$

All alleles are maintained in equilibrium if t is between 0 and $2s$. If t is negative, equilibrium is unstable; either A_1 or A_2 is eliminated according to the initial composition. If t is greater than $2s$, all alleles other than A_1 and A_2 tend to be eliminated.

It is possible for a gene that would be eliminated if only certain alleles were associated with it in a population to be maintained if others also are present. The conditions are decidedly restricted, however.

In many cases, some of which have been indicated above, there may be two or more peak values. Obviously any homozygote that is superior to all of the heterozygotes in which its allele enters is at a distinct peak. Which allele becomes fixed depends on the initial composition of the population. Any two or more alleles may form a group capable of stable equilibrium within itself but in unstable association with other such groups.

The rate of change in the mean selective value of the whole population is given approximately by the following formula, if the selective differentials are small.

$$\Delta\bar{W} = \Sigma_{i=1}^{k}[(1-q_i)(\partial\bar{W}/\partial q_i)\Delta q_i]$$

It is assumed, as before, that the frequencies of all alleles are expressed in the form $q_j = r_{ji}(1-q_i)$ in $\bar{W}$, before differentiating with respect to any frequency, q_i.

Constant Genotypic Selective Values

The treatment of selection as acting on single loci independently of all others is however highly artificial. Selective value, in a given setting, is a property of the organism as a whole and hence of its whole genotype. A gene is likely to be more favorable than an allele in some combinations, less favorable in others. The composition of the species may be thought of as located in a space of $\Sigma^n(k-1)$ dimensions, assuming n loci and a variable number (k) of alleles at each locus. If the selective value, W, for each genotype as a whole is constant, and if the frequencies at the loci are combined at random (as is approximately true under long continued random mating and small net selection coefficients of the genes), the formula for rate of change of gene frequency per generation is still

$$\Delta q_c = q_c(1-q_c)(\partial\bar{W}/\partial q_c)/2\bar{W}$$

under the same convention as before but $\bar{W}(=\Sigma Wf)$ is here a function not only of the gene frequencies at the locus in question but of the genes at all other pertinent loci.

It is again convenient to add another dimension to the geometric model: mean selective value $(\bar{W})$ of populations. The "surface" defined by $\bar{W}$ may be expected to be a very rugged one with innumerable peaks and subpeaks, corresponding to different harmonious combinations of genes.

These peaks may be of three sorts phenotypically: (a) peaks centering in different genotypes that give the same phenotype (as under the conventional multiple-factor hypothesis for quantitative variability and an intermediate optimum), (b) peaks that center in phenotypes that differ but give adaptation to the same conditions, and (c) peaks that center in adaptations to different conditions within the array of conditions to which the population has access, corresponding to the different adaptive zones and subzones of Simpson (1944).

The formula for $\triangle\overline{W}$ is the same as that above except that summation applies to all genes at all loci. The control by the gradient is obviously qualified by the term $q_c(1-q_c)/2\overline{W}$ in the formula for each $\triangle q_c$. With constant genotypic selective values, the species tends to move toward one of the peaks, though not, as carelessly stated previously, up the steepest gradient in the surface $\overline{W}$ from the point at which it is located. The peak toward which it moves is not likely to be the highest one.

Effects of Deviations from Assumed Conditions

The situation is altered more or less by various types of deviation from the idealized case to which the preceding formulae apply. If there are very large differences between the sexes in selection, special formulae must be used; but even with moderately large differences the results are almost as if the average selective coefficients for each genotype applied to both sexes as shown (with a different mathematical formulation) by Haldane (1926) (cf. Wright and Dobzhansky, 1946). If there are moderate deviations from random mating due to inbreeding, and the net selection coefficients of genes are small, the expression $\partial\overline{W}/\partial q_c$ of the preceding formulae may be replaced by $\partial(\overline{W}+F\overline{W}_I)/\partial q_c$ where $\overline{W}_I$ is the mean selective values of genotypes that are homozygous at the locus in question. The expression $\overline{W}_I$ is identical with $\overline{W}$ at loci at which there is no dominance but otherwise is different (Dobzhansky and Wright, 1941).

Selective mortality at different ages, and selective fecundity have diverse effects on observable frequencies. These must be carefully considered in analysis of concrete cases and are likely to make it necessary to go back to the basic formula.

Selection itself always produces departures from complete randomness of combination among loci if there are nonadditive interactions, but these are negligible for most purposes unless selection or linkage is strong (cf. Wright, 1942). Assortative mating, however, may produce very great departure from randomness (Wright, 1921) which requires considerable increase in complexity in its treatment. Subdivision of the population into nearly isolated strains also has effects which alter the situation drastically and are best treated under the head of intergroup selection by means of differential migration pressure. In view of a criticism by Fisher (1941) it should be emphasized that the formula for $\triangle q$ above, to which he referred, was explicitly based on the assumption of random mating of diploids and of constancy of genotypic selection coefficients. Other formulae had been used for other conditions.

Fisher (1930, 1941) himself has developed a different mode of approach for dealing with selection apart from other factors. According to his "fundamental theorem of Natural Selection": "The rate of increase in fitness of any organism at any time is equal to its genetic variance in fitness at that time." Genetic variance here refers only to the additive effects of genes. There is approximate agreement between our formulae for small selective differences and constant selective values, the difference being that Fisher's method does not introduce the term $\overline{W}$ in the denominator. The application of this theorem to cases in which there are variable genotypic selection coefficients is not, however, clear.

VARIABLE GENOTYPIC SELECTIVE VALUES

With variable genotypic selection coefficients, but otherwise the same assumptions as above and the same convention with respect to differentiation,

$$\Delta q_c = q_c(1 - q_c)\Sigma[W(\partial f/\partial q_c)]/2\overline{W}$$

where the f's are frequencies of genotypes, in general involving many loci, and the summation applies to all genotypes. The term $\Sigma[W(\partial f/\partial q_c)]$ may be written

$$[(\partial \overline{W}/\partial q_c) - (\overline{\partial W/\partial q_c})]$$

(Wright, 1942). In this, $\partial \overline{W}/\partial q_c$ measures the effect of change in the frequency of the gene in question on the average selective value of the population as a whole, while $(\overline{\partial W/\partial q_c})$ measures the average effect on the selective values of all genotypes and is of course absent if these selective values are constant.

An important case is that of competition between members of the same species such that the relative degree of competitive success of each of the genotypes is constant but the net rate of increase of the population is unaffected by changes in the relative frequencies. This situation can be represented by writing $W_c = aR_c/\overline{R}$ when a is a general constant, R_c is a constant pertaining to the particular genotype, and $\overline{R}(=\Sigma R_i f)$ is the variable average value of R. The surface of selective values for the population as a whole is constant ($\overline{W} = a$) in accordance with the hypothesis, but gene frequencies change according to the rule:

$$\Delta q_c = q_c(1 - q_c)(\partial \overline{R}/\partial q_c)/2\overline{R}$$

The surface $\overline{R}$ is not level and selection pressure tends to drive the species toward a peak value of $\overline{R}$.

If there is a combination of both kinds of selection, we can write

$$W = a[(R/\bar{R}) + s]$$
$$\bar{W} = a(1 + \bar{s})$$
$$\Delta q_c = a q_c (1 - q_c) \frac{\partial(\log \bar{R} + \bar{s})}{\partial q_c} / 2\bar{W}$$

The surfaces $\bar{W}$ and $\log \bar{R}$ may both be rugged and the peaks and valleys need not agree. The species moves toward a peak in the resultant surface $[(\log \bar{R}) + \bar{s}]$. If $\log \bar{R}$ and $\bar{s}$ are systematically opposite in sign and $\log \bar{R}$ is numerically greater, selection leads to a deterioration of the species in relation to its environment. This puts in symbolic form an interpretation of apparently orthogenetic trends toward extinction given by Julian Huxley. The special case considered by Huxley was that of selection for characters useful to males in competition but somewhat deleterious in relation to the environment.

A case that is similar in its consequences although somewhat more complicated from the mathematical standpoint is that of competition among members of the same brood. If there is no relation to success of the population and W_{11}, W_{12} and W_{22} are the selective values of A_1A_1, A_1A_2 and A_2A_2 when present in the same brood, $\bar{W} = a$

$$\Delta q_1 = a q_1 (1 - q_1) \left[\left(\frac{W_{22} - W_{12}}{W_{22} + W_{12}} \right) q_1^2 + \left(\frac{W_{22} - W_{11}}{W_{22} + 2W_{12} + W_{11}} \right) 2q_1(1 - q_1) + \left(\frac{W_{12} - W_{11}}{W_{12} + W_{11}} \right) (1 - q_1)^2 \right]$$

If there is a negative correlation between success in intrabrood competition and success in meeting the external environment, the process may lead to deterioration of the species. This case has been treated from a different mathematical viewpoint by Haldane (1924a).

Where advantages with respect to social relations and external environment are negatively correlated but closely balanced, it is possible for the opposing selection pressures to maintain alleles in equilibrium.

A genotype may have qualities that are of advantage to the species in proportion to some function of the frequency of such individuals but of no additional advantage to these individuals themselves. If this is the case, all selective values (W) are to be multiplied by the same term $[1 + \Psi(q_c)]$ in which $\Psi(q_c)$ is the function of the frequency of the genotype in question. If the W's are otherwise the same, $\bar{W} = a[1 + \overline{\Psi(q_c)}]$ with a peak or peaks corresponding to one or more values of q_c. There is, nevertheless, no tendency for q_c to change at all ($\Delta q_c = 0$). If the gene has a deleterious effect on its possessors it will tend to be eliminated in spite of its advantage to the species, still assuming random mating (cf. Haldane, 1932; Wright, 1945a).

The relative as well as the absolute selective values of genotypes may change with changes in their frequencies. It must suffice here to consider a few simple cases involving only one series of alleles. It will be assumed that the selective values of heterozygotes are always exactly intermediate between those of the corresponding homozygotes to avoid complications due to dominance or superdominance of the sorts that occur also where the selection coefficients are constant.

First consider the case in which the selective value of a homozygote rises in proportion to the frequency of its allele in the population. This may hold where the individuals of each type receive a social advantage from the presence of others of their own kind or are injured by the presence of other types. Let

$$W_{11} = a(1 + 2s_1q_1),\ W_{12} = a[1 + s_1q_1 + s_2q_2],\ W_{22} = a[1 + 2s_2q_2], \text{ etc.}$$

$$\overline{W} = a[1 + 2\Sigma_{i=1}^{k} s_i q_i^2]$$

$$\Delta q_c = aq_c[s_c q_c - \Sigma_{i=1}^{k} s_i q_i^2]/\overline{W}$$

There is unstable equilibrium (s's all positive) at the set of values

$$\hat{q}_c = (1/s_c)/\Sigma_{i=1}^{k}(1/s_i)$$

One allele or another tends to become fixed according to the initial composition of the population.

The selective advantages of genotypes may, however, decrease as the corresponding allele increases in frequency. This occurs if there is a division of labor among types to their mutual benefit, or if the population occupies a heterogeneous territory in which rare genotypes can always find a favorable niche but abundant ones are forced to live in part under unfavorable conditions. In this case the s's in the above formula are negative. There is stable equilibrium at the values $\hat{q}_c = (1/s_c)/\Sigma(1/s_i)$, exactly as if all heterozygotes had the same constant selective value and each of the homozygous types had a certain disadvantage $[W_c = a(1 - 2s_c)]$ etc. In the case with variable selection coefficients considered here, gene frequencies move in exactly the same type of surface $\overline{W}$ in the same way except that the rate is only half as great in relation to the slope as with constant coefficients (Wright and Dobzhansky, 1946).

A somewhat similar case which has been analyzed in detail in a previous paper (Wright, 1938) is that of self-sterility alleles in plants. Alleles are favored when rare and opposed when common, but as the selection depends on a relation between male gametophyte (haploid) and female sporophyte (diploid), it is more complicated mathematically than cases in which selection depends on relations among diploids, and is symmetrical or nearly so with respect to sex.

There may be cyclic movement of gene frequencies under certain limiting conditions which are probably of more mathematical than evolutionary interest. Thus if the selective values are as follows, and s is very small, there is an approach to perpetual cyclic motion although the surface of values is completely flat ($\overline{W} = a$)

$$W_{11} = a[1 + 2s(q_2 - q_3)] \qquad W_{12} = a[1 + s(q_2 - q_1)]$$
$$W_{22} = a[1 + 2s(q_3 - q_1)] \qquad W_{23} = a[1 + s(q_3 - q_2)]$$
$$W_{33} = a[1 + 2s(q_1 - q_2)] \qquad W_{31} = a[1 + s(q_1 - q_3)]$$
$$\Delta q_1 = sq_1(q_2 - q_3)$$
$$\Delta q_2 = sq_2(q_3 - q_1)$$
$$\Delta q_3 = sq_3(q_1 - q_2)$$

With the slightest departure from the limiting situation the species tends to move in a spiral in the field of gene frequencies either toward or away from a peak, if there is one, or toward or away from a depression to a point at which all Δq's are zero. Somewhat similar cases may involve more than one locus.

The formula for $\Delta\overline{W}$ is the same as with constant selection coefficients.

Net Systematic Pressure

The net effect of all of the systematic pressures on the frequency of a gene is the sum of the various components. Thus in a random breeding population of diploids, the rate of change in the frequency of a gene A_c as one of a system of simultaneous equations is approximately as follows:

$$\Delta q_c = \Sigma(u_{ci}q_i) - (\Sigma u_{ic})q_c - m(q_c - q_{c(I)}) + \frac{q_c(1-q_c)}{2\overline{W}}\left[\frac{\partial \overline{W}}{\partial q_c} - \left(\overline{\frac{\partial W}{\partial q_c}}\right)\right]$$

This type of pressure on all of the genes tends to carry the species to some position at which $\Delta q = 0$ for each of them (except in certain limiting cases referred to above). This position tends to be in the neighborhood of a peak in the surface of mean selective values in so far as due to selection related to the environment, but this is not likely to be the highest peak. Mutation and immigration pressure, if present, keep the population from attaining the exact peak. The position of stability may indeed not even be near a peak in cases in which selection depends on certain types of social relations. We have here a theory of stability of species type in spite of constant mutation, extensive variability at all times, constant action of selection, and substantial possibility of improvement if the species could arrive at certain combinations of genes already present, other than the combinations that prevail.

Accidents of Sampling

Accidents of sampling may cause fluctuation from the position of stability, but these are negligible in a large random breeding population. They become important in small populations where they are responsible for the well-known effects of inbreeding. They may be responsible for nonadaptive differentiation of small island populations but are more likely to lead to ultimate extinction in a small population, completely isolated from its kind, than to evolutionary advance (Wright, 1931 and later). These conclusions follow from the chance of fixation of a completely indifferent gene which is $1/(2N)$ where N is the effective size of population, and from the chance of fixation of a deleterious gene (selective disadvantage s in heterozygote, $2s$ in homozygote) which is $2s/(e^{4Ns} - 1)$.

NONRECURRENT MUTATIONS

Stability in a species living under constant conditions may be upset by the occurrence and establishment of a favorable mutation that has never occurred before or has previously been lost by accidents of sampling if it ever has occurred. Even a mutation with a considerable selective advantage is likely to be lost by accidents of sampling when present in only a few individuals, although it is practically certain to reach its equilibrium frequency if it ever reaches moderate frequencies in a large population. The chance of fixation of a semidominant mutation with selective advantage s in the heterozygote is $2s/(1 - e^{-4Ns})$, or approximately $2s$ as given by Haldane (1927b). A recessive mutation with selective advantage s has a chance of fixation given by Haldane as of the order of $\sqrt{s/N}$ and by Wright (1942) as $\sqrt{s/2N}$. With v the rate of recurrence of a mutation, the chance that all recurrences within a period of T generations are lost by accidents of sampling is $P = (1 - 2s)^{2NvT}$ in the former case, $P = [1 - \sqrt{s/2N}]^{2NvT}$, in the latter. The line between recurrent and nonrecurrent is necessarily arbitrary. A type of mutation with an even chance of becoming established within 1000 generations must be considered recurrent from the standpoint of a species with a life (T) of hundreds of thousands of generations; but if there is an even chance only in 100,000 generations, the mutation may be considered nonrecurrent in this case. Putting $P = ½$ in the period T generations, rate $v < 0.69/4NsT$ may be used to define nonrecurrent mutations with dominant favorable effect and $v < 0.69/\sqrt{2NsT}$ for recessives with favorable effect. Thus in a population of 10^{10} and $T = 10^4$, mutation rates above about 10^{-13} would be considered recurrent for mutations with $s = 0.01$ in the heterozygote and mutation rates above 10^{-8} in the case of recessives with $s = 0.01$.

The likelihood of disturbance of stability by nonrecurrent mutations probably increases indefinitely with increase in size of population. The number of alleles maintained in equilibrium at each locus by one of the mechanisms discussed earlier should obviously increase with increase in the size of population and thus give a broader base for new mutations. It was suggested in earlier papers (Wright, 1931 and later) that there might be a certain optimum size for evolution in the case of a homogeneous population living under constant conditions, but this probably does not give adequate weight to the likelihood of indefinitely extended allelic series.

Evolution from novel favorable mutations is obviously limited by the rate at which these occur, in so far as dependent on these mutations themselves. However, the process of establishment of a novel mutation would tend to upset the system of selective values among genes that have been maintained within the species (Fisher, 1930). The readjustments in the system required to set off to best advantage the effects of a novel favorable major mutation might be extensive. They would be less extensive with the more frequent favorable minor mutations. Even so a single novel mutation might start a chain of evolutionary changes that would continue long after the mutation itself had reached either fixation or an approximately stable frequency. Evolution by this mechanism is therefore far from being completely limited by mutation rate.

In terms of the geometric model, the establishment of a novel favorable mutation in the species adds a new dimension. If minor in effect, it may be thought of as slightly elevating the adaptive peak to which the species has been bound, thus making possible slow evolutionary advance along the established line. If, however, a novel mutation brings about a new adaptive type at a step, this means the formation of a bridge to a higher peak across what had been an impassable valley.

Changing Environment

So far we have assumed constancy of external condition. But climatic conditions change, food species and enemy species increase or decrease in numbers and change in character. A portion of the species may move to a region in which such environmental conditions are different from those hitherto encountered. Under such changes of condition, the whole system of selective values of genotypes in the species changes. Peaks in the surface of selective values may be depressed and low places elevated. With too rapid a change the location of the species in the system may be carried to such a low point that it becomes extinct, but with less extreme changes the species may move sufficiently rapidly to keep up with the changing position of the peak to which it is attached, by drawing on the store of variability provided by loci in which multiple alleles have been maintained in equilibrium. Very great changes in character may be brought about without waiting for the occurrence of any novel favorable mutation and without elimination of any established genes, merely by a shift in the equilibrium frequencies at many loci. The number of combinations of frequencies may be so great, if there are many such loci, that evolutionary adjustment to a continually changing environment may theoretically proceed indefinitely without limitation by mutation rate. However, as the system of gene frequencies changes, mutations which have been so rare as to be classified as nonrecurrent may be expected to become recurrent. While the process as a whole has the aspect of a struggle of the species to hold its own in the face of a continually deteriorating environment, rather than of evolutionary advance, there can be no doubt that a large part, perhaps the major portion of evolutionary change, is of this character (Wright, 1932).

GENOTYPIC SELECTION

In the reaction of a homogeneous species both to novel favorable mutations and to changes of environment, the course of the changes in frequency at each locus is governed largely by the net effects of change in the frequency of each gene in all of the genotypes in which it occurs. The real objective of selection is the genotype as a whole (or even a harmonious system of genotypes); but under random mating exceptionally adaptive genotypes are broken up by the reduction division immediately after they are formed, making selection by genotype impossible. Effective selection by genotype requires that these multiply as such.

This occurs most simply if there is uniparental reproduction. Selection among competing clones is indeed so effective that an array of such clones tends to be reduced to a single clone very rapidly in a homogeneous environment. Thereafter evolution can proceed only by the occurrence of favorable mutations, with no amplification through readjustments in a store of variability. If, however, there is a properly adjusted alternation of predominant uniparental reproduction with occasional crossbreeding, and sufficient isolation and local heterogeneity to mitigate the effects of selection in the uniparental phase, the situation would seem favorable for a very effective and indefinitely continuing evolutionary process based on genotypes (Wright, 1931). This system is, of course, characteristic of many species, especially ones in which there are great cyclic variations in numbers and has no doubt provided effectively for their evolution. It is less favorable in species in which population size remains relatively constant.

POPULATION STRUCTURE

This brings us to consideration of the effects of population structure in species in which there is exclusive or at least predominant biparental reproduction. The mathematical theory depends on the inbreeding coefficient F (Wright, 1943, 1946). If a species is divided into numerous local populations which are partially isolated from each other and which may become genetically differentiated from each other for any reason, the store of variability in the species as a whole is obviously much greater than in a random breeding population of the same size. Moreover the store of variability in any local population should be much greater than in a completely isolated population of its size because the postulated immigration from the rest of the species may have very much greater effects than recurrent mutation at observed rates without breaking down the basis for local differentiation. Any local population that acquires a genetic complex of exceptional adaptive value tends to increase in numbers and become a major source of immigration into neighboring populations. If this adaptation is only of local value, selection pressures may be set up against its spread that tend to isolate it and split the species, but if the adaptation is of general value it tends to transform its neighbors in this direction. There is here the basis for an extensive trial and error mechanism within the species by which the species as a whole may advance. It is to be noted that the unit of selection here is not the gene and not even the genotype as a whole but the entire system of gene frequencies of a local population.

It is important in this connection to compare the various mechanisms by which local populations may become differentiated. The most obvious is difference in the pressure of different local environments on the direction of selection. Appreciable differentiation occurs if the difference in the net selection coefficient s of a gene is greater than the immigration coefficient m (Wright, 1931). Otherwise differentiation tends to be swamped by crossbreeding. Strong differentiation of this sort belongs in the category which tends more toward splitting of the species than of promoting its evolution as a single group. Nevertheless if differentiation is not too extreme, the increase in the store of variability both locally and in the species as a whole is favorable for evolution of the species as a whole and occasionally selection directed by local conditions may lead to adaptations of general value.

Where structure is sufficiently finely divided, the effects of local accidents of sampling provide a mechanism of differentiation complementary to the preceding, since it affects all loci in which there are no important differences in selection pressure between localities. The sampling variance of gene frequencies in one generation is

$$\sigma^2_{\delta q} = q(1 - q)/2N$$

where N is the effective size of the local population (in general much smaller than its actual size).[2] The variance of the frequencies among local

[2] Random fluctuations in the systematic pressures have effects somewhat similar to those due to accidents of sampling. Formulae for the variance, $(\sigma^2_{\delta q})$ in each case, and the resulting frequency distribution $\phi(q)$ of gene frequencies, have been published (Wright, 1948) since this was written.

populations tends to increase by this amount in each generation but is counteracted by the effects of the systematic pressures toward the equilibrium frequency. The result is a frequency distribution through which the local populations tend to drift at random. If q represents a gene frequency and $(q + \delta q + \triangle q)$ its value after a random fluctuation (δq) and a systematic change $(\triangle q)$ has occurred, the frequency distribution $\phi(q)$ of the gene frequencies should be such that all moments of deviations from the mean before and after the above changes remain the same.

$$\int_0^1 \int_{-q}^{1-q} (q + \delta q + \triangle q - \bar{q})^n \phi(q) d(\delta q) dq = \int_0^1 (q - \bar{q})^n \phi(q) dq$$

For cases in which $\triangle q$ is of the same or lower order than $\sigma^2_{\delta q}$, it can be shown that this is satisfied by the equation

$$\phi(q) = (C/\sigma^2_{\delta q}) exp[2\int (\triangle q/\sigma^2_{\delta q}) dq]$$

This was demonstrated in earlier papers for the mean and second moment (Wright, 1937, 1938). The same proof can be extended at once to all higher moments. Various special cases have been derived by other methods (Wright, 1931). A method, used by Kolmogorov (1935) in confirming one of these, leads to the partial differential equation of the general case (Wright, 1945), of which the preceding equation is the solution for the case of a stable state.

$$\frac{\partial \phi(q,t)}{\partial t} = \frac{1}{2} \frac{\partial^2}{\partial q^2} [\sigma^2_{\delta q} \phi(q,t)] - \frac{\partial}{\partial q} [\triangle q \phi(q,t)]$$

I am indebted to Dr. L. J. Savage for calling to my attention that this same formula (Focker-Planck) has been used in physics in the analysis of Brownian and allied types of random movement.

In the case of an array of local populations of effective size N, all subject to the same conditions of selection of the type with constant coefficients, and each replaced to the extent m by a random sample from the total population, the rate of change of gene of gene frequency is

$$\triangle q_c = -m(q_c - \bar{q}_c) + \frac{q(1-q)}{2} \frac{\partial \log \bar{W}}{\partial q_c}$$

The distribution of frequencies of the gene A_c for a given set of frequencies at other loci is

$$\phi(q_c) = C\bar{W}^{2N} q_c^{4Nm\bar{q}_c - 1} (1 - q_c)^{4Nm(1-\bar{q}_c)-1}$$

The joint distribution for multiple alleles at a locus, still assuming that the frequencies at other loci are given, is

$$\phi(q_1 \cdots q_k) = C\bar{W}^{2N} \Pi_{i=1}^k q_i^{4Nm\bar{q}_i - 1}$$

The joint distribution for genes at all loci under the assumed conditions is of the same form except that q_i applies to any gene at any locus and there are Σk terms in the product, the summation here relating to all loci. It should be emphasized that it is assumed in these formulae that effective m is the same for all alleles, which is not necessarily the case.

Returning to the case of one pair of alleles and assuming that selection and mutation pressures are negligible, the variance of the distribution of gene frequencies is approximately (Wright, 1931)

$$\sigma_q^2 = q(1 - q)/(4Nm + 1)$$

Somewhat severe isolation is necessary to permit extensive random local differentiation. Nevertheless it can be shown that even in a large population with full continuity of interbreeding throughout its area, a slow rate of diffusion permits extensive random differentiation not only of neighborhoods but of large regions (Wright, 1943, 1946). The rate of diffusion must be such that the parents of any individual are drawn from an adult population of not more than a few dozen. If the effective population of the neighborhood (parental population) is in the hundreds random differentiation is relatively slight, while if it is in the thousands the situation is practically equivalent to random breeding, even though the species as a whole is indefinitely large (still assuming no differential selection). The possibilities of random differentiation are, however, much greater if there is merely linear continuity as along a shore line or river or if there are obstacles to interbreeding other than distance.

The continually shifting random differentiation of local populations is especially significant because it occurs at all variable loci. If there are hundreds of unfixed loci in the species, the number of substantially different genetic complexes that may arise by chance among local populations becomes indefinitely great.

This sort of differentiation cannot be expected to continue independently of selection, however, even though conditions are uniform throughout the species. Genetic complexes may be expected to arise in different localities that give adaptation to these same conditions in somewhat different ways and thus create differential selection where it had not existed before.

In certain cases such joint effects of selection and chance are exceptionally great. Any character that is affected by multiple factors with more or less additive effects and which is optimum at an intermediate grade (as is usually the case) necessarily has many peak combinations of genes. For example if capitals are used for plus factors and small letters are used for minus factors, *AABBccdd*, *AAbbCCdd*, *AAbbccDD*, *aaBBCCdd*, *aaBBccDD*, and *aabbCCDD* may determine six peaks representing the same phenotypic optimum. Stability is reached in a large random breeding population at approximate fixation of one or another of these types. On subdivision of the population into partial isolated local populations, every type may be expected to predominate somewhere. The transition from one to another may arise from slight fluctuations in the position of the optimum or else merely by accidents of sampling (Wright, 1935a, b). Such a species is obviously in a position to utilize to advantage any favorable secondary interactions of such genes with others in which local populations come to differ.

Certain conditions were referred to earlier as leading to the maintenance of two or more alleles at subequal frequencies in a random breeding population. The opposite conditions favor differentiation among partially isolated populations and hence maintenance of multiple alleles. Such differentiation is to be expected in cases in which homozygotes are superior to the corresponding heterozygotes. An important example is that of reciprocal translocations. In plants, heterozygotes for reciprocal translocations are usually semisterile and thus very strongly selected against, while the balanced homozygotes may be equally vigorous. The situation is similar though somewhat more complex in animals. The establishment of a new arrangement requires accidents of sampling in exceedingly small local populations, probably ones in which there is frequent extinction of the population and restoration from very small numbers of migrants (Wright, 1941). In this case the process obviously has more significance for ultimate splitting of the species than for its evolution as a single group.

A locus in which genes become increasingly favored as they increase in frequency has, as noted earlier, a point of unstable equilibrium. Genes with social effects favorable to their own kind come here. Recognition marks are an example. Again differentiation of local populations on such a basis tends toward splitting of the species, but less than in the case of reciprocal translocation, and in contributing to the store of variability may contribute to the evolution of the species as a whole.

The evaluation of the actual importance of the various evolutionary processes depends on the determination of values of the various coefficients. Determinations of mutation rates from a number of very different sorts of organisms (men, flies, corn plants, bacteria) indicate that genes with recurrence rates of the order of 10^{-6} per generation are not uncommon but that few have rates higher than 10^{-5}. No lower limit can of course be set. The value of m can be anything from 0 (complete isolation) to 1 (no isolation whatever), but much remains to be learned of the structure of actual populations. The observations of Dubinin, Dobzhansky, and others on the occurrence of lethals in nature have demonstrated that genes with the highest possible selection coefficient are common. Such studies as those of Dobzhansky and associates on *Drosophila pseudoobscura* indicate the existence of abundant selective differentials of high order (greater than 0.10) relating to physiological characters. These strongly indicate a shifting state of balance among selection pressures of far higher magnitude than had previously been believed to be the case. Accidents of sampling can of course be of little importance for such sets of genes except where there is isolation of very small populations. This, however, does not preclude and indeed favors the existence of a still larger number of sets of genes with much smaller selective differentials.

The general conclusion seems warranted that a certain degree of subdivision of a species into partially isolated groups provides the largest store of variability both locally and within the species as a whole, and by providing for selection in which whole genetic complexes are the objects, frees evolution most completely from dependence on rare favorable mutations and makes possible the most rapid exploitation of an ecologic opportunity (Wright, 1931 and later).

The Evolutionary Process

Under this interpretation, a continual kaleidoscopic shifting of the statistical characteristics of the local populations is to be expected within any species that occupies, not too densely, a reasonably large range. A similar but slower shifting of character is to be expected among the larger and more differentiated groups recognized as subspecies. The net result should be a gradual shifting in the characters of the species as a whole until the change becomes so great that a new species must be recognized. Subspecies on this view are only rarely incipient species.

In the occasional splitting of species to form two, the critical event seems necessarily to be complete or nearly complete interference with exchange of genes. In some cases, especially in higher plants, the primary isolation is undoubtedly genetic in origin (autopolyploidy, amphidiploidy, compound aneuploidy). In most cases, at least in higher animals, it seems however to be geographic, less frequently if ever purely ecologic (cf. Mayr). If splitting occurs by the extinction of intermediates in a chain of subspecies, it is possible for one species to become two without any current genetic change whatever in these. At the other extreme, a portion of a species which does not differ appreciably in any respect from the parent stock may become completely isolated. In the course of time the accumulation of random and selective differences in genetic composition may be expected to lead to clear-cut character differences and also to ones that prevent exchange of genes on renewed contact. The question as to the time at which two species are present instead of one is here somewhat arbitrary. The critical event however is the isolation rather than any particular genetic changes. The majority of cases are perhaps intermediate.

Similarly the multiplication of species relatively rarely leads to generic differences and still more rarely to families and higher categories by a gradual cumulative process. There seems to be a large measure of truth in the contention of Willis and Goldschmidt that evolution works down from the higher categories to the lower rather than the reverse. Nevertheless the critical event in the appearance of a higher category seems to be a major ecological opportunity rather than any sort of mutation. Such an opportunity may arise in various ways. Two extreme cases may be distinguished (1) A form, in the course of its gradual evolution, may acquire a character or character complex that happens to be of general rather than merely special significance and which thereby opens up the possibility of a relatively unexploited way of life. In many cases such an "emergence" turns on the possibility of using an organ evolved apparently for one purpose for a different purpose. (2) A form which reaches relatively unoccupied territory also has before it a major ecological opportunity independently of any character differences from its parent form. David Lack has made a most illuminating analysis of such a case in his recent study of the ground finches of the Galapagos Islands. In terms of our geometric model, the first case is one in which the genetic composition of the species manages by any means to cross a difficult valley in the surface of selective values and reaches a system of peaks higher than the single peak to which it has been bound. In the second case, the surface of selective values of the isolated portion of the species is itself changed abruptly by the elimination of competition and in many cases by changes in environmental conditions. The effect, however, is the same as in the first case; a single peak is replaced by a higher system of peaks.

Beyond a certain threshold in degree of success in the achievement of a major adaptation, improvement should be as rapid as the store of variability permits. When decisive superiority is attained or unoccupied territory is reached, a very extensive and rapid adaptive radiation should follow under the divergent selection pressures toward exploitation of the various special ecological niches, opened up by the general adaptation (in the first case) or the mere absence of competition (in the second). The most favorable conditions are those in which the population is sufficiently sparse and the territory sufficiently extensive and interrupted to permit a fine-scaled subdivision grading into complete isolation in many places. Differentiation of two forms must ordinarily start in geographically isolated regions but may be accelerated by competition if one form reinvades the territory of the other after genetic isolation has been reached. This rapid evolution in the origin and adaptive radiation of a higher category constitutes the tachytely of Simpson.

The primary adaptive radiation may be expected to be followed by secondary and tertiary radiations of progressively lower scope, and by extensive selective elimination of other higher categories especially in the case in which an original general adaptation has outmoded older forms in a great variety of ecological niches. As this process continues, the time comes when each of the successful forms has only one line of advance open to it, a single ridge in our model. There is slow and more or less orthogenetic advance, the horotely of Simpson. There may be extensive diversification of genera and species, but these are rarely of significance in starting new trends. Ultimately forms may become bound each to a single peak, with no other peak attainable and evolution virtually ceases, the bradytely of Simpson. While the sizes of the steps in such a cycle tend continually to fall off until they reach the vanishing point, a step may occasionally be larger, and on very rare occasions, much larger, initiating a new higher category and new evolutionary cycle of emergence, adaptive radiation, selective elimination of higher categories, orthogenesis and stability.

REFERENCES

Darlington, C. D., and A. A. Moffett. 1930. *J. Genet.*, *22*:129-151.

Dobzhansky, Th. 1941. *Genetics and the Origin of Species.* 2nd ed., Columbia University Press, 446 pages.

——. 1946. *Genetics, 31*:269-290.

Dobzhansky, Th., and S. Wright. 1941. *Genetics, 26*:23-51.

Dobzhansky, Th., A. M. Holz, and S. Spassky. 1942. *Genetics, 27*:463-490.

Dubinin, N. P., M. A. Heptner, Z. A. Demidova, and L. I. Djachkova. 1936. *Biol. Zhur. 5*:939-976.

Fisher, R. A. 1930. *The Genetical Theory of Natural Selection.* Oxford, The Clarendon Press, 272 pages.

——. 1937. *Annals of Eugenics, 7*:355-369.

——. 1941. *Annals of Eugenics, 11*:53-63.

Geiringer, Hilda. 1944. *Ann. Math. Stat., 15*:27-57.

——. 1948. *Genetics, 33*:548-564.

Goldschmidt, R. 1940. *The Material Basis of Evolution.* New Haven, Yale University Press, 436 pages.

Haldane, J. B. S. 1924. *Trans. Camb. Phil. Soc., 23*:19-41.

——. 1924. *Proc. Camb. Phil. Soc.* (Biol. Sci.), *1*:158-163.

——. 1926. *Proc. Camb. Phil. Soc., 23*:363-372.

——. 1927. *Proc. Camb. Phil. Soc., 23*:607-615.

——. 1927. *Proc. Camb. Phil. Soc., 23*:838-844.

——. 1932. *The Causes of Evolution.* London, Harper & Bros., 235 pages.

Hardy, G. H. 1908. *Science, 28*:49-50.

Huxley, J. S. (editor). 1940. *The New Systematics.* Oxford, Clarendon Press, 583 pages.

——. 1942. *Evolution, the Modern Synthesis.* London, Harper & Bros., 645 pages.

Kolmogorov, A. 1935. *C. R. de l'Acad. des Sciences de l'U.R.S.S., 3*(7):129-132.

Lack, D. 1947. *Darwin's Finches.* Cambridge, Cambridge University Press, 208 pages.

Mayr, E. 1942. *Systematics and the Origin of Species.* New York, Columbia University Press, 334 pages.

Morgan, C. Lloyd. 1933. *The Emergence of Novelty.* London, Williams and Norgate, 207 pages.

Osborn, H. F. 1917. *The Origin and Nature of Life.* New York, Charles Scribner's Sons, 322 pages.

Robbins, R. B. 1918. *Genetics, 3*:375-389.

Simpson, G. G. 1944. *Tempo and Mode in Evolution.* New York, Columbia University Press, 237 pages.
de Vries, Hugo. 1905. *Species and Varieties. Their Origin by Mutation.* Chicago, Open Court Publ. Co.
Weinberg, W. 1908. *Jahresheft Ver. f. vaterländische Naturkunde im Württemberg,* *64*:368-382.
—. 1909. *Zeit. ind. Abst. u. Vererbungslehre,* *1*:277-330.
—. 1910. *Arch. Rass. u. Ges. Biol.,* *7*:35-49; 169-173.
Willis, J. C. 1940. *The Course of Evolution.* Cambridge, Cambridge University Press, 207 pages.
Wright, S. 1921. *Genetics,* *6*:111-178.
—. 1921. *Amer. Nat.,* *56*:330-338.
—. 1922. *Bull. No. 1121 U.S. Dept. Agric.,* 59 pages.
—. 1929. *Amer. Nat.,* *63*:274-279.
—. 1931. *Genetics,* *16*:97-159.
—. 1932. *Proc. 6th Internat. Congress of Genetics,* *1*:356-366.
—. 1935. *Jour. Gen.,* *30*:243-256; 257-266.
—. 1937. *Proc. Nat. Acad. Sci.,* *23*:307-320.
—. 1938. *Ibid.,* *24*:253-259, 372-377.
—. 1939. *Genetics,* *24*:538-552.
—. 1941. *Amer. Nat.,* *75*:513-522.
—. 1942. *Bull. Amer. Math. Soc.,* *48*:223-246.
—. 1943. *Genetics,* *28*:114-138.
—. 1945. *Ecology,* *26*:415-419.
—. 1945. *Proc. Nat. Acad. Sci.,* *31*:382-389.
—. 1946. *Genetics,* *31*:39-59.
—. 1948. *Evolution,* *2*:279-294.
Wright, S., and Th. Dobzhansky. 1946. *Genetics,* *31*:125-156.

39
Population Structure in Evolution

Proceedings of the American Philosophical Society 93, no. 6 (1949):471–78

40
The Genetical Structure of Populations

Annals of Eugenics 15 (1951):323–54

Introduction

From the beginnings of his work on animal breeding theory and evolutionary theory Wright had strongly emphasized the importance of population structure. In the late 1940s Wright turned his attention to a synthesis of all his work on population structure. The result was a series of four papers (Wright 132, 138, 139, 140), two of which are reproduced here.

The first was delivered as a paper at the annual meeting of the American Philosophical Society, devoted that year to Natural Selection and Adaptation. This paper was directed to a sophisticated lay audience rather than professional biologists, and Wright wrote a wholly qualitative essay, distinctive for his use of compact tables to present outlines of factors in the process of evolutionary change.

The second paper, delivered in 1950 as the Galton Lecture at University College, London, during Wright's sabbatic year at the University of Edinburgh, was a quantitative presentation of his views on population structure. The strong relationship between evolution in domestic populations and evolution in nature, so important in the origins of Wright's shifting balance theory of evolution in nature, emerges clearly again in this paper. The power of Wright's *F*-statistics for elucidating the mechanisms of evolutionary change emerge more clearly in this paper than in any that he had published previously.

POPULATION STRUCTURE IN EVOLUTION[1]

SEWALL WRIGHT

Ernest D. Burton Distinguished Service Professor of Zoology, University of Chicago

(*Read April 21, 1949, in the Symposium on Natural Selection and Adaptation*)

FROM our present knowledge of genes, it seems most probable that life and evolution began with the appearance of nucleoprotein molecules endowed with the remarkable capacity for autosynthesis (table 1, I). This obviously required an environment with a richness in organic compounds which could only have been arrived at in the prior absence of life. Evolution would begin with the occurrence of mutations, duplicating as of the new type. Evolution along any single lineage would necessarily consist wholly of a succession of such mutations. From this standpoint it would appear that the prevailing direction would merely have been that toward increasing stability of the molecule. The whole evolutionary process is sometimes treated as if this were all there were to it.

TABLE 1

MAJOR STEPS IN EVOLUTION

I. *The Gene.* Origin as autocatalytic molecule.
  1. Evolution along single lineage by mutation.
  2. Evolution of population of genes.
     Mutation pressure (selection for stability).
     Selection pressure (in synthetic or predatory capacity).
     Ingression from surrounding regions.
  3. Evolution of array of local populations.
     Differential increase and dispersion.

II. *The Cell.* Equational division of assemblage of genes.
  Evolution largely as above but with greater scope.
  Chromosome aberrations supplement gene mutation.
  Selection for mitotic regularity as well as for external adaptation.

III. *The Interbreeding Species.* Cells capable of conjugation and reduction.
  1. Evolution along gene lineages by mutation.
  2. Evolution of species according to population structure.
     A. Uniparental reproduction with occasional conjugation.
        (Selection among recombinant *genotypes.*)
     B. Biparental reproduction with effective panmixia.
        (Selection according to *average effects of alleles.*)
     C. Array of partially isolated populations.
        (Interstrain selection according to *genetic complexes.*)
  3. Evolution of local array of species.

IV. *Species of Multicellular Individuals.*
  Same processes as under III but with greater scope.

The multiplying genes would soon deplete their store of natural resources: amino acids, pentose sugars, purines, pyrimidines, and energy rich substances. As Horowitz has pointed out, mutations that gave the capacity to synthesize depleted substances from precursors, arginine, for example, from citrulline, would have had much more opportunity to duplicate than ones that did not. Predatory genes, capable of breaking down others, would also be at an advantage. Thus selection would make its appearance among gene lineages and would soon take precedence over a mere trend toward stability. To understand the course taken by evolution, it becomes necessary to consider the history of the whole population of genes in a locality (table 1, I-2). The factors governing such a history are statistical: relative rates of mutation, rates of ingress from other localities, but above all the pressure of selection.

Local populations would presumably evolve into symbiotic systems. One gene might synthesize arginine in excess of its needs but be limited in its duplication by lack of methionine, unless near a gene capable of synthesizing this in excess. Different localities might arrive at somewhat different harmonious systems if sufficiently isolated to prevent swamping. If not completely isolated, useful mutations in one would be able in time to reach the others. Especially successful systems would tend to spread over wide areas but not without modification if dispersion proceeded slowly. Here we have evolution at a third level; that of the whole array of local populations, with dependence on the existence of a certain balance between isolation and dispersion (table 1, I-3).

It is evident that the scope of evolution is severely limited as long as the organism is no more than a single large molecule. It seems to require a rather large protein molecule to function

[1] This investigation was aided by a grant from the Wallace C. and Clara A. Abbott Memorial Fund of the University of Chicago.

as an enzyme in a single catalytic process. Symbiosis among such molecules is a poor substitute for permanent assemblage of all necessary metabolic capacities within one organism.

We may surmise then that the second great step in evolution was the assemblage of many genes into one structure, capable of dividing equationally after multiplication has exceeded certain physical limits (table 1, II). For our purpose it is not important whether the chromosome came first and the cell, with gene products (cytoplasm), mediating between the genes and the external environment, came later, or the reverse. In any case, the assemblage of many genes in one organism means an enormous increase in scope.

There is, however, no essential change in the mechanism of evolution. Mutations may occur at each of many loci and chromosome aberrations may occur involving many loci at once, but evolution in retrospect along any lineage is still merely a succession of mutations. There would also be evolution of local populations, guided in this case by selection for internal adaptation (e.g., perfection of mitosis) as well as by selection for external adaptation in synthetic or predatory capacity. Moreover the rate of evolution of life as a whole would be enhanced by a balance between isolation of local populations and dispersion from one to another. We have essentially the same three aspects listed under evolution of the gene.

Until recently it was thought that all bacteria and blue green algae were representative of this (or a lower) level of evolution. It now appears that in some bacteria at least (Lederberg and Tatum) and even in some viruses (Hershey; Delbruck and Bailey), the third great step has appeared, the possibility of exchange of genes, and recombination. It is indeed possible that occasional conjugation may go back to the origin of the cell and that meiosis was perfected concomitantly with mitosis. It is, however, convenient for purposes of comparison to treat these as separate steps.

With the appearance of even occasional conjugation, evolution along cell lineages virtually ceases. Individual genes still have histories which appear in retrospect as a succession of mutations but each cell lineage becomes merely a short strand in the network-like structure (in space-time) of a larger entity, the interbreeding species. This is the smallest entity above the gene which may be said to evolve (table 1, III.)

The cells with their somewhat limited scope are still the immediate objects of selection. The scope for selection is enormously increased by the appearance of multicellular organisms, which may on this account be considered the fourth great step in evolution (table 1, IV). Reproduction, however, is still cellular. The simplest entity, above the gene, that can be said to evolve is still the species.

The conjugation-reduction cycle enormously amplifies the potential amount of variability from a given number of mutations. Where 100 different mutations in a population in which there is exclusive uniparental reproduction give only 101 genotypes among which selection may operate, the same number (at different loci) in an interbreeding species provide the potentiality for 2^{100} (or more than 10^{30}) different haploid genotypes and 3^{100} (or more than 10^{47}) diploid genotypes. Taking into account the rates at which many mutations are known to occur, the estimates of the number of loci (thousands in higher organisms) and the possibility of multiple alleles at each locus, it would appear that there should be unlimited variability in any moderately numerous species. Actually, most freely interbreeding populations present a sufficiently uniform appearance to be capable of representation, adequate for many purposes, by a single type specimen. There are usually, to be sure, rare variants deviating from the type in the same way as laboratory mutants, and in some cases there is conspicuous polymorphism. These, however, do not seem to present a field of variability at all comparable to that suggested above. On the other hand, it seems to be the rule that all quantitative characters vary and that these variations are in part hereditary. If, as usually seems to be the case, these variations depend on multiple factors each of which has only a very slight effect, the balancing of the effects in individuals keeps apparent variability low even though no two individuals have the same genotype. The field of potential quantitative variability through which the species may work its way without additional mutations may however be enormous. Moreover, compound quantitative variation, beyond a certain threshold, passes into apparent qualitative variation.

In emphasizing the potentialities in a field of multiple minor factor differences, the author has never intended to deny the occasional great importance of more conspicuous unitary differences either in gene effects or chromosome patterns (Wright 1931, 1940 *a, b*). Practically all types of viable mutational change observed in the laboratory appear to have been involved in the genetic differentiation of species, judging from the results of cytogenetics and from studies of species crosses. Most of these studies, however, indicate that accumulations of vast numbers of unitary changes with individually slight effects have played the predominating role (Dobzhansky; Huxley; Sumner; Baur; Tedin; Harland; Hutchinson, Silow and Stephens; Clausen, Keck and Hiesey; Babcock, etc.).

At this stage in evolution there is a great variety of population structures with advantages and disadvantages, both with respect to current and to phylogenetic adaptation. In some forms, such as bacteria, many protozoa, rotifers, entomostraca, etc., uniparental reproduction is the rule and exchange of genes a relatively rare event. At the opposite extreme, reproduction may be practically exclusively biparental. In either of these cases, there may or may not be effective subdivision into partially isolated populations. There may only be a coarse subdivision into large subspecies, or partial isolation, if only by distance, may occur on such a fine scale that there is statistical differentiation of very small adjacent local populations.

Selection is a very effective process under prevailing uniparental reproduction, provided that conjugation and recombination occur often enough to give sufficient material for it to operate upon. The genotype as a whole, reproduced without change in the clone, is the object of selection.

Under exclusive biparental reproduction and random mating there is more variability since each individual may well have a unique genotype but going with this is the disadvantage that selection can operate only on the average effects of alleles. Genotypes, however successful they may be, are broken up at once by the reduction division. If gene effects were wholly additive with respect to selective value, this selection by mere average gene effect would be no disadvantage but it is clear that interaction effects are very common if not the rule among genes acting on the same character. Moreover the effects of genes are probably never restricted to single characters. Each is so woven into the whole developmental process that additive effects on the selective value of the organism as a whole are exceedingly unlikely. Mathematically, the array of selective values of genotypes must be thought of as a rugged one with multiple peaks (genotypes giving harmonious combinations of characters) separated by valleys. Selection in a random breeding population tends merely to bind the species to one peak, which is not likely to be the highest one.

In a subdivided species, intergroup selection operates, by means of differential growth of local populations and differential migration, on the whole genetic complex which, even more than the single genotype, is the object of selection most directly related to the success of the species as an entity.

As at the preceding stages, one must recognize a third level of action of selection in that among competing, non-interbreeding species. This competition not only has direct effects on the course of evolution within each species but, through extinction and differential rate of speciation, guides the course of evolution of life in general.

The aspect of evolution of the interbreeding species in which statistical consequences of the mechanism of heredity can most easily be analyzed is that of the random breeding population, whether this constitutes the whole species or merely a local, partially isolated race. The elementary evolutionary process here is change of gene frequency. The immediate factors in this process may be put conveniently into three categories, according to degree of determinacy (table 2) (Wright, 1948). First are the systematic pressures of mutation, selection, and immigration which are determinate in principle. Selection pressure, it may be noted, is defined to include

TABLE 2

MODES OF CHANGE OF GENE FREQUENCY (q).

I. *Immediate Change.*
   1. From systematic pressures (Δq determinate in principle.)
      a. Recurrent mutation.
      b. Intrapopulation selection.
      c. Recurrent immigration and crossbreeding.
   2. Fluctuations (δq indeterminate, $\sigma^2_{\delta q}$ and (q) determinate in principle.)
      a. In the systematic pressures.
      b. From accidents of sampling.
   3. From unique events (wholly indeterminate).
      a. Mutation favorable from first.
      b. Unique selective incident.
      c. Unique hybridization.
      d. Swamping by mass immigration.
      e. Unique reduction in numbers.

II. *Secular Change in System of Coefficients.*
   1. From causes within species (control by new peak genotype).
   2. From changes in external environment.
      a. In home territory.
      b. As that of new territory.

exhaustively all processes by which gene frequency may change in a directed fashion without physical change of the hereditary material (mutation) or introduction from without (immigration). It thus includes differential mating, differential fecundity, and differential emigration, as well as differential mortality. In the absence of a secular trend in external conditions, the systematic pressures tend to bring about a state of equilibrium rather than evolution.

Second are the effects of fluctuations in these pressures and of sampling in small populations (Wright, 1931, 1948). These cause deviations in gene frequencies which are indeterminate in direction but determinate in variance. The slight deviations in single generations are compounded by a stochastic process into wide deviations from equilibrium. The probability curve ($\varphi(q)$) describing the distribution of these deviations in single populations in the long run, or describing that among local populations subject to the same conditions, can be deduced from the systematic pressure (Δq) and the variance of the fluctuations ($\sigma^2_{\delta q}$). Thus for one pair of alleles it can be shown that $\varphi(q) = (C/\sigma^2_{\delta q})e^{\int(\Delta q/\sigma^2_{\delta q})dq}$. Formulae can be obtained for multiple alleles and for multiple loci under various assumptions.

The system of coefficients describing the determinate aspects of change of gene frequency may themselves be expected to undergo systematic changes in the course of time. It is convenient to distinguish two major categories: secular changes that trace to processes internal to the species (but not necessarily to the particular local population in question) and ones that reflect secular changes in the environment.

The modes of transformation of a species that is panmictic as a whole and one that is subdivided are compared in table 3. The strong tendency toward equilibrium in a panmictic species, assuming the population to be large, is most readily overcome by secular changes in the external environment It is possible however that a new mutation that is favorable from the first may appear and bring about adjustments in the system of other gene frequencies and thus evolutionary change under constant conditions (Fisher, 1930) but the frequency of such mutations must be very low. If the population is small, stochastic processes may occasionally lead to a new position of equilibrium. In very small completely isolated populations, such processes may, however, be expected to lead merely to random fixation of genes

TABLE 3

MODES OF TRANSFORMATION OF SPECIES

I. *Panmictic Species.*
  1. Transformation in the absence of secular environmental change.
    a. Systematic response to unique favorable mutation or hybridization.
    b. Shift to control by new adaptive peak after upset by extreme selective incident or extreme temporary reduction in numbers.
    c. Shift to control by new adaptive peak as a result of stochastic processes (small population).
    d. Degeneration from random fixation (very small population).
  2. Systematic response to selection pressure of secular environmental change.

II. *Subdivided Species* (many partially isolated local populations).
  1. Transformation in the absence of secular environmental change by local differences in rates of population growth and dispersion.
    A. Local differentiation in the absence of environmental differences.
      a. Diverse genotypes give same optimum of multiple factor character. Position of optimum fluctuates.
      b. Both homozygotes superior to heterozygote.
      c. Abundance of allele confers advantage.
      d. Responses to local non-recurrent events (favorable mutation etc.)
      e. Stochastic changes in all sufficiently neutral sets of alleles.
    B. Differentiation based on local environmental differences.
  2. Systematic response to secular environmental change, augmented by interpopulation selection.

and thus to degeneration and ultimate extinction (Wright, 1931).

In a subdivided population with varying sizes and degrees of isolation of local populations, there is much more likelihood of continuing evolution in the absence of secular changes in environment, and also greater capacity to respond rapidly to such changes, than in a homogeneous population of comparable total size. The basis is the differentiation of local populations, for which there are several distinct mechanisms, applying usually to different loci. The most important cause of differentiation is usually no doubt a difference in local environments, the response to which depends on the relation between the selection pressure at each locus and the swamping effect of immigration (Wright, 1931). There may also be differentiation at particular loci in the absence of environmental difference, from selection that favors whichever allele is locally more abundant (Wright and Dobzhansky, 1946) (e.g., genes determining recognition marks) or that favor homozygotes over heterozygotes (e.g., reciprocal translocations (Wright, 1940, 1941)). All of these have rather more tendency to split the species than to further its evolution as a single group. Fluctuations in conditions that result in fluctuations of the position of the optimum grade of quantitatively varying characters, must inevitably lead to the establishment of different genetic systems, without much character difference, among partially isolated populations (Wright, 1935). This sort of differentiation is of special importance as giving a great store of potential rather than actual variability, easily available whenever extensive systematic change is called for by changing conditions. Next a population may be jolted out of one position of near equilibrium to another by some unpredictable event. This may occur more easily in a small local population than in a large one. Finally the cumulative effects of sampling in sufficiently small and sufficiently isolated local populations may be expected to bring about differentiation at all loci at which the allelic effects are so nearly neutral that selection is not the dominating factor (Wright, 1931). The store of actual, as well as of potential, variability may obviously be expected to be enormously greater in a subdivided species than in a random breeding one of comparable size since different alleles may predominate in different localities, for any of the reasons listed.

These processes are significant for evolution of the species as a whole only as they give rise occasionally to superior *general* adaptations in particular localities. Such an event may be expected to be followed by relatively rapid growth of the population in question, greater than usual outflow of migration and the consequent grading up of adjacent populations. Linkage may be of some significance here. Within a panmictic population, all combinations tend to be formed at random in the long run, irrespective of linkage (Robbins, 1918; Geiringer, 1948). The existence of non-additive selective values (e.g., Ab and aB superior to AB and ab) interferes with this process thereby reducing apparent variability (*cf.* Mather, 1943) but in multiple factor cases in which the selection pressure on each locus is necessarily slight, the interference cannot be great (Wright, 1945). Thus incomplete linkage can rarely be important except for processes in which a few generations are significant. The hanging together of elements of an adaptive complex, during very rapid favoring selection, immediately following immigration may be such a case. Edgar Anderson has stressed this in connection with introgression from one species into another with which it occasionally hybridizes (Anderson, 1949) and it may also be of some importance in intergroup selection within a species.

The importance of population structure in the cleavage of species is more obvious than in connection with gradual transformation as a whole. Table 4 lists the probable major modes of origin of species, classified primarily according to whether the new species arises from one, or more than one, ancestral species. A secondary classification is based on the size of the population from which the new species takes its origin. This may be the whole parent species if there is merely the speciation by transformation, which has already been discussed. There may be splitting into more or less equal components either by a process of gradual transformation of subspecies into species by the development of reproductive isolation or by the extinction of intermediates in a chain of subspecies in which the ends have previously diverged so much that interbreeding does not occur on contact. The number of cases in which the extremes of a circular chain of subspecies live in somewhat different niches in the same locality without interbreeding testifies to the probable importance of this mechanism by which one species may give rise to two without any current genetic change whatever (*cf.* Osgood, 1909; Mayr, 1942).

TABLE 4

MODES OF ORIGIN OF SPECIES

I. *From Single Species.*
  1. Transformation as whole (table 3).
  2. Splitting into more or less equal populations.
     a. Subspeciation followed by selection for reproductive isolation.
     b. Subspeciation along a chain, followed by extinction of intermediates.
  3. Geographical isolation of colony, followed by divergent transformation.
  4. Partially isolating mutation followed by selection for complete reproductive isolation and by divergent transformation.
  5. Autopolyploidy followed by divergent transformation.
  6. Heteroploidy.

II. *From Hybridization.*
  1. Complete fusion of two species.
  2. Introgression into one or both parent species.
  3. Local transformation and isolation of hybrid population.
  4. Amphidiploidy.

In other cases the critical event may be the geographical isolation of a colony which does not differ from the main body of the species to any appreciable extent at the time. Speciation, however, can hardly be said to occur until divergent transformation has brought about consistent character differences and some mode of reproductive isolation that would tend to prevent interbreeding if contact should be reestablished. Such divergence may be expected to occur in the absence of appreciable environmental differences though it would of course be speeded up if differences exist.

Origin of species from single isolating mutations is probably uncommon (in the absence of hybridization). There are, however, probably cases in which autotetraploids, fertile in themselves, but producing almost sterile triploids in crosses with the parent form, have transformed into good species following the isolation (Müntzing, 1936). A partially isolating mutation, such as a reciprocal translocation with semisterile heterozygotes, may constitute a critical step toward speciation if the new pattern becomes locally fixed. This may happen relatively easily if there is the possibility of vegetative multiplication but can hardly occur under exclusive biparental reproduction except where there are many localities in which colonies of the species are continually becoming extinct to be started anew from single gravid females, or at least from very small numbers of migrants (Wright, 1940, 1942).

Hybridization may play a role in the origin of new species in various ways. It may perhaps on occasion lead to complete ultimate fusion of two apparently good species; two species become one. Anderson, as noted above, stresses the importance of introgression of small blocks of genes from one species into another, whereby, if reciprocal, two species give rise to two different ones. With sufficient local isolation, a zone of hybridization between two species with only partially overlapping ranges may lead to the formation of a hybrid species in the region of overlap: two species give rise to three. Finally, amphidiploidy (Winge, Müntzing) may undoubtedly give rise to a well differentiated new species, fertile in itself but producing sterile hybrids with its parent species. This mechanism seems to have been very important in the multiplication of species among flowering plants. Darlington and Moffett have presented cytologic evidence for the importance of heteroploidy either in an autopolyploid or an allopolyploid as an occasional mechanism producing less balanced but nevertheless viable new forms.

The adherents of the view that significant evolutionary steps arise only from single mutations with suitably great effects (Goldschmidt, 1940; Willis, 1940) lay great stress on paleontological and other evidence that seems to indicate an abrupt appearance of higher categories and relatively rapid working down from these to lower categories; and ultimately to species in which further evolution is restricted to mere multiplication of subspecies. It does indeed appear to be the case that evolution is not ordinarily a matter of subspecies tending gradually to become species, these gradually to become genera, these gradually to become families, etc. It is not, however, necessary to postulate mutations of specific, generic, familial, and ordinal rank. A species ordinarily tends to persist for very long periods in a state of equilibrium or at best, very gradual advance under ecological pressures that leave only one restricted niche in which there can be successful competition with other species. Yet such a species may all the time, especially if population structure is favorable, carry a head of steam in the form of a store of potential variability that permits very rapid change (the tachitely of Simpson (1944)) whenever any major ecologic opportunity offers (Wright, 1932, 1941*a, b*, 1945). The attainment of a general adaptation superior in a wide range of activities to any hitherto attained or an adaptation that opens up a new way of life may be the end result of a more or less gradual evolution but, once a certain threshold is passed, advance may become explosively rapid if there is a sufficient store of potential variability to draw upon. The same is the case if the species reaches unoccupied territory except that here it is not likely to arrive with a large store of potential variability. However, if a favorable population structure can develop after arrival it probably does not require much time, geologically speaking, to build up an extensive field of potential variability. Even in this case then, the ecologic situation rather than mutation rate is the limiting factor. The contrast between the extensive speciation of the Geospizidae in the Galapagos Islands, in contrast with the lack of such speciation on isolated Cocos as brought back by David Lack, seems to illustrate this point. Drastic alteration of the environment may eliminate many forms and thus give an opportunity for rapid expansion of a form that happens to be preadapted.

Perhaps the most important conclusion from statistical genetics is that neither the driving factor nor the actual limiting factor in evolution is ordinarily to be sought in the genetic situation (including the process of mutation) once the stage of the interbreeding species of multicellular organisms has been arrived at. The driving force is the universal tendency of life to expand and to seek out all opportunities. The limiting factor in each case is the ecologic pressure from other species and from the non-living environment, by which the species is ordinarily kept in place. The genetic situation in an interbreeding species with a finely divided population structure provides a virtually infinite field of potential variability which permits ready adaptation on the part of the species to fluctuating environmental conditions, ready phylogenetic adaptation to secular changes in conditions, effective exploration by trial and error processes of possibilities of breaking through the restraining pressures even under static conditions, and the rapid exploitation by similar means of any major ecologic opportunity. The result is the observed evolutionary cycle of occasional emergence of a new higher category, adaptive radiation, selective elimination of rival higher categories and return to slow more or less orthogenetic advance and mere multiplication of species, as opportunity becomes restricted to a single channel. A new break through may lead at rare intervals to a new cycle of greater or lesser scope.

The terms emergence and orthogenetic advance are used here, of course, merely as descriptive terms for two contrasting aspects of the evolutionary process not as explanations. It is the contention that the statistical consequences of the varying degrees of success of individual organisms in their efforts to live and reproduce give an adequate explanation for all known major phenomena of evolution when account is taken of the statistical effects of the known processes of heredity in populations of diverse structure.

TABLE 5

THE EVOLUTIONARY CYCLE

I. 1. *Emergence of a Higher Category* (as response to a major ecologic opportunity).
   a. Attainment of an adaptation that opens up an extensive new way of life.
   b. Arrival in territory in which many ecological niches, possible for the form, are unoccupied.
   c. Preadaptation to drastically altered environment, to which rival forms are not adapted.
2. *Adaptive Radiation* to progressively lower categories.
3. *Selective Extinction* of rival types.
4. *Apparent Orthogenesis*
   Restriction to a single possible major line of advance.
5. *Relative Stability*
   Equilibrium of genus or even species under pressures restricting form to a single ecologic niche.

II. *Rare Emergence of New Higher Category* and initiation of a new cycle at any stage of above. Degree in accordance with that of a new ecologic opportunity.

REFERENCES

ANDERSON, EDGAR. 1949. Introgressive hybridization. N. Y., John Wiley and Sons.

BABCOCK, E. B. 1949. The genus Crepis. Part I. Berkeley, Univ. of Calif. Press.

BAUR, E. 1932. Artungrenzung und Artbildung in dem Galtung Antirrhinum. *Zeit. ind. Abst. Vererb.* **63**: 256–302.

CLAUSEN, JENS, D. D. KECK, AND W. M. HIESEY. 1940. Experimental studies on the nature of species. I. Effects of varied environments on Western North American plants. *Carnegie Inst. of Washington Publ.* **520**: 1–452.

CLAUSEN, JENS, D. D. KECK, AND W. M. HIESEY. 1947. Heredity of geographically and ecologically isolated races. *Amer. Nat.* **81**: 114–133.

DARLINGTON, C. D., AND A. A. MOFFETT. 1930. Primary and secondary chromosome balance in Pyrus. *Jour. Genet.* **22**: 129–151.

DELBRÜCK, M., AND W. T. BAILEY, JR. 1946. Induced mutation in bacterial viruses. *Cold Spring Harbor Symposia on Quantitative Biology* **11**: 33–37.

DOBZHANSKY, TH. 1941. Genetics and the origin of species. N. Y., Columbia Univ. Press.

FISHER, R. A. 1922. On the dominance ratio. *Proc. Roy. Soc. Edinburgh* **42**: 321–341.

—— 1930. The genetical theory of natural selection. Oxford, Clarendon Press.

GEIRINGER, H. 1948. On the mathematics of random mating in case of different recombination values for males and females. *Genetics* **33**: 548–564.

GOLDSCHMIDT, R. 1940. The material basis of evolution. New Haven, Yale Univ. Press.

HALDANE, J. B. S. 1932. The causes of evolution. London, Harper.

HARLAND, S. C. 1936. The genetical conception of the species. *Biol. Rev.* **11**: 83–112.

HERSHEY, A. D. 1946. Spontaneous mutations in bacterial viruses. *Cold Spring Harbor Symposia on Quantitative Biology* **11**: 67–76.

HOROWITZ, N. H. 1945. On the evolution of biochemical syntheses. *Proc. Nat. Acad. Sci.* **31**: 153–157.

HUTCHINSON, J. B., R. A. SILOW, AND S. G. STEPHENS. 1947. The evolution of Gossypium. London, Oxford Univ. Press.

HUXLEY, J. S. 1942. Evolution. The modern synthesis. N. Y. and London, Harper.

LACK, D. 1947. Darwin's finches. Cambridge at the University Press.

LEDERBERG, J., AND E. L. TATUM. 1946. Novel genotypes in mixed cultures of biochemical mutants of bacteria. *Cold Spring Harbor Symposia on Quantitative Biology* **11**: 113–114.

MATHER, K. 1943. Polygenic inheritance and natural selection. *Biol. Rev.* **18**: 32–64.

MAYR, E. 1942. Systematics and the origin of species. N. Y., Columbia Univ. Press.

MÜNTZING, A. 1932. Cytogenetic investigations on synthetic Galeopsis tetrahit. *Hereditas* **16**: 105–154.

—— 1936. The evolutionary significance of autopolyploidy. *Hereditas* **21**: 263–378.

OSGOOD, W. H. 1909. Revision of the mice of the American genus Peromyscus. *North American Fauna (Bur. Biol. Survey)* **28**: 1–285.

ROBBINS, R. B. 1918. Some applications of mathematics to breeding problems. III. *Genetics* **3**: 375–389.

SIMPSON, G. G. 1944. Tempo and mode in evolution. N. Y., Columbia Univ. Press.

SUMNER, F. B. 1932. Genetic, distributional and evolutionary studies of the subspecies of deer mice (Peromyscus). *Bibl. Genet.* **9**: 1–106.

TEDIN, O. 1925. Vererbung, Variation und Systematik in den Galtung Camelina. *Hereditas* **6**: 275–386.

WILLIS, J. C. 1940. The course of evolution. Cambridge at the University Press.

WINGE, O. 1917. The chromosomes, their numbers and general importance. *C. R. Trav. Lab. Carlsberg* **13**: 131–275.

WRIGHT, S. 1931. Evolution in mendelian populations. *Genetics* **16**: 97–159.

—— 1932. The roles of mutation, inbreeding, cross breeding and selection in evolution. *Proc. 6th Internat. Congress Genetics* **1**: 356–366.

—— 1935. Evolution in populations in approximate equilibrium. *Jour. Gen.* **30**: 257–266.

—— 1940*a*. The statistical consequences of mendelian heredity in relation to speciation. *In* The New Systematics, 161–183 (ed. by J. S. Huxley), Oxford, Clarendon Press.

—— 1940*b*. Breeding structure of populations in relation to speciation. *Amer. Nat.* **74**: 232–248.

—— 1941*a*. The age and area concept extended. (A review of book by J. C. Willis.) *Ecology* **22**: 345–347.

—— 1941*b*. "The material basis of evolution" by R. Goldschmidt (review). *Scientific Monthly* **53**: 165–170.

—— 1941*c*. On the probability of fixation of reciprocal translocations. *Amer. Nat.* **25**: 513–522.

—— 1942. Statistical genetics and evolution. *Bull. Amer. Math. Soc.* **48**: 223–246.

—— 1943. Isolation by distance. *Genetics* **28**: 114–138.

—— 1945*a*. Tempo and mode in evolution: a critical review. *Ecology* **26**: 415–419.

—— 1945*b*. The differential equation of the distribution of gene frequencies. *Proc. Nat. Acad. Sci.* **31**: 382–389.

—— 1946. Isolation by distance under diverse systems of mating. *Genetics* **31**: 39–59.

—— 1948. On the roles of directed and random changes in gene frequency in the genetics of populations. *Evolution* **2**: 279–294.

WRIGHT, S., AND TH. DOBZHANSKY. 1946. Genetics of natural populations. XI. Experimental reproduction of some of the changes caused by natural selection in certain populations of Drosophila pseudoobscura. *Genetics* **31**: 125–156.

THE GENETICAL STRUCTURE OF POPULATIONS

By SEWALL WRIGHT, *University of Chicago*

Galton Lecture at University College, London, 1950

First let me acknowledge the great honour that I feel in being asked to give the Galton Lecture. The genetics of populations has two main supports, one of which traces to Galton. A geneticist cannot read *Natural Inheritance* (1889) without feeling that Galton would have welcomed Mendel's paper if it had come to his attention during his active career, and that he would have seen that it and the statistical approach which he initiated were complementary rather than antithetic. Personally, I wish to testify to the great stimulus that I had in reading *Natural Inheritance* as a student.

Random mating and inbreeding

I propose to discuss certain genetic aspects of population structure. The term is used to include such matters as numbers, composition by age and sex, and state of subdivision. The best starting point is the consideration of the situation in a large random-breeding population, in which structure in the last sense is absent.

Gene frequencies and hence genetic variability tend to remain unchanged in such a population, generation after generation, because of the persistence of genes and the symmetry of the Mendelian mechanism. This contrasts with the expected loss of half the variability in each generation (under blending heredity) which Darwin and his contemporaries recognized to be the greatest difficulty in the theory of natural selection. This principle was first expressed in general form in 1908 in independent papers by Hardy and Weinberg.

In the same year, another deduction from Mendelian principles—that self-fertilization should result in random fixation of half the heterozygous loci in each generation—was used by Shull to interpret the loss of vigour and the fixation of diverse combinations of quantitatively varying characters which he found in his selfed lines of maize.

Table 1 shows at the left the composition of a random-breeding population with respect to a pair of alleles. The array of zygotic frequencies is the square of the array of gene frequencies.

Table 1. *The frequencies of zygotes from a pair of alleles under three conditions: panmixia, an intermediate degree of inbreeding and complete fixation without change of gene frequency. The intermediate condition is expressed in three equivalent ways in terms of the inbreeding coefficient (or fixation index) F and the panmictic index P $(=1-F)$*

	Panmixia ($r_{es}=0$)	Intermediate ($r_{es}=F$)			Complete fixation ($r_{es}=1$)
		Deviation from panmixia	Panmictic and fixed components	Deviation from fixation	
Genotype	Frequency	Frequency	Frequency	Frequency	Frequency
$\mathbf{A_1A_1}$	q^2	$q^2+Fq(1-q)$	Pq^2+Fq	$q-Pq(1-q)$	q
$\mathbf{A_1a}$	$2q(1-q)$	$2q(1-q)-2Fq(1-q)$	$2Pq(1-q)$	$2Pq(1-q)$	—
aa	$(1-q)^2$	$(1-q)^2+Fq(1-q)$	$P(1-q)^2+F(1-q)$	$(1-q)-Pq(1-q)$	$(1-q)$
	1	$1+0$	$P+F=1$	$1-0$	1

Long-continued self-fertilization without selection gives the array of fixed lines, with gene frequencies unchanged, shown at the right. The other columns show an intermediate situation according to three useful points of view indicated in the headings.

It is important to note that the same coefficient, F, that measures the degree of approach toward fixation is also the Galtonian correlation coefficient r_{es} for the alleles that come together at fertilization. This is zero under random mating, 1 under complete fixation and F as the weighted average in the intermediate population.

This sort of description can be extended at once to sets of three or more alleles, not subject to selective differences, by using the principle that any group of alleles may be treated as if one, and the system can therefore be treated in a variety of ways as if it consisted of only two alleles (Table 2). The coefficient F still measures the degree of approach toward fixation, and P $(=1-F)$ measures the relative amount of heterozygosis as compared with that in the random bred population. The correlation between alleles in uniting gametes is still F, irrespective of gene frequencies or of the values assigned the alleles (Appendices A, B).

Table 2. *The frequencies of zygotes from a set of three alleles under three conditions: panmixia, an intermediate degree of inbreeding and complete fixation. The intermediate condition is expressed in terms of the inbreeding coefficient F and the panmictic index P $(=1-F)$*

	Panmixia ($r_{es}=0$)	Intermediate ($r_{es}=F$)	Complete fixation ($r_{es}=1$)
Genotype	Frequency	Frequency	Frequency
$\mathbf{A_1A_1}$	q_1^2	$Pq_1^2+Fq_1$	q_1
$\mathbf{A_1A_2}$	$2q_1q_2$	$2Pq_1q_2$	—
$\mathbf{A_1A_3}$	$2q_1q_3$	$2Pq_1q_3$	—
$\mathbf{A_2A_2}$	q_2^2	$Pq_2^2+Fq_2$	q_2
$\mathbf{A_2A_3}$	$2q_2q_3$	$2Pq_2q_3$	—
$\mathbf{A_3A_3}$	q_3^2	$Pq_3^2+Fq_3$	q_3
	1	$P+F=1$	1

Statistical properties of populations

The statistical properties of a population change with changes in zygotic composition even though gene frequencies are not changed (Appendix C). Consider first the case of exact semi-dominance. The mean is not affected, but the genetic variance of the population as a whole increases from its value under random mating ($\sigma^2_{T(0)}$) by the proportion F becoming $(1+F)\sigma^2_{T(0)}$. If the system of mating is one of subdivision into strains with internal random mating, the average variance within these falls off by the proportion F, becoming $(1-F)\sigma^2_{T(0)}$ and the variance of strain means becomes $2F\sigma^2_{T(0)}$.

If, on the other hand, there is any departure from semi-dominance, the mean shifts from its value, $m_{T(0)}$, under random mating toward a value $m_{T(1)}$ characteristic of an array of completely fixed lines and the amount of change is proportional to F. The decrease in size, fecundity and viability that are usually observed on inbreeding seem to have their basis in an association of recessiveness with deleterious effect.

In this case the variance of the population as a whole is a quadratic instead of linear function of F. Nevertheless, the coefficient F is more useful than any other single coefficient in describing the properties of the population, relative to those in a random-bred stock. It was for this reason that it was suggested that it would be the most suitable inbreeding coefficient (Wright, 1922*a*).

THE INBREEDING COEFFICIENT F

The rate of decrease of heterozygosis in systems of mating more complicated than self-fertilization was first worked out from the recurrence relation between successive generations independently by Jennings (1914) and Fish (1914) for brother-sister mating and by Jennings (1916) for some others. The present writer, who had assisted Fish in his calculations, found a simpler way of finding this quantity, the method of path coefficients, based on the correlation between uniting gametes (Wright, 1921) (Appendix A). The results, following Jennings, were expressed in this paper in terms of gene frequency $\frac{1}{2}$, but it was soon shown that the formulae applied to any gene frequency (Wright, 1922*b*). The quantity F was therefore proposed as an inbreeding coefficient giving 'the departure from the amount of homozygosis under random mating toward complete homozygosis' (Wright, 1922*a*). It has been used since as a measure of such departure relative to a specified foundation stock, not necessarily random bred.

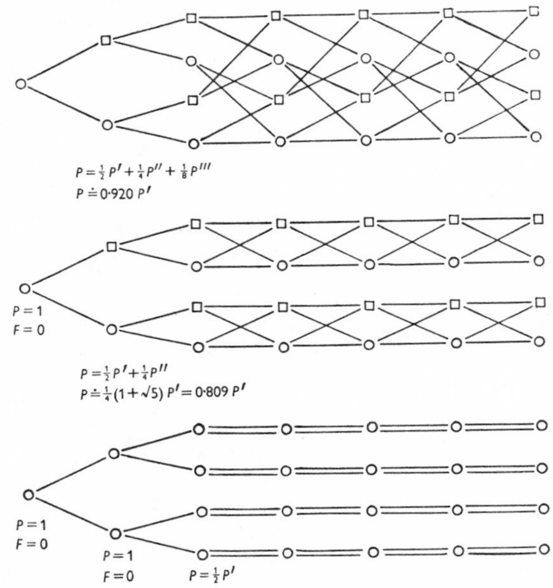

Fig. 1. Three systems of mating which agree in there being two parents and four in all more remote ancestral generations to the beginning.

The method used by Jennings and by Fish becomes too cumbersome to be practicable in dealing with systems much more complicated than mating of brother and sister. Bartlett & Haldane (1934) considerably extended its scope by using matrix algebra. The rates of decrease of heterozygosis indicated by this method have agreed in all comparable cases with those obtained by means of path coefficients (cf. Haldane, 1930, 1936, 1937, 1949; Wright, 1938). The methods are complementary. The matrix method gives a rather complete account of the history of the population in all respects. The method of path coefficients yields only one property (F) but can obtain this readily from systems which would require matrices with enormous numbers of elements.

Malécot (1948) has recently shown how the general formula for F, given by the method of path coefficients, can also be demonstrated directly from the theory of probability (cf. also Haldane, 1949).

F is not the only inbreeding coefficient that has been suggested. Pearl (1917) attempted to devise one that was independent of any theory of heredity. Unfortunately, it may have the same value under systems of mating that give the most diverse results experimentally (Fig. 1) (cf. Wright, 1923*a*, *b*). Bernstein (1930) suggested a coefficient α to describe departure from panmixia. It is identical with F. R. A. Fisher (1949) considers that attempts to set up coefficients of inbreeding are unsatisfactory for reasons which do not apply to F. He himself proposes 'an absolute measure of the amount of progress made in inbreeding' based on the matrix method. It turns out to be identical with $[-\log_e(1-F)]$ and is thus not an essentially independent coefficient.

Hierarchic structure

The method of path coefficients leads to simple general formulae for F in terms of its value in preceding generations. These are different for autosomal, sex-linked and polysomic loci (Appendix B). In the case of regular systems of mating, capable of being represented in diagrams, recurrence relations can usually be found very easily without reference to the general formulae (Wright, 1921, 1933*a*, 1938) (Appendix E). The relations between linked loci can be worked out by allied methods (Wright, 1933*b*). The effects of enforced heterozygosis (unpublished) agree with those of the matrix method as far as this has been carried (Bartlett & Haldane, 1935) and can be extended to large inbreeding populations.

The general formulae can be applied to irregular pedigrees. Fig. 2 shows the calculations from the pedigrees (not quite complete) of the bull Favourite (born 1793), on which the Shorthorn breed is essentially based, and of his son Comet (Wright, 1922*a*). The method has been applied to all of the sixty-four cows of Bates's famous Duchess strain (Wright, 1923*b*), of which Darwin

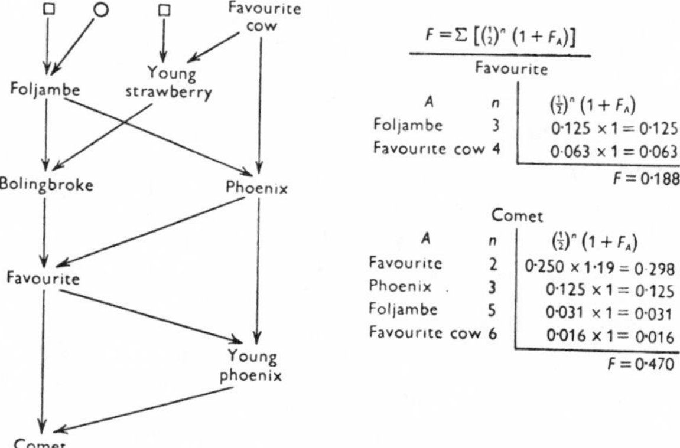

Fig. 2. Pedigree of the Shorthorn bull Comet (115) and his sire Favourite (252) and calculations of their inbreeding coefficients relative to the beginning of the Coates herd book (with minor exceptions that raise F to 0·192 for Favourite and to 0·471 for Comet).

(1868) stated: 'For thirteen years he bred most closely in-and-in; but during the next seventeen years, although he had the most exalted notion of his own stock, he thrice infused fresh blood into his herd. It is said that he did this not to improve the form of his animals, but on account of their lessened fertility.' Duchess 1 was a daughter of Comet and five generations in the straight female line from a foundation cow of the herd book. Duchesses 59 and 62 were eight generations later than Duchess 1 in the straight female line. This is about as far as one cares to go by analysis of full pedigrees. Fortunately, there is a simple sampling method (Appendix D), based on comparisons of single random lines back of sire and dam of each of a random selection of animals, which can be applied with appropriate standard errors to whole breeds (Wright & McPhee, 1925). Fig. 3 brings out the remarkably high values of F maintained by Bates throughout the whole history of the Duchess strain and the lower but still impressive figure for the breed as a whole at various periods (McPhee & Wright, 1925, 1926).

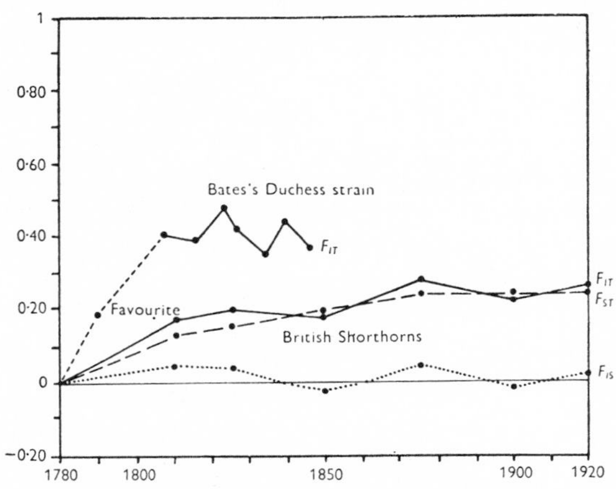

Fig. 3. The inbreeding coefficient (F_{IT}) of cows of Bates's Duchess strain and of the Shorthorn breed as a whole at various dates, relative to the beginning of the Coates herd book. The inbreeding (F_{ST}), relative to this foundation stock, that would persist if random mating were initiated, and the current inbreeding (F_{IS}) are also shown.

It has probably occurred to the reader that the coefficient of inbreeding may mean very different things in different cases. (1) There may be division of the population into completely isolated small strains, within each of which there is random mating. The inbreeding coefficient of individuals relative to the total is here due merely to the relationship of all members of the same strain and disappears at once with random mating among strains. (2) There may be frequent mating of close relatives but no permanent separation of strains. Here again random mating at once reduces the inbreeding coefficient to zero. (3) The sires used may be rather limited in number and derived from even more limited numbers of grandsires and great-grandsires. In this case there may be little apparent close inbreeding at any time, contrary to (2), and no division into strains, contrary to (1), but the value of F, relative to the foundation stock, keeps rising and cannot be much reduced by random mating. There are other possibilities. Clearly we need something more than a single value of F to give an adequate description of structure.

In the case of the pairs of random paternal and maternal lines of random Shorthorns, it was easy to find out what would happen if there were random mating merely by matching these lines at random instead of matching those back of single actual animals. It came out that in some years the artificial F was slightly greater, in others less, than the actual F. We concluded that the breed as a whole in the later years was much inbred relative to the foundation stock of about 1780 (case 3). The amount of current inbreeding must have been low in the later years. Just what it was at each period can be deduced from a seemingly rather different situation.

Suppose that a population is divided on any basis into subpopulations. The average value of P for individuals relative to their substrains is given in principle by the ratio of average actual heterozygosis within strains to that expected from internal random mating. Represent this by P_{IS} and the panmictic index for individuals relative to the total by P_{IT}. The corresponding inbreeding coefficients are F_{IS} and F_{IT}. It has been shown (Wright, 1943*a*) that the correlation between random gametes, drawn from the same subpopulation, relative to the total, is given by the formula $F_{ST}=(F_{IT}-F_{IS})/(1-F_{IS})$. This can be expressed more conveniently in terms of P's (Wright, 1948*c*): $P_{IT}=P_{IS}P_{ST}$. If there are primary subpopulations (S_1) that are themselves subdivided (S_2), $P_{IT}=P_{IS_2}P_{S_2S_1}P_{S_1T}$. This sort of analysis can be continued to any degree of hierarchic subdivision.

Returning to the Shorthorns, we may look upon the whole breed of a later period as one of an indefinitely large number of isolated strains which might conceivably have been derived from the foundation stock of 1780 by the average mating system actually followed. The inbreeding coefficient of individuals relative to the foundation stock may thus be treated as of the type of F_{IT}. That from matching random lines of ancestry is of the type F_{ST}. The amount of current inbreeding (F_{IS}) can then be estimated as above. The formula actually used for this purpose in

Table 3. *The sets of values of F_{IS}, F_{ST} and F_{IT}; P_{IS}, P_{ST} and P_{IT} under extreme types of population structure*

Total population (T) Subpopulation (S) Individual (I)	Fixation indices			Panmictic indices		
$P_{IT}=P_{IS}P_{ST}$	F_{IS}	F_{ST}	F_{IT}	P_{IS}	P_{ST}	P_{IT}
(1) Each S composed of multiple fixed lines:						
(*a*) Only slight differentiation among S's	1	0+	1	0	1−	0
(*b*) Extreme differentiation among S's	1	1−	1	0	0+	0
(2) Each S panmictic:						
(*a*) Only slight differentiation among S's	0	0+	0+	1	1−	1−
(*b*) Extreme differentiation among S's	0	1−	1−	1	0+	0+
(3) Each S a single genotype:						
(*a*) Perfect negative correlation between alleles	−1	0	−1	2	1	2
(*b*) Random combinations of alleles	−1	½	0	2	½	1
(*c*) Strong positive correlation between alleles	−1	1−	1−	2	0+	0+
(*d*) Perfect positive correlation between alleles (all genotypes homozygous)	—	1	1	—	0	0

the 1925 paper was not correct, and the published results were slightly in error. The corrected values are shown in Fig. 3 in the dotted line.

There was in general a little current inbreeding (F_{IS}) responsible for the accumulation of considerable inbreeding of the breed as a whole (F_{ST}), but at certain periods (1850, 1900) there seems to have been a little tendency toward actual outcrossing within the breed.

Lush (1943) and co-workers and others have calculated the inbreeding coefficients of many breeds of cattle, horses, sheep and swine at different periods of their histories. None of these goes back to a recorded foundation stock in which there was as much inbreeding as in the foundation Shorthorns, but all show a gradual rise in the coefficient, which indicates that the effective number at any time was of the order of 100 or at most a few hundred, a very small proportion of the total numbers registered in each generation.

F_{ST} is necessarily positive but F_{IS} and F_{IT} can be negative. Table 3 shows sets of values that apply to extreme patterns. It is evident that specification of a set of F's or P's gives a more complete description of structure than possible from any one coefficient.

Natural populations

Realistic discussion of the structure of natural populations requires simultaneous consideration of all processes by which gene frequencies may change. These may be put into categories according to the degree of determinacy. There are, first, those which tend to bring about directed changes (Δq) according to some definite function of the gene frequencies. These can be classified exhaustively into recurrent mutation, immigration and selection. Collectively these lead either toward fixation of one allele or to a state of equilibrium between two or more alleles at which $\Delta q = 0$. Second are fluctuations (δq) which are indeterminate in direction but determinate in variance ($\sigma^2_{\delta q}$). These include fluctuations about the mean values of the coefficients for the steady processes and fluctuations due to accidents of sampling in the parentage of each generation. These variances tend to be cumulative with respect to the array of possible gene frequencies and thus to lead toward random fixation. This tendency is, however, balanced by the tendency toward equilibrium due to the steady processes. The resultant is a probability distribution of frequencies of gene frequencies which applies to any one strain in the long run, or to an array of strains, subject to the same conditions, at any one time. The formula for one pair of alleles is a relatively simple function of Δq and $\sigma^2_{\delta q}$ (Wright, 1938, 1939, 1949a):

$$\phi(q) = (C/\sigma^2_{\delta q}) \exp\left[2\int (\Delta q/\sigma^2_{\delta q})dq\right].$$

It makes for clarity of thought to recognize a third category, changes in gene frequency that are best treated as indeterminate in both mean and variance; mutations that occur very rarely in the history of a species and similarly rare hybridization, unique selective incidents, etc.

Fig. 4 makes a comparison of distributions of gene frequencies that differ because of different amounts of fluctuation in a selection coefficient or because of different effective population numbers. We shall confine our attention to the latter here because of a close relation to the inbreeding coefficient F, but we note that there is a rough similarity in their effects that makes it possible to transfer general conclusions from one to the other (Wright, 1948a).

Fig. 5 shows the distribution of gene frequencies at different loci in local populations of a given effective size and subject to certain other conditions which are the same within each case.

Equilibrium at $q=0{\cdot}50$ due to equal superiority of heterozygote over both homozygotes and equal mutation rates in both directions is assumed in one case. There are only slight departures from equilibrium if the selective advantage is relatively strong, much random drift into fixation if the selective advantage is sufficiently weak. In the other case, selection tends to drive one allele into fixation, while immigration tends to reduce its frequency to 0·10. Selection tends to dominate the situation at those loci in which the selective advantage is relatively strong, while low frequencies and loss are the rule where selection advantage is weak. The following discussion is largely restricted to type or near-type alleles among which selective differences are negligible. We are concerned here with the effect of population structure on the genetic background of alleles related to quantitative variability, rather than with major differences that are dominated by selection.

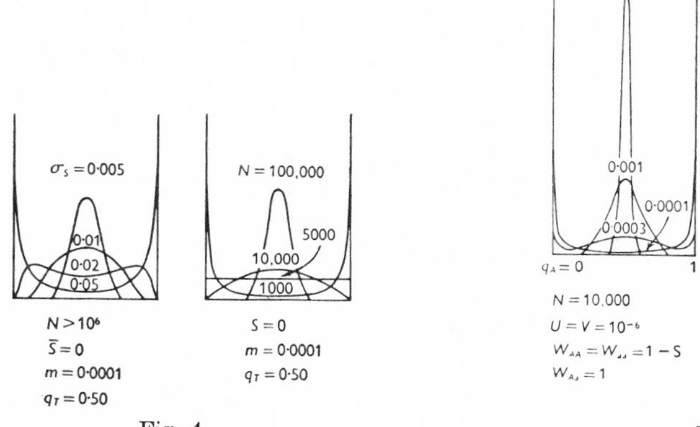

Fig. 4. Distribution of gene frequencies under two sets of conditions. In both, $\Delta q = -m(q-0{\cdot}5)$, where m is the amount of replacement by immigrants with gene frequency $q_T = 0{\cdot}50$. In those to the left, random drift is due only to fluctuations in selection, σ_s,

$$\sigma^2_{\delta q} = \sigma^2_s q^2(1-q)^2, \quad \phi(q) = Cq^{-2}\,(1-q)^{-2}\exp\{-m/[\sigma^2_s q(1-q)]\}.$$

In those to the right, fluctuations are due to small numbers, N,

$$\sigma^2_{\delta q} = q(1-q)/2N, \quad \phi(q) = C[q(1-q)]^{2Nm-1} \quad \text{(Wright, 1948}a\text{)}.$$

Fig. 5. Distributions of gene frequencies for loci which differ in the selective values of zygotes. In all cases random drift is due merely to small numbers, $N = 10{,}000$, $\sigma^2_{\delta q} = q(1-q)/20{,}000$. In those to the left, selection favours heterozygotes to varying extents (s), and there is mutation in both directions at equal rates ($u = v$), W is the selective value

$$\Delta q = -[2v + 2sq(1-q)][q - 0{\cdot}5], \quad \phi(q) = C[q(1-q)]^{-0{\cdot}96}\exp\,[40{,}000sq\,(1-q)].$$

In those to the right, selection at various rates is opposed by immigration, $m = 0{\cdot}0001$,

$$\Delta q = sq(1-q) - m(q - 0{\cdot}10), \quad \phi(q) = Cq^{-0{\cdot}6}\,(1-q)^{2{\cdot}6}\exp\,[40{,}000sq].$$

In the absence of selection, the mean gene frequency is the same, given a certain Δq, irrespective of the value of $\sigma^2_{\delta q}$. The variance of the distribution of gene frequencies is given by the formula $\sigma^2_q = q_T(1-q_T)F$ (Wright, 1943a) as well as by the formula $\sigma^2_q = \int_0^1 (q-\bar{q})^2\phi(q)dq$. These formulae connect the two modes of attack on questions of population structure. The former gives additional significance to the coefficient F as a measure of the variance of a neutral gene frequency among strains (σ_q^2), relative to the limiting value, $q_T(1-q_T)$ under complete fixation at the same gene frequency as in the total population.

THE ISLAND MODEL OF STRUCTURE

The mathematically simplest model for a heterogeneous natural population is that of an array of island populations of effective size N, largely isolated, but each replenished to the extent m from immigration, representative of the species as a whole (Wright, 1931, 1943*a*). By the method of path coefficients $F=(1-m)^2\left[\frac{1}{2N}+\left(1-\frac{1}{2N}\right)F'\right]$, where the prime indicates the preceding generation. In this case, F approaches a certain limiting value, other than 1. Putting $F=F'$

$$F=(1-m)^2/[2N-(2N-1)(1-m)^2],$$

$$F=\frac{1}{4Nm+1} \quad \text{approximately if } m \text{ is small.}$$

In the other mode of attack, we substitute $\Delta q=-m(q-q_T)$, the measure of the steady effect of immigration, and $\sigma_{\delta q}^2=q(1-q)/(2N)$, the sampling variance in one generation, in the general formula for $\phi(q)$:

$$\phi(q)=\frac{\Gamma(4Nm)}{\Gamma(4Nmq_T)\,\Gamma[4Nm(1-q_T)]}q^{4Nmq_T-1}(1-q)^{4Nm(1-q_T)-1},$$

$$\bar{q}=\int_0^1 q\phi(q)dq=q_T,$$

$$\sigma_q^2=\int_0^1(q-\bar{q})^2\phi(q)dq=\frac{q_T(1-q_T)}{4Nm+1},$$

$$F=\frac{1}{4Nm+1} \quad \text{in approximate agreement.}$$

Fig. 6 shows the distribution $\phi(q)$ for $q_T=\frac{1}{2}$ and various values of F. It is important to note that if F is as small as 0·05, the amount of differentiation in gene frequency with respect to neutral loci is considerable. Appreciable random fixation begins at $F=0{\cdot}33$.

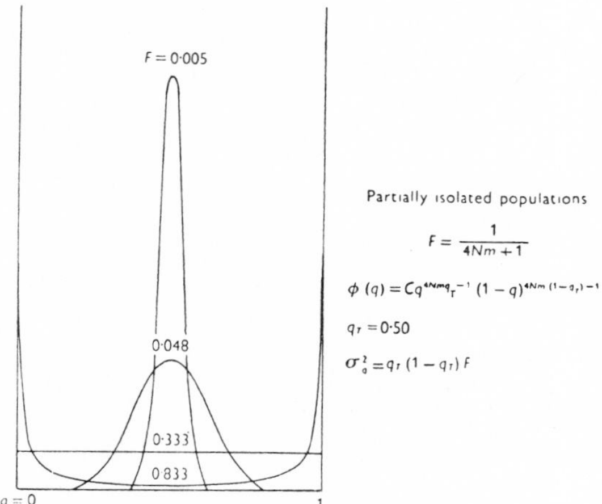

Fig. 6. The relation between variation of gene frequencies among island populations and the value of F in the total array due to the partial isolation of these. The mean gene frequency of the species, q_T, is assumed to be 0·50.

In an actual species, there may be a large central core of population in which gene frequency remains substantially constant, surrounded by small almost isolated populations with a great deal of differentiation in each element of genetic background.

Isolation by distance

Another useful model is that of a population of uniform density and a uniform and highly restricted amount of dispersal from each locality. The properties turn out to depend primarily on the size of neighbourhood, i.e. on the effective population number in an area from which the parents may be assumed to be drawn at random (Wright, 1940, 1943*a*, 1946). It appears that this is approximately the effective number in a circle of radius twice the standard deviation of the distribution of parent relative to offspring in one direction, largely irrespective of the form of this distribution curve (which seems usually to be leptokurtic; Dobzhansky & Wright, 1943, 1947; Bateman, 1947). We assume here that the variance of the distribution of grandparents is twice that of parents, and in general that the variance of ancestors of generations X is X times that of parents. These ancestors are thus drawn from an effective population of size XN_N.

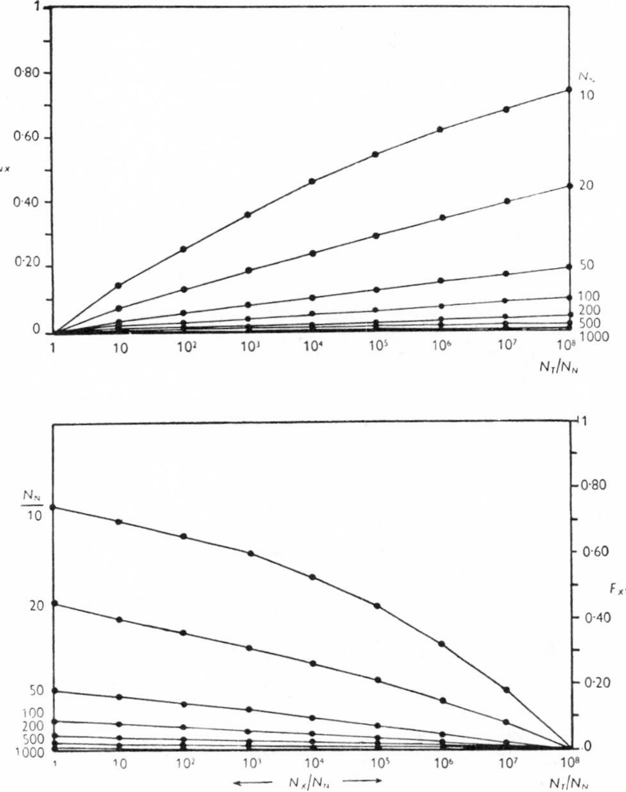

Fig. 7. Above, inbreeding coefficients (F_{NX}) of neighbourhoods of various sizes (N_N), relative to more comprehensive populations (N_X), and below, those (F_{XT}) of populations of various sizes (N_X), relative to a given total ($N_T = 10^8 N_N$) where there is continuity over an indefinitely large area. Abscissae in both: N_X/N_N.

The value of F_{NX}, measuring the inbreeding of neighbourhoods relative to a population of specified size N_X, is readily found by the method of path coefficients (Appendix F). Fig. 7 (top) shows how F_{NX} rises as N_X is taken larger. This may be interpreted, as noted, as measuring the amount of differentiation (at neutral loci) of neighbourhoods with respect to more comprehensive populations.

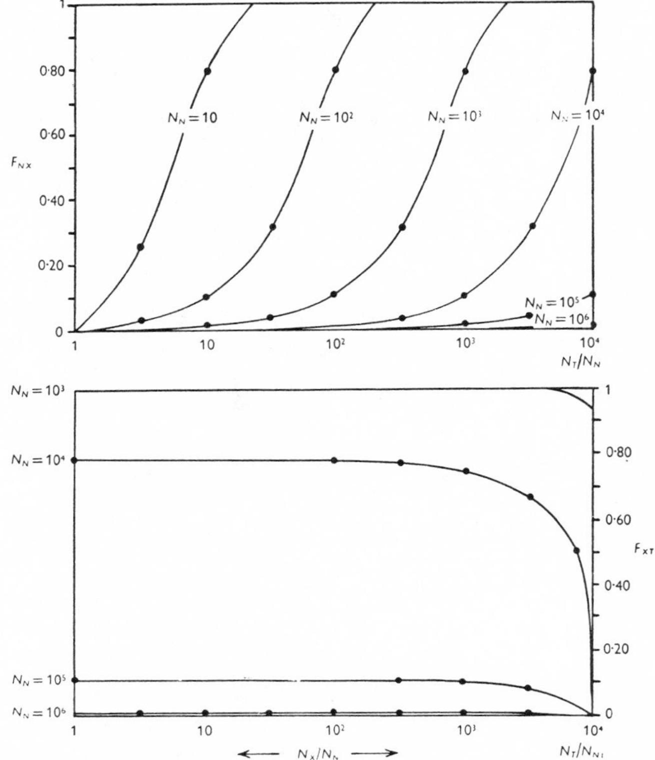

Fig. 8. Similar to Fig. 7 except that the range is along a single line instead of over an area. The total population considered, N_T, is merely $10^4 N_N$.

A better description of population structure can, however, be obtained by finding the amount of differentiation of populations of various specified sizes (N_X) within a total that is a given multiple of the neighbourhood. This depends on calculation of F_{XT} from F_{NX} and F_{NT} by the rule for hierarchic coefficients. Values are shown for increasing N_X relative to $N_T = 10^8 N_N$ in the lower part of Fig. 7.

It may be noted that differentiation of neighbourhoods carries with it considerable differentiation of much larger areas. This implies that there is a correlation between gene frequencies of neighbourhoods which falls off as their distance apart increases. Malécot (1948) has approached the problem of isolation by distance in a continuum by a different mathematical method and has made the determinations of the above correlation his primary objective. Returning to the method of attack by path coefficients, we may note that this correlation is given approximately by the ratio F_{XT}/F_{NT} for a distance of the order of $\sqrt{(N_X/N_N)}$ times the radius of a neighbourhood, and is thus the ratio of the ordinate at N_X/N_N to that at 1 in the lower part of Fig. 7.

One of the most important points that is brought out is that random local differentiation is slight under this model unless the effective number in neighbourhoods is decidedly small. Differentiation is very great if N_N is of the order of 20, not negligible if of the order of 200, but there is almost the equivalent of universal panmixia if it is as large as 1000.

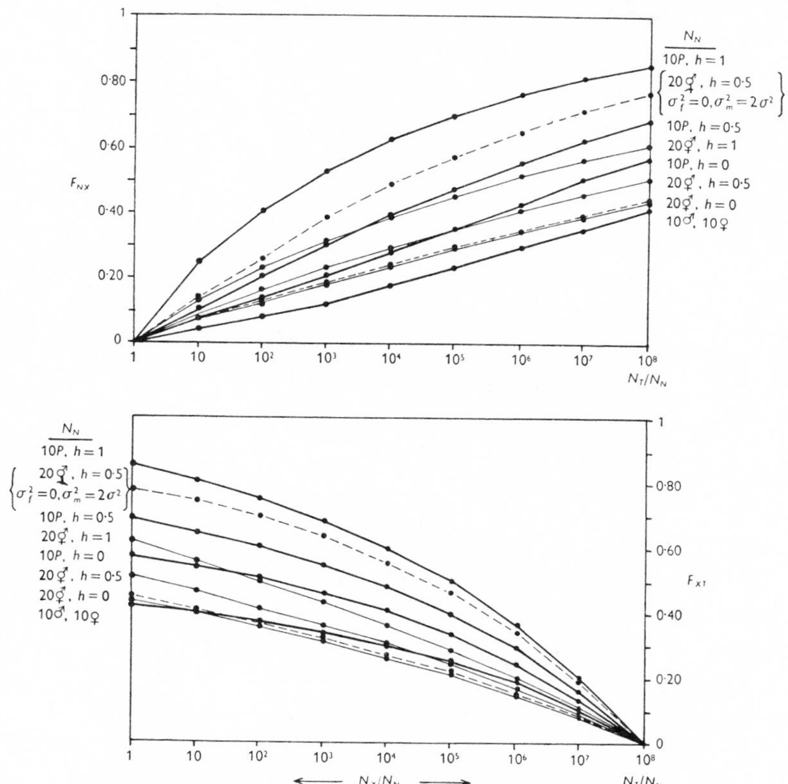

Fig. 9. Inbreeding coefficients F_{NX} and F_{XT} as in Fig. 7 but all with $N_N = 20$. Various systems of mating are compared (10♂, 10♀ mating at random, 20 monoecious individuals (20⚥)) with varying amounts of self-fertilization ($h = 0$, 0·50, 1), ten permanent pairs (10P) with varying amount of brother-sister mating ($h = 0$, 0·50, 1) and 20⚥ with no dispersal of ovules, double dispersion variance of pollen, either with merely random self-fertilization (broken line nearly same as 20⚥ $h = 0$) or with a little more than 50% self-fertilization ($r = 0{\cdot}5$, $h = 0{\cdot}525$).

In the case of differentiation along a linear range, on the other hand, the ancestors of generation X came from a population of effective size $\sqrt{X N_N}$ instead of $X N_N$. Marked differentiation occurs with relatively large neighbourhoods (Fig. 8).

The hypothesis on which all of the preceding figures are based is the arbitrary but mathematically simple one of random union of gametes from N_N monoecious individuals. There is no appreciable difference if self-fertilization is excluded ($h = 0$). The effects of other systems of mating are compared in Fig. 9 (area continuity, $N_N = 20$). The amount of differentiation increases considerably with increase in the amount of self-fertilization ($h = 0{\cdot}5$, $h = 1$). In the preceding cases equal dispersion of the sexes is assumed. If there is no appreciable dispersion of one sex, but dispersion of double the variance in the preceding cases, in the case of the other sex and if self-fertilization is no more than random, the result is practically as above, but with increased self-fertilization there is disproportionate increase in differentiation. With separate sexes and random mating (10♂, 10♀ per neighbourhood) the effect is again rather similar to that from random union of gametes of twenty monoecious organisms. If this same number of individuals is in permanent pairs (10P) instead of mating at random, differentiation is greater.

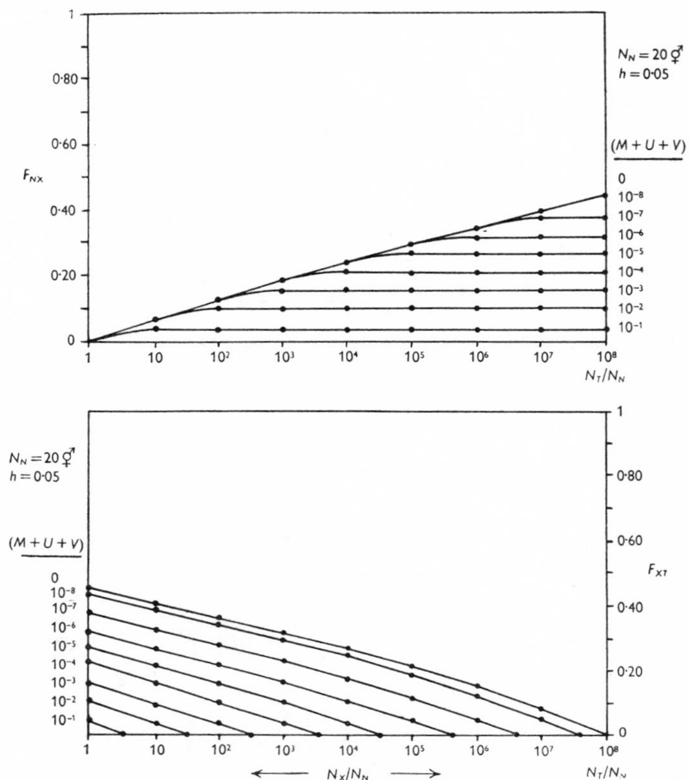

Fig. 10. Inbreeding coefficients F_{NX} and F_{XT} with $N_N = 20$, random union of gametes, but with universal dispersion (m) or reversible mutation (u, v) preventing complete fixation.

All of the cases discussed so far apply only when mutation rates are very low (less than 10^{-8} with $N_T/N_N = 10^8$), and there is no mode of occasional dispersal over the whole range of the species in one generation. It is important to consider what happens if these conditions do not hold. Fig. 10 shows calculations from random union of gametes in neighbourhoods of size 20, but mutation or universal dispersion displacing all populations to extents ranging from 10^{-1} to 10^{-8}. It may be seen that F_{NX} no longer approaches 1 asymptotically, but rather abruptly approaches a certain upper limit in each case. Correspondingly there is a definite limit to the size of population which shows any differentiation as a result of the differentiation of neighbourhoods. These are the results which can be most directly compared with those of Malécot. His final formula for the correlation as a function of distance is in a different mathematical form (involving Bessel functions) and does not involve size of neighbourhood. Empirically this correlation, as the ratio

F_{XT}/F_{NT}, seems to be almost independent of N_N in my results. I am indebted to Mr W. S. Russell for making calculations at three values of m (10^{-2}, 10^{-4}, 10^{-6}) from Malécot's formula. The results are in substantial agreement with those from F_{XT}/F_{NT}.

We may note here that if the values of F are small, systematic pressures (Δq) involving selection may be treated approximately in the same way as those involving m by using the best

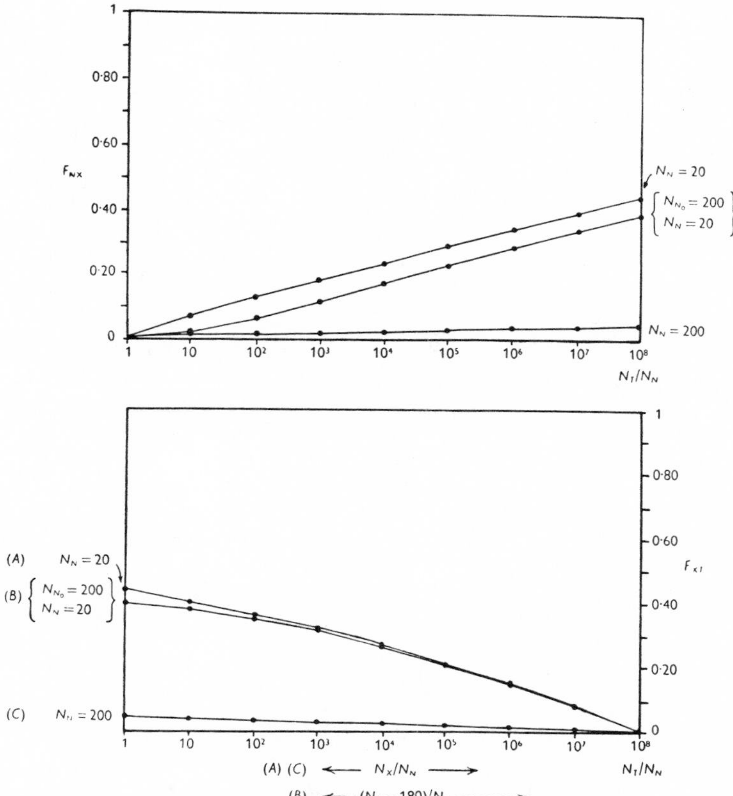

Fig. 11. Inbreeding coefficients F_{NX} and F_{XT} with random union of gametes but with three conditions with respect to numbers in neighbourhoods and to ancestral population numbers. In the highest curve in each case there is homogeneity with respect to density and dispersion and the number in a neighbourhood is 20. In the lowest curve in each case there is also homogeneity with respect to density and dispersion but N_N is 200. In the intermediate curve in each case, the neighbourhoods considered have the number 200 but because of heterogeneity in the conditions, the numbers in ancestral generations increase only by 20's.

linear expression for Δq near the equilibrium point (Wright, 1943*a*). Diversity in degree and direction of selection among localities is of course a wholly different matter and may bring about great differentiation if not overbalanced by migration (Wright, 1931, 1943*a*).

Actual populations are likely to show great heterogeneity in density and amount of dispersion. Some appreciation of the effects can be obtained by modifying the mathematical model. Assume that there are scattered relatively populous neighbourhoods with effective number MN, but that the ancestors of generation X are drawn from a population of $(M+X-1)N$ instead of

XMN or XN. The parents in such localities are thus drawn from relatively large populations, and the populations from which ancestors are drawn increase linearly with number of generations but at a low rate. Fig. 11 makes comparisons for $MN = 200$ but N only 20, with those for $N_N = 200$ and $N_N = 20$. The amount of differentiation in the heterogeneous case is much like that with N_N equal to 20 except for a lag.

It will probably be difficult to find actual cases in nature that correspond closely to these ideal cases. The primary difficulty is that data can only be obtained for cases in which there are conspicuous differences in the effects of alleles, which are cases in which selection is likely to be the dominating factor. Fig. 12, however, presents a possible case based on observed variation in flower colour (blue and white) of a small plant, *Linanthus Parryae* in an area of some 80 × 10 miles along the piedmont of the San Gabriel and San Bernardino mountains, which it occupies rather uniformly. 1258 samples of 100 plants each, distributed in a systematic way in this area, were examined by Epling & Dobzhansky (1942). The mode of inheritance is unfortunately not known, but estimates of effective size of neighbourhood on the hypothesis of no selection differed little (about 15–25) whether blue is recessive (assumed in the curves shown) or dominant or dependent on multiple factors and a threshold. Whatever the mechanism of differentiation it appears that there was marked differentiation at all levels, including not only that among six primary divisions but also of secondary and smaller divisions within the primary divisions. The curves indicate at least roughly the type of population structure discussed here (Wright, 1943*b*).

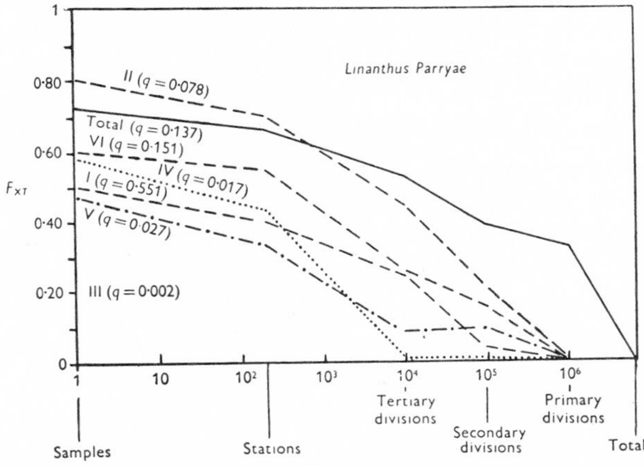

Fig. 12. Inbreeding coefficients (F_{XT}) of areas of various sizes in the total population and in the six primary subdivisions of this in *Linanthus Parryae* under the assumptions in the text.

Population structure in evolution

In considering the significance of population structure in evolution, we must distinguish two aspects of the process: multiplication of species and transformation. Partial isolation obviously tends toward the splitting off of a new species if the environmental conditions are markedly different. The thesis which I wish to stress, however, is that a fine-scaled structure of partial isolation, without marked environmental differences, presents the most favourable condition for transformation as a single species (Wright, 1931, 1932 and later).

The advantages over a panmictic population of comparable total size are of several sorts. In the first place, the subdivided population maintains more alleles at each locus and more at moderately high frequencies. In a panmictic species there is a tendency toward the establishment of one type allele at each locus. Possible favourable mutations at two or more removes (i.e. mutations of mutations) have little chance of occurrence. There are, to be sure, ways in which two or more alleles may be kept at subequal frequencies (selective advantage of heterozygotes, or advantage of mere rarity), but these are poor compared with those in a subdivided species. In the latter, different alleles may easily be maintained as the types in different places by selective advantages in relation to different conditions, as already indicated, and up to a certain point this favours adaptability of the species as a whole rather than splitting. Of more importance for transformation as a whole is local differentiation with respect to genotypes that give adaptation to the same conditions. The best adapted form in a species is usually one that is close to the average in all quantitatively varying characters. There are certain to be a great many genotypes that are of approximately this same optimal type except for pleiotropic genic effects that are of less selective significance at the time. In a panmictic population one of these genotypes inevitably gets a lead over the others and tends to become fixed. In a subdivided population, local fluctuations in conditions that shift slightly the position of the optimum insure the predominance of different genotypes of essentially the same phenotype in different places (Wright, 1935 *a*, *b*). This may indeed be brought about in sufficiently small local populations by the accumulation of sampling accidents without any fluctuations in conditions. This process, moreover, brings about differentiation in all other loci in which there are two or more type or near-type alleles with differences in selective value that are below a certain critical ratio to the reciprocal of the effective local population number (Wright, 1931). It may be well to note here that the effective population number may be expected to be much less than the apparent one for a variety of reasons (Wright, 1931, 1940).

The kaleidoscopic shiftings in the sets of gene frequencies of neighbourhoods may be expected to lead occasionally to predominance of combinations with selective advantages under the prevailing conditions that are far in excess of that indicated by the sum of the net effects of the components. Control by selection thereupon supersedes control by stochastic processes and immigration. Here we come to the second great advantage of a subdivided population over a panmictic one. Favourable combinations may spread through the species by a selective process that operates on the whole instead of merely on the separate net effects of the components.

The conditions for such a spread are more favourable in a continuum with isolation merely by distance than under the island model. In the former, centres in which a favourable combination happens to have appeared increase in population and contribute more than their share of migrants to adjacent regions. Because of the strong correlation in all gene frequencies, these adjacent regions do not need much immigration of this sort to carry them beyond the critical point at which selective increase of the combination becomes autonomous. The combination thus spreads through the species in concentric circles without any appreciable mass displacement of populations. Favourable combinations in other respects may arise at other centres and spread similarly, and the two circles of spread may overlap and cross without much interference. On the other hand, combinations that increase adaptation to the same conditions, but which differ genotypically, interfere, and the superior one ultimately displaces the other unless some aspects of both lead to a combination that is better than either, which thereupon spreads concentrically from its point of origin. With fine scaled structure, new differentiation may be expected to occur concomitantly with the spreading of combinations.

Any number of genes may contribute to a favourable new combination. Some may play an essential role and others that of minor modifiers. An important case is that of a single major mutation that has the potentiality of great advantage if it can be combined with an array of modifiers that bring its effects into harmonious adjustment with other characters. The importance of homoeotic mutations in evolution has been much stressed by Goldschmidt (1940). Such mutations usually have low penetrance and may thus be carried by a species and diffused through it at appreciable gene frequencies in spite of severe antagonistic selection when manifested in their pristine raw form. There is little chance of reaching adjustment in the face of this selection in a panmictic species. In a finely subdivided one, there is a good chance that somewhere, at some time, a combination of modifiers will be encountered that gives sufficient adjustment to reverse the direction of selection locally and thus create a centre from which the system, homoeotic mutation and modifiers can spread as a unit through the species.

Ecologic opportunity

It must be recognized that most species are held to such restricted ecologic niches by the pressure of other species that only an increase in specialization for this niche is probable even with the most favourable population structure. If, however, a more extensive ecologic opportunity is ever presented, a finely divided structure is that which is most favourable for its rapid exploitation (Wright, 1949*a*, *b*). Such an opportunity arises when migrants reach territory in which there are many unoccupied niches in which they can live. It also arises when a permanent change in conditions eliminates many species but leaves the relatively pre-adapted survivors to take over the vacant niches. Most important of all perhaps are the occasional cases in which a specialization turns out to be usable in a hitherto unoccupied way of life or to give a general advantage in many ways of life. The origin of the higher categories seems to have its basis in such ecological opportunities (Wright, 1941, 1948*b*, 1949*a*, *b*).

Evolution in general

Organic evolution is not the only sort of evolution in the sense of a process of cumulative change. When a level of intelligence was reached in an anthropoid line that made symbolic speech possible, a new evolutionary process emerged, enormously more rapid than organic evolution. The line of persistence and accumulation was that from speaker to listener, and later from writer to reader, instead of the germ line. The principles of Mendelian heredity do not apply to the evolution of culture. Nevertheless, the general qualitative conclusion would still seem to hold that this or any other evolutionary process depends on a continually shifting but never obliterated state of balance between factors of persistence and change, and that the most favourable condition for this occurs where there is a finely subdivided structure in which isolation and cross-communication keep in proper balance.

APPENDIX A

The method of path coefficients

As most of this discussion rests on the values taken by the inbreeding coefficient, F, under diverse population structures, and these values have been found by the method of path coefficients, we begin with a brief sketch of this method.

It is a method of dealing with linear systems of variables that are closed in the sense that each variable is either represented as linearly and completely determined by others in the system or is one of the ultimate factors, with indicated correlations with all other ultimate factors. The use of the method is greatly facilitated by diagrams in which arrows run to each variable that is represented as dependent from those represented as affecting it directly, and the system is completed by double-headed arrows to indicate correlations between ultimate factors, due to unknown common factors, in all cases in which a zero correlation cannot be safely postulated.

Let V_0, V_1, V_2, etc., be the variables. Express each in standard form ($X_0 = (V_0 - \bar{V}_0)/\sigma_0$, etc.). Assume that V_0 is represented as linearly and completely determined by $V_1, V_2, \ldots, V_k$:

$$X_0 = \sum_{i=1}^{k} p_{0i} X_i.$$

A coefficient p_{0i} pertaining to the indicated path of influence $V_i \rightarrow V_0$ is known as an elementary path coefficient.

If V_q is any variable in the system and there are n sets of observations:

$$r_{0q} = \frac{1}{n} \sum^{n} X_0 X_q$$

$$= \sum_{i=1}^{k} p_{0i} r_{iq} = \sum_{i=1}^{k} p_{0\overline{iq}}.$$

Any of the correlations, r_{iq}, may itself be capable of such analysis through variables by which either V_i or V_q is represented as determined. On carrying such analysis back as far as the system permits, we arrive at the basic principle that the correlation between any two variables in the system is equal to the sum of contributions (compound path coefficients such as $p_{0\overline{iq}}$ above) pertaining to the paths by which one may go from one to the other in the diagram without going forward and then back and without passing through any variable twice in the same path. The contribution of such a connecting path is the product of the path coefficients pertaining to the elementary paths along its course. Thus, if p_{acdb} indicates a compound path coefficient relating to a path of the type $a \leftarrow c \leftarrow d \rightarrow b$, its value is $p_{ac} p_{cd} p_{bd}$. The symbol $p_{a\overline{cd}b}$ indicates a connexion of the type $a \leftarrow c \longleftrightarrow d \rightarrow b$ with value $p_{ac} p_{cd} p_{bd}$ in which $p_{cd} = r_{cd}$.

If we express the correlation of a variable with itself (necessarily 1 in a closed system), we obtain a useful equation expressing complete determination:

$$r_{00} = \sum_{i=1}^{k} p_{0i} r_{0i} = \sum^{k} p_{0i}^2 + 2\Sigma p_{0i} p_{0j} r_{ij} = 1 \quad (i < j).$$

It may be noted that if a variable (e.g. V_u) is omitted, $\Sigma p_{0i} r_{0i} (= 1 - r_{0u}^2)$ is the squared multiple correlation between V_0 and the determining variables that are included, and r_{0u}^2 is the degree of determination by the residual variables. The method is identical with those of multiple regression and of factor analysis in the appropriate closed systems but is designed for algebraic use in irregular systems with intermingled known and hypothetical variables, known and unknown path coefficients and correlation coefficients, in which these methods are not applicable in the conventional ways.

The applicability of the method is greatly extended by use of the principle that a variable that is derived from another by a process of random sampling may be represented as linearly and completely determined by the latter and a hypothetical variable, 'accidents of sampling', that has a zero correlation with all variables in the system that do not involve the same actual process of sampling.

Finally, it may be noted that it makes no difference in principle, in a closed system, which variables are treated as determined by which, and which are represented as ultimate factors, but there are usually certain arrangements that are more interesting than others from the standpoint of interpretation.

APPENDIX B

General coefficients of inbreeding (Fig. 14)

In applying the method to zygotes (autosomal diploid locus) and gametes, in a diagram a system of mating (Fig. 14, left), we note first that a zygote may be considered to be linearly, completely and equally determined by the gametes that unite at fertilization. Let a be the path coefficient relating a zygote to one of the determining gametes and let F be the correlation between the latter

$$2a^2+2a^2F = 1, \quad a = \sqrt{\frac{1}{2(1+F)}}.$$

The path coefficient, b, relating a gamete to the zygote that produces it, is also the correlation coefficient as there is only one connecting path, and is equal to the correlation $(a'+a'F')$ between this gamete and a gamete that contributed to it in the preceding generation (indicated by a prime). It is assumed that no systematic changes in gene frequency due to selection, mutation or immigration intervene. Then

$$b = \sqrt{\tfrac{1}{2}(1+F')}.$$

The compound path coefficient, ba', relating gamete in one generation to one in the preceding, has the value $\frac{1}{2}$ irrespective of the system of mating, under the postulated conditions.

Application of the general principle stated in Appendix A gives at once the general formula (for diploid autosomal loci)

$$F = \Sigma[(\tfrac{1}{2})^n(1+F_A)],$$

in which n is the number of zygotes in a connecting path (including sire and dam) and F_A is the inbreeding coefficient of the common ancestor (A) to which sire and dam trace in this particular path. The summation relates to all paths by which sire and dam are connected, without tracing forward to descendants and then back and without passing through any individual twice in the same path.

This formula has been given previously (Wright, 1922*a* and in many later papers) in the form

$$F = \Sigma[(\tfrac{1}{2})^{n_s+n_d+1}(1+F_A)]$$

in which n_s and n_d are the number of *generations* from sire and dam respectively to the ancestor A. The present form leads to a simpler analogous form in the case of sex-linked inheritance than does the older one.

In the case of sex-linkage (males treated as XO, females XX), the constitution of males is completely determined by that of the egg and completely determines that of their own gametes (Fig. 14, middle). There is no contribution from a connecting path that includes males in succession:

$$F = \Sigma[(\tfrac{1}{2})^{n_f}(1+F_A)].$$

In this case, n_f is the number of females in a connecting path and the summation relates to paths which have no males in succession. F_A is treated as 0 if the connecting ancestor is a male.

The contributions of ancestral connexions in three typical cases are shown in Fig. 13.

The method has been extended to polysomic loci. In the diagram (Fig. 14, right) the many paths connecting the ultimate factors are omitted for simplicity. We define F here as the correlation between genes of the same zygote, F_D as that between genes that enter the zygote from different gametes, and F_S as that between genes that enter the zygote in the same gamete. F_S would equal F' except for the small chance, e, that two genes of the zygote trace to one in the preceding generation. In a $2k$-somic case

$$F_S = (1-e)\,F' + e,$$

$$F_D = \Sigma\left\{\left(\frac{1}{2k}\right)^n [1+(2k-1)\,F_A]\right\},$$

$$F = [(k-1)\,F_S + kF_D]/(2k-1).$$

It is to be noted that none of these general formulae for F involves gene frequency. Thus, a single coefficient characterizes the effects of a given population structure on all autosomal disomic loci, irrespective of gene frequencies in the absence of differences due to selection or mutation. The effects of immigration may usually be incorporated into the representation of population structure. Other coefficients characterize the effects on sex-linked and polysomic loci.

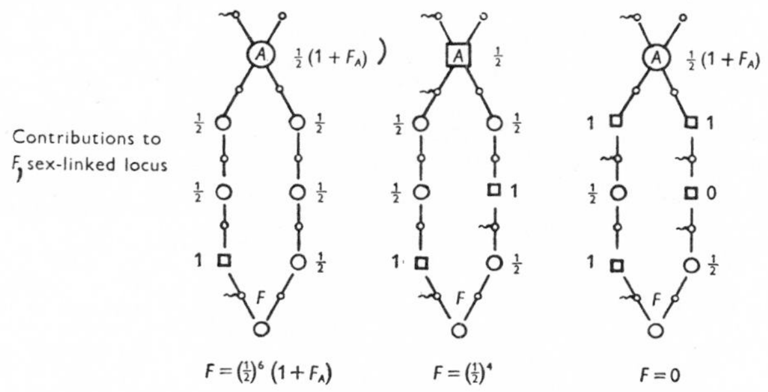

(Autosomal locus, $F = (\frac{1}{2})^7\,(1+F_A)$ in all three cases)

Fig. 13. Contribution of ancestral connexions to the value of F in three typical cases.

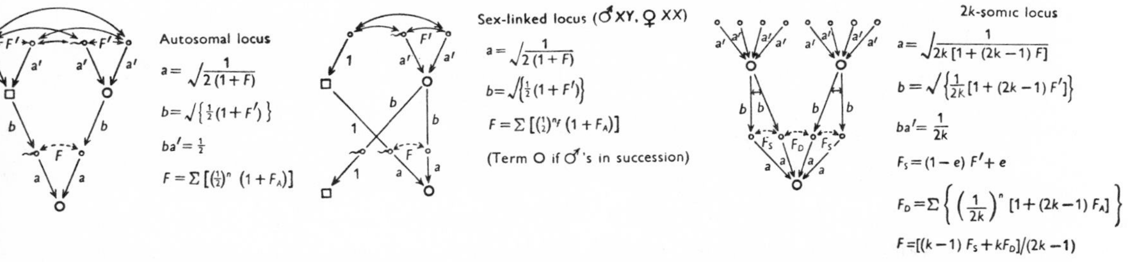

Fig. 14. Diagrams for analysis of inbreeding in the cases of autosomal loci, sex-linked loci and $2k$-somic loci. In the last case, the numerous two-headed arrows connecting the ultimate variables ~~(gametes)~~ are omitted for simplicity. The dotted two-headed arrows indicate correlations to be analysed in terms of connecting paths.

A coefficient F, as the correlation between alleles, could be found in principle for each locus subject to significant mutation rates of selective differentials. In the latter case, it would depend not only on the selection coefficient but also on gene frequency, except in so far as selection pressure can be treated as linearly related to gene frequency in the range of gene frequencies involved.

It is obvious from Table 1 that the degree of approach towards complete fixation in the case of two alleles under a system of mating that does not alter gene frequency in the total (diploid) population is measured by the same coefficient F that gives the correlation between alleles of the same zygote. The derivation of the general formulae for F in the latter sense applies as well to multiple alleles as to pairs of alleles, and it is easy to see that the other interpretation also applies to multiple alleles. The Mendelian mechanism is such that any group of alleles may be treated formally as if a single allele. Thus, if A_i and A_j in a multiple series are treated as if a single allele A_{ij} with frequency $q_{ij} = q_i + q_j$, the frequency of 'homozygotes' $A_{ij}A_{ij}$ must be $(1-F)q_{ij}^2 + Fq_{ij}$. If we subtract from this the frequencies of A_iA_i and A_jA_j which must be $(1-F)q_i^2 + Fq_i$ and $(1-F)q_j^2 + Fq_j$ respectively, recalling that F is independent of gene frequency, we find that the frequency of A_iA_j is $2(1-F)q_iq_j$ and that of all heterozygotes must be of this type. The array of zygotic frequencies can accordingly be analysed into a panmictic component $(1-F)[\sum^n q_iA_i]^2$ and a fixed component $F[\sum^n q_iA_i^2]$ as in Table 2. The correlation between alleles of the same zygote is the weighted average of the correlations within these portions (0 and 1 respectively), and is thus F irrespective of gene frequencies and of the values assigned to those alleles, as expected.

APPENDIX C

Properties of populations as related to F

The relation of the mean and variance of characters to the inbreeding coefficient, noted in the text, apply to multiple alleles as well as to pairs of alleles. We assume here that contributions of loci and of environmental factors are additive. The array of zygotic frequencies of n alleles of type A_i gene frequency q_i is

$$(1-F)[\sum^n q_iA_i]^2 + F[\Sigma q_iA_iA_i].$$

Let c_{ij} represent the contribution of A_iA_j to the character. Let $m_{T(F)}$ be the mean and $\sigma^2_{T(F)}$ the genetic component of the variance under a system of mating characterized by coefficient F. Then $m_{T(0)}$ and $\sigma^2_{T(0)}$ are the corresponding statistics under random mating ($F = 0$), and $m_{T(1)}$ and $\sigma^2_{T(1)}$ those for an array of completely fixed lines ($F = 1$), arrived at without selection:

$$\begin{aligned}
m_{T(F)} &= (1-F)\sum_{j=1}^{n}\sum_{i=1}^{n}(c_{ij}q_iq_j) + F\sum_{i=1}^{n}(c_{ii}q_i)\\
&= (1-F)\,m_{T(0)} + Fm_{T(1)}\\
&= m_{T(0)} + F[m_{T(1)} - m_{T(0)}],\\
\sigma^2_{T(F)} &= (1-F)\Sigma\Sigma(c_{ij}^2q_iq_j) + F\Sigma(c_{ii}^2q_i) - m^2_{T(F)}\\
&= (1-F)(\sigma^2_{T(0)} + m^2_{T(0)}) + F(\sigma^2_{T(1)} + m^2_{T(1)}) - [(1-F)\,m_{T(0)} + Fm_{T(1)}]^2\\
&= (1-F)\,\sigma^2_{T(0)} + F\sigma^2_{T(1)} + F(1-F)[m_{T(1)} - m_{T(0)}]^2.
\end{aligned}$$

Thus the change in mean due to the system of mating is proportional to F, but that in the total variance is a quadratic fraction of F.

The case of semi-dominance at all pertinent loci is of interest. Let c_i be contribution of allele A_i and $c_{ij} = c_i + c_j$. The mean is unaffected by the system of mating in this case:

$$\begin{aligned}
m_{T(F)} &= m_{T(0)} = m_{T(1)} = 2\sum^n c_iq_i,\\
\sigma^2_{T(0)} &= \sum_{j=1}^{n}\sum_{i=1}^{n}(c_i + c_j)^2 q_iq_j - 4(\sum^n c_iq_i)^2\\
&= 2\sum^n c_i^2q_i - 2(\sum^n c_iq_i)^2 = 2\sigma^2_{c_i},\\
\sigma^2_{T(1)} &= 4\sum^n c_i^2q_i - 4(\Sigma c_iq_i)^2 = 4\sigma^2_{c_i},\\
\sigma^2_{T(F)} &= (1-F)\,\sigma^2_{T(0)} + F\sigma^2_{T(1)}\\
&= (1+F)\,\sigma^2_{T(0)}.
\end{aligned}$$

If the inbreeding is due to isolation of a number (n) of strains (s) breeding at random within themselves, the variance within one is as follows, letting q_{is} be the frequency of A_i in this strain:

$$\sigma^2_{S(0)} = 2[\sum^n c_i^2 q_{is} - (\sum^n c_i q_{is})^2]$$

$$= 2[\sum^n c_i^2 q_{is}(1-q_{is}) - 2\Sigma c_i c_j q_{is} q_{js}] \quad (i<j).$$

But $2q_{is}(1-q_{is})$ is the amount of heterozygosis of A_i with all other alleles collectively, within the strain. The average for all strains is $2q_i(1-q_i)$ $(1-F)$. Similarly, $2q_{is}q_{js}$ is the proportion of the specific heterozygote A_iA_j within the strain, the average of which for all strains is $2q_iq_j(1-F)$. Thus, the average intra-strain variance is

$$\overline{\sigma^2_{S(0)}} = 2(1-F)[\sum^n c_i^2 q_i(1-q_i) - 2\Sigma c_i c_j q_i q_j], \quad (i<j)$$

$$= 2(1-F)[\sum^n c_i^2 q_i - (\Sigma c_i q_i)^2]$$

$$= (1-F)\,\sigma^2_{T(0)}.$$

The variance of strain means is

$$\sigma^2_{ms}(= \sigma^2_{T(F)} - \sigma^2_{S(0)}),$$

$$\sigma^2_{ms} = 2F\sigma^2_{T(0)}.$$

The genetic correlation between any relatives is readily found by the method of path coefficients assuming additive effects of genes.

In the case of autosomal diploids, the correlation between propositi Z_1 and Z_2 is given by

$$r_{Z_1Z_2} = \frac{\Sigma[(\frac{1}{2})^{n-1}(1+F_A)]}{\sqrt{\{(1+F_{Z_1})(1+F_{Z_2})\}}} \quad \text{(Wright, 1922}a\text{)}.$$

For sex-linked loci

$$r_{Z_1Z_2} = \frac{\Sigma[(\frac{1}{2})^{n_f-\frac{1}{2}n_{Zf}}(1+F_A)]}{\sqrt{\{(1+F_{Z_1})(1+F_{Z_2})\}}},$$

where n_{Zf} is the number of female propositi (0, 1 or 2). These formulae may be modified in the usual ways to take care of dominance, types of factor interaction and environmental effects.

APPENDIX D

The inbreeding coefficient of breeds

The estimation of the inbreeding coefficient of large populations on the basis of pedigrees requires use of sampling methods. The best method seems to be to take an adequate random sample of individuals and trace single *random* lines back of sire and dam (Wright & McPhee, 1925). These either show a common ancestor or they do not. No attention need be paid to the generation in which a tie occurs (autosomal F), since remoteness is exactly compensated for by increase in the number of possible ties ($2^{n_s+n_d}$) of which the one observed is representative. In m such two-line pedigrees with k observed ties,

$$F = \frac{1}{m}\sum^k [2^{n_s+n_d}(\tfrac{1}{2})^{n_s+n_d+1}(1+F_A)]$$

$$= \frac{k}{2m}(1+\overline{F}_A).$$

The value of F for the sixty-four cows of Bates's Duchess strain of Shorthorns was 0·409 from tracing of all pedigrees to the beginning of the Coates herd book. The estimate from random two-line pedigrees was $0{\cdot}422 \pm 0{\cdot}011$.

The situation is more complicated in the case of sex-linked F. F applies here only to females (XX). It is necessary to calculate the number of possible ties between lines with no successive males.

Possible contributory lines back of sire

	Ancestral generations								
Ancestor	0	1	2	3	4	5	6	7	n_s
Male	1	0	1	1	2	3	5	8	$f(n_s-1)$
Female	—	1	1	2	3	5	8	13	$f(n_s)$
Total	1	1	2	3	5	8	13	21	$f(n_s+1)$

Possible contributory lines back of dam

	Ancestral generations								
Ancestor	0	1	2	3	4	5	6	7	n_d
Male	—	1	1	2	3	5	8	13	$f(n_d)$
Female	1	1	2	3	5	8	13	21	$f(n_d+1)$
Total	1	2	3	5	8	13	21	34	$f(n_d+2)$

Of the 2^{n_s} lines tracing to ancestors of the n_sth ancestral generation back of the sire, only $f(n_s+1)$ involve no successive males, where $f(n_s)$ is the n_sth Fibonacci number starting from $f(0) = 0, f(1) = 1$, and following the rule that each is the sum of the two preceding

$$f(n) = f(n-1)+f(n-2).$$

In the case of females, the number of lines tracing to ancestors of the n_dth ancestral generation that involve no successive males is $f(n_d+2)$. If m two-line pedigrees of females show k ties and n_f is the number of females, including the dam, in a typical tie,

$$F = \frac{1}{m}\sum^{k}\left[\frac{f(n_s+1)f(n_d+2)\,(1+F_A)}{2^{n_f}}\right].$$

APPENDIX E

Regular systems of mating

In the case of regular systems of mating, the diagrammatic representation may usually be closed after a small number of ancestral generations. Consider the case of a population of monoecious individuals of constant effective number N in which fertilization is at random except for a known proportion, h, of self-fertilization, not necessarily $1/N$. Let E be the correlation between random gametes and F that between uniting gametes (autosomal diploid loci):

$$E = \frac{1}{N}b^2+\left(1-\frac{1}{N}\right)E',$$

$$F = hb^2+(1-h)E',$$

$$b^2 = \tfrac{1}{2}(1+F'),$$

$$E' = \left(\frac{F-\frac{1}{2}h(1+F')}{(1-h)}\right) = \frac{1}{N}\left(\frac{1+F''}{2}\right)+\left(1-\frac{1}{N}\right)\left(\frac{2F'-h(1+F'')}{2(1-h)}\right).$$

Solving for F:

$$F = \left(1+\frac{h}{2}-\frac{1}{N}\right)F' + \left(\frac{1}{2N}-\frac{h}{2}\right)F'' + \frac{1}{2N},$$

$$P = \left(1+\frac{h}{2}-\frac{1}{N}\right)P' + \left(\frac{1}{2N}-\frac{h}{2}\right)P''.$$

$$\text{If } h = 0, \quad P = \left(1-\frac{1}{N}\right)P' + \frac{1}{2N}P''.$$

$$\text{If } h = \frac{1}{N}, \quad P = \left(1-\frac{1}{2N}\right)P'.$$

$$\text{If } h = 1, \quad (P-\tfrac{1}{2}P') = \left(1-\frac{1}{N}\right)(P'-\tfrac{1}{2}P''), \quad P = \tfrac{1}{2}P'.$$

The rate at which heterozygosis falls off in the absence of self-fertilization can be found by equating P/P' to P'/P'' (rapidly approached in a population of constant size):

$$P = \frac{1}{2}\left[1-\frac{1}{N}+\sqrt{\left(1+\frac{1}{N^2}\right)}\right]P'$$

$$= \left[1-\frac{1}{2N}\left(1-\frac{1}{2N}\right)\right]P' \quad \text{approximately.}$$

It is evident that it makes no appreciable difference whether self-fertilization occurs at random or is excluded unless N is very small. In the other limiting case, $h = 1$, $P = \frac{1}{2}P'$ as expected.

The case of a population with separate sexes, N_p permanent pairs, and the proportion h of brother-sister mating can be worked out similarly:

$$E = \frac{1}{N_p}(ba')^2(2F'+2b'^2) + 4\left(1-\frac{1}{N_p}\right)(ba')^2E'$$

$$= \frac{1}{4N_p}(1+2F'+F'') + \left(1-\frac{1}{N_p}\right)E',$$

$$F = \tfrac{1}{4}h(1+2F'+F'') + (1-h)E',$$

$$E' = \frac{F-\frac{1}{4}h(1+2F'+F'')}{1-h} = \frac{1}{4N_p}(1+2F''+F''') + \left(1-\frac{1}{N_p}\right)\left(\frac{F'-\frac{1}{4}h(1+2F''+F''')}{1-h}\right),$$

$$F = \left(1+\frac{h}{2}-\frac{1}{N_p}\right)F' + \left(\frac{1}{2N_p}-\frac{h}{4}\right)F'' + \left(\frac{1}{4N_p}-\frac{h}{4}\right)F''' + \frac{1}{4N_p},$$

$$P = \left(1+\frac{h}{2}-\frac{1}{N_p}\right)P' + \left(\frac{1}{2N_p}-\frac{h}{4}\right)P'' + \left(\frac{1}{4N_p}-\frac{h}{4}\right)P'''.$$

$$\text{If } h = 0, \quad P = \left(1-\frac{1}{N_p}\right)P' + \frac{1}{2N_p}P'' + \frac{1}{4N_p}P'''.$$

$$\text{If } h = \frac{1}{N_p}, \quad P = \left(1-\frac{1}{2N_p}\right)P' + \frac{1}{4N_p}P''.$$

$$\text{If } h = 1, \quad (P-\tfrac{1}{2}P'-\tfrac{1}{4}P'') = \left(1-\frac{1}{N_p}\right)(P'-\tfrac{1}{2}P''-\tfrac{1}{4}P'''), \quad P = \tfrac{1}{2}P'+\tfrac{1}{4}P''.$$

It makes no appreciable difference whether brother-sister mating occurs at random or is excluded, unless N_p is very small. The rate in the former case is exactly the same as that for the same number ($N = 2N_p$) of monoecious individuals with self-fertilization excluded.

The limiting case, $h = 1$, is that of brother-sister mating with $P = \frac{1}{2}P'+\frac{1}{4}P''$, as given by Jennings (1914) (cf. Wright, 1921):

$$P \doteq \tfrac{1}{4}(1+\sqrt{5})P' = 0{\cdot}809P' \quad \text{(Wright, 1931).}$$

The simplest system, compatible with $h = 0$, is that of double-first cousin mating ($N_p = 2$)

$$P = \tfrac{1}{2}P' + \tfrac{1}{4}P'' + \tfrac{1}{8}P''' \quad \text{(Wright, 1921).}$$

On putting $P/P' = P'/P'' = P''/P'''$ to find the limiting ratio in successive generations we arrive at the equation $8X^3 - 4X^2 - 2X - 1 = 0$:

$$P = 0{\cdot}920P' \quad \text{(Wright, 1933}a\text{).}$$

The matrix method yields an equation of the 12th degree from which Fisher (1949) factors out the above cubic and thus obtains a solution ($X = 0{\cdot}91964$) in agreement with that yielded by path coefficients.

The series of results for small populations in which consanguineous mating is avoided as far as possible within each generation, make an interesting comparison (Wright, 1921):

N	P (exact)	Approximate P
1	$\tfrac{1}{2}P'$	$0{\cdot}500P'$
2	$\tfrac{1}{2}P' + \tfrac{1}{4}P''$	$0{\cdot}809P'$
4	$\tfrac{1}{2}P' + \tfrac{1}{4}P'' + \tfrac{1}{8}P'''$	$0{\cdot}920P'$
8	$\tfrac{1}{2}P' + \tfrac{1}{4}P'' + \tfrac{1}{8}P''' + \tfrac{1}{16}P^{\text{IV}}$	$0{\cdot}965P'$
16	$\tfrac{1}{2}P' + \tfrac{1}{4}P'' + \tfrac{1}{8}P''' + \tfrac{1}{16}P^{\text{IV}} + \tfrac{1}{32}P^{\text{V}}$	$0{\cdot}983P'$
N		$\left(1 - \frac{1}{4N}\right)P'$

With random mating in a population of effective size N, heterozygosis falls off by approximately $1/(2N)$ of its previous value per generation (exactly in a monoecious population). With maximum avoidance of inbreeding within each generation this rate of falling off is approximately halved.

Another important case is that of random mating in a population of N_m males, N_f females (Wright, 1931):

$$P = P' - \left(\frac{N_m + N_f}{8N_m N_f}\right)(2P' - P''),$$

$$\frac{P - P'}{P'} = -\frac{1}{2}\left(1 + \frac{N_m + N_f}{4N_m N_f}\right) + \frac{1}{2}\sqrt{\left\{1 + \left(\frac{N_m + N_f}{4N_m N_f}\right)^2\right\}}$$

$$= -\left(\frac{1}{8N_m} + \frac{1}{8N_f}\right)\left(1 - \frac{1}{8N_m} - \frac{1}{8N_f}\right) \quad \text{approximately.}$$

In the preceding cases, it has been assumed that generations are wholly distinct. Cases of overlapping generations can however be worked out. The simplest is that of alternate parent-offspring mating which was shown by Jennings (1916) to give exactly the same result as brother-sister mating. The method of path coefficients also gives the same result (Wright, 1921) as does the matrix method (Fisher, 1949).

In the case of sex-linked inheritance, N_m males N_f females (Wright, 1933a):

$$P = P' - \left(\frac{N_f + 1}{8N_f}\right)(2P' - P'') + \frac{(N_f - 1)(N_m - 1)}{8N_m N_f}(2P'' - P''').$$

Putting $P/P' = P'/P'' = P''/P'''$ and $y = \dfrac{(P - P')}{P'}$,

$$y^3 + y^2(2 + 2C_1) + y(1 + 3C_1 - 2C_2) + (C_1 - C_2) = 0,$$

where $C_1 = (N_f + 1)/8N_f$ and $C_2 = (N_f - 1)(N_m - 1)/8N_m N_f$,

$$y = -\left(\frac{2N_m + N_f}{9N_m N_f}\right), \quad \text{approximately.}$$

Following are examples of the effects of inbreeding on $2k$-somic loci (Wright, 1938). These can easily be extended. The chance that a gene may be represented twice in a gamete is ignored here for simplicity, though easily introduced if desired.

N monoecious individuals, random union of gametes:

$$P = \frac{1}{2N(2k-1)}[(6Nk-4N-2k+1)P'-(2N-2)(k-1)P''].$$

The simplest special case is that in which $N = 1$ (exclusive self-fertilization):

$$P = \left(\frac{4k-3}{4k-2}\right)P',$$

Diploid ($k = 1$) $P = \frac{1}{2}P'$,

Tetraploid ($k = 2$) $P = \frac{5}{6}P'$,

Hexaploid ($k = 3$) $P = \frac{9}{10}P'$,

Octoploid ($k = 4$) $P = \frac{13}{14}P'$.

The results for tetraploids and hexaploids agree with those arrived at by Haldane (1930) by the matrix method.

If self-fertilization is excluded we obtained

$$P = \frac{1}{2N(2k-1)}[(6Nk-4N-4k+2)P'-(2Nk-2N-4k+3)P''].$$

The simplest special case is that of sibling mating ($N = 2$):

$$P = P' - \frac{1}{8k-4}(2P'-P''),$$

$$P = \left(\frac{4k-3+\sqrt{(16k^2-16k+5)}}{8k-4}\right)P' \quad \text{approximately.}$$

	P (exact)	Approximate P
Diploid	$\frac{1}{2}P'+\frac{1}{4}P''$	$\frac{1}{4}(1+\sqrt{5})\,P' = 0{\cdot}80902P'$
Tetraploid	$(5/6)\,P'+(1/12)\,P''$	$(1/12)\,(5+\sqrt{(37)})\,P' = 0{\cdot}92356P'$
Hexaploid	$(9/10)\,P'+(1/20)\,P''$	$(1/20)\,(9+\sqrt{(101)})\,P' = 0{\cdot}95249P'$
Octoploid	$(13/14)\,P'+(1/28)\,P''$	$(1/28)\,(13+\sqrt{(197)})\,P' = 0{\cdot}96556P'$

In the case of tetraploids the matrix yields an octic equation from which Bartlett & Haldane (1934) had given 0·92356 as the pertinent root. Fisher (1949) obtains the same equation and factors out a quadratic which yields $(1/12)(5+\sqrt{(37)})$ as the pertinent root in agreement with these results.

We will not review here the somewhat more complicated cases in which the interference of inbreeding with the recombination of the linked genes has been studied (Wright, 1933*b*). General formulae for populations of N_m males, N_f females, and for the effects of restrictions on mating were considered. The results in population of 1 and 2 agreed with those from the matrix method, (Robbins, 1918*a*, *b*; Haldane & Waddington, 1931), with a few qualifications. In this case an additional correlational term must be recognized between the constitution of a gamete with respect to one locus and that with respect to the other, since the same sampling process is involved.

Bartlett & Haldane (1935) have dealt with the effects of enforced heterozygosis in a locus linked with one under consideration in populations of 1, 2 or 3 by the matrix method. These can readily be generalized for larger populations by the method of path coefficients.

APPENDIX F

Isolation by distance

Continuity over an area. Assume that there is uniform density over a large area but highly restricted dispersal in each generation, equal for male and female gametes. The variances of the distances between birthplaces of ancestors and individuals should be proportional to the number of intervening generations. Let F_{XK} be the correlation between random gametes drawn at random from a population of size XN relative to those drawn at random from a population of size KN. By the method of path coefficients

$$F_{1K} = \frac{1}{N}b^2 + 4\left(1-\frac{1}{N}\right)(ba')^2 F'_{2K} = \frac{1}{N}\left(\frac{1+F'_{1K}}{2}\right) + \left(1-\frac{1}{N}\right)F'_{2K},$$

$$F'_{2K} = \frac{1}{2N}\left(\frac{1+F''_{1K}}{2}\right) + \left(1-\frac{1}{2N}\right)F''_{3K},$$

$$F''_{3K} = \frac{1}{3N}\left(\frac{1+F'''_{1K}}{2}\right) + \left(1-\frac{1}{3N}\right)F'''_{4K}, \quad \text{etc.}$$

If a stationary state has been reached with respect to the F's, primes may be dropped. Solving for F_{1K}

$$F_{1K} = \sum_{X=1}^{K-1} t_X \Big/ \left[2N - \sum_{X=1}^{K-1} t_X\right],$$

where
$$\sum_{X=1}^{K-1} t_X = \left[1+\frac{1}{2}\left(1-\frac{1}{N}\right)+\frac{1}{3}\left(1-\frac{1}{N}\right)\left(1-\frac{1}{2N}\right)\cdots\frac{1}{K}\prod_{X=1}^{K-1}\left(1-\frac{1}{XN}\right)\right],$$

and
$$t_X = \left[\frac{(X-1)N-1}{XN}\right]t_{X-1}.$$

The inbreeding coefficient F_{1K} should approach 1 as K is increased without limit and Σt should therefore approach N. An approximate formula for Σt, given in the 1943*a* paper, was shown to do this with high precision in all cases worked out, but no general proof was given. I am indebted to Mr Alan Robertson and to Mr D. J. Hooton for two different demonstrations which lead in different cases to great simplification.

Mr Hooton's demonstration rests on the following development of the recurrence equation for t_X:

$$KNt_K = (K-1)Nt_{K-1} - t_{K-1},$$
$$(K-1)Nt_{K-1} = (K-2)Nt_{K-2} - t_{K-2},$$
$$\cdots\cdots\cdots\cdots\cdots\cdots$$
$$3Nt_3 = 2Nt_2 - t_2,$$
$$2Nt_2 = N-1.$$

Adding
$$\sum_{X=1}^{K-1} t_X = N(1-Kt_K).$$

But
$$Kt_K = \prod_{X=1}^{K-1}\left(1-\frac{1}{XN}\right) < \frac{1}{1+\frac{1}{N}\sum_{X=1}^{K-1}\frac{1}{X}}.$$

As $\Sigma\frac{1}{X}$ is divergent, $Kt_K \to 0$ and $\Sigma t \to N$.

This also follows from the expansion of $\log Kt_K$ which gives a simpler method of calculating the inbreeding coefficients than that used previously. The results agree:

$$\log Kt_K = \log \prod_{X=1}^{K-1}\left(1-\frac{1}{XN}\right)$$
$$= \sum_{X=1}^{K-1} \log\left(1-\frac{1}{XN}\right)$$
$$= -\left[\frac{1}{N}\Sigma\frac{1}{X}+\frac{1}{2N^2}\Sigma\frac{1}{X^2}+\frac{1}{3N^3}\Sigma\frac{1}{X^3}+\ldots\right]$$
$$= -\left[\frac{1}{N}[\log(K-0\cdot5)+0\cdot5772]+\frac{1}{2N^2}\left(1\cdot6449-\frac{2}{2K-1}\right)\right.$$
$$\left.+\frac{1}{3N^3}\left(1\cdot202-\frac{2}{(2K-1)^2}\right)+\frac{1}{4N^4}\left(1\cdot082-\frac{2}{(2K-1)^3}\right)\ldots\right].$$

It is convenient to express this in terms of common logarithms for purposes of calculation:

$$-\log_{10}(Kt_K) = \frac{1}{N}[\log_{10}(K-0\cdot5)+0\cdot2507]+\frac{1}{N^2}\left[0\cdot3572-\frac{0\cdot4343}{2K-1}\right]$$
$$+\frac{1}{N^3}\left[0\cdot174-\frac{0\cdot29}{(2K-1)^2}\right]+\frac{0\cdot12}{N^4}+\ldots.$$

The inbreeding coefficient may now be expressed in terms of Kt_K instead of Σt_X

$$F_{1K} = \frac{\sum^{K-1} t_X}{2N-\sum^{K-1} t_X} = \frac{1-Kt_K}{1+Kt_K}.$$

In Figs. 7–11 F_{NX} is used for the inbreeding coefficient of neighbourhoods (N_N) relative to large populations, N_X, ranging up to $10^8 N_N$ in size. $F_{NX} = F_{1X}$ under random union of gametes. It is of greater interest to find the coefficient for populations of any size N_X relative to a specified total N_T. This was previously found from the relation $F_{XT} = (F_{NT}-F_{NX})/(1-F_{NX})$ or its equivalent $P_{XT} = P_{NT}/P_{NX}$. It can also be found directly from the initial set of equations above. Returning to the symbolism used there and using S for a given value of X,

$$F_{SK} = \left[\frac{1+F_{1K}}{2}\right]\left[\frac{1}{SN}+\frac{1}{(S+1)N}\left(1-\frac{1}{SN}\right)\cdots\frac{1}{K}\prod_{X=S}^{K-1}\left(1-\frac{1}{XN}\right)\right]$$
$$= \left[\frac{1+F_{1K}}{2}\right]\left[\sum_{X=1}^{K-1} t_X-\sum_{X=1}^{S} t_X\right]\Big/\prod_{X=1}^{S-1}\left(1-\frac{1}{XN}\right)$$
$$= \left(\frac{1}{1+Kt_K}\right)\left(\frac{St_S-Kt_K}{St_S}\right),$$
$$P_{SK} = \frac{Kt_K(1+St_S)}{St_S(1+Kt_K)}.$$

Since

$$P_{1K} = \frac{2Kt_K}{1+Kt_K} \quad \text{and} \quad P_{1S} = \frac{2St_S}{1+St_S},$$
$$P_{1K} = P_{1S}P_{SK}$$

as expected.

In Figs. 7–11 F_{XT} is used for the inbreeding coefficient of populations ranging from N_N to 10^8 N_N relative to that of 10^8 N_N.

F_{XT} was interpreted in the papers 1943*a* and 1946 as the ratio of the variance (σ_q^2) of gene frequencies at neutral loci of populations of size N_X within areas of size N_T, to the limiting variance $q_T(1-q_T)$ under complete local fixation. As noted in the text, an interpretation in terms of the correlation between gene frequencies of neighbourhoods, the statistic discussed by Malécot (1948) can be given.

Let q_{N1} and q_{N2} be the gene frequencies of neighbourhoods drawn at random from the range of ancestral generations X. Their average distance apart is of the order of $\sqrt{(N_X/N_N)}\,\sigma$, where σ is the standard deviation of dispersion in one direction. Let $\sigma^2_{q(NX)}$ be the variance of the gene frequencies of such neighbourhoods.

The variances of differences between pairs is $2\sigma^2_{q(NX)}$, there being no correlation. If, however, such differences (within populations of size N_X) are averaged over a more comprehensive population of size N_T and this average is expressed in terms of the variance of neighbourhoods within this total, the correlation between the pairs, $r_{N_1N_2(XT)}$, must be taken into account:

$$2\sigma^2_{q(NX)} = 2\sigma^2_{q(NT)}(1-r_{N_1N_2(XT)}),$$

$$r_{N_1N_2(XT)} = 1-\frac{\sigma^2_{q(NX)}}{\sigma^2_{q(NT)}} = \frac{\sigma^2_{q(XT)}}{\sigma^2_{q(NT)}} = \frac{q_T(1-q_T)F_{XT}}{q_T(1-q_T)F_{NT}}$$

$$= \frac{F_{XT}}{F_{NT}}.$$

Continuity along a single line. In the case of continuity along a single line, the ancestors of generation X should be drawn from populations of $\sqrt{X}\,N_N$ instead of XN:

$$F_{1K} = \sum_1^{K-1} t_X\left[2N - \sum_1^{K-1} t_X\right],$$

$$\sum_1^{K-1} t_X = \left[1+\frac{1}{\sqrt{2}}\left(1-\frac{1}{N}\right)+\frac{1}{\sqrt{3}}\left(1-\frac{1}{N}\right)\left(1-\frac{1}{\sqrt{2}\,N}\right)\cdots\frac{1}{\sqrt{K}}\prod_{X=1}^{K-1}\left(1-\frac{1}{\sqrt{XN}}\right)\right],$$

$$t_X = \frac{N\sqrt{(X-1)}-1}{N\sqrt{X}}\,t_{X-1} \quad \text{(Wright, 1943}a\text{)}.$$

Mr Hooton notes that it can be shown in the same way as in the case of area continuity that

$$\sum_{X=1}^{K-1} t_X = N(1-\sqrt{K}\,t_K),$$

and that $\sqrt{K}\,t_K \to 0$, $\sum_{X=1}^{K-1} t_X \to N$ as K increases

$$\log(\sqrt{K}\,t_K) = \sum_{X=1}^{K-1}\log\left(1-\frac{1}{\sqrt{X}\,N}\right)$$

$$= \frac{1}{N}\Sigma\frac{1}{\sqrt{X}}+\frac{1}{2N^2}\Sigma\frac{1}{X}+\frac{1}{3N^3}\Sigma\frac{1}{X^{\frac{3}{2}}}+\frac{1}{4N^4}\Sigma\frac{1}{X^2}+\ldots.$$

Diverse systems of mating. The effects of various systems of mating where there is continuity over an area have been considered (Wright, 1946). The inbreeding coefficients for individuals, F_{1K} and for neighbourhoods, F_{NT}, are given below in terms of Kt_K instead of $\sum_1^{K-1} t_X$, the form used in the above paper.

Case 1. Monoecious individuals with the proportion h of self-fertilization. Equal dispersion of male and female gametes:

	General	$h=1/N$	$h=1$
$F_{1K}=$	$\dfrac{(N-1)-N(1-h)Kt_K}{(N-1)+N(1-h)Kt_K}$	$\dfrac{1-Kt_K}{1+Kt_K}$	1
$F_{NK}=$	$\dfrac{(N-1)(1-Kt_K)}{(N-1)+N(1-h)Kt_K}$	$\dfrac{1-Kt_K}{1+Kt_K}$	$1-Kt$

Case 2. Separate sexes with equal dispersion. N_p permanent pairs. Proportion h of brother-sister mating. Kt_K based on N_p.

	General	$h=1/N$	$h=1$
$F_{1K}=$	$\dfrac{(N_p-1)-N_p(1-h)Kt_K}{(N_p-1)+3N_p(1-h)Kt_K}$	$\dfrac{1-Kt_K}{1+3Kt_K}$	1
$F_{NK}=$	$\dfrac{(N_p-1)(1-Kt_K)}{(N_p-1)+3N_p(1-h)Kt_K}$	$\dfrac{1-Kt_K}{1+3Kt_K}$	$1-Kt_K$

Case 3. N_m males, N_f females. Equal dispersion and random mating. Kt_K based on

$$N' = 2N_m N_f/(N_m+N_f).$$

A correction is required in this case. In the formulae on pp. 47 and 48 of the 1946 paper, N_m and N_f should be multiplied by 2 to give the numbers in the parental generation, and in Tables 1 and 2 the figures for case 3 apply to $N_m = N_f = 50$, and similarly in Tables 3 and 4 they apply to $N_m = N_f = 5$.

$$F_{1K} = N_{NK} = \frac{1-Kt_K}{1+7Kt_K} \text{ approximately.}$$

Case 4. No dispersion of ovules, dispersion of pollen in normal distribution relative to distance in one direction from monoecious individuals. Size of neighbourhood N, apart from self-fertilization in excess of random. Proportion r of pollinations at random and thus $(1-r)$ excess self-fertilization and $h = (r/N)+(1-r)$ the total proportion of self-fertilization:

$$F_{NK} = \frac{\sum_{X=1}^{K-1} t_X}{N+r\left(N-\sum_{X=1}^{K-1} t_X\right)},$$

where

$$\sum_{X=1}^{K-1} t_X = \left[1+\left(\frac{1}{1+r}\right)\left(1-\frac{1}{N}\right)+\left(\frac{1}{1+2r}\right)\left(1-\frac{1}{N}\right)\left(1-\frac{1}{(1+r)N}\right)+\dots\right.$$
$$\left.+\left(\frac{1}{1+(k-1)r}\right)\prod_{X=1}^{K-1}\left(1-\frac{1}{[1+(X-1)r]N}\right)\right],$$

if $r = 1$ (self-fertilization random)

$$F_{NK} = \frac{1-Kt_K}{1+Kt_K}.$$

The result in this case is the same as with equal dispersion of ovules and pollens, provided that the variance of pollen is twice that in the latter. With excess self-fertilization, F_{NK} rises disproportionately as shown in Fig. 9.

If there is a certain amount of universal dispersion (m) or of reversible mutation at rates u and v per generation at the locus under consideration, the total amount of displacement by a random sample from the species is $(m+u+v)$. For simplicity we will use merely m. The formulae for Σt_X is as follows (Wright, 1943*a*):

$$\sum_{X=1}^{K-1} t_X = (1-m)^2+\frac{(1-m)^4}{2}\left(1-\frac{1}{N}\right)+\frac{(1-m)^6}{3}\left(1-\frac{1}{N}\right)\left(1-\frac{1}{2N}\right)+\dots\frac{(1-m)^{2K}}{K}\prod_{X=1}^{K-1}\left(1-\frac{1}{XN}\right).$$

The factor $(1-m)^{2X}$ is approximately $(1-2mX)$ if $2mX$ is small. The effect on $\sum_{X=1}^{K-1} t_X$ and hence on F_{NX} is negligible for most purposes if K is less than $1/(10m)$. If, on the other hand, X is greater than $3/m$, $(1-m)^{2X}$ becomes less than 0·0025 and contributions to $\sum^{K-1} t_X$ become negligible for most purposes. Thus the curve representing the value of F_{NX} in relation to $\log(N_X/N_N)$ follows that with $m = 0$ up to about $N_X/N_N = 1/(10m)$ and then rapidly approaches an asymptote which is less than 1. This asymptote was estimated by a rather cumbersome method, in the 1943*a* paper. Mr Robertson has arrived at a very simple formula for finding its value in the case of area continuity. This depends on writing $\frac{1}{N}\sum_{1}^{\infty} t$ as follows:

$$\frac{1}{N}\sum_{1}^{\infty} t = \left[(1-m)^2\frac{1}{N} - \frac{(1-m)^4}{2!}\frac{1}{N}\left(\frac{1}{N}-1\right) + \frac{(1-m)^6}{3!}\frac{1}{N}\left(\frac{1}{N}-1\right)\left(\frac{1}{N}-2\right)\ldots\right.$$

$$\left. = 1-[1-(1-m)^2]^{1/N}\right].$$

The numerical results in the 1943*a* paper are all in agreement.

Uniform density and amount of dispersion is a limiting case, not likely to be realized in nature. A rough model for a more typical situation may be obtained by considering individuals whose parents are drawn from a relatively large population (MN) and whose more remote ancestors are drawn from populations that increase linearly with the number of generations but by increments (N) that are only a fraction of that of the parental population:

$$F_{1K} = \frac{1}{MN}\left(\frac{1+F_{1K}}{2}\right) + \left(1-\frac{1}{MN}\right)F_{2K},$$

$$F_{2K} = \frac{1}{(M+1)N}\left(\frac{1+F_{2K}}{2}\right) + \left(1-\frac{1}{(M+1)N}\right)F_{3K}, \quad \text{etc.,}$$

$$F_{1K} = \left(\frac{1+F_{1K}}{2N}\right)\left[\frac{1}{M} + \left(\frac{1}{M+1}\right)\left(1-\frac{1}{MN}\right) + \left(\frac{1}{M+2}\right)\left(1-\frac{1}{MN}\right)\left(1-\frac{1}{(M+1)N}\right)\ldots\right.$$

$$\left. + \left(\frac{1}{M+K-1}\right)\prod_{M}^{M+K-2}\left(1-\frac{1}{XN}\right)\right],$$

$$F_{1K} = \frac{\Sigma t'}{2N-\Sigma t'},$$

$$\Sigma t' = \left[\sum_{1}^{M+K-2} t - \sum_{1}^{M-1} t\right] \Big/ \prod_{X=1}^{M-1}\left(1-\frac{1}{XN}\right)$$

$$= N[Mt_M - (M+K-1)\,t_{(M+K-1)}]/Mt_M,$$

$$F_{1K} = \frac{Mt_M - (M+K-1)\,t_{(M+K-1)}}{Mt_M + (M+K-1)\,t_{(M+K-1)}}.$$

REFERENCES

BARTLETT, M. S. & HALDANE, J. B. S. (1934). The theory of inbreeding in autotetraploids. *J. Genet.* **29**, 175–80.

BARTLETT, M. S. & HALDANE, J. B. S. (1935). The theory of inbreeding with forced heterozygosis. *J. Genet.* **31**, 327–40.

BATEMAN, A. J. (1947). Contamination in seed crops. III. Relation with isolation distance. *Heredity*, **1**, 303–36.

BERNSTEIN, F. (1930). Fortgesetzte Untersuchungen aus der Theorie der Blutgruppen. *Z. indukt. Abstamm.- u. VererbLehre*, **56**, 233–73.

DARWIN, C. (1868). *The Variation of Animals and Plants under Domestication.* London.

DOBZHANSKY, TH. & WRIGHT, S. (1943). Genetics of natural populations. X. Dispersion rates in *Drosophila pseudoobscura*. *Genetics*, **28**, 304–40.

DOBZHANSKY, TH. & WRIGHT, S. (1947). Genetics of natural populations. XV. Rate of diffusion of a mutant gene through a population of *Drosophila pseudoobscura*. *Genetics*, **32**, 303–24.

EPLING, C. & DOBZHANSKY, TH. (1942). Genetics of natural populations. VI. Microgeographical races in *Linanthus Parryae*. *Genetics*, **27**, 317–32.

FISH, H. D. (1914). On the progressive increase of homozygosis in brother-sister matings. *Amer. Nat.* **48**, 759–61.

FISHER, R. A. (1949). *The Theory of Inbreeding*, pp. 120. Edinburgh: Oliver and Boyd.

GALTON, F. (1889). *Natural Inheritance*. London: Macmillan & Co.

GOLDSCHMIDT, R. (1940). *The Material Basis of Evolution*, pp. 436. New Haven: Yale University Press.

HALDANE, J. B. S. (1930). Theoretical genetics of autopolyploids. *J. Genet.* **22**, 359–72.

HALDANE, J. B. S. (1936). The amount of heterozygosis to be expected in an approximately pure line. *J. Genet.* **32**, 375–91.

HALDANE, J. B. S. (1937). Some theoretical results of continued brother-sister mating. *J. Genet.* **34**, 265–74.

HALDANE, J. B. S. (1949). The association of characters as a result of inbreeding and linkage. *Ann. Eugen., Lond.*, **15**, 15–23.

HALDANE, J. B. S. & WADDINGTON, C. H. (1931). Inbreeding and linkage. *Genetics*, 357–74.

HARDY, G. H. (1908). Mendelian proportions in a mixed population. *Science*, **28**, 49–50.

JENNINGS, H. S. (1914). Formulae for the results of inbreeding. *Amer. Nat.* **48**, 693–6.

JENNINGS, H. S. (1916). The numerical results of diverse systems of breeding. *Genetics*, **1**, 53–89.

LUSH, J. L. (1943). *Animal Breeding Plans*, pp. 437. Ames, Iowa: Iowa State College Press.

MCPHEE, H. C. & WRIGHT, S. (1925). Mendelian analysis of the pure breeds of livestock. III. The Shorthorns. *J. Hered.* **16**, 205–15.

MCPHEE, H. C. & WRIGHT, S. (1926). Mendelian analysis of the pure breeds of livestock. IV. The British Dairy Shorthorns. *J. Hered.* **17**, 397–401.

MALÉCOT, G. (1948). *Les mathématiques de l'hérédité*, pp. 63. Paris: Masson et Cie.

PEARL, R. (1917). Studies on inbreeding. *Amer. Nat.* **51**, 545–59, 636–9.

ROBBINS, R. B. (1918*a*). Applications of mathematics to breeding problems. II. *Genetics*, **3**, 73–92.

ROBBINS, R. B. (1918*b*). Some applications of mathematics to breeding problems. III. *Genetics*, **3**, 375–89.

SHULL, G. H. (1908). The composition of a field of maize. *Rep. Amer. Breed. Ass.* **4**, 296–301.

WEINBERG, W. (1908). Über den Nachweis der Vererbung beim Menschen. *Jahreshefte d. Ver. f. vaterländische Naturkunde in Württemberg*, **64**, 368–82.

WRIGHT, S. (1921). Systems of mating. *Genetics*, **6**, 111–78.

WRIGHT, S. (1922*a*). Coefficients of inbreeding and relationship. *Amer. Nat.* **56**, 330–8.

WRIGHT, S. (1922*b*). Effects of inbreeding and crossbreeding on guinea pigs. *Bull. U.S. Dep. Agric.* no. 1121, pp. 59.

WRIGHT, S. (1923*a*). Mendelian analysis of the pure breeds of livestock. I. The measurement of inbreeding and relationship. *J. Hered.* **14**, 339–48.

WRIGHT, S. (1923*b*). Mendelian analysis of the pure breeds of livestock. II. The Duchess family of Shorthorns as bred by Thomas Bates. *J. Hered.* **14**, 405–22.

WRIGHT, S. (1931). Evolution in Mendelian population. *Genetics*, **16**, 97–159.

WRIGHT, S. (1932). The roles of mutation, inbreeding, crossbreeding and selection in evolution. *Proc. 26th Int. Congr. Genet.*, pp. 356–66.

WRIGHT, S. (1933*a*). Inbreeding and homozygosis. *Proc. Nat. Acad. Sci., Wash.*, **19**, 411–20.

WRIGHT, S. (1933*b*). Inbreeding and recombination. *Proc. Nat. Acad. Sci., Wash.*, **19**, 420–33.

WRIGHT, S. (1935*a*). The analysis of variance and the correlations between relatives with respect to deviations from an optimum. *J. Genet.* **30**, 253–6.

WRIGHT, S. (1935*b*). Evolution in populations in approximate equilibrium. *J. Genet.* **30**, 257–66.

WRIGHT, S. (1938). The distribution of gene frequencies in populations of polyploids. *Proc. Nat. Acad. Sci., Wash.*, **24**, 372–7.

WRIGHT, S. (1939). *Statistical genetics in relation to evolution. Exposés de Biométrie et de statistique biologique*, no. 802, pp. 63. Hermann et Cie, Editeurs.

WRIGHT, S. (1940). Breeding structure of populations in relation to speciation. *Amer. Nat.* **74**, 232–48.

WRIGHT, S. (1941). *The Material Basis of Evolution*, by R. Goldschmidt (Review). *Sci. Mon.* **53**, 165–70.

WRIGHT, S. (1943*a*). Isolation by distance. *Genetics*, **28**, 114–38.

WRIGHT, S. (1943*b*). An analysis of local variability of flower color in *Linanthus Parryae*. *Genetics*, **28**, 139–56.

WRIGHT, S. (1946). Isolation by distance under diverse systems of mating. *Genetics*, **31**, 39–59.

WRIGHT, S. (1948*a*). On the roles of directed and random changes in gene frequency in the genetics of populations. *Evolution*, **2**, 279–94.

WRIGHT, S. (1948*b*). Evolution, organic. *Encycl. Brit.* **8**, 915–29.

WRIGHT, S. (1948*c*). Genetics of populations. *Encycl. Brit.* **10**, 111, 111A–D, 112.

WRIGHT, S. (1949*a*). *Adaptation and Selection*. Chapter 20. Genetics, Paleontology and Evolution, pp. 365–89. Edited by G. L. Jepson, G. G. Simpson and E. Mayr.

WRIGHT, S. (1949*b*). Population structure in evolution. *Proc. Amer. Phil. Soc.* **93**, 471–8.

WRIGHT, S. & MCPHEE, H. C. (1925). An approximate method of calculating coefficients of inbreeding and relationship from livestock pedigrees. *J. Agric. Res.* **31**, 377–83.

41
Genetics and Twentieth Century Darwinism—A Review and Discussion

American Journal of Human Genetics 12, no. 3 (1960):365–72

Introduction

At the Cold Spring Harbor Symposium of 1959, dedicated to Genetics and Twentieth Century Darwinism, Ernst Mayr gave his famous introductory address "Where Are We?". The one hundredth anniversary of Darwin's *On the Origin of Species* was the perfect time for a reassessment of where evolutionary biology had gone during the century. In his address, Mayr provided a thumbnail sketch of the history of genetics in relation to evolutionary theory since 1900. Praising the contributions of Darwin, systematists, and sophisticated modern evolutionary geneticists who emphasized the interactive genome, Mayr questioned the significance of the contributions of the mathematical population geneticists (specifically Fisher, Wright, and Haldane). Mathematical population genetics, Mayr claimed, belonged to the period of "Classical Population Genetics," dominated by "beanbag genetics."

Wright disagreed vigorously with Mayr's historical interpretation of genetics in relation to evolutionary biology in the twentieth century and with Mayr's characterization of his own work as "beanbag genetics." This last point particularly rankled Wright because from the time of his thesis work he had consistently emphasized interaction effects of genes, and he had tried to integrate the consequences of genic interaction into theoretical population genetics. Thus when Wright was offered the opportunity to review the Cold Spring Harbor Symposium volume for 1959, he broke his sixteen-year-record of writing no book reviews (the last one had been of Simpson's *Tempo and Mode in Evolution*). Except for merely listing the contributors and the titles of their papers, Wright's entire review was devoted to a critique of Mayr's address.

I have examined the tension between Wright and Mayr in *SW&EB*, 477–84.

"Genetics and Twentieth Century Darwinism" A review and discussion[1]

SEWALL WRIGHT

Department of Genetics, University of Wisconsin, Madison, Wis.[2]

THIS symposium on the centenary of the publication of Darwin's "Origin of Species" gives a broad sample of current observational and experimental research that bears on evolution. The introductory address by Mayr "Where are we?" was intended to set the stage. He discussed the history of population genetics since 1900, the contributions of various lines of research and unsolved problems. As it seems to me that he has seriously misinterpreted the roles of these various lines, as well as the contributions of the one with which I am most familiar, I will single this out for special discussion later.

There followed twenty-four papers, largely on special researches. In the concluding address by Stebbins, "The synthetic approach to organic evolution", he listed seven major points on which he believed there is general agreement. He also listed a number of unsolved problems and brilliantly summarized the results of the symposium by bringing out how the various papers bore on these problems.

It is not practicable to summarize, and much less to discuss the specific contributions here. The reader may be given some idea of the scope by a list of papers not already referred to.

Dobzhansky, Th. Evolution of genes and genes in evolution.

Stubbe, Hans. Considerations on the genetical and evolutionary aspect of some mutants of *Antirrhinum, Hordeum* and *Lycopersicon.*

Buzzati-Traverso, A. A. Selection, quantitative traits and polygenic systems in evolution.

Morley, F. H. W. Natural selection in relation to ecotypic and racial differentiation in plants.

Mourant, A. E. Human blood groups and natural selection.

Lamotte, M. Polymorphism in natural populations of *Cepaea nemoralis.*

Carson, H. L. Genetic conditions which promote or retard race formation.

Schwanitz, Franz. Selection and race formation in cultivated plants.

Barnicot, N. A. Darwin's view on the evolution of human races in the light of modern research.

Sheppard, P. M. The evolution of mimicry; a problem in ecology and genetics.

Ehrendorfer, F. Differentiation-hybridization cycle and polyploidy in *Achillea.*

Coon, C. S. Race and ecology in man.

---

[1] Cold Spring Harbor Symposia on Quantitative Biology Vol. 24.

[2] Paper no. 778 from the Department of Genetics, University of Wisconsin.

Kitzmiller, J. B. and Laven, H. Speciation in mosquitoes. 1. Race formation and speciation in mosquitoes (Kitzmiller; 2. Speciation by cytoplasmic isolation (Laven); 3. Evolutionary mechanisms (both authors).
Baker, H. G. Reproductive methods as factors in speciation in flowering plants.
Wallace, Bruce. The influence of genetic systems on geographical distribution.
Kurten, B. Rates of evolution in fossil mammals.
Andrews, H. N., Jr. Evolutionary trends in early vascular plants.
Heberer, G. The descent of man and the present fossil record.
Hunt, Edward E., Jr. The continuing evolution of modern man.
Simpson, G. G. The nature and origin of supraspecific taxa.
Smith-White, S. Chromosome evolution in the Australian flora.
Rensch, B. Trends towards progress of brains and sense organs.

Mayr begins with a division of the history of the subject since 1900 into three parts:

(1) *The Mendelian Period.* "The period from 1900 to about 1920 saw a sharp cleavage, an almost bridgeless gap, between the evolution-minded naturalists on the one hand and the experimental geneticists on the other hand" (selectionist and mutationist, respectively).

(2) *Classical Population Genetics.* After referring to publications of Fisher (1930), Wright (1931) and Haldane (1932) to indicate what he is talking about, Mayr writes as follows:

"The emphasis in early population genetics was on the frequency of genes and on the control of this frequency by mutation, selection and random events. Each gene was essentially treated as an independent unit, favored or discriminated against by various causal factors. In order to permit mathematical treatment numerous simplifying assumptions had to be made, such as that of an absolute selective value of a given gene. The great contribution of this period was that it restored the prestige of natural selection, which had been rather low among the geneticists active in the early decades of the century, and that it prepared the ground for the treatment of quantitative characters. Yet this period was one of gross oversimplification. Evolutionary change was essentially presented as an input or output of genes, as the adding of certain beans to a beanbag and the withdrawing of others. This period of "beanbag genetics" was a necessary step in the development of our thinking, yet its shortcomings became obvious as a result of the work of the experimental population geneticists, the animal and plant breeders, and the population systematists, which ushered in a third era of evolutionary genetics."

(3) *The Newer Population Genetics.* Mayr describes this as follows, immediately after the above quotation:

"The next advance was characterized by an increasing emphasis on the interaction of genes. Not only individuals but even populations were no longer described atomistically as aggregates of independent genes in various frequencies, but as integrated, coadapted complexes. A gene is no longer considered to have one absolute selective value, but rather a wide range of potential values that may extend from lethality to high selective superiority, depending on genetic background and on the constellation of environmental factors. I have referred to this new mode of thinking as the genetic 'theory of relativity'. Dobzhansky's 'balance theory' of genetic variation is one of its aspects. The thinking of this newer population genetics is in considerable contrast to that of the classical population genetics and even more so to that of early Mendelism. This change of view is not always realized, even by professional geneticists who have no contact with population genetics."

"THE MENDELIAN PERIOD"

Returning to Mayr's first period my recollections of the attitudes of naturalists and geneticists toward natural selection during the latter half of it differ considerably from Mayr's statement. A very thorough and objective account of the various theories of evolution that had developed before any serious contamination by Mendelism, is given in Kellogg's "Darwinism Today" published in 1907, which I read in 1910. The impression is that of almost universal dissatisfaction with natural selection as the major principle but utter chaos with respect to substitutes. There was a bewildering array of names of theories but most of them referred to variants of Lamarckism, orthogenesis or heterogenesis. This is not surprising as there was no possibilty of settling the matter in the absence of sound knowledge of heredity. Yet, even as late as 1934, H. F. Osborn, the leading American paleontologist, probably spoke for most naturalists in holding that natural selection was inadequate.

Heterogenesis had, of course, much influence in early genetic thinking in the form of de Vries' mutation theory. Even this, however, was not as antagonistic to natural selection as the other theories. de Vries wrote in 1906 "Not withstanding all these apparently unsurmountable difficulties, Darwin discovered the great principle which rules the evolution of organisms. It is the principle of natural selection . . . It is the sieve which keeps evolution on the main line, killing all or nearly all that try to go in the other direction. By this means, natural selection is the one directing cause of the broad lines of evolution." By 1910 at least, most geneticists were much more Darwinian than de Vries and those with whom I had most contact (Castle, East) were wholly Darwinian. The development of the multiple factor theory of quantitative variability by Nilsson-Ehle, East, and Shull supplied admirably the basis for this view.

It is true, however, that most geneticists were more concerned with the kinds of mutations that are significant in evolution than with the process. Evolution is something that happens to populations and without a mathematical theory, connecting the phenomena in populations with those in individuals, there could be no very clear thinking on the subject. As to the kinds of mutations it has of course turned out that chromosome aberrations (from which de Vries was generalizing without knowing it) and both conspicuous and inconspicuous gene mutations are utilized in evolution though I think that it is generally agreed now that the last are most important in character change (Stebbins' third point).

Population genetics in a broad sense began well before 1900 in the efforts of Galton, Pearson and the rest of the Biometric school to devise tools for describing adequately the characteristics of populations, consisting of individuals each with a unique array of characters. Unfortunately for them, they went further and tried to deduce the nature of heredity from such population characters as the correlation between relatives. Shortly after 1900, experimental genetics and population genetics confronted each other in the persons of Bateson and Pearson across a seemingly "bridgeless gap."

A bridge was, however, inadvertently constructed by Pearson in 1904 in a paper in which he ruled Mendelism out because of a discrepancy between the genetic correlations expected under it and those actually observed. Yule (1906) showed that this discrepancy was based merely on a too rigid acceptance of complete dominance as a principle. Hardy, and Weinberg in 1908 extended Pearson's bridge, and Weinberg by 1910 had brought about a rather complete synthesis of the two kinds of data for homogeneous static populations. Fisher (1918) did the same from a somewhat different mathematical viewpoint. It is interesting to note that his most important extension was a more adequate treatment of factor interaction.

Another aspect of population genetics, the experimental study of the effects of inbreeding, crossbreeding and selection on laboratory populations goes back to Darwin's studies of inbreeding and crossbreeding in plants. There were several important studies of inbreeding in mammals in the latter part of the 19th century. Such experiments became more significant after 1900, when they could be given a Mendelian interpretation. Castle's experiments on inbreeding in Drosophila and on the effects of selection in hooded rats were among the most noteworthy. My own entry into population genetics, apart from serving as Prof. Castle's assistant (1912–15) in his selection experiment, came on taking charge (1915–25) of an extensive project on the effects of inbreeding and crossbreeding in guinea pigs, started in 1906 in the U. S. Bureau of Animal Industry (Wright 1922b).

Still another aspect of population genetics is exemplified by Sumner's (1918) very instructive study of the genetics of differences within and between subspecies of the wild species, *Peromyscus maniculatus*.

This brings us to Mayr's second period to which he confines mathematical population genetics in a Procrustean bed by lopping off all work before 1920 and all of the great expansion by many investigators after 1940.

"CLASSICAL POPULATION GENETICS"

The results which I obtained from the study of inbreeding and crossbreeding in guinea pigs agreed well with those of most others with other organisms and could be given a Mendelian interpretation. I was led into the mathematical aspect by the effort to develop a more general theory than that arrived at by Jennings (1916) for special cases, a theory applicable to any sort of population structure (Wright, 1921). The inbreeding coefficient (1922a,b) that was implied made it possible to generalize the Hardy-Weinberg rule to situations other than random mating. The dynamic aspects of inbreeding and assortative mating were also dealt with. The primary objective was to develop a theory on how best to combine inbreeding and crossbreeding with selection in livestock improvement. The only sort of selection considered was that directed toward an optimum, one that involves interaction in an extreme form. The mathematical analysis of this did not go far enough to be of much value until I came back to it in 1935.

Haldane (1924) made a more fruitful attack on the mathematical theory of selection by considering the case of a single gene subject to selection and mutation. This was a very important contribution to the dynamics of populations. If, however, nothing further had been developed in mathematical population genetics, Mayr's criticism would be sound. As it was, Haldane himself discussed selection under many complicating conditions including factor interaction (summarized in his 1932 book). The treatment of factor interaction was the central theme of the other two publications cited by Mayr. The situations treated in these were largely complementary.

The central theme of Fisher's book was the determination of the rate of increase of fitness (or mean selective value in my terminology) in a large homogeneous population, assuming that the effects of the genes may be involved in any sort of interaction whatever. His "fundamental theorem of natural selection" was that "the rate of increase of fitness of any organism at any time is equal to its genetic variance in fitness at that time" (in which genetic variance is defined as the additive component of the portion due to heredity). To demonstrate that progress by selection is restricted to the net effects of the genes in the combinations in which they enter is not to ignore interaction as Mayr seems to suppose. Actually Fisher's theorem holds only if the selective value of the total genotype is constant in any given environment (Wright, 1956) and under certain conditions selection may lead to decrease of mean selective value. Nevertheless the theorem covers a great deal of ground.

The objective of my papers in the 1930's was exploration of the ways in which selection may take advantage of favorable interaction effects most effectively. The most important case is that in which there are many separate peaks in the "surface" of selective values ($\overline{W}$), a case in which selection of the sort considered by Fisher merely holds the population to a single peak irrespective of whether high or low. This case includes that of selection directed toward an intermediate optimum, an almost universal situation for quantitative variability in nature, and also most cases of multiple interacting loci with pleiotropic effects. It may be noted parenthetically that Mayr states flatly that the symbol W for selective value in this theory represented an absolute value assigned to a given gene. Actually it has always been defined as applying to a *total* genotype in the system under consideration, thus involving whatever interaction effects there may be among the component genes (Wright, 1937).

The major question considered was whether there is "a mechanism by which the species may continually find its way from lower to higher peaks". One such mechanism is the production of clones since interclonal selection is obviously according to the genotype as a whole. In cases in which clones are not produced to a significant extent (as in vertebrates, most insects and many higher plants) it was concluded that the only effective mechanism was subdivision of the species into small local populations (demes), sufficiently isolated to permit more or less random differentiation with respect to alleles with minor differential effects, but not too isolated for interdemic selection (by differential growth and migration).

This process requires that there be a balance among the pressures on each of the gene frequencies so as to permit two, or better multiple, alleles at fairly high frequencies at many, or better all, of the loci that are involved. Mayr seems to imply in his beanbag analogy that it was assumed that there is typically an approach to homozygosis at all loci in the species, except for the occasional occurrence of favorable mutations that ultimately displace the previous type genes. Such an assumption was of course wholly incompatible with my theory.

In the second place, the process requires that there be a balance between the systematic pressures at each locus, tending to maintain control by the same selective peak in a given deme, and random processes that permit the gene frequencies to keep changing somewhat, and occasionally so much that the system drifts across a shallow "valley" in the surface of selective values, to come under the control of another and higher peak. Continuing evolution requires that random drift be not so great as to bring about fixation of genes. Any fixation that occurs should be that brought about by the relatively strong selection that carries the system, after crossing a valley, to the neighborhood of the new controlling peak. The course of evolution of the species as a whole is then determined by interdemic selection.

In this theory, the joint effects of random drift and intrademic selection merely supply raw material for interdemic selection. Mayr treats the role of random drift in this theory as if it were the same as the random fixation of the earlier theories of Gulick and Hagedoorn. The error here is analogous to that of treating the role of mutation in supplying raw material for Darwinian selection as if it were the same as in the mutation theories of de Vries and Goldschmidt.

"THE NEWER POPULATION GENETICS"

Mayr attributes to the newer population genetics "increasing emphasis on the interaction of genes", the discovery of "integrated coadaptive complexes", recognition of "variable selective values of genes", "Dobzhansky's balance theory of genetic variation" and in a later paragraph than that quoted earlier "the interpretation of the inheritance of quantitative characters by Mather and the development of the more sophisticated modern views on the interaction of genetic factors, on coadaptation and on genetic homeostasis".

As just noted the treatment of interaction systems was the central theme of most of "classical" population genetics. The "harmonious" system of genes that characterizes a selective peak is certainly an "integrated coadaptive complex". Such a system has by the definition of a peak, Lerner's property of genetic homeostasis and any coadaptive complex with genetic homeostasis is ipso facto located at a selective peak. Variable selective values were necessarily recognized in dealing with gene interaction. Mather's "polygenic" inheritance is basically the same thing as the multifactorial inheritance of Nilsson-Ehle, East and Shull. Mather's important contribution to the theoretical consequences of linkage is a part of mathematical population genetics. Dobzhansky's balance theory has at least elements in common with Fisher's theory of polymorphism and with one of the essential aspects of the balance theory of evolution that I developed.

What then has been the contribution of "the newer population genetics"? I think that Mayr has obscured this by his repeated treatment of mathematical population genetics and the genetics of natural populations as if they were alternative ways of making contributions of the same sort to evolutionary theory. Actually each plays a role in the synthetic theory that the other cannot possibly play. A synthesis is more than a list of contributions.

The role of the mathematical theory is that of an intermediary between the bodies of factual knowledge discovered at two levels, that of the individual and that of the population. It must deduce from the postulates at the level of the individual and from models of population structure what is to be expected in populations, and then modify its postulates and models on the basis of any discrepancies with observation and so on.

In developing the balance theory of evolution, I was trying to arrive at a judgment of the most favorable conditions for evolution under the Mendelian mechanism. No amount of mathematical analysis could prove that these conditions actually occur. The great achievement of the recent studies of natural populations has been to discover what natural populations are actually like, genetically. Dobzhansky's balance theory summarizes what he and his associates have actually found in nature and differs from abstract balance theories in concrete details that could hardly have been anticipated. I fully concur in Mayr's tribute to these tremendous achievements. Nobody, however, has understood better than Dobzhansky the reciprocal relation between concrete facts and abstract theory, as exemplified in his continual collaboration with mathematical population geneticists.

It is I think seriously misleading to divide the history of population genetics since 1900 into consecutive 20 year periods devoted respectively to the genetic mechanism, the mathematical deductions and the genetics of real populations. All of these strands have actually been developing simultaneously and in continual interaction since 1900. It is the binding together of the facts at the two levels of observation into a self-consistent theory, necessarily mathematical to a considerable extent, that constitutes the modern microevolutionary synthesis.

None of the three strands in this synthesis is complete. As we learn more of the general significance of such phenomena as the cytoplasmic heredity of Michaelis, Ephrussi and others, the activators and modulators of McClintock and Brink, paramutation of Brink, etc., these will have to be incorporated into the theory. The study of the effects of various patterns of selection in laboratory populations, in domestic animals, and cultivated plants still has far to go, and this is much more the case with the genetics of natural species. We may anticipate surprises that will require readjustments all along the line. The present symposium supplies many new data, largely at the level of actual populations, which must now be digested.

REFERENCES

FISHER, R. A. 1918. The correlation between relatives on the supposition of Mendelian inheritance. *Tr. R. Soc. Edinburgh* 52, part 2: 399–433.

FISHER, R. A. 1930. The genetical theory of natural selection. 272 pp. Oxford, Clarendon Press.

HALDANE, J. B. S. 1924. A mathematical theory of natural and artificial selection. Part 1. *Tr. Camb. Phil. Soc.* 23: 19–41.

HALDANE, J. B. S. 1932. *The causes of evolution.* New York and London; Harper and Bros.

JENNINGS, H. S. 1916. The numerical results of diverse systems of breeding. *Genetics* 1: 53–89.

KELLOGG, V. 1907. *Darwinism Today.* New York: Henry Holt & Co.

OSBORN, H. F. 1934. Aristogenesis, the creative principle in the origin of species. *Am. Natur.* 68: 193–235.

PEARSON, K. 1904. On a generalized theory of alternative inheritance with special reference to Mendel's law. *Philos. Tr. R. Soc. London* A 203: 53–86.

SUMNER, F. B. 1918. Continuous and discontinuous variations and their inheritance in Peromyscus. *Am. Natur.* 52: 177–208, 250–301, 439–454.

DE VRIES, H. 1906. Species and varieties, their origin by mutation. Chicago: The Open Court Publishing Co.

WEINBERG, W. 1910. Weiteres zur Theorie der Vererbung. *Arch. Rassenb.* 7: 35–49, 169–173.

WRIGHT, S. 1921. Systems of mating. *Genetics* 6: 111–178.

WRIGHT, S. 1922a. Coefficients of inbreeding and relationship. *Am. Natur.* 61: 330–338.

WRIGHT, S. 1922b. The effects of inbreeding and crossbreeding on guinea pigs. I, II. U. S. Dept. Agr. Bull. No. 1090, 63 pp. III. Ibid. No. 1121, 61 pp.

WRIGHT, S. 1931. Evolution in Mendelian populations. *Genetics* 14: 97–159.

WRIGHT, S. 1935a. The analysis of variance and the correlations between relatives with respect to deviations from an optimum. *J. Genet.* 30: 243–256.

WRIGHT, S. 1935b. Evolution in populations in approximate equilibrium. *J. Genet.* 30: 257–266.

WRIGHT, S. 1937. The distribution of gene frequencies in populations. *Proc. Nat. Acad. Sci.* 23: 307–320.

WRIGHT, S. 1956. Classification of the factors of evolution. *Cold Spring Harbor Symposia on Quant. Biol.* (1955) 20: 16–24D.

YULE, G. U. 1906. On the theory of inheritance of quantitative compound characters on the basis of Mendel's law—a preliminary note. *Rep. 3rd Internat. Conf. Genet.* 140–142.

42
Character Change, Speciation, and the Higher Taxa

Evolution 36, no. 3 (1982):427–43

Introduction

The fourth and last volume of Wright's *E&GP* appeared in 1978 when he was 89. But he continued, as he does to this day, his keen interest in the burning questions of evolutionary biology, his interest heightened by the increasingly frequent invocation of his own ideas (and those merely attributed to him!) in the controversies. Thus when Richard Dawkins fanned into an open flame the old argument between genic and organismic selection with his book *The Selfish Gene* (1976), Wright answered with a strong argument for organismic selection (Wright 204).

Since *E&GP* was specifically devoted to evolution in populations, Wright included little on speciation and macroevolution. Just as volume 4 of *E&GP* was in press, however, controversy over Eldredge and Gould's theory of punctuated equilibria (Eldredge and Gould 1972) was intensifying, initiating a new incarnation of the long-standing controversy over continuity in speciation and macroevolution. In reaction, Wright wrote two papers on the shifting balance theory in relation to speciation and macroevolution (Wright 205, 207); both were published when he was almost 93. The first of these, included here, Wright delivered at the famous Macroevolution Conference held at the Field Museum of Chicago in the fall of 1980. Interest in this paper was so great that Wright had more reprint requests for it than for any other paper he had ever published.

CHARACTER CHANGE, SPECIATION, AND THE HIGHER TAXA

SEWALL WRIGHT
Laboratory of Genetics, University of Wisconsin, Madison, Wisconsin 53706

Received December 9, 1980

Gould and Eldredge (1977) in their article, "Punctuated Equilibria: The Tempo and Mode of Evolution Reconsidered," took issue with the common view of phyletic gradualism. They agreed to a considerable extent with the view of Simpson, to whose book, published 36 years ago, their title referred. Simpson (1944) brought out the enormous differences among rates of evolution indicated by paleontological data: the near stasis (bradytely) of some forms for hundreds of millions of years, typical rates (horotely) (which, however, vary much among phyla) and the enormously rapid rates (tachytely) indicated especially for the periods of origin of many important groups. He used the term "quantum evolution" for evolutionary events of the sort referred to as "punctuational" by Gould and Eldredge. The latter began the abstract of their paper with the statement: "We believe that punctuational change dominates the history of life: evolution is concentrated in very rapid events of speciation (geologically instantaneous even if tolerably continuous in ecological time."

This does not contrast as much as it may seem at first sight with Simpson's assertion (p. 203) that "nine tenths of the pertinent data of paleontology fall into patterns of the phyletic mode," since he held that there might be episodes of tachytely in the phyletic evolution. There is more contrast, however, with respect to mode. Gould (1977) in his paper, "The Return of Hopeful Monsters," wrote: "I do predict that during the next decade, Goldschmidt (1940) will be largely vindicated in the new world of evolutionary biology."

In his paper, "Is a New and General Theory of Evolution Emerging?" (1980), he maintains that "Evolution is a hierarchical process with complementary but different modes of change of its three major levels, variation within populations, speciation and patterns of macroevolution."

Simpson definitely rejected Goldschmidt's (1940) thesis that speciation and the origin of the higher categories depend on types of mutation that have nothing in common with the changes that occur within species.

I am not in a position to discuss independently the data of paleontology and recognize that my field, genetics, bears directly only on microevolution, but I feel that we should explain phenomena at the higher levels as far as possible, as flowing from observed phenomena of genetics in the broad sense, including cytogenetics, before postulating wholly unknown processes. This does not bar me from accepting selection among entities at all levels of the biological hierarchy.

Historical Review

The opposed concepts, gradual and abrupt change, go back to the origins of evolutionary thought. At the beginning of the last century, Lamarck postulated a preordained ladder of life, leading from the simplest forms to man with gradually diverging branches, determined by the inheritance of characters acquired by adaptation to different conditions. E. Geoffroy St. Hilaire, on the contrary, proposed that radically different patterns of life had appeared abruptly from time to time, a view derived from the observed appearance of monstrosities. These speculations came when little was known of the actual course of evolution (cf. Nordenskiöld, 1928).

Darwin (1859), in midnineteenth century, was able to marshall the available data from all fields of biology in such a way as to win almost immediate acceptance by scientists of evolution as a fact. He also presented a theory that was at least in the spirit of physical science. He held that study of the ways in which animal and plant breeders had actually brought about striking changes by artificial selection provided insight into the evolutionary process in nature. After careful consideration (Darwin, 1868 Vol. 2) he concluded that: "Without variability, nothing can be effected: slight individual differences, however, suffice for the work and are probably the chief or sole means in the production of new species."

Thus Darwin came down strongly on the side of gradualness under natural selection though he did not wholly rule out an occasional role of the more striking changes that he called "sports."

The adequacy of Darwin's explanation of evolution was widely questioned, principally for two reasons: (1) the rapid dilution of variation, under the prevailing theory of a blending of the parental heredities, seemed to require an unbelievable efficiency of the selection process, an objection put in mathematical form by Fleeming Jenkin (1867), and (2) the difficulty of accounting for the extraordinary coadaptiveness of all parts of organisms, urged especially by St. George Mivart.

Many biologists continued to accept a perfecting principle as an essential property of life (e.g. Nägeli), a view fervantly advocated as late as 1934 by the leading paleontologist at the time, H. F. Osborn, in his doctrine of aristogenesis.

The majority, however, followed the paleotologist Cope in preferring the inheritance of the effects of use and disuse (cf. Kellogg, 1907).

At the turn of the century, de Vries proposed his mutation theory under which species appear abruptly, natural selection being relegated to the role of guiding the course of evolution beyond the species level. de Vries's theory had the merit of being based on the actual appearance of what seemed new species among plants of the American species, *Oenothera Lamarckiana,* that had escaped from cultivation in the Netherlands. It turned out later, however, that most of his supposed new species were trisomics with merely altered proportions of the elements of heredity of *O. Lamarckiana,* not transmissible, moreover, by pollen, so that the new "species" could exist only as segregants. One mutant form, *O. gigas,* turned out, however, to be a tetraploid, capable of reproducing itself and producing sterile triploid hybrids on backcrossing. It thus really did behave like a new reproductively isolated species, although differing morphologically from the parent species only very slightly. There was thus evidence that a new species might arise abruptly.

The rediscovery of Mendelian heredity in 1900 soon dissipated the first of the objections to Darwin's theory, referred to above. Yule pointed out in 1902 that segregation in the ratio 1:2:1 in F_2 of a cross persisted in randombred F_3 and in later randombred generations. Castle (1903) extended this to ratios based on other gene frequencies and Weinberg (1908) and Hardy (1908) put this in general mathematical form. Under Mendelian heredity there is no such dilution of hereditary elements as implied by the theory of blending heredity.

The study of conspicuous Mendelian difference, led most of the early geneticists to accept an attenuated form of the mutation theory: it was supposed that gene mutations with major effects were occasionally favorable and that these replaced the old type genes, one at a time. This was a form of 'punctuated' evolution but much less drastic than that of de Vries.

A few continued to accept speciation by mutation and some, notably Goldschmidt (1940), went beyond de Vries in holding that the origin of higher categories required mutations of appropriate sorts.

Castle, with an agricultural background, differed from the prevailing Mendelian view in the opposite direction. He challenged the current belief, tracing to de Vries, according to which the selection of quantitative variability could produce no permanent effects. He carried through large scale experiments with a strain of hooded rats in which he attempted to change the pattern of black and white by selection, experiments in which I was his assistant during 1912–1915. He was approaching self-black in 20 generations of plus selection, self-white in the same number of generations of minus selection, when reduced fecundity brought both lines to an end (Castle, 1916). My prior acceptance of Darwin's views on the efficacy of selection of quantitative variability was confirmed though I was somewhat disturbed by the forced termination of the experiments in spite of much heritable variability, which suggested that selection tended to have deleterious pleiotropic consequences.

Gene-Character Relations

Castle originally thought that the quantitative variability was in the piebald factor itself, a view confirmed to some extent by the appearance of a new allele by mutation in the plus series, an allele with effects intermediate between that of the allele in this series and that of the self-colored wild rats. Meanwhile, however, the multiple factor theory of quantitative variation, foreshadowed by Mendel himself and proposed by Yule in 1906, had been exhaustively demonstrated in a case in wheat by Nilsson-Ehle (1909) and more extensively in maize and *Nicotiana* by East (1910, 1916) and his associates. This provided an alternative interpretation of Castle's (1919) results which he accepted after carrying through an extensive test.

The Mendelian interpretation of natural selection requires analyses of the statistical consequences of diverse assumptions on the relations of genes to characters. Figure 1 (Wright, 1980) gives a diagrammatic illustration of a number of different assumptions. Figure 1C applies to cases in which a block of DNA (left) does not code for anything, or anything of significance, and thus has no phenotypic effect (right) that is subject to selection. Kimura (1968) and King and Jukes (1969) suggested that large portions of the protein molecules that are the primary products of DNA activity, may have no functional significance beyond filling space and that this may account for the extensive polymorphism of proteins revealed by electrophoresis in many species. This was obviously, however, not a general theory of gene-character relations.

Figure 1A represents the one-to-one relation of gene and "unit character" postulated in the prevailing Mendelian view of evolution, referred to above. The courses taken under constant selection, but diverse assumptions otherwise, were worked out systematically by Haldane in a series of papers, beginning in 1924 and summarized in 1932, following the working out of special cases by Castle (1903) and H. T. J. Norton (1915).

Such a one-to-one relation could, however, only be assumed for genes with major effects. Haldane also worked out consequences of other assumptions, including multifactorial heredity. Figure 1B represents the latter hypothesis, uncomplicated by gene interaction and pleiotropy. Fisher (1930) presented convincing reasons for holding that only minor factors are likely to be favorably selected. He accepted the occurrence of nonadditive relations between gene and character (dominance, epistacy) but showed that these merely slow down progress by selection under his assumption that species are effectively panmictic. Fisher, in contrast with Haldane, attempted to bring all evolution under a single simple formula, his "fundamental theorem of selection," "the rate of increase in fitness of any organism at any time is equal to its (additive) genetic variances in fitness at that time."

It is not wholly fair to assert that Fisher's conception of the relation of genes to characters was restricted to that of Figure 1B but he clearly tended to think in those terms. Thus his assumption (1929) that "a small selection of intensity, say 1/50,000 the magnitude of a larger one, will produce the same effect in 50,000 times the time," while true of gene character-relations of the sort represented in Figure 1B (and 1A) is not at all true of genes that modify a heterozygote and also have pleiotropic effects on the homozygotes. Fisher (1928) made the above assumption in connection with his theory that the prevailing dominance of wild type over recurrent deleterious mutations is due to specific modifiers of the rare heterozygotes, exerting a selective pressure of the order of the mutation rate, put at 10^{-6} per generation. His principal evidence for his theory was the easy modifiability of intermediate heterozygotes under strong direct selection. The latter well known phenomenon is obviously irrelevant if there are always at least slight pleiotropic effects on homozygous wild type (Wright, 1929*a*, 1977; Charlesworth, 1979).

Figure 1D represents each gene as having multiple pleiotropic effects because of interaction of its products with those of others. It may be assumed, moreover, that the effects of these interactions are not in general additive. These are inevitable consequences of the complex network of biochemical and developmental reactions that intervene between primary gene action and the ultimate effects subject to selection.

The available evidence is in harmony with this concept. My own major experimental project from the time when I was a graduate student (cf. Wright, 1968), has been the study of gene interactions in guinea pigs. Thus I have made many thousand different combinations of the genes at 11 loci that affect coat and eye color and considerable numbers of loci affecting other characters. From the first, I was fascinated by the seeming unpredictability of such combinations and have tried to devise hypothetical interaction patterns to account for them. There were also often surprising pleiotropic effects.

The latter have been even more striking in the studies of the genetics of the mouse by many workers, in which enormously more new mutations, usually deleterious, have been observed. More than 50 loci have been found that affect coat color, of which some 60% also have gross morphological effects (Searle, 1968; Silvers, 1979). On taking account of the pleiotropic effects on eumelanin and phaeomelanin, and the likelihood of other slight morphological and physiological effects, there can be little doubt that pleiotropy is universal. The situation is similar in other organisms.

My own speculations on evolution were dominated from the first by the thought that there must somehow be selection of coadaptive interaction systems as wholes. The difficulty was that under biparental heredity the reduction division breaks up combinations so rapidly in terms of geologic time, that under panmixia natural selection is capable of operating only on the average effects of genes in all combinations. Combinations of unlinked genes go halfway toward randomness per generation; with 10% recombination, half way per 7 generations; and with 1% recombination half way per 69 generations.

The Selective Topography

The nature of the field of variability available for natural selection under the various patterns of gene-character relations requires consideration.

Under the concept of one-to-one relationship (Fig. 1A), it should be possible theoretically to rank the alleles at each locus in the order of their values to the organism. Natural selection would operate under given conditions according to the courses described most systematically by Haldane as already noted.

An organism, however, is very far from being a mosaic of unit characters. The value of any gene depends in general on the array of other genes with which it is associated. This holds even under the pattern of relations of Figure 1B in which the effects of multiple loci are additive with respect to characters but usually there is an intermediate optimum close to the character mean.

Assume a group of genes *A, B, C* and *D* that contribute equally and additively to the size of some part, relative to their alleles, *a, b, c* and *d* (Fig. 2) (Wright, 1964*b*). Note that only the positive factors

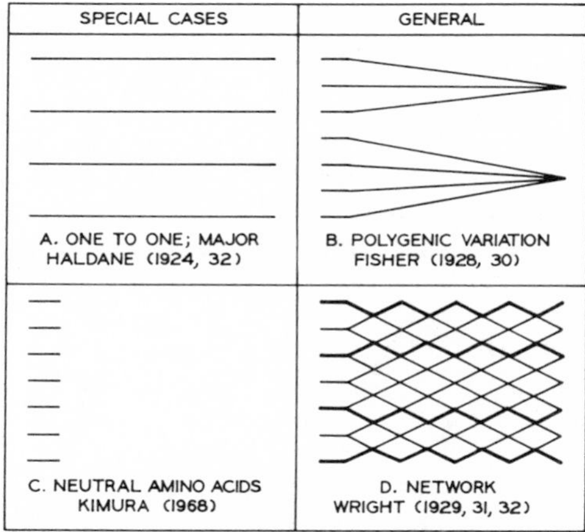

FIG. 1. Four assumptions on the relationship of genotype (left) to phenotype (right).

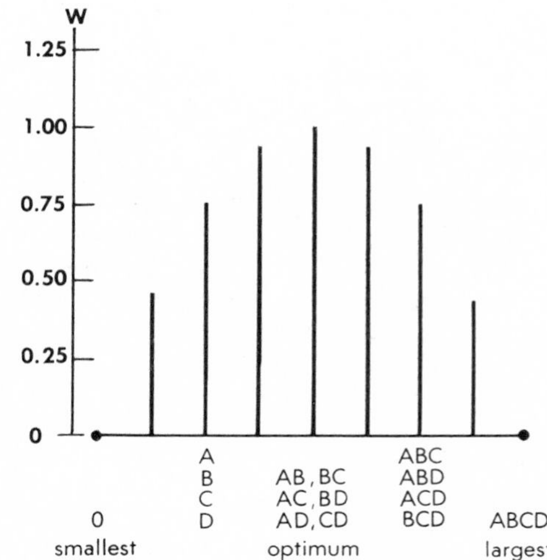

FIG. 2. Contributions to selective value of the combinations of four pairs of alleles to a quantitatively varying character, assuming equality and additivity with optimum in the middle. *AB* represents *AABBccdd*, etc. (from Wright, 1964*b*, Fig. 8).

are shown and these singly in the homozygotes. There is obviously no best allele at any locus. Fixation of any two positive genes and of the negative ones at the other two loci gives the mean and hence the optimum. There are six such optimal combinations (called selective peaks later): *AABBccdd, AAbbCCdd, AAbbccDD, aaBBCCdd, aaBBccDD* and *aabbCCDD*. The mean is also given by many other combinations such as *AaBbCcDd* but these involve heterozygosis and would give some inferior offspring.

In this case, it would make no difference which one of the six optimal types becomes established by selection. In an unfixed population, selection (according to Fisher's fundamental theory) will fix the combination to which the composition of the population is closest.

In actual cases, however, the maxima would not be equally fit, whether because of unequal gene effects or because of different pleiotropic effects. Let us assume that the effects are equal but that *A* and *B* have certain pleiotropic effects that are equal and additive. Figure 3 shows the selective values of the 16 homallelic populations on a vertical scale. With the chosen pleiotropic effect, there is one fittest type *AB*, four at the same lower level (*BC, AC, BD* and *AD*) and one lowest maximum, *CD*. The four homallelic combinations with only one positive gene and the four with three positive genes are still lower. That with all four positive genes is much lower and that with all four negative genes is the lowest of all.

Imagine now a figure with four orthogenal dimensions, one for each independent gene frequency and imagine a fifth dimension for selective values. Figure 4 shows two of the faces, including the lowest maximum *CD*, one of the four intermediate ones, *BC* and the highest one, *AB*.

The values chosen for the pleiotropic effects determine a saddle between peaks *CD* and *BC* and one between *BC* and *AB*. Arrows indicate the trajectories of populations subject to the assumed selection pressures. If it is assumed that there is recurrent mutation or a small amount of immigration from other demes, this prevents permanent fixation at any peak.

In Figure 5 the selection values of Figure 4 are shown according to a vertical scale. Selective values are also shown along

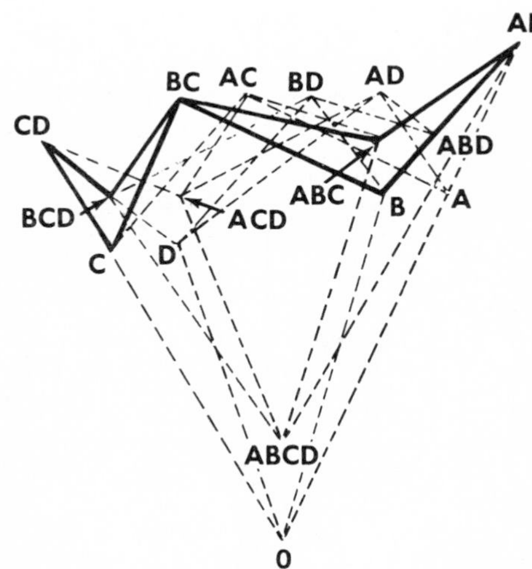

FIG. 3. Total selective values of the 16 homallelic genotypes of Fig. 2 according to the contributions indicated there, supplemented by equal semidominant pleiotropic effects of genes *A* and *B* (from Wright, 1964*b*, Fig. 9).

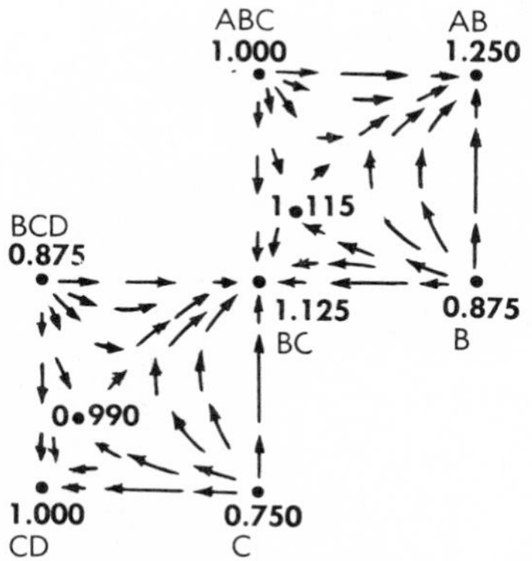

FIG. 4. Two of the surfaces of the 4-dimensional space of frequencies of genes *A, B, C* and *D* of Fig. 3. Trajectories are indicated by arrows. The selective values at the peaks, pits and saddles are given (from Wright, 1964*b*, Fig. 10).

paths from one peak to a higher one, passing through the saddle: A path directly from lowest to highest peak passes across the 4-dimensional saddle, but this is more depressed and thus presents a greater obstacle to a peak-shift than do the paths passing through the 2-dimensional saddles.

Since quantitative variability with optimum close to the mean is the usual rule for measurable characters in wild species, the foregoing model is practically universal for such characters. It has to do, however, with only small changes.

Of greater importance for major evolutionary changes are probably the unpredictable major interaction effects of genes, referred to earlier, for which Figure 1D with the full complexity expected from the network of biochemical and developmental processes, is intended to be the model. The selective topography would be correspondingly more complex.

Of special importance are interaction systems consisting of a recurrent major mutation and one or more modifiers that neutralize its inevitable deleterious side effects. A shallow saddle may lead to a great step in advance (Fig. 6).

In my 1932 paper, the first in which the concept of a selective topography was presented graphically, it was stated that "The problem of evolution, as I see it, is that of a mechanism by which the species may continually find its way from lower to higher peaks in such a field. In order that this may occur, there must be some trial-and-error mechanism on a grand scale by which the species may explore the region surrounding the small portion of the field which it occupies."

Modes of Evolution

At this point I will go back to a group of six diagrams (Fig. 7) that I used in 1932, and often later, for consideration of possible modes of evolution under Mendelian heredity. In these diagrams the multiple dimensions of gene frequency change, provided by thousands of heterallelic loci, are flattened out into two dimensions on which all genotypes are supposed to be located and the dimension of selective value is represented by contours. The innumerable selective peaks are given token representation in only two.

The three upper diagrams represent cases in which the number of individuals in an effectively random breeding population is so great that accidents of sampling can play no significant role. In Figure 7A it is assumed that recurrent

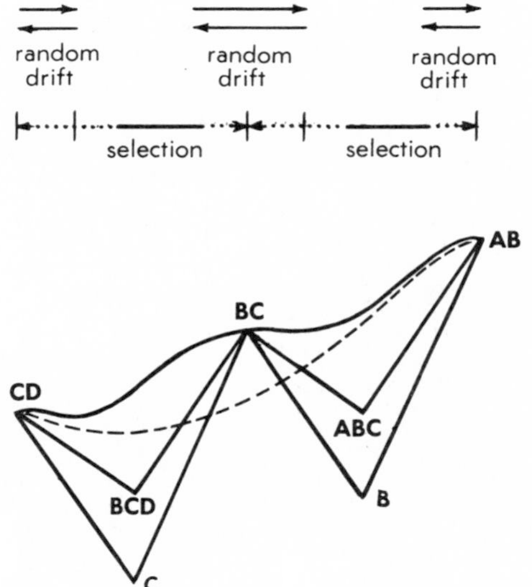

FIG. 5. Profile of trajectories from peak to peak (*CD* to *BC* to *AB*) through 2-dimensional saddles and from *CD* to *AB* through a 4-dimensional saddle and from pits to peaks (straight lines) (from Wright, 1964*b*, Fig. 11).

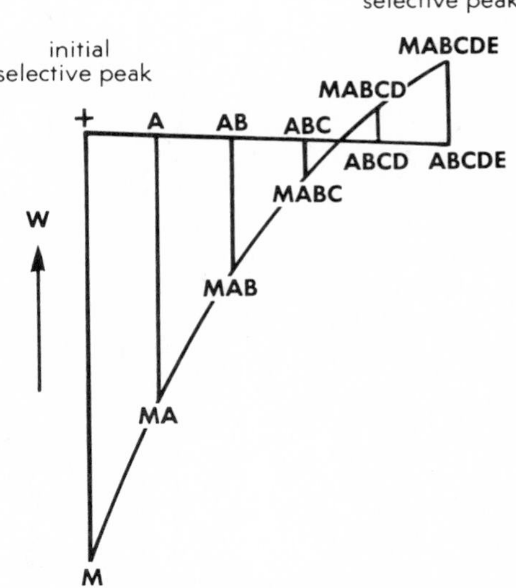

FIG. 6. Combinations of an initially deleterious, but potentially favorable, major mutation, with favorable modifiers (from Wright, 1964*b*, Fig. 13).

mutation is balanced by feeble selection. The population occupies a small portion of the field (indicated by a broken circle) about the joint equilibrium point. An increase in mutation rate would bring about an increase in the portion of the field occupied (indicated by a solid line). There is no appreciable chance, however, that this will come under control of a higher peak. A decrease in selection, such as occurs when a particular character ceases to be of value, has a similar effect. The character may wholly disappear under the mutation pressure but selection for pleiotropic effects of the genes is likely to be much more important beyond the point of selective advantage from degeneration of the character as an encumbrance (Wright, 1929*a*, 1964*a*). Mutation pressure by itself is probably of little importance for characters though it probably is for useless genes.

Figure 7B represents the case of mass selection in an effectively panmictic population, living for a long time under the same conditions. The population moves toward the selective peak, the slope of which it has reached for historic reasons, comes to occupy a region of the field about this peak and stays there without further change except in so far as novel favorable mutations create a change in the topography, an exceedingly slow process. With increased selection, the region occupied decreases.

If, however, conditions change qualitatively (Fig. 7C), the topography changes. The species tends to move with movement of the peak that it has occupied. As I noted in 1932: "Here we undoubtedly have an important evolutionary process and one which has been generally recognized. It consists largely of change without advance in adaptation. The mechanism is, however, one which shuffles the species about in the general field. Since species will be shuffled out of low peaks more easily than high ones, it should gradually find its way to the higher general regions of the field as a whole."

The process of change without advance is what Van Valen has called "the Red Queen Process." The process as a whole is that under which selection according to Fisher's fundamental theorem is most ef-

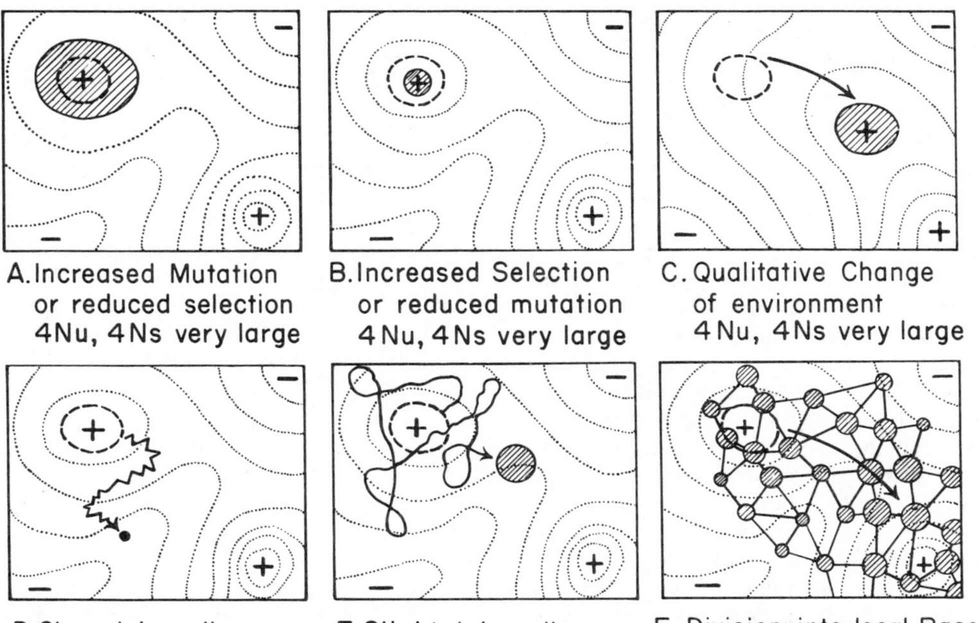

FIG. 7. Token representation of a portion of the multidimensional array of genotypes of a population with fitness contours. Field initially occupied indicated by heavy broken contour. Field occupied later indicated by crosshatched area (multiple subpopulations in F). Courses indicated in C, D, E and F by arrows. Effective population numbers, N (total), n (local); v (mutation), s (selection), m (migration) (from Wright, 1932, Fig. 4).

fective (more effective than under the unchanging conditions of Fig. 7B). It is the process which such recent authors as Maynard Smith (1975), Williams (1966) and Dawkins (1976) consider all-important.

Figure 7D refers to populations that have become so small that accidents of sampling overwhelm all but the strongest selective differences. The population wanders from the selective peak that it has occupied, moves about irregularly, decreases in variability. As I noted in 1929, (as essentially in 1931 and later): "In too small a population, there is nearly complete random fixation, little variation, little effect of selection and thus a static condition, modified occasionally by chance fixation of a new mutation, leading to degeneration and extinction."

In spite of this, most authors including Huxley (1942) and Fisher and Ford (1947) followed by many others (including textbooks published in 1979 and 1980), have attributed to me the view that fixation of nonadaptive characters by random drift was the essence of my theory.

It should be noted that I was considering the evolution of ordinary characters, not completely neutral primary gene effects such as Kimura (1968) and King and Jukes (1969) have discussed in recent years.

Figure 7E represents the case of loci with respect to which the effects of accidents of sampling and of selection are about equal within a rather small isolated population. I noted of this in 1932: "The species moves down from the extreme peak but continually wanders in the vicinity. There is some chance that it may encounter a gradient leading to another peak, shift its allegiance to this. Since it will escape relatively easily from low peaks—there is here a trial-and-error mechanism by which in time the species may work its way to the highest peak in the general field. The rate of progress is extremely slow, however, since change of gene frequency is of the order of the reciprocal of the effective population size and this reciprocal must be of the order of the mutation rate in order to meet the conditions of this case."

Finally, Figure 7F represents a species that is subdivided into local populations (demes), sufficiently small and isolated that accidents of sampling may overwhelm many weak selection pressures. There must, however, be enough diffusion that a deme that happens to acquire a favorable interaction system may transform its neighbors to the point of autonomous establishment of the same peak, and thus ultimately transform the whole species or at least that portion of it in which the new system actually is favorable. The field of variability of the species is here amplified by local differentiation, and natural selection is amplified by the selective diffusion from the superior demes.

The process within each of the demes is somewhat similar to that in Figure 7E, but is limited here only by the immigration rate which may be thousands of times the mutation rate. If there are thousands of sufficiently independent demes, the process in the species as a whole may be millions of times as effective as in Figure 7E. This is the process that I later called the shifting balance process.

I have devoted many papers (with conclusions summarized in Wright, 1969 Chapter 12) to the conditions under diverse population structures (continuous, clustered, 'island,' multiple colonies subject to frequent extinction and refounding from the superior ones) under which the process may be effective.

Going back to the model with factors *A, B, C* and *D,* the operation of the shifting balance process within the range of the species is represented in Figure 8. It is assumed that the species has been under control of the lowest of the six selective peaks (*CD*). Shifts to control by certain higher peaks, *BC, AC* and *AD* have occurred and are spreading by selective diffusion, according to their times of origin. The highest peak, *AB,* has been arrived at both within that characterization by *BC* and by overlapping of the regions controlled by *BC* and *AC*.

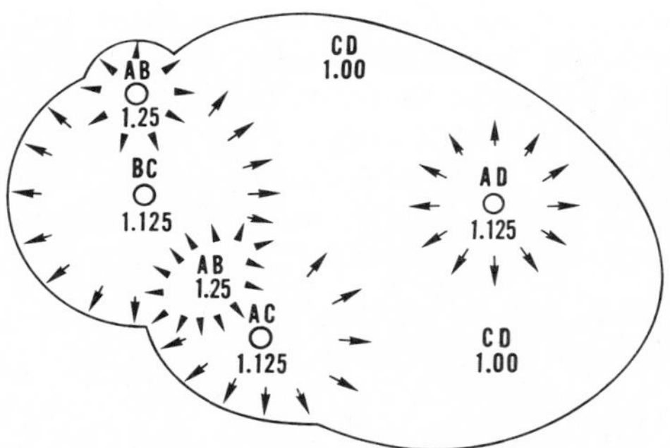

FIG. 8. Diagram of a population range, characterized initially by the lowest peak, *CD* of Fig. 4 in which intermediate peaks, *BC, AC* and *AD* have been arrived at locally and the highest peak *AB* has been arrived at from *BC* and from overlap of *BC* and *AC* (from Wright, 1964, Fig. 12).

It should be added that I suggested in 1931 and demonstrated later (Wright, 1948) that random differentiation of local populations may be due to fluctuations in the systematic pressures of selection and of amount and kind of immigration without there being small size populations. Random drift from accidents of sampling, however, affects all of the thousands of loci not subject to strong selection while fluctuations in selection would affect a more limited number.

Selection by Man in Relation to Evolutionary Theory

Darwin in "The Variation of Animals and Plants under Domestication" organized his discussion of "selection by man" under three heads: methodical, unconscious and natural, all concerned with direct selection among individuals. He did not discuss the sort of selection practiced by breeders in purchasing breeding stock from superior strains. The only references to this are in quotations. Thus in a section on "selection in ancient and semicivilized people" he quotes Virgil "as giving as strong advice as any modern agriculturist could do, carefully to select the breeding stock 'to note the tribe, the lineage, and the sire, whom to reserve for husband of the herd.'" He also quotes verses from ninth century Ireland describing a ransom demanded by Cormac:

Two pigs of the pigs of MacLir
A ram and ewe both round and red,
I brought with me from Aengus,
I brought with me a stallion, and a mare
From the beautiful stud of Manannan,
A bull and a white cow from Druim Cain.

There is clear recognition of selective diffusion from superior herds in a quotation from a Mr. Wilson to the effect that in certain districts of the Scottish Highlands: "The breeding of bulls is confined to a very limited number of persons, who by devoting their whole attention to this department are able from year to year to furnish a class of bulls which are steadily improving the general breed of the district." Darwin presented this quotation to illustrate "the great principle of division of labor" rather than that of selective diffusion from superior herds.

I do not mean to imply that Virgil, the ancient Irish author and Mr. Wilson recognized an important aspect of selection overlooked by Darwin. He undoubtedly recognized this aspect but the grading up of inferior stock by sires drawn from superior herds had no such clear analogy to anything happening in nature as did selection among individuals. It was moreover, impossible to grasp fully the qualitative difference between the merely *genic* consequences of selection among individuals and the *organismic* consequences of selection among differentiated herds, before the mechanism of heredity had come to be understood. This is still not grasped by those who urge strongly the importance of selection for coadaptation but reject the only process, interdeme selection, by which this can occur in nature under biparental reproduction.

Studies of the breeding history of the Shorthorn breed of cattle by means of a previously devised inbreeding coefficient (Wright, 1923; McPhee and Wright, 1925) led directly in 1925 to the two level shifting balance theory of evolution in nature, though publication was delayed in 1931 (abstract 1929). The general conclusions were somewhat revised in 1932 on the basis of the concept of a selective topography along the line brought out in the preceding section. The mathematical treatment in the 1931 paper dealt only with certain aspects of the simplest model. This was remedied and conclusions on various special aspects arrived at in papers in the next two decades but the general conclusions still held (cf. Wright, 1977).

These conclusions had to do with character change within species (microevolution). I had, however, accepted from the first the concept of speciation as reproductive isolation. I suggested in papers in 1940 (abstract 1938) and 1941*a* that incipient speciation from chromosomal rearrangement depends on an extreme pattern of population structure that is also very favorable for character changes by means of peak-shifts.

I tacitly treated macroevolution as merely an extension of evolution within species in my early papers but in reviews of books by Willis (Wright, 1941*b*), Goldschmidt (Wright, 1941*c*) and Simpson (Wright, 1945), all concerned especially with macroevolution, I held that the determining factor was in general an unusual ecological opportunity, not any sort of unusual mutation, an opportunity consisting of the occurrence for one reason or other, of vacant ecological niches, more or less related to that occupied by the species in question. The occupation of these in a rapid adaptive radiation constituted the origin of a new higher taxon. These concepts will be elaborated in the following sections.

Speciation

Discussion of speciation had best begin with the recognition that the term "species" has come to mean something very different, at least in principle, to neontologists from what it is possible for it to mean in practice to paleontologists. Before the beginning of the present century, there was no such divergence. The species were the kinds of organisms, distinguishable by clear-cut differences in morphology (excluding "varieties" known to occur within progenies).

It came to be recognized by such systematists as Osgood (1909) that reproductive isolation and thus incipient evolutionary branching, constituted a criterion of more significance biologically. A species came to refer to a population or group of intergrading populations, reproductively isolated from all other such populations.

Intergrading populations might differ greatly at their centers, but should be designated as merely subspecies, while ones that could be demonstrated to be reproductively isolated, even though showing no consistent morphological difference, should be considered separate species.

There are, of course, many intermediate cases, difficult to classify, but the identification of the branching points of the evolutionary process as accurately as possible came to be considered the primary objective.

This concept necessarily breaks down in the attempt to name successive species in a chain along which the morphological differences ultimately become so great that recognition of speciation becomes imperative. If change has occurred at a uniform rate, the boundaries between species are necessarily arbitrary. If, on the other hand, there have been periods of apparent stasis, separated by apparently abrupt, or at least very rapid, change, the boundaries are made most conveniently at the latter times.

Reproductive isolation may be brought about in several different ways (cf. Dobzhansky, 1970): (1) union of egg and sperm from the two populations may be prevented in various ways; (2) if fertilization occurs, normal development may fail because of imperfect cooperation of the two heredities; (3) chromosome rearrangements may interfere with normal meiosis, leading to more or less aneuploidy and thus incipient separation; and (4) there may be nucleo-cytoplasmic incompatibility which I will not discuss further.

Under the first head, mating may be prevented by geographic isolation, by difference in the breeding seasons, by difference in ethological patterns such as courtship, by ecological difference within the same region or by anatomical incompatability. Finally there may be biochemical bars to fertilization.

Under the second head, faulty development or low fecundity may occur in F_1 but if F_1 is normal, the irregular proportions of the components of the two heredities in F_2 and later may lead to enough faulty development or low fecundity to bring about at least incipient speciation.

Under both of these heads, the difference may be due either to a single locus or may require passage of a threshold after accumulation of slight effects at many loci. There is no essential difference from microevolutionary changes except in that there is some contribution to reproductive isolation.

Chromosomal differences are conveniently divided into balanced and unbalanced. The latter includes duplications and losses either of whole chromosomes or of portions down to single loci. The unlikelihood of speciation involving whole chromosomes has already been touched on in connection with de Vries' *Oenothera* mutations.

Deficiencies of any extent are usually lethal when homozygous and thus also have little or no significance for evolution. Small duplications on the other hand, are usually viable even as homozygotes. Typically they have effects similar to those of gene mutations and are transmitted as if Mendelian genes. Many of them probably become inactivated ultimately. Those that persist tend to become differentiated and increase the stock of loci. They are undoubtedly important in evolution but not with respect to speciation.

This brings us to the balanced chromosome changes. Here, as noted earlier, we encounter a real abrupt species-forming mechanism, the duplication of the entire genome. If occurring in a hybrid between species whose chromosomes have become so differentiated that none of them pair in meiosis, amphidiploids are formed, capable of breeding true to the hybrid phenotype and reproductively isolated from both parent species because of the sterility of the triploids resulting from backcrosses. There is no doubt that a great many species of plants, though few of animals, had this origin. On the other hand, the autotetraploids formed from doubling the genome of a single species differ little from the parent in phenotype. They can reproduce their type and produce sterile triploids in backcrosses but the sets of four chromosomes cause irregularities in meiosis. They are, no doubt, responsible for many new species but not as many as amphidiploids. Allotetraploids with pairing of some but not all chromosomes resemble full amphidiploids in originating a morphologically new species but resemble autotetraploids in having some irregularity in meiosis.

Balanced chromosome rearrangements include inversions and translocations of various sorts. Paracentric inversions, at least in *Drosophila* and some grasshoppers, suffer very little reduction in fecundity. They are important in evolution by permitting segregation of chromosomes, or at least large blocks, as wholes. They permit adaptation to different ecological niches and thus increase the adaptability of the species, but are not involved in speciation.

Pericentric inversions and translocations typically lead to significant reduction of the fecundity of heterokaryons, 50% in the case of a typical reciprocal translocation, and thus to incipient speciation. They may be important in speciation for this reason even though the character changes (from position effects) are slight or absent.

According to M. J. D. White (1978): "Over 90 percent (and perhaps over 98 percent) of all speciation events are accompanied by karyotypic changes and—in the majority of these cases, the structural chromosomal rearrangements have played a primary role in initiating divergence."

On the other hand, there are many cases in which a chain of intergrading subspecies has returned on itself in a circle and it is found that the overlapping populations coexist without interbreeding as if distinct species (Osgood, 1909; Mayr, 1963). Extinction of the intermediates would elevate them to this rank without any change whatever in themselves. These cases indicate that not all speciation depends on chromosomal rearrangement.

Returning to the latter, there is a problem as to how a new chromosomal rearrangement can pass the barrier imposed by strong selection against the heterokaryons. White (1978) and Hedrick (1981) lean to meiotic drive from asymmetrical segregation, as the explanation.

Fixation by accidents of sampling is, however, another possibility. It can occur only in extremely small colonies (Wright, 1940, 1941*a*). The most favorable situation is that in a region in which there are numerous colonies, subject to frequent extinction and refounding by stray fertilized females from the more successful ones. Since this is also a situation especially favorable to peak-shifts, there should be a strong correlation between speciation by the above process and favorable character changes.

Studies of the differences between closely related species have made it clear that reproductive isolation usually involves several processes (review by Dobzhansky, 1970). It is clear that incipient reproductive isolation, whatever its cause, tends to be made complete, in one way or other, by natural selection against individuals involved in hybridization.

The general conclusion of this section is that speciation may take place in a great many different ways that have nothing in common except the promotion of reproductive isolation and hence branching of the evolutionary process. The effects of speciating events on other characters range from zero where the event is merely the extinction of intermediates in a chain of subspecies, to abrupt and great in the case of amphidiploidy, but in general the effects on characters are at the microevolutionary level.

The Higher Categories

The only aspect of the evolution of the higher categories of which the actual genetics is fairly clear, is the course of substitution in the amino acid sequences of several proteins over long periods of geologic time (Zuckerkandl and Pauling, 1965). The rates of substitution separating hemoglobin β of man from the globin of the lamprey (1.3 per site per billion years), hemoglobins α or β of several mammals from α of a bony fish (carp), and either α or β of several mammals with each other, are surprisingly uniform (grand average 1.1 per site per billion years) (Kimura, 1969).

This is largely true of substitutions in the amino acid sequences of cytochrome-c throughout eukaryotic evolution since the separation of the higher plants, insects, lower chordates, reptiles; birds, and mammals from the fungi some 1.3 billion years ago and the separation of the higher groups from each other (Fitch and Markowitz, 1970).

The rates in the nonessential portions of any given protein have been so uniform that Kimura has held that natural selection must be ruled out and has proposed that the substitutions are the cumulative consequences of accidents of sampling. While most others advocate mass selection (e.g., Ayala, 1975) and I have proposed the shifting balance process (Wright, 1978), the near uniformity of rate over a period encompassing the origins of kingdoms and phyla implies continuity of the evolutionary process.

The evolutionary potentialities of most species are probably restricted to very slow, gradual progress in adaptation to the single ecological niche which they occupy, by the occupancy of all related niches by other species. From time to time, however, an opportunity is presented for adaptive radiation into other niches. Such ecological opportunities are, it has seemed to me, much the most important cause of the origin of higher categories (Wright, 1941*a*, 1941*b*, 1945, 1949). The occurrence of any particular kind of mutation (usually recurrent) is of little significance in the absence of such an opportunity.

Such an opportunity arises on entry of the species into a region in which niches, related to its own, are unoccupied. The remarkable adaptive radiation of families of marsupials in Australia and of families of several primitive orders of mammals in South America, both in the early Cenozoic, are examples. At a lower level is the adaptive radiation of the Geospizidae from a species of ground finch that reached the Galapagos Islands (that attracted Darwin's attention) and that of the Drepanididae from a species of bird (honeycreeper) that reached the Hawaiian Islands.

A similar opportunity is presented to the survivors of a catastrophe that has caused extensive extinction of species. The extraordinary adaptive radiation of the orders and families of mammals in the Paleocene and Eocene following the worldwide extinction of the dinosaurs and many other forms at the end of the Mesozoic is an example. The mammals had existed as a small group for some hundred million years without much differentiation before this opportunity occurred.

Of special importance, however, are the cases in which evolution along a restricted line happens to lead to an adaptation that turns out to open up an extensive new way of life. The evolution of a motile, tadpole-like larva of a certain type of sessile echinoderm seems to have opened the way, by means of neoteny, to the active life of the chordates other than tunicates. Later modification of the first gill arch made possible an effective jaw, leading to the adaptive radiation of the fishes. Modification of the ventral fins to permit locomotion on land and of the swim bladder for respiration in the air, enabled crossopterygian fishes to move from one drying pool to another, a peripheral niche for a fish, but this opened up life on land and the adaptive radiation of the amphibia. Similarly certain specializing adaptations later led to the adaptive radiations of the reptiles and birds.

A species presented with an opportunity to invade an unoccupied niche would be able to use mutations with more drastic effects than the quantitative variants used before and to do so sometimes in spite of rather unfavorable side effects because of the absence of competition. These may in a sense be considered to be the species mutations postulated by Goldschmidt but it is the ecological opportunity, not mere occurrence of this sort of mutation, that leads the way.

It should be added that much more drastic mutations may be utilized and fixation may be much more rapid if the mutation is recurrent and comes to be associated with otherwise neutral modifiers that remove its inevitable deleterious side effects. In all three of the book reviews referred to earlier (Wright 1941*b*, 1941*c*, 1945), it was noted that the origin of a higher taxon is expected to be an "explosively" rapid process under the shifting balance theory. This has been reiterated in many later papers.

The reorganization required for the origin of the highest categories may seem so great that only "hopeful monsters" will do. Here, however, we must consider the size and complexity of the organisms. Such changes would probably have been impossible except in an organism of very small size and simple anatomy. I have recorded more than 100,000 newborn guinea pigs and have seen many hundreds of monsters of diverse sorts (Wright, 1960) but none were remotely "hopeful," all having died shortly after birth if not earlier. Yet among nine specimens of a small trematode (*Microphallus opacus*) about 1.5 mm long, of which I made serial sections in my first research project (Wright, 1912), one was highly abnormal in form and had two large ovaries instead of only one. It would probably have been considered a monster if it had been a large complicated organism, but it was apparently flourishing as well as the others before it was fixed.

It would hardly be possible for a typical clam to be derived from a typical snail by mutation or a succession of mutations but it is not unreasonable to suppose that a small protomollusk with a single simple shell, produced a mutant in which this was divided laterally into two and that the pelecypods evolved from this mutant type. All of the classes of mollusks could reasonably have arisen in somewhat this way. All of the phyla of multicellular invertebrates and most of their classes probably arose from small rather simple forms, most of them not long, in geologic terms, after the origin of the eukaryotic cell.

The pattern of evolution thereafter was probably one of very gradual orthogenetic progress along many lines with occasional appearance of ecological opportunities of the sorts discussed above, followed by explosively rapid divergence of species in exploiting these. The group could be considered to constitute a new genus unless secondary divergencies raised the level of the group to the family and tertiary divergencies, perhaps, raised it to the level of a new order, all within no more than a moment in geologic time. At some level, major divergence ceases, and the species settle down with only minor differentiation of new species to slow orthogenetic progress (Wright, 1949).

Summary

The implications of the shifting balance theory with respect to the course of evolution agree in the main with the pattern indicated by the fossil record, according to Simpson in 1944 (cf. Wright, 1945), reiterated by Gould and Eldredge (1977) in their statement quoted in our introduction: "Punctuated change dominates the history of life: evolution is concentrated in very rapid events of speciation (geologically instantaneous even if tolerably continuous in ecological time." I would, however, substitute the phrase "of character change" for "of speciation." Character change and speciation (in the sense of reproductive isolation) are wholly different phenomena genetically, even though closely correlated in occurrence.

There is agreement only with the first but not the last part of the sentence also quoted in our introduction, from Gould (1980): "Evolution is a hierarchic process with complementary but different modes of change of its three leading varieties: within species, speciation and patterns of macroevolution." The shifting balance process is a two-level one (selection among individuals and among differentiated local populations), but no difference is assumed in the rates of minor and major mutation during the phases of near-stasis and rapid change. The interpretation of these phases under the shifting balance theory is in terms of differences in ecological opportunity. Speciation tends to accompany rapid change both because each of these processes tends to bring about the other and because speciation from chromosome rearrangement and peak-shifts is favored by the same population structure (numerous small colonies subject to frequent extinction and refounding by stray individuals from the more flourishing colonies).

According to the shifting balance theory the determining factor for rapid change, and the origin of a new higher taxon that usually accompanies such change, is the presence of one or more vacant ecological niches, whether from entrance of the species into relatively unexploited territory or from its survival after a catastrophe has eliminated other species occupying related niches, or from gradual attainment of an adaptation that opens up a new way of life.

We consider first the course of evolution of a species restricted to a single niche (because of occupancy of all related niches by other species) and living for a long time under relatively unchanging conditions. If its population density is great or there is a wide dispersion of offspring, a state of near-stasis should soon be reached as it comes to occupy the most available selective peak. It cannot move down from this against natural selection to reach a saddle leading to a higher selective peak.

If on the other hand, the species occupies a wide range but in part at least only sparsely and with restricted dispersion, the operation of the shifting balance process leads to gradually improving adaptation of the species as a whole, by means of successive minor peak-shifts and selective diffusion from them.

Still with only a single niche but under continuously changing conditions, there is continual readaptation largely of the treadmill sort, change without progress. A very gradual improvement is, however, to be expected as the species is shuffled into the higher general regions of the selective topography. This occurs even in populations that are effectively panmictic. In such a population, the rate of change is approximately according to Fisher's fundamental theorem except as qualified by frequency dependent selection or linkage disequilibrium. If, however, population structure permits significant operation of the shifting balance process, readaptation is facilitated by minor peak-shifts, not allowed for in Fisher's theory.

In cases in which new ecological niches become available in any of the ways referred to, their occupation may require allelic substitutions with major effects. Such substitution may occur in spite of imperfect adaptation and the inevitable deleterious effects of any major change, because of the absence of competition, but is greatly facilitated if population structure is favorable to peak-shifts, involving the gene in question and one or more nearly neutral modifiers that tend to eliminate the more deleterious of the side effects. Such major peak-shifts are most likely if the major mutation is recurrent and thus becomes available, sooner or later, wherever a favorable modifier or a pair of such modifiers, reaches sufficiently high frequencies for crossing of a saddle that pulls the major mutation to occupancy of the higher peak.

Such occupancy tends to give incipient reproductive isolation, followed by full speciation under selection against hybridization. On the other hand, speciation may come first, because of geographical isolation of a portion of the species or local establishment of a chromosome rearrangement or other cause, and be followed by occupancy of the new niche by means of a major peak-shift. In either case, the decisive peak-shift occurs within a species and is likely to be accompanied by minor peak-shifts that improve adaptation.

The occupancy of new niches, accompanied by speciation, constitutes the origin of a new higher taxon. This may merely be a new genus but if there is extensive adaptive radiation into new niches, the array may constitute a new family, new order, or very rarely a new class or phylum.

It may seem that mutations with impossibly drastic effects would be required for the origins of the higher of the taxa. Such origins, however, probably all occurred from species, the individuals of which were so small and simple in their anatomies that mutational changes, that would be complex in a large form, were not actually very complex.

The final conclusion is that the evolutionary processes indicated by the fossil record can be interpreted by the shifting balance theory without invoking any causes unknown to genetics or ecology.

Literature Cited

Ayala, F. J. 1975. Scientific hypotheses, natural selection and the neutrality theory of protein evolution, p. 19–42. *In* F. M. Solzanao (ed.), The Role of Natural Selection in Human Evolution. North Holland Publ. Co., Amsterdam.

Castle, W. E. 1903. The laws of Galton and Mendel and some laws governing race improvement by selection. Proc. Amer. Acad. Arts Sci. 39:233–242.

———. 1916. Further studies of piebald rats and selection with observations on gametic coupling. Carnegie Inst. Washington. Publ. no. 241, p. 163–190.

———. 1919. Piebald rats and selection: a correction. Amer. Natur. 53:370–376.

Charlesworth, B. 1979. Evidence against Fisher's theory of dominance. Nature 78:848–849.

Darwin, C. 1859. The Origin of Species by Means of Natural Selection. 6th ed. John Murray, London. D. Appleton and Co., New York. 1910.

———. 1868. The Variation of Animals and Plants Under Domestication. 2nd ed. John Murray, London. D. Appleton, London. 1883.

Dawkins, R. 1976. The Selfish Gene. New York Univ. Press, N.Y.

Dobzhansky, Th. 1970. Genetics and the Evolutionary Process. Columbia Univ. Press, N.Y.

East, E. M. 1910. A Mendelian interpretation of variation that is apparently continuous. Amer. Natur. 44:65–82.

———. 1916. Studies on size inheritance in Nicotiana. Genetics 1:164–176.

Fisher, R. A. 1928. The possible modification of the responses of the wildtype to recurrent mutations. Amer. Natur. 63:115–126.

———. 1929. The evolution of dominance: a reply to Professor Sewall Wright. Amer. Natur. 63:553–556.

———. 1930. The Genetical Theory of Natural Selection 2nd ed. Clarendon Press, Oxford. Dover, Publ. 1958.

Fisher, R. A., and E. B. Ford. 1947. The spread of a gene in natural conditions in a colony of the moth *Panaxia dominula*. Heredity 1:143–174.

Fitch, W. M., and E. Markowitz. 1970. An improved method for determining codon variability in a gene and its applicability to the rate of fixation of mutations in evolution. Biochem. Genet. 4:579–593.

Goldschmidt, R. 1940. The Material Basis of Evolution. Yale Univ. Press, New Haven.

Gould, S. J. 1977. The return of hopeful monsters. Nat. Hist. Magazine, June–July 1977:22–30.

———. 1980. Is a new and general theory of evolution emerging? Paleontology 6:119–130.

Gould, S. J., and N. Eldredge. 1977. Punctuated equilibria: the tempo and mode of evolution reconsidered. Paleontology 3:115–151.

Haldane, J. B. S. 1924. A mathematical theory of natural and artificial selection. Part I. Cambridge Phil. Soc. 23:19–41.

———. 1932. The Causes of Evolution. Harper and Brothers, Publ., N.Y.

Hardy, G. H. 1908. Mendelian proportions in a mixed population. Science 28:49–50.

Hedrick, P. W. 1981. The establishment of chromosomal variants. Evolution 35:322–332.

Huxley, J. S. 1942. Evolution, the Modern Synthesis. Harper and Brothers, Publ., N.Y.

Jenkin, F. 1867. Origin of species. North British Rev. 46:277–318.

Kellogg, V. L. 1907. Darwinism Today. Henry Holt and Co., N.Y.

Kimura, M. 1968. Evolutionary rate at the molecular level. Nature 217:624–626.

———. 1969. The rate of molecular evolution considered from the standpoint of population genetics. Proc. Nat. Acad. Sci. USA 63:1181–1188.

King, J. L., and T. H. Jukes. 1969. Non-Darwinian evolution. Science 164:788–798.

Maynard Smith, J. 1975. The Theory of Evolution. Penguin, London.

Mayr, E. 1963. Animal Speciation and Evolution. Harvard Univ. Press, Cambridge.

Nilsson-Ehle, H. 1909. Kreuzungsuntersuchungen an Häfer und Weizen. Lunds Univ. Aarskr. NF 5:2:1–22.

NORDENSKIÖLD, E. 1928. The History of Biology. Translated by L. B. Eyre. Tudor Publ. Co., N.Y.

NORTON, H. T. J. 1915. *In* R. C. Punnett (ed.), Mimicry in Butterflies. Cambridge Univ. Press, Cambridge.

OSBORN, H. F. 1934. Aristogenesis, the creative principle in the origin of species. Amer. Natur. 68:193–235.

OSGOOD, W. H. 1909. Revision of the mice of the American genus *Peromyscus,* North American Fauna Bull. No. 28 Washington, D.C. Bur. Biol. Survey, USDA.

SEARLE, A. G. 1968. Comparative Genetics and Coat Colour in Mammals. Academic Press, N.Y.

SILVERS, W. K. 1979. The Coat Colors of Mice: a Model for Mammalian Gene Action and Interaction. Springer-Verlag, N.Y.

SIMPSON, G. G. 1944. Tempo and Mode in Evolution. Columbia Univ. Press, N.Y.

VRIES, H. DE. 1901–1903. Die Mutations theorie, 2 Vol. Veit Co., Leipzig.

WEINBERG, W. 1908. Ueber den Nachweis der Vererbung beim Menschen. Jahresheft Ver. Naturforsch. F. Württemberg 64:368–382.

WHITE, M. J. D. 1978. Modes of Speciation. W. H. Freeman and Co., San Francisco.

WILLIAMS, G. C. 1966. Adaptation and Natural Selection. A Critique of Evolutionary Thought. Princeton Univ. Press, Princeton.

WRIGHT, S. 1912. Notes on the anatomy of the trematode, *Microphallus opacus*. Trans. Amer. Micr. Soc. 31:167–175.

———. 1929*a*. Fisher's theory of dominance. Amer. Natur. 63:274–279.

———. 1929*b*. Evolution in a Mendelian population. Anat. Rec. 44:287.

———. 1931. Evolution in Mendelian populations. Genetics 16:97–159.

———. 1932. The roles of mutation, inbreeding, crossbreeding and selection in evolution. Proc. VI Internat. Genet. Cong. 1:356–366.

———. 1940. Breeding structure of populations in relation to speciation. Amer. Natur. 74:232–248.

———. 1941*a*. On the probability of fixation of reciprocal translocations. Amer. Natur. 75:513–522.

———. 1941*b*. The "Age and Area" concept extended. (Review of book by J. C. Willis) Ecology 22:345–347.

———. 1941*c*. The material basis of evolution (Review of book by R. Goldschmidt) Sci. Monthly 53:165–170.

———. 1945. Tempo and mode in evolution: a critical review. Ecology 28:415–419.

———. 1948. On the roles of directed and random changes in gene frequency in the genetics of populations. Evolution 2:279–294.

———. 1949. Population structure in evolution. Proc. Amer. Phil. Assoc. 93:471–478.

———. 1960. The genetics of vital characters of the guinea pig. J. Cell Comp. Physiol. 56 Suppl. 1:123–151.

———. 1964*a*. Pleiotropy in the evolution of structural reduction and of dominance. Amer. Natur. 98:65–69.

———. 1964*b*. Stochastic processes in evolution, p. 199–241. *In* John Garland (ed.), Stochastic Models in Medicine and Biology. Univ. Wisconsin Press, Madison.

———. 1968. Evolution and the Genetics of Populations, Vol. 1. Genetic and Biometric Foundations. Univ. Chicago Press, Chicago.

———. 1969. Evolution and the Genetics of Populations, Vol. 2. The Theory of Gene Frequencies. Univ. Chicago Press, Chicago.

———. 1977. Evolution and the Genetics of Populations, Vol. 3. Experimental Results and Evolutionary Deductions. Univ. Chicago Press, Chicago.

———. 1978. Evolution and the Genetics of Populations, Vol. 4. Variability Within and Among Natural Populations. Univ. Chicago Press, Chicago.

———. 1980. Genic and organismic selection. Evolution 34:825–843.

YULE, G. U. 1902. Mendel's laws and their probable relation to interracial heredity. New Phytol. 1:192–207, 222–238.

———. 1906. On the theory of inheritance of quantitative compound characters and the basis of Mendel's laws: a preliminary note. Proc. III Internat. Cong. Genet. p. 140–142.

ZUCKERKANDL, E., AND L. PAULING. 1965. Evolutionary divergence and convergence in progeins, p. 97–166. *In* V. Bryson and H. J. Vogel (eds.), Evolving Genes and Proteins. Academic Press, N.Y.

Corresponding Editor: D. J. Futuyma

References Cited in Introductions

(Extensive bibliographies can be found in *SW&EB* and *E&GP*.)

Crow, J. F., and M. Kimura. 1970. An introduction to population genetics theory. New York: Harper and Row. Crow and Kimura dedicated this book to Wright.

Dawkins, R. 1976. *The selfish gene*. New York: Oxford University Press.

Dobzhansky, T. 1937. *Genetics and the origin of species*. New York: Columbia University Press.

Dobzhansky, T., and C. Epling. 1942. Genetics of natural populations, VI: Microgeographic races in *Linanthus parryae*. *Genetics* 27:317–32.

Eldredge, N., and S. J. Gould. 1972. Punctuated equilibria: An alternative to phyletic gradualism. In *Models in paleobiology*, ed. T. J. M. Schopf, 82–115. San Francisco: W. H. Freeman.

Fisher, R. A. 1922. On the dominance ratio. *Proceedings of the Royal Society of Edinburgh* 42:321–41.

———. 1928a. The possible modification of the response of the wild type to recurrent mutation. *American Naturalist* 62:115–26.

———. 1928b. Two further notes on the origin of dominance. *American Naturalist* 62:571–74.

———. 1929. The evolution of dominance: A reply to Professor Sewall Wright. *American Naturalist* 63:553–56.

———. 1930. *The genetical theory of natural selection*. Oxford: Oxford University Press. 2d ed., 1958. New York: Dover.

———. 1934. Professor Wright on the theory of dominance. *American Naturalist* 68:370–74.

Fisher, R. A., and E. B. Ford. 1947. The spread of a gene in natural conditions in a colony of the moth *Panaxia dominula*. *Heredity* 1:143–74.

———. 1950. The "Sewall Wright effect." *Heredity* 4:117–19.

Ford, E. B. 1975. *Ecological genetics*. 4th ed. London: Chapman and Hall.

Goldschmidt, R. B. 1940. *The material basis of evolution*. New Haven: Yale University Press. Reprinted 1982, with introduction by S. Rachootin.

Gould, S. J. 1980. G. G. Simpson, paleontology, and the modern synthesis. In *The evolutionary synthesis*, ed. E. Mayr and W. Provine, 153–71. Cambridge: Harvard University Press.

———. 1982. Introduction. In the reprint of Dobzhansky 1937, xvii–xli. New York: Columbia University Press.

———. 1983. The hardening of the modern synthesis. In *Dimensions of Darwinism*, ed. M. Grene, 71–93. New York: Cambridge University Press.

Haldane, J. B. S. 1932. *The causes of evolution*. London: Longmans.

Huxley, J. S. 1940. *The new systematics*. Oxford: Oxford University Press.

Kimura, M., and J. F. Crow. 1964. The number of alleles that can be maintained in a finite population. *Genetics* 49:725–38.

Lewontin, R. C., J. A. Moore, B. Wallace, and W. B. Provine, eds. 1981. *Dobzhansky's genetics of natural populations*. New York: Columbia University Press.

Mayr, E. 1940. Speciation phenomena in birds. *American Naturalist* 74:249–78.

———. 1942. *Systematics and the origin of species*. New York: Columbia University Press.

———. 1954. Change of genetic environment and evolution. In *Evolution as a process*, ed. J. Huxley, A. C. Hardy, and E. B. Ford, 157–180. London: Allen and Unwin.

———. 1959. Where are we? *Cold Spring Harbor Symposia on Quantitative Biology* 24:1–14.

Mayr, E., and W. B. Provine, eds. 1980. *The evolutionary synthesis*. Cambridge: Harvard University Press.

Provine, W. B. 1971. *Origins of theoretical population genetics*. Chicago: University of Chicago Press.

Simpson, G. G. 1944. *Tempo and mode in evolution*. New York: Columbia University Press.

Willis, J. C. 1940. *The course of evolution by differentiation or divergent mutation rather than by selection*. Cambridge: Cambridge University Press.

Wilson, D. S. 1980. *The natural selection of populations and communities*. Menlo Park, Calif.: Benjamin/Cummings.

———. 1983. The theory of kin selection. *Annual Review of Ecology and Systematics* 13:23–55.

Publications by Sewall Wright

*(Those marked by an asterisk * are reproduced in this volume)*

1. 1912 Notes on the anatomy of the trematode, *Microphallus opacus*. *Transactions of the American Microscopical Society* 31:167–75.
2. 1914 Duplicate genes. *American Naturalist* 48:638–39.
3. 1915 The albino series of allelomorphs in guinea pigs. *American Naturalist* 49:140–48.
4. Two color mutations of rats which show partial coupling. With W. E. Castle. *Science* 42:193–95.
5. 1916 An intensive study of the inheritance of color and of other coat characters in guinea pigs with especial reference to graded variation. In Carnegie Institution of Washington Publication No. 241,59–160.
6. 1917 On the probable error of Mendelian class frequencies. *American Naturalist* 51:373–75.
7. The average correlation within subgroups of a population. *Journal of the Washington Academy of Science* 7:532–35.
8. Color inheritance in mammals, I. *Journal of Heredity* 8:224–35.
9. II: The mouse. *Journal of Heredity* 8:373–78.
10. III: The rat. *Journal of Heredity* 8:426–30.
11. IV: The rabbit. *Journal of Heredity* 8:473–75.
12. V: The guinea pig. *Journal of Heredity* 8:476–80.
13. VI: Cattle. *Journal of Heredity* 8:521–27.
14. VII: The horse. *Journal of Heredity* 8:561–64.
15. 1918 VIII: Swine. *Journal of Heredity* 9:33–38.
16. IX: The dog. *Journal of Heredity* 9:89–90.
17. X: The cat. *Journal of Heredity* 9:139–44.
18. XI: Man. *Journal of Heredity* 9:227–40.
19. Pigmentation in guinea pig hair. With H. R. Hunt. *Journal of Heredity* 9:178–81.
20. On the nature of size factors. *Genetics* 3:367–74.
21. 1919 Scientific principles applied to breeding. *Breeders Gazette* 75:401–2.
22. 1920 The relative importance of heredity and environment in determining the piebald pattern of guinea pigs. *Proceedings of the National Academy of Science* 6:320–32.
23. *Principles of livestock breeding*. Bulletin No. 905. U.S. Department of Agriculture.
24. 1921 Correlation and causation. *Journal of Agricultural Research* 20:557–85.
25. Factors in the resistance of guinea pigs to tuberculosis with especial regard to inbreeding and heredity. With Paul A. Lewis. *American Naturalist* 55:20–50.
26. Review of *The origin and development of the nervous system,* by C. M. Child. *Journal of Heredity* 12:72–75.
27. Systems of mating, I: The biometric relation between parent and offspring. *Genetics* 6:111–23.

28. II: The effects of inbreeding on the genetic composition of a population. *Genetics* 6:124–43.
29. III: Assortative mating based on somatic resemblance. *Genetics* 6:144–61.
30. IV: The effects of selection. *Genetics* 6:162–66.
31. V: General considerations. *Genetics* 6:168–78.
*32. 1922 Coefficients of inbreeding and relationship. *American Naturalist* 56:330–38.
33. The effects of inbreeding and crossbreeding on guinea pigs, I: Decline in vigor. In Bulletin No. 1090, 1–36. U.S. Department of Agriculture.
34. II: Differentiation among inbred families. Bulletin No. 1090, 37–63. U.S. Department of Agriculture.
35. III: Crosses between highly inbred families. Bulletin No. 1121. U.S. Department of Agriculture.
36. 1923 Two new color factors of the guinea pig. *American Naturalist* 57:42–51.
37. The theory of path coefficients: A reply to Niles' criticism. *Genetics* 8:239–55.
*38. Mendelian analysis of the pure breeds of livestock, I: The measurement of inbreeding and relationship. *Journal of Heredity* 14:339–48.
*39. II: The Duchess family of shorthorns as bred by Thomas Bates. *Journal of Heredity* 14:405–22.
40. The relation between piebald and tortoise shell color pattern in guinea pigs. *Anatomical Record* 23:393.
41. Factors which determine otocephaly in guinea pigs. With O. N. Eaton. *Journal of Agricultural Research* 26:161–82.
42. 1925 Corn and hog correlations. Bulletin No. 1300. U.S. Department of Agriculture.
43. The factors of the albino series of guinea pigs and their effects on black and yellow pigmentation. *Genetics* 10:223–60.
*44. Mendelian analysis of the pure breeds of livestock, III: The shorthorns. With H. C. McPhee. *Journal of Heredity* 16:205–15.
45. An approximate method of calculating coefficients of inbreeding and relationship from livestock pedigrees. With H. C. McPhee. *Journal of Agricultural Research* 31:377–83.
46. 1926 A frequency curve adapted to variation in percentage occurrence. *Journal of the American Statistical Association* 21: 161–78.
47. Mutational mosaic coat patterns of the guinea pig. With O. N. Eaton. *Genetics* 11:333–51.
48. Effects of age of parents on characteristics of the guinea pig. *American Naturalist* 60:552–59.
49. Mendelian analysis of the pure breeds of livestock, IV: The British dairy shorthorns. With H. C. McPhee. *Journal of Heredity* 17:397–401.
50. Review of *The biology of population growth,* by Raymond Pearl, and *The Natural Increase of Mankind,* by J. Shirley Sweeney. *Journal of the American Statistical Association* 21:493–97.
51. 1927 Transplantation and individuality differentials in inbred families of guinea pigs. With Leo Loeb. *American Journal of Pathology* 3:251–85.
52. The effects in combination of the major color-factors of the guinea pig. *Genetics* 12:530–69.

53. 1928 Review of *The rate of living,* by Raymond Pearl. *Journal of the American Statistical Association* 23:336–39.
54. An eight-factor cross in the guinea pig. *Genetics* 13:508–31.
55. 1929 The persistence of differentiation among inbred families of guinea pigs. With O. N. Eaton. Technical Bulletin No. 103. U.S. Department of Agriculture.
*56. Fisher's theory of dominance. *American Naturalist* 63:274–79.
*57. The evolution of dominance. *American Naturalist* 63:556–61.
58. The dominance of bar over infra bar in *Drosophila. American Naturalist* 63:479–80.
*59. Evolution in a Mendelian population. *Anatomical Record* 44: 287.
*60. 1930 Review of *The genetical theory of natural selection,* by R. A. Fisher. *Journal of Heredity* 21:349–56.
*61. 1931 Statistical theory of evolution. *Journal of the American Statistical Association* 26, suppl., 201–8.
62. Statistical methods in biology. *Journal of the American Statistical Association* 26, suppl., 155–163.
63. Review of *The measurement of man,* by J. A. Harris, C. M. Jackson, D .C Paterson, and R. E. Scammon. *Journal of the American Statistical Association* 26:358–60.
*64. Evolution in Mendelian populations. *Genetics* 16:97–159.
65. On the genetics of number of digits of the guinea pig. *Anatomical Record* 51:115.
66. 1932 On the evaluation of dairy sires. *Proceedings of the American Society for Animal Production.*
67. Complementary factors for eye color in Drosophila. *American Naturalist* 66:282–83.
68. General, group and special size factors. *Genetics* 17:603–19.
69. Hereditary variations of the guinea pig. *Proceedings of the Sixth International Congress of Genetics* 2:247–49.
*70. The roles of mutation, inbreeding, crossbreeding and selection in evolution. *Proceedings of the Sixth International Congress of Genetics* 1:356–66.
71. 1933 Inbreeding and homozygosis. *Proceedings of the National Academy of Science* 19:411–20.
72. Inbreeding and recombination. *Proceedings of the National Academy of Science* 19:420–33.
73. Review of *Order of birth, parent-age, and intelligence,* by L. L. Thurstone and R. L. Jenkins. *Journal of Heredity* 24: 193–94.
74. Review of *Some recent researches in the theory of statistics and actuarial science,* by J. F. Steffensen. *Journal of Heredity* 24:364–66.
*75. 1934 Physiological and evolutionary theories of dominance. *American Naturalist* 68:25–53.
76. The genetics of growth. *Proceedings of the American Society for Animal Production* (1933): 233–37.
77. Types of subnormal development of the head from inbred strains of guinea pigs and their bearing on the classification and interpretation of vertebrate monsters With K. Wagner. *American Journal of Anatomy* 54:383–447.
78. On the genetics of submornal development of the head (otocephaly) in the guinea pig. *Genetics* 19:471–505.
79. Polydactylous guinea pigs: Two types respectively heterozygous

and homozygous in the same mutant gene. *Journal of Heredity* 25:359–62.

80. An analysis of variability in number of digits in an inbred strain of guinea pigs. *Genetics* 19:506–36.

81. The results of crosses between inbred strains of guinea pigs differing in number of digits. *Genetics* 19:537–51.

82. Genetics of abnormal growth in the guinea pig. *Cold Spring Harbor Symposium on Quantitative Biology* 2:137–47.

*83. Professor Fisher on the theory of dominance. *American Naturalist* 68:562–65.

84. The method of path coefficients. *Annals of Mathematical Statistics* 5:161–215.

85. 1935 A mutation of the guinea pig, tending to restore the pentadactyl foot when heterozygous, producing a monstrosity when homozygous. *Genetics* 20:84–107.

*86. The analysis of variance and the correlations between relatives with respect to deviations from an optimum. *Journal of Genetics* 30:243–56.

*87. Evolution in populations in approximate equilibrium. *Journal of Genetics* 30:257–66.

88. The emergency of novelty: A review of Lloyd Morgan's "emergent" theory of evolution. *Journal of Heredity* 26: 369–73.

89. On the genetics of rosette pattern in guinea pigs. *Genetica* 17:547–60.

90. 1936 On the genetics of the spotted pattern of the guinea pig. With Herman B. Chase. *Genetics* 21:758–87.

*91. 1937 The distribution of gene frequencies in populations. *Science* 85:504.

*92. The distribution of gene frequencies in populations. *Proceedings of the National Academy of Science* 23:307–20.

93. The hereditary factor in abnormal development. *Proceedings of the Institute of Medicine*. vol. 11, November 15, 1937.

*94. 1938 Size of population and breeding structure in relation to evolution. *Science* 87:430–31.

*95. The distribution of gene frequencies under irreversible mutation. *Proceedings of the National Academy of Science* 24: 253–59.

*96. The distribution of gene frequencies in populations of polyploids. *Proceedings of the National Academy of Science* 24: 372–77.

*97. 1939 The distribution of self-sterility alleles in populations. *Genetics* 24:538–52.

98. Genetic principles governing the rate of progress of livestock breeding. *Proceedings of the American Society for Animal Production* (1939):18–26.

*99. *Statistical genetics in relation to evolution: Actualités scientifiques et industrielles*. 802. Exposés de Biometrie et de la statistique biologique, XIII. Paris: Hermann & Cie.

*100. 1940 Breeding structure of populations in relation to speciation. *American Naturalist* 74:232–48.

*101. The statistical consequences of Mendelian heredity in relation to speciation. In *The new systematics,* ed. Julian S. Huxley, 161–83. Oxford: Oxford University Press.

102. 1941 A quantitative study of the interactions of the major colour factors of the guinea pig. *Proceedings of the Seventh International Genetics Congress* (1939):319–29.

103. Genetics of natural populations, V: Relations between mutation rate and accumulation of lethals in populations of *Drosophila pseudoobscura*. With T. Dobzhansky. *Genetics* 26:23–51.

104. Review of *A philosophy of science*, by W. J. Werkmeister. *The American Biology Teacher* 3:276–78.

*105. Review of *The "Age and Area" concept extended*, by J. C. Willis. *Ecology* 22:345–47.

*106. Review of *The material basis of evolution*, by R. Goldschmidt. *The Scientific Monthly* 53:165–70.

107. The physiology of the gene. *Physiological Reviews* 21:487–527.

108. Tests for linkage in the guinea pig. *Genetics* 26:650–69.

109. On the probability of fixation of reciprocal translocations. *American Naturalist* 75:513–22.

110. 1942 Genetics of natural populations, VII: The allelism of lethals in the third chromosome of *Drosophila pseudoobscura*. With T. Dobzhansky and W. Hovanitz. *Genetics* 27:363–94.

111. The physiological genetics of coat color of the guinea pig. *Biological Symposia* 6:337–55.

*112. Statistical genetics and evolution. *Bulletin of the American Mathematical Society* 48:223–46.

113. Comment on "Mating customs in North Carolina," by Florence C. Dudley and William Allen. *Journal of Heredity* 33:333–34.

*114. 1943 Isolation by distance. *Genetics* 28:114-38.

*115. Analysis of local variability of flower color in *Linanthus parryae*. *Genetics* 28:139–56.

116. Genetics of natural populations, X: Dispersion rates in *Drosophila pseudoobscura*. With T. Dobzhansky. *Genetics* 28:304–40.

117. 1945 Physiological aspects of genetics. *Annual Reviews of Physiology* 5:75–106.

118. Genes as physiological agents. General considerations. *American Naturalist* 79:289–303.

*119. *Tempo and mode in evolution*: A critical review. *Ecology* 26:415–19.

*120. 1946 The differential equation of the distribution of gene frequencies. *Proceedings of the National Academy of Science* 31: 383–89.

*121. Isolation by distance under diverse systems of mating. *Genetics* 31:39–59.

122. Genetics of natural populations, XII: Experimental reproduction of some of the changes caused by natural selection in certain populations of *Drosophila pseudoobscura*. With T. Dobzhansky. *Genetics* 31:125–156.

123. 1947 On the genetics of several types of silvering in the guinea pig. *Genetics* 32:115–41.

124. Genetics of natural populations, XV: Rate of diffusion of a mutant gene through a population of *Drosophila pseudoobscura*. With T. Dobzhansky. *Genetics* 32:303–24.

*125. 1948 On the roles of directed and random changes in gene frequency in the genetics of populations. *Evolution* 2:279–94.

*126. Evolution, organic. *Encyclopaedia Britannica*. 14th ed. rev. 8:915–29.

*127. Genetics of populations. *Encyclopaedia Britannica*. 10:111–112.

*128. 1949 Adaptation and selection. In *Genetics, paleontology, and evolution,* ed. G.L. Jepson, G. G. Simpson, and E. Mayr, 365–89. Princeton: Princeton University Press.

129. Colorimetric determination of the amounts of melanin in the hair of diverse genotypes of the guinea pig. With Zora I. Braddock. *Genetics* 34:223–44.

130. Estimates of the amounts of melanin in the hair of diverse genotypes of the guinea pig from transformation of empirical grades. *Genetics* 34:245–71.

131. Differentiation of strains of guinea pigs under inbreeding. *Proceedings of the First National Cancer Conference,* 13–27.

*132. Population structure in evolution. *Proceedings of the American Philosophical Society* 93:471–78.

133. On the genetics of hair direction in the guinea pig, I: Variability in the patterns found in combinations of the R and M loci. *Journal of Experimental Zoology* 112:303–24.

134. II: Evidences for a new dominant gene, Star, and tests for linkage with eleven other loci. *Journal of Experimental Zoology* 112:325–40.

135. 1949 Dogma or opportunism. Commentary on "The Russian purge of Genetics": History of the conflict. *Bulletin of Atomic Scientists* 5:141–42.

136. 1950 III: Interaction between the processes due to loci R and St. *Journal of Experimental Zoology* 113:33–64.

137. Discussion on population genetics and radiation. *Journal of Cell and Comparative Physiology* 35:187–210.

138. Genetical structure of populations. *Nature* 166:247–53.

139. Population structure as a factor in evolution. In *Moderne Biologie: Festschrift für Hans Nachtsheim,* 274–87. Berlin: F. W. Peter.

*140. 1951 The genetical structure of populations. *Annals of Eugenics* 15:323–54.

*141. Fisher and Ford on "the Sewall Wright effect." *American Scientist* 39:452–58, 479.

142. 1952 The genetics of quantitative variability. Agricultural Research Council, Quantitative Inheritance. London: Her Majesty's Stationery Office (1) 1952, pp. 5–41. Reprinted in *Yearbook of Physical Anthropology,* 1952, 159–95.

143. The theoretical variance within and among subdivisions of a population that is in a steady state. *Genetics* 27:312–21.

144. 1953 Gene and organism. *American Naturalist* 87:5–18.

145. The interpretation of multivariate systems. In *Statistics and mathematics in biology,* ed. O. Kempthorne, T. A. Bancroft, J. W. Gowen and J. L. Lush, 11–33. Ames: Iowa State College Press.

146. 1954 Experimental studies of the distribution of gene frequencies in very small populations of *Drosophila melanogaster*. With W. E. Kerr. I: Forked. *Evolution* 8:172–77.

147. II: Bar. *Evolution* 8:225–40.

148. III: Aristapedia and spineless. *Evolution* 8:293–302.

149. Summary of patterns of mammalian gene action. *Journal of the National Cancer Institute* 15:837–51.

150. 1955 Discussion on responses of populations to radiation. Conference on genetics held at Argonne Nat. Lab., November 19–20, 1954, 59–62. AEC Division of Biology and Medicine Washington, D.C.

151. 1956 Modes of selection. *American Naturalist* 90:5–24.

152. Classification of the factors of evolution. *Cold Spring Harbor Symposium on Quantitative Biology* (1955) 20:16–24.

153. 1959 Genetics, the gene, and the hierarchy of biological sciences. *Proceedings of the Tenth International Congress of Genetics* 1:475–89. Also in *Science* 130:959–65.

154. On the genetics of silvering in the guinea pig with especial reference to interaction and linkage. *Genetics* 44:383.

155. Silvering (si) and diminution (dm) of coat color of the guinea pig and male sterility of the white or near-white combination of these. *Genetics* 44:563–90.

156. A quantitative study of variations in intensity of genotypes of the guinea pig at birth. *Genetics* 44:1001–26.

157. Qualitative differences among the colors of the guinea pig due to diverse genotypes. *Journal of Experimental Zoology* 142:75–114.

158. Physiological genetics, ecology of populations, and natural selection. *Perspectives in Biology and Medicine* 3:107–31. Also in *Evolution after Darwin,* ed. Sol Tax, 1:429–55. Chicago: University of Chicago Press, 1960.

159. 1960 On the appraisal of genetic effects of radiation in man. In *The Biological Effects of Atomic Radiation: Summary Reports 1960,* 18–24. National Academy of Science, National Research Council.

160. On the number of self-incompatibility alleles maintained in equilibrium by a given mutation rate in a population of given size: A re-examination. *Biometrics* 16:61–85.

161. Path coefficients and path regression: Alternative or complementary concepts. *Biometrics* 16:189–202.

162. The treatment of reciprocal interaction, with or without lag, in path analysis. *Biometrics* 16:423–45.

163. Residual variability of intensity of coat color in the guinea pig. *Genetics* 45:583–612.

164. Thomas Park: President elect. *Science* 131:502–3.

165. Postnatal changes in the intensity of coat color in diverse genotypes of the guinea pig. *Genetics* 45:1503–29.

166. The genetics of vital characters of the guinea pig. *Journal of Cell and Comparative Physiology* 56, suppl. 1, 123–51.

*167. Genetics and twentieth century Darwinism: A review and discussion. *American Journal of Human Genetics* 12:365–72.

168. 1962 Review of *The statistical processes of evolutionary theory,* by P. A. P. Moran. *American Scientist*, December 1962, p. 460A.

169. 1963 Discussion of "Systems of mating in mammalian genetics," by E. L. Green and D. F. Doolittle. In *Methodology in mammalian genetics,* ed. W. J. Burdette, 42–53. San Francisco: Holden-Day.

170. Genic interaction. In *Methodology in mammalian genetics,* ed. W. J. Burdette, 159–88. San Francisco: Holden-Day.

171. William Ernest Castle, 1867–1962. *Genetics* 48:1–5.

172. Plant and animal improvement in the presence of multiple selective peaks. In *Statistical genetics in plant breeding,* ed. W. D. Hanson and H. F. Robinson, 116–22. National Academy of Science, National Research Council.

173. Selection toward an optimum and linkage disequilibrium. *Proceedings of the Eleventh International Congress of Genetics* (The Hague) 1:147.

174. 1964 Pleiotropy in the evolution of structural reduction and of dominance. *American Naturalist* 98:65–69.

175. Biology and the philosophy of science. *Monist* 48:265–90. Also in *The Hartshorne festschrift: Process and divinity,* ed. W. R.

Freese and E. Freeman, 101–25. La Salle, Ill: Open Court Publishing.

176. Stochastic processes in evolution. In *Stochastic models in medicine and biology*, ed. John Gurland, 199–242. Madison: University of Wisconsin Press.

177. 1965 Factor interaction and linkage in evolution. *Proceedings of the Royal Society of London* B 162:80–104.

178. The distribution of self-incompatibility alleles in populations. *Evolution* 18:609–19.

179. The interpretation of population structure by F-statistics with special regard to systems of mating. *Evolution* 19: 355–420.

180. Dr. Wilhelmina Key. *Journal of Heredity* 56:195–96.

181. 1966 Polyallelic random drift in relation to evolution. *Proceedings of the National Academy of Science* 55:1074–81.

182. Mendel's ratios. In *The origin of genetics: A Mendel source book,* ed. Curt Stern and Eva A. Sherwood, 173–79. San Francisco: W. H. Freeman.

183. 1967 Comments on the preliminary working papers of Eden and Waddington. In *Mathematical challenges to the neo-Darwinian interpretation of evolution,* Wistar Symposium monograph no. 5, 117–20.

184. The foundations of population genetics. In *Heritage from Mendel,* ed. R. Alexander Brink, 245–63. Madison: University of Wisconsin Press.

185. "Surfaces" of selective value. *Proceedings of the National Academy of Science* 102:81–84.

186. 1968 Dispersion of *Drosophila melanogaster. American Naturalist* 102:81–84.

187. Contributions to genetics by J. B. S. Haldane. In *Haldane memorial volume,* ed. K. R. Dronamraju. Baltimore: Johns Hopkins University Press.

188. *Evolution and the genetics of populations, vol. 1: Genetic and biometric foundations*. Chicago: University of Chicago Press.

189. 1969 Haldane's contributions to population and evolutionary genetics. *Proceedings of the Twelfth International Congress of Genetics* 2:445–51.

190. Deviations from random combination in the optimum model. *Proceedings of the Twelfth International Congress of Genetics* 2:150.

191. Deviations from random combination in the optimum model. *Japanese Journal of Genetics* 44, suppl. 1,152–59.

192. The theoretical course of directional selection. *American Naturalist* 103:561–74.

193. *Evolution and the genetics of populations, vol. 2: The theory of gene frequencies*. Chicago: University of Chicago Press.

194. 1970 Random drift and the shifting balance theory of evolution. In *Mathematical topics in population genetics,* ed. K. Kojima. Heidelberg: Springer-Verlag.

195. 1971 Evolution, organic. Rev. *Encyclopedia Britannica*.

196. 1973 The origin of the F-statistics for describing the genetic aspects of population structure. In *Genetic structure of populations,* ed. N. E. Morton, 3–25. Honolulu: University of Hawaii Press.

197. 1975 Panpsychism and science. In *Mind in nature,* ed. J. E. Cobb, Jr. and D. R. Griffin. Washington, D.C.: University Press of America.

198. 1977 *Evolution and the genetics of populations, vol. 3: Experimental results and evolutionary deductions*. Chicago: University of Chicago Press.

199. Modes of evolutionary change of characters. In *Proceedings of the International Conference on Quantitative Genetics*, pp. 681–701. Ames: Iowa State University Press.

200. 1978 *Evolution and the genetics of populations, vol. 4: Variability within and among natural populations*. Chicago: University of Chicago Press.

*201. The relation of livestock breeding to theories of evolution. *Journal of Animal Science* 46:1192–1200.

202. The application of path analysis to etiology. In *Genetic Epidemiology*, ed. N. E. Morton. New York: Academic Press.

203. Review of *Modes of speciation*, by Michael J. D. White. *Paleobiology* 4:373–79.

204. 1980 Genic and organismic selection. *Evolution* 34:825–43.

*205. 1982 Character change, speciation, and the higher taxa. *Evolution* 36:427–43.

206. Dobzhansky's genetics of natural populations. *Evolution* 36:1102–6.

207. The shifting balance theory and macroevolution. *Annual Review of Genetics* 16:1–19.

208. 1983 On "Path analysis in genetic epidemiology: A critique." *American Journal of Human Genetics* 35:757–68.

209. 1984 The first Meckel oration: On the causes of morphological differences in a population of guinea pigs. *American Journal of Medical Genetics* 18:591–616.

210. Diverse uses of path analysis. In *Human population genetics*, ed. A. Chakravarti, 1–34. New York: Van Nostrand Reinhold.

Biology/History of science

After a seventy-five-year career, Sewall Wright is one of the most influential evolutionary biologists of the twentieth century. Many of his early, seminal papers are now available only in this book, which presents all of Wright's published papers on evolution up to 1950 and a few published later. William B. Provine's introductions include pertinent references to related portions of Provine's scientific biography of Wright, *Sewall Wright and Evolutionary Biology,* and Wright's own four-volume masterwork, *Evolution and the Genetics of Populations.* By comparing the papers in this volume with the corresponding topics in the larger work, it is possible to determine the respects in which Wright extended, changed, or remained constant in his ideas over a period of sixty years.

Wright's shifting-balance theory of evolution, first conceived in 1925, has proved enormously useful in modern evolutionary biology. Wright's international prestige has never been higher than it is currently, and the time is ripe for a rereading of his papers. They are not only historically important for understanding the period of the "evolutionary synthesis" of the 1930s and 1940s, but continue to be stimulating and useful to working evolutionary biologists today.

SEWALL WRIGHT is the Ernest D. Burton Distinguished Service Professor Emeritus of zoology at the University of Chicago and professor emeritus of genetics at the University of Wisconsin. He has received the National Medal of Science, the Darwin Medal of the Royal Society, the Balzan Prize, and many other honors during his extraordinary career. His four-volume work, *Evolution and the Genetics of Populations,* is also published by the University of Chicago Press.

The University of Chicago Press

ISBN 0-226-91054-

REDWOOD
LIBRARY
NEWPORT
R.I.